Organic Semiconductors

Part B

Organic Semiconductors

Part B

Felix Gutmann
*School of Chemistry
Macquarie University
Sydney, New South Wales
Australia*

Hendrik Keyzer
*Chemistry Department
California State University, Los Angeles
5151 State University Drive
Los Angeles, California 90032*

Lawrence E. Lyons
*School of Chemistry
Brisbane, Queensland
Australia*

With a Chapter on Organic Semi-Metals by
Robert B. Somoano
*Jet Propulsion Laboratory
Pasadena, California 91103*

ROBERT E. KRIEGER PUBLISHING COMPANY
MALABAR, FLORIDA
1983

Original Edition, Part B 1983
Part A 1981 Reprint w/corrections

Printed and Published by
**ROBERT E. KRIEGER PUBLISHING COMPANY, INC.
KRIEGER DRIVE
MALABAR, FLORIDA 32950**

Copyright © 1983 Robert E. Krieger Publishing Co., Inc.

All rights reserved. No reproduction in any form of this book, in whole or in part (except for brief quotation in critical articles or reviews), may be made without written authorization from the publisher.

Printed in the United States of America

To Irene, Mary and Alison

PER ARDUA AD ASTRA

Foreword

In the sixteen years which have elapsed since the publication of "Organic Semiconductors" by Gutmann and Lyons a great deal has occurred in the field, and our understanding of electronic conduction in organic solids has moved from the position it once occupied as a special subject to one which impinges on a wide variety of molecular problems. Therefore, the appearance by the original authors, with the addition of one collaborator in Part B, of what amounts to an entirely new book, as well as an update of the older edition, represents a great contribution to the literature not only of the special subject but of many others related to it.

Part B of the new volume is an entirely new update of the subjects as they were initially described; Part B (Chapters 12-25) adds another fifteen chapters differing completely from those of the original volume. This is mentioned simply to indicate the enormous increase in breadth which has taken place in this field. Materials that received only a cursory mention within a chapter in the first edition are now described in special chapters on theory, for example, and another one on charge-transfer complexes.

The other major subject which has reached a certain stage of maturity, and which is expanded here, is one-dimensional conductors. Finally, there exists the more speculative work in the application of all these compounds to biology.

The present volume will continue to be the leading reference work not only in electrical conductivity of organic solids but in all related topics as well.

Melvin Calvin
University Professor of Chemistry
University of California, Berkeley
November 1982

Preface

Since the appearance of the first edition of *Organic Semiconductors* an enormous amount of work has been published and several other books have appeared on this topic.

This book purports to be an exhaustive compendium of further research results in the field of the electrical organic solid state from 1965 to date. The definitive tradition begun with *Organic Semiconductors, Part A,* is followed. Some hypotheses advanced in the earlier work have been modified, and the original theories are evaluated in the light of new data. Major advances in the organic solid state of living matter are treated in more detail. A considerable number of organic materials previously mentioned only in terms of potential have now entered the market place. These are competing successfully with traditional materials, and in some cases exceeding early promise. Data on new organic semiconductors have burgeoned. These have been compiled in extensive tabular form.

This book examines critically the extensions of now conventional solid state theory, as well as new theories. Concepts of the generation of electric charge and the transport of electrical energy at the microscopic level in organic substances, receive the close attention which placed *Organic Semiconductors, Part A,* in high acientific and technological demand.

While the basic structure of volume A has to a certain extent been maintained, developments in different areas, *e.g.,* biological materials, have mandated new chapters. The new topic of organic semi-metals, which appeared since volume A was written, is dealt with in a chapter—Quasi-One-Dimensional Organic Conductors—kindly contributed by Dr. R. B. Somoano.

PREFACE

The volume of publication is increasing now at such a rate that we thought it advisable to add an annotated and selected bibliography covering mainly work published between late 1980 up to about October 1982. This information appears as a Post-scriptum.

"Organic Semiconductors, Part B" begins with Chapter 12. All Figures and Tables in the text are preceded with the chapter number, except those called Addenda Tables in Chapter 24. The latter Tables are preceded by numerals and end with letters or letters and numerals combined. Items from "Organic Semiconductors, Part A" are indicated by chapter and section number as well as the appropriate page number wherever possible. For convenience a list of all Tables in Part A and Part B is given on page 630.

We are indebted to the following publishers for permission to reproduce figures and/or tables appear in the text:

Academic Press for Figures 13.3, 13.4, 15.2 and 15.13.

Akademie Verlag DDR Germany for Figures 15.19, 15.20, 16.18, 16.19, 16.20.

The American Chemical Society for Figures 12.5, 13.1, 17.4, 17.5, 19.9, Tables 6.8F and 8.7F and condensed Note on p. 232.

The American Institute of Physics for Figures 12.2, 12.4A, 12.8, 12.20, 15.6, 16.9, 16.12, 17.3, 19.8, 22.1 and Tables 15.21, 15.22, 16.6, 4.7D, 6.6B, 6.17A.

The American Physical Society for Figures 15.3, 15.4, 15.5, 15.7, 15.8, 15.15, 20.3 and Tables 15.1, 15.9, 15.10, 15.11.

The Australia and New Zealand Association for the Advancement of Science for Figures 12.17, 22.5, 22.7, 22.8.

The Editor of the Australian Journal of Chemistry (Commonwealth Scientific and Industrial Research Organization) for Figures 12.1, 12.4B, 14.1, 15.23, 15.24, 15.25, 22.6 and Tables 14.4, 14.5, 19.2, 19.3, 22.5 and 6.9A.

B A G Brunner Verlag A.G. for Tables 8.21A, B and C.

The Chemical Society of Japan for Figures 12.6, 12.7, 14.2, 18.2, 22.9, 22.10, 22.3 and Tables 12.1, 8.6B, 8.6C.

The Chemical Society (London) for Figure 12.3 and for Table 5.4A.

The Electrochemical Society for Figure 15.11 and for Table 6.8E.

Friedrich Vieweg & Sohne Verlag for Figures 12.5, 12.12, 13.1, 17.4, 17.5, 19.9, and Tables 6.8F and 8.7F.

Gauthier-Villars (C. R. Academic Francaise) for Figures 16.16 and 16.17.

Gauthier-Villars and the Editor of the Revue de Chimie Minerale for Figure 15.27.

Gordon and Breach Science Publishers Ltd., for Figure 15.12, and Tables 14.1, 14.2, 15.6 and 15.7.

The Institute for Molecular Science, Japan, for Table 6.8B.

The Institute of Physics, England, for Figure 16.9.

John Wiley & Sons Inc. for Figures 12.16, 12.18, 15.25, 16.5 and Tables 6.8C, 6.8D and 8.7D.

The Editor of the Journal of Biological Physics for Figure 16.4, and for Tables 8.7D, 8.7E, 8.14B, 10.2C, 10.4A.

MacMillan Publishers Ltd. (Nature, London) for Table 10.2A.

The National Academy of Science U.S.A. for Figure 18.3.

North Holland-Elsevier Publ. Co. for Figures 12.11, 12.15, 13.5, 15.9, 15.10, 16.2, 16.21, 17.1, 17.6, 17.7, 17.8, 17.9, 19.1, and for Tables 19.1 and 9.1A.

Pergamon Press for Figures 12.11, 12.15, 13.5, 15.9, 15.10, 16.2, 16.11, 16.21, 17.1, 17.6, 17.7, 17.8, 17.9, and Table 19.1.

Plenum Press for Table 22.3.

Springer Verlag for Figure 15.26 and Table 6.19A.

Society for Polymer Science, Japan, for Tables 8.20A and 10.3A.

Taylor and Francis Ltd., for Figures 15.1 and 15.16.

VEB-Deutscher Verlag für Grundstoffindustrie for Figure 16.8.

Verlag Chemie for Figures 12.10, 16.13 and Table 22.6.

Contents

Chapter 12 Experimental and Methodology   1

- 12.1   Dark Conductance, 1
- 12.2   ac Methods, 3
- 12.3   Cyclotron Resonance, 4
- 12.3   Drift Mobility, 5
- 12.4   Dielectric Properties, 6
- 12.5   Hall Effect, 9
- 12.6   Thermo-electric Effect, 12
- 12.7   Electron Affinity, 13
- 12.8   Ionization Potentials, 16
- 12.9   Contact Potentials, 21
- 12.10   Photo Effects, 22
- 12.11   Catalysis, 25
- 12.12   Titration Apparatus and Methodology, 26
- 12.13   Microwave Methods, 29
- 12.14   Other Contact-less Methods, 32
- 12.15   Cell Arrangements Using Metallic Contacts, 32
- 12.16   Electrolytic Contacts, 33
- 12.17   Surfaces, 37
- 12.18   Effect of Gases, Adsorption and Chemisorption, 39
- 12.19   Noise Measurements, 42
- 12.20   Purification, 44
- 12.21   Compacted Powders. Annealing. Pressure, 47
- 12.22   Thin Films, 49
- 12.23   Liquids and Melts, 51

Chapter 13 Excited States 61

- 13.1 Introduction, 61
- 13.2 Excitons. Polarons, 63
- 13.3 Exciton Migration, 65
- 13.4 Exciton Reactions, 74
- 13.5 Bi-excitons and Exciton Clusters, 75
- 13.6 Exciplexes, 77
- 13.7 Excimers, 77
- 13.8 Surface Excitons. Polaritons, 79
- 13.9 Dipole-dipole (Förster) Transfer, 80
- 13.10 Other Collective Localized Excitations, 82

Chapter 14 Ionized States 89

- 14.1 Introduction, 89
- 14.2 Electron Affinities, 89
- 14.3 Ionization Potentials, 94
- 14.4 Carrier Generation, 97
- 14.5 Polarization Energies, 102

Chapter 15 Theories of Charge Transfer 109

- 15.1 Introduction, 109
- 15.2 Energy Band Model. Delocalized Transport, 113
- 15.3 The Transition Regime between Coherent Wave Propagation and Transfer *via* Localized States. The Munn-Siebrand Formalism, 118
  Linear Interaction with Lattice Modes, 120
  Quadratic Interaction with Molecular Modes, 121
- 15.4 Hopping Transfer *vs.* Band Model; further discussion, 122
  Pressure Dependence of Mobility, 127
- 15.5 Hopping, 130
- 15.6 Tunnelling, 141
- 15.7 Tunnelling *vs.* Hopping Transfer, 145
  Christov Characteristic Temperature, 145
- 15.8 Hindered Rotation, 148
- 15.9 Ligand Mediated Charge Transfer, 149
- 15.10 Trap-Limited Transport, 152
- 15.11 Hall Mobilities, 159
- 15.12 Contact-Limited Conductivity, 161

CONTENTS

15.13 Energy Correlations, 171
  The Compensation Law, 171
  Electrochemical Energy Gap Correlations, 178
  Catalytic Activity and Energy Gap, 182
  Other Correlations, 183

Chapter 16 Review of Published Data   193

16.1 Introduction, 193
16.2 Molecular Solids, 194
16.3 Carrier Mobilities in Molecular Solids, 194
16.4 Long-Chain Compounds and Polymers, 196
  Highly Conducting Polymers, 207
16.5 Nomadic Polarization Effects, 209
16.6 Dyes, 214
16.7 Doping and Sensitization, 219
  Spectral Sensitization, 223
  Catalysis, 228
16.8 Doped Polymers, 228
16.9 Liquids and Glasses, 231
16.10 Seebeck Coefficients, 236
16.11 Semiconducting TCNQ Free Radical Salts, 243

Chapter 17 Charge Transfer Complexes   259

17.1 Introduction, 259
17.2 Donicity and Electron Acceptance Numbers, 262
17.3 Weak Charge Transfer Complexes, 264
17.4 Strong Charge Transfer Complexes, 266
17.5 Activated Charge Transfer Complexes, 269
17.6 Intramolecular (Auto-) Complexes, 272
17.7 Ternary and Inclusion Adducts, 274
  Solvation, 278
17.8 Surface and Micellar Complexes, 280
  Micellar and Colloidal Transfer Complexes, 281
17.9 Proton Transfer Complexes, 283
17.10 Polymeric Charge Transfer Complexes, 286
17.11 Identification of Charge Transfer Complexes, 288
  Spectroscopy and Other Optical Methods, 288
  Conductivity Titrations, 291
  Dipole Moments, 297
  Permittivity Titrations, 298
  Other Dielectric Methods, 299

Other Electrical and Magnetic Methods, 300
Electron Spin Resonance, 300
Nuclear Magnetic Resonance, 300
Nuclear Quadrupole Resonance, 301
Magnetic Susceptibilities, 301
Electrochemical Methods. Polarography and Voltammetry, 301

Chapter 18 Biological Materials, 319

- 18.1 Introduction, 319
- 18.2 Conduction Data, 319
- 18.3 Conduction and Biological Structure, 320
- 18.4 Hydration of Proteins, 321
- 18.5 Non-ionic Conduction in Troteins, 323
- 18.6 Dielectric Measurements on Protein-Complexes, 324
- 18.7 Cytochrome Oxidase, 325
- 18.8 Methylglyoxal Complexes, 325
- 18.9 Methylglyoxal-Ascorbic Acid-Protein Interaction, 327
- 18.10 Integrated Biological Systems. Photobiology, 327
- 18.11 Nerve Conduction, 331
- 18.12 Bone, 336
- 18.13 Cooperative Phenomena, 336

Chapter 19 Structure, 343

- 19.1 Introduction, 343
- 19.2 Anisotropy, 344
- 19.3 Crystalline Transitions, 345
- 19.4 Amorphous Systems, 345
  Electron Transfer in Amorphous Solids, 347
- 19.5 Intramolecular Changes, 361
- 19.6 Doping and Structure, 363
- 19.7 Charge Transfer Complexes, 363
- 19.8 Pressure, 366
- 19.9 Intercalation Compounds, 371
- 19.10 Layered Materials and Thin Films, 371

Chapter 20 Space Charge Effects, 377

- 20.1 Space Charge Limited Currents, 377
- 20.2 Glow Curves, 384
- 20.3 Ferro-, Pyro- and Piezo-electric Effects, 385

20.4   Electrets, 387
20.5   Plasmas, 391

Chapter 21 Quasi-One-Dimensional Organic Conductors, 397

Chapter 22 Photo Effects, 419

22.1   Introduction, 419
22.2   Contacts, 420
22.3   Carrier Generation, 422
22.4   Photoconductivity, 430
22.5   Photovoltaic Effects, 437
22.6   Photoemission, 449
22.7   Other Photo Effects, 453
        Luminescence, 453
        Photomagnetic Effects, 453
        Dye-Lasers, 453
        Photocatalysis, 454

Chapter 23 Retrospect, Outlook and Speculations. Summary, 259

23.1   Retrospect, 459
23.2   Outlook and Speculations, 460
        Devices, 460
        Electro-Photography, 462
        Energy Conversion and Storage, 462
        Catalysis, 464
        Biological, 464

Chapter 24 Addenda Tables, 471

        List of Tables in Parts A and B, 471
        Abbreviations Used in Addenda Tables, 474

Chapter 25 Post-scriptum, 648

Author Index, 675

Subject Index, 709

12

Experimental and Methodology

This Chapter updates Chapters 2 and 3. In addition, ref. 274 in Chapter 12 lists several recent reviews on this subject.

12.1 Dark Conductance

Montgomery[1] has devised a method to measure the electrical resistivity of anisotropic crystals, based on some theoretical work by Logan et al.[2] A rectangular prism with edges in the principal crystal directions is prepared with electrodes on the corners of one face. Voltage-current ratios for opposite pairs of electrodes permit calculation of components of the resistivity tensor. The method can use small samples, and is best suited to materials describable by two or three tensor components.

An electrochemical method to derive the value of the energy gap in eq. 2.14 has been developed by Lyons and is fully discussed in Section 15.13.

Application of four-electrode probes to monoclinic crystals is discussed by McMullan.[3] A four-electrode cell for conductivity measurements on small, fragile, needle-like crystals has been developed by Coleman;[4] the sample rests on a bed of four parallel fine wires acting as the current and voltage leads. The crystal itself is not attached to its substrate but rests on a teflon block free to expand thermally.

The frequently employed van der Pauw 4-probe method (Section 3.7) which uses 4 randomly placed contacts around or on a solid slab of dimensions larger than the electrode spacing, is said[5] to be liable to excessively

EXPERIMENTAL AND METHODOLOGY

low conductivity values if there exists a radial carrier concentration gradient and/or the equivalent of a radial *p-n* junction.

The conductance and conductivity usually referred to is that at constant pressure with the sample allowed to expand freely; the conductivity at constant volume σ_v can be evaluated from the compressibility and coefficient of thermal expansion of the compound under test. σ_v is said to be of value in studying anharmonicity effects.[26]

In high resistance samples, the current becomes extremely minute requiring extreme care in its measurement. Moreover, they also result in long capacitive, instrumental time constants which may reach several minutes and thus make such work rather time consuming. A tested[266] cell for measurements on anthracene flakes is shown in Fig. 12.1.

When a voltage is applied to a crystal, a steady-state current is attained only after sufficient time is allowed for the time-dependent, or "polarization", currents to decrease to zero. Two methods are used to measure the steady-state current-voltage response:

(a) from a series of current-temperature responses at different voltages; and

(b) at a constant temperature, by varying the voltage and measuring the current for each voltage.

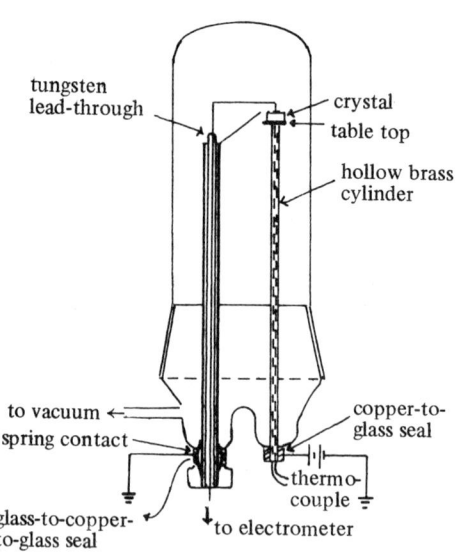

Fig. 12.1 High resistance dark conduction cell, incorporating glass-to-metal seals. After Johnston and Lyons.[266]

ac METHODS

The first method is usually inaccurate because between some heating and cooling cycles, the current irreversibly decreases. This effect might be associated with the electrodes. In method (b) above, only small voltage increments should be employed so that a steady state current is attained without excessive waiting.[269] In a log (current) vs. voltage plot, a linear relation is found to hold between log i and $v^{1/2}$ (see Sections 15.10 and 15.12) which permits extrapolation and computation of the sample resistance at constant voltage.

An interesting size effect has recently been reported[272] on TCNQ complex salts (not semi-metallic) in that conductance discontinuities occur at a certain temperature involving a drop of conductance by a factor as large as 300. This effect is confined to very small single crystals, and must be related to a dimerization of the columnar structure of these crystals. In larger specimens, however, no such discontinuity is observed[272] which appears to be confined to the cooling cycle only; the conductivity rises continuously and monotonically upon reheating. The matter is still unresolved and similar size effects night well occur elsewhere, requiring additional study.

12.2 ac Methods

A four point probe system for ac conductance measurements has been developed by Philips;[6] it involves driven coaxial shields and a controlled current source, covering the range of 10^{-8} to 10^{-5} Siemens and 20-200 Hz.

The conventional bridge methods often given rise to earthing problems which are quite serious when measuring high impedance samples. A "pseudo-bridge" developed by Borchardt and Holland[7] circumvents these problems by arranging four impedances around an operational amplifier similar to a bridge but permitting the direct earthing of source and output, i.e., also of sample and standard, without the use of coupling transformers.

A special cell allowing capacitance determinations on minute specimens has been proposed by Lal and Pahwa;[8] stray capacitances are eliminated.

The very low, sub-audio frequency region is now receiving more attention and a special bridge covering the interval of 0.01 to 1000 Hz, connecting the sample in an either 3- or 4-terminal configuration, has been developed by Lakes and Harper.[9]

In the case of a disordered, especially amorphous solid the a.c. conductivity σ_{ac} is primarily determined by the number of pairs of localized energy states having a dielectric relaxation time τ not below $1/\omega$, the period of the applied electric field.[10] In general, there will exist a whole range of τ values corresponding[10] to different hopping distances of the carrier. The interaction between the carrier and the solid matrix tends to increase all relaxation times so that σ_{ac} will be determined by pairs of energy states with a reduced separation distance. If these sites are randomly distributed, then σ_{ac} will be but little affected. However, if the interaction is so strong that τ no longer remains comparable to $1/\omega$, then σ_{ac} will tend towards zero. The effective ac conductivity of disordered systems is also discussed by Beeby and Hayes.[14] The frequency dependent impedance of organic semi-metals (see Chapter 21) is discussed by Rice and Bernasconi.[278]

Defining θ as the ratio of the density of states per kT near the Fermi level to the density of electrons in the conduction band, it can be shown[12] that the zero frequency limit of the parallel capacitance is proportional to (density of states per unit volume and energy near the Fermi level)$^{1/2}$, reducing to the classical space charge capacitance, as long as θ is small. For larger values of θ considerable corrections become necessary.[12] The theory of small ac signal response of solids and liquids with mobile charges subject to recombination is discussed by MacDonald and Franceschetti.[13] The ac regime of two-phase systems in which one phase conducts considerably better than the other, leads to frequency dependence of both the resistive and the capacitive impedance components.[11]

Cells and electrodes for conductimetric titrations are discussed in Section 12.12.

Cyclotron Resonance

A carrier moving in a magnetic field undergoes a circular motion at a frequency ω_c given by (eq. 12.1), where H is the magnetic field strength and m^* the effective mass of the carrier.

$$\omega_c = \frac{eH}{m^*c} \qquad (12.1)$$

The technique is practicable only at cryoscopic, usually liquid He, temperatures; it is very useful for investigating the band structure and carrier-scattering properties in covalent, ionic and metallic crystals. Burland and Konzelmann[15] have measured the effective mass of holes in the ab plane of anthracene crystals. The measured value of $11m_e$, where m_e is the free electron mass, is in good agreement with band structure calculations and demonstrates that, at least at 2 K holes in anthracene travel in bands. The question of band versus hopping motion at higher temperatures remains open.[16] Direct determination of the effective mass of charge carrier is, however, a significant development, in that values obtained may serve as valuable calibration points for band structure calculations.[16]

12.3 Drift Mobility

A typical arrangement for its measurement employed successfully by Sonnonstine and Hermann[17] is described below: the experimental technique used to determine drift mobilities devised by Kepler and LeBlanc,[265] involves direct measurement of transit times of photoinjected carriers. The light source was a Xenon Corporation S-130B flashtube and a model 465A micropulser. Quartz, coated with a tin tin-oxide layer, served as the illuminated electrode. While the back electrode was conductive silver paste painted on the nonilluminated surface on the sample. A weak phosphor-bronze spring clip held the sample in place and also provided electrical contact to the silver paste. The sample temperature was varied from 203 to 353 K by means of blowing either hot or cold dry nitrogen into the cell, and was monitored with a copper-constantan thermocouple imbedded in an anthracene crystal mounted about 3 mm from the sample. The temperature was allowed to stabilize for a few minutes at each value before drift mobility measurements were made.

This "time-of-flight" method has also been applied to mobility studies in a doped host matrix[18] and plotting the transit pulses on a log scale. The required photogenerated sheet of charge is created by a say, 5 nsec pulse of say, 3371 Å light from a pulsed laser. At this wavelength the dopant strongly absorbs, whereas the host (polymer) is essentially transparent, so that the exciting light is exclusively absorbed by the dopant molecule. If the duration of the light pulse is kept much shorter than typical current transients, then it becomes possible to study separately transport and photogeneration processes. To avoid space charge effects the light intensity must be kept low and only the first transient following a standard resting procedure is analyzed. A xerographic discharge also, under certain circumstances, reflects the photogeneration process.[19]

The theory of the time-of-flight method has been critically discussed by Dodelet and Freeman.[20] Scher and Montroll[21] have applied the method to mobility measurements of films: the sample is sandwiched between two plane parallel contacts, one of which is semi-transparent. The time constant of the system, given by RC (R being the sample resistance and C its capacitance, as arranged) must be kept much below the flash duration t. Again, log (current) is plotted vs. log t. The carrier transit time is indicated by a sharp change in the slope of the plot, i.e., of $d(\log I)/d\log t$.

A surface type cell for such measurements has been developed[22] (see Fig. 12.2), and applied to, e.g., azulene and anthracene by Young et al.[23] In the xerographic arrangement a charge is deposited on the sample surface e.g., from a corona discharge, and the potential decay after cessation of the

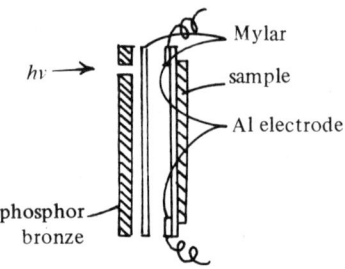

Fig. 12.2 Surface-type drift mobility measurement cell. After Young et al.[23]

charging cycle, is measured.[23] The decay does not reach zero again; this is due to charges becoming trapped within the solid.[25]

It has now become possible[27] to time-resolve these photo-generated transients to 800 psec; in anthracene the three main transient effects, viz., photogeneration, polaron formation, and free (hot) carrier motion, take place within less than the above time interval.[27]

The carriers may be injected[28] rather than photo-generated: A low energy electron beam injects electrons into a specimen film, and the dissipation of the injected charge is measured. In many injection experiments the injected charge may not be dissipated by drift through the specimen but locally neutralized by the natural leakage conductivity. It is therefore essential to employ extremely thin films only.[29] The method has been applied to n-terphenyl[30] and to polyvinylacetate[31] and other materials; it is of special value to mobility studies in liquids; see Section 16.9 and Chapter 15.

Theoretical aspects of mobility measurements as well as the interpretation of the resulting plots are also discussed in Chapter 15.

12.4 Dielectric Properties

While the permittivity is an important parameter of all semiconductors, even for organic polymers with a chain of conjugation it has not been extensively investigated. Knowledge of the dielectric constant may resolve many problems of semiconductor physics, and particularly in field stimulated electron emission.[32]

One of the difficulties in dielectric studies is determining whether the observed dielectric relaxation is a Debye or Maxwell-Wagner process.[34] In a Debye process,[33] the dielectric dispersion is due to the rotation of dipolar molecules in the applied field and hence is characteristic of the molecular structure of the adsorbate. In contrast, Maxwell-Wagner processes[33] are caused by inherent heterogeneity of the dielectric and originate in the differences in conductivities. Both processes give a similar dependence of the dielectric constant and loss on frequency.

Dielectric dispersion may arise from (i) interfacial polarizations involving electrode or intercrystalline interfaces, (ii) relaxations of the charge-transfer dipole moment, or (iii) activated hopping of electrons over potential energy barriers. Several experimental factors indicate that the dispersion represents a bulk rather than interfacial phenomenon. The reproducibility from sample to sample supports this conclusion, as does its temperature dependence since dielectric loss peaks resulting from interfacial effects are usually not temperature sensitive.[35]

The dielectric loss function $\chi(\omega)$ defined as[36]

$$\sigma(\omega) = \omega\chi(\omega)\alpha \int_0^\infty \frac{\omega^2}{1 + \omega^2/\nu^2} g(\nu) d\nu \qquad (12.1B)$$

where $n < 1$ over a wide frequency range and $g(\nu)$ is the function describing the spectrum of relaxation frequencies ν. This physically involves (1) a relaxing center at which relaxation is assisted by the diffusional arrival of movable defects, and (2) by variable range hopping conductivity (Chapter 15). Process (2) is dominated by low frequency modes. It then leads to a frequency dependence of the ac conductivity of the form $\sigma(\omega)\alpha\omega^n$ with $n < 0.8$. Process (1) predominates at high frequencies, especially above the characteristic Debye relaxation frequency. It results in a $\omega^{1/2}$ proportionality for the conductivity.

The permittivity of many organic solids drops reversibly[37] with increasing pressure, in contrast to the conductivity which, due to plastic deformation, tends to exhibit a hysteresis effect.[37]

A very elegant method to determine the effective permittivity of small single crystals at zero frequency, has been developed by Johnston and Lyons.[267] It is based upon two measurements of photoconduction transients produced by light flashes: one with the front transparent electrode in direct contact with the crystal and the other also transparent separated by a small distance. For details of the method, the original publication should be consulted.[267]

A dynamic bridge method suitable for the measurement of the very small capacitive component in an essentially resistive impedance is described by Tick and Johnson.[38]

A system involving a monopole probe suitable also for *in vivo* measurements, has been developed by Toler and Seals.[39] It consists of a signal source, a network analyzer and a short monopole antenna, the impedance of which in the dielectric is compared with its impedance in air. Permittivity as well as conductivity are determined simultaneously, at frequencies ranging from 10-100 MHz.

A microwave method (see also Section 12.15) to measure the permittivity of highly conducting organic metals using a dielectric resonance is described by Khanna et al.[40] The dielectric resonator is analogous to a resonant metal cavity in that highly reflecting dielectric walls serve to confine waves in the dielectric. The size of the resonator is reduced by $(\epsilon)^{\frac{1}{2}}$. Therefore, if a solid possesses a high dielectric constant, it can be placed directly inside a standard microwave waveguide and will absorb power at its characteristic resonance frequencies.

The theory of time-dependent polarization is discussed and reviewed in considerable details by Mracek;[41] the effect has been applied to a new method of chemical analysis, viz., Dielectric Time Domain Spectroscopy.[42]

Theories of complex permittivity are discussed by Hill.[43] For amorphous semiconductors, Hindley has shown[44] that the temperature dependence of the permittivity is related to that of the energy gap. Jonscher proposes[45] to plot not the real and imaginary components of the complex permittivity as in the usual Cole-Cole plot, (see p. 68, Part A) but rather to plot the components of the complex admittance in order to evaluate the relaxation behavior of the dielectric. However, even the conventional Cole-Cole plot gives a great deal of information as to the dielectric relaxation times and the association, or polymerization, of a specimen, as illustrated by studies of thiazine derivatives:[46] glass transition temperatures, supercooling as well as dispersion processes are all derivable from such studies.

Dielectric methods have been applied e.g., to study the ejection of counterions during the thermal "fusion" of Na-DNA solutions.[47] The study of charge transfer complexes such as DNA-dye complexes[240] made considerable use of dielectric methods.[48]

A typical Cole-Cole plot for the perylene-chloranil charge-transfer complex[35] is shown in Fig. 12.3. In such plots, the components of the complex permittivity

$$\epsilon = \epsilon' - i'\epsilon'' \qquad (12.2A)$$

are plotted against each other.

The Debye theory of dispersion and absorption in a dielectric lead to a complex relative permittivity

$$\epsilon_r = \epsilon_\infty + \{(\epsilon_s - \epsilon_\infty)/[1 + (i\omega\tau)^{1-\alpha}]\} \qquad (12.2B)$$

where α is a parameter determining the width of the distribution of relaxation times about the most probably relaxation time τ, ω is the angular frequency, and ϵ_s and ϵ_∞ are respectively the limiting low and high frequency relative

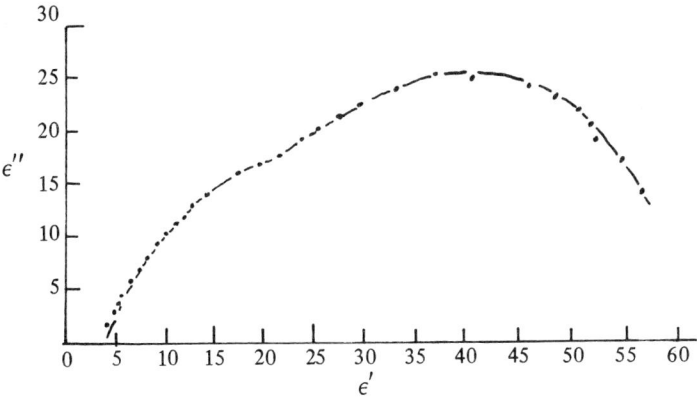

Fig. 12.3 Plot of ϵ' against ϵ'' at 316.5 K for the frequency range 5×10^{-3} to 10^5 Hz. The value for ϵ' increases with decreasing frequency. After Bone and Pethig.[35]

permittivity values. With reference to Fig. 12.3, the dominant and minor dispersion components have distribution parameter values of $\alpha = 0.32$ and $\alpha = 0$, respectively.

In the present case,[35] the total dielectric dispersion can be represented by two distinct arcs indicating the presence of two separate relaxation processes. Assuming that both processes involve a Debye-type dipolar relaxation, the total dispersion can be separated into two distinct components with relaxation time activation energies of 0.68 and 0.71 eV, for the dominant and minor components, respectively.

Several recent reviews on dielectric methods are available.[49] Aspects of this topic are also discussed in Section 17.11.

12.5 The Hall Effect

The following summary is based upon a discussion of this topic by Pethig:[50]

In an applied magnetic field, the force on the carriers, as well as the scattering, depends on their velocities, so that the effect becomes determined by an average over the total velocity range of the carriers. This average may not be the same as that for the current. The Hall coefficient must then be redefined as

$$R = r/nq \qquad (12.3)$$

where r is the scattering parameter and is a function of the applied magnetic field, the scattering mechanisms effective, and the shape of the energy surfaces. The Hall mobility μ_H is then related to the drift mobility μ by

$$\mu_H = r\mu \tag{12.4}$$

for small magnetic fields and spherical energy surfaces, r has the value $3\pi/8$ and $315\pi/512$ for lattice and ionized impurity scattering respectively. The value of r decreases and approaches unity as the magnetic field increases in magnitude. An anisotropy factor, which tends to reduce the value of r, must be taken into account for nonspherical energy surfaces, which therefore increases and tends to unity as the magnetic field increases. Mobilities can be ascertained only to within the uncertainty of the value for r, which ranges from about 0.9 to about 1.8.

For two or more kinds of dominant charge carriers the Hall mobility becomes an average mobility, but not an obviously meaningful one.

Re-entrant energy surfaces can produce anomalous Hall effect results for the sign and magnitude of the Hall mobility and in general the Hall mobility will be less than the drift mobility. If the energy bands are very narrow, the band states may be almost equally populated by charge carriers and the major electronic contributions are not necessarily from energy levels situated at the appropriate valence or conduction band edges. For electrons, for example, the negative contributions to the effective mass may be more dominant that the positive contributions, resulting in a reversal of the conventional Lorentz force producing the Hall effect. In this way electrons can be mistaken as positive holes.[50]

Any effect which causes the scattering of the conduction electrons by the scattering centers—whether atoms or molecules—to become asymmetric gives rise to an anomalous Hall voltage. The presence of centers exhibiting a magnetic moment is a frequent source of the anomalous Hall effect;[53] the phenomenon is specially pronounced in amorphous solids where practically every molecule, or atom, acts as a scattering center. Even traces of ferromagnetic impurities are liable to produce anomalous Hall voltages which completely swamp the normal effect.

Refined potential distribution calculations for thin solid films have been reported by DeMey mainly for the ac Hall technique.[54] Correction terms to allow for the finite size of the sample as well as for its surface properties are given by Manifacier,[55] while the theory of the Hall effect in the hopping region of a disordered solid is treated by Boettger and Bryskin.[56]

The interpretation of Hall Effect data obtained from measurements on polycrystalline solids, and *a fortiori*, on polycrystalline films, is still a matter of dispute: Volger concludes[57] that the method yields correct values of carrier concentration within the crystallites but not of carrier mobility; Bube and coworkers[58] have suggested that Hall data in such materials are determined by the intercrystalline phase, while Lipskis et al.[59] state that the carrier concentration is correctly determined by the Hall Effect if $b\beta < 1$ holds; here, $\beta = L_2/L_1$ and $b = \mu_2/\mu_1$ where L_1 and L_2 refer to the dimension of the crystallite and of the intercrystallite materials, respectively, and μ_1 and μ_2 stand for the carrier mobilities in these two phases. They point out that the usual method of representing the real film in terms of equivalent circuits cannot yield more than rough approximations and limits of applicability, at best.[59]

Measurement of the Hall voltage gives accurate results only for single crystal specimens of relatively low resistivity and large mobility values. To minimize the effect of the current contacts, the length of the sample should preferably be of the order of ten times its width.[50]

The voltage between the Hall voltage probes may not be the Hall voltage alone. Other galvanomagnetic and thermomagnetic effects (Nernst, Rhighi-Leduc and Ettingshausen effect) can produce voltages between the Hall probes. A thermoelectric voltage due to a transverse thermal gradient and an IR voltage drop due to misalignment of the Hall probes may also be present. All of these additional voltages, except that caused by the Ettingshausen effect, are eliminated by the averaging of four measurements taken for both polarities of the sample current and magnetic field.[50] The Ettingshausen effect is negligible for materials in which a high thermal conductivity is primarily due to lattice conductivity, or in which the thermoelectric power is small.

A practical method to eliminate non-Hall interferences except the Ettingshausen (which is also proportional to the cross product between magnetic induction and current, like the Hall effect) employs an alternating current as well as an alternating magnetic field, but of different frequencies. A signal of the frequency difference then results which can be isolated by means of a lock-in amplifier. The sensitivity of the system is about 10^{-10} V or better.

Hall effect measurements on amorphous solids are discussed by Colson et al.[24]; Friedman and Pollak treat[71] the Hall mobility in disordered systems exhibiting a hopping-type conductivity. They find that the Hall coefficient depends on the ratio between the mean interstitial spacing and the extent of the localized wave function. Dispersion in local site energies lowers the apparent Hall mobility (see discussion in Section 15.11).

Apparatus for ac Hall measurements on high impedance, low mobility materials, for which the ac method is superior to the simpler dc method, has been devised by several authors.[60] A non-uniform magnetic field is claimed to improve the sensitivity of the method.[61] Instead of measuring the Hall voltage, it is sometimes of advantage to measure the Hall current.[50]

The electrical contacts at the specimen ends are used to "short-out" the Hall field opposing the Lorentz force, and the resulting transverse Hall current is measured. The required geometry of the specimen is such that its width be at least 10 X its length. The main advantage is that with a guard ring, surface conduction effects can be separated.

The method has been used to determine charge carrier mobilities in anthracene single crystals.[51] The magnitude of the Hall currents were of the order 10^{-15} A corresponding to mobility values of around 1 cm^2/Vsec.[51] The technique has been further developed[52] to allow Hall current measurements on samples of arbitrary shape.

A simple Hall probe for measurements using a stationary sample and an alternating magnetic field has been designed by Cottles and Hermann;[62] it allows nulling spurious induced voltages due to local magnetic interferences.

Techniques for Hall measurements on 1-dimensional semimetals are described by Voronyuk.[63]

For a discussion of microwave Hall techniques, see Section 3.7. Theoretical aspects and significance of Hall effect data are discussed in Section 4.4 and Section 15.11.

Hall effect measurements are of special importance in studies of the electric transport properties of disordered materials and of organic semi-metals.[64] A numerical simulation of the effect in terms of a cubic array of resistances has been attained by Webmann et al.[65]

12.6 Thermoelectric Effects

The advantage of thermoelectric effect data is that the thermoelectric power is a zero current transport measurement which is not affected by breaks in a molecular chain or similar arrangement, as long as most of the temperature gradient occurs across the unbroken regions.[66]

Beni[66] has developed a theory for the thermoelectric effect in a narrow band Hubbard chain such as in one of the TCNQ charge transfer salts (see Section 16.11) for an arbitrary number of electrons per site. If this characteristic electron density is below 2/3, it is shown that the sign of the thermoelectric power, and thus the Seebeck coefficient, is negative at all temperatures. Above 2/3 it is small and negative only above a certain characteristic, below that temperature the sign and slope change. For a half-filled band the thermoelectric power tends towards zero because of the electron-hole symmetry.[68] The theory now has been extended also for a lattice-electron "gas" model.[69] For the same type of solids, Shchegolev gives the following equation for the Seebeck coefficient[70] S:

$$S = \frac{k}{e}\left\{\ln\left[1 + \exp(E/kT)\right] - \left[\frac{\frac{E}{kT}\exp(E/kT)}{1 + \exp(E/kT)}\right]\right\} \qquad (12.5)$$

where k is Boltzmann's constant, T the temperature in K, and E the energy gap or activation energy of conductivity.

Apparatus for the measurement of Seebeck coefficients, specially suited to biopolymers, is described by Eley et al.[67]

The methodology of Seebeck coefficient determinations on (polymeric) fibers and films is discussed by Grady and Hersh.[72] A 2-electrode method employing a slowly varying alternating field, especially suitable for small, fragile, needle-like samples, has been developed by Chaikin and Kwak:[73] it is suitable for the temperature region of 1.2 to 400 K and it requires a temperature differential of only 0.5 K.

The interpretation of thermoelectric effect data in polycrystalline films is discussed by Lipskis et al.[59]

12.7 Electron Affinity

The adiabatic electron affinity of a molecule is of an importance comparable with that of bond dissociation energies and ionization potentials, and is necessary for the description of electron transfer between molecules. A number of experimental and theoretical approaches have been used to measure electron affinities,[74] but the agreement generally has been poor.

Photodetachment experiments,[75] in which an electron of a gaseous negative ion M_{gas}^- is detached by a photon of energy $h\nu$ (eq. 12.6), provide a relatively direct means of determining electron affinities:

$$M_{gas}^- + h\nu \rightarrow M_{gas} + e^- \qquad (12.6)$$

The energy equation is

$$h\nu = A_G + E_k + \Delta E \qquad (12.7)$$

where A_G is the electron affinity of M_{gas}, E_k is the kinetic energy of the liberated electron, and ΔE accounts for a possible difference in the internal energies of the product neutral molecule M_{gas} and the absorbing negative ion. The threshold energy $h\nu_0$ of the cross-section corresponds to $E_k = 0$, and in many cases $\Delta E = 0$ at threshold.

The method may involve the use of a laser;[81] negative ions are extracted from an electric discharge formed into a beam of selected mass which crosses an intense laser beam at the inlet of a hemispherical electron energy analyzer. Photodetached electrons pass through the analyzer and are counted. Varying the analyzer energy yields a spectrum of electron counts vs. kinetic energy. A reference is required; oxygen serves well, having an electron affinity of 1.465 ± 0.005 eV.[76] The apparatus is shown, as a typical example,[77] in Fig. 12.4-A.

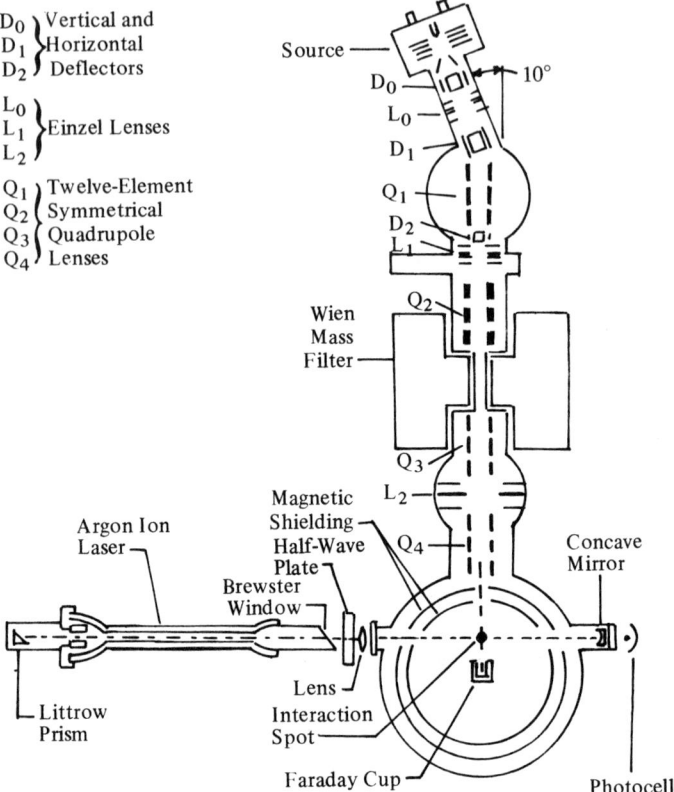

Fig. 12.4A A complete view of a photodetachment apparatus. After Celotta et al.[77]

Negative ions are produced in a hot cathode gas discharge source, accelerated to an energy of 680 eV, focused, mass analyzed by a Wien mass filter, refocused and pass through an interaction region, and finally collected in a Faraday cup. Beam currents of the mass selected species are typically in the range of $10^{-10} - 10^{-8}$ A.

An apparatus for the measurement of electron affinities using an electron capture technique is described by Lyons et al.[264] (See Fig. 12.4B). This is a variant of what sometimes is called the magnetron method.[80] Table 6.8B lists values obtained by this technique, while Table 6.18 lists values obtained by Lyons and Palmer[75] with their variant of the photodetachment method.

The methodology of the electron capture (Lovelock, cf. Section 6.8) technique has been discussed in detail by Lyons et al.[264] Their experimental arrangement[264] is shown in Fig. 12.4-B.

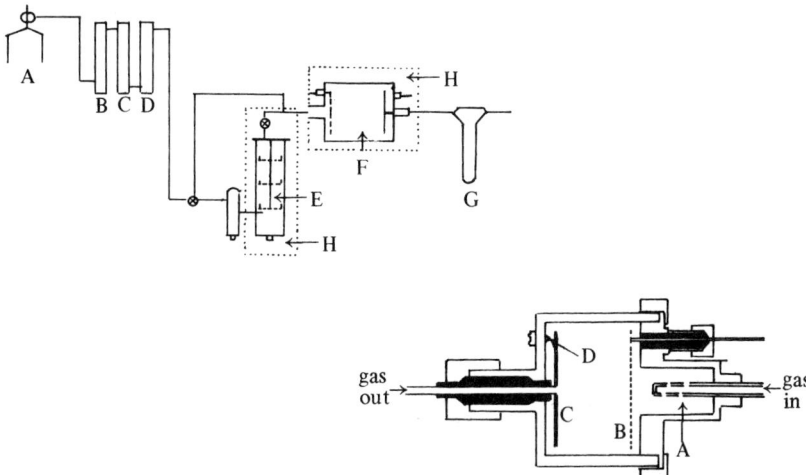

Fig. 12.4B Flow diagram for gaseous electron capture method. After Lyons et al.[264]

Changing the acceptor in a series of charge transfer complexes having the same donor, permits the electron affinity in the solid, A_c, to be estimated from changes in the maximum of the charge transfer absorption band if A_c of one acceptor is known. The method neglects changes in the polarization energy as well as possible entropy changes consequent upon changing the acceptor; it is discussed in Section 2.12 p. 92, and 6.9 p. 356.

A linear relation has been shown to hold between[78] electron affinity and the free energy ΔG^0 of trapped electrons in a solvent such as p-difluorobenzene, in the reaction

$$e_s^- + C_6H_4F_2 \leftrightarrow C_6H_4F_2^- \tag{12.8}$$

It would only be linear for molecules of approximately the same size so that the polarization energies would be comparable. (See Fig. 12.5). This is illustrated in a study of 29 donors using diazodiquinone as the acceptor.[79] (See Fig. 12.6).

Photoemission of electrons or holes into an insulator from a metal electrode can be employed for obtaining information on the electronic energy levels in the material. This technique has been used[82] for solids and was introduced for the study of electronic levels in non-polar liquids by Holroyd and Allen.[83]

The experiment consists of the determination of the work function of a suitable metal electrode *in vacuo* and in the liquid. The work function in the liquid ϕ_1 differs from that *in vacuo* ϕ_{vac} by V_0, the energy necessary to bring an electron from the vacuum into the liquid.

16   EXPERIMENTAL AND METHODOLOGY

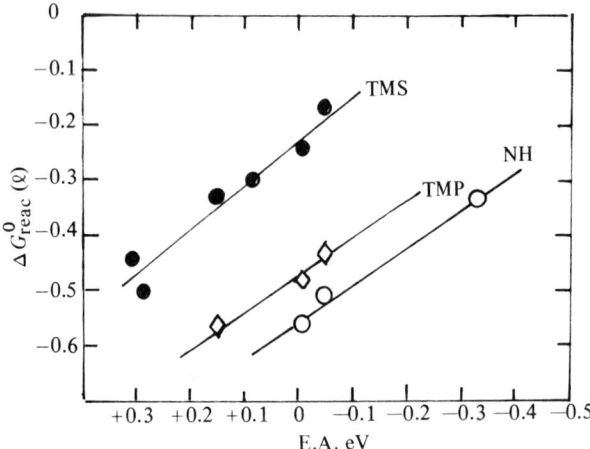

Fig. 12.5 Free energy of reaction vs. electron affinity: (●) triphenylene, phenanthrene, naphthalene, biphenyl, styrene, and α-methylstyrene in tetramethylsilane (◊) naphthalene, styrene, and α-methylstyrene in 2,2,4-trimethylpentane, (○) styrene, α-methylstyrene, and p-difluorobenzene in n-hexane. After Holroyd et al.[78]

$$\phi_1 = \phi_{vac} + V_0 \qquad (12.9)$$

Negative V_0 values mean that energy is released and thus represent the electron affinity of the liquid.

Numerical values of electron affinities will be found in Tables 6.8A, 6.8C, 6.10A, 6.17 and 6.19. A list of absolute values of electron affinities has been given by Chen and Wentworth.[84] The subject is further discussed in Section 2.12, 6.8, 6.9, and Section 14.2.

12.8 Ionization Potentials

The basics of photoemission spectroscopy are discussed in Section 2.13, 6.6 and 6.7. The photoemission spectra are usually obtained by means of retarding potentials applied in a form of Faraday cage. The observables are the values of the photocurrent and of the retarding potential, I and V, with wavelength or frequency as the parameter.

The energy distribution curves (EDC's) of photoemitted electrons are obtained by differentiating the I-V characteristic curves. The spectral dependence of the quantum yield (SDQY) of the photoelectrons is obtained by dividing the total photocurrent by the photon number. The total photocurrent is measured by applying enough accelerating voltage (+10 V) between the collector and the emitter of the analyzer. During the operation, the retarding

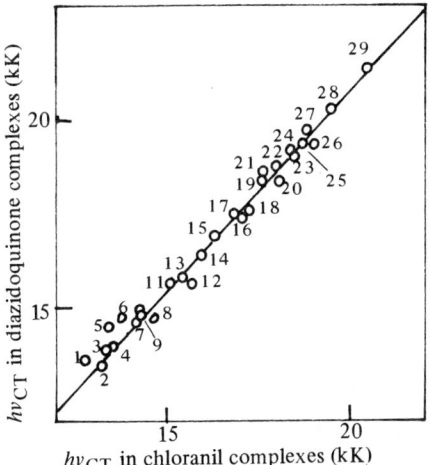

Fig. 12.6 Plots of wave numbers of maxima of the charge-transfer absorptions in the diazidoquinone complexes against those in the corresponding p-chloranil complexes. Point designation corresponds to the numbering of the donors.
(1) phenothiazine, (2) o-tolidine, (3) phenoxazine, (4) perylene, (5) 1,5-diaminonaphthalene, (6) dimethyl-p-toluidine, (7) dichloro-o-tolidine, (8) benzidine, (9) dibromo-o-tolidine, (10) N-methylphenothiazine, (11) dimethylaniline, (12) phenyl-naphthylamine, (13) diphenylamine, (14) anthracene, (15) pyrene, (16) phenoxathiin, (17) o-anisidine, (18) 1,2-benzanthracene, (19) 4,4'-dimethoxybiphenyl, (20) m-toluidine, (21) o-toluidine, (22) p-dimethoxybenzene, (23) p-chloroanline, (24) 1-naphthol, (25) aniline, (26) thianthrene, (27) 2-naphthol, (28) m-chloroaniline, and (29) triphenylene. Solvent: dichloroethane. After Koizumi and Matsunaga.[79]

potential is swept from minus voltage to plus. The potential when the peak takes off from the base line is the stopping voltage, V_0 and the potential when the peak falls down to the base line is the saturation voltage.

There are several methods to obtain the ionization potentials.[88] One method is from the EDC's. With the values of stopping voltage, V_0, and saturation voltage, V_2, the ionization potential is given as

$$I_p = h\nu - e(V_s - V_0) \tag{12.10}$$

where $h\nu$ is the incident photon energy and e is the electron charge. When the EDC's have tails, V_0 and V_s are determined by approximating the edges of the EDC's by straight lines.

Another method is from the quantum yield measurement. According to Lyons and Morris[85] the threshold of ionization, E_{th}, is given by the photon energy at which the quantum yield is 10^{-9} electron/photon.

Yet another method is from the relation between the quantum yield Y

and the incident photon energy $h\nu$. Near the threshold energy, the quantum yield is given as[86]

$$Y^{1/3} \propto (h\nu - E_{\text{th}}) \qquad (12.11)$$

The first method has some ambiguities when it is difficult to determine V_0 and V_s precisely. The determination of V_0 in the high-photon-energy region is also difficult, because high-energy electrons lose energy by scattering due to nonuniformity in the sample and/or intrinsic secondary effects. Therefore V_0 and V_s values at 7.75 eV photon energy are more reliable.

The semi-empirical relationship eq. (12.11), often referred to as the cube power law, is illustrated in Fig. 12.7,[86] for several compounds (see also discussion in Section 14.4 and 22.6).

A formally similar relationship between carrier yield and light intensity has been reported by Schlotter, using intense laser generated light pulses and measuring the resulting photocurrent transient.[87]

Gas phase ionization potentials of amino acids and of other large organic molecules have been determined from photoelectron spectra,[89] the method has also been applied to ion radical salts.[101]

The energy of excess electrons in a condensed matrix has been measured by comparing the ionization potential of guest molecules in the medium as against *in vacuo*;[90] tryptophan or TMPD$^+$ are suitable guest species. The gas phase uv-photoelectron spectrum of tryptophane has been thus determined by Seki and Inokuchi[91] yielding an I_g value of 7.3 ± 0.1 eV.

Changing the donor in a series of charge transfer complexes, the acceptor remaining the same, permits the ionization potentials of donors to be estimated from changes in the maximum of the charge transfer absorption band, if I_c of one donor is known. This method neglects changes in the polarization energy and entropy upon changing the donor. Matters may become complicated by the appearance of multiple absorption peaks; their separation will vary if there is a partial bond formation in the excited state of the complex so that, in effect, two excited states coexist. However, if the separation is below kT, only one charge transfer emission peak should be observable.[92] Other effects, though, may also be involved such as multiple accepting orbitals or steric changes.[93]

As has already been pointed out in Section 6.7, p. 351, measured photoionization potentials are really surface values; the question thus arises, exactly where are the photoelectrons generated?[280] The distance measured from the surface is termed the escape depth or electron attenuation length. Its determination gives useful information for the analysis of photoemission processes,[95] particularly the transport step. The attenuation lengths of low energy electrons (less than 10 eV for ultraviolet photoemission) of many

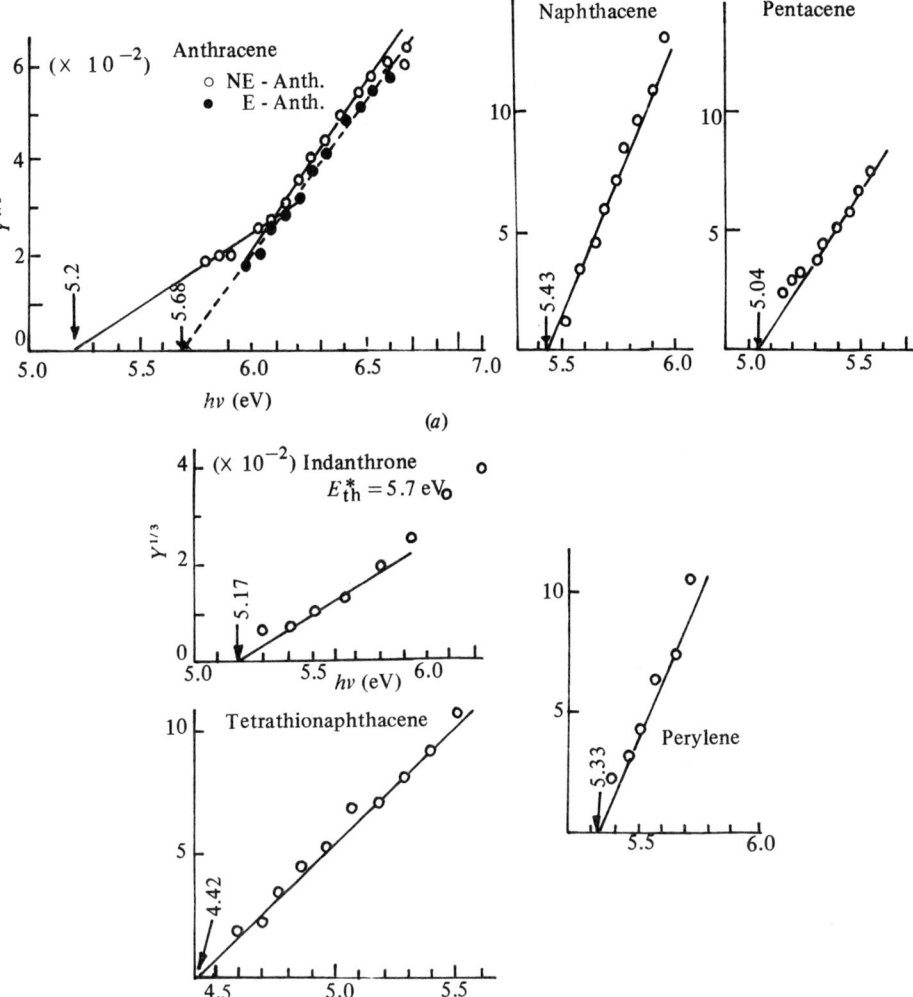

Fig. 12.7 Cube power-law fit for the photoelectric yield near the threshold. After Kochi et al.[86]

materials vary from several angstroms to more than 50 Å.[96] This suggests that in some cases the photoelectrons are produced in the first few surface layers of the solid and in other cases in the bulk of the solid.

The electron-attenuation lengths of several inorganic materials have been studied extensively, but only a few investigations have been made on organic materials.[102] Pong and Smith[97] have found the electron-attenuation length of copper phthalocyanine to be 11 Å in thin films (less than 40 Å thick) and about 150 Å in thick films (more than 50 Å). Berry[98] and Chang and Berry.[99] report that the films at 77 K varied from several angstroms to about a hundred angstroms. Belkind et al.[100] criticized the work of Pong and Smith on the grounds that 11 Å in the thin copper phthalocyanine film is too small and the effects may be due to pin-holes in the film. They obtain 140 Å as the escape depth for a naphthacene film on a gold electron emitter (electron energy being about 1 eV). Hino et al.[94] measured the low energy (<3 eV) electron-attenuation lengths of pentacene and perylene evaporated polycrystalline films, deposited onto copper iodide (CuI) film as electron source. The escape depths of electrons were 75 ± 10 Å for pentacene and 800 ± 80 Å for perylene.

Fig. 12.8 illustrates the experimental apparatus. The main vacuum chamber is attached to a vacuum ultraviolet monochromator. Monochromatic light is introduced into the chamber through the exit slit and also the LiF window. The incident light irradiates the cuprous iodide (CuI) and aromatic hydrocarbon laminated specimen deposited on the LiF (25 mm diameter and 2 mm thick) substrate. Whilst the electron current emitted from the specimen and also the transmittance of the incident light are measured, the furnace, from which the samples are evaporated, is removed from the light path. The intensity of the transmitted light is measured by a photomultiplier.

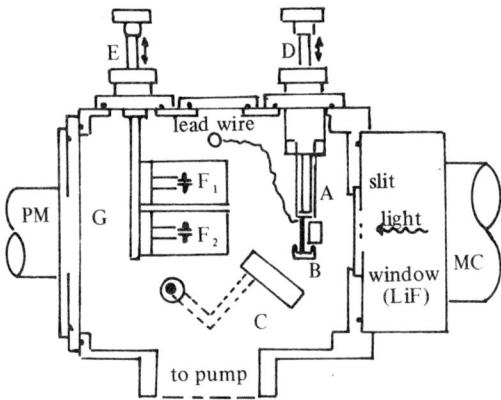

Fig. 12.8 Apparatus for the measurement of escape depths A: LiF substrate; B: quartz oscillator; C: electron collector; D: manipulator for movement of the substrate; E: manipulator for movement of the furnace; F_1 and F_2: furnaces; G: 1,12-benzoperylene wave-length converter; MC: a half-meter Seya-Namioka-type monochromator; PM: photomultiplier. After Hino et al.[94]

IONIZATION POTENTIALS

The most difficult problem in the measurement of the electron escape depth is the high electric resistance of organic solids. A gold film is frequently used to coat the substrate to provide electrical contact. However, the quantum yield (ejected electrons/incident photon) of gold[104] below 12 eV of $h\nu$ is equal to or smaller than that of most aromatic hydrocarbons.[102] One can overcome this problem by using high photoelectric efficiency CuI as a coating on the substrate.[102]

Numerical data of ionization potentials[102] are collected in Tables 6.6A, 6.6B, 6.7A, 6.19 and 6.21. Further discussion will be found in Section 14.3.

12.9 Contact Potentials

The appearance of a contact potential (see Section 2.14, p. 95) is a consequence of the requirement that the Fermi levels across any interface in thermodynamic equilibrium must equalize. The resulting band bending causes a potential barrier across the interface, the potential across the barrier being the contact potential.

The effect is necessarily closely associated with surface conditions which are discussed in Section 12.17. It was first formulated as Coehn's Rule in 1898[105] — a substance with a higher permittivity in contact with one of lower permittivity charges up positively relative to the latter. Of course, neither a continuous current nor external work can be obtained from a contact potential difference which refers to an equilibrium condition only. It can only be measured so that this equilibrium is not violated.

A suggestion by Langmuir[106] for conversion of the contact potential to a readily measurable electromotive force has been taken up by Frumkin[107] and has recently received attention in the study of contact potentials in electrochemistry:[108] plates of two different solids are first put into direct contact and then separated, acquiring electric charges giving rise to an electric field in the intervening space, driven by the contact potential. An ionized gas is now introduced into the space between the two solids until the electric field disappears by short circuiting *via* the gas. The contact potential then reappears between the two solids and can be measured; the energy now comes from introducing and maintaining the ionized gas.

The charge transfer in metal-polymer contacts has recently been proposed as a new method for the study of surface states, termed Contact Charge Spectroscopy.[109] The polymer is charged to a saturation value by repeated contacting/separation cycles with a metal plate; this is said[109] to require about 1000 cycles. The charges transferred in each cycle are supposed to be additive. The metal is said to inject carriers into certain states of the polymer with which states metals other than the one employed cannot comnumicate.[109] For electron transfer the electron affinity of the polymer needs to be close to the Fermi level of the metal. It is suggested that in this way the energy distribution of localized surface states in the polymer may be explored.[109] The method however, is still controversial; other authors[110] claim that in every repeated contact new areas of the two solids are involved and that the change depends only on the most recent contact.

12.10 Photo Effects

Experimental methodology of this topic is discussed in Sections 2.16 to 2.20, 8.11 and 8.12, in 10.4 and 11.2 as well as in Chapter 22. Here it is only intended to supplement these discussions with a few comments and selected references.[111]

Multichannel spectrometers are now available in which a separate detector is associated with each individual slit position instead of a single detector being used to scan the spectrum. The signal to noise ratio in a system using N detectors is then improved by a factor of $N^{1/2}$, the so-called Felgett advantage. Alternatively, the time domain response may be transformed into a frequency-domain response by Fourier transformation, or employing a Hadamard transform[111] in which only a portion of the slit positions is scanned at any one time. A further improvement may be attained by replacing the usual incoherent monochromator with a coherent tunable laser source the frequency of which is used as a reference signal.

In photo-electron spectroscopy, arriving electrons in a vidicon type detector, similar to a TV camera, may be dispersed so as to strike different regions of the screen and this results effectively in an multichannel system. Several dyes have been studied in this way (see Section 16.6).

The above techniques may, of course, also be applied to other, non-optical techniques such as NMR.[114]

In spatially Resolved Picosecond Spectroscopy[110] spatial oscillation in the initial density or excited states is produced by the interferences of pico-second excitation beams crossed in the samples. This results in an oscillatory susceptibility, i.e., a diffraction grating. A variable delay probe pulse is Bragg defracted by the grating. Excited state transport reduces the grating diffraction efficiency permitting a direct examination of the transport, e.g., in p-terphenyl.

The use of x-rays with energies exceeding the K-edge can be made to yield a great deal of relevant information: the short range order can be deduced from the backscattering effect of the surrounding atomic arrangement on the ejected 1s electron. The shift of the K absorption edge can provide a measure of charge transfer. Structural information is derived from the oscillatory part of the extended x-ray absorption (EXAFS). Short range order as obtained by Fourier transformation of the experimental data, allows average coordination numbers and neighbor distances to be determined.[136]

Optical reflectance spectroscopy is a valuable tool for the study of anisotropic crystals[112] though by no means confined to such.[113] Its relation to plasma effects are discussed in Section 10.7, Chapter 19, and to exciton dynamics[263] in Section 13.2. A typical infrared reflection spectrum of TEA(TCNQ)$_2$ anisotropic single crystals is shown in Fig. 12.9.[116] Irradiation is by polarized light along the three orthogonal directions 1,2,3 coinciding with the principal axes of the conductivity tensor of these materials. In spectrum R_1, with the direction of polarization along the columns, several intense reflection maxima are obtained. Application of an electric field perpendicular to the surface alters the reflection spectrum; this is referred to as Electroreflectance.[117]

In view of its possibilities as a solar energy converter, the photovoltaic effect has received a great deal of attention (see Section 22.5).

Photoconductivity transient methods are discussed in connection with their evaluation and interpretation in Section 12.3.

Chemiluminescence, light emitted from a chemical reaction, was first observed in fireflies, luminescent bacteria and marine organisms. In 1727, Swift described a fictional, 16-year effort in one laboratory to isolate sunbeams from cucumbers.[118] The phenomenon is a virtually universal effect in oxidizable organics. Several reviews are available.[119] A typical apparatus

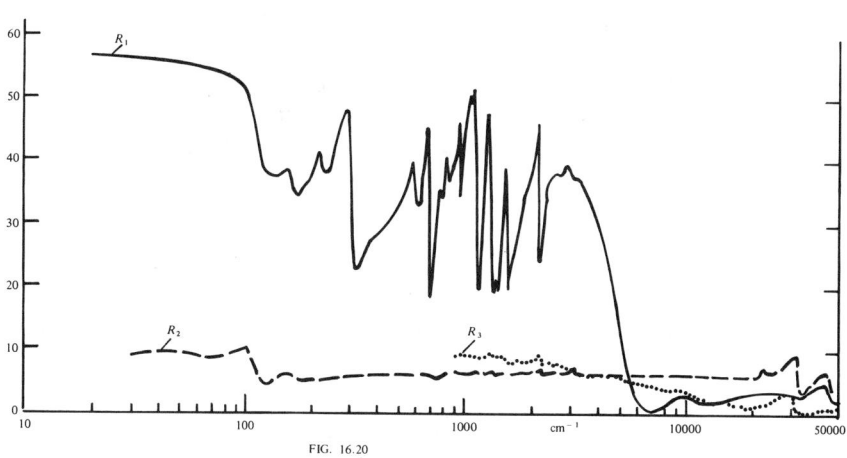

Fig. 12.9 Details of reflectance infrared spectra of TEA(TCNQ)$_2$(cm^{-1}). After Farges.[116]

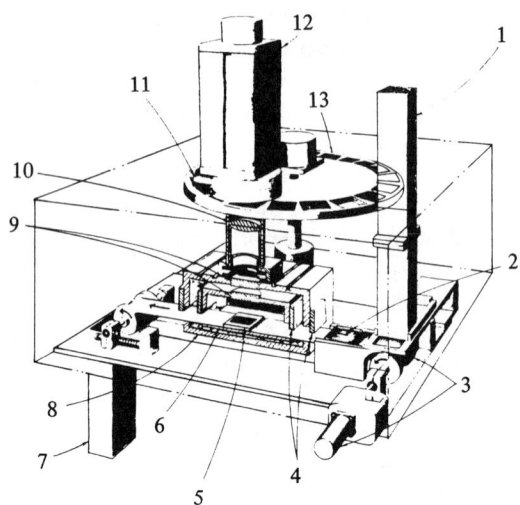

Fig. 12.10 Apparatus constructed at Battelle-Columbia (Ohio, USA) for measuring chemiluminescence. 1. sample bank and charger; 2. liquid sample; 3. sample conveyor and drive; 4. automatic doors; 5. solid samples (external dimensions ca. 50×50 mm^2); 6. sample oven; 7. sample bank; 8. water-cooled oven shield; 9. quartz windows; 10. collector lens; 11. manually operated shutter; 12. photomultiplier; 13. filter wheel with 20 filters. After Mendenhall.[119]

for studies of the effect (also at elevated temperatures) is shown in Fig. 12.10.

The apparatus depicted in Fig. 12.10 is particularly suited for examination of thin sheets of material at elevated temperatures. The samples mounted on glass or aluminum backing onto 35-mm aluminum slide holders to ensure a constant surface area exposed to the photomultiplier, are carried on a conveyer belt into an inner heated compartment. A second compartment surrounding the first is water-cooled to avoid heating the entire apparatus, which is enclosed with ¼ inch aluminum panels. The temperature, atmosphere, and position of the filter wheel in the apparatus are controlled from an adjacent room where the recording facilities are also located.[119]

When emission from a sample is low ($10 - 10^3$ counts/sec), a correction is made by subtracting the counting rate observed from the oven without the samples. Noise in the associated electronics, small sources of emission in the oven compartment, and cosmic radiation contribute to this background, which is temperature-dependent but rarely exceeds 10^3 counts/sec with this apparatus.

The chemiluminescence peaks obey an Arrhenius type relationship, permitting the calculation of an activation energy.[119] Intensity-time curves are highly characteristic and may be employed to check composition and identity of apparently identical specimens. Several physical properties may be correlated with the phenomenon.[119] A variant of this technique is electro-generated chemiluminescence in which an electric field is applied to the system electrochemically.[120] It involves the generation of excited radiation-emitting species associated with free radical or ion-radical processes.[120]

Techniques to measure the optical polarization of luminescence have been developed.[137]

Several other photoeffects are of relevance to organic semiconduction phenomena: if the compound contains a sufficient concentration of ferro- or ferrimagnetic ions, Faraday rotation of infrared and/or visible light may be observable. The effect involves the rotation of the plane of polarization of linearly polarized light passing through the magnetized crystal.[121]

Double refraction due to application of an electric field, and changes in the refractive index due the field, i.e., the Kerr effect, have been applied to the study of optically compounds and complexes containing an optically active component.[122] The ac Kerr effect involves an alternating electric field; it has been employed for studies of liquids such as xylene, toluene and others.[124] Instrumentation is based upon a 2-channel signal averager to obtain simultaneously two birefringence signals from two Kerr cells in series, one serving as a reference standard and the other containing the unknown sample.[124]

However, e.g., solid polymers are not very well suited to Kerr electro-optical methods because of complications arising from mechanically, (say, strain) induced birefringence effects. Polymeric liquids such as polypropyleneglycol having dipolar groups rigidly contained within their backbone, have thus been studied.[268]

Circular dichroism and rotational dispersion, i.e., the Cotton effect, has been employed to study asymmetric compounds and their complexes.[123]

The photo-dielectric effect, i.e., changes in the permittivity upon illumination in an electric field, has been used to study excited states in semiconductors.[125]

12.11 Catalysis

Some recent developments in this field, involving and relating specifically to organic semiconductors, are discussed in Sections 2.7, 3.8 and 15.12; practical processes based thereon are dealt with in Section 22. The subject has been reviewed by several authors;[126] Catalysis on organic solids and specially on phthalocyanines is treated in detail by Kopf and Steinbach[127] and by others.[134]

The catalytic activity of several charge transfer complexes has been demonstrated for a variety of reactions, such as for the condensation reaction between CO and H_2 (Fischer-Tropsch synthesis[128]), for the 1,2-butene isomerization,[128] for the Rosenmund-von Braun nitrile synthesis,[130] and related reactions,[130] and polymerization reactions.[160] The complexes employed were alkali-metal-metal-phthalocyanine for Fischer-Tropsch syntheses,[128] violanthrene-iodine for the butene isomerization,[129] Cu(I)Cl-pyridine for the nitrile syntheses,[130] and maleic-anhydride-amide complexes for polymerizations.[130] The latter required additional energy in the form of illumination, but needed no initiator.[131]

Halogen exchange reactions were observed in the reactions between bromoferrocene and Cu(I)halide-pyridine complexes.[132] The mechanisms of these reactions appear to resemble those for the hydrogen exchange reactions.

Electrocatalysis *via* charge transfer complexes has been recently reviewed.[133]

The catalytic activities of organics and, especially, of charge transfer complexes for the ortho-para hydrogen conversion and for the hydrogen deuterium exchange reaction have been extensively studied.[135]

Two different reaction paths, one through the chemisorption process and the other *via* the exchange process, were observed. With simultaneous measurements of the rate of hydrogen adsorption and also of the time course of ortho-para hydrogen conversion on the catalysts, a kinetic analysis, based on the two processes supports the chemisorption process to be predominant in every reaction over the complexes in question. Among the charge transfer complexes thus investigated, there are[133] aromatic hydrocarbons complexed with Na, and other alkali metals, phthalocyanine-Li, and Na-organic acceptor complexes, anthracene-trinitrobenzene, phthalocyanine-Na, violanthrene-B-Cs, and Ba-napthacene complexes.

Structural prerequisites have been critically examined.[138] Photocatalysis through excitation of an adsorbate, usually a dye, has been studied;[139] the excited dye molecule injects an electron into the conduction band of the substrate. A similar mechanism appears to be active at a semiconductor-solution interface[160] where charge transfer is governed by the position of the electron donor level of the excited dye relative to the conduction band edge of the substrate.[160]

12.12 Titration Apparatus and Methodology

Conductivity and/or permittivity titrations[141] (see Section 17.10) may be carried out in a cell shown in Fig. 12.11,[141] which cell is basically a modified

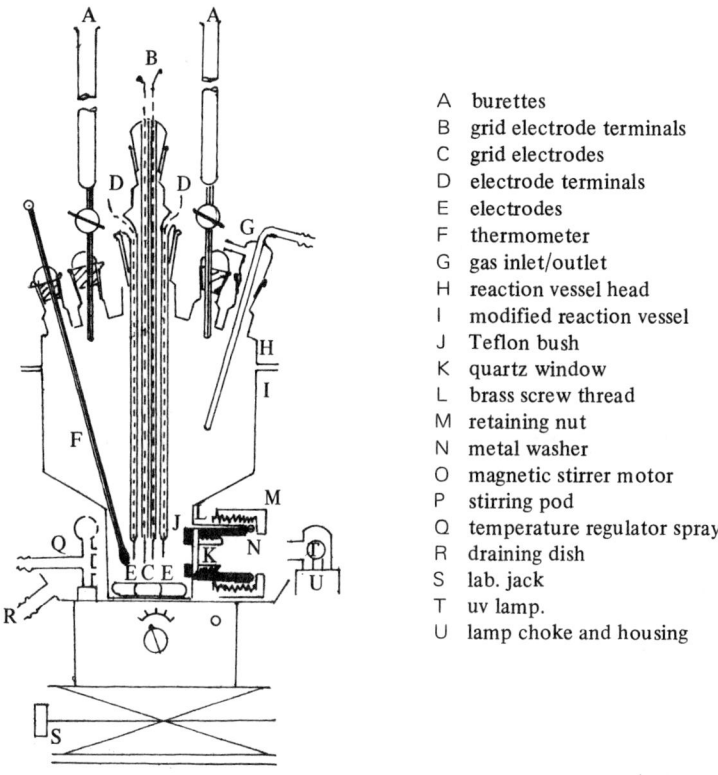

Fig. 12.11 Conductivity titration cell. After Gutmann and Keyzer.[141]

reaction vessel and reaction head. Titrations can be done in a protective atmosphere; the temperature of the cell may be controlled by spraying thermostated water on it. A quartz window at electrode level permits study of conductivity changes upon uv irradiation of the solution.[142] Separate current and voltage electrodes are provided, allowing measurements with direct current as well as studies of the effect of superimposing direct voltage while measuring the conductivity with an alternating voltage. The design may be much simplified by omitting the special features, and by replacing the water spray with a water jacket. The changes in conductivity are followed with a suitable bridge. Titrations are carried out by adding approximately 2 ml aliquots at 3 to 5 min intervals, the period required for the conductivity to attain a stable value. Continuous, remote magnetic control stirring is carried out for 2-4 min; the remaining period is used for bridge adjustment and reading. The whole assembly must be electrostatically shielded. Gold

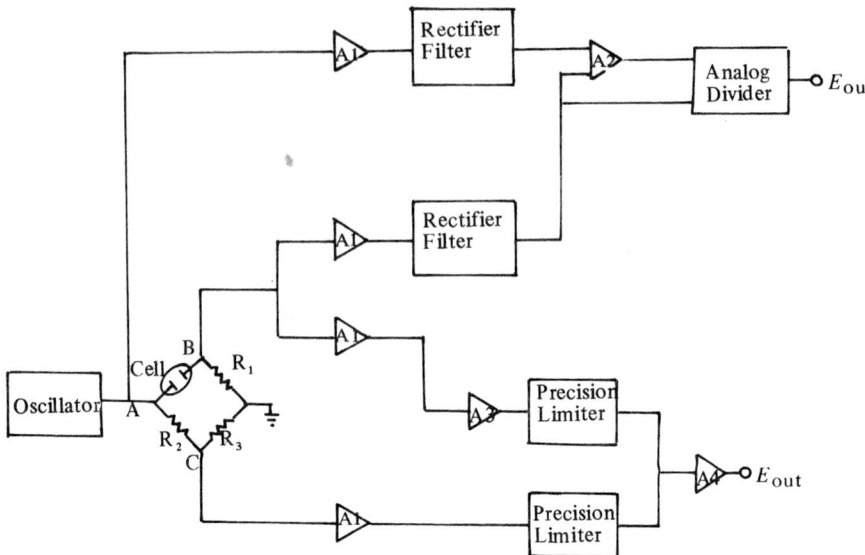

Fig. 12.12 Block diagram of conductometric device. Amplifiers: A1, buffer; A2, difference amplifier; A3, inverter; and A4, signal amplifier. After Doane and Stock.[145]

electrodes are sometimes preferable to platinum because the latter often proves troublesome by promoting unwanted side reactions by virtue of their high catalytic activity. Alternative designs have been published by Levi et al.[143] An ultrasonic technique for the dispersion of otherwise insoluble compounds so as to permit measurements on such systems, has been developed by Keyzer.[144]

A conductometric device that allows simultaneous ac bridge balance, without regard to capacitance effects, and generation of a signal proportional to solution conductance has been constructed by Doane and Stock.[145] Its block diagram is shown in Fig. 12.12.

The signal from the bridge at point B is inverted by A3 and then fed into a precision limiter, which simulates ideal diode behaviour. A similar limiter accepts the signal from point C. Position half-cycles are thus removed from each signal. The two resultants are added by A4 and this combination signal is applied to the vertical input of an oscilloscope with a horizontal scan rate of 100 kHz. Two overlapping rasters appear. The bridge is balanced by changing R_3 until the bright-lit tops of the rasters coincide. Any capacitance changes merely shift the half-cycle across the screen, leaving the amplitudes essentially unaffected. The analog signal is obtained from the voltage at point B, and E, the voltage applied to the other electrode at A. Provided that essentially no current is drawn from B, any output is directly proportional to solution conductance. Phasing problems are removed by rectifying and smoothing both E_2 and E.[145]

12.13 Microwave Methods

The statement made on p. 196, Part A, that high frequency methods are inapplicable to organics, can no longer be upheld in the light of later technical developments.[164] Thus, microwave methods, following their first employment* for mobility measurements by Trukhan[147] and by Eley,[198] have been devised and applied successfully to measurements[149] especially on organic semi-metals (see Chapter 21). The microwave cavity perturbation technique developed by Buravov and Shchegolev[150] places the sample in the maximum field position of the cavity. The shift in cavity frequency (Δf) and change in half-width at half-maximum ($\frac{1}{2}\Delta$) resulting from the insertion of the sample is measured. The field inside the specimen is given by

$$E_i = E_0/[1 + n(\epsilon - 1)], \qquad (12.12)$$

where E_0 is the unperturbed electric field in the cavity, $\epsilon = \epsilon_1 - i\epsilon_2$ the complex dielectric constant, and n the depolarization factor. Eq. 12.12 is valid only when the skin depth δ is greater than the sample thickness t,

$$\delta > t \qquad (12.13)$$

For higher conductivities, this does not hold since the electric field falls exponentially inside the sample. However, the depolarization factor can be estimated for an ellipsoid of revolution,

$$n = \frac{bc}{a^2}\left[\ln\left(\frac{4a}{b+c}\right) - 1\right] \qquad (12.14)$$

where a, b, and c are the semiaxes of the ellipsoid.

Eq. (12.14) is claimed[152] to agree with experiment to better than ±50%. The change in half-width and the frequency shift are given by[150]

$$\frac{1}{2}\Delta_0 = \frac{\frac{1}{2}\Delta}{f_0} = \frac{\alpha\epsilon_2}{[1 + n(\epsilon_1 - 1)]^2 + (n\epsilon_2)^2}, \qquad (12.15)$$

$$\Delta f_0 = \frac{\Delta f}{f_0} = \frac{\alpha\{(\epsilon_1 - 1)[1 + n(\epsilon_1 - 1)] + n\epsilon_2^2\}}{[1 + n(\epsilon_1 - 1)]^2 + (n\epsilon_2)^2}, \qquad (12.16)$$

---

*The first Hall effect measurements at microwave frequencies were reported by Cooke, who worked on covalent and relatively high mobility solids.[151]

where the filling factor $\alpha = \frac{1}{2}V_s E_{max}^2 / \int V_c |E_0|^2 dV$ (for a TE_{101} rectangular cavity $\alpha = 2V_s/V_c$), E_{max} is the magnitude of the electric field at the sample, V_s the sample volume, V_c the volume of the cavity, and f_0 its central frequency). Therefore,

$$\epsilon_1 - 1 = \frac{1}{n}\left(\frac{\Delta f_0 [(\alpha/n) - \Delta f_0] - (\frac{1}{2}\Delta_0)^2}{(\frac{1}{2}\Delta_0)^2 + [(\alpha/n) - \Delta f_0]^2}\right) \quad (12.17)$$

$$\epsilon_2 = \frac{\alpha}{n^2}\left(\frac{\frac{1}{2}\Delta_0}{(\frac{1}{2}\Delta_0)^2 + [(\alpha/n) - \Delta f_0]^2}\right) \quad (12.18)$$

When the conductivity is large, $\epsilon_2 \gg \epsilon_1 \gg 1$, the shift is completely determined by sample geometry so that

$$\Delta f_0|_\infty = \alpha/n, \quad (12.19)$$

which provides an independent experimental measure of the depolarization factor.[150]

The theory has been further developed by Cohen et al[153] who have solved the cavity perturbation problem for the surface impedance region where the above analysis fails. They obtain for the conductivity

$$\sigma^{\frac{1}{2}} = \left(\frac{9\pi^2}{2^5}\right)\frac{\beta^2}{\alpha}\left(\frac{b}{c}\right)\left(\frac{f^{3/2}}{\Delta}\right) \quad (12.20)$$

where β is the experimentally observed fractional shift in resonance frequency

$$\frac{f - f_0}{f_0} = \frac{\alpha}{2},$$

α is the filling factor $= 2V_c$; V_s being the volume of the sample and V_c that of the cavity. The depolarization factor

$$\eta = \left(\frac{b}{a}\right)^2\left[\ln\frac{2a}{b} - 1\right]$$

and $\Delta/2$ is again defined as the change in cavity half-width at half power, divided by f, c is the velocity of light; $\omega = 2\pi f$. The sample is dimensionally approximated by an ellipsoid (see eq. 12.14). In practice, it takes the form of a cylinder of length $2a$ and of cross-sectional diameter $2b$.

A suitable sample holder employed by Khanna et al.[152] consisted of a slotted styrofoam pellet inserted into a thin-wall polystyrene tube connected to a gear arrangement which allowed turning of the sample from outside. The pellet was trimmed so that there was no detectable shift on rotation with no sample in the cavity. With the sample needle axis perpendicular to the cavity E field, the depolarization factor is of order unity, resulting in no observable shift and minimal loss of Q.

The method has been applied,[154] *inter alia*, to the measurement of the transport properties of 1-dimensional semimetals (see Chapter 21). Especially if the skin depth, or the penetration distance, is small, these methods essentially measure a surface impedance which may differ from the bulk impedance (resistance) of the sample, because of the coupling of conduction electrons to localized surface states. An enhanced surface impedance due to this effect, involving localized electronic states in the oxide layer next to the metal, has been reported by Halbritter.[155]

Sayed and Westgate[156] point out that the Trukhan as well as the Eley techniques use an impedance matching transformer so that, effectively, it is the Q of the cavity plus transformer which is being measured. This may introduce errors up to a factor of five.[156] A technique developed by Sayed and Westgate is claimed[156] to be free of this error as well as the difficulties involved in calibrating the cavity by means of reference samples of known, but relatively high mobility. Their method is claimed[156] to have a detection limit of 0.07 cm^2/V sec.

Hardy and Berlinsky[159] have devised a method whereby the sample is located so as to act as the central conductor in a coaxial transmission cavity exhibiting a Q-factor of

$$Q = \frac{2b}{\delta} \ln \frac{R}{b} \qquad (12.21)$$

where R is the resonator radius and b that of the sample. The skin (penetration) depth δ is given by

$$\delta = \frac{c}{2\pi\omega\sigma} \qquad (12.22A)$$

where c is the velocity of light *in vacuo* and ω the angular frequency of the applied electric field and σ the surface conductivity; $1/\sigma\delta$ thus is the effective surface impedance.

Another variant, devised for the study of non-linear electron transport phenomena in TTF:TCNQ[160], uses a very low value of applied field to minimize Joule heating; the non-linearities of the specimen are used to mix the fundamental and its second harmonic, giving rise to a dc voltage appearing between the ends of the filamentary sample. The method is based upon prior work by Schneider and Seeger.[161]

The application of microwave Hall measurements to highly conducting organic semimetals *via* the Faraday rotation of the plane of polarization of an incident electromagnetic wave is discussed by Ong and Portis.[157] Reviews of microwave Hall and conductivity methodology have been published by several authors.[158]

12.14 Other Contact-less Methods

The problems associated with providing electrically invisible, or at least ohmic, contacts (see Sections 3.7, 6.10 p. 373, 12.15 and 15.12) has prompted development of several methods, apart from the microwave arrangements discussed in Section 12.13.

In an arrangement proposed by Brenner and Slight,[162] current is supplied to the solid sample under test for the purpose of electrolysis, *via* two electron beams or glow discharges, the sample forming the septum between an anodic and a cathodic compartment. While this necessarily results in carrier injection, it nevertheless should be a promising, and interesting, method for measurements on small, high resistivity, specimens.

Long has used[163] a somewhat related method to measure the conductance of very small solid samples: he employs the electron beam in a commercially available electron scanning microscope to provide a moveable current source to map the surface potential distribution on a crystal face; Ag paint provides a return path to ground and secures the sample.

The conductivity of very small filaments or needles can be determined by comparing the energy distribution of field-emitted ions originating from the sample with that originating from a Pt wire.[164]

The surface potential decay in naphthalene has been studied by Campos and Giacometti using a negative corona discharge to contact the sample.[25] It has also been suggested that the electric field-induced changes in surface impedance can be used to study surface states and surface properties.[165]

An apparatus to measure surface charge densities with a precision of 1%, range 5×10^{-11} to 10^{-6} C/cm^2 has been developed by Sessler and West.[166]

12.15 Cell Arrangements Using Metallic Contacts

Carrier transfer through and over contact potential barriers[167] has been extensively studied and here we can only refer to some work in this field. For further discussion of carrier injection from electrodes see Sections 3.7, 6.10 p. 374, 12.15 and 15.12.

Metal contacts of Pb, Sn, Ag, Cu, Ti, Ni, Mo and W covered with a contaminant film have been studied by Tamai;[168] currents through the film are considered to involve Schottky (thermionic) transfer rather than Fowler-Nordheim (field) emission of carriers: the contact resistance drops with increasing temperature as to be expected from the Schottky theory. The contact resistance is lowest if the voltage drop across the contact potential barrier equals the barrier height. See also discussion in Section 15.12.

Chen finds[165] that the dielectric relaxation time τ of a dielectric film covering the electrode determines whether the contact will be ohmic or not; if the carrier transit time in the applied field E remains $<\tau$ the contact is able to supply a space charge limited current and behaves ohmically; if this condition is not met, then the contact is non-ohmic. The model involves two parameters: the injection current I, assumed to relate to E by

$$I = kE^p \tag{12.22B}$$

where k is a constant of proportionality and p a parameter to be obtained from step voltage vs. current measurements.

Other studies of film-covered contacts[170] suggest that electron transfer through such contacts is by inelastic tunneling (see Section 15.6) and/or phonon-assisted hopping (see Section 15.7) and primarily involves the presence of localized electronic states.

Fabish et al. have suggested[109] that a metal-to-polymer contact involves charge injection *via* localized polymer energy states which happen to be located near the Fermi Level of the metal. It is not clear whether these energy states are extrinsic or intrinsic; also the role of the probably incomplete nature of the metal-to-polymer contact remains to be clarified. Cottrell et al.[110] have queried these results. One difficulty in such experiments is the frequently obscure prehistory of the polymer involving mechanical and other stresses, locally oxidized regions, traces of plasticiser and the like (see also the discussion in Section 12.9). A method to alleviate or even to avoid the film barrier by introducing metals such as Cu from a contact directly into a film has been proposed.[171]

A complete cell design for dark conductance, photoconductivity and photovoltaic studies for solid discs and flakes is described by Rosseinsky et al.[172]

12.16 Electrolytic Contacts[173]

In order to observe carrier injection into crystals from an electrolyte, it is necessary that either (i) the RedOx potential of the electrolyte be of similar energy to the crystal conduction or valence band energies or (ii) that an amount of external energy (possibly in the form of singlet or triplet excitons) be available to aid injection.[174] Figure 12.13 aids in determining approximately the energy relationships between the crystal and the electrolyte.[174] These considerations are equilibrium thermodynamic considerations only. Kinetic effects, not considered here, may also be important in determining

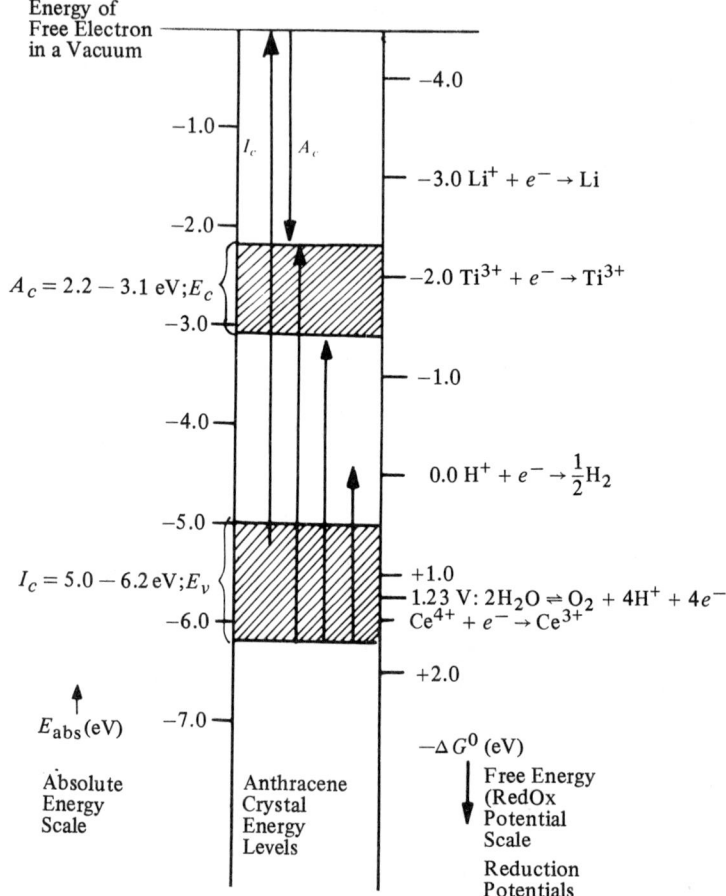

Fig. 12.13 The relationship of the absolute energy scale to the electrochemical ReDox potential scale. After Lyons.[174]

whether or not carrier injection will occur. The reference state is the standard electrode potential of the normal hydrogen electrode. (Reduction potentials are used). Noyes[175] and Lohmann[176] have calculated the standard free energy change for the half-reaction of the normal hydrogen electrode

$$H^+ + e^-_{(\infty)} \rightleftarrows \frac{1}{2}H_2 \qquad (12\text{-}23)$$

to be -4.5 eV on the absolute energy scale; $e^-_{(\infty)}$ represents an electron in a vacuum). Thus, Fig. 12.13 allows comparison of the RedOx potential scale with the absolute energy scale.

The position of the Fermi level in the semiconductor, which causes the conduction band edge energy E_{cb}, in the surface and in the bulk to have equal energy, defines the flat-band potential E_{fb} which thus can be determined from electrochemical measurements,

$$E_{cb} = eE_{fb} - kT \ln D/S \qquad (12\text{-}24)$$

where e is the electronic charge, D the donor density and S the effective density of states at the bottom of the conduction band. E_{fb} necessarily is a function of the pH of the electrolyte.

Lohmann[177] studies the hole injection from an electrolytic contact into a typical molecular crystal, vis., I_2. At the interface I_2/solution of iodine radical cations a positive space charge develops from which an applied electric field can inject holes into the solid. This should be the case whenever a molecular crystal interfaces with a solution of its radical ions. Thus, e.g., holes may be injected into naphthalene single crystals from a RedOx couple[178] such as Ag^{++}/Ag^+ in 7.5 N HNO_3 or Ce^{4+}/Ce^{3+} in 15 N H_2SO_4; electron injection results from a Na/K eutectic which is liquid at room temperatures;[178] the current then is space charge limited. Liquid electrodes, also resulting in space charge limited currents, are also reported[179] with p-chloranil single crystals; the authors claim[179] that the current does not depend on the carrier injecting electrode.

In the case of anthracene, Lohmann and Mehl[180] report that high voltage pulses yield current pulses of about 200 Å/cm^2: the surface of the crystal becomes degenerate and (electronically) metal-like. The hole injection rate does not appear to depend on the proton partial pressure at the anthracene surface.[181] They deduce[180] that a strongly oxidizing or reducing system in contact with the solid electrode bends the top of the valency band and the bottom of the conduction band within the surface region into close proximity of the Fermi level. The initially insulating surface thus becomes degenerate and exchange current densities result[180] which are of the same order as those obtained with metal electrodes. Since the electron exchange current[182] depends primarily on the overlap integral between the donating and accepting species one may expect that the electron exchange between an absorbed reduced layer and oxidized bulk species is large.[183]

Water has been used to discharge holes at anthracene crystal electrodes.[184] In Mehl and Hale's work,[184] the nature of the species injecting the holes into the front side of the electrode was varied so that the RedOx potential of the injecting couple changed. Saturation currents depended on the relation of its RedOx potential to the potential of the valence band of anthracene on the

Fig. 12.14 *p*-anilinobenzenesulphonic acid (1) and oxidation product (2).

RedOx scale. The back surface was assumed to discharge holes spontaneously. The back contact played no role in the kinetics of current flow. The use of aqueous contacts in the presence of oxygen results in the oxidation of anthracene to anthraquinone and other products with resultant degradation of the crystal (back) surface. Back surface reactions are difficult to observe because the saturation current on the insulator is limited by the front surface. Lyons and coworkers report[183] that, with *p*-anilinobenzenesulphonic acid (1) in the back electrode, all discharged holes oxidized, and so oxidation either of the anthracene crystal or of water does not then occur. The standard RedOx potential of the oxidation of (1) to (2) in Fig. 12.14 is +0.85 V (N.H.E.) in 10^{-3} mole 1^{-1} sulphuric acid. The standard RedOx potential of the water reaction

$$\frac{1}{2}O_2 + 2H^+ + 2e \rightleftarrows H_2O \tag{12.25}$$

at pH 3 is +1.05 V (N.H.E.). Energetically the holes reaching the back surface have sufficient energy to oxidize either the water or the compound (1),[186] with water possibly being expected to be oxidized preferentially since the usual assumption is that, for all electrolyte contacts, water forms the primary contact layer to the crystal.

For high injection efficiency, de-oxygenation and high stirring rates are essential. This is ascribed[183] to the existence of a reaction layer of small but finite thickness between the crystal and the oxidizing species.

All evacuable cell arrangement incorporating electrolytic contacts has been developed[183] by Lyons and McGregor.

Electron transfer reaction at film-covered electrodes have been studied[187] theoretically by Dogonadze et al., especially as the exchange current density is affected; for the important case of impurity assisted resonance tunneling it is shown that impurities located at the center of the film dominate the charge transfer.[187]

The mean neutralization distance measured from the electrode-solution interface, is reported to be a few 10 Å if the electron passes the surface barrier by a tunneling mechanism.[270] Electrodes with covalently attached monolayers have been studied by Diaz and Kanazawa.[188] Thus, the initially formed structures of alkylsilyl derivatives on metal oxide surfaces contain some unreacted labile alkoxide groups. The final product may have some polymeric nature. When the electroactive species are attached to the surface *via* linear alkylsilyl derivatives, they appear to exist in various chemically different environments. Interconversion of the electroactive moieties among the various environments is slow and depends on the folding behaviour of the linear alkylsilyl derivatives. The net effect is that RedOx reactions become electrochemically quasi-reversible. This behaviour can influence the rate of charge transfer. When the monolayer contains an electroactive center, this may act as a mediator, facilitating electron transfer to an electroactive species in the solution.

12.17 Surfaces

Surface structures and surface states are discussed, *inter alia,* in Section 1.10, 3.8, 10.1, 10.4, 12.9, and 12.11. Several reviews on this very extensive topic are available.[189]

Because the surface marks a discontinuity in an otherwise infinite lattice, new wave functions are allowed; this leads to energy states localized at the surface. If these states occur at the same energy as existing energy levels of the bulk they become surface resonances and are not easily distinguishable from bulk states. If, however, they occur in a bulk band gap, they are easily identified and, can play an important part in the overall electronic properties of interfaces.

Should a chemical reaction take place at the surface, then non-equilibrium, excess carriers may be generated in the bulk as well as in the surface region of the solid. This appears to have been studied so far only in the case of covalent semiconductors such as Ge and CdS;[190] it should be interesting to explore this for molecular crystals.

Since the surface molecules are less strongly bonded than those in the bulk of the solid, surface diffusion is facilitated. If a clean surface is formed with fine grooves in it, the atoms or molecules at the crests will tend to migrate into the valleys to reduce the surface area of the specimen. The driving force is the surface energy; the potential barrier to be surmounted is the (activation) energy Q_s needed to move one surface molecule (atom) to a neighboring site. Q_s generally is slightly below that for bulk diffusion and about half to one third of the energy of vaporization, and about 0.1 to 0.3 of the adsorption energy, depending on the adsorbate. Surface molecules should thus exhibit melting points below those for the bulk solid though there is no real experimental evidence for this effect.

The surface of a crystal is not just the end of a perfect crystal. Rather, the atoms and charges arrange themselves differently from the bulk material: a surface, like a defect within the bulk of the crystal, breaks the fundamental periodicity. Localized configurations—like atoms, molecules, defects, and surfaces—present problems for calculations that depend on crystal periodicity. One solution, is to match bulk pseudopotential solutions to surface-state-like solutions.[191]

The Fourier expansion for the pseudopotential in terms of reciprocal lattice vectors $G = m2\pi/a$, where m is an integer and a the (one-dimensional) lattice spacing, is given by[191]

$$V(r) = \sum_G V(G)S(G)e^{iGr} \tag{12.26}$$

The coefficients $V(G)$ are called "form factors" of the potential; $S(G)$ is the "structure factor" that locates the atoms in the unit cell; it can be determined by X-ray or neutron diffraction analysis. For a discussion of pseudopotential techniques and theory, the literature should be consulted.[191]

Another approach is to retain the periodicity by introducing the concept of "supercells." A supercell is a large "unit" cell of a crystal that can contain many lattice sites of an ordinary crystal. The structure factor, $S(G)$ of eq. (12.26) accounts for the positions of the atomic cores, and one can use it to describe supercells with any arrangement of atoms. To study surfaces one can use a supercell that has the regular crystal periodicity in the x- and y-directions, but has slabs of finite thickness with space between them in the z-direction. The "crystal" consists of an infinite number of slabs of material stacked in the z-direction, with space to isolate them. The new geometry retains the periodicity but the much larger unit cell introduces new reciprocal lattice vectors, G, and requires extrapolation of the $V(\bar{q})$ curves, $\bar{q}$ being the wave vector.[191] It then appears that the surface "dangling bonds" become linked, the surface "heals" and smoothes out. The surface disturbance of the periodicity extends only a few layers into the crystal; it is likely that impurities can migrate into the crystal through these channels.

Surface states would not show up in the total charge density unless they had amplitudes comparable with bulk states. It is possible to search for these states by comparing the electronic states (using a density-of-states function) at the surface layer with those of the bulk. Analysis shows that there is a prominent surface state, whose energy lies in the forbidden semiconductor gap region.[191] This state has been called the "dangling bond" surface state. Other surface states appear in the second or third layer and are sometimes referred to as "back-bond states."[191] More complicated systems may arise.

Metastable activated surface states have been shown to revert to stable states by electron capture; the (thermal) activation is said[192] to form an intermediate step in the sequence. The electron capture cross-section is reported to be about 10^{-16} cm^2.

There are many powerful methods, mainly of an electronic nature, which are applicable to surface studies, like Auger, LEED, ESCA, and others,[273] which are beyond the scope of this book. An interesting method specifically aimed at studying surface carrier trapping and changes in the local surface potential of semiconductors, based on measurements of the capacitive photocurrent upon illuminating the sample with a laser beam, is due to Zuev and Popov.[193]

Very sensitive inelastic-electron-tunneling spectroscopy (IETS), gives information on the vibrational structure of organic monolayers on surfaces. Electrochemical methods for surface studies are reviewed by Gerischer et al.[195]

Chemisorption processes which are associated with charge transfer across the interface, can be studied with great accuracy, even when only submonolayer amounts are adsorbed. The optical reflectance of metal electrodes measures sensitively overlayers and the electric charge on the surface, both of which can be controlled by the voltage applied to the electrochemical cell.

12.18 Effect of Gases. Adsorption and Chemisorption

There is now a vast literature[196] on the way in which gases adsorb on clean surfaces, involving physisorption, chemisorption and chemical reaction. Sometimes the adsorbed gases form superstructures with very large periodicities. It seems unlikely that this can be due to long-range interactions between the gas atoms themselves: it is more probably due to some type of energy resonance produced in the solid substrate.

Chemisorption of gases generally constitutes the intermediate stage of catalysis. The catalyst must not be too reactive, since it will at once produce reaction products. On the other hand it must not be too indifferent: it must be able to polarize the adsorbate so that it presents just the right degree of activity when another gas is admitted to the system. In the vicinity of a surface exists an inhomogeneous, static electric field arising from the array of ions or from peculiar spatial distribution of the electrons of the solid.

The adsorbed atoms are polarized by the surface electric field and acquire dipole moments, which result in a decrease of the work function of the surface on which the atoms are adsorbed. The polarization causes lowering of the energy of the system, given by

$$\Delta E = -\frac{1}{2}\alpha F^2, \qquad (12.27)$$

where α is the polarizability, or the dipole polarizability of the atom and F is the electric field strength at the center of the atom. The polarization energy ΔE contributes to the interaction energy between the surface and the atom, it is part of the adsorption energy. In the theory of intermolecular forces, it is called the energy arising from the induction effect.

This matter has been treated in detail by Nakamura and Tatekawi.[198] The discussion of chemisorption which follows is based upon that by Ahmed.[199]

Let E be a gas species capable of being adsorbed on the same semiconductor surface in the three forms E^0, E^-, and E^+, i.e., neutral, acceptor and donor forms, respectively. In electronic equilibrium on the surface, a fraction of the total number of acceptor levels (A) will be occupied by electrons and a fraction of the total number of donor levels (D), will be depleted of electrons. Thus, of the total number of particles N of the given species chemisorbed on a unit area, a certain number N^0 will be in a state of weak bonding to the surface, while a number N^- will be strongly acceptor bonded to the surface and a number N^+ will be strongly donor bonded to it. Thus,

$$\eta^0 = \frac{N^+}{N}, \qquad \eta^1 = \frac{N^0}{N}, \qquad \eta^2 = \frac{N^-}{N}, \qquad (12.28A)$$

One has

$$\eta^0 + \eta^1 + \eta^2 = 1 \qquad (12.28B)$$

The densities of particles sorbed, or bonded, to the surface are such that η^0, η^1, and η^2 characterize the relative contents of the positive, neutral and negative forms of chemisorption at equilibrium, respectively. The probability that the chemisorbed particle will be in a particular state (characterized by a certain type of bond with the surface), or the average relative lifetime of the chemisorbed particle in the corresponding states, is thus described.

Ahmed[199] then obtains (see Fig. 12.15) the variation of the relative content of the positive, neutral, and negative (η^0, η^1 and η^2 respectively) forms of chemisorption with the change in Fermi level position ϵ^-_s. As the distance of the Fermi level from the valence band is increased, the relative content of particles bound to the surface by a donor bond decreases. The value of η^1 which characterizes the relative content of the weak form of chemisorption evidently passes through a maximum as the Fermi level is steadily displaced.

The semiconductor surface becomes charged as a consequence of strong chemisorption. Hence, the degree and nature of the surface charge arising in chemisorption depends not only on the nature of the chemisorbed particle

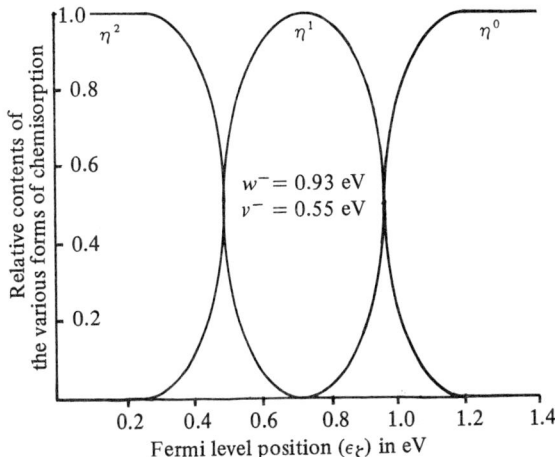

Fig. 12.15 The relative content of the various forms of chemisorption versus Fermi level position. η^2, η^1, η^0 refer to the negatively, positively charged and neutral forms, respectively. After Ahmed.[199]

and the degree of surface coverage, but also on the position of the Fermi level, i.e., on the state of the system as a whole.

When the adsorbate is capable of being adsorbed in three forms, neutral, singly negatively charged and doubly negatively charged then it can be shown[199] that, as the distance of the Fermi level from the bottom of the conduction band and its proximity to the midgap increases, the relative content of the doubly ionized particles decreases monotonically, and the relative content of the singly negatively charged particles increases to a maximum value. In this region the negatively charged form of chemisorption is the predominant form. On displacing the Fermi level even further, the relative content of the neutral form starts to increase monotonically and becomes the dominant one.

Some species, notably ions, are specifically adsorbed; this alters the surface properties by lowering the work function and, specially if there are vacant d-electronic bonds, these are affected.[197]

The effect of the presence of O_2 on the conductivity of compactions has been studied[200] by Berets and Smith who find that oxygen partial pressure and apparent conductivity are inversely related. Exposing (semi-metallic) polypyrrole to NH_3 gas at room temperature drops the conductivity by a factor of ten; acceptor-doped polyacetylenes exhibit like behavior.[201] The polypyrrole interaction appears to involve physical adsorption only because the effect is reversible upon pumping off the gap. The polyacetylenes, however, are said[201] to react with NH_3 at the carbonium sites along the polymeric chains. Exposing insulating, undoped, N-methylpyrrole to Br_2 vapor rises the conductivity twenty-fold; this is probably due to the formation of a (surface-) charge transfer complex.

A high electron affinity gas has been shown[202] to raise the conductivity of 9-nitroanthracene exponentially with its vapor pressure while adsorption of a gas of low ionization potential has the converse effect.

The effect of oxygen on the dark conductivity of metal-free phthalocyanine has been studied in detail.[203] Surface interactions of gases with organic solids have been reviewed by Margulis and Boguslavskii.[204]

The nature of the gaseous ambient also affects measured values of the ionization potential, probably because of chemisorption (see Section 14.3).

As mentioned in Section 12.17 a powerful tool for the study of adsorbed molecules is IETS, in which the energy distribution of the electron tunneling current through such structures is measured.[279]

12.19 Noise Measurements

Charge fluctuation phenomena have been extensively studied in connection with semiconductor (inorganic) devices,[205] and at contacts,[205] in solution[206] and in membranes.[207] Only isolated studies appear to have been made on organic liquids such as pure acetone.[208] Noise studies on charge transfer complexes have recently been proposed by Farges and Gutmann,[209] who suggest that they might yield a method for the direct measurement of recombination phenomena. Such, of course, need not be confined to charge transfer complex formation; the study of the stochastic processes involved in carrier recombination in various photoeffects, but especially in photovoltaic and photoconductive effects, should prove valuable. Noise arising from fluctuations in the ionization state of impurities in inorganic semiconductors has been reported[210] as well as noise arising from electrochemical reactions[211] especially in primary cells.[212]

The noise arising in an ideal resistor, and in the absence of any irreversible processes, is white,[213] i.e., it is frequency independent and given as Johnson noise in terms of the Nyquist eq. (12.29)

$$[V^2(t)] = 4kTR\Delta f \qquad (12.29)$$

where $[V^2(t)]$ refers to the mean value of the instantaneous terminal voltages as a function of time t, T is the resistance of the resistor, k Boltzmann's constant, and T absolute temperature. Δf stands for the pass band of the measuring apparatus.

Since one may define a "noise mobility" μ_n:

$$\mu_n \equiv \frac{e\tau_0}{m} \tag{12.30}$$

in terms of the molecular relaxation time τ_0 and the carrier mass m, e being the electronic charge, noise measurements in terms of an equivalent noise conductivity allow determination of either m or μ_n because the relaxation time is readily obtainable from dielectric measurements. The noise conductivity is the reciprocal of that resistivity of a resistance entering in the Nyquist eq. (12.29) which produces the same value of $[V^2(t)]$. Thus, one measures the noise voltage produced by the unknown and then replaces it by an "ideal" resistance—a good resistance decade box serves quite well if it uses metallic resistors only—which is adjusted until the same value of noise voltage is obtained. The Nyquist eq. (12.29) holds for the equilibrium conditions only; it is derivable from equilibrium thermodynamics. As soon as current flows, or if any non-equilibrium processes take place either in the bulk of the sample, or, quite likely, at the contacts, excess noise will be observed[213] which usually obeys an f^{-n} relationship[216] with n often equal to unity. In liquids, under conditions of current flow and especially in ionic solutions, an f^{-1} noise component has been reported, claimed[214] to be due to the stochastic nature of the ionic diffusion. This is also a promising field which ought to be explored.

Noise voltages are usually measured by means of a noise spectrograph as illustrated schematically in Fig. 12.16.

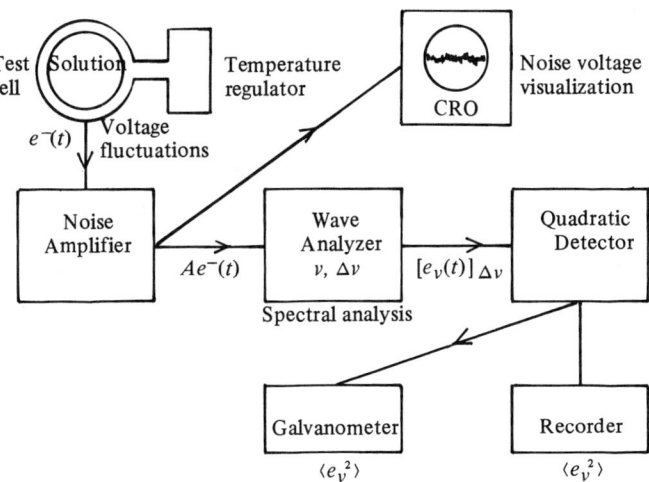

Fig. 12.16 Schematic diagram of a noise spectrograph. After Vasilescu et al.[215]

12.20 Purification

A new zone melting apparatus said to be specially suited for the purification of organic solids, has been devised by Ishizuka.[217] He points out that very precise heat control is essential because any local temperature changes cause changes in the widths of the molten zones, resulting in irregular impurity transfer and inefficient refining.

Zone refining of anthracene has been further perfected.[218] A commercial zone refiner is described by Munits.[219] Boguslavaskii and Lozhkin point out[220] that anthracene contains a great deal of dissolved oxygen which cannot be removed by zone refining. Dislocation-free single crystals of semiconductors are said[221] to result from an elaboration of the classical procedure: an after-heater, or a reflecting screen of length not less than the length of the molten zone is located at a distance not less than $R/4$ above or below the solid-liquid interface, where R is the radius of the growing crystal. This lowers the axial temperature gradient below the critical value above which dislocations and other defects are likely to be formed. For the Si(111) surface and $R = 2.5$ cm, this gradient is $-53.1°$ C/cm.

Zone refined coronene is said[222] to contain no disorder, and thus is free of dislocations. Zone refining of organic compounds, and in particular of biphenylstilbene, is discussed in detail by Noyima et al.[223]

McLevige et al. use a chemical etch to remove successive surface layers of a solid in order to determine the electrically active impurity distribution in the crystal;[224] the effectiveness of impurity removal is followed by Hall effect measurements.

For controlled doping, as well as for synthesis of crystalline charge transfer complexes, it is sometimes of advantage to co-sublime two compounds. An apparatus which has been successfully employed[225] for this purpose is shown in Fig. 12.17.

Affinity chromatography[226] offers selective isolation and purification of a desired macromolecule from a mixture. The method takes advantage of the inherent functional specificity of binding by a biological macromolecule: a specific enzyme inhibitor, for example, is covalently attached to a gel column whose pores are sufficiently large to admit the enzyme of interest. Enzyme added to the column will then bind to the column-bound inhibitor while other unrelated proteins will pass unaffected. To remove the column-bound enzyme, it suffices to elute with a potent soluble enzyme inhibitor, which detaches the desired enzyme from the column.[226]

Another powerful method for the separation of ampholytes, especially proteins, is isoelectric focussing.[227] Every protein or other ampholyte has an isoelectric point, pI, which is a pH value at which it may be dissolved to

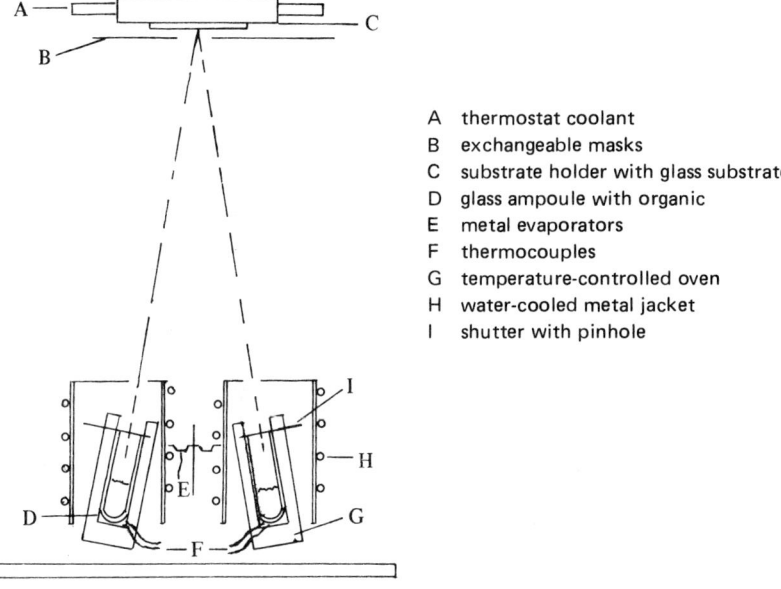

Fig. 12.17 Schematic diagram of a cosublimation apparatus with an evacuated chamber. After Lyons.[225]

yield a net charge of zero. If the protein is added to a solution with a higher pH it loses protons and becomes negatively charged. Conversely, if the protein is in surroundings with a lower pH it will capture protons and become positively charged. If one arranges a buffer with a varying pH, that is, increasing from one end to the other, a pH-gradient is obtained. Such a gradient has always the lowest pH value at the anode and the highest pH at the cathode.

When a sample of proteins with isoelectric points within the pH of the gradient is added, the protein molecules will obtain different charges. The charge of every individual protein will be determined by its pK and the pH corresponding to its position. Applying a voltage results in each molecule migrating toward the pH value where it is isoelectric, that is, where its net charge is zero. The proteins are thus exactly focussed at the point where the pH is equal to the pK.

This process is illustrated in Fig. 12.18, which shows how three proteins, marked (pI_1), (pI_2) and (pI_3) are focussed in a column. Diffusion counteracts the electrofocussing. As the diffusion coefficient is inversely proportional to the molecular weight, better focusing is obtained with higher molecular weight proteins. Proteins with as small a difference as 0.02 pH units at the isoelectric point have been separated.

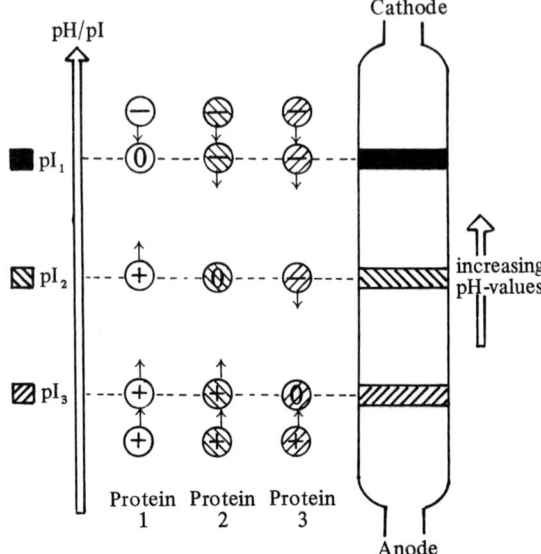

Fig. 12.18 Three proteins, marked pI_1, pI_2, pI_3, schematically electrofocussed in a column. Each of the three protein components is negatively charged above its respective point and positively charged beneath. For that reason it will move in the electric field until the position in the column is reached where the pH of the pH-gradient is the same as the isoelectric point for that particular protein. After Haglund.[227]

An adaptation of the well-known Cottrell filter for gas purification to electrostatically purity insulating liquids is described[228] by Kok et al.

In the context of purification, it has been pointed out by Pohl that, as far as highly conducting polymers are concerned,[229] high levels of electronic conduction are associated with the long-range delocalization of electronic orbitals of the polymeric molecule, which is only possible in some highly conjugated (e.g., polyacene) polymers. These *must*, moreover, be of a ladder or sheet-type structure. Only relatively few are known at present. Comparing this paucity of electronically effective compounds with the several million other available compounds which could serve as impurities but which are "bland" or ineffective electronically, one appreciates that the usual problem of electronically active impurities in covalent or ionic semiconductors is reduced by orders of magnitude when dealing with conductive polymers. Only the rare, highly and specially conjugated (known as eka-conjugated) types of polymer compounds having a *greater* degree of specific electronic conduction than the host polymer can increase the conductivity of a given conductive polymer sample.[229] However, ionic impurities may also exist which may considerably affect the conduction properties of the host polymer.

Considerable care must be taken to avoid this type of impurity in the molecular solids.

The presence of ionic impurities must be suspected if the conductivity is time dependent or tends to drop rather than to increase with increasing pressure.[229]

Ionized centers also are more effective scattering centers than are magnetic impurities though such become significant at low temperatures and are likely to affect the thermoelectric effects.[230]

Impurity effects in TCNQ adducts are reported on by Kamaras et al.[276] The effect of impurities on phase transitions caused by relatively weak electron–electron interactions in 3-dimensional arrays of linear conducting chains–the model for organic semimetals (see Chapter 21) is discussed by Abrahams[277] who finds that impurities tend to destroy the correlations between electrons on different chains, and thus cause 3-dimensional phase transitions.

12.21 Compacted Powders. Annealing. Pressure.

A stainless steel cell permitting the simultaneous measurement of conductance and thermo-emf of amorphous and polycrystalline samples is described[231] by Hadek; it is suitable for ambient gas pressures from 0.001 torr vacuum to 10^4 At pressure, and for temperatures ranging from 40 to above 300 K. It allows the sample resistivity to vary between 10^{-4} and 10^{11} ohm cm; a two-electrode method is employed.

A four-electrode cell, using the van der Pauw method already described, has been designed[232] by Cahen, based on a design by Hottman and Pohl.[233] It employs compactions in a controlled atmosphere and covers the temperature range 77 to 400 K. Both designs allow measurements under high pressure using Bridgeman anvils.

A four-electrode cell suitable for measurements on liquids, has been described by Greatorex and Stokes.[234]

The role of surface, interfacial and spurious space charge effects in conductivity determinations is discussed[235] by Kramer and van Ruyen.

An apparatus for the annealing of compactions at high temperature and in the presence of a strong electric field is described[236] by Pandey; its use is of advantage not only in order to relieve residual mechanical strains but also to remove spurious space charges which otherwise may prove very troublesome, see also a discussion[237] by Nakada and Ishihara. Annealing of polymeric films is reported to reduce the apparent conductivity because of increasing crystallinity.[238] Annealing of (horn) keratin has been shown[239] to raise the dielectric relaxation time and the resistivity. *In vacuo* annealing of

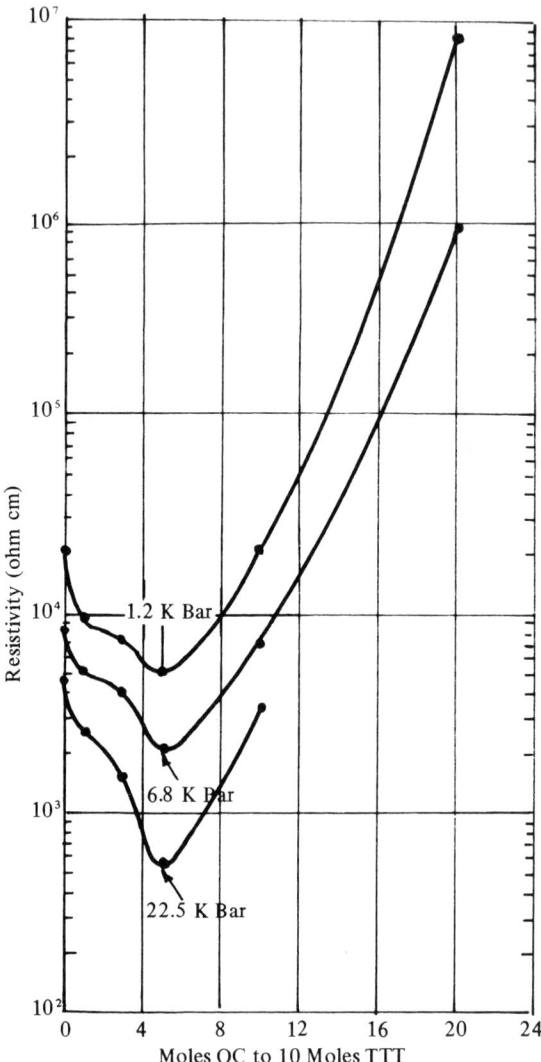

Fig. 12.19 Effect of OC additive concentration and preparation pressure on the room temperature resistivity of TTT. After Krikorian and Sneed.[243]

polynaphthalene is reported[240] to increase the trap concentration due to an increase in the size of the polymeric spherulites.

The dependence of measured resistivity and of Seebeck coefficient values on the preparation pressure of compactions is illustrated in Figs. 12.19 and 12.20, using the charge transfer complex between tetrathiotetracene (TTT) and o-chloranil (OC) as an example.

THIN FILMS

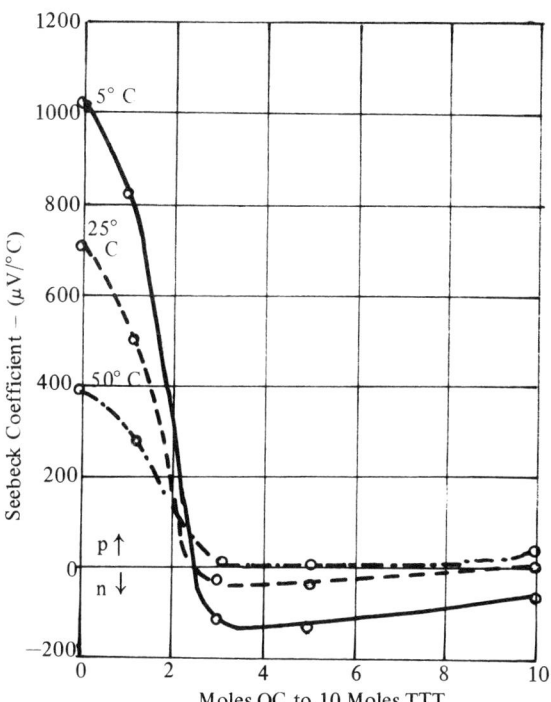

Fig. 12.20 Effect of OC concentration on the Seebeck coefficient of TTT, (preparation pressure 1.2 kbar). After Krikorian and Sneed.[243]

The results obtained on single crystals and on compactions of e.g., TCNQ complexes, have also been compared by Lane et al.[244]

Coleman has developed[245] at least a qualitative method to evaluate the intrinsic conductivity of compactions by eliminating, to some extent, interparticle contact resistances and, in the case of anisotropic solids, the low conductivity contributions to the conductivity tensor. This voltage-shortened compaction (VSC),[245] uses a four-electrode method but coats the region between the potential contacts with a thin layer of silver paint, thus partially shorting the voltage leads, but also forming a microcrystalline layer embedded in a conductive matrix. Though the resistance values at any one temperatures have no meaning, their temperature dependence is significant.[245] The conductance of a mixture of conducting and of insulating spheres has been studied theoretically by Ottavi et al.[275]

12.21 Thin Films

Charge transport through thin films is discussed in Chapter 15, especially in Section 15.5 and 15.6; such transfer is thought to be mainly due to phonon-assisted hopping plus a contribution from tunneling;[247] carrier injection from the contacts (see Section 15.12) has also been observed, e.g., in thin films of polycrystalline tetracene and of p-quaterphenyl.[248] A thin film may be epitaxially grown on a crystalline substrate; the crystal lattice of the substrate surface acts as an atomic template for the crystal growth of the overlayer film.

When certain crystalline materials are deposited on a flat amorphous surface, the result is a "textured" polycrystalline film: the deposited film consists of small crystalline grains with a particular crystal plane parallel to the substrate surface, but with the other two lattice directions randomly oriented from one grain to the next. The absence of long range order in the overlayer reflects the absence of order in the substrate. If one now artificially imposes long-range order by producing a relief pattern of grating lines with the right periodicity and profile on the amorphous substrate, the individual grains will tend to align themselves to produce a crystalline film with a single orientation. The periodicity of the grating may exceed atomic or molecular dimensions; it need to be only less than the characteristic grain size of the film to be deposited.[249] The method should be equally applicable to organic compounds though this has not yet been done. Crossed molecular beam vapor condensation techniques have been used[250] for the preparation of thin films of charge transfer complexes and other adducts.

Film deposition in an ultrahigh vacuum chamber, by means of an electron beam evaporator has been reported.[251] Structure and crystallinity of thin films have been studied by several authors.[252] The absorption spectra of perylene or of coronene evaporated films indicate an amorphous structure at low temperatures, gradually changing to polycrystalline with rising substrate temperature.[253] Tetracene films formed at 180 K are reported to exhibit a stable, noncrystalline and monomeric structure with a low trap concentration, while those formed at lower temperatures are said[254] to be mixtures of monomers and dimers with the concentration of localized trapping centers increasing with lower temperatures. The valency band is said[254] to split into localized states below 130 K.

Aging, or annealing, of thin polycrystalline films of p-terphenyl is said[246] to cause the disappearance of shallow traps which are replaced by deeper traps. The enormous differences, and uncertainties, which are liable to arise from different experimental arrangements are illustrated in Table 12.1. Surface-type and sandwich-type refer to films deposited by vacuum sublimation in cells of the kind referred to (cf., Section 3.5); "compressed" means compaction of a polycrystalline sample at 290 kg/cm^2. A recent comparison of resistivities measured with these two types of cells in a large number of metal-diphthalocyanines[255] agrees with Table 12.1.

Table 12.1 The Electrical Resistivity of Tetrathiotetracene. After Inokuchi et al.[241]

Specimen	Resistivity ρ (ohm cm) from $\rho = \rho_0 \exp(E/2kT)$ $T = 288$ K	Activation energy E (eV)	Electrode
Surface type	1.3×10^6	0.57	Al
Sandwich-type	4.2×10^{10}	0.41	Ag
Compressed	8.3×10^3	0.44	Cu
Matsunaga's[242]	10^4	0.46	–

12.23 Liquids and Melts

The methodology of conductance measurements on liquids has been discussed by Huang and Freeman.[256] A four electrode cell for such determinations allowing either constant voltage or constant current measurements is described[257] by Tamamushi and Takahashi, and a cell suitable for conductance measurements on liquids as well as on solids placed in a resonant cavity has been developed[258] by Andreev et al. It operates at 10^{10} Hz and allows measurements over a temperature range of 300 to 1000 K.

A measuring cell for liquid conductances has been patented by Zellweger.[259] Polarization effects in liquid conductimetry are discussed by Artamonov et al.[260] Microwave methods have been applied to liquid samples by Hallenga,[261] and by Loon and Finsy[262] who use it to measure the components of the complex permittivity, thus obtaining values for both conductance and capacitance.

Experimental aspects for impedance measurements on liquids are discussed[271] in detail by Barthel et al.; they also give a design permitting titrations and *in situ* purification.

References

1. H. C. Montgomery, *J. Appl. Phys.*, **42**, 2971 (1971).
2. B. F. Logan, *J. Appl. Phys.*, **42**, 2975 (1971).
3. J. T. McMullan, *Phys. Stat. Solidi*, **A4**, K-181 (1971).
4. L. B. Coleman, *Rev. Sci. Inst.*, **46**, 1125 (1975).
5. R. D. Westbrook, *J. Electrochem. Soc.*, **121**, 1212 (1974).
6. T. E. Philips et al., *Rev. Sci. Inst.*, **50**, 263 (1979).
7. I. G. Morchardt and L. R. Holland, *Rev. Sci. Inst.*, **46**, 67 (1975).

8. K. Lal and D. R. Pahwa, *Rev. Sci. Inst.*, **42**, 534 (1971).
9. R. S. Lakes and R. A. Harper, *Rev. Sci. Inst.*, **46**, 1583 (1975).
10. W. A. Philips, *Phil. Mag.*, **34**, 983 (1976); M. Vodenicharova and I. Amov, Bulgar. *J. Phys.*, **1**, 22 (1974); Sh. Ikeno, *Polymer J.*, **10**, 123, 231 (1978).
11. J. L. Jacquemin and G. Bordu, *Phys. Stat. Solidi*, **A-45**, 579 (1978).
12. R. and M. Meaudre, *Phys. Stat. Solidi*, **A-37**, 633 (1976).
13. J. R. MacDonald and D. R. Franceschetti, *J. Chem. Phys.*, **68**, 1614 (1978).
14. J. L. Beeby and T. M. Hayes, *Solid State Phys.*, **4**, 1757 (1971).
15. D. M. Burland and U. Konzelmann, *J. Chem. Phys.*, **67**, 319 (1977).
16. J. D. Williams, *Adv. Phys. Organic Chem.*, **16**, 159 (1978).
17. T. J. Sonnonstine and A. M. Hermann, *J. Chem. Phys.*, **60**, 1335 (1974).
18. G. Pfister, *Phys. Rev. Lett.*, **33**, 1474 (1974).
19. J. Mort et al., *J. Appl. Phys.*, **43**, 2285 (1972).
20. J. P. Dodelet and G. R. Freeman, *Canad. J. Chem.*, **55**, 2264 (1977).
21. H. Scher and E. Montroll, *Phys. Rev.*, **B-12**, 2455 (1975).
22. W. E. Spear, *J. Non-Cryst. Solids*, **1**, 197 (1969); Y. Kamura, *Chem. Lett*, 301 (1974).
23. R. H. Young et al., *J. Chem. Phys.*, **70**, 443 (1979).
24. R. Colson et al., in "Amorphous and Liquid Semicond. Proc. 7th Int. Conf.," 1977; W. E. Spears, ed., University of Edinburgh, Scotland, Cent. Indust. Consult. Liaison Bur.
25. M. Campos and J. A. Giacometti, *Appl. Phys. Lett.*, **32**, 794 (1978).
26. R. R. Cooper, *Phys. Rev.*, **B-19**, 2404 (1979); R. H. Friend et al., *Phys. (Paris) Lett.*, **39**, 134 (1978); M. Weger et al., *J. Chem. Phys.*, **71**, 3916 (1979).
27. L. B. Schein et al., *J. Chem. Phys.*, **71**, 3189 (1979).
28. W. Tantraporn, *J. Appl. Phys.*, **39**, 2012 (1968).
29. J. Hirsch and E. M. Martin, Kodak Ltd. Horrow, England Internal Report, 1971.
30. E. L. Frankevich and E. I. Balabanov, *Sov. Phys.-Solid State*, **7**, 570 (1965).
31. A. V. Vannikov, *Sov. Phys.-Solid State*, **9**, 1068 (1967).
32. Sh. Ikeno, *Polymer J.*, **10**, 123, 231 (1978).
33. P. Debye, "Polar Molecules," The Chemical Catalogue Co., N.Y., (1929), p. 89; C. P. Smyth, "Dielectric Behaviour and Structure," McGraw Hill, N.Y., (1955); C. G. Koops, *Phys. Rev.*, **83**, 121 (1951).
34. E. M. McCafferty, *J. Phys. Chem.*, **82**, 2044 (1978).
35. S. Bone and R. Pethig, *J. Chem. Soc. Faraday Trans.*, I. **74**, 720 (1978).
36. P. V. Giaguinta et al., *J. Chem. Phys.*, **68**, 2621 (1978).
37. A. V. Maksimychev et al., *Dokl. Akad. Nauk*, SSSR, **241**, 141 (1978).
38. P. A. Tick and D. Johnson, *Rev. Sci. Inst.*, **44**, 798 (1973).
39. J. Toler and J. Saels, DHEW (NIOSH) Publ. (U.S.), **77-176** (1977), via *Chem. Abst.*, **88**, 34029y (1978).
40. S. K. Khanna et al., *Solid State Commun.*, **16**, 667 (1975).
41. J. Mracek, *Acta Technica CSAV*, The Czechoslovak Academy of Sci., No. 4, 402 (1972).
42. B. Gestblom and E. Noreland, *J. Phys. Chem.*, **81**, 782 (1977); M. J. C. V. Gemert, *Philips Res. Repts.*, **28**, 530 (1973).
43. N. E. Hill, *J. Phys. Chem. Solid State Phys.*, **5**, 415 (1972).
44. N. K. Hindley, *Solid State Commun.*, **9**, 987 (1971).
45. A. K. Jonscher, *J. Mater. Sci.*, **13**, 553 (1978).
46. F. Gutmann and H. Keyzer, *Electrochim. Acta*, **13**, 693 (1968); *J. Chem. Phys.*, **50**, 550 (1969).

REFERENCES

47. H. Grassi and D. Vasilescu, *Biopolymers*, **10**, 1543 (1971).
48. S. Bone and R. Pethig, *Trans. Faraday Soc.*, 720 (1978); J. P. Farges and F. Gutmann, in "Modern Aspects of Electrochemistry;" J. O'M. Bockris and B. E. Conway, eds. **12**, 267 (1978).
49. H. A. Pohl, *J. Non-cryst. Solids*, **22**, 291 (1976); Y. Wadda, *Dielectr. and Relax. Molecul. Processes*, **3**, 143 (1977); A. K. Jonscher, *Nature* (London), **267**, 673 (1977); M. Davies, *Acta Phys. Polon.*, **A-50**, 241 (1976); M. Evans, *J. Chem. Soc. Faraday II*, **10**, 1442 (1979).
50. R. Pethig, *J. Biol. Phys.*, **1**, 198 (1973).
51. R. Pethig and K. Morgan, *Nature* (London), **214**, 266 (1967); K. Morgan and R. Pethig, *J. Materials Sci.*, **6**, 179 (1971).
52. V. N. Dobrovolskii and A. N. Kroleveta, *Fiz. Tekh. Poluprovodn.*, **12**, 2129 (1978).
53. G. Bergmann, *Phys. Rev.*, **B-19**, 3933 (1979); *Physics Today*, **32**, 25 (1979).
54. A. M. Hermann and J. S. Ham., *Rev. Sci. Inst.*, **36**, 1553 (1965); G. Perluzzo and J. Yahia, *Canad. J. Phys.*, **50**, 1379 (1972); G. De Mey, *Solid State Electronics*, **16**, 995 (1973); **17**, 977 (1974).
55. J. C. Manifacier, *Phys. Rev.*, **17**, 3926 (1978); J. C. Manifacier and K. K. Henisch, *ibid.*, 2640.
56. H. Boettger and V. Bryskin, *Phys. Stat. Sol.*, **B-80**, 569 (1977).
57. J. Volger, *Phys. Rev.*, **79**, 1023 (1950).
58. R. H. Bube, *Appl. Phys. Lett.*, **13**, 136 (1968); **14**, 84 (1969); G. H. Blount et al., *J. Appl. Phys.*, **41**, 2190 (1970).
59. K. Lipskis et al., *Phys. Stat. Sol.*, (a)-4, K-217 (1971).
60. A. Rembaum et al., *J. Phys. Chem.*, **73**, 513 (1969); A. R. Blythe and P. G. Wright, *Phys. Lett.*, **34a**, 1 (1971); D. Zosel et al., *Phys. Stat. Solidi*, **38**, 183 (1970); J. P. Farges et al., *ibid.*, **37**, 745 (1970); G. Perluzzo and J. Yahia, ref. 54.
61. G. Ya. Kolbasov and V. A. Tyagai, *Fiz. Tekh. Poluprov.*, **6**, 946 (1972).
62. V. M. Cottles and A. M. Hermann, *Rev. Sci. Inst.*, **44**, 334 (1973).
63. P. I. Voronyuk and O. I. Danilevich, *Fiz. Elektron.*, (Lvov), **13**, 33 (1976).
64. M. H. Cohen and J. Jortner, *Phys. Rev. Lett.*, **30**, 699 (1973); M. H. Cohen and J. Jortner, *Phys. Rev.*, **A-10**, 978 (1974); J. Jortner and M. H. Cohen, *J. Chem. Phys.*, **58**, 5170 (1973); M. H. Cohen and J. Jortner, "Proceedings of the Fifth International Conference on Amorphous and Liquid Semiconductors," Garmisch, Taylor and Francis, London, 1974, p. 167; M. H. Cohen and J. Jortner, *J. Phys.* (Paris), **35**, C4-345 (1974); I. Webman, J. Jortner and M. H. Cohen, *Phys. Rev.*, **B-13**, 713 (1976); L. Friedmann and N. F. Mott, *J. Non-Cryst. Solids*, **7**, 103 (1972).
65. I. Webmann et al., *Phys. Rev.*, **B-15**, 1936 (1978).
66. P. H. Chaikin et al., *Phys. Rev. Lett.*, **31**, 601 (1973).
67. D. D. Eley et al., *J. Phys.*, **E10**, 1220 (1977).
68. G. Beni, C. F. Coll, *Phys. Rev.*, **B-11**, 573 (1974), and references therein.
69. S. M. Girvin and G. D. Mahan, *Phys. Rev.*, **B-19**, 1302 (1979).
70. I. F. Shchegolev, *Phys. Stat. Solidi*, **A-12**, 9 (1972).
71. L. Friedman and M. Pollak, *Phil. Mag.*, **B-38**, 173 (1978).
72. P. L. Grady and S. P. Hersh, *Rev. Sci. Inst.*, **46**, 20 (1975).
73. P. M. Chaikin and J. F. Kwak, *Rev. Sci. Inst.*, **46**, 218 (1975).
74. L. G. Christophorou, "Atomic and Molecular Radiation Physics," John Wiley, London, 1971, pp. 521-584; C. A. McDowell, in "Physical Chemistry, An Advanced Treatise," Vol. 3, D. Henderson, ed., Academic Press, London, 1969, pp. 495-536; G. Briegleb, *Angew. Chem.*, (Int. Ed. Engl.), **3**, 617 (1964).

75. L. E. Lyons and L. D. Palmer, *Aust. J. Chem.*, **29**, 1919 (1976); *Int. J. Mass Spectrometry and Ion Physics*, **16**, 431 (1975); *Chem. Phys. Lett.*, **21**, 442 (1973).
76. H. Hotop et al., *J. Chem. Phys.*, **58**, 2373 (1973).
77. R. J. Celotta et al., *J. Chem. Phys.*, **60**, 1740 (1974).
78. R. A. Holroyd et al., *J. Phys. Chem.*, **83**, 435 (1979); R. A. Holroyd, *Ber. Bunsen Ges. Phys. Chem.*, **81**, 298 (1977).
79. S. Koizumi and Y. Matsunaga, *Bull. Chem. Soc. Japan*, **43**, 3010 (1970).
80. cf. H. S. W. Massey, "Negative Ions" in "Advanced Atomic and Molecular Physics," **15**, Dec., 1979, D. R. Bates and B. Bederson, eds. Academic Press, N.Y., 1979; F. M. Page and G. C. Goode, "Negative Ions and the Magnetron," Wiley-Interscience, N.Y., 1969.
81. P. C. Engelking et al., *J. Chem. Phys.*, **69**, 1826 (1978); M. Siegel et al., *Phys. Rev.*, **A-6**, 607 (1972); R. Celotta et al., *ibid.*, p. 631.
82. J. M. Caywood, *Mol. Cryst. Liq. Cryst.*, **12**, 1 (1970); J. Mort and A. I. Lakatos, *J. Non-Cryst. Solids*, **4**, 117 (1970); R. Williams and J. Dresner, *J. Chem. Phys.*, **46**, 2133 (1967); R. Williams, *Phys. Rev.*, **140**, 569 (1965); A. M. Goodman, *Phys. Rev.*, **144**, 588 (1966).
83. R. A. Holroyd and M. Allen, *J. Chem. Phys.*, **54**, 5014 (1971); see also W. Tauchert et al., *Canad. J. Chem.*, **55**, 1860 (1977).
84. E. C. M. Chen and W. E. Wentworth, *J. Chem. Phys.*, **63**, 3183 (1975).
85. L. E. Lyons and G. C. Morris, *J. Chem. Soc.*, 5182 (1960).
86. K. Kimura et al., *J. Am. Chem. Soc.*, **100**, 6564 (1978), and references cited therein; M. Kochi et al., *Bull. Chem. Soc. Japan*, **43**, 2690 (1970).
87. P. Schlotter et al., *Phys. Stat. Solidi*, **B-81**, 521 (1970).
88. K. Seki et al., *Bull. Chem. Soc. Japan*, **47**, 1608 (1974); S. Hino et al., *ibid.*, **48**, 1133 (1974); *Chem. Phys. Lett.*, **36**, 335 (1975).
89. T. P. Debies and J. W. Rabelais, *J. Electron Spectrosc.*, **3**, 315 (1974); L. Klasinc, *ibid.*, **8**, 161 (1976).
90. A. Bernas et al., *Chem. Phys. Lett.*, **17**, 439 (1972); **30**, 383 (1975); S. Noda et al., *J. Phys. Chem.*, **79**, 2666 (1975); J. Moan, *Chem. Phys. Lett.*, **18**, 446 (1973).
91. K. Seki and H. Inokuchi, *Chem. Plys, Lett.*, **65**, 158 (1979).
92. J. Bagchi and M. Chowdhury, *J. Phys. Chem.*, **83**, 629 (1979).
93. J. W. Verhoven et al., *Tetrahedron*, **25**, 3395 (1969); R. A. Mackay et al., *J. Amer. Chem. Soc.*, **93**, 5026 (1971); E. M. Kosower et al., *ibid.*, **94**, 986 (1972); see also ref. 92.
94. S. Hino et al., *Chem. Phys. Lett.*, **37**, 494 (1976).
95. D. Haarer, *Chem. Phys. Lett.*, **27**, 91 (1974); **31**, 192 (1975); I. Lindau and W. E. Spicer, *J. Electron Spectrosc.*, **3**, 409 (1974); C. R. Brundle, *Surface Sci.*, **48**, 99 (1975).
96. N. Karl and J. Ziegler, *Chem. Phys. Lett.*, **32**, 438 (1975); S. Hino and H. Inokuchi, *J. Chem. Phys.*, **70**, 1142 (1979).
97. W. Pong and J. A. Smith, *J. Appl. Phys.*, **44**, 174 (1973).
98. W. Berry, *J. Electrochem. Soc.*, **118**, 597 (1971).
99. Y. C. Chang and W. B. Berry, *J. Chem. Phys.*, **61**, 2727 (1974).
100. P. I. Belkind et al., Elect. Prop. Org. Solids Conf., Karpacz, Hungary, 1974.
101. H. Kawamura and H. Inokuchi, *Bull. Chem. Soc. Japan*, **45**, 710 (1972).
102. S. Hino and H. Inokuchi, *J. Chem. Phys.*, **70**, 1142 (1979).
103. S. F. Lin et al., *Phys. Rev.*, **B-12**, 4184 (1975); P. Nielsen et al., *Solid State Comm.*, **17**, 1067 (1975).

REFERENCES

104. W. F. Krolikowski and W. E. Spicer, *Phys. Rev.*, **B-1**, 478 (1970).
105. C. F. Gallo and W. L. Lama, *J. Electrostat*, **2**, 145 (1976).
106. I. Langmuir, *Trans. Amer. Electrochem. Soc.*, **29**, 125 (1916).
107. A. N. Frumkin and A. V. Gorodetskaya, *Z. Phys. Chem.*, **136**, 451 (1928).
108. A. Matsuda, *J. Res. Inst. Catalysis*, Hokkaido Univ., **27**, 31 (1979).
109. T. J. Fabish et al., *J. Appl. Phys.*, **47**, 930,940 (1976); T. J. Fabish and C. B. Duke, *J. Appl. Phys.*, **48**, 4256 (1977); C. B. Duke and T. J. Fabish, *J. Appl. Phys.*, **49**, 315 (1978); cf., also conducting oxide electrodes, R. J. Jajinski, *J. Electrochem. Soc.*, **125**, 1619 (1978).
110. M. D. Fayer, *Bull. Amer. Phys. Soc.*, **24**, 326 (1979); G. A. Cottrell et al., *J. Appl. Phys.*, **50**, 374 (1979).
111. A. G. Marshall and M. B. Comisarow, "Multichannel Methods in Spectroscopy" in "Transformations in Chemistry," P. R. Griffith, ed., Plenum, N.Y., 1978; J. L. Koenig, *Appl. Spectroscopy*, **29**, 293 (1975); M. B. Comisarow and A. G. MacNeil, *J. Chem. Phys.*, **62**, 293 (1975).
112. R. W. Frei and J. D. MacNeil, "Diffuse Reflectance Spectroscopy," CRC Press, Boca Raton, Fla., 1973; M. Tanaka, *Bull. Chem. Soc. Japan*, **50**, 2881 (1977); A. Brau et al., *Phys. Stat. Solidi*, **62B**, 615 (1974); A. Brau, Ph.d. Thesis, U. of Nice (France), (1976); F. Wooten, "Optical Properties of Solids," Academic, N.Y., 1972.
113. R. B. Somoano et al., *Phys. Rev.*, **B-17**, 2853, (1978); B. Weber et al., *Phys. Semicond. Proc. Int. Conf.* 13th, F. G. Funi, ed., Nth-Holland, Amsterdam 1976, p. 349; Y. Matsunaga, *Bull. Chem. Soc. Japan*, **45**, 770 (1972); W. Gabes and D. J. Stufkens, *Spectrochem. Acta*, **A30**, 1835 (1974); V. K. Malyutenko and A. I. Liptuga, *Phys. Stat. Solidi*, **A51**, K-137 (1975); A. A. Bright et al., *Phys. Rev. Lett.*, **34**, 206 (1975).
114. H. Hill and R. Freeman, "Introduction to Fourier Transform NMR," Varian Assoc., Palo Alto, Calif., 1970.
115. H. Meier and W. Albrecht, *Ber. Bunsen Ges. Phys. Chem.*, **73**, 86 (1969).
116. J. P. Farges, D. Sc. Thesis. Univ. of Nice (France) Dec. 1974.
117. C. E. Okeke, Yeshiva Univ. Bronx, N.Y., Dissertation, 1977.
118. J. Swift, "Gulliver's Travels, Voyage to Laputa," Signet Classics, New Amsterdam Library, New York, 1960, p. 197.
119. G. D. Mendenhall, *Angew. Chemie*, **89**, 220 (1977); F. McCapra, *Acc. Chem. Res.*, **9**, 201 (1976); V. Ya. Shlyapinktokh, *Russ. Chem. Revs.*, **35**, 292 (1966); W. R. Seitz et al., "Chemiluminescence and Bioluminescence," Plenum, N.Y., 1973.
120. K. Balcerowicz and J. Slavinski, *Acta Phys. Polonica*, **A39**, 237 (1971).
121. P. F. Bongers, *IEEE Trans. Magnetics*, **MAG-5**, 472 (1969).
122. G. Briegleb et al., *Ber. Bunsen Ges. Phys. Chem.*, **76**, 101 (1972); Z. Croitorou, *Props. Dielect.*, **6**, 103 (1965); P. Durand and R. Fournie, "Dielect. Materials Meas. and Applic. Conf. IEE.," Publ. No. 67, IEE London 1970, p. 142; M. S. Beevers et al., *J. Chem. Soc. Faraday, II*, **72**, 1482 (1976); **73**, 458 (1977); J. Crossley et al., *J. Chem. Soc. Faraday II*, **73**, 1651,1906 (1979).
123. J. P. Carrion et al., *Helv. Chem. Acta*, **51**, 459 (1967); T. B. Landar et al., *C. R. Acad. Sci.*, (Paris), **271**, 1201 (1971); A. I. Saitt and A. D. Wrixon, *Chem. Commun.*, **1969**, 1184; M. Hatano et al., *Bull. Chem. Soc. Japan*, **46**, 3698 (1973).
124. J. D. Ellis and J. P. Llewellyn, *J. Phys.*, **E-10**, 1249 (1977).
125. E. E. Godik and A. I. Kuznetsov, *Izv. Akad. Nauk. SSR. Ser. Piz.*, **42**, 1206 (1978).
126. H. Inokuchi, *Disc. Faraday Soc.*, **51**, 183 (1971); H. Hidefumi, *Kagaku (Kyoto)*, **32**, 420 (1977); T. Okuhara and K. Tanaka, *Catalyst.*, **19**, 429 (1977); R. Ohnishi, *Catalyst*, **20**, 392 (1978); S. R. Morrison, *Chem. Tech.*, **7**, 570 (1977).

127. H. Kopf and F. Steinbach, eds. "Katalyse an Phthalocyaninen" Stuttgart (W. Ger), 1973.
128. K. Tamaru et al., Jap. Pat. 72 08 284, March 9, 1972.
129. M. Tsuda et al., *J. Catalysis,* 11, 81 (1968).
130. M. Sato et al., *Bull. Chem. Soc. Japan,* 43, 2972 (1970).
131. G. Gaylord, *Nuova Chem.,* 49, 81 (1973); G. P. Belov et al., *ibid.,* 48, 73 (1972); E. Tsuchida et al., Nippon Kagaku Kaishi, 2416 (1972).
132. M. Sato et al., *Bull. Chem. Soc. Japan,* 42, 1976 (1969).
133. J. P. Farges and F. Gutmann, "Mod. Aspects of Electrochemistry," J. O'M. Bockris and B. E. Conway, eds., Plenum Press, New York, 1980, p. 380.
134. L. Y. Johansson et al., *Electrochem. Acta,* 18, 255 (1973); J. Beck, *Ber. Bunsen Ges. Phys. Chem.,* 77, 353 (1973).
135. K. Kimura and H. Inokuchi, *J. Catalysis,* 29, 49 (1973).
136. H. Morawitz et al., IBM Res. Lab. San Jose, Calif. Priv. Commun. (1979).
137. K. Ohno and H. Inokuchi, *Chem. Phys. Lett.,* 33, 585 (1975).
138. K. Tanaka, *J. Crystallog. Soc. Japan,* 20, 150 (1978).
139. T. Takizawa et al., *J. Phys. Chem.,* 82, 1391 (1978); T. Watanabe et al., *ibid.,* 81, 1845 (1977).
140. H. G. and F. Willig, *Top. Curr. Chem.,* 61, 31 (1976).
141. F. Gutmann, *J. Sci. Indust. Res.,* 26, 19 (1967); F. Gutmann and H. Keyzer, *Electrochem. Acta,* 11, 555, 1163 (1966).
142. I. R. Forrest et al., *Aggressologie,* 7, 147 (1966).
143. L. Levi et al., *Zavod. Lab.,* 43, 659 (1977).
144. D. Beltran, S. Chan and H. Keyzer in "Bioelectrochemistry Proc. US-Australia Joint Seminar on Bioelectrochemistry." Pasadena, Calif., 1979, H. Keyzer and F. Gutmann, eds., Plenum Press, N.Y., (1980).
145. L. M. Doane and J. T. Stock, *Anal. Chem.,* 50, 1891 (1978).
146. M. Jaworski and Z. Romaczewski, *Lect. Notes Phys.,* 65, (Org. Conduct. Semiconduct.) 409 (1977).
147. E. M. Trukhan, *Pribory Tekh. Eksper.,* 1965, 198; *Biofizika,* 11, 142 (1966); 15, 1052 (1970).
148. D. D. Eley and R. Pethig, *J. Bioenergetics,* 1, 109 (1970); 2, 39 (1971); 3, 271 (1972); "Conduction in Low Mobility Materials," Taylor and Francis, London, 1971.
149. Yu. G. Arapov and A. B. Davydov, *Defektoskopiya,* 63 (1978); M. Godlewski, *Phys. Stat. Solidi,* A-51, K141 (1979).
150. L. J. Buravov and I. F. Shchegolev, *Prib. Tek. Eksp.,* 2, 171 (1971).
151. S. P. Cooke, *Phys. Rev.,* 74, 701 (1948).
152. S. K. Khanna et al., *Phys. Rev.,* B10, 2208 (1974).
153. M. Cohen et al., *Solid State Commun.,* 17, 367 (1975).
154. S. K. Khanna et al., *Solid State Commun.,* 18, 1405 (1976).
155. J. Halbritter, *Z. Physik,* B-31, 19 (1978).
156. M. M. Sayed and C. R. Westgate, *Rev. Sci. Inst.,* 46, 1074 (1975).
157. N. P. Ong and A. M. Portis, *Phys. Rev.,* B15, 1782 (1977).
158. B. Molnar and T. A. Kennedy, *J. Electrochem. Soc.,* 125, 1318 (1978); R. Pethig, *J. Biol. Phys.,* 1, 193 (1973); cf. also ref. 149.
159. W. N. Hardy et al., *Bull Amer. Phys., Soc.* 20, 466 (1975).
160. K. Seeger and W. Maurer, *Solid State Commun.,* 27, 603 (1978).
161. W. Schneider and K. Seeger, *Appl. Phys. Lett.,* 8, 133 (1966); K. Hess and K. Seeger, *Z. Physik,* 218, 431 (1969); 237, 252 (1970).
162. A. Brenner and J. L. Sligh, *J. Electrochem. Soc.,* 117, 602 (1970).

163. J. P. Long, Rpt. C00-1198-1185 (1977) avail. NTIS.
164. N. Ernst and J. H. Block, Forschungsber. Wehrtech., (Bundes-Minist. Verteidigung, W. Germany) 1976, p. 75.
165. S. G. Kalishnikov and V. I. Fedosov, *Fiz. Tekh. Poluprovodn.*, 12, 1154 (1978).
166. G. M. Sessler and J. E. West, *Rev. Sci. Inst.*, 42, 15 (1971).
167. T. E. Hartmann et al., *J. Appl. Phys.*, 37, 2488 (1966); A. C. Lilly and J. R. McDowell, *ibid.*, 39, 141 (1968); M. Stuart, *Brit. J. Appl. Phys.*, 18, 1637 (1967); D. L. Pulfrey et al., *J. Appl. Phys.*, 41, 2838 (1970); M. Kryszewski and A. Szymanski, *J. Polymer Sci. D;* Macromolec. Revs., 4, 245 (1970); W. R. Runyan, "Semiconductor Measurements and Instrumentation," McGraw-Hill, N.Y., 1976; I. Chen, *Solid State Commun.*, 26, 359 (1978); K. P. Charle and F. Willig, *Chem. Phys. Lett.*, 57, 253 (1978); V. P. Shuvaev et al., *Elektrokhimiya*, 14, 667 (1978).
168. T. Tamai,"Proc. Int. Symp. Contam. Control," 4th, (1978) p. 221, Inst. Environment. Sci., Mt. Prospect, Ill., (1978).
169. I. Chen, *Solid State Commun.*, 26, 359 (1978).
170. K. Doblhofer and J. Ulstrup, *J. Phys. (Paris) Colloq.*, 59 (1977).
171. C. A. Hogarth and T. Igbal, *Thin Solid Films*, 51, 245 (1978).
172. D. R. Rosseinsky et al., *J. Phys.*, E-10, 1236 (1977).
173. H. Gerischer, *J. Electroanal. Interfacial Electrochem.*, 82, 133 (1977); R. R. Dogonadze et al., *Electrochem. Acta*, 22, 967 (1977); S. G. Louie et al., *ibid.*, 13, 790 (1976); H. Gerischer, in "Physical Chemistry—An Advanced Treatise," H. Eyring et al., eds., Academic Press, N.Y., 1970, pp. 488 ff.; H. Gerischer and F. Willig, *Top. Curr. Chem.*, 61, 31 (1976); T. C. McGill, *J. Vac. Sci. Technol.*, 11, 6 (1974).
174. L. E. Lyons, *Priv. Commun.*, (1977).
175. W. A. Noyes, *Pure Appl. Chem.*, 9, 461 (1964).
176. F. Lohmann, *Surface Sci.*, 14, 431 (1969); Z. Natusforschg. 229, 843 (1967).
177. F. Lohmann, *J. Phys. Chem. Solids*, 29, 1693 (1968).
178. F. Lohmann and W. Mehl, *J. Chem. Phys.*, 50, 500 (1969).
179. K. Pigon and J. Sworakowski, *Acta Phys. Polon.*, 32, 329 (1967).
180. F. Lohmann and W. Mehl, *Ber. Bunsen Ges. Phys. Chem.*, 71, 493 (1967).
181. F. Lohmann and W. Mehl, *Electrochem. Acta*, 13, 1469 (1968).
182. P. P. Schmidt and H. B. Mark Jr., *J. Chem. Phys.*, 58, 4290 (1973); R. A. Marcus, *J. Chem. Phys.*, 43, 679 (1965); R. R. Dogonadze et al., *Russ. J. Phys. Chem.*, 38, 652 (1964); V. G. Levich, *Adv. Electrochem. Electrochem. Eng.*, 4, 249 (1964).
183. L. E. Lyons and K. G. McGregor, *Aust. J. Chem.*, 29, 21 1401 (1976).
184. H. Kallmann and M. Pope, *J. Chem. Phys.*, 32, 300 (1959); W. Mehl et al., *J. Electrochem. Soc.*, 113, 1166 (1966); W. Mehl and J. M. Hale, *Adv. Electrochem. Electrochem. Eng.*, 6, 399 (1967); L. I. Boguslavskii and B. T. Lozhkin, *Electrochem. Acta*, 17, 1007 (1972).
185. R. C. Jarnagin et al., *J. Chem. Phys.*, 39, 573 (1963).
186. "U.V. Atlas of Organic Compounds," J3/1,2, Vol. 5, Butterworth, London 1971.
187. D. N. Goswami, *Indian J. Biochem. Biophys.*, 14, 372 (1977).
188. A. F. Kiaz and K. K. Kanazawa, *IBM J. Res. Develop.*, 23, 316 (1979).
189. D. Tabor, *Physics Bulletin*, the Institute of Physics (London), 29, 521 (1978); e.g., M. Prutton, "Surface Physics," Oxford Univ. Press, 1975; G. A. Samorjai, "Principles of Surface Chemistry," Prentice-Hall, London, 1972; J. M. Blakely, "Introduction to the Properties of Crystal Surfaces," Pergamon Press, Oxford, 1973; A. Clark, "The Chemisorption Bond," Academic Press, N.Y., 1974; A. E. Morgan and H. W. Werner, *Phys. Sci.*, 18, 451 (1978).
190. Yu. I. Tyurin and U. V. Styrov, *Fiz. Tekh. Poluprovodn.*, 11, 2157 (1977).

191. M. L. Cohen, *Physics Today*, **32**, 40 (1979); M. L. Cohen et al., *Solid State Phys.*, **24**, 37 (1970); C. Kittel, "Introduction to Solid State Physics" 5th ed. Wiley, N.Y., 1976; T. Nakamura, *J. Res. Inst. Catal. Hokkaido Univ.*, **26**, 145 (1978).

192. J. Lagowski et al., *J. Appl. Phys.*, **48**, 3566 (1977).

193. V. A. Zuev and V. G. Popov, *Kvantovaya Elektron.*, (Kiev), **12**, 53 (1977).

194. M. G. Simonsen et al., *J. Chem. Phys.*, **61**, 3789 (1974).

195. H. Gerischer et al., *Adv. Phys.*, **27**, 437 (1978).

196. M. Haroniec and W. Rudzinski, *J. Res. Inst. Catalysis*, Hokkaido Univ., **25**, 197 (1977); R. A. Alberty and F. Daniels, "Physical Chemistry," Wiley, N.Y., 1979; cf. also refs. 189 and 197.

197. B. E. Conway, in "Electrochemistry – the Last Thirty and the Next Thirty Years," H. Bloom and F. Gutmann, eds., Plenum Press, N.Y., 1977, p. 183.

198. T. Nakamura and M. Tatewaki, *J. Res. Inst. Catalysis* Hokkaido Univ., **25**, 159 (1977).

199. L. I. Ahmed, *J. Phys. Chem. Solids*, **29**, 1653 (1968).

200. D. J. Berets and D. S. Smith, *Trans. Faraday Soc.*, **64**, 823 (1968).

201. K. Kanazawa, A. F. Diaz et al., *IBM Research Lab.*, San Jose Calif., Private Commun. (1979).

202. K. M. Jain et al., *Indian J. Phys.*, **52-A**, 543 (1978).

203. N. I. Ionescu and P. Banyai, *Rev. Roumaine de Chimie*, **23**, 1023 (1978).

204. V. B. Margulis and L. I. Boguslavskii, *Proc. Conf. Org. Semicond. Izd. Akad. Nauk Latv. SSR, Riga*, 1968; *Elektrokhimiya*, **3**, 329 (1967); *Kinetika and Kataliz.*, **9**, 211 (1967).

205. J. Brophy, "Basic Electronics for Scientists," McGraw-Hill, N.Y., 2nd Ed., 1972; V. D. Ziel, "Fluctuation Phenomena in Semiconductors," Butterworth, London, 1959; N. M. Hosseini and B. K. Jones, *Phys. Stat. Solidi*, **A40**, K185 (1977).

206. D. Vasilescu et al., *Electrochim. Acta*, **19**, 181 (1974); H. Kranck et al., *ibid.*, **23**, 891 (1978); V. A. Tyagai, *ibid.*, **18**, 229 (1973); G. Blanc et al., *ibid.*, **20**, 599, 687 (1975); Commun. 27th Reunion S.I.E. CNRS, Paris, Zurich, Sept. 1976; *J. Electroanal. Interfacial Electrochem.*, **75**, 97 (1976).

207. L. J. DeFelice and D. R. Firth, *IEEE Trans. Biomed. Eng.*, **18**, 339 (1971); L. J. DeFelice and J. P. O. M. Michaelides, *J. Membrane Biol.*, **9**, 261 (1972); H. Fishman, *Biophys. Soc. Abst.*, **119A**, (1972).

208. W. F. Pickard, *Nature* (London), **201**, 283 (1964).

209. J. P. Farges and F. Gutmann in "Modern Aspects of Electrochemistry," J. O'M. Bockris and B. E. Conway, eds., **13**, 361 (1979), Plenum Press, N.Y., 1979.

210. L. J. Giacoletto, *Proc. IEEE*, **49**, 921 (1961).

211. F. Cardon, *Physica*, **57**, 390 (1972); F. N. Hooge, *Phys. Lett.*, **33A**, 169 (1970); N. M. Hosseini, ref. 205; G. P. Vasileu et al., *Pisma Zh. Tekh. Fiz.*, **2**, 604 (1976); G. Blanc et al., ref. 206.

212. K. J. Euler, *Naturwissenschaften*, **58**, 621 (1971); G. C. Barker, *J. Electroanal. Chem. Interfacial Electrochem.*, **39**, 48 (1972); cf. also refs. 206 and 209.

213. F. N. Hooge, *Phys. Lett.*, **33a**, 169 (1970), ref. also 211.

214. F. N. Hooge, *Phys. Lett.*, **29A**, 139 (1969); **33A**, 169 (1970); J. N. Loreitye and A. M. M. Hoppenbrouwers, *Philips. Res. Rep.*, **26**, 29 (1971); M. Weissman and G. Feher, *J. Chem. Phys.*, **63**, 586 (1975); K. Klason and J. Kubaj, *J. Appl. Phys.*, **47**, 1970 (1976).

215. D. Vasilescu et al., *Biopolymers*, **12**, 341 (1973).

216. G. Y. Kolbasov and V. A. Tyagai, *Fiz. Tekh. Poluprov.*, **6**, 946 (1972).

217. Y. Ishizuka, *Bull. Chem. Soc. Japan*, **50**, 563 (1977); cf. also N. I. Wakayama

et al., *ibid.,* **46**, 2277 (1973).
218. M. V. Kurik and I. I. Poberezhets, *Zh. Fiz. Khim.,* **52**, 798 (1978).
219. I. N. Munits et al., German Pat., 2,635,936 Feb. 16, 1978.
220. L. I. Buguslavskii and B. T. Lozhkin, *Surface Sci.,* **38**, 413 (1973); M. Zander, *Angew. Chem.,* **72**, 513 (1960).
221. W. Geil and K. Schmugge, (East) German Pat., 124,148 Feb. 9, 1977.
222. G. S. Pawley, *J. Phys. Chem. Solids,* (in print 1979/1980).
223. H. Noyima et al., *Bull. Chem. Soc. Japan,* **51**, 2513 (1978).
224. W. V. McLevige, *J. Phys.,* **10**, 335 (1977).
225. L. E. Lyons, *Search,* **7**, 341 (1976).
226. P. Cuatrecasas, *J. Biol. Chem.,* **245**, 3059 (1970).
227. H. Haglund, *Methods of Biochem. Analysis,* **19**, 1 (1971).
228. J. A. Kok et al., *Appl. Sci. Res.,* **17**, 461 (1967).
229. H. A. Pohl, *J. Polymer Sci.,* **17C**, 13 (1967); J. H. T. Kho and H. A. Pohl, *ibid.,* A-1, **7**, 139 (1969); H. A. Pohl, in "Electronic Aspects of Biochemistry," B. Pullman, ed., Academic Press, N.Y., 1964, p. 121.
230. J. Kossut, *Phys. Stat. Solidi,* **B-78**, 537 (1976).
231. V. Hadek, *Rev. Sci. Inst.,* **42**, 393 (1971).
232. D. Cahen et al., *Rev. Sci. Inst.,* **44**, 1567 (1973).
233. S. D. Hottman and H. A. Pohl, *Rev. Sci. Inst.,* **42**, 387 (1971).
234. D. Greatorex and G. Stokes, *Sch. Sci. Rev.,* **59**, 337 (1977).
235. P. Kramer and L. J. van Ruyen, *Solid State Electron.,* **20**, 1011 (1977).
236. R. K. Pandey, *Rev. Sci. Inst.,* **44**, 907 (1973).
237. I. Nakada and Y. Ishihara, *J. Phys. Soc. Japan,* **19**, 695 (1964).
238. C. L. Gupta, *Indian J. Pure. Appl. Phys.,* **15**, 684 (1977).
239. J. E. Algie, *Colloid Polymer Sci.,* **257**, 1 (1979).
240. L. A. Berkovich and A. Fomin, SB. ASPIR. RAB.–Kasan. Gos. Univ. Techn. Nauk., **1976**, 3, 12.
241. H. Inokuchi et al., *Bull. Chem. Soc. Japan,* **40**, 2695 (1967).
242. Y. Matsunaga, *J. Chem. Phys.,* **42**, 2248 (1965).
243. E. Krikorian and R. J. Sneed, *J. Appl. Phys.,* **40**, 2306 (1969).
244. J. E. Lane et al., *J. Chem. Phys.,* **69**, 3981 (1978).
245. L. B. Coleman, *Rev. Sci. Inst.,* **49**, 60 (1978).
246. J. Swiatek, *Pr. Nauk Inst. Chem. Org. Fiz. Politech.,* Wroclaw, **16**, 299 (1978).
247. K. Doblhofer and J. Ulstrup, *J. de Phys. (Paris) Colloq.,* (1977) (5) 49; M. Careem and A. K. Jonscher, *Phil. Mag.,* **35**, 1489, 1503 (1977).
248. W. Wlodarski and A. Lipinski, *Pr. Nauk Inst. Chem. Org. Fiz. Politech.,* Wroclaw (Poland), **16**, 317 (1978).
249. H. I. Smith and D. C. Flanders, *Appl. Phys. Lett.,* **32**, 349 (1978); M. W. Geis et al., "Proc. 15th Symp. Electron. Ion and Photon Beam Bechnol.," Boston, Mass., 1979, in press; M. W. Geis et al., *Appl. Phys. Lett.,* **35**, 71 (1979).
250. P. C. Li, J. P. Devlin and H. A. Pohl, *J. Phys. Chem.,* **76**, 1026 (1972); J. Stanby et al., *ibid.,* **70**, 2011 (1966).
251. C. A. Crider et al., *Nucl. Instrum. Method,* **149**, 701 (1978).
252. B. M. Abdurakchamno et al., *Thin Solid Films,* **37**, 1 (1976).
253. Y. Kamura et al., *Bull. Chem. Soc. Japan,* **49**, 418 (1976).
254. H. Eiermann and W. Hofberger, *Pr. Nauk Inst. Chem. Org. Fiz. Politech.,* Wroclaw, **16**, 103 (1978).
255. M. I. Fedorov et al., *Izv. Vyssh. Uchebn. Zaved/Fiz.,* **21**, 158 (1978).
256. S. S. S. Huang and G. R. Freeman, *Canad. J. Chem.,* **55**, 2264 (1977).

257. R. Tamamushi and K. Takahashi, *J. Electroanal. Chem. Interfacial Electrochem.*, **50**, 277 (1974).
258. A. A. Andreev et al., *Zavoid. Lab.*, **43**, 192 (1977).
259. U. Zellweger, Brit. Pat., 1,568,887; 26 April (1978).
260. B. P. Zrtamonov et al., *Zh. Fiz. Khim.*, **51**, 2868 (1977).
261. K. Hallenga, *Rev. Sci. Inst.*, **46**, 1691 (1975).
262. R. V. Loon and R. Finsy, *Rev. Sci. Inst.*, **44**, 1204 (1973).
263. G. C. Morris and M. G. Sceats, *Chem. Phys.*, **1**, (1973); M. G. Sceats and G. C. Morris, *Phys. Stat. Solidi*, **14A**, 643 (1972).
264. L. E. Lyons et al., *Aust. J. Chem.*, **21**, 853 (1968).
265. R. G. Kepler, *Phys. Rev.*, **119**, 1226 (1960); O. H. LeBlanc, *ibid.*, **33**, 626 (1960); **37**, 916 (1962).
266. G. R. Johnston and L. E. Lyons, *Aust. J. Chem.*, **23**, 2187 (1970).
267. G. R. Johnston and L. E. Lyons, *Phys. Stat. Solidi*, **37**, K-75 (1970).
268. J. Crossley et al., *J. Chem. Soc. Faraday II*, **73**, 1651, 1906 (1977).
269. C. R. Johnston, *Chem. Phys. Lett.*, **3**, 699 (1969).
270. A. Calusaru and A. Moldavan, *Electrochim. Acta*, **17**, 1299 (1972).
271. J. Barthel et al., in "Modern Aspects of Electrochemistry," J. O'M. Bockris and B. E. Conway, eds. Plenum Press, N.Y., **13**, 1 (1979).
272. H. Grassi et al., *Phys. Stat. Sol.*, **A55**, K 179 (1979); J. P. Farges et al., *ibid.*, 89.
273. e.g., reviews of "Photo-Electron Spectroscopy," J. Augustynski and L. Balsenc, in "Modern Aspects of Electrochemistry," J. O'M. Bockris and B. E. Conway, eds., **13**, 251 (1979); B. G. Baker, *ibid.*, **10**, (1975); "Auger Spectroscopy," J. Augustynski and L L. Balsenc, loc. cit., B. Baker, loc. cit.
274. Z. G. Soos, *Ann. Rev. Phys. Chem.*, **25**, 121 (1974); G. D. Stucky et al., *Ann. Rev. Mater. Sci.*, **7**, 301 (1977); A. J. Berlinsky, *Contemp. Phys.*, **17**, 331 (1976); L. N. Bulaevskii, *Sov. Phys.–Usp.*, **18**, 131 (1976); G. A. Toombs, *Phys. Rep.*, **40C**, 181 (1978); J. M. Perlstein, *Angew. Chem. Int. Ed.* (Engl.), **16**, 519 (1977); R. H. Friend and D. Jerome, *J. Phys. C: Solid State Phys.*, **12**, 1441 (1979); H. Meier "Organic Semiconductors," Monographs in Modern Chem., Vol. 2, H. F. Ebel, ed., Verlag Chemie, 1974; Z. G. Soos and D. J. Klein, "Organic Molecular Crystals: Charge Transfer Complexes," in "Treatise on Solid State Chemistry," Vol. 3, Ch. 9, 679-767, N. B. Hannay ed., Plenum Press, N.Y., 1976; R. G. Kepler, "Organic Molecular Crystals: Anthracene," in "Treatise on Solid State Chemistry," Vol. 3, Ch. 8, 615-678, N. B. Hannay, ed., Plenum Press, N.Y., 1976; J. J. Andre, A. Bieber and F. Gautier, *Annales de Physique*, (Paris), **1**, 145-256 (1976); Z. G. Soos, *J. Chem. Educ.*, **55**, 546-552 (1978).
275. H. Ottavi et al., *J. Phys.*, **C-11**, (1978).
276. K. Kamaras et al., *J. Phys.*, **C-10**, L423 (1977).
277. E. Abrahams et al., *J. Low Temp. Phys.*, **32**, 673 (1978).
278. M. J. Rice and J. Bernasconi, *Phys. Lett.*, **38A**, 277 (1972).
279. M. G. Simonsen et al., *J. Chem. Phys.*, **61**, 3789 (1974); J. M. Clark and R. V. Coleman, *Proc. Natl. Acad. Sci. USA*, **73**, 1598 (1976).
280. e.g., the ionization potential of the topmost molecule at crystalline surface in contact with an aqueous electrolyte is 0.4 eV lower than its vacuum value. F. Willig and G. Scherer, *Chem. Phys. Lett.*, **58**, 128 (1978).

13

Excited States

13.1 Introduction

This chapter intends to supplement discussions given, mainly, in Chapter 5, but also in Chapter 4. In addition to the excitons and polarons treated in these chapters, there now have been defined several other types of localized excited states which will be discussed in the following sections. Several reviews on the subject are available.[1] The literature on this topic is truly enormous and, indeed, daunting. Much of it refers to luminescence phenomena. Discussion here will be confined to a selection of reports which are directly concerned with (semi-)conductivity.

Injection of sufficient energy into a molecular compound generally results in an excited singlet state. Usually, the energy is supplied in the form of radiation, but not necessarily so; it may also be furnished chemically.[2] The excited singlet may relax according to several possible paths shown in Table 13.1.

The decay mechanisms involving an activator, or guest (impurity) molecule are of special interest. Collision of the exciton with such a center results in fluorescence, or in recombination of the electron-hole pair *via* energy-losing collisions with the center plus thermalization of any excess energy. When the exciton is localized in a shallow trap such as an impurity center, its binding energy is of the order or even below the width of the exciton band itself.[3]

The excited singlet state is highly interacting, mainly by dipole-dipole interactions, and thus results in an energy transfer though not necessarily in a

Table 13.1 Relaxation Processes of Excited Molecules. After Seanor.[5]

A molecule in the excited state may lose its energy by any of the following processes:

Singlet excited state:

Intersystem crossing to triplet level

Direct relaxation to ground state with fluorescence.

Non-radiative decay to ground state $\tau = 10^{-13}$ sec

Formation of excited multimers

Direct formation of charge carriers

Indirect formation of charge carriers

Decomposition of photoproducts

Two exciton processes → charge carriers of photoproducts

Transfer of energy to another molecule of the same or different species

Triplet state:

Relaxation to ground state with emission of light-phosphorescence. $\tau = 10^{-4} - 10$ sec

Non-radiative relaxation to ground state

Formation of multimers

Decomposition → photoproducts

Two exciton processes
→ charge carriers
→ photoproducts
→ new excited singlet state (triplet-triplet or triplet-singlet processes)

Transfer of energy to another molecule of the same or different species

charge transfer, over distances of the order of 100 Å. Its lifetime typically is 25 nsec (see also Section 13.4).

Far larger distances occur for energy transfer *via* excited triplet states because their interaction with the lattice is mainly *via* much weaker exchange interactions; the triplet state also has a far longer lifetime.

In donor-acceptor complexes the excited state can be more polar and more closely bonded than the ground state even in solution; thus the donor-acceptor distance in the (weak) ethylene:Cl_2 complex drops from 2.26 to 1.97 Å upon excitation.[4] However, complexes between one polar and one non-polar component show a reduction in the polarization and thus of the permittivity. In aqueous solution, molecules excited to the singlet state may give rise to hydrates while dimers, or excimers, arise from transitions between a molecule in the triplet state and a ground state molecule. Such transitions are illustrated for pyrimidine[44] in Fig. 13.1.

Other reactions involving excited states will be discussed in Section 13.3. Excited states of charge transfer complexes are discussed in Section 17.5.

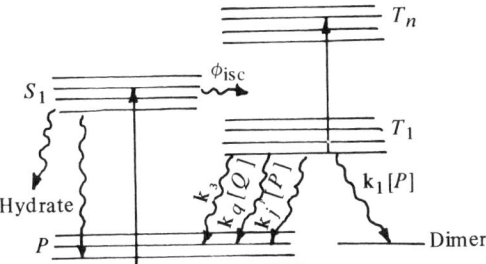

Fig. 13.1 Model for excited-state reactions of pyrimidines. After Whillans and Johns.[44] The rate constants for triplet reactions are as follows: k_3, spontaneous deactivation to the solvent; k_q, bimolecular quenching by Q; k_1', self-quenching by the pyrimidine not leading to product formation; and k_1, self-quenching by the pyrimidine in the formation of photodimers.

13.2 Excitons. Polarons

An exciton[7] is a mobile but localized non-conducting excited state, in other words, an energy packet forming a quasi-particle obeying Bose-Einstein statistics: a boson. Depending on the strength of the phonon-exciton interaction, its motion may be described either in terms of a band model, i.e., by coherent propagation within an energy band, viz., the exciton band, or else in terms of a hopping model, i.e., by transitions between more or less localized states. These, however, need not be nearest neighbors. There appears to be no definite experimental method to decide between these models. For a further discussion see Section 5.8.

The binding energy E of an exciton to an ionized or neutral impurity is related[15] to the ionization energy I of the corresponding defect structure by an equation of the form

$$E = A + BI \tag{13.1}$$

where A and B are functions of the effective mass ratio of electrons and associated holes. The binding energy of Wannier (large radius) excitons is of the order[55] of 0.05 eV. The coupling of Frenkel (small radius) excitons and of ionic excited states is discussed by Petelenz.[56] The energy of the ground state of Wannier excitons has been calculated[57] for the case of a polar matrix containing free charge carriers. The binding of an exciton in an indirect gap semiconductor can exceed the band gap.[58] This renders the system unstable towards spontaneous formation of excitons, thus modifying the ground state. It is referred to as an excitonic insulator and has a spatial period equal to the reciprocal of the wave vector off-set between the conduction band minimum and the valence band maximum.[58] The ground state may be viewed as a low-density condensate of bosons (Wannier excitons) with short-range repulsion due to the exclusion principle operating between the electrons and between the holes and exhibiting a remarkable similarity[58] to the ground state of a superconductor though there are important differences.

Single photon excitation of anthracene or tetracene in the near *uv* leads predominantly to the formation of neutral Frenkel excitons. These singlet excitons can interact with electrodes or impurities to give rise to carriers (holes). If the exciton density is sufficiently high, carriers of both signs are produced simultaneously (intrinsically) without the assistance of the electrode (or an impurity) by an exciton-exciton annihilation mechanism.

In anthracene, there exists a "hot" energy band with an absorption intensity obeying an Arrhenius form of temperature relation; the activation energy E is about 0.13 eV.[61]

Most organic molecules with π-electrons have transitions with oscillator strengths ≥ 0.1. In a crystal, these transitions give rise to broad exciton bands with energies determined mainly by lattice sums of point dipole-dipole interactions. Ground state vibrational excitons commence to populate that band as the temperature is raised above a certain critical value. The vibrational excitons may then be further optically excited into the first exciton band. A typical exciton energy level diagram is shown in Fig. 13.2. Mobile triplet excitons in charge transfer complexes persist in anthracene:TNB 1:1 adducts[25] down to 1.2 K.

Charge transfer excitons have been directly observed[36] in intramolecular

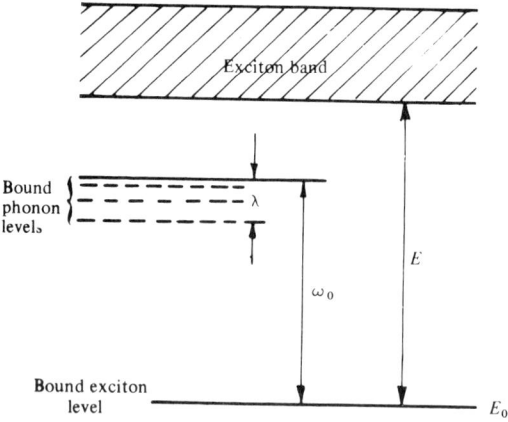

Fig. 13.2 Typical exciton energy level diagram. After Rashba and Zimin.[3] ω_0 = exciton-phonon coupling constant, ω = phonon frequency, $\lambda \equiv \omega_0 - \omega$.

(auto-) complexes of e.g., dichloroanthracene by the modulation of spectral reflectivity by an applied electric field. Their experimental life time in TCNQ salts is reported[37] to be of the order of 10^{-4} sec. The diffusion length of triplet excitons in tetracene is reported[38] as about 5000 Å. Other results,[39] from alternating current harmonics, for liquid benzene are about 1000 Å.

The extraordinarily long life time is associated with the length of the mean free path L. Recombination of the exciton can only occur[40] if the electron and hole approach each other to a distance considerably below the Onsager radius $r_0 = 2e^2/3kT\epsilon$, so that their excess energy may be lost in many deactivating collisions. Diffusion theory does not apply within r_0.[40] For a polar crystal, say, $\epsilon = 15, r_0 = 25$ Å.

Triplet excitons are reported[45] to be generated by the fission of 2-phonon excited singlet states; further photo-ionization then yields free carriers exhibiting an electron-hole recombination rate constant of about 10^{-6} cm^{3}sec^{-1}. Charge carriers, free or bound to centers, generally should be regarded as polarons. See Section 4.5. A polaron consists of the charge carrier and the distortion of the lattice induced by the carrier itself. One can distinguish large and small polarons. For the former the distortion of the lattice, induced around a charge carrier, extends over distances larger than, and for the latter over distances smaller than the lattice constant. Energetically, for the cases of large and small polarons, half the bandwidth is larger and smaller, respectively, than the maximum polaron binding energy. This latter quantity is the energy gained by an infinitely slow carrier (zero bandwidth) due to polarization and distortion induced in the lattice by the carrier itself.

Mott[59] stresses the difference between a "dielectric" polaron and a small, self-trapped polaron. For the formation of the latter, a potential barrier must be penetrated or surmounted causing a time-delay in the self-trapping. This situation does not apply to dielectric polarons, and thus furnishes a criterion to distinguish between the two types. Extensive reviews of polaron theory are available.[43]

13.3 Exciton Migration

Exciton dynamics in molecular crystals is a problem characterized by many subtle complicating effects. Exciton transport phenomena are complicated by exciton trapping, exciton ionization, radiationless decay processes, exciton-exciton annihilation and other difficulties. Two limiting cases can be described for exciton transfer: the random walk and the coherent motion models. In the strong scattering model, the mean free path is of the order of the lattice spacing, wherein the localized states are dominant and the exciton motion can be described in terms of a diffusive random walk model.[65] In the co-

EXCITED STATES

herent motion model, the mean free path is considerably larger than the lattice spacing and the exciton motion can be described in terms of an exciton-band model. Exciton transfer may involve:

Trapping	Radiationless Decay
Ionization	Exciton-Exciton Annihilation
Higher Exciton States	Radiative Decay yielding a photon
Exciton-Exciton Interactions yielding a free carrier	Internal Conversion Processes yielding a phonon

The theory of exciton transfer is discussed by several authors[18] whose original papers should be consulted.

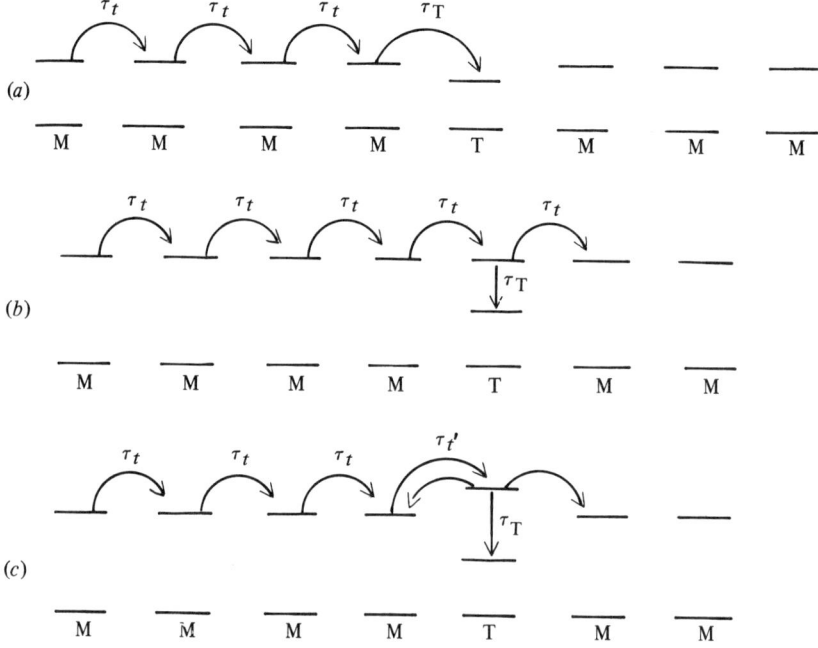

Fig. 13.3 Mechanisms for the trapping of excitation energy from an ensemble of molecules M by a trapping molecule T; (a) when the first excited state of T is lower than that of the transferring molecules M; (b) when the first excited states of M and T are equal but T has, in addition, an excited state *below* the first excited state which is absent in M; (c) when the first excited state of T is *higher* than that of M and T has in addition an excited state below the first excited state of M which is absent in M. In cases (b) and (c) efficient trapping can occur when the trapping time τ_T is far less than the transfer time τ_t. After Sybesma.[117]

In exciton trapping, the excitation becomes fixed on one molecule *via* a localized excited state. Trapping may involve an excited singlet state of the acceptor, which is lower in energy than that of the donor molecule (see Fig. 13.3*a*). This situation is likely to occur in photosynthesis *via* chlorophyll *a*; the energy difference there is about 0.05 eV. Alternatively, as shown in Fig. 13.3*b*, the acceptor may have another, lower, excited state which is accessible *via* its singlet level which remains the same as that of the donor. The trapping probability is governed by the ratio between the trapping time τ_t, of the excited localized state, to the time constant of the exciton transfer τ_f. If $\tau_f \ll \tau_t$, then a high population density of trapped excitons may result even if the first excited state accessible to the exciton derived from the donor is at a higher level than the first excited singlet (see Fig. 13.3*c*). This situation appears to prevail in many photosynthetic bacteria.

Trapping of excitons very often results in a lowering of the $\pi\pi^*$ and $n\pi^*$ levels and thus facilitates the formation of activated complexes. The relative position of these two levels is greatly affected by the environment, especially by the polarity of a solvent which may be present; in highly polar solvents their position may even become inverted. This effect bears on the frequently observed solvent effects in activated complexes (see Section 17.5 and 17.6).

Other most important primary processes governing exciton transport are[62] exciton-phonon scattering events, which can be studied spectroscopically, especially by reflection spectroscopy,[63] (see Section 12.10). In anthracene, where the exciton is weakly coupled with the intermolecular phonons but strongly coupled to a small degree with the intramolecular phonons,[63] crystals of size more than 1 μ may be considered semi-infinite for obtaining these reflection spectra. Impurities remaining after usual ultrapurification processes do not appear[63] to affect the results. Two limiting types of exciton in doped molecular crystals can be distinguished:[64] (1) An impurity exciton state when the energy band associated with the predominant resonance transfer of excitation *via* impurity molecules splits out from the host crystal band; (2) the mixed exciton state when the impurity introduction is followed only by some modification of the initial exciton band. Even for Frenkel excitons with their small radii of elementary excitation, one can expect the mixed exciton state (2) when the excitation energy difference between the impurity and host molecules is substantially less than the exciton band width.

As Hochstrasser has pointed out,[118] the radiative properties of excitons may be considered to fall in between two extreme cases: they may be treated in terms of polaritons (see Section 13.8) or in terms of impurity states involving coherent polarization of the matrix. The difference, however, is really one of degree only.[118]

In the random walk model, the excitons randomly move in the lattice which is assumed to contain extended trapping regions. Noting that the energy transfer rate must be proportional to the rate at which the exciton samples previously unsampled sites, a general expression for the energy transfer rate $k(t)$, derived from this model[65] is:

$$k(t) = N'_a C(A) t_H^{-1} \{1 + [2C(A)/(2\pi\sigma^2)^{3/2}](t_H/t)^{1/2}\} \quad (13.2)$$

$$\tau_s^0/\tau_s = 1 + C(A)\tau_s^0 k(\infty), \quad (13.3)$$

where terms of the order of t^{-1} and higher have been neglected and the emission probability per step has been assumed to be small. Two new parameters are σ, which depends on the hopping probabilities for hops of different lengths and directions on a specific lattice, and $C(A)$, which is called the capacity of the random walk and, in the limit of many hops, gives the number of new sites sampled by the exciton on each hop. In general, $(2\sigma^2)$ will be close to 1 for an isotropic random walk and closer to $\frac{1}{2}$ for an anisotropic walk.[65] N'_a stands for the fraction of host molecules replaced by dopant (activator) molecules, t for time, t_H for the average time for each hop, and τ^0/τ_s for the relative host fluorescence decay time.

It is equivalent to treat the excitons as point particles and consider a finite trapping region around each activator or to treat the activators as point traps and associate a finite region of sampled sites (equal in size and shape to the trapping region) around each exciton. The parameter A in $C(A)$ gives the number of sites in the sampling (or trapping) region other than the one on which the exciton (or activator) is located. The capacities for a random walk on a simple cubic lattice have been calculated as a function of the size of the trapping region, shape of the trapping region, anisotropy of the random walk, and effects of greater than nearest-neighbor steps.[65] As compared with the capacity for the usual isotropic nearest-neighbor walk with point traps, the capacity is increased by about a factor of 5 by considering a spherical trapping region consisting of all sites within two lattice spacings and is further increased by a factor of 2 if the trapping region contains the same number of points but is linear. An anisotropic random walk tends to decrease the capacity. Including a distribution of hopping lengths consistent with dipole-dipole resonance interaction increases both $(2\sigma^2)$ and $C(A)$ by slightly less than a factor of two.

The magnitude of the capacity is directly related to[65] the size and shape of the trapping region which depends on two things: the strength of the interaction and the amount of distortion of the host lattice due to the presence of the activator. The former can be characterized by the critical interaction distance, R_0, defined as the distance over which energy is transferred at the rate equal to the rate of de-excitation by all other mechanisms.[65]

For dipole-dipole resonance interaction with random dipole orientations, this is expressed by

$$R_0 = [(5.86 \times 10^{-25} \phi_s / n^4 \tilde{\nu}_{sa}^4) \int g_s(\tilde{\nu}) G_a(\tilde{\nu}) d\tilde{\nu}]^{1/6} \quad (13.4)$$

where ϕ_s is the quantum efficiency of the sensitizer, n is the refractive index of the host, $g_s(\tilde{\nu})$ and $G_a(\tilde{\nu})$ are the spectral distribution functions of the sensitizer emission and activator absorption spectra, respectively, and $\tilde{\nu}_{sa}$ is the mean wave number in the region of spectral overlap.

Powell[65] obtains a value of about 1.5×10^{-10} for the room temperature hopping time in pyrene-doped napthalene, and a value of R_0 of about 25 Å. This is slightly larger than the value of $R_0 = 23$ Å found for anthracene in napthalene and much greater than the value of $R_0 \approx 12$ Å determined for tetracene in napthalene.[66] Since the capacities are about the same for anthracene and tetracene and slightly smaller for pyrene, this seems to indicate that lattice distortion is the dominant factor in determining the size of the trapping region.

A random walk model has also been applied[6] to describe the transfer and decay of a small Frenkel-type (see Part A, p. 290) exciton. The diffusion length of single excitons[9] perpendicular to the (ab) plane of anthracene has been found[8] to be 598 ± 51 Å; doping with acridine lowers this to 322 ± 80 Å; this distance depends strongly on crystal purity and also on surface conditions. In single crystals of napthalene, exciton diffusion is said[10] to be almost a two-dimensional process: the anisotropy ratio exceeds 2000.

To differentiate between coherent and diffusion limited triplet Frenkel exciton migration in molecular crystals one must specify both the coherence time associated with the wave vector $\mathbf{k}$ and the correlation time associated with the experimental approach used.[75]

The experimental correlation time is of the order of the reciprocal frequency of the applied field. If the lifetime of an exciton $\mathbf{k}$ state is much longer than this, excitons associated with individual $\mathbf{k}$ states may be investigated by the applied field. A complete description of excitron migration in the Frenkel limit requires the electronic states, the phonon states, and phonon-exciton coupling all to be considered in terms of the crystal states. It is the latter interaction that determines the primary mechanism responsible for the electronic energy transfer in solids at both high and low temperatures.[76] At low temperature the density of the phonon states becomes so small that phonon scattering becomes less frequent than energy migration between lattice sites. Once the time between scattering events approaches the lifetime of the excited electron state, a Frenkel exciton can be thought of as a delocalized excitation propagating coherently[79] as a wave packet with a velocity characteristic of its energy and the linear combination of crystal $\mathbf{k}$

states which describe the wave packet: this group velocity is given by[75]

$$V_g(\mathbf{k}) = (2\pi/\hbar)(d\epsilon/d\mathbf{k}) \qquad (13.5)$$

For a one-dimensional crystal,

$$\epsilon(\mathbf{k}) = E^0 + 2\beta \cos ka \qquad (13.6)$$

where $\epsilon(\mathbf{k})$ is the band dispersion associated with translational equivalent interactions along a direction $\vec{a}$. E^0 is the electronic energy of the localized excited state while β is the effective nearest neighbor intermolecular interaction. In a stochastic model the distance, $\ell(\mathbf{k})$, along which an exciton moves coherently, without changing direction or velocity, is given by:

$$\ell(\mathbf{k}) = V_g(\mathbf{k}) \cdot \tau(\mathbf{k}) \qquad (13.7)$$

where $\tau(\mathbf{k})$ is the lifetime of the coherent state, $\ell(\mathbf{k})$ is a mean free path and $\tau(\mathbf{k})$ corresponds to a correlation time for the wave vector state $\mathbf{k}$.[75] Excitons can propagate in the crystal at a variety of velocities and distances depending upon the $\mathbf{k}$ states populated:[75] the group velocity is zero at the top and bottom of the band ($\mathbf{k} = 0$ and $\pm \pi/a$) but $8\pi\beta/\hbar$ at the center of the band ($\mathbf{k} = \pm \pi/2a$). The extent to which these states contribute to the propagation of electronic energy is determined by the population distribution in the band. For a thermal distribution, the number of excitons, $N(\mathbf{k})$, propagating with a velocity, $V_g(\mathbf{k})$, at a given temperature is proportional to the density of states $\rho(\epsilon)$ times the Boltzmann factor:[75]

$$N(k) = \frac{\rho(\epsilon)\exp(-\epsilon(\mathbf{k})/kT)}{\int \rho(\epsilon)\exp(-\epsilon(\mathbf{k})/kT)} \qquad (13.8)$$

It results that[75] $\mathbf{k}$ states in the center of the band can have velocities $10^6 - 10^7$ times those associated with random walk migration, for bands between 1 and 10 cm^{-1}. The coherence length can approach macroscopic dimensions if phonon-exciton scattering is weak (i.e., $\tau(\mathbf{k})$ is long) and the excited states are long-lived (e.g., triplet states). This is only achieved at very low temperatures where the distribution of phonon states approaches the $T \to 0$ limit. At intermediate temperatures the principal limitation of $\tau(\mathbf{k})$ is phonon-exciton scattering. In such cases an exciton, initially in a $\mathbf{k}$ state, scatters to other $\mathbf{k}'$ states *via* phonon interactions in a time short compared to its lifetime, but in a time long compared to intermolecular exchange. As a result[75] the coherence time is shortened, the mean-free path or coherence length is reduced, and the individual $\mathbf{k}$ states acquire a width $\Gamma(\mathbf{k})$, given by the reciprocal of the coherence lifetime of the individual $\mathbf{k}$ states. $\Gamma(\mathbf{k})$ is given by:[75]

$$\Gamma(\mathbf{k}) \equiv (\tau(\mathbf{k}))^{-1} = \sum_{\mathbf{k}'} (\tau_{\mathbf{k}\mathbf{k}'})^{-1} \qquad (13.9)$$

where $\tau_{\mathbf{k}\mathbf{k}'}$ is the probability of an exciton initially in the $\mathbf{k}^{th}$ state, scattering *via* phonon-exciton interactions to a final state $\mathbf{k}'$.

Attention must be given to the relationship between the correlation time associated with exciton migration and the time scale of the particular experimental approach being employed. If the experimental correlation time, which is of the order of the reciprocal of the radiation field, is much shorter than $\tau(\mathbf{k})$ (as is the case for optical absorption), only manifestations of the coherent model are apparent from the data.[75] Conversely, when the experimental correlation time is longer than $\tau(\mathbf{k})$ for all $\mathbf{k}$, only the random walk processes are displayed. A reliable measure of phenomena connecting coherent migration and diffusion limited migration, such as phonon-exciton scattering, $V_g(\mathbf{k})$ and $\ell(\mathbf{k})$, can only be determined when[75] the experimental correlation time is of the order of $\tau(\mathbf{k})$.

However, any real crystal will also contain impurity and other trapping states. In such a crystal, where the energy separation between the impurity (trap) states and the exciton ($\mathbf{k} = 0$) states can be spectroscopically measured, the temperature dependent emission from the band states should be a sensitive function of the energy dispersion of the band.[75] Exciton migration between trap sites so as to equilibrate thermally the excitation within the lifetime of the state depends upon the average velocity of exciton migration which in turn is related to the extent of coherence. For a one-dimensional crystal the partition function z for band and trap states can be written as:[75]

$$z = 1 + e^{-\Delta/kT} + \sum_{k = \pi/na}^{(n-1)(\pi/an)} 2e^{-[\Delta - 2\beta(1 - \cos ka)]/kT} \qquad (13.10)$$

The zero of energy is taken at the energy of the trap which is Δ below the $\mathbf{k} = 0$ state of the band. The dispersion is restricted to a nearest neighbor with an intermolecular interaction β along a translation direction $\hat{a}$. The leading and second terms in the partition function characterize the Boltzmann factor for the trap level and the nondegenerate exciton $\mathbf{k} = 0$ state from which emission occurs. The remaining doubly degenerate terms include the non-$\mathbf{k} = 0$ states. The exciton states and trap states are assumed to be in thermal equilibrium. The trap probability is taken to be $P_{\text{trap}} = 1/z$, and hence the intensity of trap emission is given by $I_{\text{trap}} = 1/z$. For all this to be valid, the system must reach equilibrium within the lifetimes of the excited states, implying that the average excitation migrates between many trap sites *via* the exciton states in a time less than the lifetime of the excitation. In the coherent model the average velocity at a temperature T for the exciton is given by

$$\langle V_g \rangle = \frac{2\beta a}{\hbar} \frac{\sum_k \sin(ka) e^{-(2\beta\cos ka)/kT}}{\sum_k e^{-(2\beta\cos ka)/kT}} \tag{13.11}$$

For a 1 cm^{-1} band at 4.2 K, $\langle V_g \rangle$ = 2500 cm/sec while in the hopping model the median velocity,

$$\bar{V}_{\text{hop}} = \frac{3}{8}\left(\frac{48}{\hbar}\right)^{\frac{1}{2}} a,$$

is a factor of 10^6 smaller. When the impurity density becomes low, $\bar{V}_{\text{hop}}$ is too small to allow the sample to come into thermal equilibrium within the lifetime of the states. On the other hand, coherent migration is more than adequate to ensure equilibrium. The latter description has been found to give excellent agreement with experiment at low temperatures and agrees well with the results of the spin resonance experiments[77] for two crystals which have been studied and which can be considered models for one dimensional exciton migration, viz., 1,2,5,5-tetrachlorobenzene[77] and 1,4-dibromonaphthalene.[78]

Studies by Yoshihara et al.[83] on monomolecular and bimolecular recombination of excitons in several crystals are summarized in Table 13.2. In excitons, recombination *via* the collision of two excitons is the diffusion governed encounter, rather than the subsequent energy transfer step which appears to be rate determining.[83] For such processes, the rate constant δ should be[84] proportional to $T^{-\frac{1}{2}}$ assuming the exciton to move coherently in an energy band. This appears to hold for anthracene. Thermally activated exciton transfer appears to hold for the naphthalene-tetracyanobenzene complex, between 100-297 K, suggesting the applicability of a hopping model. At low temperatures, again, a band model appears to be preferable. Results for pyrene and naphthalene also indicate a mixed regime (see Table 13.2).

Trap-to-trap[66] long range triplet exciton transfer may occur by tunneling, i.e., a super-exchange type interaction through the host exciton band,[67] the direct Förster-Dexter type interaction[68] being of small importance for such long range transfer. The concept of exciton percolation[68] describes free exciton flow through the impurity. Due to the increased trap-to-trap migration length in a long-lived triplet, the effective percolation for the triplet can occur at a much lower concentration than the static percolation which defines the concentration at which the impurity density-of-states becomes a quasi-continuum.[69]

Exciton percolation appears[70] to be highly dependent on the trap-depth, showing the importance of the indirect, i.e., superexchange interaction through the host exciton band. Using this approach, for trap-to-trap ("virtual

Table 13.2 Rate Constants for Monomolecular and for Bimolecular Exciton Recombination.

After K. Yoshihara et al.[83]

α and γ are the rate constants for unimolecular and for bimolecular recombination, respectively. The temperature dependence of γ is obtained from fluorescence decay times τ at temperatures T. D_3 stands for the diffusion coefficient.

Compounds	Temperature Dependence of γ on D	ϵ (M^{-1} cm^{-1})	τ (nsec)	γ (cm^3 sec^{-1})
Anthracene	$1/\sqrt{T}$ (5 ~ 250 K)	9000 Monomer	18	5×10^{-11} (300 K)
Naphthalene –TCNB	$1/\sqrt{T}$ (5 ~ 50 K) $\exp(-\Delta E/kT)$ (~100 ~ 300 K)	2380 CT$^{(2)}$	18 19	3×10^{-13} (5 K) 6×10^{-13} (300 K)
Pyrene	constant (5 ~ 100 K) $\exp(-\Delta E/kT)^{(2)}$ (~120 ~ 300 K)	---- Excimer	182 106	2×10^{-15} (5 K) 9×10^{-15} (300 K)
Naphthalene$^{(1)}$	$1/\sqrt{T}$ (6 ~ 100 K)	290 Monomer	--	-----

ϵ: Molar extinction coefficient at the first absorption band in solution, CT: Charge-transfer band, $\Delta E \sim 160$ cm^{-1}, $\Delta E \sim 330$ cm^{-1}. $\Delta E \sim 440$ cm^{-1} was obtained using host-guest systems.$^{(2)}$

$^{(1)}$A. Hammer and H. C. Wolf, *Molec. Cryst.*, 4, 191 (1968).
$^{(2)}$W. Klopffer, H. Bauser, F. Dolezalek, and G. Naundorf, *Molec. Cryst. Liq. Cryst.*, 16, 229 (1972).

band") transfer one obtains good agreement with experiment without adjustable parameters.[70] In "long-range" percolation, a long-range bond[71] is defined as a succession of near-neighbor bonds. Using solely the known interchange equivalent[72] nearest-neighbor exciton interactions again, one gets agreement with experiment. Due to the finite lifetime of the triplet state, the excitation can move by trap-to-trap migration only[70] through a certain number of trap sites. The separation between trap sites is the tunnel length.[70] The effective tunnel length for doped naphthalenes is about 6 to 7 nearest neighbor bonds in succession and is closely related to the tunneling (or hop) time t. Also related is the tunneling (superexchange) energy interaction, which in this case is only of the order of 10 to 10^3 Hz, compared to exchange interactions of the order of 1 cm^{-1} (3×10^{10} Hz). This gives impurity exciton conduction bandwidths of 10 to 10^3 Hz.

Triplet exciton diffusion in anthracene has been claimed to be mainly in the (ab) plane; trapping occurs at dislocations due to e.g., plastic deformation which generate traps of a depth of 0.3 eV.[17] The coherence time of excitons in naphthalene has been found to be limited to 10-50 psec, because of phonon scattering.[11] The kinetics of energy transfer from a single crystal of tetracene host to guest molecules of pentacene have been studied by Campillo.[12] The theory of phonon scattering of excitons in molecular crystals such as naphthalene and anthracene is treated by Dissado and Brillante.[13] The ESR spectra of triplet excitons in single crystal pyrene have been studied by Bizzaro et al.[14]

Triplet exciton diffusion in molten naphthalene, anthracene and pyrene has been studied by Baessler[16] by a photoconductivity arrangement using a sandwich cell. Holes are injected into the matrix by dissociation of excitons striking the anode. The diffusion coefficient of naphthalene is reported[16] as $6 \pm 2 \times 10^{-2}$ cm^2/sec while that of anthracene is 1/100 of that value. Inoue et al.[80] report a value of 15 Å for the critical transfer distance for triplet-triplet exciton transfer in naphthalene and of 13 Å for phenanthrene.

In disordered lattices, theory[73] indicates that the decay time of the excitation, $P_0(t)$, tends exponentially towards zero above the percolation threshold P_c, while below same it tends towards a constant value. In the intermediate region, $P_0(t)$ appears to tend to be diffusion limited as in linear chains.[74]

The interaction of exciton transfer with proton relaxation in organic solids has been studied by Schwarzer and Haken.[81] The theory of polaron motion at low temperature is discussed by Kartheuser et al.[82]

13.4 Exciton Reactions

Probably the most important exciton reaction is that at a metallic contact giving rise to photoconductivity. This matter is discussed in Section 2.19, 6.11, 8.11, and in Chapter 22.

In this effect, only those exciton reactions are registered which lead to formation of a charge carrier. From quantum yield studies for hole production by singlet excitons in anthracene crystals it is found[85] that the rate constant for singlet exciton oxidation decreases in an exponential fashion with reaction distance:

$$k_{CT} = k_{CT,0} \exp(-d/d_0) \tag{13.12}$$

where $k_{CT,0} = 3 \times 10^{13}$ sec^{-1} and $d_0 = 2.2 \pm 0.3$ Å. The quantum yield allows for the probability that the carrier generated inside the crystal escapes geminate recombination with its image charge (see Section 6.11).

For reduction of triplet excitons in a chloranil crystal[86] $k_{CT,0} = 4 \times 10^{13}$ sec^{-1} and $d_0 = 1.2 \pm 0.1$ Å. The functional dependence $k_{CT}(d)$ indicates that charge carrier formation occurs independently of energy transfer. It must be due to a charge transfer process, in the course of which the excited molecule ejects an electron or a hole to empty metal states across a potential barrier. d_0 is characteristic of the surface barrier formed either by a fatty acid surface layer or the last molecular layer adjacent to the interface. $k_{CT,0}$ represents the rate constant at which an excited state located at the very interface dissociates. At a normal anthracene-aluminum contact approximately 20 percent of the singlet excitons striking the interface dissociate at a rate constant $k_{CT} = 6 \times 10^{10}$ sec^{-1}. For triplet excitons dissociation is the dominant decay channel. The rate at which excitons are quenched at the surface is usually controlled by diffusion of the excitons towards the active surface rather than by the reaction itself. From the solution of the diffusion equation under steady-state conditions it follows[97] that the total number of excitons disappearing per unit time and area is

$$N_{ex} = I_0 \cdot \frac{l_a}{(l_a + l_d)} \cdot \frac{1}{1 + \left(\frac{l_d}{c\tau_\infty \sum_i k_i}\right)} \quad (13.13)$$

I_0 is the incident photon flux, l_d and τ_∞ denote diffusion length and intrinsic lifetime of the excitons. Σk_i can be expressed as $k_q + k_{CT}$, where k_q refers to exciton quenching by energy transfer or induced intersystem crossing and k_{CT} to exciton dissociation *via* charge transfer. $\dot{N}_{ex}$ can be obtained from an experiment counting the total number of exciton quenching events at the surface, e.g., fluorescence quenching studies.[88]

For singlet and triplet excitons in crystalline anthracene $l_a/(c\tau_\infty)$ has the values[89] 4×10^9 and 4×10^5 sec^{-1}, respectively, for triplet excitons in a p-chloranil crystal $l_a/(c\tau_\infty) \approx 10^7$ sec^{-1}. Chemical reactions involving excitons play a major role in catalysis (see Section 12.11 and 23.2). The subject has been reviewed.[90]

13.5 Bi-excitons and Exciton Clusters

Two excitons may interact forming virtually an excitonic molecule, or bi-exciton.[24] These are produced e.g., by the (simultaneous) absorption of two light quanta and may form a "cold gas," which is called that because there is virtually no interaction between its constituents.[27] Recombination is strongly affected by polariton (see Section 13.8) interactions.[27] Exciton cluster-states in benzene have been studied by Le Sar and Koppelman;[28] these are pseudo-localized states in the middle of the energy band formed from extended states. The generalized theory of these and related entities is discussed by Sakoda.[29] Singlet bi-excitons formed *via* absorption of 2 quanta in anthracene are said to be quenched by triplet excitons.[30]

The binding energy W of the bi-exciton is given by[91]

$$W = E - 2E_{exc} \qquad (13.14)$$

where E is the total energy of the bi-exciton as calculated e.g., by the variational method, and E_{exc} refers to the energy of a single exciton. This assumes that the effective masses of the constituent electrons and holes are equal. W depends strongly on the phonon-electron coupling constant as well as on the energy ζ of longitudinal optical phonons:

$$W = f(\zeta); \qquad \zeta = \frac{\epsilon_0}{\epsilon_\infty - 1}; \qquad \zeta = \left(\frac{2\hbar\omega}{R_{ex}}\right)^{1/2} \qquad (13.15)$$

R_{ex} refers to the Rydberg constant for the exciton and ϵ_0 and ϵ_∞ stand for the permittivity at low and at optical frequencies, respectively. For values of ζ below 2, W is a monotonically decreasing function of ζ;[91] as ζ increases, the bi-exciton becomes less stable.[91]

A bi-exciton can also result from the interaction of two coupled polarons of the Wannier exciton type.[92]

Exciton-exciton interactions and their dissociation in the electric field of a charged defect have been studied by Petelenz.[23]

Conduction electrons may become[94] bound to the electronic excitation and form an electron-exciton complex. This complex is necessarily unstable with respect to auto-ionization and can decay by returning the excess electron to the conduction band. In the case of elastic scattering, the auto-ionizing electron returns to the conduction band with the same quantum numbers as before the reaction. Therefore, it has a high probability of occupying the same quantum state at a nearby equivalent site: the auto-ionizing electron is likely to be recaptured by the same resonant process in a nearby molecule and tends to propagate through the lattice by a continuous creation and annihilation of an exciton. The creation of an exciton causes the electron to occupy the resonant level E_R lying below the vacuum level.[94] If E_R is fairly independent of exciton energy, it becomes possible to align the energy levels of the solid with the transmission maxima by increasing the energy scale by a constant factor E_R. In molecular solids, electronic excitation occurs at nearly the same energy as in the gas, and alignment with the gas-phase transitions is also possible.[94] The energy shifts of 2.25 and 2.70 eV for benzene and pyridene, respectively, are therefore an experimental determination of E_R.

The methodology involves[95] a well-collimated electron beam directed towards the target as a thin (about 100 Å) film, and measuring the current as a function of electron energy. Such high resolution measurements have been made with benzene, fluorobenzene, pyridine, benzaldehyde, thiophene and furan.[94]

Multi-exciton complexes in semiconductors are said[93] to be associated with zero energy oscillations. The free complexes are unstable[93] but may become stabilized if formed on a fixed impurity center. These complexes are reported[93] to retain their individuality and to become transformed into wider electron/hole systems.

For a theoretical treatment of double and multiple excitations, see several papers by Hush and associates.[119]

The thermodynamics of excitons in semiconductors are treated by Combescot;[19] he concludes that this entity gives rise to a "liquid" and may evaporate into a "gas." Electron-hole "droplets" in silicon and germanium are well known[20] to occur at temperatures of a few K, where they form a metallic phase consequent to a transition from a neutral excitonic "gas" to a metallic liquid,[20] if the density of excitons is high enough.[26]

Most of the experiments have been done with germanium, because the drops therein are relatively long-lived, but they have also been found in phosphides etc, though short-lived. Their diffusion in the absence of external forces is below about 10^{-9} cm^2/sec, non-thermal phonons, the so-called phonon wind[22] can cause them to move much faster.[21] These phenomena should also be investigated in organic solids and in molecular crystals. There are many references and reviews on this topic.[96]

13.6 Exciplexes

This term refers to excited complexes, the main discussion on this topic will be found in Section 17.5. Several reviews on this subject are available.[97]

Triple exciplexes of the form $(D_2A)^*$ or $(DA_2)^*$ arise, e.g., in dicyanobenzene-naphthalene complexes;[98] this is said[99] to occur especially at high donor or acceptor concentrations. The exciplex decay time in doped polyvinylcarbazole is reported to be proportional to the square of an applied electric field.[100]

Exciplex formation in anthracene as well as photodimerization has been studied by Costa and Melo.[116]

13.7 Excimers

An excimer is a metastable dimer formed between an excited molecule and one in its ground state. In polymers, they may result from an intramolecular reaction between adjacent molecules on the chain, or can be formed intermolecularly. Their role in charge transfer complexes is discussed in Section 17.5. It is interesting to note that dinaphthylamine is unique in that it forms both an intramolecular excimer and exciplex.[120]

In e.g., anthracene and iminophenanthrene, excimeric regions behave as defects[31] causing broad, red-shifted emission bands. They appear to have much longer relaxation times than the monomeric, or unassociated, more perfect regions.[31] The potential energy barrier between an excimer and the excited monomeric state in tetracene is said to be only 0.07 eV, over the temperature range[32] from 85 to 300 K. This value compares to an activation energy of 0.035 eV for the excimer formation in single crystal perylene;[104] it is reported that no activation energy is required[104] for the excitation of a monomer defect.

The excitation energy may be transferred from a given excimer to a neighboring pair of monomers, resulting in excimer diffusion and energy transfer. This may occur by one of the following processes.[33]

1. the excimer dissociates into an excited monomer plus a molecule in its ground state. The energy of the excited monomer is then transferred to one molecule of an adjacent pair of molecules which then forms another excimer by interacting with its neighboring ground-state-molecule, or

2. the excimer expands to the equilibrium spacing of a pair of molecules in their ground state and transfers its energy to a neighboring dimer, or

3. a dimer in its ground state contracts to the excimer equilibrium spacing and receives the excitation energy from a neighboring excimer having the same spacing.

Process (1) is unlikely at room temperature because of insufficient thermal energy being available; (2) is considered the most likely.[34]

Excimers play a role in exciton diffusion: thus, Frand and Harrah,[41] in a detailed study of excimers in polymers, show that excitons move by thermally activated hopping to preformed excimer sites, though Yokoyama et al. argue[42] that the trapping of triplet excitons at excimer-forming sites is less important than it is for singlet excitons.

Excimer-to-excimer energy transfer, probably by a hopping mechanism in alternative (1) has been reported[101] e.g., in pyrene; in anthracene it appears that the transfer also involves molecular movement,[102] perhaps rotation (see Section 15.8).

The lifetime of excimer (emission) decay at 90 K is 160 ± 5 nsec[103] and is insensitive to impurities. The decay time for a pure crystal at 300 K is 57 ± 4 nsec; this time is shorter (41 ± 3 nsec) in "impure" crystals. These exhibit, in addition, a second and more rapid decay component probably

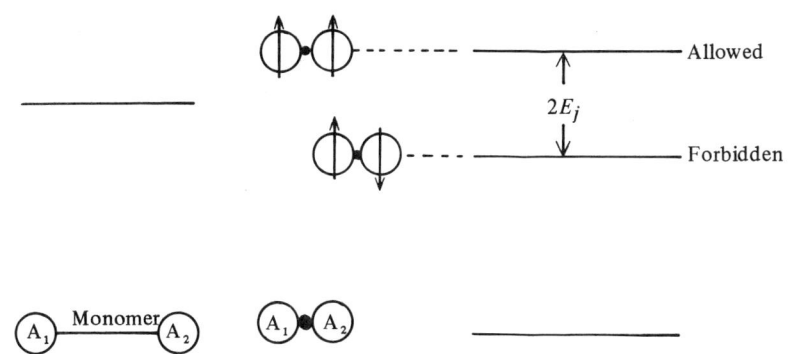

Fig. 13.4 Energy levels of an excited dimer. The arrows point to the direction of the transition dipole. In the illustrated case, the transition dipole is perpendicular to the dimer axis. E_j is the dipole interaction energy. After Sybesma.[117]

due to impurity emission. The lifetime of the excimer is also short in samples with a high defect population. Thus, it appears that at 300 K but not at 90 K, energy is transferred *via* impurity molecules and defect sites.

Figure 13.4 illustrates the energy level diagram for a dimer whose transition dipole moments are perpendicular to the axis through the center of each dipole. In this case, the allowed state is the upper antibonding one where the transition dipoles are parallel to each other (in phase). The lower state is forbidden: since the dimensions of the dipole are small compared to the wavelength of the exciting light, the molecules must be in the same region of the radiation field. The phase of the electromagnetic wave is, therefore, the same throughout that region, and the generated dipoles must be in phase with each other. If the transition dipoles are aligned with the dimer axis the lower (bonding) state would be the one in which the transition dipoles are in phase and would, hence, be the allowed one. In an oblique arrangement both dipoles have perpendicular and parallel components and both levels are allowed. The energy gap between the levels is twice the dipole interaction energy.

13.8 Surface Excitons. Polaritons.

Excitons are liable to form at surfaces, part or all of the excitation energy coming from the surface energy (see Section 12.17). Such have been observed in anthracene,[46] in tetracene[47] and in many mixed molecular crystals.[48] Their binding energy is of the order of 0.1 eV;[49] these surface excitons are closely related to surface charge transfer complexes (see Section 17.8).

The surface polarization charge at a surface or interface gives rise to an electromagnetic wave travelling along that surface; a localized surface state arises from the coupling of a photon with the electronic and vibronic transitions in the crystal.[121] The polariton therefore is a moveable excited surface polarization.[54,122] Vibrationless, non-vibronic polaritons have also been observed;[54] below 100 K, the polariton propagates in e.g., anthracene along its b-axis in the (ab) face by moving coherently over distances of several 10-100 Å; above 100 K the polarization is rapidly scattered by lattice phonons leading to what has been called "bulk polaritons."[54] From reflectance spectra, the existence of three localized surface states in anthracene has been deduced.[60]

Polaritons may arise from the coupling of an incident photon to the electric dipole polarization created by surface excitons; they are non-radiative. Their associated electromagnetic wave remains confined to the surface, or interface, by electric or magnetic polarization interactions. Their electric field decays exponentially on either side of the interface or surface, and the field vector is polarized in the sagittal plane; the real part of the dielectric permittivity may assume negative apparent values.[50]

Polaritons in mixed molecular crystals have been studied *inter alia* by Ueba and Ichimura[105] who use anthracene-perdeuterated anthracene as the model system. Surface plasmons, i.e., polariton dispersion have been investigated by several authors.[106] They tend to arise especially if the bulk excitation takes the form of Wannier excitons and are associated with macroscopic electromagnetic surface fields decaying exponentially in amplitude with distance from the surface.[106] Surface condition, e.g., roughness, affects the dispersion because the polaritons are eigenstates of the surface (or interface).

13.9 Dipole-Dipole (Förster) Transfer

The basics of this mechanism are discussed in Section 5.12, p. 309. This mode of transfer appears to be of considerable relevance to energy transfer processes in photosynthesis where the singlet excitation energy following the absorption of a quantum of light is transferred from molecule to molecule until the energy reaches the photosynthetic reaction center.[112] A four-state model of Förster transfer is described in Fig. 13.5. The system consists of molecules A and B and the surrounding solvent medium. The first symbol in parentheses following the letter designating the molecule (*i.e.,* A or B) denotes the populated electronic state of the molecule and the second symbol denotes the electronic state for which the intramolecular geometry and solvent arrangement are appropriate. Thus, $A(S_1, S_0)$ means that molecule A is in its first excited electronic state, S_1, with an intramolecular geometry and

DIPOLE-DIPOLE (FÖRSTER) TRANSFER

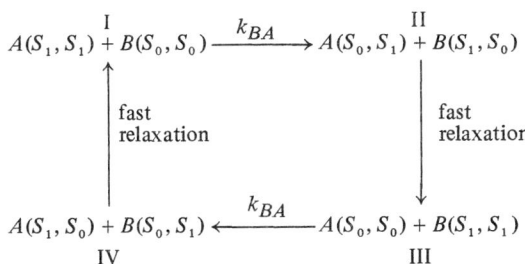

Fig. 13.5 A four state diagram representing Förster transfer between molecules A and B. See text for a description of the symbols. After Shipman and Housman.[113]

solvent arrangement appropriate for its ground electronic state, S_0. The Förster transfer rate constants are k_{AB} and k_{BA}, Fig. 13.5. The other two rate constants, designated 'fast relaxation,' are the rate constants for thermal equilibration to a Boltzmann population of vibrational levels; this relaxation should be of the order 10^{12} sec^{-1} at 25°C.

In the range of applicability of the Förster transfer mechanism, the equilibration rate constant is much faster than both k_{AB} and k_{BA}, and therefore the equilibrium populations of II and IV are quite small, while the ratio of the populations of I and III is k_{BA}/k_{AB}.

The Förster transfer rate is directly proportional to the square of the interaction energy between the transition densities on molecules A and B. For intermolecular spacings large compared to the extent of the transition density on a molecule, the dipole-dipole term in a multipole expansion dominates the interaction energy.

The limits of applicability of the dipole-dipole approximation have been considered by Chang.[114] It has been assumed[113] that the electronic excitation is localized completely on one molecule or the other. For sufficiently strong transition density coupling, the electronic excitation will be delocalized extending over more than one molecule. Since the transition density interaction falls off as R^{-3}, this is a more serious problem when the molecules are close together than when they are far apart. At close contact between molecules with significant overlap between the molecular orbitals, the charge transfer states corresponding to the transfer of an electron from A to B or *vice versa* can mix with the locally excited states and change the character of the excited states involved. Since the overlap falls off exponentially with separation, the mixing-in of charge transfer character is only important for short distances. The model appears to hold for separation distances of the order of 25 Å such as arise in photosynthetic arrangements of chlorophylls.[113]

13.10 Other Collective Localized Excitations

These are of primary importance to the study of amorphous systems, though by no means exclusively so. (See discussion in Section 19.4.)

Collective excitations called plasmons[51] associated with plasma effects and often with dipoles imbedded in a non-polar medium, may be probed by infrared spectroscopy; they are of importance also in some crystalline media.[35] In narrow band semiconductors, a resonance effect between the width of the energy gap and the plasma frequency may cause very fast electron-hole recombination.

Plasmons in crystalline media and in molecular crystals especially have been reviewed.[107] Their anisotropic behavior is reported to be a good guide to the reduced dimensionality as e.g., in TTF/TCNQ complexes.[108]

Similar collective plasmalike oscillations, which can also be called polarization waves, exist in systems of neutral molecules with permanent dipole moments which are free to precess and interact *via* long range dipole-dipole interactions. Such modes have been studied theoretically in bulk dipolar liquids.

Dispersion relations for interfacial dipolar plasmons in liquid mono- and multilayer structures have been studied by Banville et al.[52] They use dry layers of fatty acids in which the terminal methyl group is halogenated.

In polymers, excitations called "solitons,"[109] are kinks moving along the polymer chain, separating two domains, to the left and right of the kink, whose single-double bond alternative patterns are out-of-phase with one another. At low doping levels it is energetically more economical for a carrier to bind to a soliton kink than to enter the conduction (or valence) band.

The soliton thus acts as a trap. Some workers have proposed that movement of charged solitons, perhaps by diffusion, is the dominant contribution to the conductivity of lightly doped, or somewhat impure, organic polymers. However, the matter is still highly controversial.[110]

Since solitons may arise in non-linear dispersive media[115] as particle-like stable waves, their importance transcends that of the subject of this volume; they play a role in non-linear dissipative transmission lines and lattice shock waves to mention just a few. Formally, the soliton is a solitary wave solution of a non-linear term; physically, solitons are highly localized entities. A two-dimensional soliton is sometimes referred to as a vorton.

Highly mobile defects may arise from neutral free radicals especially in undoped polyacetylene films.[53]

Local rearrangement modes, so-called softons, as well as entities produced by their association with one electron, a softaron, or two electrons, a bi-softaron, are discussed in the context of amorphous solids (Section 19.4).

References

1. A. S. Davydov, "Theory of Molecular Excitons," Plenum Press, N.Y., 1971; G. Klein, "Radiationless Decay, Fission and Fusion of Excitons in Irradiated Molecular Crystals," CRN–CNPA–77–4 Rept. 1977; Avail. Inis. M. D. Galanin, *Acta Phys. Chem.*, 23, 83 (1977); E. A. Salkov, *Visn. Akad. Nauk. UKR. SSR*, 1977, 25, (Reviews Russian work); M. P. Lisitsa, *Zh. Prikl. Spektrosk.*, 27, 589 (1977); E. C. Lim, "Excited States," Academic, N.Y., 1979, Vol. 4 and preceeding volumes; W. Hanke, *Adv. Phys.*, 27, 287 (1978); H. Meier et al., *Chem. Eng. Tech.*, 51 (6), 653 (1979); C. G. Morris, "Mol. Rate Process," Paper Symposium (1979), p. F2, RACI, Melbourne; R. G. Ulbrich, *Solid State Electronics*, 21, 51 (1978).

2. H. H. Seliger and J. P. Haman, *Jerusalem Quantum Chem. Biochem.*, 10, 345 (1977).

3. E. I. Rashba and A. B. Zimin, "Proc. 6th Molec. Cryst. Symp.," Elmau (Ger.) 1973, p. 20.

4. L. Fredin and B. Nelander, *J. Molec. Struct.*, 16, 205 (1973); B. Nelander, *Theor. Chem. Acta* (Berlin), 25, 283 (1972).

5. D. A. Seanor, in "Polymer Science," A. D. Jenkins, ed., Nth. Holland, 1972, Ch. 17.

6. W. Spannring and H. Bässler, *Chem. Phys.*, 25, 325 (1977).

7. R. Nuyts and P. Phariseau, *Physics*, A-90, 260 (1978); J. Klaffer and Y. Jortner, *J. Chem. Phys.*, 68, 1513 (1978); D. T. Clark, *Prog. Theoret. Org. Chem.*, 2, 426 (1977); See also Ref. 1; D. C. Reynolds and Th. C. Collins, *Excitons*, Academic, New York, 1980.

8. D. Donati and J. O. Williams, *Mol. Cryst. Liq. Cryst.*, 44, 23 (1978).

9. H. Bässler, *Electrochim. Acta*, 13, 1497 (1968).

10. L. Altwegg et al., *Phys. Rev.*, B-14, 1963 (1976).

11. P. Argyrakis and R. Kopelman, *J. Chem. Phys.*, 66, 3301 (1977).

12. A. C. Campillo, *Chem. Phys. Lett.*, 52, 11 (1977).

13. L. A. Dissado and A. Brillante, *J. Chem. Soc. Faraday Trans.*, II, 73, 1262 (1977).

14. W. Bizzaro et al., *Chem. Phys. Lett.*, 53, 49 (1978); *Phys. Stat. Solidi*, B-84, 27 (1977).

15. H. Atzmüller and V. Schröder, *Phys. Stat. Solidi*, B-89, 349 (1978).

16. H. Bässler, *J. Chem. Phys.*, 49, 5198 (1968).

17. A. Sasaki and S. Hayakawa, *Japan J. Appl. Phys.*, 17, 283 (1978); D. D. Dlott et al., 67, 3808 (1977).

18. R. W. Munn and R. Silbey, *J. Chem. Phys.*, 68, 2439 (1978); A. Elli, *Phys. Rev.*, B-16, 5443, (1977); B. J. Mulder, *Philips Res. Repts.*, 23, 128 (1968); R. Silbey, *Ann. Revs. Phys. Chem.*, 29, 203 (1978).

19. M. Combescot, *Phys. Stat. Solidi*, B-86, 349 (1978).

20. H. Combescot and P. Nozieret, *J. Phys.*, C-5, 2301 (1972); Ya. E. Pokrovskii, *Phys. Stat. Solidi*, A-11, 385 (1972); R. N. Silver, *Phys. Lett.*, 44-H, 61, (1973); J. Hensel et al., *Phys. Rev. Lett.*, 31, 386 (1973); J. McGroddy et al., *Solid State Commun.*, 13, 1801 (1973); V. Marello et al., *Phys. Rev. Lett.*, 31, 583 (1973).

21. J. C. Hensel and R. C. Dynes, *Phys. Rev. Lett.*, 39, 969 (1977); J. C. Hensel, *Bull. Amer. Phys. Soc.*, 13, 421 (1978); M. Greenstein and J. P. Wolfe, *Phys. Rev. Lett.*, 41, 715 (1978); J. Doehler and J. M. Worlock, *Solid State Comm.*, 27, 229 (1978); *Phys. Rev. Lett.*, 41, 980 (1978).

22. Editorial, *Physics Today*, 26, 17 Dec. 1973; *ibid.*, 32, 17 Feb. 1979; Y. E. Pokrovskii, *Phys. Stat. Solidi*, A-11, 385 (1972); C. D. Jeffries, *Science*, 189, 955 (1975).

23. P. Petelenz, *Acta. Phys. Polon.*, A-53, 177 (1978); *Inst. Fiz. Jag.* (Krakow), 918-C (1976).

24. A. R. Hassan, *Solid State Commun.*, **25**, 817 (1978); F. Mouross and H. Buettner, *Phys. Stat. Solidi*, **B94**, 107 (1979).

25. W. Steudle et al., *Chem. Phys. Lett.*, **54**, 461 (1978).

26. R. M. Rice et al., *Solid State Phys.*, **32**, 1 (1977); E. Hanamura and H. Haug, *Phys. Rep.*, **33C**, 209 (1977); J. H. Rose and M. B. Shore, *Phys. Rev.*, **17**, 1884 (1978); R. K. Kalis and P. Vashishta, *ibid.*, 2653; A. Mandliu and C. K. Kittel, *ibid.*, 2685.

27. E. Ostertag et al., *Phys. Stat. Solidi*, **B-84**, 673 (1977).

28. R. Le Sar and R. Koppelman, *Chem. Phys.*, **29**, 289 (1978).

29. Sh. Sakoda, *J. Phys. Soc. Japan*, **44**, 211 (1978).

30. M. E. Michel-Bayerle et al., *Phys. Stat. Solidi*, **B-85**, 45, (1978).

31. J. D. Williams et al., "Laser Chem. Proc. Conf." M. A. West, ed., Elsevier, Amsterdam, 1977, p. 426.

32. D. D. Kolendritskii et al., *Opt. Spektrosk.*, **44**, 281 (1978).

33. W. Klöpffer et al., *Molec. Cryst. Liq. Cryst.*, **16**, 229 (1972).

34. Z. Lebovitz et al., *J. Chem. Phys.*, **69**, 647 (1978).

35. E. Tosatti, "Plasmons in Crystalline Media," in "Interaction Radiat. Condensed Matter Lect." Internatl. Winter Coll. 1976, publ. 1977, Internatl. Atomic Energy A., Vienna, Austria, 1977.

36. S. C. Abbi and D. M. Hanson, *J. Chem. Phys.*, **60**, 319 (1974).

37. F. Devreux et al., *Phys. Lett.*, **A-46**, 49 (1973).

38. V. A. Terentev, *Zh. Fix. Khim*, **47**, 2083 (1973).

39. F. Gutmann, *Electrochim. Acta*, **11**, 1099 (1966).

40. R. C. Hughes, *J. Chem. Phys.*, **58**, 2212 (1973).

41. G. W. Frank and L. A. Harrah, *J. Chem. Phys.*, **61**, 1526 (1974). cf. also G. W. Frank, *ibid.*, 2015.

42. M. Yokoyama et al., *Chem. Lett.*, 509, (1973).

43. J. Appel, *Solid State Physics*, **21**, 193 (1968), F. Seitz et al., eds., Academic Press, N.Y., 1968; J. J. Nettel and S. Heinekamp, *J. Phys. Chem. Solids*, **40**, 597 (1979); K. Okamoto and S. Takeda, *J. Phys. Soc. Japan*, **37**, 333 (1974).

44. D. W. Whillans and H. E. Johns, *J. Am. Chem. Soc.*, **93**, 1358 (1971); cf., also H. E. Johns, "Proc. 23rd Internatl. Cong. Pure and Applied Chem.", Boston, Mass., 1971, 8, 11 (1971), Butterworth, London, 1971.

45. P. Schlotter et al., *Phys. Stat. Solidi*, **B-81**, 521 (1977).

46. M. R. Philpott and J. M. Turlet, *J. Chem. Phys.*, **64**, 3852 (1976); J. M. Turlet et al., *J. Luminesc.*, **18-19**, 1, 47 (1979).

47. M. R. Philpott and J. M. Turlet, *J. Chem. Phys.*, **64**, 4260 (1976).

48. H. Ueba and S. Ichimura, *J. Chem. Phys.*, **70**, 1745 (1979); *J. Phys. Soc. Japan*, **41**, 1974 (1976).

49. R. Del Sole and E. Tosatti, *Solid State Comm.*, **22**, 307 (1977).

50. M. R. Philpott and J. D. Swalen, *J. Chem. Phys.*, **69**, 2912 (1978).

51. R. Brout and P. Carruthers, *Lectures in the Many-Body Problem*, Wiley, N.Y., 1963; *J. Chem. Phys.*, **66**, 3664 (1977); T. Miyakawa, *Solid State Comm.*, **25**, 133 (1978); M. Fukui et al., *J. Phys. Chem. Solids*, **40**, 523 (1979); W. H. Weber and C. E. Eagen, *Bull. Amer. Chem. Soc.*, **24** (3), 441 (1979).

52. B. M. Banville, *J. Chem. Phys.*, **66**, 3664 (1977).

53. I. B. Goldenberg et al., *J. Chem. Phys.*, **70**, 1132 (1979).

54. K. Tomioka, *J. Chem. Phys.*, **66**, 2984 (1977); M. R. Philpott and J. D. Swalen, *J. Chem. Phys.*, **69**, 2912 (1978); R. T. Holm and E. D. Palik, *Phys. Rev.*, **B-17**, 2673 (1978); G. Borster et al., *Solid State Phys.*, **74**, 107 (1974); K. L. Kliewer and R. Fuchs,

Adv. Chem. Phys., **27**, 355 (1974); A. Otto, *Adv. Solid State Phys.* (Festkörperprobleme), **14**, 1 (1974); H. Räther, *Phys. Thin Films,* **9**, 145 (1977).

55. V. H. Smith and P. Petelenz, *Phys. Rev.*, **B-17**, 3253 (1978); J. Pollman and H. Buettner, *Phys. Rev.*, **B16**, 4480 (1977).

56. Yu. P. Petelenz, *Pr. Nauk Inst. Chem. Org. Fiz. Politech. Wroclaw*, **16**, 257 (1978).

57. K. K. Bajaj and C. Aldrich, *Bull. Amer. Phys. Soc.*, **24**, 337 (1979).

58. R. E. Amritkar and N. Kumar, *Solid State Commun.*, **26**, 627 (1978); G. Hanamura and H. Hang, *Phys. Repts.*, **33**, 209 (1977).

59. N. F. Mott and A. M. Stoneham, *J. Phys.*, **C-10**, 3391 (1977).

60. M. Turcel and M. R. Philpott, *J. Chem. Phys.*, **62**, 2777 (1975); *J. Chem. Phys. Lett.*, **35**, 92 (1975); V. V. Bondar and M. V. Kurik, *UKR. Fiz. Zh.*, **22**, 1922 (1977).

61. A. Matsui and Y. Ishii, *J. Phys. Soc. Japan*, **23**, 581 (1967).

62. A. S. Davydov, Ref. 1.

63. M. G. Sceats and G. C. Morris, *Phys. Stat. Solidi*, **14A**, 643 (1972); G. C. Morris and M. G. Sceats, *J. Chem. Phys.*, **16**, 375 (1974); *Chem. Phys.*, **3**, 332, 342 (1974).

64. V. L. Broude, "6th Molecular Crystal Symp. Proc.," Elmau; W. Germany, 1973, p. 49.

65. Z. G. Soos and R. C. Powell, *Phys. Rev.*, **B-6**, 4035 (1972); R. C. Powell, *J. Chem. Phys.*, **58**, 920 (1973); see also D. D. Dlott et al., *J. Chem. Phys.*, **67**, 3808 (1977); and R. Keiper and R. Schuchardt, *Phys. Stat. Solidi*, **B-85**, 155 (1978).

66. M. A. El-Sayed, M. T. Wauk and G. W. Robinson, *Mol. Phys.*, **5**, 205 (1962); G. C. Nieman and G. W. Robinson, *J. Chem. Phys.*, **37**, 2150 (1962); G. W. Robinson and R. P. Frosh, *J. Chem. Phys.*, **38**, 1187 (1963); H. Sternlicht, G. C. Nieman and G. W. Robinson, *J. Chem. Phys.*, **38**, 1326 (1963); G. F. Hatch and G. C. Nieman, *J. Chem. Phys.*, **48**, 4116 (1968); S. D. Colson and G. W. Robinson, *J. Chem. Phys.*, **48**, 2550 (1968); R. Kopelmann, *Rec. Chem. Progr.*, **31**, 211 (1970); R. Kopelmann et al., *Chem. Phys.*, **19**, 413 (1977); H. K. Hong and R. Kopelmann, *J. Chem. Phys.*, **55**, 724 (1971).

67. D. L. Dexter, *J. Chem. Phys.*, **21**, 836 (1953); Th. Förster, *Annal. Phys.*, Ser. 6, Vol. 2, **55**, 55 (1948), English translation, by R. S. Knox, Univ. of Rochester, N.Y., 1974.

68. R. Kopelman, E. M. Monberg, F. W. Ochs and P. N. Prasad, *J. Chem. Phys.*, **62**, 292 (1975); R. Kopelman, *Topics in Applied Physics*, Vol. 15: "Radiationless Processes in Molecules and Condensed Phases," F. K. Fong, ed., Springer, Berlin, 1976; J. Hoshen and R. Kopelman, *J. Chem. Phys.*, **65**, 2817 (1976); R. Kopelman, *J. Phys. Chem.*, **80**, 2191 (1976).

69. H. K. Hong and R. Kopelman, *J. Chem. Phys.*, **55**, 5380 (1971).

70. R. Kopelman et al., *Chem. Phys.*, **19**, 413 (1977).

71. J. Hoshen, E. M. Monberg and R. Kopelman, unpublished; see also J. Hoshen and R. Kopelman, *Phys. Rev.*, **B-14**, 3438 (1976).

72. D. M. Hanson, *J. Chem. Phys.*, **52**, 3409 (1970); D. M. Hanson and G. W. Robinson, *J. Chem. Phys.*, **43**, 4174 (1965); T. L. Muchnik, R. E. Turner and S. D. Colson, *Chem. Phys. Lett.*, **42**, 570 (1976).

73. J. Heinrichs and N. Kumar, *Bull. Amer. Phys. Soc.*, **24**, 308 (1975).

74. S. Alexander et al., *Phys. Rev.*, **B-17**, 4311 (1978).

75. C. B. Harris, "6th Molecular Crystal Symp.," Elmau, Proc., W. Germany, 1973, p. 29.

76. M. Grover and R. Silbuy, *J. Chem. Phys.*, **54**, 4843 (1971); R. W. Munn and W. Siebrand, *ibid.*, **52**, 47 (1970).

77. A. H. Francis and C. B. Harris, *Chem. Phys. Lett.*, **9**, 188 (1971).

78. R. M. Hochstrasser and J. D. Whiteman, *J. Chem. Phys.*, **56**, 5945 (1972).
79. R. Silbey, "Coherent Exciton Transfer in Solids," *Ann. Revs. Phys. Chem.*, **29**, 473 (1978).
80. A. Inoue et al., *Bull. Chem. Soc. Japan*, **51**, 345 (1978).
81. E. Schwarzer and H. Haken, *Phys. Stat. Solidi*, **B-84**, 253 (1977).
82. E. Kartheuser et al., *Phys. Rev.*, **B-19**, 546 (1978).
83. K. Yoshihara et al., "6th Molec. Cryst. Symp. Proc.," Elmau, W. Germany, 1973, p. 16.
84. V. M. Agranovich and Yu. V. Konobeev, *Phys. Stat. Solidi*, **27**, 435 (1968).
85. H. Killesreiter and H. Bässler, *Chem. Phys. Lett.*, **11**, 411 (1971).
86. H. Killesreiter and H. Bässler, *Phys. Stat. Solidi*, **B-51**, 657 (1972).
87. B. J. Mulder, *Philips Res. Repts. Suppl.*, **4**, (1967).
88. H. Kallman et al., *Phys. Stat. Solidi*, **B-44**, 813 (1971).
89. H. Killesreiter and R. Braun, *Phys. Stat. Solidi*, **B-48**, 201 (1971).
90. H. Blume and H. Guesten, "Chemical Processes of Electronically Excited Molecules," *Ultravioletstrahlen*, J. Kiefer, ed., De Gruiter, Berlin, 1977, p. 34; D. L. King and D. W. Setser, *Ann. Rev. Phys. Chem.*, **29**, 407 (1978); J. P. Farges and F. Gutmann, *Modern Aspects of Electrochemistry*, J. O'M Bockris and B. E. Conway, eds., Plenum, N.Y., **13**, 361 (1980).
91. J. Adamowski and S. Bednarek, *Pr. Inst. Fiz. Pan.*, **74**, 144 (1977).
92. J. Pollmann and H. Büttner, *Phys. Rev.*, **B-16**, 4480 (1977).
93. Z. A. Insepov and G. E. Norman, *Zh. Exsp. Teor. Fiz.*, **73**, 1517 (1977).
94. L. Sanche, *Chem. Phys. Lett.*, **65**, 61 (1979).
95. L. Sanche and G. J. Schulz, *Phys. Rev.*, **A6**, 69 (1972); K. Hiraoka and W. H. Hamill, *J. Chem. Phys.*, **57**, 3870 (1972); *ibid.*, **59**, 5749 (1973).
96. E. A. Andryushin and A. P. Silin, *Fiz. Nizk. Temp., (Kiev)*, **3**, 1365 (1977); D. S. Pan et al., *Phys. Rev.*, **17B**, 3284, 3297 (1978); see also refs. 22.
97. *Molecular Association*, R. Foster, ed., **2**, (1979) Academic, N.Y., 1979; P. Fröhlich and E. L. Wehry "The Study of Excited State Complexes (Exciplexes)," *Modern Fluorescence Spectroscopy*, Vol. **2**, 319, E. L. Wehry, ed., Plenum, N.Y., 1976; H. Leonhardt and A. Weller, *Ber. Bunsen Ges. Phys. Chem.*, **61**, 791 (1963); M. Gordon and W. R. Ware, *Exciplex*, Academic, N.Y., 1975.
98. H. Beens and A. Weller, *Chem. Phys. Lett.*, **2**, 82 (1968); J. Beens et al., *J. Chem. Phys.*, **47**, 1183 (1967); N. Tyutyulkov et al., *Theoret. Chem. Acta*, (Berlin) **20**, 385 (1971).
99. T. Mimura and M. Itoh, *Bull. Chem. Soc. Japan*, **50**, 1739 (1977); *J. Am. Chem. Soc.*, **98**, 1095 (1976); T. Mimura et al., *Bull. Chem. Soc. Japan*, **50**, 1665 (1977).
100. M. Yokoyama et al., *Chem. Phys. Lett.*, **34**, 597 (1975); P. Petelenz, *ibid.*, **65**, 579 (1979).
101. N.Y.C. Chu et al., *J. Chem. Phys.*, **55**, 3069 (1971); W. Klöpffer et al., *Molec. Cryst. Liq. Cryst.*, **16**, 229 (1972).
102. Z. Ludmer, Ph. D. Thesis, Weizmann Inst. of Sci., Israel, (1973).
103. M. D. Cohen and Z. Ludmer, "6th Molec. Cryst. Symp. Proc.," **106**, W. Germany, 1973, p. 106.
104. A. Inoue et al., *Bull. Chem. Soc. Japan*, **45**, 720 (1972).
105. H. Ueba and Sh. Ichimura, *J. Chem. Phys.*, **70**, 1745 (1979).
106. D. G. Hall and A. J. Braundmeier Jr., *Phys. Rev.*, **B-17**, 3808 (1978); J. Laguis and B. Fischer, *ibid.*, 3814.
107. E. Tosatti, "Interactions of Radiation and Condensed Matter, Lect. Int. Winter Colloq.," 1976 (Publ. 1977), **1**, 281 (1977), Interntl. Atomic Energy, Comm. Vienna

(Austria); A. S. Davydov, *Fiz. Mol. (Kiev)*, **4**, 98 (1977).

108. L. M. Kahn et al., *Phys. Rev.*, **B-17**, 4600 (1978).

109. W. P. Su et al., *Phys. Rev. Lett.*, **42**. 1678 (1979); M. Rice et al., *Phys. Lett.*, **71A**, 152 (1979); M. B. Fogel et al., *Phys. Rev. Lett.*, **36**, 1411 (1976); M. J. Rice et al., *ibid.*, 432 (1976).

110. J. F. Rabolt et al., *J. Chem. Phys.*, **71**, 4614 (1979); Th. C. Clarke et al., *J. Chem. Soc. Chem. Commun.*, **(1972)** 332.

111. R. Lobo et al., *J. Chem. Phys.*, **59**, 5992 (1973).

112. *Bioenergetics of Photosynthesis*, Govindjee, ed., Academic Press, N.Y., 1975; *Primary Process of Photosynthesis*, J. Barber, ed., Elsevier, N.Y., 1977.

113. L. L. Shipman and D. L. Housman, *Photochem. Photobiol.*, **29**, 1163 (1979).

114. J. C. Chang, *J. Chem. Phys.*, **67**, 3901 (1977).

115. K. Lonngren and A. Scott, eds., *Solitons in Action*, Academic Press, N.Y., 1978.

116. S. M. De B. Costa and E. C. C. Melo, *J. Chem. Soc. Faraday*, II, **76**, 1 (1980).

117. C. Sybesma, *An Introduction to Biophysics*, Academic Press, N.Y., 1977.

118. R. M. Hochstrasser and G. R. Meredith, *J. Luminesc.*, **18-19**, Pt. 1. 32 (1978).

119. P. R. Taylor et al., *J. Chem. Phys.*, **69**, 1971 (1978); *Chem. Phys. Lett.*, **41**, 444 (1976).

120. R. S. Davidson and T. D. Whelan, *Chem. Phys. Lett.*, **56**, 54 (1978).

121. e.g., G. Borstel and H. J. Falge, *Phys. Stat. Solidi*, **B83**, 11 (1977).

122. Alternative definitions of a polariton are given as, "A renormalized photon surrounded by a cloud of negative and positive charge pairs," or "A combined system of excitations," J. Hopfield, *J. Phys. Soc., Japan*, Suppl., **21**, 77 **(1966)**; A. Stahl, *Phys. Stat. Solidi*, **B94**, 221 (1979).

14

Ionized States

14.1 Introduction

Discussion in this Chapter will be mainly confined to some recent developments in the fields of electron-affinity and ionization potentials. The methodology and aspects related to it are treated in Section 12.7 and 12.8. Extensive tables and numerical values of electron affinities are given in Tables 6.8A, 6.8C, 6.8D, 6.8E, 6.8F, 6.9A, 6.10A, 6.17A, 6.18A, 6.20A and of ionization potentials in Tables 6.6A1, 6.6A2, 6.6B, 6.7A1, 6.7A2, 6.8B, 6.19A, 14.1 and 14.2. Ionization potentials and electron affinities of dyes are also discussed in Section 16.6.

14.2 Electron Affinities

The A_G values derived from charge-transfer spectra or from polarographic reduction potentials are subject to considerable uncertainty on an absolute scale.[1] An example of this uncertainty is the range of values obtained for tetracyanoethylene from charge-transfer spectra. The values[1] for the estimated electron affinity are: 1.50, 1.53, 1.7, 1.80, 1.9, 2.2 and 2.9 eV. The value 2.2 eV, relative to the p-chloranil value of 1.37 eV reported[1] by Briegleb in his Table 6, seems to be in error. By Briegleb's procedure we obtain 1.73 eV as the affinity of $C_2(CN)_4$.

Surface ionization studies also have yielded results for A_G: the magnetron triode method[2] gave 2.88 ± 0.05 eV and another surface ionization technique[1] gave 1.7 ± 0.3 eV. While surface ionization has yielded accurate electron affinities for atoms, particularly the halogens, it has been less successful for molecules. Differences between the molecular electron affinities from the magnetron surface ionization method and those from photodetachment experiments often exceed 0.5 eV. A closer investigation of the processes involved in surface ionization apparently is needed before molecular electron affinities can be determined reliably by this method.

Determination of the threshold energy of photodetachment offers what appears to be a superior method to obtain consistent and comparable data for the electron affinities of organic material. The methodology is discussed in Section 12.7. The wave mechanical description of photodetachment provides the basis for determining the relative probabilities of transitions from the various vibrational states of the ion to those of the molecule. Time-dependent perturbation theory shows that the cross-section for photodetachment $\sigma(h\nu)$ can be expressed in terms of the dipole matrix element R connecting the initial (negative ion) and the final (neutral molecule + electron) states, and the density of continuum states, which is proportional to $E_K^{1/2}$:

$$\sigma(h\nu) = \text{constant} \cdot E_K^{1/2} R^2 \tag{14.1}$$

Molecular photodetachment transitions are governed[3] by the Franck-Condon principle, which allows the dipole matrix element for transitions from the vibrational state v'' of the ion to the v' state of the molecule to be expressed as

$$\bar{R}(v',v'') = \bar{R}_e \int \psi_{v'} \psi_{v''} d\tau_v \tag{14.2}$$

$\bar{R}_e$ is an averaged value of the electronic transition moment, and $\int \psi_{v'} \psi_{v''} d\tau_v$ is the overlap integral for the vibrational wavefunctions $\psi_{v''}$ and $\psi_{v'}$ of the ion and molecule respectively.

With $\bar{R}_e$ treated as constant, eq. (14.2) has been very useful for the analysis of the relative intensities of photodetachment transitions between specific vibrational levels of diatomic systems such as O_2^{-1} and OH^{-1}. For a complex system, detailed computation of the overlap integrals is not possible. However, if it can be shown that the potential energy-nuclear coordinate hypersurfaces of the neutral molecule and its negative ion are identical (apart from their absolute energy), then the orthogonality properties of the vibrational wavefunctions require[1] that $R(v',v'') = 0$ unless $\Delta v = v' - v'' = 0$ for all $3N-6$ normal modes of vibration.

While eq. (14.1) and (14.2) can give the relative intensities of photodetachment transitions between specific vibrational states, they do not give, without further development, an explicit expression for the way in which the cross-section varies with photon energy near threshold. The energy dependence near threshold is of great importance for the extraction of the electron affinity from the experimentally obtained plots of relative photo-detachment cross-section vs. photon energy. For atomic and diatomic negative ions, theory predicts[4] that the cross-section near threshold has a functional dependence determined by the angular momentum of the photodetached electron. For H^-, O_2^- and C_2^-, the cross-section near threshold obeys

$$\sigma = \text{constant} \cdot \nu E_\kappa^{3/2} \qquad (14.3)$$

resulting in a zero slope at threshold. On the other hand, the cross-sections for C^-, O^-, OH^-, CH^-, SH^-, SeH^- and SO^- are required to rise very steeply from threshold according to

$$\sigma = \text{constant} \cdot \nu E_\kappa^{1/2} \qquad (14.4)$$

allowing experimental data to be extrapolated to yield accurate threshold energies.

For non-linear polyatomic systems, no theoretical prediction of the cross-section has yet been achieved. This leads to some reduction in the accuracy of the electron affinity. However, only a small error is caused by not knowing the cross-section shape near threshold: the experimental cross-section[1] in the energy range to 0.5 eV above threshold is approximately linear, and extrapolating this region linearly gives a threshold energy which is very nearly equal to that obtained from the

$$\sigma = \text{constant} \cdot \nu E_\kappa^{3/2} \qquad (14.5)$$

extrapolation of data close to threshold. Both values are only about 0.02 eV greater than the electron affinity of hydrogen.[4]

Under certain simplifying assumptions,[1] a synthetic cross-section may be constructed, as shown in Fig. 14.1 for tetracyanoethylene.[1] Simple harmonic oscillator wavefunctions were used to calculate the overlap integrals. The bond force constants were estimated by linear interpolation of the single- and double-bond values[5] to the calculated bond orders.[6] When the constructed curve is compared with the experimental cross-section, it should be remembered that the low signal-to-noise ratios of the photo-detachment data and the relatively large monochromator bandwidth may prevent resolution of any vibrational structure which may exist. For the purpose of positioning the experimental curve on the energy axis of Fig. 14.1, establishing the electron affinity of the molecule, it should be remembered also that for the energy range within 1 eV of the photodetachment threshold, the correspondence of the experimental and constructed curves should be best near threshold.

For more details, the original publication[1] should be consulted. Uncertainty is much reduced if attention is limited to relative electron affinities, i.e., the

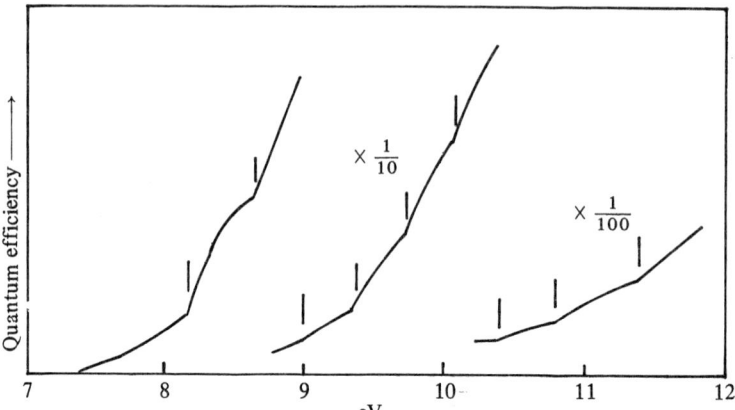

Fig. 14.1 (a) The photodetachment cross-section constructed with a simplified diatomic model for tetracyanoethylene. The steps are labelled by the vibrational quantum numbers (v', v'') of the final and initial states. (b) The convolution for the constructed cross-section and a triangular monochromator bandpass function of 22 nm full bandwidth at half maximum. (c) The experimental cross-section, assuming $A_G = 2.3$ eV. After Lyons and Palmer.[1]

difference in electron affinity of the molecule studied and that of a chosen reference molecule. The relative affinities are probably accurate to ±0.1 or 0.2 eV. For example, if charge-transfer spectra are used to compare the electron affinities of tetracyanoethylene, p-chloranil and p-benzoquinone, then the electron affinities of molecule (i) relative to that of molecule (j), $_i\Delta_j [\equiv A_G(i) - A_G(j)]$, are respectively,

$$_2\Delta_1 = -0.3 \text{ eV}, \qquad _3\Delta_1 = -0.9 \text{ eV}, \qquad _3\Delta_2 = -0.6 \text{ eV},$$

or, from different data

$$_2\Delta_1 = -0.4 \text{ eV}, \qquad _3\Delta_1 = -1.1 \text{ eV}, \qquad _3\Delta_2 = -0.6 \text{ eV}.$$

Taking the absolute electron affinity of tetracyanoethylene as 2.3 ± 0.3 eV from the work described in the cited paper, it follows that A_G of p-chloranil is 2.0 ± 0.4 eV and of p-benzoquinone 1.4 ± 0.4 eV. In a similar fashion the other absolute values in Table 6.9A have been obtained from relative electron affinities derived from charge transfer spectra.[1] For p-benzoquinone previous estimates of A_G have been 2.0,[1] 1.34 ± 0.09,[2] 2.08,[7] 1.89 (+0.2, −0.3).[7] The result of Cooper et al.[7] is therefore consistent with the results on tetracyanoethylene as quoted.[1]

Extensive tables of the electron affinities of 176 organic compounds have been published by Kampars and Neilands.[8] Theirs are relative values based upon that of chloranil taken as 2.45 ± 0.15 eV. These authors[8] state that the electron affinity A_g of a compound may be obtained from the additional formula:

$$A_g = A_g^0 + 0.9 C \Sigma A_g^{sub} \tag{14.6}$$

where C is a coefficient determined by the sensitivity of a system to the introduction of a substituent; the A_g^{sub} values of several substituents are listed in Table 6.18. A_g^0 refers to the parent compound. Any extension in the π-electron system of an acceptor is said[8] to lower A_g so that one can conclude that the lowest vacant molecular orbital is largely localized in that part of the molecule which is directly linked to the hetero-atom.[8]

Other group contributions obtained by the electron attachment method,[12] are listed in Table 6.8D.

For comparison purposes, the electron affinities of several elements[9] are listed in Table 6.17A.

A recent list of polarographically obtained[10] electron affinities is given in Table 6.8E.

The electron affinities of methiodides, constituting a series of closely related single-ring cationic acceptors were studied[11] in reference to the correlation of the $CT\bar{\nu}_{max}$ with the electron affinity of the cation. The donor is changed to see the effect of the ionization potential. These compounds exhibit multiple absorption maxima, commonly ascribed to the final donor states $^2P_{1/2}$ and $^2P_{3/2}$. The separation between the two maxima should be fixed if the final donor state is a free atom but will vary if there is partial bond formation in the CT excited state. The separation between the two maxima is a function of the acceptor cation and thus of the degree of $D:A$ interaction in the excited CT state. The authors report[11] that good correlation exists after allowing for solvent interaction effects. The separation of the two absorption maxima is said[11] to be associated with an interaction of the iodine atom with the donor in the excited charge transfer state. Their numerical results are reproduced in Table 6.8F.

Extensive computer calculations recently carried out by Åsbrink et al.[13] suggest that the electron affinity of naphthalene is not 0.2 eV as frequently quoted (see Table 6.8A) but 0.00 eV.

The electron affinities of unsaturated, uncharged π-electron acceptors of the type $C_w N_x O_y F_z$, where fluorine is taken as representative of halogens, has been studied by Hammond.[16] It appears that a high degree of unsaturation is a prerequisite for a high electron affinity.[16] On the basis of the above reasoning, Koizumi and Matsunaga[14] conclude that the azido group, N_3, is just as good as electron acceptor as is p-chloranil.

Beitz and Miller[17] have obtained values of electron affinities polarographically as well as from survival probability measurements of electrons injected, and subsequently trapped, in 2-methyltetrahydrofuran glass at 77 K. Their data are shown in Table 6.20A.

14.3 Ionization Potentials

Assuming the ionization cross-section to be (as theoretically expected) a step function for monochromatic incident light, the derivative of the experimental curve should reproduce the energy distribution of the incident radiation and thus hould show one or more often overlapping peaks. A sharp and well-defined onset of ionization indicates that the energy involved is the adiabatic ionization energy: the vertical one to be obtained at the steepest point of the curve. Vertical and adiabatic ionization energies differ, as discussed in Section 6.6. A few typical comparative values are given in Table 14.1. Asymmetry and peak broadening are indicative of auto-ionization. (See discussion in Section 6.6 and 6.7).

A fairly typical photo-ionization (PIE) curve is shown in Fig. 14.2 for anthracene; the ionization potentials so deduced[34] are listed in Table 14.2. Table 14.2 shows the locations of the breaks in comparison with the PE bands.[43] Most breaks are in good agreement with the PE data, though the former might be somewhat affected by the difference between the adiabatic and vertical ionization potentials concerned. However, some bands corresponding to breaks are missing in the PE spectrum. For example, there is a clear break at 8.19 eV between two π-ionization potentials of 7.40 and 8.65 eV. This cannot be attributed to the excitation of the vibrations of the molecular ions, since the energy interval is too large. The effect has been ascribed to these bands involving at least a partial resonance.[43]

Table 14.1 Gas Phase Ionization Energies. After Batley and Lyons.[43]

Compound	Vertical ionization energy (eV)	Adiabatic ionization energy (eV)
Anthracene	7.42	7.42
Phenothiazine	7.38	<7.0
N,N,N',N'-Tetramethyl-p-phenylenediamine	6.47; $\geqslant$7.35	$\leqslant$6.25; $\leqslant$7.00
N,N,N',N'-Tetramethylbenzidine	6.55; $\geqslant$7.36	$\leqslant$6.40; $\leqslant$7.04
Perylene	7.12	6.92

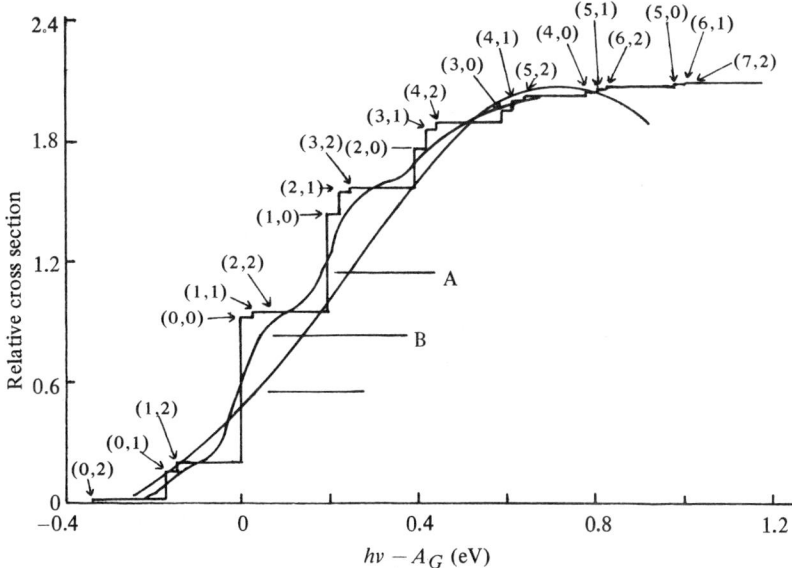

Fig. 14.2 Photoionization efficiency curve of anthracene (breaks are indicated by lines). After Aihara and Inokuchi.[34]

Table 14.2 Ionization Potentials of Anthracene. After Batley and Lyons.[43]

PIE curve (a)	PE spectrum (b)
7.40 (eV)	7.40 (eV)
8.19	--
8.65	8.52
8.98	9.16
9.35	--
9.72	--
10.07	10.13
10.39	10.21
10.78	10.7
11.73	11.3

(a) PIE curve: values obtained from photoionization efficiency data (cf., Fig. 14.2).
(b) PE spectrum: Photo-Electron spectrum.

Ionization energies are affected by ambient gases, probably because of chemisorption. Thus, the photo-electric threshold energy for benzonitrile-rubidium was found[36] to be 5.05 ± 0.02 eV but dropped to 4.95 ± 0.02 eV upon chemisorption of hydrogen.[36] The probing depth of photoelectron spectroscopy depends upon both the electron escape depth and the light penetration depth (the inverse of the absorption coefficient). In most cases, the former is usually shorter than the latter. Therefore, the determination of the escape depth is necessary for the interpretation of the photo-electron energy distribution curves (EDC's).[37] If the escape depth is very short (around 1 nm), the observed EDC's reflect the electronic structure of the solid surface, and if it is long (about 10 nm or more), they are the reflection of the electronic structure of the bulk as well as of that surface. Particularly, in ultraviolet photoelectron spectroscopy, the escape depth changes drastically in accordance with the change of the kinetic energy of the photoelectrons.

Electron mean free paths for metals and inorganic semiconductors are compiled in several reviews.[38] Hot electron scattering lengths of molecular crystals have been studied to a lesser extent. Pong and Smith report[39] the electron escape depth of Cu-phthalocyanine to be 1.1 nm in its thin film (<4 nm) and to be 15 nm in its thick region (>5 nm). They concluded that 1.1 nm is the true escape depth of Cu-phthalocyanine, because the longer length (15 nm) is due to the effect of light absorption. This conclusion was criticized by Belkind et al.[40] on the ground that the 1.1 nm value is too small and may be due to the effect of its coverage. These authors obtained 14 nm as the escape depth for a naphthacene film on gold when the electron energy was about 1 eV above the vacuum level.[40] The electron escape depths of aliphatic compounds at 77 K were reported[41] to be from a few tenths nanometer to about 10 nm. Studies on TTF-TCNQ and its related compounds have yielded values of 1 nm or less for electrons of energy less than 10 eV above the Fermi level of each compound.[42] The energy dependence of the electron escape of naphthacene and perylene deposited films has been measured by an overlayer technique.[37] Energy distribution curves of transmitted electrons through deposited organic layers were obtained[37] by differentiating the I-V characteristic curves, and the electron escape depths classified by their kinetic energies were obtained from the above results. Presently obtained[37] escape depth is between 8 and 15 nm for naphthacene, and between 30 and 37 nm for perylene. They differ from previous results. The matter awaits further classification. Numerical data are listed in Tables 6.6A, 6.7A, 6.19A, and 14.2; recent values of vertical ionization potential of upper occupied π-orbitals of several compounds are listed by Auiyami et al.[56]

14.4 Carrier Generation

Free carrier generation upon illumination can be considered as a three step process:[29]

(1) photon absorption results in a quasi-free electron; this in itself is a complex process involving the separation of an electron-hole pair;
(2) transport to the electrode and
(3) surmounting the surface potential barrier.

The ionization process (1) may occur by direct ionization or, more likely in molecular crystals, *via* a more or less highly excited molecular exciton, a process sometimes referred to as auto-ionization.[30] The carrier must obtain sufficient energy to avoid geminate recombination with the positive hole which its generation leaves behind. In the migration process, the electron is likely to lose energy by a variety of scattering, trapping and phonon interaction processes. At high photon energies, secondary electrons may be formed which produce secondary peaks in the photoemission spectra. However, such secondary peaks may also be associated with the auto-ionization reactions for which the following sequence has been proposed:[31]

$$M \xrightarrow{h\nu} M^{**} \to M^* \to M^+ + e^- \\ \searrow M^{+*} + e^- \qquad (14.7)$$

Here, M is the ground state of the molecule, M^{**} a high energy excitonic state, M^* a lower energy auto-ionizing state, M^+ the resulting positive ion in its ground state and M^{+*} same in an excited state. For a detailed discussion of these phenomena in connection with the observed quantum yield, a recent review[30] by Inokuchi and coworkers should be consulted. The photoionization of adsorbates has been discussed by Wendin and Ohno.[59]

The essence of the widely accepted auto-ionization model[57] of intrinsic photoconductivity may be summarized, following Lyons,[58] as follows: one-photon absorption gives rise to a state which either degrades or forms an ion-pair which then is separated into two oppositely charged species under the combined action of thermal energy and an electric field. The threshold of intrinsic conductivity does not necessarily mark E_G but denotes that excitation energy which can produce an ion-pair which can be dissociated under the conditions prevailing (see also discussion in Section 15.1). In the primary photoelectric process, absorption of a light quantum must yield an energy of $2 E_g$, where E_g is an energy gap—not necessarily that appearing in conventional, *Ge* or *Si*, band theory. An electron-hole pair then results. For a discussion of E_g, see Section 15.12. The energy of the lowest available conduction band, usually very narrow, can be written[30] from the Lyons equation (8.11), thus:

$$E'_g = 2E_{th} - I_g - A_g \qquad (14.8)$$

where E'_g is the gap energy involved, E_{th} the photo-emission threshold and I_g and A_g the ionization energy and electron affinity, respectively, of the free molecules.

Often, though not always—e.g., not in p-terphenyl—one can distinguish 3 distinct regions in the spectral dependence of the quantum yield: for low energy quanta, near one and about within 1 eV of the threshold, the quantum efficiency rises steeply with increasing $h\nu$. This is followed by an intermediate region, probably involving exciton interactions and characterized by a rather gradual increase in quantum yield, and a final high energy—say above 9.5 eV—region characterized by a monotonic rise in the quantum yield. The cube law eq. 12.11 holds in the first-mentioned interval, and has been derived by Kochi et al.[32] from the assumption that the quantum efficiency is then governed by the energy distribution of photoelectrons arriving at the contact and by their transition probability through the interfacial region. The intermediate region appears to be due to changes in the effective absorption coefficient in the surface layer exposed to the light flux. If the escape depth $L(E)$ of the electrons (see Section 12.8 and 14.3), sometimes referred to as the attenuation length, is much shorter than the extinction length of the incident light beam, then only radiation absorbed within a very thin interfacial layer contributes to photoemission. Batley[33] has analyzed the problem of photoemission in these circumstances where photoemission has to compete against other processes, probably in the main, geminate recombination. His results are summarized in Table 14.3, where DI stands for direct ionization and AI for auto-ionization processes.

In the third, high energy, region, either emission from deeper levels within the solid and/or secondary electrons produced by a variety of processes, appear to contribute.[30] The matter is still unresolved. For the yield of carriers as a function E, a straight line should result, with a slope equal to $e^3/2\epsilon k^2 T^2$. Hughes[26] has shown this theory to be applicable to carriers by X-rays in anthracene, and others[27] have shown that it holds for photogenerated carriers.

When the mean free path of both carriers is less than r_0, recombination becomes highly probable. Under these conditions the relative drift velocities v_d of an electron and a hole a distance r apart is $(\mu_h + \mu_e)(e/\epsilon r^2)$, where μ_e and μ_h are the electron and hole drift mobilities, respectively. The rate of influx of electrons v_D into a sphere of arbitrary radius r drawn around a

Table 14.3 Expected Behavior of Quantum Yield $Y(h\nu)$.* After Batley.[33]

Scattering condition	Dominant mechanism of ionization		Expected behavior of $Y(h\nu)$ when a_t decreases
$a_t L \gg 1$	DI	$Y \sim A_1 \int_{a_t}^{a_{DI}} dE$	increases
	AI	$Y \sim A_2 \int_{a_t}^{a_{AI}k} dE$	nearly constant
$a_t L \ll 1$	DI	$Y \sim A_3 \int a_{DI} L dE$	gradually increases
	AI	$Y \sim A_4 \cdot a_{AI} kL$ $\sim A_4 \cdot a_t kL$	decreases

*$a_{DI}(h\nu, E)$ is the effective absorption coefficient to form quasi-free electrons with energy E, a_{AI} is the absorption coefficient of the autoionizing exciton, $k(h\nu, E)$ is the autoionizing probability to form quasi-free electrons with the energy E and A_i includes all factors other than the change in a_{DI}, a_{AI}, and a_t. In the last column it is assumed that $\int a_{DI} dE$ gradually increases with the photon energy. Note that $a_t \sim a_{AI}$ when autoionization dominates.

hole, and thus the recombination rate constant is

$$v_D = \frac{e}{r^2 \epsilon}(\mu_h + \mu_e) \qquad (14.9)$$

Taking $\mu_e = \mu_h = 1$ cm^2/V sec, a recombination rate of 1.2×10^{-6} cm^3 sec^{-1} results, in good agreement with experiment.[28] Thus, r_0 is the distance at which the coulombic mutual potential energy of the electron-hole pair equals the thermal energy kT. The escape probability for an initial separation distance r is then given by $\exp(-r_0/r)$. The number n of carrier pairs then generated against geminate recombination is

$$n \cong A(T)[1 + Ee^3/2\epsilon k^2 T^2] \qquad (14.10)$$

where the function $A(T)$ is independent of the applied field E but depends on the initial carrier distribution. ϵ is the permittivity of the matrix which the carriers "see."

The escape time τ_0 for the carriers to separate sufficiently to avoid geminate recombination follows from the diffusion equation

$$n_0 = [D\tau_0]^{1/2} \qquad (14.11)$$

where D is a diffusion constant related to the mobility μ by the Einstein relation[21]

$$\mu = eD/kT \qquad (14.12)$$

so that, for anthracene, taking $r_0 = 50$ Å and assuming a mobility of 0.4 cm^2/V-sec: $\tau_0 = 2.4 \times 10^{-11}$ sec. Thus, geminate recombination requires only about 24 picoseconds. Jortner et al.[22] also proposing a type of auto-ionization mechanism for carrier formation, arrive at a value of about 20 psec for anthracene. These authors[22] suggest that the observed low quantum yield, which is about 10^{-4} for anthracene, is due to the competitive interaction between ionization and other radiationless recombination processes.

Karl and Sommer[24] have pointed out that electrons and holes may interact for relatively long time intervals if the applied electric field is low; recombination will then be favored, and the photo-current component due to the effect is strongly field dependent.

Two-photon absorption experiments by Kepler[25] suggest that the photoelectron first transits into a broad band of energy bands but spends very little time therein before undergoing other reactions, with geminate recombination predominating.

Carrier generation in anthracene[23] under illumination appears to go via a singlet exciton;[18] this may be shown directly by employing photons of energy in excess of about 4 eV or from the singlet-singlet interaction[19] which seems to be the most probable cause of the 3533 Å absorption in anthracene. As proposed by Onsager, the electrons thus produced in a higher energy state, eventually thermalize, i.e., come into thermal equilibrium with the lattice via phonon interactions, reaching the Onsager[20] distance r_0 from their counterpart hole.

Intrinsic photogeneration of carriers specially in anthracene is discussed in detail in Section 6.11. More recent investigations have attempted to clarify the primary process in the formation of free charge carriers. There has been no direct evidence for the presence of a charge transfer state in the crystal, and no direct optical transition to the conduction band has been observed in the one-photon absorption spectrum.

Batt, Braun and Hornig,[44] using blocking contacts, found that up to 2×10^6 V m^{-1}, the intrinsic photocurrent varied linearly with field, and the slope/intercept ratio of the current-field plot was in near quantitative agreement with that predicted by the Onsager theory of geminate recombination. They reported $r_0 = 8.5$ nm at photon energies of 4.2 to 5.1 eV and $r_0 = 13.8$ nm at 5.8 to 5.9 eV. Geacintov and Pope,[45] using water, aluminum and blocking contact, found good agreement in absolute photoconduction efficiency between blocking and aluminum contacts, but for water contacts the efficiency was an order of magnitude larger. Also, they extended the current-field relationships to higher fields than did Batt, Braun and Hornig,[44] and found poor agreement with the Poole-Frenkel and Schottky effects and also with the Onsager geminate recombination theory. They assumed $r_0 = 8.3$ nm.

Chance and Braun,[46] using blocking electrodes, confirmed that for fields up to 2.5×10^6 V m^{-1}, the intrinsic photocurrents are explained by the Onsager model. They explained the deviation, at fields below 10^5 V m^{-1}, of the photoconduction efficiency, from that predicted by the Onsager theory, in terms of diffusion controlled recombination of free and trapped carriers. Their absolute quantum yield agreed well with the experiments of Geacintov and Pope using blocking contacts. Measurements of the activation energy of photoconduction, yielded $r_0 = 5$ nm for photon energies of 4.4 to 5.2 eV, and $r_0 = 6.7$ nm for 5.4 to 6.2 eV ($\epsilon_r = 3.2$).[47]

From a detailed study, Lyons and Milne[48] conclude that in anthracene two ion-pair states appear to be formed: between 3.9 and 4.2 eV with an r_0 of 1.8 nm and between 4.5 and 5.0 eV with an r_0 of 2.5 nm. In the region of 4.2 to 4.5 eV, r_0 undergoes a smooth transition.

The spectral dependence of ϕ_1, the one-photon ion-pair production efficiency, has a peak at 4.2 ± 0.15 eV with a value of 0.1, agreeing with the spectral dependence of $\bar{\eta}$ (two-photon ion-pair yield). The two-photon absorption cross-section σ_2^F has a peak at 4.1 eV which appears only in the crystal spectrum and not in the solution spectrum. Thus, the peak in the two-photon absorption spectrum may be assigned to absorption to an ion pair state. However, this assignment contradicts theoretical calculations of σ_2^F for a transition to an ion pair state. A suitable explanation of the high intensity of σ_2^F at 4.1 eV is needed. In the process of singlet exciton fission into two triplet excitons, the lowest ion pair state is not the fissionable singlet.

They appear also to have resolved previously reported anomalous results of the intrinsic photoconductivity of anthracene with aqueous electrodes, by employing 1 M Na_2SO_3 aqueous contacts instead. Values of steady state photoconduction quantum yield thus obtained are reported to agree with those measuring other types of contacts as well as pulse techniques.[48] The observed currents strongly dependent *inter alia,* on

(1) the wavelength and intensity of the incident light;
(2) the magnitude and polarity of the applied field;
(3) the state of the illuminated surface;
(4) the ambient atmosphere; and
(5) the temperature of the crystal.

Direct one-photon, intrinsic photoconduction has been shown[49] to be possible only with light of energies greater than 3.85 eV (320 nm). At low light intensities, any carrier generation with light of lower energy (i.e., $\lambda > 320$ nm) must occur by an "extrinsic" process. The most common of these generation processes is exciton dissociation at the surface of the crystal.[50,51] Such a mechanism has been invoked[51] for photogeneration by light in the singlet absorption region (340 nm $< \lambda <$ 410 nm), as well as for weakly absorbed light (415 $< \lambda <$ 455 nm) at low intensities. Multi-photonic generation processes have been reported for anthracene using high-intensity light at a number of wavelengths. With long-wavelength light, Strome[52] reported generation of free carriers by a direct, two-photon excitation. Similar two-photon and three-photon absorptions have been established from fluorescence studies.[51] Two-photon processes involving excitons have also been found. The proposed exciton-exciton generation with weakly absorbed light has been criticized,[51] in that the results were better approximated by singlet-exciton photoionization.[53]

14.5 Polarization Energies

The methods or calculating polarization energies, discussed in Section 6.4, have been refined and extended by Batley et al.[54] For details, the original papers should be consulted; numerical results for several hydrocarbons as well as for some charge transfer complexes are reproduced[54] in Table 14.4 and 14.5. I_g and I_c refer to the gaseous (isolated molecule) and crystal ionization energies; it is seen that the hydrocarbons yield reasonable agreement between these values and the $I_g - I_c$ difference obtained from the Lyons equation. The complexes, however, tend to exhibit considerable discrepancies. Computations have also shown[54] that the polarization energies for positive and negative charges are quite different in these complexes. The ion-quadrupole terms were neglected, but the errors thus introduced are estimated[54] to be of the order of 0.1 eV.

Capek[55] has studied the effects of introducing dynamic self-consistency into polarization theory. He finds that the correction to the polarization energy obtained from the Lyons model (see Section 6.4) is about half the bandwidth, numerically <0.1 eV.

Though unimportant at moderately high field strengths where the free carriers are rapidly removed from the recombination region, free-trapped carrier recombination can dramatically affect the low field quantum yield.[57]

Chance and Braun,[57] using blocking electrodes and a pulsed photoconductivity technique on single crystals of anthracene, have confirmed that up to 2.5×10^6 V/m, the Onsager theory of geminate recombination holds. Some deviations at higher fields are explained in terms of the diffusion

Table 14.4 Calculated Polarization Energies (eV) compared with $I_G - I_c$.

P_I, Ion-dipole sum; point charge. P_D, Dipole-dipole sum; point charge. ΔP_I, the change in P_I resulting from the distribution of charge over the ion (for $n = 3$). After Batley et al.[54]

Hydrocarbon	Ion-Dipole Sums (P_I)				P_D	ΔP_I	$P_I + P_D + \Delta P_I$	$P_I + P_D$	$I_G - I_c$
	$n = 3$	6	10	∞					
Naphthalene	1·720	1·850	1·909	2·01	−0·529	−0·0791	1·40	1·48	1·38
Anthracene	2·053	2·185	2·246	2·35	−0·717	−0·2542	1·38	1·63	1·45
Tetracene	2·515	2·648	2·707	2·81	−0·965	−0·6670	1·18	1·85	1·57
Pentacene	3·047	3·207	3·280	3·40	−1·414	−1·1125	0·88	1·99	1·60
Phenanthrene	1·919	2·046	2·103	2·21	−0·533	−0·2712	1·40	1·67	1·40
1,2-Benzanthracene	2·113	2·246	2·304	2·41	−0·701	−0·5341	1·18	1·71	1·88
Pyrene	2·090	2·200	2·249	2·34	−0·559			1·78	1·77
Perylene	2·328	2·435	2·483	2·58	−1·219			1·36	1·52
Triphenylene	1·814	1·950	2·009	2·12	−0·584			1·54	

Table 14.5 Calculated Polarization Energies (eV) for Molecular Complex Crystals (Point Charge Approximation) After Batley et al.[54]

Complex		$n = 3$	P_I 6	10	∞	P_D	$P_I + P_D$	$I_G - I_c$
Anthracene-trinitrobenzene	P^+	2·442	2·519	2·554	2·62	−0·50	2·12	1·57
	P^-	3·936	4·014	4·049	4·11	−0·36	3·75	
Anthracene-pyromellitic dianhydride	P^+	2·123	2·250	2·309	2·41	−0·19	2·22	2·16
	P^-	2·558	2·687	2·746	2·85	−0·28	2·57	
Perylene-pyromellitic dianhydride	P^+	2·004	2·101	2·145	2·24	−0·42	1·82	1·76
	P^-	3·087	3·184	3·228	3·31	−0·64	2·66	
Perylene-p-fluoranil	P^+	2·025	2·133	2·184	2·27	−0·47	1·80	
	P^-	3·428	3·537	3·587	3·67	−0·77	2·80	
Pyrene-tetracyanoethylene	P^+	1·668	1·762	1·805	1·88	−0·30	1·58	
	P^-	2·289	2·379	2·421	2·50	−0·45	2·05	

This work has been further extended by Lyons and Milne[57] who studied the steady state one-photon intrinsic photogeneration in anthracene single crystals and report[57] that the Onsager model remains applicable for aqueous, blocking and metal contacts involving pulsed as well as steady state conditions. Assuming a delta function to be applicable for the initial separation of the photo-generated charges against recombination, r_0, two ion pair states appear to be formed in anthracene; for light quanta of 3.9 to 4.2 eV, r_0 results as 18 Å while for quanta between 4.5 and 5 eV it comes to 25 Å; the intermediate region exhibits a smooth transition regime. The ion-pair production efficiency of photon absorption, equal to 0.1, occurs at a photon energy of about 4.2 eV which is close to the peak of the ion-pair yield consequent upon the absorption of two photons. The authors conclude[57] that in the process of singlet exciton fission into two triplet excitons, the lowest ion-pair state is not the fissionable singlet.

The highest polarization energy for any organic solid so far reported appears to be 3.0 eV, which value has been found[60] to hold for hexaiodobenzene; its ionization potentials are 8.58 for the gaseous and 5.6 eV for the solid state (these are adiabatic energies). Halogen anils, in general, are said to act as strong electron acceptors in charge transfer complexes.

References

1. L. E. Lyons and L. D. Palmer, *Aust. J. Chem.*, **29**, 1919 (1976).
2. See Table 6.8C.
3. L. M. Branscomb, *Atomic and Molecular Processes*, D. R. Bates, ed., Academic Press, London, 1962, p. 100; R. C. Cellota et al., *Phys. Rev.*, A6, 631 (1972).
4. B. W. Steiner, *Case Studies in Atomic Collision Physics*, E. W. McDaniel and M. R. C. McDowell, eds., Nth. Holland, Amsterdam, 1972, **2**, 484.
5. G. Herzberg, *Molecular Spectra and Molecular Structure*, Van Nostrand, N.Y., 1945, **2**, 193.
6. M. F. Rettig and M. Wing, *Inorg. Chem.*, **8**, 2685 (1969).
7. T. L. Kunii and H. Kuroda, *Theor. Chim. Acta*, **11**, 97 (1968); C. D. Cooper et al., *J. Chem. Phys.*, **63**, 2752 (1975).
8. V. Kampars and D. Neilands, *Russ. Chem. Revs.*, **46**, 503(1977); USP. Kim., **46**, 945 (1977).
9. H. Hotop and R. A. Bennett, *J. Chem. Phys.*, **58**, 2373 (1973).
10. J. E. Kuder et al., *J. Electrochem. Soc.*, **125**, 1750 (1978).
11. S. Bagchi and M. Chowdhury, *J. Phys. Chem.*, **83**, 629 (1979).
12. For a Review, see H. S. W. Massey, "Negative Ions," *Adv. Atomic and Molecular Phys.*, **15**, D. R. Bates and B. Bederson, eds., 1979.
13. L. Åsbrink et al., *Z. Naturforsch.*, **33A**, 172 (1978).
14. S. Koizumi and Y. Matsunaga, *Bull. Chem. Soc. Japan*, **43**, 3010 (1970).
15. R. A. Holroyd et al., *J. Phys. Chem.*, **83**, 435 (1979).
16. P. R. Hammond, *J. Chem. Soc.*, A-1968, 145, Nature (London), **206**, 891 (1965).
17. J. V. Beitz and J. R. Miller, *J. Chem. Phys.*, **71**, 4579 (1979).

18. R. R. Chance and C. L. Braun, *J. Chem. Phys.*, **64**, 3573 (1976).
19. C. L. Braun, *Phys. Rev. Lett.*, **21**, 215 (1968).
20. D. M. Pai and R. C. Enck, *Phys. Rev.*, **B-11**, 5163 (1975); J. C. Knights and E. A. Davies, *J. Phys. Chem. Solids*, **35**, 543 (1974); H. Sand and A. Mozunder, *J. Chem. Phys.*, **66**, 689 (1977); L. Onsager, *Phys. Rev.*, **54**, 554 (1938).
21. N. Karl et al., *Z. Naturforsch.*, **A-25**, 382 (1970).
22. J. Jortner and M. Bixon, *Mol. Cryst.*, **9**, 213 (1969).
23. L. B. Schein et al., *J. Chem. Phys.*, **71**, 3189 (1979).
24. N. Karl and G. Sommer, *Phys. Stat. Solidi*, **6**, 231 (1971).
25. R. G. Kepler, *Phys. Rev.*, **B-9**, 4468 (1974).
26. R. C. Hughes, *J. Chem. Phys.*, **55**, 5442 (1971).
27. N. E. Geacintov and M. Pope, "Proc. Internatl. Conf. Photoconduct. Conduct., 3rd Conf.," p. 293, E. M. Dell, ed., Pergamon, Oxford, 1969; R. H. Batt et al., *Appl. Opt. Suppl.*, **3**, 20 (1969).
28. R. G. Kepler and F. N. Coppage, *Phys. Rev.*, **151**, 610 (1966).
29. A. I. Belkind et al., "Preprints of Elect. Prop. of Organic Solids Conf." Karpacz, Poland, 1974; D. M. Hanson, *Crit. Revs. Solid State Sci.*, **13**, 243 (1973).
30. K. Seki et al., *Bull. Chem. Soc. Japan*, **49**, 904 (1976); cf., also, T. I. Quickenden and G. K. Yim-Tan, *Electrochim. Acta*, **24**, 143 (1979); W. J. Albery and M. D. Archer, *ibid*, **21**, 1155 (1976).
31. A. Zagrubskii and F. I. Vilesov, *Fiz. Tverd. Tela*, **13**, 2300 (1971); *Sov. Phys. Solid State*, **13**, 927 (1972); "Uspechi Fotoniki," F. I. Vilesov, ed., Leningrad Univ., 1974, p. 109.
32. M. Kochi et al., *Bull. Chem. Soc. Japan*, **40**, 531 (1967); **43**, 2690 (1970).
33. M. Batley, Ph. D. Thesis, Univ. of Sydney, 1966.
34. J. P. Aihara and H. Inokuchi, *Chem. Lett.*, **1973**, 421.
35. A. Fulton and L. E. Lyons, *Aust. J. Chem.*, **21**, 873 (1968).
36. H. Kawamura and H. Inokuchi, *Bull. Chem. Soc. Japan*, **45**, 710 (1972).
37. S. Hino and H. Inokuchi, *J. Chem. Phys.*, **70**, 1142 (1979).
38. I. Lindau and W. E. Spicer, *J. Electron. Spectrosc.*, **3**, 409 (1974); C. J. Powell, *Surf. Sci.*, **44**, 29 (1974); C. R. Brundle, *Surf. Sci.*, **48**, 99 (1975).
39. W. Pong and J. A. Smith, *J. Appl. Phys.*, **44**, 174 (1973).
40. A. I. Belkind, S. B. Aleksandrov, and V. V. Grechov, "Electrical Properties of Organic Solids Conference," Karpacz, 1974.
41. W. B. Berry, *J. Electrochem. Soc.*, **118**, 597 (1971); Y. C. Chang and W. B. Berry, *J. Chem. Phys.*, **61**, 2727 (1974).
42. S. F. Lin, W. E. Spicer, and B. H. Schechtman, *Phys. Rev.*, **B-12**, 4184 (1975); P. Nielsen, D. J. Sandman, and A. J. Epstein, *Solid State Commun.*, **17**, 1067 (1975).
43. M. Batley and L. E. Lyons, *Molec. Cryst.*, **3**, 357 (1968).
44. R. H. Batt, C. L. Braun, and J. F. Hornig, *J. Chem. Phys.*, **49**, 1967 (1968); R. H. Batt, C. L. Braun, and J. F. Hornig, *Appl. Opt. Suppl. Electrophotog.*, **20**, (1969).
45. N. E. Geacintov and M. Pope, "Proc. 3rd. Int. Conf. Photoconduct.," Stanford, 289, 1969.
46. R. R. Chance and C. L. Braun, *J. Chem. Phys.*, **59**, 2269 (1973).
47. C. L. Braun, Session for Discussion, Seventh Molecular Crystals Symposium, Nikko, Japan, September 1975; R. R. Chance and C. L. Braun, *J. Chem. Phys.*, **64**, 3573 (1976).
48. L. E. Lyons and K. A. Milne, *J. Chem. Phys.*, **65**, 1474 (1976); Preprints, 7th. Molec. Cryst. Symp. Nikko, Japan, Sept. 1975, p. 125; K. A. Milne and L. E. Lyons, *Aust. J. Chem.*, **31**, 699 (1978).

REFERENCES

49. N. Geacintov and M. Pope, *J. Chem. Phys.*, **45**, 3884 (1966); see also Ref. 45.
50. V. V. Eremenko and V. S. Medvedev, *Soviet Phys. Solid State*, **2**, 1426 (1961).
51. G. R. Johnston and L. E. Lyons, *Aust. J. Chem.*, **23**, 1571 (1970).
52. F. C. Strome, *Phys. Rev. Lett.*, **20**, 3 (1968).
53. E. Courtens et al., *Phys. Rev.*, **156**, 948 (1948).
54. M. Batley et al., *Aust. J. Chem.*, **23**, 2397 (1970).
55. V. Capek, *Czech. J. Phys.*, **B-29**, 439 (1979).
56. I. Akiyama et al., *J. Phys. Chem.*, **83**, 2997 (1979).
57. R. R. Chance and C. L. Braun, *J. Chem. Phys.*, **59**, 2269 (1973); **64**, 3573 (1976); L. E. Lyons and K. A. Milne, **65**, 1474 (1976).
58. L. E. Lyons, *Priv. Commun. (1980).*
59. e.g., G. Wendin and M. Ohno, *Physica Scripta*, **14**, 148 (1976).
60. N. Sato et al., *J. Phys. Soc. Japan*, **48**, 1254 (1980); cf., also, I. Shirotani et al., *J. Solid State Chem.*, **18**, 47 (1978).

15

Theories of Charge Transfer

15.1 Introduction

For dc conductivity to exist, two requirements must be fulfilled: free carriers must enter and leave the external circuit *via* the contacts, and these carriers which have entered the system must in some way transfer their charges to the other contact so as to leave it again. This finds expression in the conductivity equation, eq. (1.1), Section 1.1.

During the last 10 to 15 years attention has centered on the mechanisms of charge transfer rather than on the generation of free charges in a solid or liquid; that topic is discussed in Section 14.4. Such charges may be present intrinsically, or extrinsically by virtue of impurities or defects, or else they may be injected from the outside, say photoelectrically or by electron bombardment. The following discussion will be concerned with charge transport theories and thus with (carrier) mobility.

Such transfer may take place by several means:

(1) As coherent transfer in a conduction band which, in an organic solid, is likely to be narrow, i.e., comparable to the thermal energy kT. This mechanism is similar to that in say *Ge*. However, the band model used for inorganic semiconductors is fundamentally inapplicable to organic solids; the band model is essentially a one-electron approximation, whereas electronic polarization is a many-electron phenomenon. Electronic polarization certainly exists in organic solids. The (inorganic) band model should be replaced by the analogous (but different) auto-ionization model,[132] though an analogy

exists with respect to the energy gap between this and the band model. The same symbol E_G is used in both models.

The auto-ionization model[5] (of intrinsic photoconductivity) proposes that one-photon absorption gives rise to a state which either degrades or else forms an ion-pair which is then separated into two oppositely charged species under the combined action of thermal energy and an electric field (see also Sections 14.4 and 16.6). The threshold of intrinsic conductivity does not necessarily mark E_G but denotes that excitation energy which can produce an ion-pair which can be dissociated under the conditions prevailing. It is known experimentally that an ion-pair formed after excitation requires an activation energy to dissociate it.

The energy gap E_G in an organic semiconductor may be defined as the energy difference between the ground state of the crystal and the state with an excess electron and an excess hole, the two charged species being sufficiently separated from each other for coulombic interaction between them to be negligible, i.e., by about 15 nm in a typical case (see Section 14.4).

Unlike the situation in the typical inorganic semiconductor, such as *Ge* or *Se*, the energy gap in many organic crystals (e.g., anthracene) is invisible in the absorption spectrum of the crystal. Furthermore, most observations of the thermal activation energies of the dark electrical conductivity of crystals of the anthracene type do not yield the value of E_G, again in strong contrast to the situation with the typical inorganic semiconductors in the absence of doping. Many claims in literature that an observed thermal activation energy of dark conductivity of organic solids did mark an energy gap are clearly wrong.

(2) By phonon-assisted hopping over the potential barrier separating two adjacent molecular, or ionic, sites. This again results in a drift velocity component in the direction of the applied field.

(3) By quantum mechanical tunneling through, not hopping over, the potential barrier; the transition takes place between exactly matching energy levels on either side of the barrier.

(4) By long or variable-range hopping, or percolation, meaning non-coherent charge transfer between non-adjacent molecular or ionic sites; this replaces the sharp energy band edges by slowly decaying tails extending well into the energy gap, and arising from the very disorder characterizing amorphous systems, (see also Section 19.4).

(5) By transferring not the free charge carrier as such but rather its energy in form of an exciton, usually a polaron. This process requires a mechanism for the exciton to be dissociated at the carrier exit electrode so as to yield again a free carrier which can enter the external circuit.

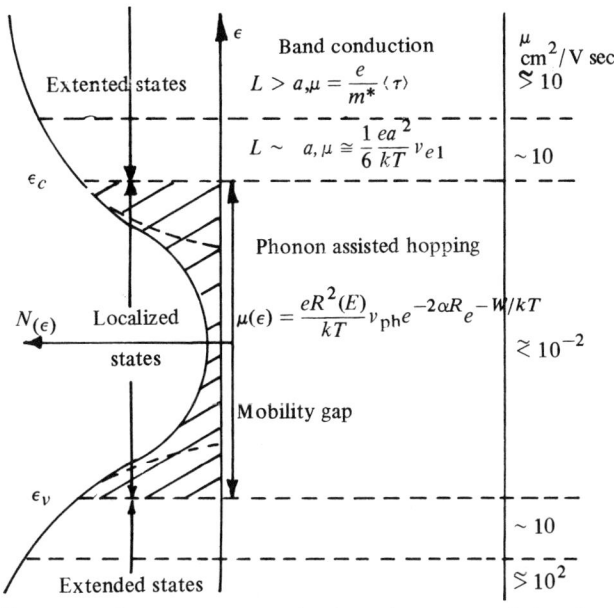

Fig. 15.1 Summary of transport properties in various parts of the electronic state distribution of an amorphous solid. The graph shows a schematic density of state distribution $N(\epsilon)$. Localized states are shaded. After Spear.[46]

Process (4), percolation, is discussed in more detail when dealing with amorphous systems in Section 19.4.

Figure 15.1 summarizes different modes of transport associated with the various regions of the state distribution.[46] The figure shows schematically the density of the state distribution; the energy is plotted as ordinate. In high-lying extended electron states, well within the conduction band, the effect of random potentials and intermolecular distance fluctuations on charge transport is but slight; the mean free path of the carrier, L, is appreciably longer than the mean intermolecular spacing α and a spacing theory based on Bloch wavefunctions and only occasional scattering (see Chapter 4) remains applicable. The mobility μ is relatively high.

The onset of localization is marked by a critical energy ϵ_c when the effects associated with the lack of long range order commence to dominate the charge transport. The mean free path of the carrier and the coherence length of the wave function describing the carrier, approach the intermolecular spacing α. Below ϵ_c in the diagram, transport cannot any longer be considered as motion in an energy band with only occasional scattering. Cohen[47] pointed out that in this border region, conduction is essentially a diffusive process akin to Brownian motion: the electron jumps from site to site with an atomic, or molecular, frequency ν_{el}, but without any thermal activation; ν_{el} is an electronic frequency of the order of 10^{15} sec^{-1}. The estimated drift mobility is about 10 cm^2/Vsec or less. A detailed analysis of the transport in the ϵ_c region is based on a random phase model[48] in which the extended wave functions are represented as linear combinations of atomic, or molecular, wave functions having no phase correlation between site and site.

The fairly sudden transition below ϵ_c, at the onset of localization, is of fundamental importance in the theory of the non-crystalline state and is discussed in Section 19.4. States below ϵ_c thus are localized in that relevant physical quantity averaged over these states vanish in a rigid lattice. At temperatures above 0 K charge transport through the localized state distribution is possible, but requires phonon assistance. The mobility $\mu(\epsilon)$ for the resulting thermally activated hopping regime is shown in Fig. 15.1. The quantity $\exp(-2\alpha R)$ describes the overlap of the wavefunctions between adjacent hopping sites, while the term $\nu_{ph} \exp(-W/kT)$ refers to the probability that the localized electron transfers, or hops, to another site at an energy W above the originating one. Hopping mobilities, at room temperatures, of the order of 0.01 cm^2/Vsec or less result so that in the vicinity of ϵ_c and ϵ_ν the mobility drops by several orders of magnitude.

The thermal equilibrium densities of carriers in the extended states may become so large at elevated temperatures that conduction *via* these states becomes dominant. At lower temperatures, the occupation probability at ϵ_c (for electrons) and ϵ_ν (for holes) drops rapidly so that transport *via* the localized state distribution takes over though at reduced mobility values. As T is still further lowered, this type of hopping transfer eventually approaches the Fermi level to within a few times kT. Then, one enters a variable range hopping regime.[48] This arises, because at sufficiently low temperatures nearest neighbor hops become increasingly unlikely while longer range transitions to more distant sites for which W is small, now enter the picture. Thus, in nearest-neighbor hopping there are many but short distance hops while under variable range hopping there are few but long distance hops. Physically, this may be explained by remembering that the amplitudes of thermal vibrations of the lattice molecules become smaller so that there is more opportunity for the hopping carrier to encounter a favorable situation in which it may cover large distances before becoming, again, temporarily trapped. Analysis of the variable hopping regime at the Fermi level yields the relation

$$\sigma \propto \exp(-B/T^{1/4}) \qquad (15.1)$$

for temperature dependence of the conductivity σ. The topic of different modes of mobility regimes has been reviewed by several authors[2] (see also Section 15.5, and for mobility data, Section 16.3).

15.2 Energy Band Model. Delocalized Transport

In carrier drift *via* a conduction band, the mobility is determined by scattering. A great many different scattering centers have been identified, such as defects of various kinds, impurities, phonons, etc. (see Chapter 4). Some of these may also trap the carrier; trapping is discussed in Section 10.2, 15.10 and 20.1.

Only shallow traps affect the mobility; deep traps fill quickly and since their depth, in terms of energy, exceeds kT considerably, they remain filled though they may affect the mobility indirectly by acting as scattering centers.

The interaction between a charged center, say an ionized impurity, and a free carrier, may alter the effective value of the local permittivity and thus, in turn, the local electric potential field involved in the scattering process.[45] This effect is usually neglected, though it has been studied of late. The matter remains to be clarified.[45]

A recent energy level diagram[4] for the "classical," best-studied moelcular crystal, viz., anthracene, is shown in Fig. 15.2. I_g is the ionization energy

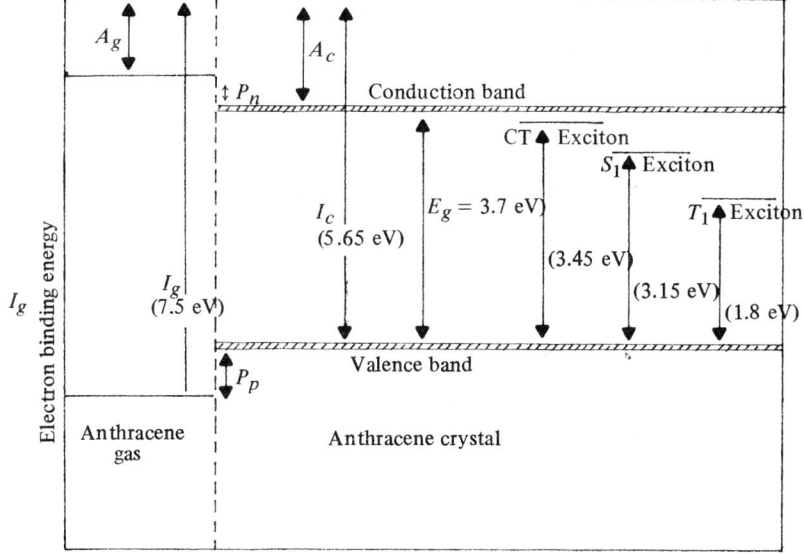

Fig. 15.2 Schematic diagram of energy levels in anthracene. After Williams.[4]

in the gas phase, and I_c that of the crystal, while A_g and A_c stand for the electron affinities of the 2 phases. P_n and P_p are the polarization energies for an electron and a hole, respectively. E_g stands for the energy gap which, according to the Lyons equation (see Part A, p. 470, eq. (8.11).

$$E_g = I_g - A_g - (P_p + P_n) = I_c - A_c \tag{15.2}$$

results in 3.7 eV. Exciton levels for triplet T_1, singlet S_1 and charge transfer (CT) states are also indicated. If there is sufficient overlap of molecular wave functions to give rise to energy bands, these are likely to be quite narrow, corresponding to very high effective masses of the carriers. However, if the band width J_0 becomes very narrow, i.e., if $J_0 \leqslant kT$, then the energy states even at the top of the "band" will be occupied, with negative masses, and the concept of an effective mass will no longer be meaningful.[5] The conductivity should then become proportional to the inverse temperature.[5] $\sigma \propto j_0/kT$. This contradicts experiment because in nearly all cases the conductivity varies exponentially with inverse temperature. Very narrow bands mean very low mobilities and high effective masses.

In the band model, the scattering cross-section presented to the electron wave by a scattering center should increase with the temperature T so that $\mu \propto T^{-n}$ where the value of n depends on the details of the dominant scattering mechanism(s) and the theory employed (Chapter 4, especially Section 4.5 and 4.6). It should range from 1 to 2. The model is claimed to account for experimentally determined mobility components in naphthalene[6] in all crystallographic directions except for electrons in the a and b directions; these

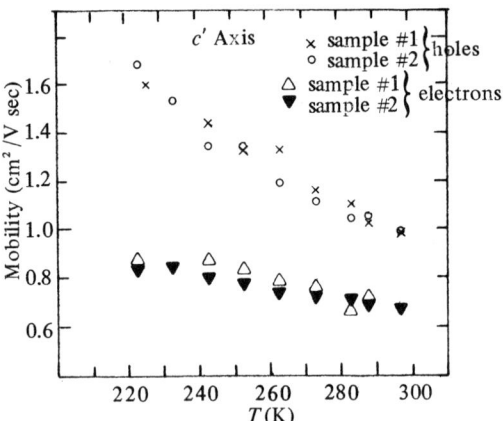

Fig. 15.3 Mobility vs. temperature for electrons and holes in the $\vec{c}'$ direction; Naphthalene. After Mey and Hermann.[6]

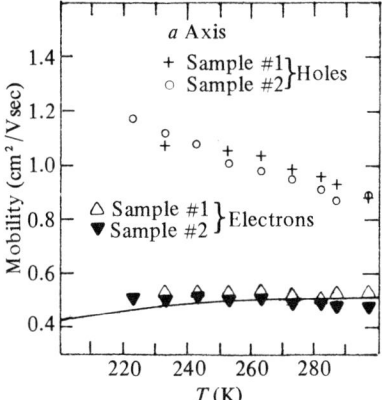

Fig. 15.4 Mobility *vs.* temperature for electrons and holes in the $\vec{a}$ direction; Naphthalene. After Mey and Hermann.[6]

mobilities are found to be essentially temperature dependent[218] and hence not compatible with band theory as seen from Figs. 15.3, 15.4 and 15.5, taken from a study by Mey and Hermann;[6] another report by Sumi[218] agrees with these data. This author considers the field dependence of the mobility as well as its behavior when transiting the Debye temperature θ_D of the solid: as the crystal is cooled below θ_D, the mobility rises abruptly; in this

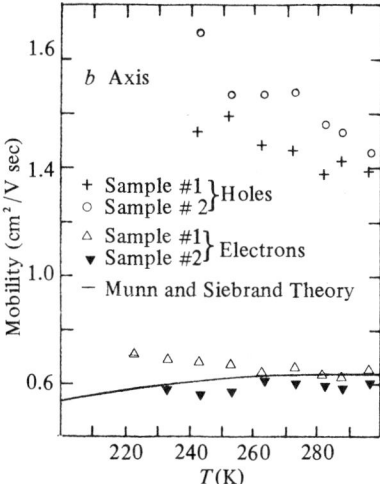

Fig. 15.5 Mobility vs. temperature for electrons and holes in the $\vec{b}$ direction; Naphthalene. The solid curve is the Munn-Siebrand prediction.[7] After Mey and Hermann.[6]

region the drift mobility also exhibits maxima and minima as long as the applied electric field is kept below 10^5 V/cm. At θ_D, the mobility saturates.[218] It is concluded[218] that below θ_D there is an increasing electron transfer as the temperature drops further below θ_D—without phonon participation: charge transfer occurs not by phonon-assisted hopping between the crystallographic planes but by coherent motion in a 2-dimensional energy band within the plane.

Figure 15.3 displays theoretical predictions derived from the Munn-Siebrand formalism[7] (see Section 15.3). The fit is seen to be reasonably good. Thus, it may be concluded[8] that the electron-phonon interaction in this case is stronger than that compatible with the band model. This incompatibility has also been demonstrated for anthracene.[8]

For perdeuterated anthracene crystals, however, a trap-free mobility varying with temperature as

$$\mu_0(T) = \text{Const.}\ T^{-1.7} \qquad (15.3)$$

has been calculated,[22] in good agreement with the experimental value of the exponent of -1.8 and quite compatible with the band model. Also in the highly conducting polyacetylenes, with their extensive π-electron delocalization, the band model appears to hold.[9] Such solids are discussed in Chapter 21 and in Section 17.10.

Burshtein and Williams have reported carrier mobilities for durene (1,2,4,5-tetramethylbenzene).[10] In the crystal ab plane both hole and electron mobilities are isotropic (5 cm^2 V^{-1}sec^{-1} and 8 cm^2 V^{-1}sec^{-1} respectively at room temperature) with a $T^{-2.5}$ temperature dependence. In the c^* direction only hole mobilities have been measured (0.15 cm^2 V^{-1}sec^{-1} at room temperature) which showed a $T^{-2.8}$ temperature dependence. These results[10] indicate that charge-carrier transport in durene agrees with the band model. The mobilities are larger than those measured for the iso-structural anthracene probably because of the changed electron exchange interactions following chemical substitution around the aromatic ring.

It must be realized that, in a given temperature interval, conductivity may be due not to any one mechanism but to several with different ones contributing to a varying extent at different temperatures. Moreover, within the framework of any one transfer mode, the mobility of the carrier is determined by several physical and chemical variables which, again, will contribute to the charge transfer to a varying degree as the temperature is changed. Frequently one or the other dominates, but at any one temperature several processes may contribute to a comparable extent. Thus, Mey and Hermann in a careful study of the mobilities in single crystals of naphthalene conclude[6] that the transfer mechanism appears to be intermediate between drift in an energy band and phonon-assisted hopping. In phenanthrene, coherent transport prevails[3] in the a and c' directions while the mobility along the b axis is dominated by phonon assisted hopping.[3] Sakia et al.[20] in studies of the ac conductivity $\sigma_\sim$ of thin films of Cu-phthalocyanine and of other aromatic compounds[21] find that

$$\sigma_\sim = \sigma_0 e^{-E/kT} + \text{Const.} \, \omega^n e^{-E(\omega T)/kT} \qquad (15.4)$$

where $n = f(T)$, and $E(\omega T)$ is an apparent activation energy for nearest neighbor hopping.[21] Thus, $\sigma_\sim$ appears to be the sum of a classical coherent carrier motion term plus a strongly frequency-dependent hopping term.

In benzophenone, too, contributions from both thermally activated hopping and small polaron band modes have been proposed.[33]

In, e.g., some TCNQ adducts, conductivity, though not intrinsic, is said[216] to be due to both electrons and holes to a comparable extent, with different activation energies for the mobility. The mobility ratio μ_+/μ_- is reported[216] to drop with increasing pressure.

In a discussion akin to that centering on the relation between carrier mean free path and lattice spacing, (cf., pp. 40 and 268–the well-known Ioffe objection) Morton has pointed out[70] a further difficulty in the application of the classical band model to narrow band semiconductors, also arising from the Uncertainty Principle: when the uncertainty energy $\Delta\epsilon$ exceeds the range of excited energies kT the standard equilibrium distribution functions for electrons and phonons will be inappropriate. The usual assumption of statistical mechanics, namely that the particles are weakly interacting (equivalent to $t \to \infty$), does not apply, and for the electrons the spread of excited states will be of order $\Delta\epsilon$ rather than kT. For the phonons the spread of energies will still lie within the usual range allowed by the Debye cut-off procedure, and the lattice heat capacity, for example, will be little changed by the reduction in the free life time t appearing in the Heisenberg Uncertainty Relation.

The uncertainty energy $\Delta\epsilon$ is given by

$$\Delta\epsilon \approx h/t = hv/L \qquad (15.5)$$

where v is the velocity of the electron, of the order of the usual Fermi velocity, say 10^5 or 10^6 msec^{-1} for d- or s-electrons, respectively. From free-electron theory, which is assumed to apply approximately, the resistivity ρ is given by

$$\rho = \frac{m^* a^3}{N_a e^2 t} \sim \frac{10^{-21} m^*}{N_a t m} \qquad (15.6)$$

where m^* and m are the effective carrier mass and the electron mass respectively, N_a is the electron per atom ratio for the unfilled sub-bands at the Fermi surface and e is the electronic charge. The resistivity ρ is the total, including any scattering by lattice defects. a stands for the lattice spacing and L for the mean free path of the electron. For a simple metal, Morton has calculated[70] the orders of magnitude of the resistivity and the uncertainty energy, assuming that the mean free path is of the order of a few Å. It is generally assumed that this represents a lower limit for the mean free path dominated by impurity scattering. For the shortest mean free paths likely to be observed, the uncertainty in energy is of the order of the full bandwidth of the electrons.[70] The bandwidth energy may be taken to represent an upper limit to the uncertainty energy, and correspondingly there will be a lower limit to the relaxation time t. The total resistivity will exhibit a tendency to saturate at a value of the order of the maximum observed residual resistivity. Conversely, due to the reduced interaction, the lattice thermal conductivity at low temperatures, limited by electron scattering, will be unusually large for samples with residual resistivity comparable with the maximum attainable.

In the semi-classical theory of the electron-phonon interaction the electron is a localized particle with dimension less that the phonon wavelength. To construct an electron wave-packet even with this dimension it is necessary to superpose states from outside the usual range of thermally excited states. The corresponding uncertainty in energy is of the order[70] kTv/s for scattering by typical phonons at temperatures below the Debye temperature, where s is the velocity of sound. This figure may be comparable with the bandwidth, and one may assume the mean free path will not then readily reduce further; that is, the scattering efficiency will reduce.

While the above discussion[70] is meant to apply primarily to metals, it should also hold *mutatis mutandis,* for the class of well-conducting, but narrow-band organics; this matter deserves further study and consideration.

15.3 The Transition Regime between Coherent Wave Propagation and Transfer via Localized States. The Munn-Siebrand Formalism.[11]

The Munn-Siebrand[11] approach includes the effect of intramolecular vibrations upon transport and may yield smaller bandwidths, and thus heavier effective carrier masses, than those resulting from rigid crystal calculations, though without altering the band-like properties. This concept appears to be, so far, a satisfactory and widely accepted model for the understanding of charge carrier transport in molecular organic crystals.

The basic mechanism of electronic transport depends on the nature of the electron exchange and the electron-phonon interactions. In molecular crystals the former are much smaller than in inorganic solids like Ge or Si where the electrons behave as quasi-free particles only occasionally scattered by phonons. The electron-phonon interactions, however, are very similar. Thus, the latter may become dominant and the electrons then behave as quasi-localized particles which only occasionally transfer to neighboring sites under the influence of electron exchange interactions. The electron mobility in the resulting hopping regime is thermally activated, i.e., increasing with temperature. Munn and Siebrand[11] have pointed out that there are two limiting modes of transfer in the hopping regime.[11] In the slow-electron limit the electron exchange energies are small compared with phonon dispersion energies and the transfer of the electron between adjacent molecules is the rate-determining step, as usually assumed. In the slow-phonon limit the electron exchange energies are large compared with phonon dispersion energies, and since the electron is strongly coupled to the phonons its rate of transfer is limited by the rate of phonon transfer. Finally, the electronic transport depends not only on the strength of the electron-phonon interaction but also on its nature whether the interaction is best regarded as affecting principally intermolecular vibrations of rigid molecules or intramolecular vibrations of molecules in an otherwise static lattice, and whether the interaction is linear or quadratic in the phonon coordinates.

Complications arise in calculating the electron-phonon interactions because of their reciprocity:[11] the electron is affected by the vibrational modes and they in turn are affected by the electrons. If the transport is coherent, the vibrations destroy the (local) crystal symmetry, causing local variations in the electron exchange integrals so that the energy band states lose coherence and the electron has a finite free lifetime.

If transport is by hopping, the electron affects the vibrational modes in its vicinity, resulting in a polaron. This entity may transit by hopping, or by tunneling (see Section 15.6), to its nearest neighbor only if the vibrational state of that neighbor is favorable.

Munn and Siebrand then proceed to calculate the conductivity σ in the intermediate region between these two charge transfer regimes.

The dc conductivity may be written[11]

$$\sigma = i \int_{-\infty}^{0} t \langle [j(t), j(0)] \rangle dt. \tag{15.7A}$$

$j(t)$ is the current density operator in the Heisenberg representation at time t with the volume of the system taken as unity; the angular brackets denote a thermodynamic average in zero electric field; the square bracket is a commutator. By means of standard transformations, eq. (15.7A) can be written as[10,16]

$$\sigma = \frac{1}{2}\beta \int_{-\infty}^{\infty} \langle j(t)j(0)\rangle dt = \lim_{w\to 0} [\phi(w)/2w] \quad (15.7B)$$

where $\beta = 1/kT$, where $\phi(w)$ is the spectral density for the Green function

$$F(t) = T\langle j(t)j(0)\rangle$$

in which the operator T orders the earlier time on the right.

$j(t)$ is the rate of change of the polarization P which is given by

$$P = e\sum_n r_n a_n^+ a_n \quad (15.8)$$

so that

$$j(t) = \frac{dp}{dt} = -i[p\mathcal{H}] \quad (15.9)$$

a_n^+ creates an electron in a state localized about molecule n at position r_n. $\mathcal{H}$ is the total Hamiltonian, a sum of independent electron and phonon terms plus the electron-phonon interaction. For simplicity we take a linear chain with nearest-neighbor interactions, when

$$\mathcal{H} = \sum_n [\epsilon_0 a_n^+ a_n - Ja_n^+(a_{n+1} + a_{n-1})] + \sum_q \omega_q b_q^+ b_q + \mathcal{H}_{ep}, \quad (15.10)$$

where ϵ_0 is the energy of the localized electron state, J is the exchange integral at the mean molecular separation, and b_q^+ creates a phonon of frequency ω_q in mode q (units with $\hbar = 1$ are used). The term $\mathcal{H}_{ep}$ governs σ through the thermodynamic average over $\mathcal{H}$, and also through a term in j unless p and $\mathcal{H}_{ep}$ commute.

Linear Interaction with Lattice Modes

The results described in this subsection are essentially those due to Gosar and Choi[12] and to Gosar and Vilfau.[13] The electron-phonon interaction causing local variations in the exchange integrals is, for the present simplified model

$$\mathcal{H}_{ep} = \sum_{n,q} K_q(b_q + b_{-q}^+)a_n^+(a_{n+1} + a_{n-1}) \quad (15.11)$$

Since the exchange integrals change very rapidly with the intermolecular separation, the difference between r_n and the mean molecular position can be neglected so that

$$j = -ied\sum_n [J - \sum_q K_q(b_q + b_{-q}^+)] a_n^+(a_{n+1} - a_{n-1}) \quad (15.12)$$

where d is the mean molecular separation. Because $\mathcal{H}_{ep}$ does not commute with p, there is a phonon-assisted part of the current. The conductivity is calculated from eq. (15.7) by assuming that the correlation between electrons and phonons can be decoupled into a product of independent electron and phonon correlations. Then σ depends on

$$g(t) = \langle a_n^+(t) a_{n+1}(t) a_{n+1}^+(0) a_n(0) \rangle \tag{15.13}$$

which is assumed to have the Gaussian form,

$$g(t) = (n/N)\exp(-\alpha^2 t^2/4\pi) \tag{15.14}$$

where n is the number of excess electrons and N the number of molecules in the crystal. The parameter α is determined by equating coefficients of t^2 in the expansions of eqs. (15.10) and (15.11). The major part of the resulting conductivity is

$$\sigma = (2ne^2\beta d^2/\alpha)[J^2 + \sum_q K_q^2 (2\nu_q + 1)] \tag{15.15}$$

where ν_q is the mean number of phonons in mode q. The drift mobility $\mu^D = \sigma/ne$ is directly obtainable from eq. (15.12).

These calculations do not assume any transport model. They show that the effective transfer integral increases with temperature through ν_q, opposing the decrease to be expected through α.[11]

Quadratic Interaction with Molecular Modes

The results in this subsection are based on the electron-phonon interaction[11]

$$\mathcal{H}_{ep} = -(\omega_2^2/4\omega_0)(b_n + b_n^+)^2 a_n^+ a_n \tag{15.16}$$

where b_n^+ creates a molecular phonon of frequency ω_0 at site n. Since $\mathcal{H}_{ep}$ commutes with p, there is no phonon-assisted current, so that

$$j = -ied\sum_n J a_n^+ (a_{n+1} - a_{n-1}) \tag{15.17}$$

Thus, the conductivity depends purely on a two-particle correlation function, but averaged over the full $\mathcal{H}$.

The Fourier expansion of $F(t)$ in the complex frequency plane is expanded as a perturbation series in $\mathcal{H}_{ep}$, and the series is approximated by an infinite subset of terms involving particularly simple phonon correlations. This subset is resummed exactly, and the conductivity is calculated from it. Because of the infinite summation the results should be appropriate to both band and hopping conduction. The conductivity can be written as

$$\sigma = ne^2\beta d^2 J^2 \gamma \tag{15.18}$$

where γ consists of a number of terms, each a complicated function of the model parameters.

To evaluate eq. (15.18) numerically, one has to make additional approximations. Within the limits where first-order perturbation theory is valid[11] one may treat the intermediate regions in a qualitative way. These limits will be expressed in terms of the four parameters of the model, namely, the electron exchange integral J, the electron-phonon interaction ω^2, the phonon frequency ω_0 through the dispersion relation $\omega_q^2 = \omega_1^2 \cos q$.

For very narrow bands, such that $\omega_2^2 \gg \omega_1^2 \gg J\omega_0$, transport takes place *via* hopping. Except at very low temperatures, the electron exchange is the rate-determining step (slow-electron hopping). The leading term of γ in eq. (15.18) becomes proportional to $\omega_0 \omega_1^{-2}$ and σ is found to increase slightly with temperature if the carrier density is constant. The drift mobility μ^D shows the same behavior.

If one now increases J, e.g., by rotating an anisotropic crystal, such that $\omega_2^2 \gg J\omega_0 \gg \omega_1^2$, the leading term of γ tends to become proportional to $\omega_0^{-2} \omega_1^4 J^{-3}$ so that σ and thus μ^D vary as $\omega_0^{-2} \omega_1^4 J^{-1}$. The overall transport mechanism still corresponds to hopping but the rate-determining step is now phonon exchange (slow-phonon hopping). The temperature dependence remains essentially the same as before.

A further increase in J will ultimately lead to $J\omega_0 \gg \omega_2^2 \gg \omega_1^2$, at which point coherent transport becomes dominant. The leading term of γ becomes proportional to $\omega_0 \omega_1^{-4} \omega_2^2$ and μ^D decreases rapidly (roughly exponentially) with temperature.

The way in which these three limiting regions may be joined in a crystal like anthracene is schematically depicted in Fig. 15.6. Slow-electron hopping may prevail for electrons in anthracene moving in the c' direction, as indicated by temperature, pressure, and isotope effects.[14] Slow-phonon hopping is barely realizable in anthracene. The actual transport mechanism in all other directions is expected to be intermediate between slow-phonon hopping and coherent transport. The $\mu^D(J)$ curves depicted in Fig. 15.6 are based on the listed parameter values together with the band structure mobility components in anthracene. Figure 15.6 also shows the general behavior of the Hall effect for this model.[14] The product of Hall and drift mobility, which is independent of the direction of the electric field has been plotted[11] rather than the field dependent μ_H.

15.4 Hopping Transfer vs. Band Model; further discussion

Schein[18] has pointed out that in molecular crystals exciton effects are strong (exciton binding energies are approximately half the one-electron "bandgap") resulting in optical absorption curves dominated by the effects of electron-electron interactions and phonon emission and absorption. In fact, the "conduction band" has not been identified definitively in any molec-

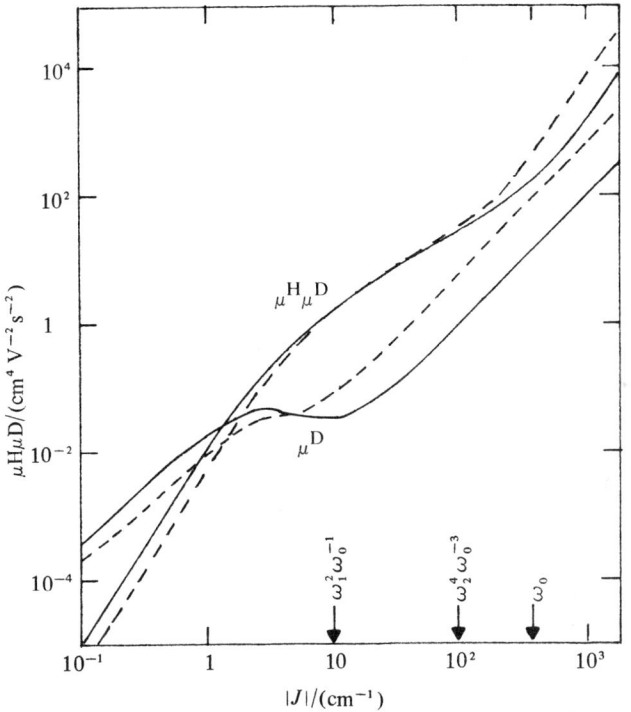

Fig. 15.6 Drift and Hall mobilities of charge carriers in anthracene as a function of the electronic bandwidth for $T = 300$ K (solid curves) and 200 K (broken curves). The curves, which represent a schematic solution to eq. (15.18) and its counterpart for the Hall effect, fit all the unambiguously observed mobility components of electrons and holes. After Munn and Siebrand.[11]

ular crystal by optical measurements. Only internal-photoemission measurements have been used[18] to determine bandwidths in molecular crystals; however, doubts have been expressed concerning the interpretation of these measurements.

Schein argues[18] that the weak temperature dependence observed appears to be inconsistent with narrow energy band theory based on either a single or a combination of scattering mechanisms. Independently, Forster[19] has shown that e.g., carbon-filled polymers do not conduct by any single classical conduction mechanism: band conduction as well as hopping transfer plus, perhaps, ionic diffusion may occur.[19] In addition to many internal parameters he emphasises that the observed conductivity will depend on the conditions of the experiment, especially the contacts. However, at or near room temperature, hopping is probably dominant.[19]

In narrow band theory (see eq. 4.56b, Part A, p. 268) the mobility μ is given by

$$\mu = \frac{e\tau}{kT}\frac{J^2 a^2}{\hbar^2}$$

Then, the temperature dependence of μ can be attributed to the temperature dependence of a, J, and τ.

The temperature dependence of the lattice constant is

$$a = a_0(1 + \beta T) \qquad (15.20)$$

where

$$\beta \cong 10^{-4} K^{-1} \qquad (15.21)$$

which is negligibly small.[89] However, as the lattice expands, the electron overlap integral decreases. Delacote[90] has shown this to be small for anthracene, 5% in 200 K. Another source is the modification to the bandwidth due to the formation of polarons. The effective (polaron) bandwidth[58] is We^{-S}, where S is a vibrational overlap integral which is exponential in the phonon occupation number and therefore in T. In either case as T increases J (the effective bandwidth) decreases; the temperature dependence of J is at least

$$J \propto T^{-p}, \qquad p > 0. \qquad (15.22)$$

It remains to establish the temperature dependence of the relaxation time τ for various scattering mechanisms. In acoustic phonon scattering, the cross-section a_s^2 is linear[91,92] in T and therefore the inverse of the mean free path (the product of a_s^2 and the number of scattering events per unit volume N).

$$Na_s^2 \propto T \qquad (15.23)$$

The relaxation time therefore follows as

$$\bar{\tau} = Na_s^2 v^{-1} \propto v^{-1} T^{-1} \qquad (15.24)$$

Thus, for "classical" wide band, semiconductors, eq. 4.27, Part A, p. 258 yields

$$\mu \propto T^{-3/2} \qquad (15.25)$$

In the narrow band case, and for the time being neglecting the temperature dependence of J, it is shown in Part A, p. 259 that

$$\mu \propto T^{-2} \qquad (15.26)$$

The difference temperature dependence of eqs. (15.25) and (15.26) is directly determined by the temperature dependence of ν_i for wide- and narrow-band semiconductors.

This result is shown[18] in Table 15.1 along with results obtained similarily for two-phonon acoustic, Coulomb, neutral impurity, one- and two-phonon optical, and dislocation scattering.

It is seen that an at least T^{-1} dependence of μ results. If phonons are involved, an additional temperature dependence is obtained in order to account for the thermal occupancy of the phonon states. Inclusion of the temperature dependence of J only further raises the temperature dependence so that one can write, within the confines of narrow band theory; (see also discussion in Part A, pp. 264-268).

$$\mu \propto T^{-(1+m)}, \qquad m > 0 \qquad (15.27)$$

Eq. (15.27) is obviously inconsistent with experimental data of electron mobilities as displayed in Table 15.1: therefore, it must be concluded that band theory does not hold in these systems. Hopping theories, in which the carrier is treated as localized, appear to be a more appropriate approach.[18]

Detailed studies by Maruyama et al.[23] on highly pure perylene also indicate that for this compound a hopping rather than a band model is applicable. In the course of this study, Maruyama has also computed intermolecular resonance integrals for perylene which values are reproduced in Table 4.3A.

The rather flat nature of the higher temperature region of $\mu(T)$ has been ascribed[24] to anharmonicity of the lattice vibrations; at low temperatures coherent transport has to be considered in addition to the hopping mode.[24]

Sumi,[218] in order to explain the almost temperature independent electron mobility observed in anthracene in the c' direction[140] between about 77 and 500 K, and also naphthalene in a much more restricted temperature interval,[141] introduces the idea of a two-channel, anisotropic, hopping mobility. One operates *via* the usual transfer integral of the crystal lattice containing scattering centers, while the other channel operates *via* the transfer integral involving rotational motion of the molecules. The latter is said[218] to be dominant at high temperatures while the first channel results in mobility within the ab plane.[140] Electronic motion within the ab plane is supposed to be coherent (band model).

Mobility through the first channel increases as the electron-phonon scattering within the (a,b) plane becomes less frequent below the Debye temperature, while mobility through the second channel is suppressed in this temperature region, since rotational vibrations are not excited enough (see also discussion in Section 15.8). Accordingly, the steep rise of the mobility observed below about 100 − 130 K in naphthalene is ascribed to the crossover between the two hopping channels. The calculated (a,b) mobility increases

Table 15.1 Temperature dependence of μ (excluding temperature dependence of J) After Schein.[18]

Scattering	a_s^2	$I_R = (Na_s^2 v)^{-1}$	"Classical" μ	Narrow band μ ($W \lesssim kT$)
Acoustic				
one phonon	T	$(vT)^{-1}$	$T^{-3/2}$	T^{-2}
two phonon	T^2	$(vT^2)^{-1}$		T^{-3}
Coulomb	v^{-4}	v^3	$T^{3/2}$	T^{-1}
Neutral impurity	v^{-1}	v^0	T^0	T^{-1}
	v^0	v^{-1}		
Optical				
one phonon	$T^{-1}[\exp(hv/kT) - 1]$	$(T/v)[\exp(hv/kT) - 1]$	$T^{1/2}[\exp(hv/kT) - 1]$	$[\exp(hv/kT) - 1]$
two phonon	$T^{-1}[\exp(hv/kT) - 1]^2$	$(T/v)[\exp(hv/kT) - 1]^2$		$[\exp(hv/kT) - 1]^2$
Dislocation	v	v^{-1}	$T^{-1/2}$	T^{-1}

with decreasing temperature obeying approximately $T^{-1} - T^{-2}$ around room temperature, consistent with observation.[141] The observed anisotropy in magnitude of the electron mobility was reproduced with the use of a reasonable strength for coupling with rotational vibrations. Thus the model[140] also explains why the observed c' mobility is comparable in magnitude to the (a,b) mobility, even though the conduction band width in the c' direction is very narrow compared to that along the (a,b) plane.

A temperature independent mobility has also been reported[212] for naphthalene above 100 K. It has been suggested[213] that this effect is due to transfer between energy states which are localized by virtue of the disorder produced by the local thermal vibrations in the crystal. The matter is still not resolved.

Pressure Dependence of Mobility

This is intended to complement the discussion in Part A, p. 250 ff with a few more recent data, results and references (see also Section 19.8).

Munn and Siebrand studied[54] effects of pressure on the drift-mobility components using values of the linear compressibilities calcualted from the measured elastic constants.[55] Elnahwy et al. have measured[56] the linear compressibilities of crystalline anthracene using neutron-diffraction techniques. Their results show that the linear compressibilities are anisotropic, contrary to the assumption of Katz et al. (cf., Part A, p. 244); the values of these compressibilities are significantly different from those used by Munn and Siebrand.

Elnahwy et al.[56] have calculated the intermolecular-resonance integrals for an excess electron and an excess hole as functions of pressure in a band model similar to that used by Katz et al., but using the Hoyland and Goodman coefficients;[57] and including two- and three-center integrals. These intermolecular-resonance integrals were then used to calculate the band structures for the carrier and their dependence on pressure. Drift-mobility components were then calculated in the constant-free-path and in the constant relaxation-time approximations, as functions of pressure.

Elnahwy's et al. results[56] are displayed in Tables 15.2, 15.3, 15.4 and 15.5.

Drift mobility components for different pressures and for electrons as well as for holes are also listed in Tables 15.4 and 15.5. Data are given for both the constant free time τ (see Part A, p. 248) and for the constant free path λ approximation (see Part A, p. 254).

Table 15.4 compres calculated values of the anisotropy of the drift mobility μ_d with experimental values. Table 15.5 likewise compares the ratio of μ_d at 3 kbar to that at atmospheric pressure with experimental data.

Table 15.2 Intermolecular Resonance Integrals for an Excess Hole and An Excess Electron in Units of 10^{-4} eV at Different Pressures. After Elnawhy et al.[56]

Intermolecular resonance integrals	Hole pressure (kbar)				Electron pressure (kbar)			
	0.00	2.00	4.00	6.00	0.00	2.00	4.00	6.00
e_2	−0.495	−0.439	−0.315	−0.153	0.274	0.729	1.275	1.957
e_3	−125.9	−135.1	−145.1	−155.7	74.25	79.95	86.11	92.74
e_4	−0.129	−0.148	−0.171	−0.192	−0.025	−0.028	−0.031	−0.035
e_5	0.273	0.306	0.368	0.431	0.215	0.219	0.217	0.230
e_6	−3.230	−3.838	−4.573	−4.651	−0.489	−0.628	−0.804	−1.195
e_7	0.073	0.081	0.001	0.112	0.027	0.022	0.013	0.009
e_9	−103.3	−107.9	−112.3	−116.8	−124.5	−133.2	−142.4	−152.1
e_{10}	28.96	33.17	37.87	43.13	1.591	2.041	2.565	3.138

Table 15.3 Bandwidths vs. Pressure in Units of 10^{-4} eV. After Elnawhy et al.[56]

	Hole				Electron			
	pressure (kbar)				pressure (kbar)			
Direction	0.00	2.00	4.00	6.00	0.00	2.00	4.00	6.00
a^{-1}_{lower}	309	312	314	311	493	526	562	600
a^{-1}_{upper}	286	286	282	279	491	523	557	592
b^{-1}_{lower}	799	840	879	918	203	215	227	239
b^{-1}_{upper}	249	283	321	364	789	844	904	967
c^{-1}_{lower}	218	247	282	324	11.7	16.5	22.2	27.9
c^{-1}_{upper}	246	283	324	366	13.8	16.1	18.9	22.3

Although the constant-free-path approximation yields good agreement between the calculated anisotropy of the drift mobility at atmospheric pressure and experiment it predicts[56] essentially no change of the mobility with pressure in the $\vec{a}$ direction for holes, contrary to the experimentally observed increase. The constant-relaxation-time approximation predicts that the drift mobility of holes increases in all three directions, in qualitative agreement

Table 15.4 Comparison of the Theoretical and Experimental Anisotropy of the Drift Mobility of Anthracene at Atmospheric Pressure. After Elnawhy et al.[56]

	Two- and three-center integrals		Experimental value	
	Constant τ	Constant λ	Kepler[58]	Kajiwara et al.[60]
	Holes			
μ_a	1.00	1.00	1	
μ_b	2.1	2.0	2.0	
$\mu_{c'}$	0.45	0.54	0.8	
	Electrons			
μ_a	1.00	1.00	1.0	1.0
μ_b	0.66	0.64	0.59	0.64
$\mu_{c'}$	0.001	0.002	0.23	0.28

Table 15.5 Comparison of the Theoretical and Experimental Dependence of the Drift Mobility of Anthracene under Pressure at 3 kbar. After Elnawhy et al.[56]

	Two- and Three-center integrals		Experimental value	
	Constant τ	Constant λ	Kepler[58]	Kajiwara et al.[60]
Holes				
μ_a/μ_{a0}	1.06	0.98	1.4	
μ_b/μ_{b0}	1.16	1.075	1.4	
$\mu_{c'}/\mu_{c'0}$	1.41	1.26	1.4	
Electrons				
μ_a/μ_{a0}	1.14	1.07	1.4	1.33
μ_b/μ_{b0}	1.135	1.055	1.3	1.26
$\mu_{c'}/\mu_{c'0}$	. . .	. . .	1.0	1.0

with experiment. In fact, this approximation reproduces the exact experimental value of the percentage change of the drift mobility of holes in the $\vec{c}$ direction. The fractional changes in the drift mobility of holes and of electrons, in the ab plane at 3 kbar are smaller than experimental data, but closer to these than values obtained from the constant free path model. Thus, the constant-relaxation-time approximation is superior to constant-free-path approximation.[56] Both models, however, fail to describe the mobility of electrons in the $\vec{c}$ direction.[56] The conduction in this direction seems to be more likely due to a hopping model, a conclusion[56] also suggested by the increase of this mobility component with temperature. By comparing the experimental values of the drift-mobility components in the ab plane with the values calculated in the constant-relaxation-time approximation one finds that $\tau_0 = 2.3 \times 10^{-14}$ sec for holes and for electrons.[56] Therefore, the uncertainty in energy given by $\hbar/\tau_0$ is 0.027 eV, comparable with the bandwidths in the ab plane. The model is therefore internally consistent. Anisotropy effects are discussed in the frame-work of percolation theory[221] (Chapter 19).

15.5 Hopping

Probably the first conclusive experimental evidence for thermally activated hopping has been adduced by Gutmann et al.[25] working with 4 different polymeric and monomeric TCNQ complexes in the temperature range of

83 to 250 K. A large body of further evidence for this transport has since accumulated, out of which a few representative examples can be quoted: in poly(2-vinylpyridine),[26] polycrystalline thin films of coronene,[27] polyvinylchloride,[28] perylene,[23] many liquids such as benzene,[29] poly(arylenevinylenes),[30] polyvinylcarbazole[31] and many other polymers[32] and TNCQ complexes.[53]

Electrons can hop iso-energetically to another site only if the energy levels on both sides of the intervening potential barrier coincide; this is made possible by thermal fluctuations. Mott has pointed out[34] that this does not lead to a single activation energy ΔE at low temperatures but to an apparently decreasing activation energy as the temperature drops. The hopping frequency between the two sites has the form

$$\nu = \nu_0 e^{-\alpha R} e^{-\Delta E/kT} \quad (15.28A)$$

where R is the separation between the sites and α is determined by the decay rate of the wave function with distance,[35] so that $\exp(-\alpha R)$ represents the overlaps of the wave function between sites. ν_0 represents a lattice (phonon) frequency. At low temperatures the amplitudes of thermal fluctuations drop so that hops over large R's become relatively, as a proportion of all hops, more probable. If an electron hops a distance below R, then the number of states within the range dE is

$$(4\pi R^3/3)N(E)dE \quad (15.28B)$$

where $N(E)$ is the density of states. The average spacing between energy levels will then be $\Delta E = 3/[\pi 4 R^3 N(E)]$ and this will be the hopping activation energy.

The jump frequency given in eq. (15.28A) can be minimized so that

$$R^4 = \frac{Q}{8\pi\alpha N(E)kT} \quad (15.29)$$

If this gives a value of R less than the average distance between centers R_0 then the activation energy in $\Delta E = 3/4\pi R_0^3 N(E)$, independent of temperature. At low temperature, when $R > R_0$, the jump frequency, and thus the mobility, varies with temperature as $\exp[-(T_0/T)^{1/4}]$.

This relation has been the subject of much discussion.[35] Experimentally it has been found that it correctly predicts the temperature dependence of the conductivity of a number of materials but the $\exp[-(T_0/T)^{1/4}]$ dependence of the conductivity appears to extend to temperatures too high for this model[34] to hold. Some workers[36] have employed percolation theory (see Section 19.4) to extend Mott's theory to higher temperatures. Emin[37] has studied the jump rate assuming that all multiphonon hops contribute to the mobility; he finds that this hopping rate involves a non-activated temperature dependence similar to $\exp-(T/T_0)^{1/4}$, for temperatures below the Debye characteristic temperature T_{Deb}. The parameter T_0 is inversely proportional to the density of localized states.[64] Chen has supplied[226] a theory for thermally stimulated currents in hopping systems.

The theory of Sumi[140] (discussed in Section 15.4) also involves an attempt to clarify this issue. His theory predicts[140] that the drift velocity in the c' direction of highly anisotropic crystals such as anthracene or naphthalene should saturate with electric field F when the energy difference between neighboring (a,b) planes exceeds the rotational-vibration energy $\hbar\omega_2$ and the temperature is moderately low, $T \leqslant \hbar\omega_2/k$ (~180 K). The tendency towards saturation should be detectable at about room temperature.

At temperatures below T_{Deb} the c' drift velocity should show marked deviation from a simple linear relationship with the field strength, even for applied fields such that cFa is smaller than $\hbar\omega_2$. Here, it should be sensitive to defect scattering as, for example, by impurities. These phenomena should be more pronounced in naphthalene than in anthracene.

An alternative approach proposed by Bernasconi[135] suggests that a log (σ/σ_0) vs. $(T_0/T)^{-1/4}$ as well as a $(T_0/T)^{-1/2}$ relation can be simulated by a nearest-neighbor hopping model having an appropriate distribution of activation energies. In the disorder network type of analysis in this model, there is, for each temperature, a percolation along an optimal set of paths. This leads to a particular mean activation energy at each temperature. Studies by Saha, Abbi and Pohl[136] on polymers of the polyacene quinone radical type appears to support this reasoning.

Thus, one must distinguish[125] between (classical) nearest neighboring hopping which is a thermally activated transit of a carrier from one site to an adjacent one, and variable range hopping[126] to a more distant center. The latter is characterized by the disorder parameter α, eq. (15.28) which may be defined as the reciprocal of the localization length characterizing the spatial extension of the wave function. In some intermediate cases, the observed activation energy will be the composite of two individual activation energies. There exists a critical temperature making the transition between the two regimes,[123,124] which, however, is difficult to observe and locate experimentally.[123]

Let the ratio of carrier donating to carrier accepting sites N_A be denoted by K. Then

$$\text{For } K \ll 1 \quad T_{\text{crit}} = \frac{e^2 N_A^{2/3} K^{1/3}}{K\alpha\epsilon}$$

$$K > 0.5 \quad T_{\text{crit}} = \frac{e^2 N_A^{2/3}}{K\alpha\epsilon} \quad (15.30)$$

Here, ϵ is the permittivity of the medium, and e the electronic charge. Physically, at high temperatures the carrier does many hops over very small distances, while at very low temperatures, it does very few hops but over relatively large distances. The matter is further discussed when dealing with amorphous systems (see Section 19.4).

Mobility is often measured by the "time-of-flight" method discussed in Section 2.3, Part A, p. 59 ff and in Section 13.3. Ideally, for one type of carrier having a single-valued mobility, and in the absence of trapping effects, the current pulse observed should have a rectilinear edge, falling off vertically to zero at a certain instant of time. This, of course, is never found: (1) the capacitance of the system tends to distort the sharp edge of the pulse into a logarithmic decay curve, which effect, however, is small in a well-designed experiment, and also readily calculable; (2) and more importantly, carrier diffusion results in a probability distribution of the arrival times and in a "tail" developing in the pulse-shape observed.

In the Gaussian case the time dependence of the trace can be described in terms of the mean displacement ℓ from the illuminated surface and the dispersion σ which characterizes the spread of the charge sheet about the mean (see Fig. 15.7). The current induced by the drifting charge sheet is

$$I = eq \cdot \mu_d \cdot E/L \tag{15.31}$$

where L is the sample thickness, e is the electronic charge, μ_d is the drift mobility and q the number of carriers injected. In the absence of deep trapping, i.e., if $\mu_d E \tau \gg L$ where τ is the deep trapping life-time, the transient current remains constant and independent of the spreading σ about the mean ℓ.

When the leading edge of the carrier space charge reaches the counter electrode, the current starts to decay so that the width of its decay curve is a measure of σ at that time. The current peak, due to the arrival of the charge peak at the counter electrode, is determined by the transit time t_T of the space charge. Hence, for a carrier packet spreading according to Gaussian statistics, $(\sigma/\ell)_{t_T} \propto t_T^{-1/2}$ i.e., the current pulse will sharpen with increasing transit time when plotted in units of t_T (by lowering the external field, for instance).[40]

In the dispersive non-Gaussian case,[40] the carrier packet is not expected to grow symmetrically about its mean position. Immediately after the carrier-producing light flash some carriers will rapidly move out of the generation region due to a rare succession of short event times. As time evolves, an increasing number of carriers will suffer an event immobilizing them for times of the order of the observation time, t_T. Under extreme non-Gaussian conditions, the distribution of the carrier packet grows asymmetrically, resulting in a leading edge penetrating deep into the bulk, while the maximum of the charge density moves only slowly out of the generation

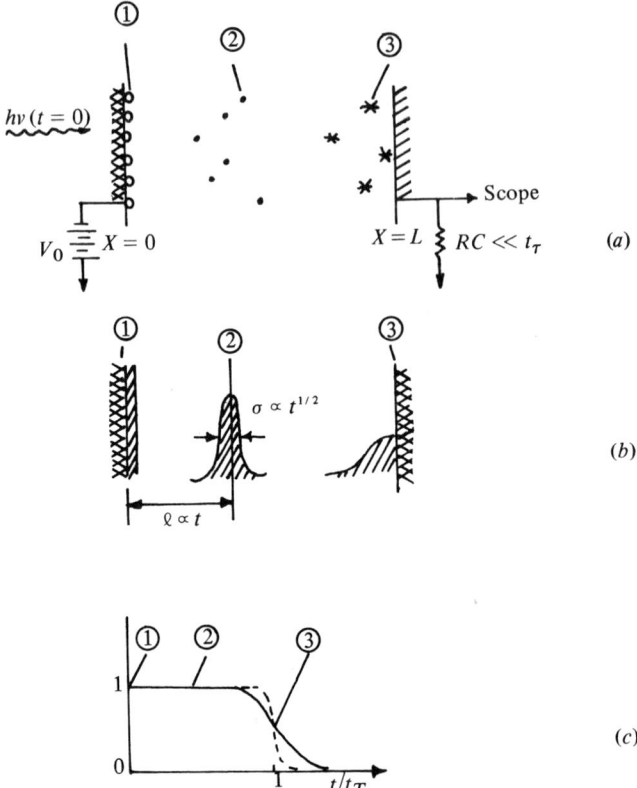

Fig. 15.7 Schematic representation of carrier propagation under Gaussian conditions. (a) Position of representative carriers in the sample bulk at $t = 0$ (○), $t < t_T$ (●) and $t \sim t_T$ (★). (b) Charge distribution in sample bulk at $t = 0$, $t < t_T$, and $t \sim t_T$. (c) Current pulse in external circuit induced by charge displacement. Units normalized to t_T and $i_T = i(t_T)$. The dashed line represents the transient current for lower applied bias field, i.e., longer transit time. After Pfister and Scher.[40]

region (see Fig. 15.8). For such asymmetric carrier propagation, the spread and the mean position have the same time dependence, hence $\sigma/\ell =$ const. Thus the shape of the transient current is independent of the transit time when plotted in units of t_T, a feature which has been termed "universality of the current shape."[40]

The mean drift velocity v_d of the moving space charge decreases with time because the number of carriers immobilized for a time of the order of t_T increases for a sufficiently wide distribution of individual events, or charge arrivals. Then, given $v_d = f(t)$, the transit time defined as $t_T \equiv L/v_d$, rises

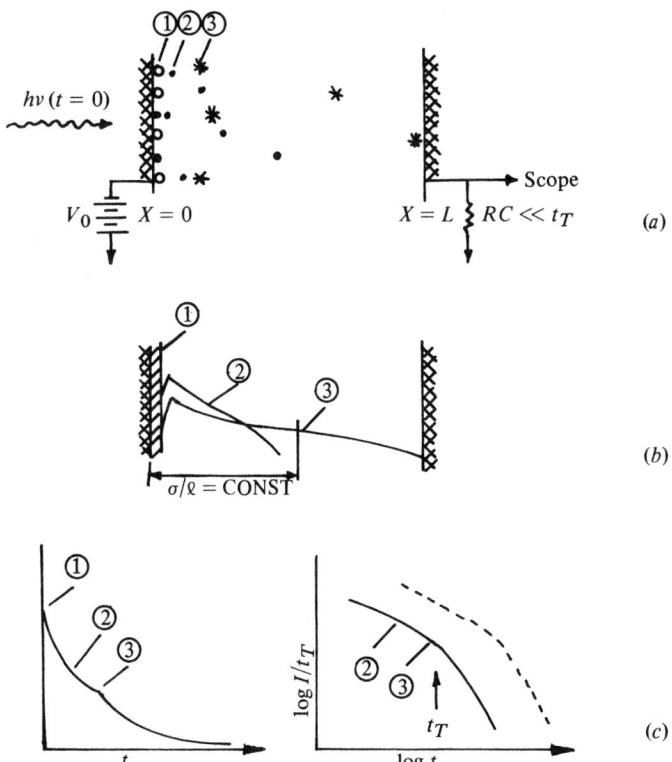

Fig. 15.8 Schematic representation of carrier propagation under ideal non-Gaussian conditions. (a) Position of representative carriers in the sample bulk at $t = 0$ (○), $t < t_T$ (●) and $t \sim t_T$ (★). (b) Charge distribution in sample bulk at $t = 0$, $t < t_T$ and $t \sim t_T$. (c) Current pulse in external circuit induced by charge displacement in linear units (left) and logarithmic units (right). The dashed line represents transient current for a lower applied bias field, i.e., longer transit time. After Pfister and Scher.[40]

faster than the simple proportionality to the sample thickness, L allows. Therefore, a thickness-dependent drift mobility μ_d results:

$$\mu_d = L/t_T E \tag{15.32}$$

This behavior contrasts sharply with the Gaussian case where the drift mobility remains a well-defined, intrinsic and sample thickness-independent quantity.

The general result obtained for an algebraic distribution function

$$\psi(t) \propto t^{-(1+\alpha)}$$

is summarized in the following equations[40]

$$I(t) \sim \begin{Bmatrix} t^{-(1-\alpha)}, & t < t_T \\ t^{-(1+\alpha)}, & t > t_T \end{Bmatrix}$$
$$t_T \sim \left(\frac{L}{l(E)}\right)^{1/\alpha} \exp(\Delta/kT),$$
(15.33)

where the disorder parameter α, $0 < \alpha < 1$, is determined by the time dependence of the distribution function. Following eq. (15.33), the rate of current decay increases at a characteristic time t_T ('transit time'). At this time the rate of carrier loss at the collecting electrode begins to dominate the rate of temporary carrier immobilization in the bulk of the sample. Thus, t_T represent the arrival time of the leading edge of the space charge at the collecting electrode. Equation (15.33) indicates a superlinear relation between sample thickness and t_T. The shape of transient current $I(t)$ and the field and thickness dependence of the transit time t_T are correlated by the disorder parameter α. The smaller α, the stronger the (E/L) dependence of t_T and the more dispersive the shape $I(t)$. At constant temperature, α is roughly constant, hence the current shape displays the scaling property α/ℓ constant. The sum of the power exponents describing the time dependence of $I(t)$ at times shorter and longer than t_T equals -2 and is therefore independent of the actual disorder and underlying transport mechanism.[40]

Equation (15.33) strictly applies to dispersive transport which can be characterized by an algebraic time dependence of the distribution function $\psi(t)$. More complicated $\psi(t)$ expressions may be necessary to explain the experimental data over broad experimental ranges.

In addition to non-algebraic distribution functions, deviations from the current transient curves described by eq. (15.33) may arise if the spreading of the travelling space sharge is distorted by a non-uniform field due e.g., to other space charges and/or surface effects. For a discussion of the mathematical details, the original publications[40] should be consulted. The theory, often referred to as the CTR model (Continuous Time Random Walk) is formally equivalent to trap-controlled transport (see Section 15.10).

Non-Gaussian transport has been observed in a broad range of disordered solids, both inorganic and organic. The latter offer experimental advantages in that the states involved in charge transfer are controllable in the sample preparation. Hopping transport may be identified by the exponential dependence of the transit time on the average spacing between localized (acceptor) sites which can be adjusted by chemical doping.[42] Several organic disordered solids such as carbazole polymers,[49] doped polymers,[50] organic glasses[51] have been investigated using the above theory. Other examples are naphthalene[97] and doped polyacetylene.[52]

Mort et al.[41] have carried out detailed studies of charge transport in a polycarbonate polymer doped with N-isopropyl carbazole, NIPC, as a function of NIPC concentration. In his system strong evidence exists that the only transport mechanism is hopping among localized states. Studies by Gill,[43] centered on trinitrofluorenone (TNF) doped polyesters, show that the transport of electrons is associated with hopping between TNF molecules and is controllable by varying the dopant concentration[43] (cf., also Section 16.7).

Figure 15.9 shows a composite plot of transient currents in logarithmic scales for a 1:0.3 weight ratio polycarbonate:NIPC sample. Transient responses could only be detected when the illuminated electrode was biased positively and thus indicate hole transport. Currents for different bias fields were shifted along the logarithmic current and time axes until superposition resulted. The units of both scales are relative. According to eq. (15.33), the current transient should decay first as $t^{-(1-\alpha)}$ and then as $t^{-(1+\alpha)}$.

The transition between the two slopes occurs when the fastest carriers encounter the absorbing substrate boundary and defines this time, t_T. The reduced time axis in the Fig. 15.9 is in terms of t_T and the various data refer to transit pulses measured on the same sample for bias voltages ranging from 150 V to 1400 V. As indicated, t_T over this voltage range varies from 640 msec to 3.6 msec, thus showing that t_T is a convenient measure of the charge transfer process. The figure also demonstrates that the current pulse

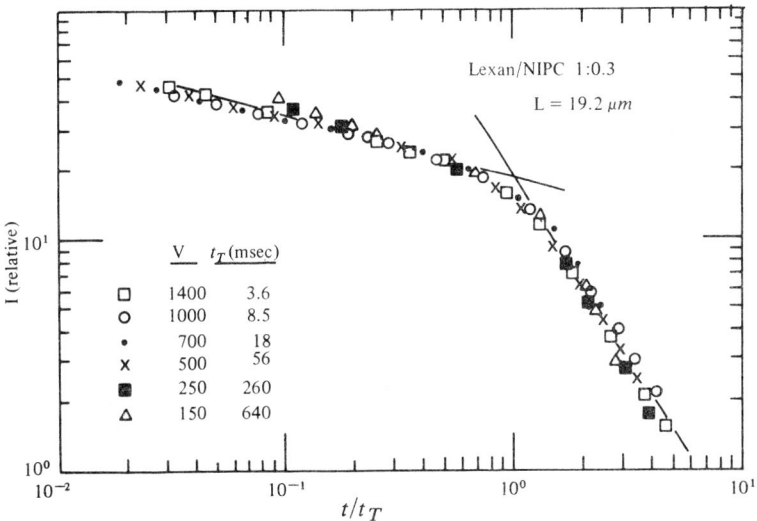

Fig. 15.9 Composite plot of transient currents for a 1:0.3 weight ratio polycarbonate: N-isopropylcarbazole system. Lexan is a trade name for the polymer. For discussion, see text. After Mort et al.[41]

shape as such remains unaffected by bias voltage changes if plotted on a log-log scale. This invariance, which has also been obtained in polyvinylcarbazole (PVK)[44] is inconsistent with either diffusion, a statistical variation in trapping events or thermal release from a distribution of traps.[44] This unusual aspect of transient pulses is, however, a major prediction of the Scher-Lax-Montroll stochastic hopping theory[40] and in itself, this agreement between the observed and predicted universality of the transit pulse shape with respect to the time scale is strong evidence that hole transport in polycarbonate:NIPC occurs *via* hopping. In addition, the sum of two slopes in the current transient is very close to 2 as predicted. Hopping occurs between the dopant molecules

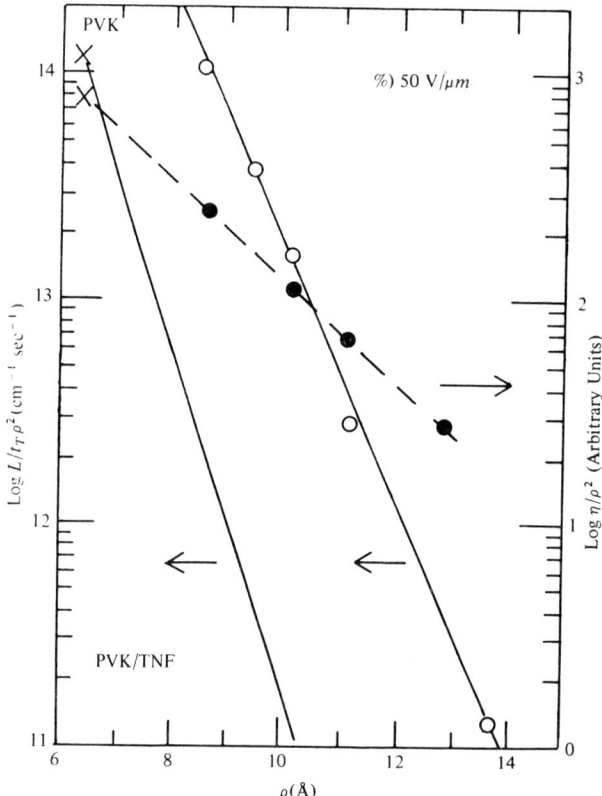

Fig. 15.10 Polycarbonate: N-isopropyl-carbazole system. For discussion, see text. The ordinates represent the relative quantum efficiency of the incident light/ρ^2 in arbitrary units as a function of ρ, the average distance between dopant molecules, as per the dashed line. The other lines refer to the ordinate units as stated, viz., log $(L/t_T\rho^2)$. After Mort et al.[41]

and is controlled by overlap between the wave functions of charge localized at the hopping sites and an Arrhenius temperature term if the hopping is phonon-assisted. Hence,

$$\frac{L}{t_T} \sim \rho^2 \exp\left(-\frac{2\rho}{\rho_0}\right)\exp\left(-\frac{\Delta}{kT}\right) \quad (15.34)$$

where ρ is the mean distance between dopant molecules, ρ_0 is the localization radius of the charge at the hopping site, and Δ refers to an activation energy. L is the film thickness, as determined e.g., from capacitance measurements. For hopping transport, a plot of $\log(L/t_T\rho^2)$ vs. ρ should yield a straight line with slope $-2/\rho_0$. Figure 15.10 shows such a plot[41] for the polycarbonate:NIPC system where the L/t_T values were measured at room temperature and a fixed field of 50 V/μm. The data strongly support a transport mechanism involving hops between neighboring NIPC molecules.

For comparison, Gill's data[43] for hole transfer in PVK:trinitrofluorenone (PVK:TNF) are also displayed. In that system, the spacings between uncomplexed carbazole molecules increase with higher loadings of TNF. The localization radii determined from the slopes are ρ_0(NIPC) $\approx$ 1.5 Å and ρ_0(PVK) $\approx$ 1.2 Å. The similar values of ρ_0 for PVK, where the carbazole molecules are chemically bonded to the vinyl backbone of the polymer, and the polycarbonate:NIPC, where no bonding occurs, suggest that hole transport in PVK can be viewed[41] as hopping in a random or disordered array of carbazole molecules. The vinyl backbone of the polymer plays no primary role in the transport process but merely fixed the carbazole chromophores in space. As in the case of PVK, hole transport in the polycarbonate:NIPC system is thermally activated and carrier velocities have a similar superlinear field dependence.[43,44]

The activation energy Δ is about 0.1 eV smaller for the polycarbonate:NIPC then for PVK. This difference in Δ appears to account for the ratio in t_T values for the two materials at a given value of ρ. The concentration dependence of the quantum efficiency, as in the case of the charge transport, shows that overlap between neighboring dopant molecules plays an important role in the photogeneration process. From the slope, a value of $\sim$3.0 Å for the localization radius results.

As another, and final example, reference is made to the careful studies by Kuder et al.[71] on fluorenone doped PVK (polyvinylcarbazole). Four different regimes in the photoinduced discharge curves (PIDC) can be distinguished, as shown[71] in Fig. 15.11 which is a schematic representation of various PIDC curves observed for 40% acceptor loadings of fluorene derivatives in PVK. Although there must be gradual changes between the regions defined, four different modes of transport are apparent: (i) mobility limited, (ii)

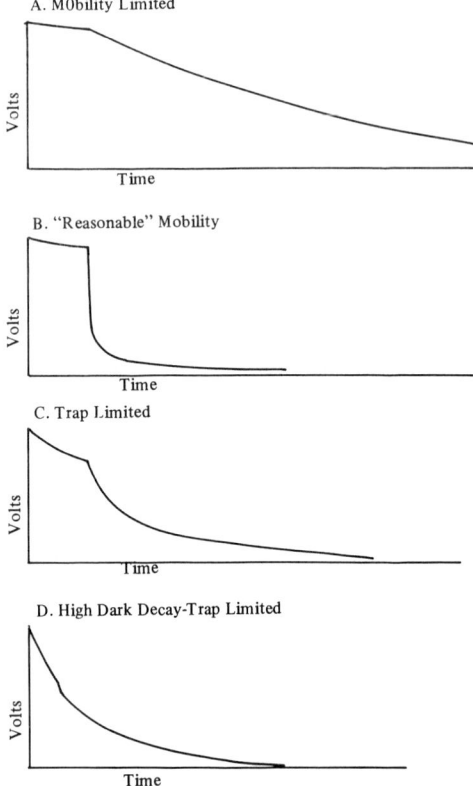

Fig. 15.11 Qualitative PIDC curves for PVK/acceptor mixtures. Horizontal axis = 20 mm/sec. 1 cm marker shown in A. Vertical scale dependent on dark decay. A and B, 100 V/cm, C, 50 V/cm; and D, 20 V/cm. After Kuder et al.[71]

"reasonable" mobility characteristics, (iii) trap limited, and (iv) high dark decay.

These definitions of mobility are qualitative in nature and depend on the time scale of measurement. For these results reasonable mobility is defined as light-induced discharge to 50% V_0 in 0.1 sec. Mobility limited is defined as low dark decay-low light-induced discharge rate (Fig. 15.11a); trap limited is defined as increasing dark decay (Fig. 15.11c) low light-induced discharge; and high dark decay trap-limited is defined as low charge acceptance due to dark decay and no light induced transport; (Fig. 15.11d). It is interesting to note[71] that the four regions indicated in (Fig. 15.11) correlate quantitatively with electron affinity A_g values for fluorenone derivatives: the first (Fig. 15.11a), with less than about 2.0 eV, the second (Fig. 15.11b) with

$2.0 < A_g < 2.2$ eV the third (Fig. 15.11c) with $2.2 < A_g \leq 2.4$ eV and the fourth, (Fig. 15.11d), with $A_g > 2.4$ eV.

Kivelson[61] has recently re-examined the validity of the Scher-Lax-Montroll theory here discussed and finds that their single state scheme underestimates the role of carrier diffusion because sites where the carrier may be trapped for long periods of time—in order words, deep traps—tend to be sites which are difficult to reach. The interpretation of time-of-flight experiments specially in disordered semiconductors remains to be re-examined.

Hopping theory in one dimension has been treated by several authors.[222]

15.6 Tunnelling

The physics of tunnelling[69] are discussed in Chapter 8, Part A, pp. 421 ff. The tunnelling mobility μ_0 may be written in the form[68]

$$\mu_0 = \frac{\pi e}{\hbar kT} \Sigma (r_j - r_i)(r_j - r_i)[w^2(i,j)/\alpha(i,j)] \quad (15.35)$$

where r_i and r_j are the position vectors of molecules i and j and $w(i,j)$ denotes a transfer integral operating between i and j. $\alpha(i,j)$ denotes the fluctuation of $w(i,j)$: a decrease in the distance between adjacent—say dimeric—molecules causes fluctuations Δw of the values of the transfer integrals—say between molecules 1 and 8 in Fig. 15.12—which fluctuations may be calculated from the equation[68]

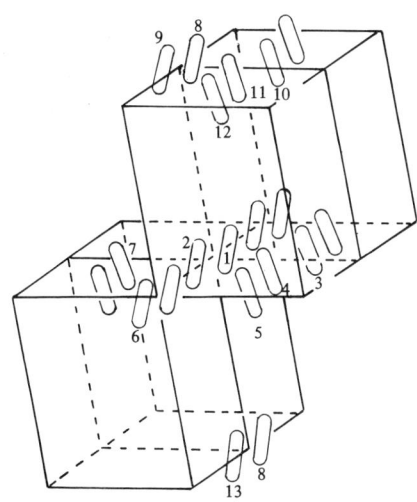

Fig. 15.12 Schematic representation of the perylene unit cell showing the numbering of the molecules. After Maruyama et al.[23]

$$\Delta w = \frac{dw}{dr}\Delta r \tag{15.36}$$

Transfer integrals in the c' direction for perylene are compared with those for anthracene in Table 15.7. Thus, in perylene, fluctuations in the 1-8 transfer integrals are significant.[23]

An excess electron placed on a dimer may be temporarily localized through the formation of an excimer like state between the two molecules; in perylene the excimer state is located at about 0.4 eV below the mutually independent molecular excited states.[67] The relative displacement between dimeric molecules due to excimer formation has been estimated[67] as about 0.2 Å. The same order of stabilization in energy and in displacement could be assumed to occur in an "excess electron dimer" produced in a perylene crystal, resulting in fluctuations in the value of the transfer integrals.

Taking Δr in eq. 15.36 as 0.2 Å and neglecting fluctuations due to lattice phonons which are expected to be much smaller,[23] $|\Delta\omega|$ is found to be 0.893×10^{-4} eV. With this value, and 3.57×10^{-4} eV for $w(i,j)$ in eq. (15.35)–taking $j = 8$–a value of 145 cm^2/V sec is obtained for the electron tunnelling mobility in the c' direction of perylene at 296 K.

In order to obtain the experimental drift mobility, the tunnelling mobility thus obtained has to be multiplied by an Arrhenius type term, $\exp(-E/kt)$, where E refers to an activation energy. The drift mobility then results as about 0.015 cm^{2}V sec for electrons at 296 K, inserting the experimentally observed[23] value for E, viz., 0.2 eV. This mobility values agrees fairly well

Table 15.6 Intermolecular Resonance Integrals for Perylene in Units of 10^{-4} eV. After Maruyama et al.[23]

Molecule number	Electron	Hole
2	−58.9	−25.7
3	19.6	17.8
4	9.72	9.15
5	19.6	17.8
6	−11.4	0.82
7	9.73	9.15
8	−3.57	2.52
9	0.01	0.03
10	−0.04	0.13
11	0.84	1.17
12	0.01	0.11
13	0.13	0.17

TUNNELLING 143

with the measured drift mobility of 0.017 cm^2/V sec.

This is a case of direct (elastic) tunnelling; however there also may occur another tunnelling mode for carrier transport in organic systems, viz., resonance tunnelling *via* localized states.[72] Charge may thus be transferred over several molecular spacings by a percolation type of transfer. The effect is of special importance in cases involving space charge layers—depletion or accumulation layers—arising at contacts, interfaces, grain boundaries and similar discontinuities involving energy barriers; it is also of interest in amorphous solids where the band edges are but ill-defined and where localized states exist which give rise to a resonance tunnelling current.

In this context, resonance tunnelling *via* donor states shall be considered; other types of localized states behave in a like fashion. Consider[73] a one-dimensional model of the barrier, thus neglecting the contribution of electrons with nonvanishing momentum parallel to the interphase. In the space charge region the energy E, of the donor states is also dependent on the distance x from the interface:

$$E_1(x) = E_1^b + E_c(x) - E_c^b \tag{15.37}$$

where E_1^b is the corresponding bulk value. Since $E_c(x)$, the potential energy of the space charge region is a monotonically decreasing function of x, there is a one-to-one correspondence between the energy E, of intermediate states and their distance x from the interface. One can write the contribution of the states situated at $x = \ell$ to the total resonant current in the form:

$$\delta j_+(\ell) = C[1 - n(E_1)] R(E_1 en) \theta(E_1 - E_1^b) D(E_1 en) \tag{15.38}$$

where C is a collection of constants containing the concentration of the reactants and of the resonant states, n is the Fermi-Dirac distribution, R is the resonant tunnelling probability, and D is the probability to find the heavy particle subsystem on the reaction hypersurface.[73] The localized states can serve as intermediate states for elastic resonant tunnelling only if their energy is greater than the bulk value of the lower edge of the conduction

Table 15.7 Electron Transfer Integrals in the c' Direction of Perylene Single Crystals, Compared with those in an Anthracene Crystal. Units of 10^{-4} eV. After Maruyama et al.[23]

Molecule-pairs	Perylene	Anthracene
1-8	3.57	0.38
1-11	0.84	0.67
1-13	0.13	0.38

band. It is thus necessary to introduce a factor $\theta(E_1 - E_c^h)$ into eq. (15.38) where θ denotes the Heaviside step function.

The tunnelling probability R for resonant tunnelling can be written[73]

$$R = \frac{R_1 R_2}{R_1 + R_2} \qquad (15.39)$$

R_1 and R_2 are the tunnelling probabilities from one side of the barrier to the resonant state and from the latter to the conduction band they are given by:

$$R_1 = \exp\left[-2\int_0^\ell k(x)dx\right], \qquad R_2 = \exp\left[-2\int_\ell^{l_g} k(x)dx\right] \qquad (15.40)$$

where

$$k(x) = \frac{|2m^*|}{\hbar^2}[E_c(x) - E_1(\ell)]^{1/2}$$

is the propagation constant in the barrier and l_0 is the classical turning point determined by $E_c(l_0)$. The width of the resonant state has been neglected. The total resonant current is obtained by integrating eq. (15.38);

$$j(\eta) = \int_0^L \delta j(\ell\eta)\delta\ell \qquad (15.41)$$

where L is determined by the relation: $E_1(L) = E_c^b$. As the band bending becomes smaller with decreasing potential, the number of intermediate states which can participate in elastic resonant tunnelling becomes also smaller. When the band bending is less than $E_c^b - E_1^b$ no intermediate states are available, and the resonant tunnelling current vanishes. This is a feature peculiar to resonant tunnelling through space charge barriers.

Resonant tunnelling has been suggested[74] to provide a way for a carrier to be released even from a deep trap. The effect has also been shown to contribute to the observed mobility by allowing the carrier to surmount grain barriers in polycrystalline solid films.[75]

It is well known[76] that reactions of the form

$$\left.\begin{array}{c}D^- + A \rightleftharpoons D + A^- \\ D^- + A^+ \rightleftharpoons D + A \text{ or } (A^*)\end{array}\right\} \qquad (15.42)$$

(see Section 15.9), where the donor D may be a physical trap, can occur even if D and A are separated by distances of up to 50 Å. This is referred to as long range tunnelling and involves a percolation type of transfer (see Section 15.7). It is important not only in molecular solids[77] but especially also in biological structures.[78] The reverse process to long-range transfer has also been observed during the absorption of incident light.[80]

Most studies of long range electron transfer processes are based on excess electrons injected e.g., by beta radiation into a solid, or often glassy though sometimes liquid, matrix. Thus, Beitz and Miller[134] investigated tunnelling reactions in 2-methyl-tetrahydrofuran glasses at 77 K between trapped electrons and 48 different types of organic acceptors. Tunnelling rates reported range from 10^{-6} to 100 $\sec^{-1}$ over distance of 15 to 40 Å. While obviously relevant to conductivity, and especially to mobility studies, there appears to have been very few, if any, attempts made to look closely into such correlations.

In many cases, the problem arises of the relative probability of the electron going over or tunnelling through a barrier, in other words, electron transfer by hopping vs. tunnelling.

15.7 Tunnelling vs. Hopping Transfer

Christov Characteristic Temperature[81]

At the absolute zero of temperature, the hopping probability is zero; any change transfer then must occur by tunnelling. As the temperature is raised, the tunnelling probability, which is virtually temperature independent, remains substantially constant while the hopping probability, being a thermally activated process, increases exponentially. Thus, a characteristic temperature T_k is reached at which both tunnelling and hopping probabilities are equal and the number of carriers passing over the barrier, the hopping current, equals the tunnelling current. This problem has been treated by Christov[81] who gives equations for symmetrical as well as asymmetrical barriers corresponding to the Eckart,[82] the parabolic, and an arbitrary potential energy function. He then shows that this characteristic temperature, which we shall term the Christov characteristic temperature T_k, is mainly determined by the rate of curvature atop the barrier and only to a very minor degree by its asymmetry and by its exact shape. For an arbitrary but smooth potential energy barrier having a curvature L_m at its top where the translational coordinate x has the value x_m, the characteristic temperature, T_k, for a carrier of mass m results as[81]

$$T_k \cong \frac{h\sqrt{L_m}}{2\pi k\sqrt{12m}}; \qquad L_m = \left[\frac{\partial^2 V}{\partial x^2}\right]_{x=x_m} \qquad (15.43)$$

h and k being Planck's and Boltzmann's constants, respectively.

146   THEORIES OF CHARGE TRANSFER

While thus T_k is seen to depend critically on the shape of the barrier at its top, one may assume that the actual shape will fall somewhere between two extremes: the rather steep Eckart barrier[82] and the rather flat parabolic barrier. The Christov characteristic temperatures for these are:[81]

(symmetric Eckart):

$$T_k = 0.28 \frac{hE_0^{1/2}}{kl(2m)^{1/2}}, \quad (15.44)$$

(symmetric or asymmetric parabolic):

$$T_k = \frac{hE_0^{1/2}}{(12)^{1/2}\pi kl(2m)^{1/2}}, \quad (15.45)$$

where l is the half-width of the barrier of height E_0. A fuller discussion of the Christov theory is given in that part of Section 15.12 dealing with contact phenomena.

However, in order for the electron to be transferred rather than to oscillate the potential energy function must be asymmetrical; the acceptor must provide a deeper trap than the donor. The asymmetry may be provided by an externally applied electric field of sufficient magnitude (see discussion in Section 15.10). The transfer itself is iso-energetic so that any excess energy of the transiting carrier must be taken up as vibration and/or excitation.

For a 1 eV barrier 20 Å wide, T_k has a value of between 4000 K for the Eckart barrier and 1300 K for a parabolic barrier.[83] A resonance condition is tactily assumed to prevail in that there exists an exact, though even short-lived, alignment condition between the highest occupied and the lowest vacant energy levels on both sides of the barrier. Also, this highly simplified and perhaps somewhat naive treatment takes no account of the spin unpairing which may have to precede the tunnelling, or of any phonon interactions. However, the resulting Christov characteristic temperatures are so high that one must expect electron transport at or near room temperatures under these conditions to occur predominantly by tunnelling (see also discussion in Section 15.8).

This is not so for proton tunnelling:[84] assuming a 1 eV barrier and a Christov characteristic temperature of 400 K, a barrier width $2l$ of 4.6 Å results for the case of an Eckart barrier and 1.5 Å for a parabolic barrier. Thus, proton tunnelling will depend critically on details of the shape of the potential barrier.

Seki[85] has proposed a model of fluctuating hopping sites which provides a mathematical description of transport similar to that of Bagley whose model is[86] based on hopping between localized states over potential barriers. Seki also includes a probability factor due to tunnelling or overlap or individual molecular wave functions, which has the form E^{-R/R_0}. The equation proposed by Seki is

$$\mu F = (\lambda_0/\tau_0)\exp(-R/R_0)\exp(-U_0/kT)\sinh(e\lambda_0 F/2kT) \qquad (15.46)$$

where μ is the mobility; F, the electric field; R, the intermolecular distance between exchange sites; τ_0, λ_0, and R_0 are constants of a specific system with units of time and characteristic distances, respectively; e if the electron charge, U_0 the thermal barrier to transport, and kT is the thermal energy of the system, R_0 can be related to the electron affinity A_c or the ionization potential I_c by[87]

$$\frac{I_c}{R_0} = \left(\frac{2}{\hbar}\right)\sqrt{2mE_i} \qquad (15.47)$$

where m is the mass of the carrier and E_i is identified with either A_c or I_c depending on the carrier sign, i.e., electron or holes. Experiments by Kuder et al.[71] on fluorenone doped polyvinylcarbazole gave results in good agreement with this reasoning (see also Section 15.5).

Brocklehurst[88] has pointed out a serious difficulty inherent in a long range tunnelling model: The simplest tunnelling model involves penetration through a one-dimensional rectangular barrier. For an electron with energy $E < V$, the barrier height, the transmission rate, W, is given by

$$W = \nu\exp(-2bR), \qquad b^2 = 2m(V-E)/\hbar^2 \qquad (15.48)$$

If the electron collides with the barrier at a rate, ν, of 10^{15} sec^{-1}, $V-E$ is 1 eV, R the barrier width is 40 Å, then $W = 1.58 \times 10^{-3}$ sec^{-1}, for $R = 20$ Å, $W = 1.26 \times 10^6$ sec^{-1}. Long range tunnelling is readily accounted for, but this would suggest that there will be no differences between acceptors if their electron affinity is great enough, and that the only relevant property of the donor is the value of E. In practice, huge variations are found as e.g., in Kuder's[71] experiments.

An elegant method for the study of charge transfer, by tunnelling or otherwise, involves conductivity measurements of Langmuir-Blodgett (mono-) layers. The film parameters then are accurately known and macroscopic current data may be readily related to microscopic properties.[219] Thus, Yamamoto has studied[220] the conductivity of 3, 5, 7, and 9 layer films in a direction perpendicular to the plane of the film. The conductivity was of the same order as that of (solid) paraffin, viz., 10^{-16} to 10^{-17} (ohm-cm)$^{-1}$. The films consisted of fatty acids doped with all-trans β-carotene, the molecules of which lie parallel to the plane of the film and are aggregated. The conductivity is said[220] to drop with increasing fatty acid chain length in the order Ba-stearate, -arachidate, -behenate, and, above 0.3 V rises exponentially with applied voltage. Charge transport is said[220] to be due to (long range) tunnelling. In cholesteryl-laureate[62] and in stearate multilayer films conductivity is reported to be due to thermally activated hopping[63] though other workers report that in monolayers of stearic acid deposited by the Langmuir-Blodgett technique tunnelling prevails.[127] Above 11 layers, the electrical properties of the multilayer systems closely approach those of the bulk, in which charge transport appears to be limited by carrier injection from the contact[127] (see Section 15.12). Suzi[227] has recently demonstrated an inverse square root temperature relationship of conductivity for Langmuir films, which suggests a hopping regime *via* interface states.

Though several other Langmuir layers such as stearate multilayers[63] and layers of cholesteryl-laureate[62] have been studied, much more work needs to be done in this promising field.

Inelastic tunnelling of electrons has been employed for adsorption studies on complete and incomplete monolayers, especially for investigations of the adsorbate orientation.[149]

15.8 Hindered Rotation

It has been proposed[83] that the actual charge transfer from molecule to molecule is by tunnelling, but the rate determining precondition for tunnelling, at least in the case of relatively large and complex molecules, is a sterically favorable alignment between an "active donor site" on one molecule and an "active receptor site" on another, neighboring, molecule. Such alignment will occur by a process of hindered rotation involving the crossing of a rotational energy barrier and it is the activation energy involved in this process which appears as the thermal activation energy of the conductivity. The primary event is charge injection from the contacts resulting in the formation of a negative ion at the cathode and a positive ion at the hole injecting anode. Depending on the relative probability of an electron to move from the anion to a neighboring neutral molecule, against the probability of electron-transfer from a neutral molecule to a positive ion, the material will exhibit n-type or p-type conduction. No translational motion of the ions is involved, only rotation which need not, and probably rarely does, involve more than a portion of the molecule, or ion.

In a large complex molecule, the intramolecular electron distribution will fluctuate with frequencies far in excess of the relatively slow molecular rotation; it is thus the mutual alignment of such two resonating energy levels which precedes the actual tunnelling and which is the rate determining process. Barrier tunnelling has been invoked[98] to account for the onset of internal rotation below the melting point in e.g., benzene, anthracene, and others. The frequencies involved are of the order of 10^{11} to 10^{13} Hz; even higher frequencies may be involved where only a vibration-rotation interaction is required in order to produce the necessary resonance condition.

The charge transfer thus is effectively a bimolecular reaction;[83] it requires the sterically favorable orientation of two molecules relative to each other while the energy barrier separating them is temporarily lowered (Franck-Condon Principle). The tunnelling takes place from a thermally excited state, i.e., *via* a small polaron, as illustrated in Fig. 15.13. In this Figure, the abscissae are in arbitrary units of distance while the ordinates represent energy. Diagram (*a*) represents a linear lattice of diatomic (for the sake of simplicity) molecules. A fairly substantial thermal distortion may occur, which brings the electron energy level of the occupied site into momentary coincidence with an empty one on a neighboring site, as illustrated in (*b*). The minimum vibrational energy required to achieve such a coincidence configuration is $\frac{1}{2}W_p$. In this situation it is then very likely that the charge carrier will tunnel, or hop to the new site.[46]

W_p here is the (excess) polarization energy of the distorted molecule which energy is dissipated after transfer, leaving the carrier trapped again at the second site. The lattice relaxation time is determined by the reciprocal of the parameter describing the bandwidth of the phonon energy spectrum. If its dispersion is small, then the electron may transfer to yet another site before the distortion associated with the previous coincidence has been fully dissipated.[46] One would expect a correlation between the thermal activation energy of electrical conductivity, E_{cond} and that of the dielectric relaxation time E_{diel}. Since vibronic energy levels are separated by about 0.2 eV, about that much energy would be required for the requisite spin unpairing prior to tunnelling, assuming a sterically favorable orientation. Thus, one would expect E_{cond} to be about 0.2 eV in excess of E_{diel}. Table 15.8 offers such a comparison; comparable data are very scarce because materials of dielectric interest have only rarely been studied from the point of view of semiconductivity, and *vice versa*.[83]

In tetrafluoroethylene films, it appears that a dipole orientation process along the lines discussed in this Section, is the main mechanism of conduction in the intermediate temperature range and at frequencies below about 100 kHz.[211]

15.9 Ligand Mediated Charge Transfer

Electron transfer from a donor D to an acceptor A molecule site is facilitated, and the mean free path of the carrier is increased, if an intermediate site exists linking D and A. Such effects are frequent, and of crucial im-

Table 15.8 Comparison of Conduction and Dielectric Activation Energies. The Activation Energies are in eV and refer to E/kT. After Gutmann.[83]

Substance	$E_{cond.}$	$E_{diel.}$
Liquids:		
Benzene	0.42	0.61
Toluene	0.41	0.7
1-Nitronaphthalene	0.27	0.093
Solids:		
Chlorpromazine	1.5	1.4
Promethazine	1.5	1.37
Trifluoperazine	1.4	1.4
Polyethylene-terephthalate	1.65	0.66 to 0.87 / 0.78
Albumen	1.1	0.8
Polyacrylonitrile	0.85	(1.7) & 0.61
Nylon	1 to 2	2.1 & 0.75
Gelatin	1.5	1.14 to 1.3
Collagen	1.36	1.1

portance, in biological systems such as oxidative phosphorylation, photosynthesis, etc., (cf. Section 8.8, Chapter 18 and 20).

A theoretical treatment has recently been given by Fischer et al.[93] and by Karl;[94] electron transfer in biological transfer chains is discussed in detail by van Heuvelen.[95]

Very often, a metal atom is involved as the intermediate;[96] this has been studied[130] for triple-decker multi-electron compounds such as bis(cyclopentadienylcobalt) cyclo-octa-tetracene.

The covalent bonding usually existing between the metal ion and its partner(s) involves considerable electron delocalization[95] in ligand orbitals and the resulting complexes exhibit but little change in their geometrical arrangements upon a change in valency. The transfer scheme may be described thus:

$$M_D^- - L - M_A \rightleftharpoons M_D - L - M_A^- \qquad (15.49)$$

Metal bridged charge transfer complexes are discussed in Sections 17.7 and 19.9.

Again, the Franck-Condon principle requires that an at least transitory energy level coincidence exists between donor and acceptor, as shown in Fig. 15.13. Since the electron (or hole) rests on the donating site until a

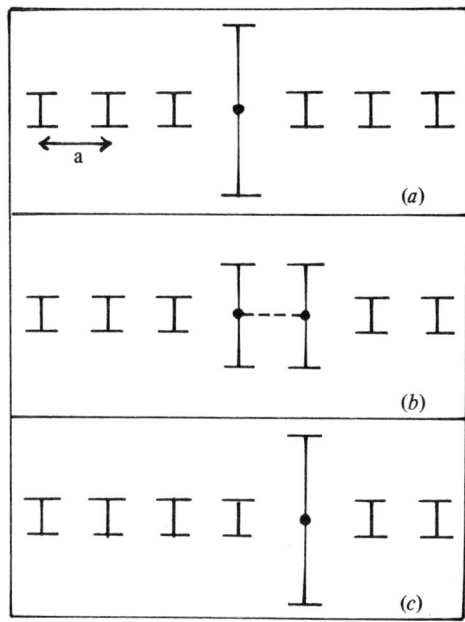

Fig. 15.13 Small polaron transport in a linear lattice of diatomic molecules. (a) self-trapped electron and molecular distortion at initial site; (b) transient coincidence of energy levels of initial and of final site; (c) the electron has tunnelled to the final site. After Emin.[100]

resonance condition arises allowing it to transfer, the D site may be considered a trapping site. In many cases, also, the molecular charge transport envisaged above may compete with a chemical reaction

$$M_D^- \to P \qquad (15.50)$$

in which the product P acts as a deep trap in which the carrier remains immobilized. Thus, for efficient transport, the residence time of the carrier on the donating site must be less than the time constant of the chemical reaction giving rise to P.

If there is a distribution of electron donating and electron accepting sites in an inert host matrix, charge transfer *via* a percolation, diffusion-like mechanism (see Section 19.4) may occur;[138] when the concentration of donors reaches a critical value, energy is transferred to the acceptors over long distances. The critical donor concentration is interpreted as that concentration above which the donors form a single infinite "cluster," i.e., as the percolation concentration.

The use of the percolation concept is still controversial. Borrowing from magnetic theory, Bouchez et al.[139] discuss electron transfer between 2 hydrogen atoms on different molecules *via* a third heavy intervening atom by a "super-exchange" interaction.

Lyons et al.[143] have studied heterogeneous electron transfer, i.e., transfer between different molecular species, in solution. For molecules of similar electron affinity the rate of transfer can be obtained from ESR line broadening; thus that between anthracene anions and pyrene molecules was found[143] to vary between 1.0×10^8 and 3.5×10^8 ℓ mole^{-1} sec^{-1} within an excess pyrene concentration range of 0.8×10^{-3} to 7.5×10^{-3} M.

Electrical conductivity involving maleonitrile dithiol as the ligand has been reviewed.[131]

Several authors[128] have used the Schottky model (see Fig. 15.18, discussed in Section 15.12) in studies of the charge transfer from a donor to an acceptor site. Memming and Moellers[129] suggest that the Schottky potential barrier model only holds for fully ionized donor states. Localized electronic states within a space charge barrier are reported to act as intermediates in elastic resonant tunnelling.[133]

15.10 Trap-Limited Transport

Irrespective of the actual transport mechanism—whether (a) by coherent wave motion in a conduction band or (b) by hopping, or (c) by tunnelling—in many cases the mobility is determined by trapping/detrapping processes. Aspects of trapping effects, where the band model applies, are dealt with in Part A, Sections 2.3, p. 67; 4.5, p. 257; 10.2, p. 566, as well as in Section 20.1.

Application of an external electric field increases the free time of a carrier trapped near or at a defect because of the lowering of the potential barrier hindering its escape. This is usually treated in terms of the Poole-Frenkel effect,[35] illustrated in Fig. 15.14. The combined effects of an applied electric field and coulombic attraction between the members of an ion pair cause an energy barrier to the separation of the charges in the ion pair. In the absence of a field, let ψ_0 be the energy barrier to separation. When a constant field is applied

$$\psi_E = \psi_0 - (e^3 E/\pi\epsilon)^{1/2} \qquad (15.51)$$

where ψ_E is the energy barrier to separation, at field E; e, the electronic charge; and ϵ the permittivity.

If diffusion of the charges within the coulombic potential well is neglected, and the current is determined by the rate of thermal excitation over the barrier,

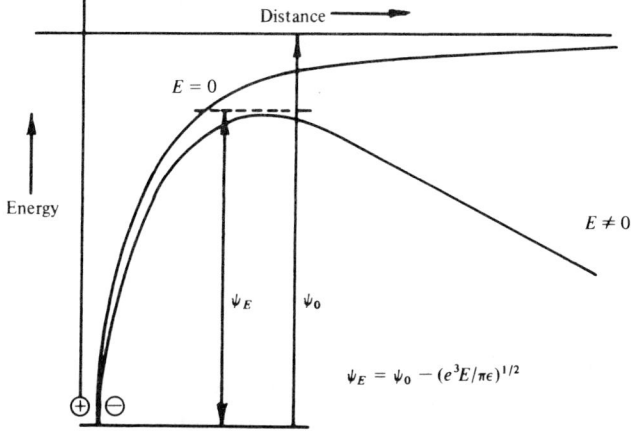

Fig. 15.14 The Poole-Frenkel Effect. After Milne.[225]

$$j = A \exp(-\psi_E/kT)$$
$$= A \exp\{-[\psi_0 - (e^3E/\pi\epsilon)^{1/2}]/kT\} \quad (15.52)$$
$$= j_0 \exp\{(e^3E/\pi\epsilon)^{1/2}/kT\}$$

where j_0 is the current at zero applied field. Equation (15.52) predicts an $E^{1/2}$ dependence of log (current) on the applied field E, which agrees with an empirical equation for the field dependence of the mobility, proposed by Gill[49] and found to hold for both electrons and holes in doped PVK polymeric films:

$$\mu = \mu_0 e^{-(E_0 - \beta E^{1/2})/kT_{\text{eff}}} \quad (15.53)$$

where

$$1/T_{\text{eff}} = 1/T - 1/T_0 \quad (15.54)$$

μ_0, E_0 and T_0 being constants. The value of β, for electrons as well as for holes, is 4.08×10^{-5} eV (m/V)$^{1/2}$, which compares with[35] the experimentally determined 2.72×10^{-5} eV (m/V)$^{1/2}$. However, Gill[49] argues against the general applicability of this model because it is based upon an energy band model and involves a high concentration of charged trapping sites which appears to be unlikely in an inorganic solid.

A conductivity, or mobility, relation of log (conductivity) vs. $E^{1/2}$, however, is by no means uncommon, especially in polymeric films[122] though it is usually considered as a contact phenomenon associated with carrier emission from the contact into the solid. This matter is discussed in Section 15.12.

154 THEORIES OF CHARGE TRANSFER

Simmons[99] discusses in detail 2 two-trap-level models illustrated in Fig. 15.15. He assumes that the trap levels are discrete. The upper trap of density N_1 and energy E_1 is located in the upper-half of the band gap ($E_1 > E_i$), while the lower trap (N_2, E_2) is located in the lower-half ($E_2 < E_i$) of the band gap; E_1 and E_2 are such that

$$(E_1 + E_2)/2 > E_i, \qquad (15.55)$$

where E_i is the intrinsic Fermi energy.

He assumes for algebraic expediency[99] that the upper traps are neutral when empty and the lower traps are neutral when filled, and that the energy of the top of the valence band, E_v, is equal to zero energy, and thus the bottom of the conduction band, E_c, is equal to E_g, the energy of the band gap. He also initially assumed that E_F, the Fermi energy, is sufficiently far from either trap level so that Boltzmann statistics may be used.

Introducing:

$$T_{c,1} = \frac{E_g - E_1}{k \ln(N_c/N_1)} \quad \text{and} \quad T_{v,2} = \frac{E_2}{k \ln(N_v/N_2)} \qquad (15.56)$$

where N_c and N_v are the effective densities of states in the conduction band and in the valency band, respectively, k being Boltzmann's constant, his results may be summarized in Tables 15.9 and 15.10.

For $T < (T_{v,2}), T_{c,1}$ the trapped carrier concentrations considerably exceed those of the free carriers. Therefore,[99] for an appropriate tunnelling or hopping mobility μ_h, the conductivity σ_v associated with the trapped carriers can exceed the conductivity σ associated with free carriers:

$$\mu_h(n_t + p_t) = 2\mu_h n_t \gg \mu(n + p) \qquad (15.57)$$

as long as the mobilities for electrons and holes differ not too much. At low temperature $p_t = n_t \gg n,p$, so that the conductivity is tunnel-hopping

Table 15.9 Trap Model Results. After Simmons[99]

Temperature range	Conductivity type	Activation energy
$T < T_t$	n	$E_g - E_{F0}$
$T_t < T < T_{v,2}$	p	$E_{F0} > E_g - E_{F0}$
$T_{v,2} < T < T_{c,1}$	p	$E_1/2 < E_{F0}$
$T > T_{c,1}$	intrinsic	$E_g/2 > E_1/2$

E_F is the Fermi level at operating temperature and E_{F0} at 0 K. E_g refers to the energy gap and E_1 to the trap energy. $T_{c,1}$ and $T_{v,2}$ are defined in eq. (15.56).

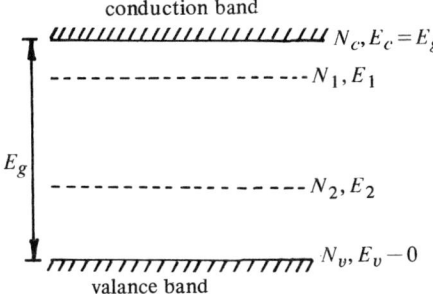

Fig. 15.15 Simmons two-trap model. After Simons.[99]

limited, with an activation energy of $(E_1 - E_2)/2$. In the range

$$T_{v2} < T < T_{c1}, \qquad n_t = p,$$

so since $\mu \gg \mu_h$, the conductivity is due to free-carrier conduction and is p-type with an activation energy of $E_1/2$. Above T_{c1}, $n = p$, the conductivity is intrinsic with an activation energy of $E_g/2$.

For further details and for an extension of the discussion to the multi-trap case, the original study should be consulted.[99]

As mentioned before, purely formally, multiple trapping and the Scher-Montroll-Pfister "Continuous Time Random Walk" theory[40] (see Section 15.5) are equivalent. For the case of coherent transport in an energy band, the theory[40] may be summarized thus: introducing a disorder parameter α (cf., eq. (15.28) and ff) and $\psi(t)$, the (transit) time distribution function, consider conventional multiple trapping of a carrier moving in an energy band state, into various localized states: then, the most significant stochastic variable is the release rate W_i which is a single site quantity.[40]

The carrier moves an average distance $\mu\tau E$ before it is trapped in a particular level with a probability ξ_i. Thus, the lattice constant in the direction of the

Table 15.10 Trap Model Results. After Simmons.[99]

Temperature range	Conductivity type	Activation energy
$T < T_{c,1}$	n	$E_g - E_{F0}$
$T_{v,2} > T > T_{c,1}$	n	$(E_g - E_1)/2 < E_g - E_{F0}$
$T > T_{c,1}$	intrinsic	$E_g/2 > (E_g - E_1)/2$

E_F is the Fermi level at operating temperature and E_{F0} at 0 K. E_g refers to the energy gap and E_1 to the trap energy. $T_{c,1}$ and $T_{v,2}$ are defined in eq. (15.56).

field a_0 is set equal to $\mu\tau E$ and $\psi(t)$ is a weighted sum of the probability per unit time to be released from one of the levels to the band,

$$\psi(t) = \sum_i \xi_i W_i \exp(-W_i t). \tag{15.58}$$

If one uses a density of states with a finite width (i.e., a maximum trap energy), one could generate a spectrum of rates

$$p(W) = c W^{\alpha-1} \exp(-W_l/W). \tag{15.59}$$

this gives rise to

$$\psi(t) = 2c(W_l/t)^{(\alpha+1)/2} K_{\alpha+1}[2(W_l t)^{1/2}], \tag{15.60}$$

where W_l is a minimum release rate which corresponds to the largest trap energy Δ_1. If the density of states is peaked around an energy Δ, the transit time could vary as $t_T \propto \exp(\Delta/kT)$ and $t_T W_l$ can increase with increasing T.

For $W_l t_T \ll 1$ one has dispersive transport and for $W_l t_T \gg 1$ non-dispersive or Gaussian transport. In addition, if there is a temperature independent contribution to α, the $\psi(t)$ can describe a Gaussian to non-Gaussian transition with decreasing temperature, with a weakly T-dependent α, and no change in the activation energy at the transition. $\psi(t)$ in eq. (15.60) has an algebraic form only for $W_1 t \ll 1$; a serious limiting condition because the theory requires the employment of $\psi(t)$ in an algebraic form.[40]

In time-dependent hopping or tunnelling amongst a random distribution of sites, in traversing the disordered matrix, a carrier experiences a wide variety of environmental conditions.

The distribution of hopping times, due to the variation in site separations and site energy fluctuations, over the entire medium is folded into the hopping time distribution to leave a single site, $\psi(t)$.[40] For hopping, the spectral form of $\psi(t)$ is different from the coherent motion plus multiple trapping case:[40] then, one may assume a superposition of independent release rates (weighted by the probability of the carrier being in the level) because there is only one way to leave each state, so that the weighting of a specific 'release' rate $W\exp(-Wt)$ must include the probability of whether the carrier is still in the state ($\phi(t)$). Hence, the physical interpretation of $\{\xi_i, w_i\}$ in eq. (15.58) for a $\psi(t)$ applied to a hopping problem is significantly modified. One can have a carrier hopping through a material and experiencing fluctuations in the energy level of the localized site, as well as the dispersion in intersite separation. The fluctuation in energy levels will tend to increase the dispersion in hopping times and add temperature dependence to the effective α, as shown above. If the carrier is interacting with hopping sites corresponding to distinct sets of energy levels, then the fluctuations can be discrete, as opposed to disorder-induced energy fluctuations in, e.g., impurity hopping conduction in semiconductors.[101] Consider two sets of states, and designate the sites as h and t, where the density N_h is much larger than N_t. The carrier can hop from h to h with an activation energy Δ_h and from t to h with Δ_t. If $\Delta_t > \Delta_h$ then the typical hop time τ_t from t to h is much greater than the typical hop time τ_h from h to h. This situation is trap-controlled hopping because there is a discrete separation between the spectrum of τ_h and τ_t. There are two activation energies in this hopping case; the transit time activation can range from Δ_h to Δ_t depending on the relative densities N_h/N_t. For $N_h \gg N_t$ there will be a large number of hopping paths that do not pass through a t site; hence, most of the fastest carriers (which determine t_T) will not encounter a hop with Δ_t. As N_t increases, the number of paths that do not contain a t site decreases and the activation energy Δ of t_T tends to Δ_t.

In trap-free hopping, the dispersion as expressed by α and the displacement between events are correlated. This does not hold for trap-controlled hopping, where the number of events as well as the dispersion of event times may be independently varied. The rate-limiting steps are the release times from a set of isolated trapping sites (density N_t), while the dispersion of the release times is determined by the local distribution of hopping sites (density N_h) around the trap, as well as any fluctuation in the trap energy.

Pollak[102] has used a percolation method (see also Section 19.4) to study the time-dependent hopping problem in a random medium. Sites are grouped into clusters in which all the intersite transition rates $W_{ij}(r)$ are less than some limiting value $W_{ij}(r_m)$ and then assumes that these clusters may be connected with a single spatial link of separation r_m. This reduces the time-dependent problem to motion along a one-dimensional chain with a series of limiting steps each with the same transition rate.

Pfister and Scher remark[40] that it is not surprising that essentially dispersion-free transport would follow from such a mode; Pollak[102] may have implicitly pointed to some limitations in the extent of dispersion due solely to positional disorder, but his treatment of the problem, at this point, does not substantially demonstrate this limitation.[35,40] His basic conclusion is that, for a carrier, if a site is hard to leave, it is hard to enter. Therefore, the hopping carrier avoids all sites which are difficult to enter. This apparent difficulty can be overcome if the hopping is among fluctuating energy levels, Pfister and Scher[40] view is that the anomalous dispersion can be caused by the relatively few long hopping times on the time scale set by the fastest carriers. In other words, most of the carriers must experience a wide dispersion of a statistically small sampling of long hops in order to have non-Gaussian spreading. Of course, the physically isolated site will be avoided but that type of site is not necessary to cause the accumulative spreading in time that has been observed. The issue remains to be further clarified.

The role of structural traps introduced into anthracene crystals by mechanical deformation has been investigated[103] by studying electron beam induced drift mobility transients; the electron beam replacing the more frequently used beam of light as the means of carrier injection. Such trapping centers, associated with basal dislocations in the (001) plane and with a stress induced polymorph of anthracene[104] provide trapping centers about 1.0 eV deep for electrons but do not affect the hole trapping properties. Dimers and anthracene molecules are said[105] to be responsible for these deep traps.

Litvinenko[106] reports that the conductivity of anthracene rises exponentially with increasing density of dislocations, which effect he considers to be due to either an increase in the concentration of acceptor levels available for a type of (ligand) mediated charge transfer (see Section 15.9) or else to the dislocation lines acting as channels of higher local conductivity. This matter, too, remains unresolved.

It is often problematic whether or not an observed mobility is the microscopic mobility.[210] If many shallow traps are present, the drift mobility, as measured, may be reduced. In such cases, the Hall mobility is generally closer to the microscopic mobility than the directly measured drift mobility. The topic of drift vs. microscopic mobility is discussed in Section 2.3 and 15.12. For the trapping case, the observed mobility μ_{obs} may be written

$$\mu_{obs} = \mu_L P + \mu_F (1 - P) \tag{15.61}$$

where μ_L refers to the carrier mobility in a localized state, μ_F to the free carrier (microscopic) mobility, and P to the probability of carrier localization. If the trap is shallow, the discussion of Section 4.5 p. 257 is applicable. If the trap is deep, then $\mu_L = 0$. The mean free path distance which the carrier travels between two successive trapping events defines the range of the carrier.[8]

Generally, this is of the order of tens of Å though several cases of very long range have been reported: e.g., in thin films of Mylar (a polyester) it is[107] about 600 Å in a field of 800 kV/cm. In highly purified PVK it is reported[107] to exceed the enormous value of 10^4 Å. The range of carriers has also been discussed repeatedly in this chapter. A table of microscopic and effective mobilities values is reproduced in Table 4.7D (see Section 16.8).

Shallow traps are reported by Mierzejewski et al.[142] to dominate the mobility in single crystals of TCNE below 245 K; above that temperature the mobility is said to be proportional[142] to $T^{-3.5}$.

A phenomenological theory of double injection into an insulator with traps has been derived under the injection plasma approximation, and confirmed by an exact calculation for a particular case.[161]

Given sufficient information about trapping, detrapping and recombination parameters, the theory allows prediction of the $j(V,I,d)$ equation for a given system. By assuming that recombination radiation can cause photodetrapping, the frequently observed failure of the scaling law relating current j to sample thickness d, V being the applied voltage:

$$j/d = f\left(\frac{V}{d^2}\right) \quad (15.62)$$

can be explained.[161] Difficulties arise in applying the theory containing two sets of traps for one type of carriers, so that some act as trapping and others as recombination centers. The main results of this somewhat involved theory[161] may be summarized assuming that the mobilities, thermal velocities and recombination capture cross-sections for electrons and holes are not too different, and writing n_f for the free electron concentration and p_f for that of free holes: if $n_f > p_f$ the expected current-voltage equations are

$$j \propto V^{[(3q-r+2)/(q-r+2)]} \quad (15.63a)$$

or

$$j \propto V^{[(3q-r+2)/\{q(1-v)+(r-2)(u-1)\}]} \quad (15.63b)$$

where $n_f \propto n^q I^u, p_f \propto n^r I^v$. If $p_f > n_f$, then q,u must be interchanged with r,v.

15.11 Hall Mobilities

The methodology of the Hall effect is discussed in Section 2.5 and 4.4. In many cases, an anomalous Hall voltage is observed which may even yield the wrong sign for the majority carriers as ascertained from e.g., thermotric or Dember voltages, or temperature independent apparent mobility values.

These anomalies have been treated theoretically by Friedman and Pollak[109] on the basis of a random phase model: the Hall voltage then arises from a magnetic quantum mechanical interference involving three or more sites which are momentarily resonant, having attained temporarily identical energy states. For long mean free paths, the theory reduces to the classical Hall effect. If μ_D is the drift mobility, the Hall mobility μ_H then results as

$$\mu_H = \mu_D kT/j \quad (15.64)$$

where j is the interaction integral associated with the extended states and estimated to be of the order of a few eV for amorphous covalent solids.[109] The theory accounts well for the low value of μ_H as well as for its temperature independence. The sign of the Hall effect is found always to be negative, irrespective of the sign of the majority carrier. The theory is based on a hopping type conduction and applies percolation theory to the disordered systems considered. Thus, μ_H becomes a function of the ratio between the mean intersite spacing involved in the hopping transfer and the size of the localized wave function. It is the disorder in the local site energies which lowers the Hall voltage. Pohl[108] has pointed out that the Hall effect is much affected by the possible presence of two types of carriers, holes and electrons concommitantly.[108] Since the concentrations of carriers in organic materials is often not well known, the Hall data must be treated with circumspection.

At present it would appear[108] that the Hall effect might well be "normal" in long chain eka-conjugated polymers, since the carrier drift within the long chains may be treated semiclassically. The tunnelling or hopping to different sites which is important in the sense of the Friedman atomic-site model[109] is then of major importance only in interchain transitions.

Pohl and Burnay[110] consider the question whether the Friedman theory[109] is applicable to quasi-one or quasi-two dimensional solids such as polymers: will such be "Friedman-normal" or "Friedman-anomalous"? Will their long range and relatively high mobility permit a quasi-classical response for the carriers and a correspondingly large Hall voltage? Their results on several polymers, especially on metal-free polyphthalocyanine,[110] indicate that these behave in a "Friedman anomalous" manner. The Seebeck coefficient for the phthalocyanine indicated p-type conductivity while the Hall voltage was negative, as in the case of n-type conductivity. In other cases, the Hall voltage was too small to be measured. Yamaguchi's studies on graphitic carbons[111] also indicate an anomalous Hall effect, and so do data reported by Suzuki et al. on polyacrylonitrile films.[118]

Other results, however, (cf., Part A, p. 255, Section 4.4 and Table 4.8A) indicate that at least in some cases the Hall effect may behave normally.

Results reported by Spielberg et al.[112] on single crystals of anthracene indicate that roughly correct ratios of drift to Hall mobility, as well as the correct sign of the Hall constant, may be obtained from band theory, but are critically dependent on the wave functions used in the calculation of the transfer integrals (see Table 15.11).

Schaadt and Williams also report[117] a normal Hall behavior for electrons in anthracene. However, more recent studies by Korn et al.[116] and by Elnahwy et al.[56] on anthracene indicate an anomalous Hall effect.

Granted a hopping model of charge transfer, Emin[121] finds that the value of the measured Hall coefficient depends on the nature and relative orientation of the local orbitals between which the carrier transits. The hopping of an excess electron between anti-bonding orbitals in an odd-membered ring is said to yield an anomalous Hall coeffcient suggesting p-type conductivity. A like transfer of a positive hole would indicate n-type behavior.[121]

Nagaev and Sokolova find that the Hall coefficient becomes more and more anomalous as the width of the energy gap decreases;[119] this is difficult to accept because it would mean that, in the limiting case of an ideal metal where the energy gap vanishes, the Hall coefficient would be highly anomalous, in contradiction to well known experimental results.

In anisotropic solids, the shorting effect of the current leads must be allowed for.[214] If L/W is the length to width ratio of the anisotropic sample, the corresponding ratio for the isotropic equivalent[214] is L'/W with

$$L/L' = \sqrt{\sigma_1/\sigma_2}.$$

Then, μ_H (observed) = $\alpha\mu_H$ (true), in which α is a known function of L'/W for isotropic samples.[214] In a sample of TEA(TCNQ)$_2$, for which $\sqrt{\sigma_1/\sigma_2} = 13$,[215] $L/W = 2.5$ and $L'/W = 0.2$, the shorting effect is considerable, corresponding to $\alpha = 0.15$.

Equations relating carrier mobilities and Hall voltages are discussed in detail by Swart and Campbell.[120] The important problems associated with anomalous Hall voltages remain far from clarified and deserve further and detailed study. Meton and Gerard have listed[228] data for inorganic acids, bases and salts which may be of interest in this context.

15.12 Contact-Limited Conductivity

This Section is concerned with cases where the current through the sample is governed by contact phenomena rather than by the sample bulk conductivity. Aspects of this topic are dealt with in Part A, Sections 3.7, 3.7D, 6.10, and 10.1; pp. 185, 373, 456, and 559, respectively.

Ideally, a contact[209] should be neutral or "invisible," i.e., neither injecting excess carriers not blocking their discharge, i.e., rectifying. The ideal contact also should be ohmic so that the contact resistance necessarily introduced by it should not falsify the current-voltage characteristic of the sample by contact non-linearities.

An ohmic contact must satisfy two conditions. When the applied field is sufficiently low any excess injected carrier is neutralized before it completes

Table 15.11 Ratio of the Hall to the Drift Mobilities for Naphthalene as Calculated from the Wave Functions of the Indicated Investigators. After Spielberg et al.[112]

Magnetic field parallel to	Holes			Electrons			References
	a	b	c'	a	b	c'	
	−15.8	1.1	−15.8	−9.9	1.5	−10.7	113
	−1.3	0.08	−1.3	0.70	1.3	−1.9	114
	−0.21	0.75	−0.2	0.82	0.87	−0.095	115

transit. The carrier density, n, is unperturbed by the contact and only then can a dc current be interpreted as $I = \sigma E$, where $\sigma = en\mu$. μ is the microscopic mobility. The second condition is that when the field is high enough that any excess injected carrier completes its transit before it is neutralized, the ohmic contact provides an unlimited carrier reservoir to supply a space-charge-limited current (SCLC). The transition between the high and low field behavior occurs at fields for which the transit time $t_T = L/E$ approaches the dielectric relaxation time $\tau = \epsilon_0 \epsilon \rho$, where ρ is the unperturbed bulk resistivity and ϵ the permittivity.

The behavior of a contact may be explored[145] by applying a voltage step to the system, resulting in a current transient such as shown in Fig. 15.16 and usually displayed on a log/log plot.

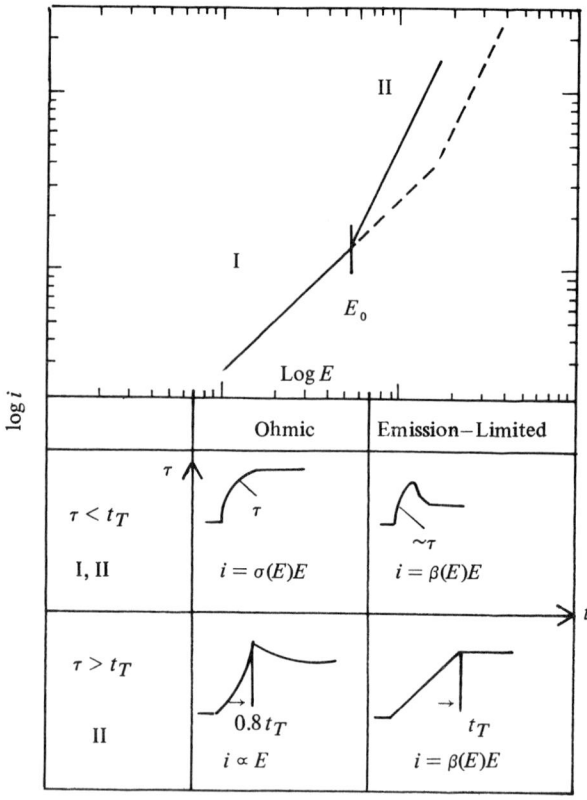

Fig. 15.16 Schematic of current response curves to a voltage step applied to ohmic and to emission-limited contacts. After Pfister and Scher.[145]

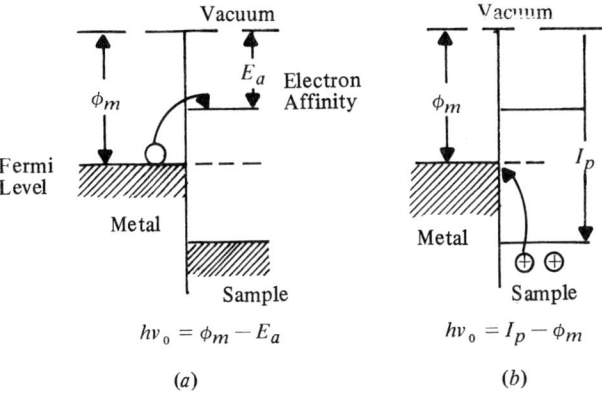

Fig. 15.17 Photo-injection from a metal contact into a semiconductor. (a) electrons; (b) holes. The quantum condition which must be fulfilled for injection to occur is also stated. $h\nu_0$ is the photo-emission threshold.

If the current is ohmic, then the time required to establish equilibrium is determined by the dielectric relaxation time, or the RC time constant of the measurement circuit, whichever is the greater. If for $E > E_0$ the dc current shows the thickness dependence characteristic for space-charge limited currents, then the current response following a field step $E > E_0$ should exhibit the cusp well known from transient space-charge-limited currents. However, unlike the photo-induced transient current which approaches zero (i.e., dark level) for $t > t_T$, the dark current now should be asymptotic to the space charge limited dark current value. However, failures of the thickness dependence relation, i.e., of the scaling law, are often observed as discussed later in this Section when dealing with double injection. If neither of the current responses illustrated in Fig. 15.16 for ohmic conditions are observed, one must conclude that the contact is emission limited: for t_T it requires a transit time to establish an (emission limited) dc current and therefore the response becomes a ramp of width t_T.

Photo-injection is discussed in more detail in Section 15.1. The relation of between photo-injection and the ionization potential I_p and the electron affinity E_a of the sample and the work function ϕ_m of the contact is shown is an idealized fashion in Fig. 15.17; band bending is neglected, see Fig. 3.46, in Part A, p. 189.

For the analysis of current vs. voltage curves, the well-known expressions[150] for the Poole-Frenkel (see Section 15.10) and the Schottky effect (see eq. 15.65 and 15.66) are usually employed.

The current voltage plot in Fig. 15.16 consists of a linear part where the current is proportional to the field E, which itself is proportional to the applied voltage; this is the low field region where $E < E_0$. Above a certain critical field E_0 the relation becomes superlinear: $i = E^n$.

If the superlinearity is due to space charge limitations, then the scaling law holds in that E_0 rises in proportion to the sample thickness L; see Section 10.1. Thus, changing the sample thickness provides a means to check the ohmic behavior of the contact and to ascertain whether the superlinearity is indeed due to the presence of space charges (see the dashed line in Fig. 15.16). If the contact is ohmic and the current for $E > E_0$ is not space-charge-limited, the I-E curve reflects a field-dependent conductivity, $\sigma(E)$. If the contact is emission-limited, the dc current level is determined by the rate at which the contact can supply carriers. Then, quantitatively, one may employ a combination of the Schottky and Poole-Frenkel expression:

$$I = I_0 \exp\{(\alpha' E^{1/2} - \phi)/kT\} \quad \text{and} \quad I = I_0 \exp\{(\alpha E^{1/2} - \phi)/kT\} \quad (15.65)$$

where I is the current, I_0 a constant, E the applied electric field, k Boltzmann's constant, T the temperature and ϕ the difference in work functions across the contact potential barrier. The Poole-Frenkel parameter α and Schottky parameter α' are given by

$$\alpha = (e^3/4\pi\epsilon\epsilon_0 d)^{1/2} \quad \text{and} \quad \alpha' = \frac{1}{2}\alpha, \quad (15.66)$$

where e is the electronic charge, ϵ_0 is the permittivity of free space, ϵ is the high-frequency relative permittivity of the sample, and d is the sample thickness. The Schottky effect is concerned with injection of charge from the electrodes into the samples over a potential barrier, the height of which is field-dependent. In the Poole-Frenkel effect the current is considered to be due to field-assisted thermal excitation of electrons[146] from traps into the conduction band of the sample; ϕ is the ionization potential I_p/ϵ.

This "classical" treatment has been refined and extended by Christov.[147] The following, necessarily greatly abbreviated, synopsis will be based upon his work, which does not impose any restrictions concerning the band structure of the solid sample, and takes account also of processes occurring at both the cathodic and the anodic (metal) contact. Field emission then is essentially a tunnelling problem, unless the temperature approaches the Christov characteristic temperature[81] at which hopping over the top of the barrier becomes important (see discussion in Section 15.7).

In the free electron approximation, the transition probability W for the crossing of a potential barrier can be written[147]

$$W(E, p_y, p_z) = \frac{1}{1 + \exp[H(E, p_y, p_z)]} \quad (15.67a)$$

with

$$H(E, p_y, p_z) = -\frac{4\pi i}{h} \int_{x_1}^{x_2} p_x \, dx \tag{15.67b}$$

where p_x, p_y, p_z are the components of the electron's momentum p related to its total energy E through the quadratic expression

$$\frac{p_x^2 + p_y^2 + p_z^2}{2m} = E - V(x) \tag{15.68}$$

assuming that the potential energy depends only on x according to one-dimensional barrier representation. $V(x)$ represents the potential barrier of height V_m. To investigate the energy distribution of the emitted, or injected, electrons; the exponent in eq. (15.67) is expanded in a series

$$H(E, p_y, p_z) = b - c(E - E_0) + f(E - E_0)^2 + \cdots \tag{15.69}$$

about a suitable energy value E_0, where the coefficients b, c, f depend on the barrier parameters and on the band structure of the insulator. The same expansion applies for $H(E_x)$ in the case of free electron approximation. In practice, one needs only the first two or three terms of eq. (15.69).

It is convenient to divide the whole energy range (from $E = 0$ to ∞) into three intervals, whose limits are determined by the quantities

$$\mu' = \mu + 2kT, \qquad E' = V_m - kT_k \tag{15.70}$$

where T_k is the Christov characteristic temperature, see Section 15.7, eqs. (15.43), (15.44), and (15.45); and repeated below in the same notation:

$$T_k \cong \frac{h}{2\pi^2 k} \sqrt{\frac{L_m}{m_c}} \qquad L_m = -\left(\frac{\partial^2 V}{\partial x^2}\right)_{x=x_m} \tag{15.43}$$

It is useful to introduce another characteristic temperature

$$T_c \equiv \frac{1}{kc} \tag{15.71}$$

where c is the corresponding coefficient of the expansion (15.69) about the Fermi level ($E_0 = \mu$). Both T_k and T_c have well defined physical meanings as will be explained below.

In terms of the quantities μ' and E' of eq. (15.70) one may divide the entire range of energy variation between 0 and ∞ into three intervals

$$\begin{array}{lll} \text{lower range } A_1 & \text{middle range } A_2 & \text{upper range } B \\ 0 < E < \mu' & \mu' < E < E' & E > E' \end{array} \tag{15.72}$$

In each of these ranges either $W(E)$ or $f(E)$ or both, can be expressed in a simplified form. Thus, in the ranges A_1 and A_2 ($E<E'$) the general formula (15.67) for the transition probability becomes

$$W(E,p_y,p_z) = \exp[-H(E,p_y,p_z)], \qquad (15.73)$$

while in the ranges A_2 and B ($E > \mu'$) the Fermi-Dirac function may be replaced by the classical energy distribution

$$f(E) = \exp[-(E-\mu)/kT]. \qquad (15.74)$$

In the lower and middle energy ranges A_1 and A_2 ($E < E'$) the exponent H in eq. (15.73) can be further simplified by taking into account that it has a minimum (hence W has a maximum) at $P_x = p$ so that only electrons moving nearly normal to the metal surface can tunnel through the barrier. The function

$$F(E) = \frac{2}{h^3}f(E)\int\int W(E,p_y,p_z)dp_y dp_z \qquad (15.75)$$

represents the total energy distribution of the electrons injected from the cathode (M_1).

For large inter-electrode spacings and sufficiently high applied voltages, the current density for emission from the cathode may be obtained by integrating the energy distribution over the entire energy range:[147]

$$I = e\int_0^\infty F(E)dE = \frac{2e}{h^2}\int_0^\infty f(E)dE \int\int W(E,p_y,p_z)dp_y dp_z \qquad (15.76)$$

where e is the electron charge.

The total current integral eq. (15.65) can be represented as a sum of three integrals over the energy ranges A_1 and A_2 and B,

$$I = I_1' + I_2' + I'' = e\int_0^{\mu'} P(E)dE + e\int_{\mu'}^E P(E)dE + e\int_{E''}^\infty P(E)dE \qquad (15.77)$$

Not all the three current components make important contributions to the total current, so that in most cases eq. (15.77) reduces to two or only one term: first,[147] an "extended" field emission region is defined by the condition $T < \frac{5}{6}(T_c)$ so that the maximum of the energy distribution falls within the lower energy range $0 < E_m < \mu'$. In that case, the current component I' in eq. (15.77) becomes dominant so that

$$I = Q_1(T/T_c)B_1'\exp(-b), \qquad Q_1(T/T_c) = \frac{\pi(T/T_c)}{\sin[\pi(T/T_c)]} \qquad (15.78)$$

for the total current is obtained as a good approximation. If $T < T_c/4$ one has $Q < 1.12$ which means that the current becomes practically independent of the temperature; i.e., in the region of 'pure' field emission. Further, an intermediate emission region exists for $\frac{5}{6}T_c < T < T_k/2$ in which the peak of the distribution falls in the middle energy range ($\mu' < E_m < E'$) and, in general, the currents have to be computed using eq. (15.77). The total current may be approximated by

$$I = B_2' \exp\left(-\left[b_m + \frac{E_{m-\mu}}{kT} - \frac{\{c_m - (1/kT)\}^2}{4f_m}\right]\right) \quad (15.79)$$

Finally,[147] an "extended" thermionic emission region may be defined valid for $T > T_k/2$ for which the energy distribution maximum falls within the upper energy range $E_m > E$. The current components I'' in eq. (15.77) predominate in that region so that the total current results as

$$I = \kappa(T_k/T)\exp[-(V_m - \mu)/kT], \quad \kappa = \frac{(\pi/2)(T_k/T)}{\sin[(\pi/2)(T_k/T)]} \quad (15.80)$$

This is a good approximation for $T > \frac{2}{3}T_k$, in which region $E_m > V_{max}$. If $T > 2T_k$, practically all emitted electrons cross over the barrier, defining a region of "pure" thermionic emission, $T > 2T_k$, Q is close to unity and the current follows in good approximation as

$$I = Q\frac{4\pi mek^2}{h^3}\exp\left(-\frac{\phi_s}{kT}\right)\exp\left(\frac{1}{kT}\sqrt{\frac{e^3 E}{4\pi\epsilon_r\epsilon_0}}\right) \quad (15.81)$$

For small inter-electrode spacings and relatively low values of applied voltage electron, emission in both directions has to be taken into account from the anode as well as from the cathode. Then the resulting current density can be represented by the difference

$$I_{res} = I_I - I_{II} = e\int_0^\infty [F_{12}(E) - F_{21}(E)]dE = e\int_0^\infty [F_1(E) - F_2(E)]dE \quad (15.82)$$

of the two opposite current density I_I and I_{II}, where F_1 and F_2 are given by eq. (15.75) with corresponding indices for each of the opposite current directions.

Thus, I_I and I_{II} are obtained for any inter-electrode distance and applied voltage. For small voltages, however, Ohm's law can be readily obtained from eq. (15.82) if $W(E)$, at constant P_y and P_z, is also constant. For the field and thermionic regions this relation can be obtained directly by using

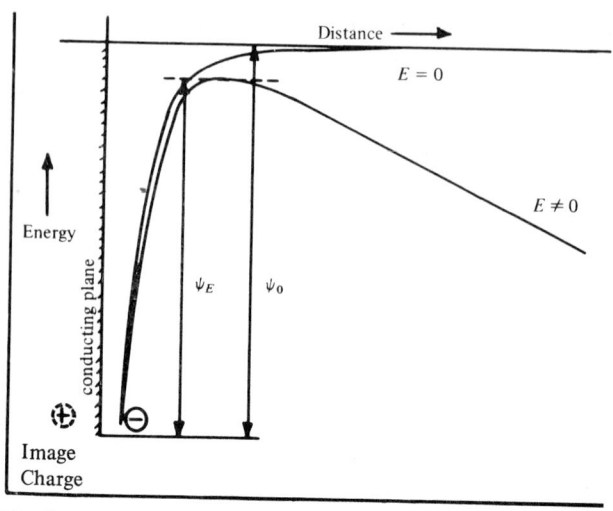

Fig. 15.18 The Schottky Effect. E refers to the electric field and ψ_0 to the energy barrier at zero field while ψ_E stands for that of the applied field E. After Milne.[225]

the approximation expressions for $F(E)$ instead of the general formula (15.75). Ohm's law has been derived previously in the quasi-free electron approximation by means of the normal energy distribution both for field emission[148] and for thermionic emission[149] through thin insulating layers.

Particularly at higher temperatures, the potential barrier function $V(x)$ in eq. (15.68) may be approximated by the Schottky potential function

$$V(x) = -E\frac{e^2}{4\epsilon x} - eFx \qquad (15.83)$$

where, again, ϵ is the permittivity and e the electronic charge. This is illustrated in Fig. 15.18. In the Schottky model, a carrier experiences a coulombic potential due to the presence of an image charge in a nearby conducting plane. Diffusion of the carrier is neglected.

For the Schottky potential barrier, the Christov characteristic temperature, eq. (15.43) takes the form

$$T_k = \frac{hq^{1/4}\epsilon_r^{1/4}}{\pi^2 k m^{1/2}} E^{3/4}, \qquad (15.84)$$

where h and k are the Planck and Boltzmann constants, ϵ_r the relative dielectric permittivity of the sample, E the applied field V/d, d being the thickness of the sample, and m the effective mass of the carrier.

The Schottky barrier model is a reasonable approximation[149] if $\phi/ev < 1$

where ϕ is the work function of the contact metal. The field dependence of the Schottky function may be written

$$\psi(E) = \psi_0 - (e^3 D/4\pi\epsilon)^{\frac{1}{2}} \qquad (15.85)$$

A Schottky interfacial barrier arises at a contact because, in order to equilibrate the Fermi levels across a metal-semiconductor contact in thermodynamic equilibrium, carriers must be transferred into or out of surface states within several 100 Å of the interface. The resulting band bending creates the Schottky barrier, i.e., the energy difference from the Fermi level in the metal bulk to the conduction band minimum at the interface. As discussed in Part A, Sections 2.12 and 3.7, pp. 93, 188, it should equal the difference in work functions. However, it appears that[155] this has to be modified by introducing a coefficient S characteristic of the semiconductor and called the Index of Interface Behavior, and determined by resonances derived from "dangling bonds" on the semiconductor surface such as are discussed in Part A, Section 3.8, pg. 213. The above theoretical treatment[158] has so far centered on inorganic semiconductors but should be extended to (organic) molecular crystals.

Band bending and field penetration have been calculated[160] for near surface layers for semiconductors of simple geometries up to surface fields of a few times 0.1 V/Å, such as are met in many surface and interfacial effects. It then appears that the surface potential arising from band bending amounts[160] to a few V. The field penetrates about 10 Å to 200 Å into the surface region, depending on the conductance type, assuming a positive field, and a plane parallel sample. The electronic surface properties thus will be considerably affected by the field penetration and the photon absorption edge will be red-shifted.

Figure 15.19 compares experimental data obtained by Vodenicharova[151] on polyquinoline-carboxylic acid films. While Fig. 15.20 illustrates the log I vs. $E^{\frac{1}{2}}$ relationship resulting from the Schottky barrier and expressed in eq. (15.81). It is seen that deviations from the classical Schottky behavior[150] becomes evident at higher fields. Similar results have been obtained for polysilazane films,[152] for polyvinylcarbazole,[158] for its complexes,[158] as well as for several other systems.[153]

Several authors[156] have applied the Schottky model to electron transfer in the bulk of the solid, viz., between electron donating and accepting sites, (see Section 15.9) as well as to contact phenomena.[157]

Energy barriers arising at contacts play a major role in the photovoltaic effect and are discussed in Part A, Sections 8.12, pg. 516; 8.13, pg. 517; 10.4, pg. 595; as well as in Section 22.5. Schottky barriers are often invoked especially in photovoltaic cells based on phthalocyanine films.[159]

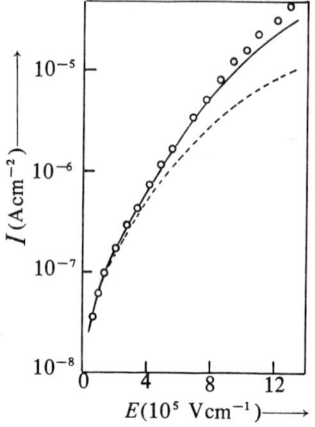

Fig. 15.19 A comparison of the experimental data (o) with theory, polyquinoline-carboxylic acid film. The solid line shows values calculated from Christov's theory,[147] while the dashed line refers to values calculated from the classical Richardson-Schottky equation for thermionic emission, which results from the Christov theory by setting the Quantum Correction Factor κ in eq. 15.80 to yield $q = 1$. After Vodenicharova.[151]

To explain the observation of Schottky emission in some systems, Bonham and Lyons[162] propose that the initial charge transfer occurs from the electrolyte to a deep energy state on the crystal surface. This could be a physical defect on the surface or perhaps an oxidized anthracene molecule. Anthracene is well-known to undergo a variety of reactions if exposed to light and air, and it is reasonable to expect reaction products on the surface to affect injection mechanisms strongly. It is not necessary for the carrier to be trapped in this way right at the surface. The trapping center need only lie sufficiently close to the surface for the carrier's image force in the solution to be significant.[162]

The role of minority carrier injection is discussed by several authors.[154] The minority carriers move into the opposite direction from the majority carriers. However, since the latter diffuse also against the applied field, the effect becomes important only at low current densities.[154] Based on a random

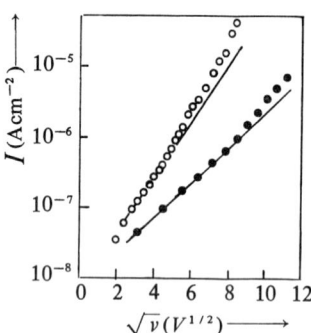

Fig. 15.20 Current voltage characteristics at $T = 297$ K of two samples with thickness of the polymer layer 5.8×10^{-4} cm (o), and 1.4×10.4 cm (•). It is seen that all experimental points which deviate from the linear ln I vs. $V^{1/2}$ relationship are within the extended region of thermionic emission where $T_k/T \leq 3/2$. After Vodenicharova.[151]

walk model to describe the transport and decay of Frenkel excitons near a metal contact interface, Spannring and Bässler[224] report that in the case of anthracene, most of the energy transfer events involve the second, third and fourth molecules lattice plane from the interface. Charge transfer leading to photogenerated carriers is said[224] to proceed mainly from the second layer.

15.13 Energy Correlations

The Compensation Law

Results on electrical conduction in organic crystals often have been reported[163] as giving rise to the "compensation law"

$$\log \sigma_0 = \alpha E + \beta \qquad (15.86)$$

where α and β are constants: σ_0 and E occur in the conductivity equation

$$\frac{I(T)}{F} = \sigma_0 \exp\left(-\frac{E}{kT}\right) \qquad (15.87)$$

where $I(T)$ is the measured current density at temperature T for an applied field F, σ_0 and E are determined as the intercept, at $(1/T) \to 0$, and the slope, of the linear plot of $\log I(T)/F$ vs. $(1/T)$.

Correlations between I, σ_0 and E of the type of eq. (15.86) have been obtained[163] for many different substances and also for different experiments on the one system either with different samples of the one system or with the one sample measured on various occasions.

In connection with the kinetics of chemical reactions Exner[164] pointed out that for relations similar to eq. (15.86) the quantities analogous to E and $\log \sigma_0$ are intrinsically related through the method of their calculation. Any deviation in E is automatically reflected and magnified in the value of $\log \sigma_0$, when measurements of $I(T)$ are confined to a restricted range of temperature and when the range of measurable values of $I(T)$ is itself restricted. Under these circumstances a relation such as eq. (15.86) must be observed.

An excellent example of such apparent compensation occurs in currents through anthracene single crystals with guarded electrodes of silver paste. A correlation coefficient of 0.998 was found between $\log \sigma_0$ and E, in a plot of 18 points obtained with ultra-pure anthracene.[165]

For inorganic semiconductors, eq. (15.86) is referred to as the Meyer-Neldel rule;[166] for these compounds it does certainly appear to hold.

Rosenberg et al.[167] introduced a characteristic temperature T_0 as a further constant into the expression for the temperature dependence of conductivity. The existence of T_0 was derived from the common intersection of straight

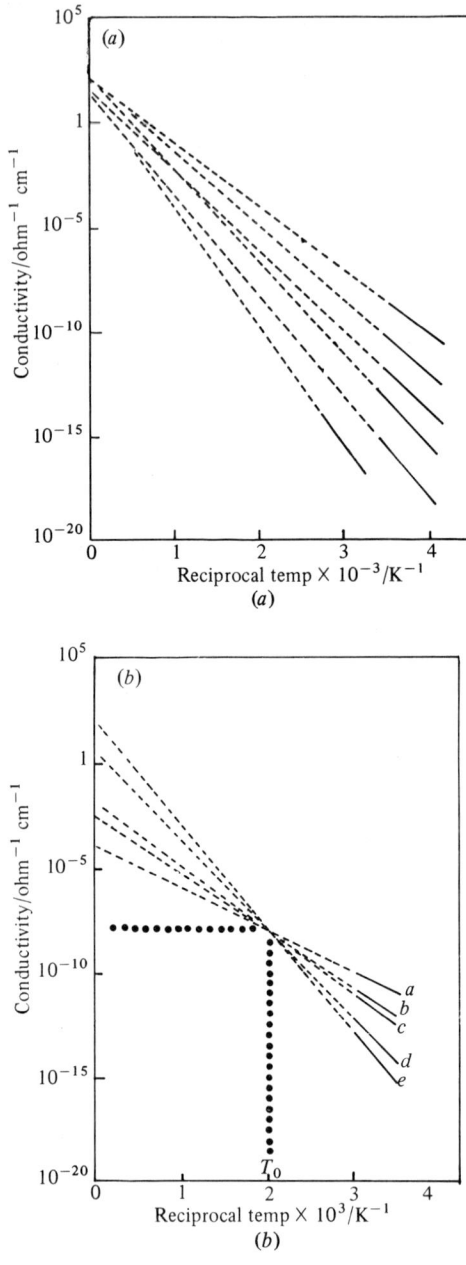

Fig. 15.21 A set of activation energy curves for (a) haemoglobin crystals; (b) oxidized cholesterol extrapolated to $1/T = 0$. Each straight line refers to a specific hydration state in (a) and to a specific donor-acceptor complex in (b). After Rosenberg et al.[167]

lines for a set of semiconductors in a plot of I/T against $\log \sigma$, which is an alternative graphic illustration of the compensation phenomenon. T_0 is the slope of the compensation curve in the graph E against $\log \sigma$. It can attain values within very broad limits (∞, 272 K) for the substances investigated[167] (proteins, hemoglobin, cholesterol, retinals). The authors[167] varied the activation energy of conductivity by different methods, e.g., by forming weak complexes, hydration, or by using different isomers of the same substance. They observed that T_0 does not depend on the method of the change of activation energy and proposed therefore a modified equation

$$\sigma_{(T)} = \sigma_0' \exp(E/2kT_0)\exp(-E/2kT) \tag{15.88}$$

The matter is illustrated in Fig. 15.21. Diagram (b) shows the $\log \sigma$ vs. $10^3/T$ plots for oxidized cholesterol[167] which are seen to intersect in a focal point having coordinates σ_0^1 and T_0. Diagram (a) shows data for haemoglobin from which it is seen that σ_0 extrapolates to $10^3/T = 0$, i.e., $T \to \infty$.

Another method to demonstrate the compensation law consists in plotting $\log \sigma$ at constant temperature against the activation energy E. The slope of the resulting straight line equals $-1/kT_0$ while its intercept with the conductivity axis yields σ_0, as illustrated in Fig. 15.22.

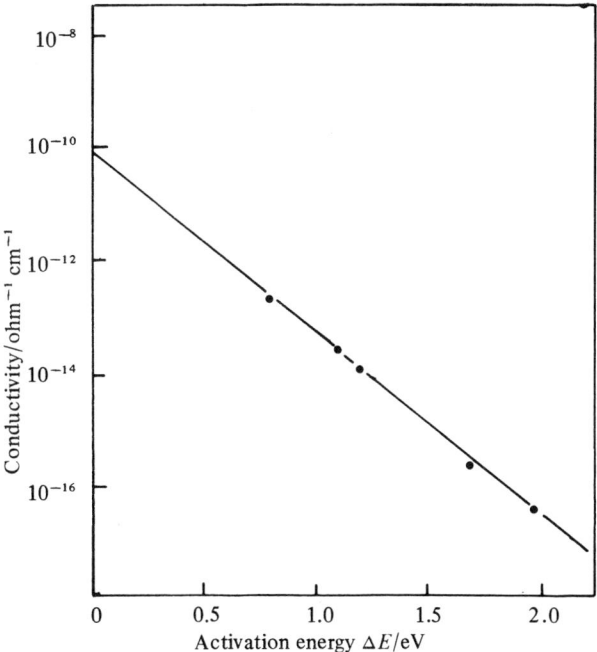

Fig. 15.22 Alternative plot of data in 15.21(b); conductivity at a constant temperature is plotted against the activation energy. After Rosenberg et al.[167]

A further plot[165] for ultrapure anthracene single crystals, and for that of lesser purity and also for doped anthracene is displayed in Fig. 15.23. Other examples where the law has been reported to hold are indophenine polymers,[168] phenolformaldehyde resins[169] and mixed organic single crystals.[170]

Kemeny and Rosenberg[171] ascribe the effect to the tunnelling of small polarons of effective mass not more than about 100 electron masses and leading to a relation between T_0 and the Debye temperature.

If the conduction band, assuming electronic conductivity, tails a finite distance towards the valency band ϵ into the energy gap, as is the case for many organic solids exhibiting a distribution of traps of different energies, and if the position of the Fermi level is governed by a fixed dominant hole level deeper inside the gap, then:[172]

$$n = n_0^1 \exp\left[\frac{(E_m - E_F)}{kT_c}\right] \exp\left[\frac{-(E_m - E_F)}{kT}\right] \quad (15.89)$$

where E_F is the position of the Fermi level. Comparison shows that $T_0 \equiv T_c$ and $\Delta E \equiv (E_m - E_F)$.

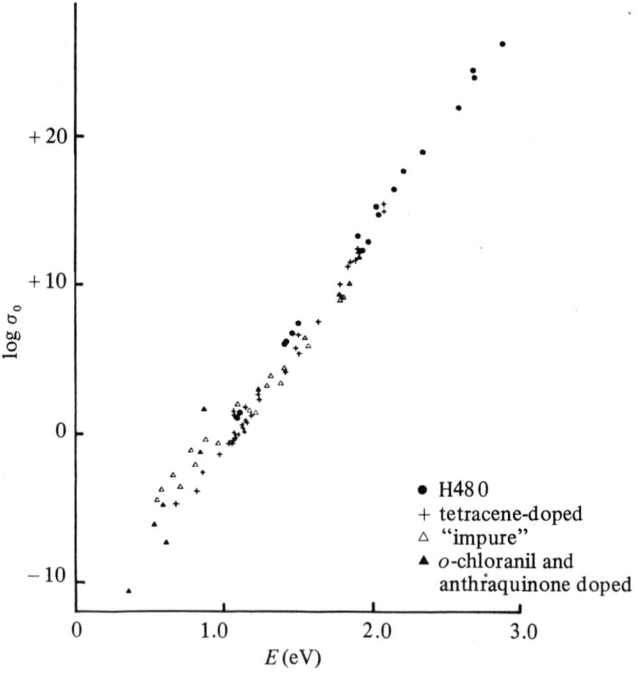

Fig. 15.23 Log σ_0 vs. E plots for several samples of anthracene. H-480 refers to an ultrapure specimen. After Johnson and Lyons.[165]

ENERGY CORRELATIONS

Support for this model comes from measurements of space-charge limited currents in anthracene where a correlation has been found[173] between the total density of traps H and the distribution parameter T_c which, again, has been ascribed to a constant trap distribution.[174]

A similar hypothesis has been proposed[175] by Lee who also derives the compensation law from the assumption that the carrier concentration tails off exponentially into the gap so that the majority carrier concentration $n(E)$ is given by

$$n(E) = \frac{N}{kT_c} e^{-E/kT_c} \qquad (15.90)$$

where, again, T_c is a characteristic temperature of a rather empirical character; N is the total density of energy states.

However, the linear relationship between $\log \sigma_0$ and E may originate solely from the calculation of these parameters.[165]

If a true linear free-energy relationship (LFER) does hold for dark conduction, then, for each particular E values, unique values of σ_0 and $\sigma(T)$ must exist, defining the system precisely. The observed E value does not necessarily define the other parameters if the LFER is simply "apparent." A true LFER demands a point of intersection for the Arrhenius plots of the experimental data.[165] The temperature at this intersection point, T_0, is a characteristic temperature for the system. The constants α and β from eq. (15.86) are then given by

$$\alpha = (2.303 \, kT_0)^{-1}, \qquad \beta = \log \sigma(T_0) \qquad (15.91)$$

While it is obvious that $\log \sigma_0$ and E are linearly related, in order to test whether σ_0 and E are physically related, plots of $\log \sigma(T)$ against E and $\log \sigma(T_1)$ against $\log \sigma(T_2)$ are required. Figure 15.24 shows such a plot for ultrapure anthracene samples at 345 K;[165] while one could construct some type of correlation into that plot, its results does certainly not support the assumption of any physical basis for the compensation law in high resistivity, pure organic solids.

However, further clarification can be obtained from mixing experiments[177] because mixing is the simplest possible method of influencing conductivity without chemical interactions. It should therefore be possible to calculate the conductivity from basic physical properties of the components without introducing additional constants.[177] A simple mixing rule was derived by Lichtenecker[176]

$$\sigma_r = \sigma_1^p \sigma_2^q \qquad (15.92)$$

176 THEORIES OF CHARGE TRANSFER

where σ_r is the resulting conductivity; σ_1 and σ_2 are the specific conductivities of the respective components; and p and q are volume proportions of the mixture so that $p + q = 1$. Substituting for σ_1 and σ_2:

$$\sigma_r = \sigma_{01}{}^p \sigma_{02}{}^q \exp[-(pE_1 + qE_2)/kT] \quad (15.93)$$

where σ_{01} and σ_{02} pre-exponential factors of the components and E_1, E_2 the respective activation energies of conductivity. Denoting $\sigma_{01}{}^p \sigma_{02}{}^q = \sigma_{0r}$, $pE_1 + qE_2 = E_r$:

$$\sigma_r = \sigma_{0r} \exp(-E_r/kT) \quad (15.94)$$

or

$$\log \sigma_r = \log \sigma_{0r} - 0 \cdot 4343 E_r/kT$$

This relation can be separated into two parametric equations[177]

$$\begin{aligned} \log \sigma_r &= -KE_r + e \\ \log \sigma_{0r} &= -K'E_r + e \end{aligned} \quad (15.95)$$

which are the equations of the compensation effect.

Introducing into the eq. (15.95) the values for σ_r, σ_{0r} and E_r as they follow

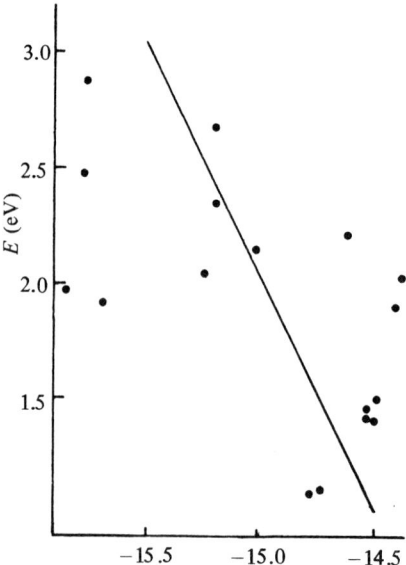

Fig. 15.24 E vs. log σ plots for ultrapure anthracene samples at 345 K. After Johnson and Lyons.[165]

from the Lichtenecker's logarithmic mixing rule and obtaining expressions with the pairs ($p = 1, q = 0$) and ($p = 0, q = 1$) (as the equations are valid for any pair of $p + q = 1$):

$$K = (\log\sigma_1 - \log\sigma_2)/(E_2 - E_1)$$
$$K' = (\log\sigma_{01} - \log\sigma_{02})/(E_2 - E_1) \quad (15.96)$$
$$c = (E_2\log\sigma_1 - E_1\log\sigma_2)/(E_2 - E_1)$$

This formalism for the mechanical mixtures of two semiconductors shows that each member of the series obeys the compensation relation. The slope of the line in the plot $1/T$ against log (activation energy) can attain the value selected in advance. Figure 15.25 illustrates the results of such mixing experiments[177] which support a physical basis for the compensation law, at least for inhomogeneous systems;[177]

Systems exhibiting the compensation effect, may be "mixtures" between an "unperturbed" phase and one affected by some sort of treatment because of this phase being more readily accessible to it: water vapor, complexation, state of aggregation or film formation, etc.[177] The resulting conductivity and its activation energy are thus the result of mixing the original unperturbed substance with the "influenced" portion. The characteristic temperature T_0 is, in this case, not a characteristic of a single substance, but of two substances, viz., that of the basic component and either of its most conductive combination with the complex-forming agent, or of its most insulating state due to the treatment, or to a spontaneous change (e.g., escape of iodine from the perylene-iodine complex). In certain cases this perturbed component

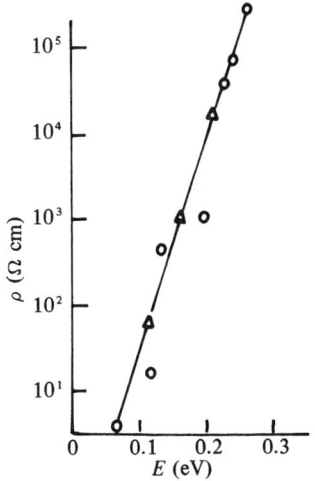

Fig. 15.25 Plot of log ρ against E of the mixture of 4,4'-bis(dimethylamino) diphenylamine iodide radical salt and the iodine complex of N,N'-bis(4-aminophenyl)-p-phenylenediamine. ○ experimental values; △ theoretical values. After Ulbert.[177]

may be prepared or isolated from the "mixture" but frequently it exists only as a surface state, and its existence can be derived only from the existence of the compensation effect.

The compensation law has been shown to hold for tetrathiotetracene complexes with mole ratio as the parameter;[178] for different pressures very nearly parallel straight lines result.[178] It is also reported[179] to hold for ionic conductors such as Li^+ in LiF with conductivity replaced by diffusion. This, if confirmed, would suggest a deeper reason than purely electronic effects, to be involved. In fact, Yatsimirskii[180] has proposed that it is an entropy effect which causes it. These matters should be further studied and clarified.

Electrochemical Energy Gap Correlations

Lyons derived[181] numerical values for the electron affinities of a number of aromatic hydrocarbons from listed reduction potentials. Although those first numerical results were too high for the absolute values of A_G, the adoption of Hush's value[182] for the free energy of the reaction

$$Hg(\ell) + Cl^-(sat.,aq.) \to \frac{1}{2}Hg_2Cl_2(s) + e^-(g) \quad (15.97)$$

which took account of a surface potential previously neglected,[185] together with allowance for a liquid junction potential previously assumed negligible, was shown[184] to yield values of A_G in reasonable agreement with experiment.[185] The same approach has been employed by Matsen[186] as well as by several other authors.[187]

Similar considerations allow the calculation[188] of the ionization potential (of the isolated molecules) I_g from oxidation potentials as well as of the electron affinity A_g from reduction potentials. The solvation energy of the ion results from Hush's modification[182] of the Born equation though there is uncertainty as to the constancy of the solvation energy even over a set of related molecules, as well as its value; it is usually about 1.5 eV. Then, the liquid junction potential which is of the order of about 0.25 V and is also not accurately known, enters the calculations of I_g and A_g based on

$$\begin{aligned} I_G &= eE_{1/2}^{ox} - S^+ + W - kT\ln Q^+ + E_{\ell j} \\ A_G &= eE_{1/2}^{red} + S^- + W - kT\ln Q^- + E_{\ell j} \end{aligned} \quad (15.98)$$

where $E_{1/2}^{red}$, $E_{1/2}^{ox}$ are the observed reduction, oxidation potentials [vs. SCE (aq.)] for a given solvent, S^+, S^- are the free energy of solvation of the ion relative to that of the neutral molecule, Q^+, Q^- are the ratios of the partition functions of the reduced and the oxidized species in the electron transfer process, $E_{\ell j}$ is the liquid junction potential in the cell used, W is the free energy change in the reaction, e is the charge of the electron.

Equations (15.98) imply[189] that these potentials refer to reversible con-electron reactions. Departures from this will be neglected.

In a solution, the energy of the reaction

$$2M \to M^+ + M^- \tag{15.99}$$

is given fairly closely by $E_{\frac{1}{2}}^{ox} - E_{\frac{1}{2}}^{red}$, provided that similar solvents and conditions are used in determining the two potentials,[25] from (15.98):

$$eE_{\frac{1}{2}}^{ox} - eE_{\frac{1}{2}}^{red} = I_G - A_G + S^+ + S^- \tag{15.100}$$

where $kT\ln Q^{+(-)}$ has been neglected. If this term is determined chiefly by the electron spin term and equals 0.02 eV the error so introduced is about 0.04 eV.

Lyons[189] deduces from eq. (15.100) an expression for the energy gap, E_g, for a given medium (solvent) as follows.

To calculate E_G for an organic solid one asks what difference to the quantity $E_{\frac{1}{2}}^{ox} - E_{\frac{1}{2}}^{red}$ will be made by changing the permittivity ϵ_r of the medium from the value it has in the solvent to its value in the organic solid.[25]

The free energy of solvation of a single ion depends on ϵ_n proportional to the ratio $(1 - 1/\epsilon_{n,1})/(1 - 1/\epsilon_{n,2})$ (see Section 6.4 p. 345). Thus:

$$E_{G,2} = (E_{\frac{1}{2},1}^{ox} - E_{\frac{1}{2},1}^{red}) + (S_2^+ + S_2^-)\left[1 - \frac{1 - 1/\epsilon_{r,2}}{1 - 1/\epsilon_{r,1}}\right] \tag{15.101}$$

From the original Born equation, $S_2^+ = S_2^-$. The term in brackets is <1; being often 0.2 and 0.4.[25]

Reduction and oxidation potentials can be determined with good accuracy. The uncertainty in the solvation energy term is reduced because it is multiplied by a factor <1, it is also usually less in magnitude than $E_{\frac{1}{2}}^{ox} - E_{\frac{1}{2}}^{red}$. A 20% error in the solvation energy thus produces a much smaller (ca. 5%) error in $E_{G,2}$. Errors in W and E_{Qj} do not effect the value of $E_{G,2}$.

Use of eq. (15.101) requires the value of $\epsilon_{r,2}$. In only a few cases have permittivity values been accurately measured[190] for solids of interest in organic semiconductor studies, and even then there is the uncertainty of the error in using the average value of the three principal terms. This disadvantage is offset[184] to some extent by the nature of the term $(1 - 1/\epsilon_{r,2})$ in which $\epsilon_{r,2}$ appears. Taking tetracene as an example, a change in $\epsilon_{r,2}$ from 4 to 3.5 makes a difference in the calculated E_G of 0.09 eV, which certainly is significant.

However, with all these limitations, the method does offer not only a readily accessible check on energy gap values derived from conductivity/conductance measurements, but also can act as a valuable screening method since electro-chemical potential measurements are far less time-consuming than solid state measurements. Table 15.12 lists results for E_g and $E_{\frac{1}{2}}$ thus

Table 15.12 Calculated Energy Gaps E_g in Organic Solids, and Related Quantities (eV). After Lyons.[189, 2285]*

	$eE^{ox}_{1/2}$ [a]	$-eE^{red}_{1/2}$ [b]	S^- [c]	$\epsilon_{r,2}$ [d]	E_G
Anthracene	1.34; 1.37	1.96; 1.99	1.77	3.2 (3.5)	4.37 4.27
1,2-Benzanthracene	1.44; 1.42	2.00; 2.04	1.67	(4)	4.25
Biphenyl	1.98	2.70; 2.58	1.82	(2.5)	7.01
Chrysene	1.65^r	2.30; 2.28	1.60	(4)	4.70
Naphthalene	1.84;^{i} 1.84; 1.81	2.50	1.92	2.85	5.61
Pentacene	0.82	1.34; 1.37	1.58	(4)	2.9
Perylene	1.06^R	1.67; 1.67	1.61	(4)	3.50
Phenanthrene	1.83;^{i} 1.73	2.46; 2.43	1.75	(3)	5.33
Pyrene	1.36;^{R} 1.36	2.11; 2.07	1.71	(3.5)	4.37
Tetracene	1.04	1.58; 1.65	2.66	(3.5) (4.0)	3.55 3.46
Cyanine dyes *(See Footnote)*					
A	1.21	1.48	1.45	(4)	3.04
B	0.66	1.05	1.43	(4)	2.05
C	0.66	1.07	1.41	(4)	2.07
D	0.67	1.08	1.39	(4)	2.08
E	0.41	0.95	1.40	(4)	1.70
F	0.21	0.77	1.38	(4)	1.31
G	0.88^i	1.13	1.42	(4)	2.35
H	0.52	1.10	1.38	(4)	1.95
J	0.72	1.12	1.42	(4)	2.18
K	0.62	1.07	1.42	(4)	2.24
L	0.73^i	1.14	1.33	(4)	2.18
Acriflavin	>1.10	1.05	1.60	(4)	≥2.53
Capri blue	>1.10	0.36^j	1.49	(4)	≥1.82
Eosin	0.99^{ij}	1.05	1.38	(4)	2.37
Erythrosin	0.92^{ij}	1.05	1.34	(4)	2.29

*This reference is to be found in the Table References at the end of the book

Table 15.12 (*Continued*)

	$eE_{1/2}^{ox}$ [a]	$-eE_{1/2}^{red}$ [b]	S^- [c]	$\epsilon_{r,2}$ [d]	E_G
Fluorescein	0.91	1.22	1.49	(4)	2.49
Methylene blue	>1.10	0.35[j]	1.49	(4)	≥1.81
Phenosafranine	>1.10	0.65	1.49	(4)	≥2.11
Pinacryptol green	>1.10	0.63[i]	1.49	(4)	≥2.09
Thionine	>1.10	0.39[j]	1.60	(4)	≥1.87
Chlorophyll a	0.57	1.15	1.11	(4)	2.2

[a] vs. aq. SCE; solvent CH_3CN
[b] vs. aq. SCE; solvent CH_3CN or dimethyl formamide
[c] S^- calculated for hydrocarbons by Batley,[184] for dyes by Gouverneur et al.,[191] for chlorophyll a, S^+ is assumed equal to S^-.
[d] values not in parentheses are from Ref. 190
[i] electrochemically irreversible
[j] 2-electron reduction or oxidation
[R] electrochemically reversible

Cyanine dyes

Dye	n	X	Y	R_1	R	A
A	0	S	S	H	C_2H_5	Br
B	1	S	S	H	C_2H_5	Br
C	1	S	S	CH_3	C_2H_5	Br
D	1	S	S	H	C_2H_5	Br
E	2	S	S	H	C_2H_5	I
F	3	S	S	H	C_2H_5	ClO_4
G	0	−CH:CH−	−CH:CH−	H	C_2H_5	I
H	1	−CH:CH−	−CH:CH−	H	C_2H_5	Br
K	1	−CH:CH−	−CH:CH−	H	C_2H_5	Br
L	1	S	S	H	C_6H_5	I

Table 15.12 (*Continued*)

J is
[structure: 1,1'-diethyl-2,4'-cyanine iodide cation with I⁻]

Oxidation potentials, $E_{1/2}^{ox}$: hydrocarbons, Ref. 191; dyes, Ref. 191, chlorophyll a, Ref. 192. The value for pentacene has been derived from the linear plot of oxidation potential vs. the coefficient of the highest occupied molecular orbital in the series of linear polyacenes, benzene to tetracene extended to pentacene.[193] Reduction potentials, $E_{1/2}^{red}$: Ref. 194 and references therein.

electrochemically obtained. Both for dyes and for aromatic hydrocarbons, the energy gap comes close to the excited singlet state of the crystal and of the isolated molecule.[189]

The thermal activation energy of photoconductivity has been related by Vijh[296] to the polarographic RedOx (half-wave) potentials; both quantities are said to be linearly related.

Naphthalene, anthracene, pyrene, perylene, coronene and diphenyl are said to exhibit this relationship which is indeed an independent confirmation of the Lyons theory discussed above.[189]

The energy of the optical charge transfer absorption band $h\nu_{ct}$ for π-π complexes has also been derived[195] from the RedOx potentials of the donors and acceptors, disregarding, however, the coulombic term, steric hindrance and solvent contributions. Reasonably good agreement for 20 organic acceptor compounds (with hexamethylbenzene always as the donor) between the acceptor reduction potential and $h\nu_{ct}$ has been obtained by Williams.[197] The electrochemistry of charge transfer complexes has been reviewed.[196]

Catalytic Activity and Energy Gap

This correlation is of course basic to the modern electronic theory of catalysis[198] as devised, e.g., by Volkenshtein and his school. (See Section 2.7, 12.11, and discussion in Ref. 196.)

However, the only study referring directly to the relation between energy gap and catalytic activity is that by Hanke[199] who reports that the catalytic activity of monomeric and polymeric phthalocyanines relates inversely to the gap energy; a wide energy gap material exhibiting low catalytic activity

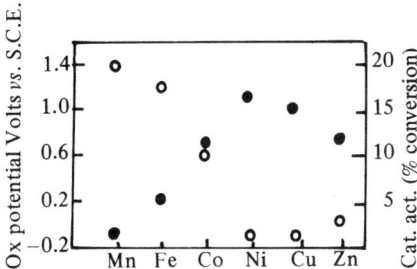

Fig. 15.26 Oxidation potentials and catalytic activities of several metal-phthalocyanines as a function of the central metal atom. ○, Oxidation potential; ●, catalytic activity. After Manassen.[200]

as measured by the decomposition of gaseous formic acid. The reaction constants obey an Arrhenius relation in respect to $1/T$ so that the resulting activation energy must be related to that of dark conductance, or resistance.

The catalytic activity of phthalocyanines has also been related to their oxidation potential[200] as illustrated in Fig. 15.26. It is seen that the lowest potential values go hand in hand with the highest catalytic activity. In the light of Lyons' recent work[189] discussed before in this section, these results are readily understandable. Other references[230] on the catalysis of phthalocyanines should be consulted.

Other Correlations

Vijh[201] has correlated the energy gap in inorganic semiconductors such as metallic oxides with the bond energy, the heat of formation and the free energy of sublimation per atom, of the compounds. Ruppel et al. have suggested[202] that the energy gap for inorganic compounds should fall between 1 and 2 times the heat of formation per mole. Vijh[201] has shown that it is not the molar heat of formation, but that per atom equivalent, i.e., the molar heat of formation/total number of valencies—whether electron donating or accepting—of the compound, ΔH_F, which enter into such relationships Δ. Then, $E_g/2 \cong \Delta H_F$. Since the analysis is based upon considerations of lattice energy, Vijh[201] himself points out that it is not applicable to molecular solids such as anthracene. Indeed, attempts to relate the energy gap with colligative properties such as heat of formation, bond energy, heat of sublimation, etc., only result in scatter diagrams.

For charge transfer complexes, however, V. Gutmann[203] has demonstrated that a correlation between the energy of the charge transfer optical absorption

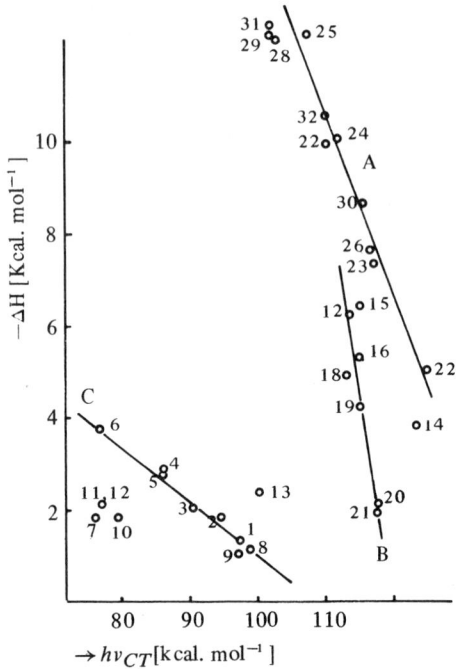

Fig. 15.27 Correlation between the enthalphy of complex formation, ΔH, and the charge transfer energy (optical absorption) $h\nu_{CT}$, of several electron donors with iodine as the acceptor. From V. Gutmann.[203]

1. Benzene
2. Toluene
3. o-Xylene
4. Mesitylene
5. Durene
6. Hexamethylbenzene
7. Hexaethylbenzene
8. Chlorobenzene
9. o-Dichlorobenzene
10. Naphthalene
11. 1-Methylnaphthalene
12. 2-Methylnaphthalene
13. Cyclohexene
14. Propyleneoxide
15. Trimethyleneoxide
16. Tetrahydrofuran
17. 2-Methyltetrahydrofuran
18. Tetrahydropyran
19. Diethylether
20. Ethanol
21. Methanol
22. Ammonia
23. Methylamine
24. Dimethylamine
25. Trimethylamin
26. Ethylamine
27. Diethylamine
28. Triethylamine
29. Tri-n-propylamine
30. n-Butylamine
31. Tri-n-butylamine
32. Piperidine

band peak, $h\nu_{ct}$ which is closely related to E_g^v, and the enthalpy for complexation ΔH does hold. This is illustrated in Fig. 15.27.

There is some positive evidence[204] that E_g may be related to the cohesive energy density of a solid, which is defined as[205]

$$E_{\text{coh}} \equiv \frac{H_{\text{vap}} - RT}{v} \cong \left(\frac{n^2 - 1}{n^2 + 1}\right) \qquad (15.102$$

where H_{vap} is the heat of vaporization, R the gas constant, T the temperature in K, and v the molar volume in the approximation given, n is the refractive index and the parameter C is a constant for any given chemical type of compounds.[205] Plots are a set of parallel straight lines though at this stage no definite conclusions may as yet be proposed.

Shchegolev[207] discussed the inter-relation between the bond energies of TCNQ salts and their thermoelectric, Seebeck, coefficients.

Wheland[223] has correlated conductivities of organic charge transfer complexes with heavy atom effects and redox potentials. He found[223] high conductivity for complexes derived from acceptors with electron affinities E_A between 0.02 and 0.35 eV, and from donors with ionization potentials (E_I) between 0.1 and 0.4 eV. He also found a close match between ReDox $(E_E - E_D) \approx 0.25$ eV and molecular dimensions.

References

1. L. B. Schein and A. R. McGhie, *Phys. Rev.*, **B-26**, 1631 (1978); W. Mey and A. M. Hermann, *Phys. Rev.*, **B-7**, 1652 (1973).
2. S. D. Druger, *Organic Molecular Photophysics*, **2**, 313 ff. J. Birks, ed., Wiley, N.Y., (1975); J. O. Williams, "Electrical Conduction in Organic Solids," *Adv. Phys. Organic Chem.*, **16**, 159 (1978); R. G. Kepler, *Properties of Solid Polymeric Materials*, J. Schultz, ed., Academic Press, N.Y., **B-10**, 637 ff, 1977; J. T. Yardley, *Introduction to Molecular Energy Transfer*, Academic Press, N.Y., 1980.
3. T. J. Sonnonstine and A. M. Hermann, *J. Chem. Phys.*, **60**, 1335 (1974).
4. J. O. Williams, in Ref. 2.
5. K. Heikes, in "Proc. Buhl Internatl. Conf. on Materials," Pittsburgh, Pa., p. 2, (1963), ed., E. R. Schatz, Gordon & Breach, N.Y., 1964; S. I. Kubarev and I. D. Mikhailov, *Teor. Eksper. Khim*, **1**, 279 (1965), Translation No. 2, p. 148.
6. W. Mey and A. M. Hermann, *Phys. Rev.*, **B-7**, 1652 (1973).
7. R. W. Munn and W. Siebrand, *Chem. Phys. Lett.*, **3**, 655 (1969); *J. Chem. Phys.*, **52**, 47, 6391 (1970); R. W. Munn and R. Silbey, *J. Chem. Phys.*, **68**, 2439 (1978); see also Ref. 17.
8. R. G. Kepler, in Ref. 2.
9. P. M. Grant and I. P. Batra, paper no. 12, "Sympos. on the Structure and Prop. of Highly Conduct. Polymers and Graphite," IBM, San Jose, Calif. (1979).
10. Z. Burshtein and O. F. Williams, *Phys. Rev.*, **B-15**, 5769 (1977).
11. R. W. Munn and W. Siebrand, *J. Chem. Phys.*, **52**, 6391 (1970); **53**, 3393 (1970); *Disc. Faraday Soc.*, No. 51, 17 (1971); see also Ref. 7.
12. P. Gosar and S. I. Choi, *Phys. Rev.*, **150**, 529 (1966).
13. P. Gosar and I. Vilfau, *Mol. Phys.*, **18**, 49 (1970).
14. R. W. Munn, et al., *J. Chem. Phys.*, **52**, 6442 (1970); R. M. Glaser and R. S. Berry, *J. Chem. Phys.*, **44**, 3797 (1966).

16. V. Ambegaokar, *Astrophysics and the Many Body Problem*, Benjamin, N.Y., 1963, p. 375.
17. R. W. Munn and W. Siebrand, *J. Chem. Phys.*, **53**, 3343 (1970).
18. L. B. Schein, *Phys. Rev.*, **B-15**, 1024 (1977); *Phys. Rev. Lett.*, **40**, 197 (1978); see also L. B. Schein and A. R. McGhie, *Phys. Rev.*, **B-20**, 1631 (1978).
19. E. O. Forster, *IEEE Trans.*, **PAS-90**, 913 (1971).
20. Y. Sakai et al., *Bull. Chem. Soc. Japan*, **47**, 1886 (1974); cf., also A. R. Foweraker and B. R. Jennings, *Spectrochim. Acta*, **31-A**, 1065 (1975).
21. Y. Sakai, *Japan J. Phys.*, **12**, 1463, 1952 (1973).
22. W. Mey et al., *J. Chem. Phys.*, **58**, 2542 (1973).
23. Y. Maruyama et al., *Molec. Cryst. Liq. Cryst.*, **20**, 373 (1973).
24. Sh. Efrima and H. Metiu, *Phys. Chem. Lett.*, **60**, 226 (1979).
25. F. Gutmann et al., *Nature* (London), **221**, 1237 (1969).
26. C. B. Duke and R. J. Meyer, *Bull. Amer. Phys. Soc.*, **24**, 309 (1979).
27. J. Swiatek, *Phys. Stat. Solidi*, **A-38**, 285 (1976).
28. H. Ranicar et al., *Aust. J. Phys.*, **24**, 325 (1971).
29. R. G. Romanets et al., *Elektron. Obrab. Mater.*, 44 (1976).
30. A. Abu-Bakr et al., *Eur. Polymer. J.*, **13**, 799 (1977).
31. G. Pfister et al., *Phys. Rev. Lett.*, **40**, 659 (1978).
32. e.g., T. J. Lewis, Ann. Rep. Conf. Electr. Insul. Dielect. Plenum, 1976, Publ. 1978, p. 533; H. A. Pohl, *J. Biol. Phys.*, **2**, 113 (1974); A. A. Dulov, *Usp. Khim*, **35**, 1853 (1966); V. S. Mylnikov, *ibid.*, **37**, 78 (1968); I. S. Schopov and C. Vodenicharov, *J. Macromolec. Sci.*, **A-4**, 1627 (1970).
33. J. B. Webb and D. R. Williams, *J. Phys.*, **C-11**, 3245 (1978).
34. N. F. Mott, *Phil. Mag.*, **19**, 835 (1969); see also Ref. 208.
35. R. G. Kepler, *Properties of Solid Polymeric Materials*, J. M. Schultz ed., **B-10**, (1977), Academic Press, N.Y., 1977, p. 637; H. Bottger and V. Bryskin, *Phys. Stat. Solidi*, **B-78**, 415 (1976).
36. M. Pollak et al., *Phys. Rev. Lett.*, **30**, 586 (1973); V. Ambeagokar et al., *Phys. Rev.*, **B-4**, 2612 (1971).
37. D. Emin, *Phys. Rev. Lett.*, **32**, 303 (1974).
38. H. Scher, in "Proc. 5th Int. Conf. Amorph. Liq. Semicond.," J. Stuke and W. Brenig, eds., **1**, 135 (1973).
39. M. D. Silver et al., *J. Non-Cryst. Solids*, **B-10**, 773 (1972).
40. G. Pfister, *Phys. Rev. Lett.*, **33**, 1474 (1974); H. Scher, *Bull. Amer. Phys. Soc.*, **24**, 308 (1979); H. Scher and E. Montroll, *Phys. Rev.*, **B-12**, 2455 (1975); G. Pfister and H. Scher, *Adv. Phys.*, **27**, 747 (1978).
41. J. Mort et al., *Solid State Commun.*, **18**, 693 (1976).
42. H. Hoegl, *J. Phys. Chem.*, **69**, 755 (1965).
43. W. D. Gill, "Proc. 5th Int. Conf. Amorph. and Liquid Semicond.," 901, Garmisch-Partenkirchen, Taylor and Francis, London, 1974, p. 901.
44. J. Mort and A. I. Lakatos, *J. Non-Cryst. Solids*, **4**, 117 (1970).
45. P. Csavinszky, *Phys. Rev.*, **B-14**, 1649 (1976); L. M. Richardson and L. M. Scarfone, *ibid.*, **B-19**, 925 (1979).
46. W. E. Spear, *Adv. In Phys.*, **23**, 513 (1974); H. Sumi, *Solid State Commun.*, **28**, 309 (1979); **29**, 495 (1979); *Chem. Phys.*, **70**, 3775 (1979); L. B. Schein et al., *Phys. Rev. Lett.*, **40**, 197 (1978).
47. M. H. Cohen, *J. Non-Cryst. Solids*, **4**, 391 (1970).
48. N. F. Mott, *Phil. Mag.*, **22**, 7 (1970); see also Ref. 209; cf also D. Redfield, *Advances in Physics*, **24**, 563 (1975).

REFERENCES

49. W. D. Gill, *J. Appl. Phys.*, **43**, 5033 (1972); G. Pfister and C. H. Griffiths, *Phys. Rev. Lett.*, **40**, 659 (1978).
50. E. W. Montroll and H. Scher, *Stat. Phys.*, **9**, 101 (1973); G. Pfister et al., *Phys. Rev. Lett.*, **37**, 1360 (1976); G. Pfister, *Phys. Rev.*, **B-16**, 3673 (1977).
51. W. D. Gill, "5th Int. Conf. Amorph. and Liquid Semicond.," Garmisch-Partenkirchen, J. Stuke and W. Brenig, eds., Taylor and Francis, London, 1973, p. 901.
52. C. K. Chiang et al., *Phys. Rev. Lett.*, **39**, 1098 (1977).
53. A. J. Epstein et al., *Solid State Commun.*, **23**, 355 (1977).
54. R. W. Munn and W. Siebrand, *Phys. Rev.*, **B-2**, 3435 (1970).
55. T. Danno and H. Inokuchi, *Bull. Chem. Soc. Japan*, **41**, 1783 (1968).
56. S. Elnahwy et al., *J. Chem. Phys.*, in press, (1979); *Phys. Rev.*, **19**, 1108 (1979).
57. J. R. Hoyland and L. Goodman, *J. Chem. Phys.*, **36**, 12 (1962).
58. R. G. Kepler, *Phys. Rev.*, **119**, 1226 (1960); T. Holstein, *Ann. Phys. (N.Y.)*, **8**, 325, 434 (1959).
60. T. Kajiwara et al., *Bull. Chem. Soc. Japan*, **40**, 1055 (1967).
61. H. Kivelson, *Bull. Amer. Phys. Soc.*, **24**, 308 (1978).
62. M. Hociuc et al., *Rev. Roum. Phys.*, **23**, 221 (1978).
63. M. Sugi et al., *Chem. Phys. Lett.*, **45**, 163 (1977).
64. Boon-Keng Teo et al., *J. Am. Chem. Soc.*, **99**, 4862 (1977); cf., also A. P. Ginsberg et al., *Inorg. Chem.*, **15**, 514 (1976).
65. J. M. Caywood, *Mol. Cryst. Liq. Cryst.*, **12**, 1 (1970).
66. H. J. Gaehas and C. Willig, *Chem. Phys. Lett.*, **32**, 300 (1975).
67. J. B. Birks, *Photophysics of Aromatic Molecules,* Wiley Interscience, N.Y. and London, 1970, p. 331.
68. P. Gosar and S. I. Choi, *Phys. Rev.*, **150**, 529 (1966).
69. M. K. Konkin et al., *Solid State Commun.*, **26**, 49 (1978); P. Lu and C. K. Chan, *Nuovo Cimento. Soc. Inter. Fiz.*, **22**, 410 (1978); S. A. Rice and M. M. Pilling, *Progr. React. Kinetics*, **9**, 93 (1978); R. Memming and F. Mollers, *Ber. Bunsen Ges. Phys. Chem.*, **77**, 960 (1973).
70. N. Morton, *Solid State Commun.*, **26**, 973 (1978).
71. J. E. Kuder et al., *J. Electrochem. Soc.*, **125**, 1750 (1978); see also C. B. Duke in "Proc. 3rd Symp. Electrode Processes," The Electrochemical Soc., Princeton, N.Y., S. Bruckenstein et al., eds., 1979, p. 15.
72. G. M. Parker and C. A. Mead, *Appl. Phys. Lett.*, **14**, 21 (1969); J. C. Penn et al., *Phys. Rev.*, **B-5**, 768 (1977); J. W. Gadzuk and E. W. Plummer, *Rev. Mol. Phys.*, **45**, 487 (1973).
73. W. Schmickler, *Ber. Bunsen Ges. Phys. Chem.*, **82**, 477 (1978).
74. G. Sarrabayrouse et al., *J. Phys. (Paris)*, **38**, 1443 (1977).
75. A. Kubovy and M. Janda, *Phys. Stat. Solidi*, **A-40**, 225 (1977).
76. S. A. Rice and M. J. Pilling, In ref. 69; W. M. Bartczak et al., *Curr. Top. Radiat. Res. Q.*, **11**, 307 (1977).
77. B. Brocklehurst, *Chem. Phys. Lett.*, **39**, 61 (1976); see also ref. 76; K. I. Zamaraev and R. F. Khairutinov, *Russ. Chem. Revs.*, **47**, 518 (1978).
78. D. DeVault et al., *Nature (London)*, **215**, 642 (1967); F. Gutmann, *ibid*, **219**, 1359 (1968); J. Jortner, *J. Chem. Phys.*, **48**, 60 (1976); Ling Y. Wei, *Bull. Math. Biophys.*, **29**, 411 (1967); J. R. Brocklehurst, *Phys. Bull.*, **31**, 54 (1980); C. B. Duke, *Tunnelling in Biological Systems,* B. Chance, ed., Academic Press, N.Y., 1979, 31 ff; W. F. Libby, *Ann. Rev. Phys. Chem.*, **28**, (1977).
79. B. Brocklehurst, *J. Phys. Chem.*, **83**, 536 (1979); K. I. Zamaraev and R. F.

Khairu'dinov, *Russ. Chem. Revs.*, **47**, 518 (1978); J. R. Miller, *J. Phys. Chem.*, **82**, 767 (1978).
80. M. J. Potasek and J. J. Hopfield, *Proc. Natl. Acad. Sci. USA*, **74**, 3817 (1977).
81. S. G. Christov, *Ann. Physik. Leipzig*, **VII-12**, 20 (1963); **VII-15**, 87 (1965); *Disc. Faraday Soc.*, **39**, 60, 259, 263 (1965); *Phys. Stat. Solidi*, **42**, 583 (1970); **7-A**, 371 (1971); *Contemp. Phys.*, **13**, 199 (1972).
82. C. Eckert, *Phys. Rev.*, **35**, 1303 (1970).
83. F. Gutmann, *Japan. J. Appl. Phys.*, **8**, 1417 (1969); see also C. B. Duke, Ref. 71.
84. F. Gutmann, in Ref. 78.
85. H. Seki, *Amorphous and Liquid Semiconductors*, J. Studke and W. Brenig, eds., Taylor and Francis, London, (1974), p. 1015.
86. B. G. Bagley, *Solid State. Commun.*, **8**, 345 (1970).
87. W. D. Gill, *J. App. Phys.*, **43**, 5033 (1973).
88. B. Brocklehurst, Ref. 79.
89. A. I. Kitaigorodsky, *Molecular Crystals and Molecules*, Academic Press, N.Y., 1973.
90. G. Delacote, *Mol. Cryst.*, **5**, 309 (1969).
91. W. Shockley and J. Bardeen, *Phys. Rev.*, **77**, 407 (1949); **80**, 72 (1950).
92. R. A. Smith, *Semiconductors*, Cambridge U.P., London, 1959.
93. S. F. Fischer et al., *Bull. Amer. Phys. Soc.*, **24**, 346 (1979).
94. N. Karl, *Pr. Nauk. Inst. Chem. Org. Fiz. Politech. Wroclaw*, **16**, 43 (1978).
95. A. Van Heuvelen, *J. Biol. Phys.*, **1**, 215 (1973).
96. R. J. P. Williams, *Current Topics in Bioenergetics*, **3**, 79 (1969).
97. M. Campus and J. A. Giacometti, *Appl. Phys. Lett.*, **32**, 794 (1978).
98. R. L. Dexter and W. B. Fowler, *J. Chem. Phys.*, **47**, 1378 (1967).
99. J. G. Simmons, *Phys. Rev.*, **B-15**, 964 (1977); cf., also Ref. 217.
100. D. Emin, *Electronic and Structural Properties of Semiconductors*, P. G. LeComber and J. Mort, eds., Academic Press, London, 1974.
101. H. Scher and M. Lax, *Phys. Rev.*, **B-7**, 4491, 4502 (1973).
102. M. Pollak, *Phil. Mag.*, **36**, 1157 (1977).
103. F. C. Aris et al., *Solid State Comm.*, **12**, 913 (1973).
104. G. M. Parkinson et al., "8th Molec. Cryst. Sympos.," Sta. Barbara, Calif., (1977).
105. J. O. Williams, in Ref. 2.
106. V. Yu. Litvinenko, *Kristallografiya*, **21**, 850 (1976).
107. R. C. Hughes, in "Electrophotography," D. R. White, ed., Soc. Photogr. Sci. and Eng., Wash., D.C., 1974, p. 147.
108. H. A. Pohl, *Biol. Phys.*, **2**, 113 (1974).
109. L. Friedman and M. Pollak, *Phil. Mag.*, **B-38**, 173 (1978); L. Friedman, "Conduction in Low Mobility Materials," Report of the Eilat (Israel) Conf. 1971, M. Pollak, ed.
110. S. G. Burnay and H. A. Pohl, in press, *J. Non-Cryst. Solidi*, (1979).
111. T. Yanaguchi, *Carbon*, **1**, 47, 535 (1964); **2**, 95 (1965).
112. D. H. Spielberg et al., *Phys. Rev.*, **B-3**, 2012 (1971).
113. G. D. Thakton et al., *J. Phys. Chem.*, **66**, 2461 (1962).
114. J. I. Katz et al., *J. Chem. Phys.*, **39**, 1683 (1963).
115. R. Silbey et al., *J. Chem. Phys.*, **42**, 733 (1965), **43**, 2925 (1965).
116. A. Korn et al., *Phys. Rev.*, **186**, 938 (1969).
117. M. Schaadt and D. R. Williams, *Phys. Stat. Solidi*, **39**, 223 (1970).
118. M. Suzuki et al., *Jap. J. Appl. Phys.*, **14**, 741 (1975).
119. E. L. Nagaev and E. B. Sokolova, *Fiz. Tverd. Tela (Leningrad)*, **19**, 732 (1977).

REFERENCES

120. P. L. Swart and C. K. Campbell, *Thin Solid Films*, **44**, 83 (1977).
121. J. Emin, *Phil. Mag.*, **35**, 1189 (1977).
122. M. Vodenicharova, *Phys. Stat. Solidi*, **28A**, 263 (1975); C. M. Vodenicharov et al., *C. R. Acad. Sci. Bulgar.*, **24**, 1309 (1971).
123. B. I. Schklovskii, *Soviet Phys. Semicond.*, **6**, 1053 (1973).
124. M. Pollak, *J. Non-Cryst. Solids*, 8-10, 486 (1972); **11**, 1 (1972).
125. B. I. Shklovskii et al., *J.E.T.P. Lett.*, **14**, 233 (1971); *Phys. Stat. Solidi*, **50**, 45, (1972).
126. A. J. Grant and E. A. Davis, *Solid State Commun.*, **15**, 563 (1974); R. M. Hill, *Phys. Stat. Solidi*, **A-34**, 601 (1976).
127. M. A. Careem and R. M. Hill, *Thin Solid Films*, **51**, 363 (1978).
128. G. M. Parker and C. A. Mead, *Appl. Phys. Lett.*, **14**, 21 (1969); J. C. Penn et al., *Phys. Rev.*, **B-5**, 768 (1977); J. W. Gadzok and E. W. Plummer, *Rev. Mol. Phys.*, **45**, 487 (1973).
129. R. Memming and F. Mollers, *Ber. Bunsen. Ger. Phys. Chem.*, **77**, 960 (1973).
130. J. Mordclewski and W. E. Geiger, *J. Am. Chem. Soc.*, **100**, 7429 (1978).
131. C. C. Isett and E. A. Perez-Albuerne, *Res. Disc.*, **158**, 59 (1977).
132. N. Geacintov and M. Pope, *J. Chem. Phys.*, **45**, 3884 (1966); **50**, 814 (1969); R. J. Batt, C. L. Braun and J. F. Hornig, *Appl. Opt. Suppl.*, **3**, 20 (1969); N. Karl and G. Sommer, *Phys. Status Solidi*, **A-6**, 231 (1971); R. R. Chance and C. L. Braun, *J. Chem. Phys.*, **59**, 2269 (1973); **64**, 3573 (1976); L. E. Lyons and K. A. Milne, *J. Chem. Phys.*, **65**, 1474 (1976).
133. J. W. Gadzur and E. W. Plummer, *Rev. Mol. Phys.*, **45**, 487 (1973).
134. J. V. Beitz and J. R. Miller, *J. Chem. Phys.*, **71**, 45, 79 (1979).
135. J. Bernasconi, *Phys. Rev.*, **B-7**, 2252 (1973).
136. K. Saha, S. C. Abbi and H. A. Pohl, *J. Non-Cryst. Solids*, **21**, 117 (1976); **22**, 291 (1976).
137. S. D. Colson et al., *J. Chem. Phys.*, **19**, 4941 (1977); R. Kopelman et al., *J. Chem. Phys.*, **19**, 413 (1977); A. Argyrakis and R. Kopelman, *J. Theoret. Biol.*, **73**, 205 (1978); C. Swenberg et al., *Photochem. Photobiol.*, **24**, 201 (1976).
138. T. Keyes and S. Pratt, *Chem. Phys. Lett.*, **65**, 100 (1979).
139. P. Bouchez et al., *Chem. Phys. Lett.*, **65**, 212 (1979).
140. H. Sumi, *J. Chem. Phys.*, **70**, 3775 (1979); **71**, 3403 (1979); see also Ref. 218.
141. W. Warta, Diplomarbeit, Univ. Stuttgart (1978).
142. A. Mierzejewski et al., *Pr. Nauk. Inst. Chem. Org. Fiz. Politech. Warclaw*, **16**, 245 (1978).
143. L. E. Lyons et al., *Aust. J. Chem.*, **21**, 2789 (1968).
144. P. K. Hansma, *Phys. Rep. (Phys. Lett. C)*, **30**, 145 (1977); J. D. Langan and P. K. Hansma, *Surface Sci.*, **52**, 211 (1975); N. M. D. Brown et al., *J. Chem. Soc. Faraday II*, **1**, (1979).
145. G. Pfister and H. Scher, *Adv. in Phys.*, **27**, 747 (1978); H. Scher et al., *J. Appl. Phys.*, **42**, 3939 (1971).
146. e.g., R. Stratton, *J. Phys. Chem. Solids*, **23**, 1177 (1962); R. Stratton et al., *ibid.*, **27**, 1599 (1960); J. Antula, *Phys. Stat. Solidi*, **24**, 89 (1967); A. Sommerford and H. Bethe, *Handbuch d. Phys.*, 24/II, Springer, Berlin, 1933.
147. S. G. Christov, *Contemp. Phys.*, **13**, 199 (1972); *Surface Sci.*, **70**, 32 (1978); *Phys. Stat. Solidi*, **42**, 583 (1970); **7A**, 371 (1971).
148. J. Frenkel, *Phys. Rev.*, **36**, 1604 (1930); R. Stratton, *J. Phys. Chem. Solids*, **23**, 1177 (1962).

149. S. G. Christov, *Phys. Stat. Solidi,* **32,** 509 (1969); *Surface Sci.,* **70,** 32 (1978); C. Vodenicharov and S. G. Christov, *Solid State Electronics,* **15,** 933 (1972).

150. e.g., C. Kittel, *Introduction to Solid State Physics,* **266,** 2nd ed., Wiley, N.Y., 1956.

151. M. Vodenicharova, *Phys. Stat. Solidi,* **28A,** 263 (1975); C. R. Acad. Bulgare des Sciences, **24,** 1309 (1979).

152. J. Tyczkowski et al., *Thin Solid Films,* **55,** 253 (1978).

153. W. Schmickler, *Ber. Bunsen Ges. Phys. Chem.,* **82,** 477 (1978).

154. J. C. Manifacier and H. K. Henisch, *Phys. Rev.,* **B-17,** 2640, 2648 (1978); F. Koch, *Surf. Sci.,* **80,** 110 (1979); P. Popescu and H. K. Henisch, *Phys. Rev.,* **B-11,** 1563 (1975); **B-14,** 517 (1976).

155. E. J. Mele and J. D. Joannopoulos, *Phys. Rev.,* **B-17,** 1528 (1978).

156. G. Lengyel, *J. Appl. Phys.,* **37,** 807 (1966); J. H. Ranicar et al., *Austral. J. Phys.,* **24,** 325 (1971); M. K. Konkin et al., *Solid State Commun.,* **26,** 949 (1978); G. Sarrabayrouse et al., *J. Phys. (Paris),* **38,** 1443 (1977); A. Kubovy and M. Janda, *Phys. Stat. Solidi,* **A-40,** 225 (1977).

157. J. C. Inkson, *J. Phys.,* **C-5,** 2599 (1972); **C-6,** 1350 (1973); S. G. Louie et al., *Phys. Rev.,* **B-13,** 2461 (1976); **B-15,** 2154 (1977); L. J. Brillson, *Phys. Rev. Lett.,* **38,** 245 (1977); J. E. Rowe et al., *ibid.,* **35,** 1471 (1975); K. C. Pandey, *J. Vac. Sci. Technol,* **15,** 440 (1978); D. R. Hamann, *Surface Sci.,* **68,** 167 (1977).

158. W. J. Gill, *J. Appl. Phys.,* **43,** 5033 (1972).

159. e.g., Fu-Ren Fan and L. R. Faulkner, *J. Chem. Phys.,* **69,** 3334, 3341 (1978).

160. T. T. Tsong, *Bull. Amer. Phys. Soc.,* **24,** 279 (1979).

161. J. S. Bonham, *Aust. J. Chem.,* **28,** 1631 (1975).

162. J. S. Bonham and L. E. Lyons, *Aust. J. Chem.,* **26,** 489 (1973).

163. G. R. Johnston and L. E. Lyons, *Phys. Stat. Solidi,* **37,** K43 (1970); K. Ulbert, *Aust. J. Chem.,* **23,** 1347 (1970); D. D. Eley, *J. Polym. Sci.,* (C), **17,** 73 (1967); K. Ulbert, *ibid.,* **22,** 881 (1969); E. Krikorian, U.S. Air Force Rept. AF CRL-68-0384 (1968); D. D. Eley et al., *Trans. Faraday Soc.,* **64,** 1513 (1968); V. Madek, *Solid State Commun.,* **6,** 337 (1968).

164. O. Exner, *Coll. Czech. Chem. Commun.,* **29,** 1094 (1964).

165. G. R. Johnston and L. E. Lyons, *Aust. J. Chem.,* **23,** 2187 (1970); see also G. R. Johnston and L. E. Lyons, Ref. 163.

166. W. Meyer and H. Neldel, *Z. Tech. Phys.,* **1937,** 558.

167. B. Rosenberg et al., *J. Chem. Phys.,* **49,** 4108 (1968).

168. I. Schopov and C. Vodenicharov, *J. Macromolec. Sci.,* **A-14,** 1627 (1970).

169. W. Bucker and A. Herspring, Tech. Hochschule, Aachen, W. Germany, Priv. Commun. (1978).

170. M. Hisa et al., *J. Phys. Soc. Japan,* **21,** 23 (1971).

171. G. Kemeny and B. Rosenberg, *J. Chem. Phys.,* **52,** 4151 (1970); **53,** 34, 49, (1970).

172. G. G. Roberts, *Transfer and Storage of Energy by Molecules,* G. M. Burnett et al., eds., Wiley, N.Y., 1974, p. 153.

173. G. P. Owen et al., *J. Chem. Soc. Faraday Trans. II,* **70,** 853 (1974).

174. S. Nespurek and E. A. Silinsh, *Phys. Stat. Solidi,* **A-34,** 747 (1976).

175. K. O. Lee, *Phys. Stat. Solidi,* **A-47,** K47 (1978).

176. K. Z. Lichtenecker, *Z. Elektrochem.,* 1934, 11.

177. K. Ulbert, Ref. 163.

178. E. Krikorian and R. J. Sneed, *J. Appl. Phys.,* **40,** 2306 (1969).

179. T. Dosfale and R. J. Brook, *J. Mater. Sci.,* **13,** 167 (1978).

REFERENCES

180. V. K. Yatsimirskii, *Teor. Eksp. Khim.*, **12**, 566 (1976).
181. L. E. Lyons, *Nature* (London), **166**, 193 (1950); M. Batley, L. Johnston and L. E. Lyons, *Aust. J. Chem.*, **23**, 2397 (1970).
182. N. S. Hush, *Aust. J. Sci. Res.*, **A-1**, 480 (1948).
183. M. Latimer et al., *J. Chem. Phys.*, **7**, 108 (1939).
184. M. Batley, Ph. D. Thesis, Univ. of Sydney (1966).
185. W. E. Wentworth, E. Chen and J. E. Lovelock, *J. Phys. Chem.*, **70**, 445 (1966); L. E. Lyons, G. C. Morris and L. J. Warren, *Aust. J. Chem.*, **21**, 853 (1968).
186. F. A. Matsen, *J. Chem. Phys.*, **24**, 602 (1956).
187. e.g., G. J. Hoijtink and J. Van Schooten, *Rec. Trav. Chim. Pays-bas*, **73**, 355 (1954); K. M. C. Davis, P. R. Hammond and M. E. Peover, *Trans. Faraday Soc.*, **61**, 1516 (1965); A. Streitwieser Jr., *Molecular Orbital Theory for Organic Chemists*, Wiley, New York, 1961; M. E. Peover, *Nature*, **191**, 703 (1961); *Trans. Far. Soc.*, **58**, 1656 and 2370 (1962); *Trans. Faraday Soc.*, **60**, 417 (1964); *J. Chem. Soc.*, 4540 (1962); O. Chalvet and I. Jano, *Compt. Rend.*, **259**, 1857 (1964).
188. G. J. Goistink, *Rec. Trav. Chim. Pays-Bas*, **77**, 555 (1958), cf., also Ref. 184.
189. L. E. Lyons, Unpublished work.
190. R. W. Munn, J. R. Nicholson, H. P. Schwab and D. F. Williams, *J. Chem. Phys.*, **58**, 2873 (1973); R. W. Munn and D. R. Williams, *J. Chem. Phys.*, **59**, 1742-1746 (1973).
191. C. Gouverneur et al., *Electrochim. Acta*, **19**, 215 (1974).
192. J. Harbour and G. Tollin, *Photochem. Photobiol.*, **19**, 72 (1974).
193. J. Koutecky, *Z. Phys. Chem.*, Frankfurt/Main, **52**, 8 (1967).
194. T. Imura et al., *Photochem. Photobiol.*, **22**, 129 (1975); A. Streitwieser, *Molecular Orbital Theory for Organic Chemists*, Wiley, N.Y., 1961.
195. J. Bendig and D. Kreysig, *Z. Chemie*, **18**, 33 (1978).
196. see J.-P. Farges and F. Gutmann, *Modern Aspects of Electrochemistry*, J. O'M. Bockris and B. E. Conway, eds., Plenum Press, N.Y., **12**, 267 (1977); **13**, 361 (1979).
197. J. O. Williams, *J. Chem. Soc.*, in print (1979).
198. Th. Volkenshtein, "4th Intern. Cong. Catalysis," Moscow, 1968, Ya. T. Eidus, ed., Nauk. Moscow, 1970; *The Electronic Theory of Catalysis*, Pergamon, Oxford, 1963; D. Schuhmann, ed., "Prop. Electriques des Interfaces Charges," Moscow, Paris, 178; W. Broslow, *Science of Materials*, Wiley, N.Y., 1978; K. I. Zamaraev and R. F. Khairutdinov, M. E. Volpin, eds., *Soviet. Sci. Revs.*, **B-2**, (1980); K. Seeger and W. Maurer, *Solid State Commun.*, **27**, 603 (1978); T. Kanizawa et al., *J. Phys. Chem.*, **82**, 1391 (1978).
199. W. Hanke, *Z. Anorg. Allgem. Chem.*, **347**, 67 (1966).
200. J. Manassen, *Fortschr. Chem. Forschg.*, **25**, 1 (1972); W. Hanke, *Z. Chem.*, **9**, 1 (1969); see also ref. 196.
201. A. K. Vijh, *Electrochim. Acta*, **14**, 921 (1969); *J. Phys. Chem. Solidi*, **29**, 2233 (1968); *J. Mater. Sci.*, **10**, 123 (1975); *J. Electrochem. Soc.*, **117**, 173C, (1970).
202. W. Ruppel et al., *Helv. Phys. Acta*, **30**, 238 (1957).
203. V. Gutmann and V. Mayer, *Rev. de Chimie Minerale*, 8, 429 (1971).
204. F. Gutmann and H. Keyzer, *J. Res. Inst. Catal.*, Hokkaido Univ., **28**, 199 (1981).
205. D. D. Lawson and J. D. Ingham, *Nature* (London), **223**, 614 (1969); D. D. Lawson et al., "Proc. 1st World Hydrogen Energy Conf.," **2**, 913 (1976); A. F. M. Barton, *Chem. Rev.*, **75**, 731 (1975).
206. A. K. Vijh, *J. Chim. Phys. Phys-Chim. Biol.*, **72**, 245 (1975).
207. I. F. Shchegolev, *Phys. Stat. Solidi*, **A-12**, 9 (1972).
208. N. F. Mott and E. A. Davis, *Electronic Processes in Non-Cyrstalline Materials*, Clarendon, Oxford Univ. Press, 1979.

209. Several reviews and books have been published dealing with this topic, cf., E. H. Rhoderick, Clarendon, Oxford Univ. Press, 1980.
210. Yu. A. Berlin et al., *J. Chem. Phys.*, **69**, 2401 (1978).
211. U. Hetzler and E. Kay, *J. Appl. Phys.*, **49**, 5617 (1978).
212. L. B. Schein and A. R. McGhie, *Phys. Rev.*, **B-70**, 1631 (1979); L. E. Schein et al., ref. 40.
213. S. Efrima et al., *J. Chem. Phys.*, **69**, 5113 (1978); *Chem. Phys. Lett.*, **60**, 226 (1979); cf., also Ref. 212.
214. H. R. Zeller and A. Beck, *J. Phys. Chem. Solids*, **35**, 77 (1974); E. H. Putley, *The Hall Effect and Related Phenomena*, Butterworth, London, 1960; J. P. Farges and A. Brau, *Phys. Stat. Solidi*, in press (1980).
215. A. Brau and J. P. Farges, *Phys. Stat. Solidi*, **B-61**, 257 (1974).
216. A. A. Adkhamov et al., *Dokl. Akad. Nauk. Tadzh. SSR*, **20**, 20 (1977).
217. F. W. Schmidlin, *Solid State Commun.*, **22**, 451 (1977); W. D. Lakin et al., *Phys. Rev.*, **B-15**, 5834 (1977).
218. H. Sumi, *Solid State Commun.*, **29**, 495 (1979); *28*, 309 (1978); see also ref. 140.
219. H. Kuhn et al., *Techniques of Chemistry*, Vol. 1, pt. 3B, p. 577, A. Weissberger and B. W. Rossitor, eds., Wiley, N.Y., 1972; *J. Appl. Phys.*, **42**, 4390 (1971); B. Mann et al., *Chem. Phys. Lett.*, 8, 82 (1971); H. Bucher et al., *Molec. Cryst.*, 2, 199 (1967).
220. N. Yamamoto et al., *Bull. Chem. Soc. Japan*, **51**, 1714, 3462 (1978).
221. B. I. Shklovskii, *Fiz. Tekh. Poluprovodn.*, **11**, 2135 (1977).
222. e.g., A. P. Fedder, *Phys. Rev.*, **B-17**, 2098 (1978).
223. R. C. Wheland, *J. Am. Chem. Soc.*, **98**, 3926 (1976).
224. W. Spannring and H. Bässler, *J. Chem. Phys.*, **25**, 325 (1977).
225. K. A. Milne, Ph.D. Thesis, Univ. Queensland, 1976.
226. I. Chem, *J. Appl. Phys.*, **47**, 2988 (1976).
227. M. Suzi et al., *Chem. Phys. Lett.*, **45**, 163 (1973).
228. M. Meton, and P. Gerard, *Chem. Phys. Lett.*, **44**, 582 (1976).
229. C. L. Chien and C. R. Westgate, *The Hall Effect and Its Applications*, Plenum, New York, 1980.
230. R. Onishi, *Catalyst*, **20**, 392 (1978); H. Kopf and F. Steinbach, "Katalyse an Phthalocyaninen," Stuttgart, 1973.

16

Review of Published Data

16.1 Introduction

This chapter is intended to update parts of Chapter 8. However, the emphasis of on-going research during the last decade or so, has tended to shift away from conductivity pure and simple to more specific studies of *e.g.*, mobility (see Chapter 15), apart from the highly conducting organic semimetals which will be treated in Chapter 21. The volume of studies devoted to charge transfer complexes has become so large that this matter, too, is now the subject of a separate chapter, *viz.*, Chapter 17.

Likewise, biological materials have become the subject of extensive studies and are discussed separately in Chapter 18.

Semiconducting adducts based on TCNQ, in contradistinction to semimetallic ones, are discussed in Section 16.11 in this chapter.

Mobilities, especially, have now been determined for many compounds; theoretical aspects are treated in Chapter 15 while tables of numerical values of mobility are collected in Section 16.3 of this chapter.

The increasing application of organic semiconductors to electro-photography has made it advisable to deal separately with this matter (see Section 16.7 for a discussion of doping and sensitization). Doped polymers are dealt with in Section 16.8.

The enormous values of permittivity attainable in certain polymers by the mechanism of nomadic polarization, in view of its practical possibilities, have also led to this topic being dealt with separately in Section 16.5.

194 REVIEW OF PUBLISHED DATA

Numerical data on several physical quantities relevant to the subject in this book will also be found listed in conjunction with the topic being discussed, see Index and lists of Tables on pages 471 and 630.

References to the Addenda Tables in Chapter 24 are found on page 635 and are given in four digit numbers commencing with 2001.

16.2 Molecular Solids

Experimental data for molecular crystals, metal-free as well as metal containing organic compounds, are listed in Table 8.2A and B. It has been realized that little is to be gained from data on still more high resistivity solids, often of questionable purity and under ill-defined experimental conditions. Thus, only 27 entries are found in that table; insufficient to apply any statistics for further evaluation. Some structural aspects of studies of metal-organics are treated in Section 19.7, see also Chapter 17.

The best data for aromatic hydrocarbons, which, generally, are those of the highest values for both resistivity and gap (activation) energy, have values of the latter close to the excited singlet state of the isolated molecule, and of the crystal. The values are so large that no intrinsic conductivity can be expected to be measurable at or slightly above room temperatures. Measured conductivities thus must be ascribed to either impurities or, most likely, to charge-injection from the contact(s); it is mobility which is of primary interest in these solids, see Chapter 15. The conductivity of metal-organic compounds has been reviewed.[101]

16.3 Carrier Mobilities in Molecular Solids

Mobility data for anthracene are collected in Table 4.7A and those for naphthalene in Table 4.7B. These two compounds, and especially anthracene, may be considered as the archetype of organic molecular solids and great efforts have been devoted to their purification and sample preparation in general.

The mean value for 8 entries for the drift mobility of electrons at 25° C in anthracene is 0.45 cm^2/Vsec; in a direction parallel to the c' axis, the mean of 3 entries is 0.34 cm^2/Vsec. It is interesting to note that these values agree quite well with those obtained for the perdeuterated compound.

The mean of 6 entries for holes, likewise, is 0.84 cm^2/Vsec, again in agreement with values obtained from perdeuterated samples. There is only one entry for the drift mobility of holes parallel to the c'-axis, $viz.$, 0.05 to 0.1 cm^2/Vsec.

Values reported for Hall mobilities are, generally, about an order of magnitude above the drift mobilities; the only values for magneto-resistance are extremely low and the reasons for this should be further investigated. It is interesting to note that a reputedly trapfree specimen is reported to have a hole mobility only about 50% higher than one containing shallow traps; one would expect the discrepancy to be much higher. The drift mobilities at 25° C for electrons in naphthalene are, qualitatively, rather similar to those found in anthracene; the anisotropy, as expected, is much below that observed in anthracene and the consistency of the data is seen to be quite good, though there are fewer observations. The average electron mobility in all 3 crystallographic directions is 0.59 cm^2/Vsec while that for holes averages at 0.97 cm^2/Vsec. The Hall mobility for electrons in a direction parallel to the a-axis is said to be normal and agrees with the drift mobility; for holes it is also reported to be normal but exhibiting quite a low value of 0.3 cm^2/Vsec. All other Hall mobilities appear to be anomalous.

Mobilities for other solids are listed in Tables 4.8A1 and 4.8A2, ranging from an unbelievably low 10^{-12} cm^2/Vsec for polyisoprene to a high of 14,700 for the semi-metal TTF:TCNQ, as expected. Many compounds exhibit anisotropic mobilities, *e.g.,* benzophenone or anthraquinone. The majority carriers are electrons in just about as many cases as for those reported to be *p*-type. Taking what appear to be the most reliable single crystal data on TCNE, durene, benzophenone, perylene, phenanthrene, anthraquinone, azulene and tetrabenzoperylene and averaging the anisotropy, this yields a mean electron mobility of 0.8 and a mean hole mobility of 0.95 cm^2/Vsec. The effect of controlled addition of traps has been studied in phenanthrene and it is seen from the table that the introduction of shallow traps causes the mobility to drop to one-half to one-quarter of the reputedly trap-free values. It is regrettable that only 11 data are available for the temperature dependence of the mobility; in most cases it is said to be exponential with an activation energy of the order of 0.5 eV.

For comparison purposes, Table 4.7C reproduces mobility values collected by Schein.[2210] He lists measured microscopic mobilities (by time of flight measurement) for organic and inorganic molecular crystals, including the carrier sign (electron or holes), orientation in the crystal, room-temperature mobility, temperature dependence of μ, the temperature range for the temperature dependence was determined. Inspection of this table reveals the following trend: independent of the particular material, $\mu \approx 1$ cm^2/Vsec (within an order of magnitude) and is weakly temperature dependent. Excluded from this table are those measurements of the drift mobility which were demonstrated to be determined by impurities or measurements in which the temperature dependence or the impurity dependence was not reported, making it impossible to determine whether the microscopic mobility has been observed. For a discussion of these results, see Section 15.4.

Mobilities in liquids and melts are discussed in Section 16.9. For a full discussion of mobility theories, see Chapter 15. Minimum resistance values of phthalocyanines under pressure are tabulated in Table 9.1A.

16.4 Long-Chain Compounds and Polymers

Several reviews and books on polymeric semiconductors are available[1] so that the present discussion can be confined to some selected results of special significant and interest.

Measurements have now been carried out on a wide range of solids synthethized so as to create as many conjugated double bonds as possible. Generally, these solids are infusible, insoluble and black or high colored, but only rarely are they well characterized. Though conductivities as high as about 10 (ohm-cm)$^{-1}$ have been reported, no significant regularities which can be interpreted can be identified. Presumably there are too many complicating processes present such as impurity trapping and amorphous material effects and these dominate any molecular effects.[2] In general, the conductivity is found to vary with temperatures as $e^{-\Delta E/kT}$ with widely varying values reported for ΔE. A compilation of conductivity data on polymers is shown in Table 8.7C, supplementing Table 8.7A. Table 8.7B, dealing with PAQR polymers, is supplemented in Tables 8.7D and 8.7E. Thermoelectric power data for PAQR polymers are given in Table 8.14B.

Table 8.7C, which is by no means claimed to be exhaustive, in view of the large bulk of data published, lists 53 highly incomplete entries; even the sign of the majority carrier has been reported in only 23 cases. Of these 10 are electrons and 13 are holes. The resistivities range from a rather doubtful 10^{18} ohmcm for teflon to a nearly metallic 0.0005 in polyacetylenes, with the energy gap having values from 3.04 eV in PVC down to 0.01 eV in polyphthalocyanines. Mobilities, when given, are seen to be very low. In many cases, the conductivity is likely to be at least partially ionic rather than electronic, e.g., in PVC this is probably so.

PVK (polyvinylcarbazole) in an oxygen ambient under low values of applied electric field is predominantly an ionic and not an electronic conductor.[32] The activation energies are said[32] to be 0.8 eV for the ionic and 0.4 eV for the electronic component. In order to reconcile these two statements, it is necessary to assume that the pre-exponential factor of the electronic component is at least a few tens of times as large as that of the ionic component.

More than one conduction mechanism has been observed in several polymers[13,17] (see Section 15.2 and 15.4) without, however, ascribing one of them to ionic transport. PVC, especially when plasticized, showed appreciable ionic conductivity at temperatures above 300 K.[19]

Poly(ethylene-terephthalate) is also reported[55] to exhibit ionic conductivity plus an electronic component involving the Poole-Frenkel effect (see Section 15.12).

PVK is, in many respects, a model compound for the study of conductivity phenomena in polymers. It is a good photoconductor and has found electrophotographic application. A large amount of information on PVK is available.[4] Its energy diagram[5] is shown in Fig. 16.1, though it does not show any of the many trapping levels which are of crucial importance in the electrical behavior of polymers, even if there is little detailed knowledge of their energetics. Molecular orbital calculations of the electronic structure of PVK have been carried out by Hattori and Wada.[56]

The variation of hole mobility[7] in PVK is illustrated in Fig. 16.2 as a function of temperature, with the applied electric field as parameter;[7] the behavior is typical for simple polymers. In the experiment,[7] positive holes were generated directly at the surface of the polymer film or injected into the polymer from a thin layer of selenium. The mobility was potential-dependent and varied exponentially with temperatures, as shown in Fig. 16.2.

Charge transfer appears to be by thermally activated hopping as in dipyridilium polymers,[40] in poly(arylene-vinylenes),[37] in polyindophenines,[16] and many others. It does seem that this is the predominant mode of charge transfer in polymers as a species; because of local variations in the spatial arrangement of the polymeric chains in a solid, one would expect[242] that nearly all the low energy states for electrons, holes as well as for excitons are localized states. In polymers, trapping can occur at particular molecular sites and at chain-folds. Consider the motion of a charge carrier along a randomly oriented polymer chain shown in Fig. 16.3. If there is a conduction

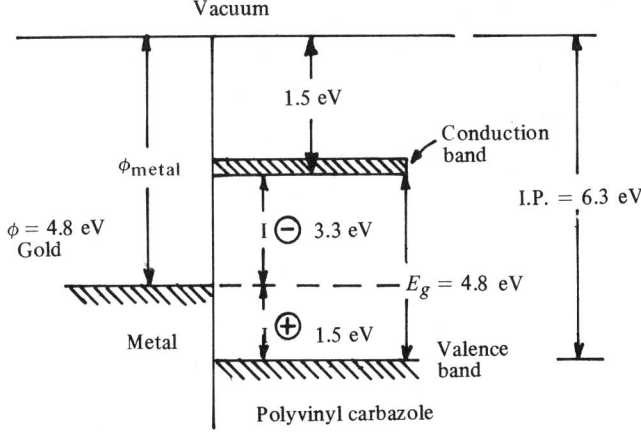

Fig. 16.1 Energy levels in poly-N-vinylcarbazole. After Seanor.[5]

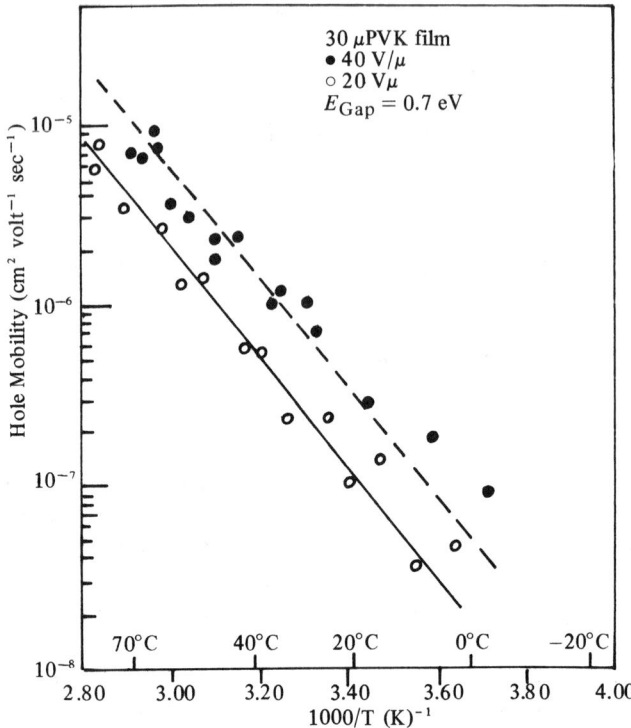

Fig. 16.2 Variation of hole mobility with temperature in poly-N-vinylcarbazole. After Regensburger.[7]

band in which the charge carrier can move, then it will move along until it reaches the point σ at which the chain is no longer aligned with the electric field.[5] At this point, it will not have any momentum in directions perpendicular to the electric field nor can momentum perpendicular to the field be supplied by any other source than the random fluctuation of thermal energy. It cannot be accelerated around the bend to the point C, because there is no component of electric field in this direction.[5] Therefore, the point B is effectively a localized state. The charge carrier will stop there, the medium around it will polarize, and considerable thermal energy will be required to excite it out of the potential well into the next chain at D. It is quite possible that the depth of the potential well will be a time-dependent function related to the rotational or vibrational modes of the polymer. Some close correspondence between the temperature dependence of dielectric constant and conductivity might, therefore, be expected. There does not seem to exist any report on the presence, or absence of such a correlation, though Pohl and coworkers[8]

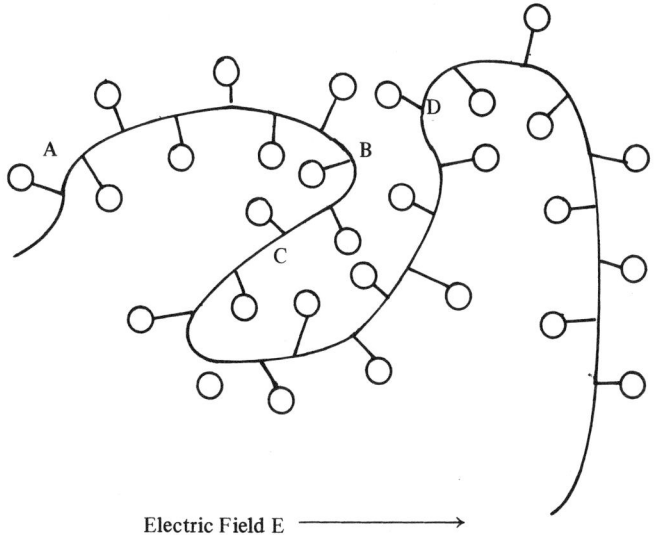

Electric Field E ⟶

Fig. 16.3 Randomly oriented polymer chain. After Seanor.[5]

report that the permittivity of polymers exhibiting nomadic polarization (see Section 16.5) does vary exponentially with temperature with an activation energy which is close to that for electrical conductivity.

An exacting study of polyquinolinecarboxylic acid[9] has shown that the permittivity ϵ increases almost linearly with temperatures with a temperature coefficient between 0.0013 and 0.0021 K^{-1}. At temperatures above 340 K the increase slows down, and thereafter, for most of the samples, an insignificant decrease of ϵ is observed with further rise in temperature. Phenol-formaldehyde resins show only a very weak temperature dependence of their permittivity.[10] after heat treatment this becomes virtually temperature independent.[10]

Dielectric measurement offers a powerful tool for the study of polymeric systems[70] (see Section 12.4). Referring to eq. 12.2, for a Cole-Cole distribution of relaxation times, the variation of the dielectric loss factor ϵ'' with frequency is given by

$$\epsilon''(\omega) = \frac{(\epsilon_s - \epsilon_\infty)(\omega\tau)^n \sin(n\pi/2)}{1 + 2(\omega\tau)^n \cos(n\pi/2) + \omega t)^{2n}} \quad (16.1)$$

where $n = 1 - \alpha$, α being defined in eq. 12.2B. The conductivity $\sigma(\omega)$ is given by

$$\sigma(\omega) = \omega\epsilon_0 \epsilon'' \quad (16.2)$$

Upon combining eq. (16.1) and (16.2), an expression is obtained showing that the log-log plot of conductivity vs. frequency will have a slope tending toward the value $(2 - \alpha)$ at low frequencies, and towards the value of α at high frequencies:[71]

$$\sigma(\omega) = \frac{\epsilon_0(\epsilon_s - \epsilon_\infty)\omega^{2-\alpha}\tau^{1-\alpha}\sin[(1-\alpha)\pi/2]}{1 + 2(\omega\tau)^{1-\alpha}\cos[(1-\alpha)\pi/2] + (\omega\tau)^{2-2\alpha}} \quad (16.3)$$

The conductivity thus should vary with frequency slower than linearly at high frequencies, i.e., above the peak of the relaxation time spectrum and super-linearly below the same.

For many polymers, the dielectric absorption frequency peak ν_{max}, obtained from plots of the dielectric loss tangent $\tan\delta$ vs. temperature, obeys eq. (16.4) so that plots of $\log(\nu_{max}/T)$ vs. $1/T$ are linear:

$$\frac{\nu_{max}}{T} = \frac{k}{h}\exp-\frac{\Delta H^* - T\Delta S^*}{kT} \quad (16.4)$$

This equation has been derived by Pohl[72] from the assumption that intrachain transport dominates. Here ΔH^* and ΔS^* are the activation enthalpy and entropy, respectively, referring to the carrier transport within the polymeric chain.

Cole-Cole plots (see Section 12.4) for many polymers support a two-component mechanism of charge transport: intrachain and/or interchain transfer being rate determining. In the first case, the range of relaxation times obtained increases with temperature while for the second case considerable deviations from the circular arc shape are observed, indicating a wider spread of relaxation times.[70]

Charge transport in the model of Fig. 16.3 is closely related to the interrupted strand model of Bernasconi[57] (see Section 15.5), discussed by Alexander et al.[58] A band model then applies to intramolecular transport while interchain transfer probably is by hopping, though other modes of transfer cannot be ruled out, see Chapter 15; such a two-step process has been stipulated, e.g., by Ranicar and Fleming for PVC.[59] In polyethylene, likewise, Yoshino et al.[60] propose a fast intramolecular migration plus a slow intermolecular one. The mobility then is limited by space charge and trapping of carriers at sites in the molecular structure.[61] Further support for this model comes from the findings of Hetzler and Kay[62] on tetrafluoroethylene films who also invoke a two-mechanism process of charge transfer: Williams[30] discusses the reasons why polymeric systems, as a rule, fail to exhibit high conductivities; in a polyene chain, $-(CH:CH)_n-$, the π electrons are delocalized tending to equalize all the $C-C$ bond lengths. Such a chain, therefore, for n sufficiently large, should behave as a one-dimensional conductor. Thus, as n rises, the activation energy E should drop (cf., Section

9.6) following[30]

$$E = \frac{19(N+1)}{N^2} \tag{16.5}$$

becoming of the order of kT for $n \gtrsim 1000$, when metallic conductivity would be expected. Inspection of Tables 8.7A, B and C, shows that this does not occur, though metallic conductivity has been observed in, e.g., the polypyrroles or in polyacetylenes.[28] However, this is due to quite a different mechanism (see Chapter 21).

Williams reasons[30] that the main causes for this poor conductivity include large intermolecular barriers, bond alternation, the presence of defects which reduce the chain length and also facilitate semiconduction by a hopping motion of the defect, and chain rotation leading to an interruption of conjugation and a reduction of coplanarity of substituents.[29] The energy and structure of some n-alkanes and polyethylene has been studied in detail by Ueno et al.[252] A possible primary carrier generation process might involve chain rotation leading to severance of double bonds[30] (see Section 15.8).

It is well known that greater chain flexibility raises the conductivity of polymers; dielectric relaxation processes in polymers are dominated by segmental motion.[36] However, the introduction of groups likely to cause greater flexibility, such as methylene links, into the conjugated chain tends to destroy the intramolecular conjugation. For instance, single $-CH_2-$ groups inserted into the chain do increase the crystallinity and result in a higher conductivity because of easier intermolecular charge transfer. However, longer links such as $-CH_2CH_2-$ lower the conductivity in spite of an improvement in the crystallinity, or degree of order. The failure of eq. (16.1) shows that it is intermolecular charge transfer which governs the conductance process.

Charge transfer between chains may occur either by tunneling or, more likely at room temperature and above, by thermally activated hopping;[41] see Fig. 16.4.

Defects in the chain, or strand, such as vacancies or interstitials impede intramolecular as well as intermolecular transfer.[14] Such interruptions may exhibit a reflection coefficient near to unity and thus transform the conduction electron states from coherent Bloch wave states into Bloch standing wave states, resulting in a non-conducting ground state.

The argument is supported by the finding of two sets of traps of polystyrene;[11] only the lower energy trap appears to be related to the state of order of the solid.[11] Three localized electronic levels were found in the band gap of the polymer.[11] In linear organic compounds such as TTF/TCNQ, strong lateral elastic interactions between adjacent chains have been invoked[12] to explain the electrical behavior of the solid.

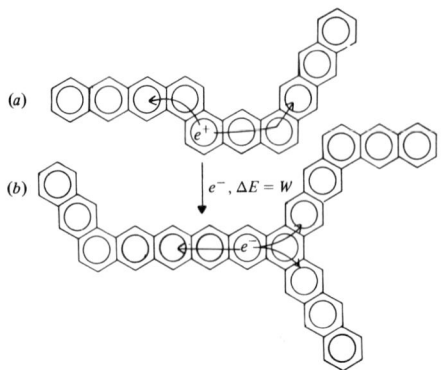

Fig. 16.4 Diagram of thermally or field-induced charge separation between polymer molecules (a) and (b) of unlike or similar size. After Pohl.[8]

A detailed study of the electrical conduction in several commercial polymers, principally in polystyrene,[13] using pulsed electron bombardment so that these penetrated fully the thickness of the polymeric film, has shown[13] that the resulting current consists of two components: a short-lived, "prompt" component obeying first-order kinetics, and a long-lived current following second-order kinetics. The decay of the latter, long-lived, current involves a bimolecular recombination lifetime. The first, prompt, component is said[13] to be associated with intramolecular carrier motion, with a mobility in excess of 0.01 $cm^2/Vsec$. The other component is said[13] to be associated with intermolecular carrier transport, with a mobility, at least in the kinked chain polymers employed, governed by the hopping probability. In polythene at least, additional carrier localization as in Fig. 16.3 appears to occur.

Support for these ideas comes from further and more recent work[17] on polyethylene, teflon, polyester, and polystyrene: the transient charging currents appear[17] to be the sum of two components with time constants differing by at least a factor of ten. This, however, is interpreted in terms of carrier injection from the contact as responsible for the shorter time constant while the longer one is thought[17] to be associated with carrier migration. These matters require further study and clarification.

The charge transport in polytetrafluoroethylene films is said[62] to be by hopping at low temperatures and/or high frequencies, while at high temperatures contributions from charge emission (probably from the contact) via the Poole-Frenkel effect become dominant. At intermediate temperatures, both mechanisms, activated by dipole orientation processes, contribute.

A study[19] of the electrical conductivity of PVC (polyvinylchloride) employing both "pure" films as well as industrial preparations containing additives has shown[19] that the current carriers in both pure and impure PVC below 300 K are probably electrons, injected from the metal contacts though thermal ionization of impurities may also contribute; the electron mobility in pure PVC is of the order of 0.01 cm^2/Vsec. In both pure and impure PVC, transport is said[19] to be governed by trapping-detrapping processes and best described in terms of a hopping model.

Polymers built up of 1,3,4-oxadiazole-2,5-diyl and vinylene, respectively, ethynylene groups prepared[17] by polycondensation of fumaramide, respectively, acetylenedicarboxamide with hydrazine sulfate in polyphosphoric acid exhibit the properties of conjugated polymers—dark color, semiconductivity, and paramagnetism. Comparison[17] of these polymers with each other and with poly-p-xylylidene suggests that the trans double bond is a better building unit for semiconductive conjugated polymers than the triple bond and that the 1,3,4 oxadiazole-2,5 diyl group is superior in that respect to the 1,4 phenylene group.[17]

Polymalonitrile above 435 K is said[33] to be an intrinsic semiconductor with an activation energy (on E/kT) of 0.86 eV. At 380 K, the conductivity, which is n-type, appears to be dominated by traps; one electron trapping level is said[33] to be 0.34 eV below the edge of the conduction band, the trap concentration being about 5×10^{16} cm^{-3}.

The trap density in polynaphthylene is reported[34] to increase upon annealing due to an increase in the size of the polymeric spherulites.

A saturation effect in radiation-induced conductivity of poly(ethyleneterephthalate) in fields of $10^6 - 10^8$ V/m has been observed by Maeda et al.[26]

The conductivity in many polymers appears to be dominated by carrier injection from the contact *via* Schottky emission processes;[30] this has been shown to hold for polysilazane films,[35] for polyquinoline-carboxylic acid[9] and several others[17] (see also Section 15.12). The important subject of polymeric liquid crystals[38] is considered to be outside the scope of this book.

Polypyrimidines are reported to have resistivities of $10^{12} - 10^{14}$ ohm cm at room temperatures,[21] while polyazopyrimidines have much lower resistivities, ranging from 10^9 to 10^{10} ohm cm at 25° C; this is paralleled by a drop in the activation energies from 0.56-2.29 eV to values between 0.76 to 1.33 eV in the polyazo polymer.[21]

TTF (Tetrathiafulvalene) treated with metals such as Ni, Mn, Co or with acetylacetonate yields dark, insoluble and infusible polymers with a phthalocyanine-like structure.[22] Their resistivity is said[22] to be about 1.6×10^4 ohm cm, or higher, at 300 K; conductivity from thermo-emf measurements appears to be p-type.

Polyimides exhibit two breaks in their log (conductance) *vs.* $1/T$ relation, one at about 200° C and one at 340° C or above.[23] However, insertion of p-phenylene groups between the pyromellimide and benzimidazole rings lowers the first mentioned discontinuity by about 20° compared to polymers containing m-phenylenediamine groups.[23] While this is ascribed to a drop in the conjugation, the matter is not at all clear.

Relationships between the monomer structure and the conductivity as well as other structural features have been studied by Graovac et al.[24] Other workers have looked for correlations between the thermal and the electrical conductivity of metal-containing polymers.[25]

Since structural transitions very often show up on a log (conductance) vs. $1/T$ diagram (see Section 9.3) as a change of slope because of the different activation energy involved, the glass transition temperature T_g of polymers* likewise can, in many cases, thus be deduced.[53] This is shown in Fig. 16.5 for an elastomeric ionene-TCNQ complex, where it can be readily confirmed from mechanical measurements.[27]

Such a correlation between T_g and the electrical properties of a polymer is to be expected because T_g depends on the coulombic interactions within the system.[63] The change, however, need not always be so drastic as to affect the log σ vs. $1/T$ slope, though sometimes the dominant mechanism of charge transfer changes at this temperature, as is said to be the case in polystyrene.[64] In this polymer, conductivity above the glass transition temperature at 95-100° C is governed by space charges and trapping/detrapping. Actual transport again is proposed to be by hopping,[64] and is associated with segmental and/or chain motion; a slow decay of the current due to the filling of traps is observed above T_g but absent below T_g. T_g may also be determined from dielectric relaxation times[65] and from glow curves (see Section 20.1, 20.2 and 10.3). The latter method has been applied e.g., to study multiple transitions in polypropylene occurring at 39 and at 68° C in undrawn and at 28 to 55° C in the drawn polymer.[66] T_g in polycarbonate has likewise been found[67] to depend on mechanical deformation and treatment. A great

---

*The glass transition is a thermodynamic process common to all amorphous solids. When an amorphous solid is heated into its glass transition region, the dynamic states of its constituent molecules, its structural and thermodynamic parameters undergo characteristic changes reflecting the onset of new configurational degrees of freedom. One always observes[221] on heating a sample through its T_g region, a change in the temperature dependencies of electron and hole drift mobilities and of photogeneration efficiency. Whereas these processes are activated below T_g, they become essentially temperature independent on heating through T_g and maintain the temperature independence over a range of temperatures above T_g. Mobility values and the absolute generation efficiency remain stable in this range even under isothermal conditions, changing only as the sample begins to crystallize. Measurements on molecularly doped amorphous polymers suggest that a diminution in transport activation energy over a range of temperatures above T_g could indeed to a feature common to many amorphous solids in which transport proceeds via hopping.

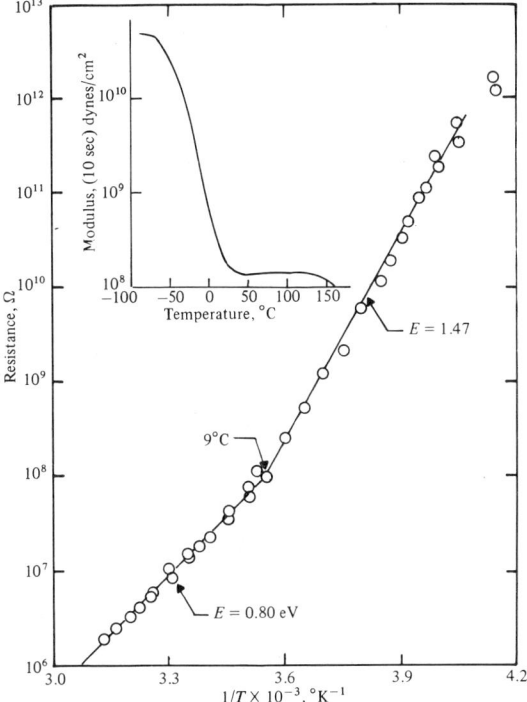

Fig. 16.5 Comparison of torsion modulus with electrical resistivity as a function of temperature for ionene elastomers. After Rembaum et al.[27]

deal of literature, is of course, available on glass transitions and related effects.[68] Several T_g values calculated from glow curves (see Section 10.3, 20.1 and 20.2) are tabulated in Table 8.20A; interfacial polarization effects associated with structural transitions are discussed in Section 20.4

Somoano et al.[54] consider a polymer which may exist in a rubbery and/or glassy state: The conductivity, σ, in such a model is given by:

$$\sigma = n_0 e^{-E_n/kT} q\mu_0 e^{-E_\mu/kT} = 1/\rho \qquad (16.6)$$

where n_0 and μ_0 are the charge density and the mobility at infinite temperatures, respectively, q is the charge of the carrier, and E_n and E_μ are the charge density and mobility activation energies, respectively. In the rubbery state (RS), the molecular chains are in motion involving continual changes of configuration. In this sense, one may picture the rubbery state as a "loose lattice" in which the lattice constraint forces, which hold the elements of each molecular chain together, are very weak, thereby enabling the molecular chains to easily alter their configurations. With the configuration constantly changing, there is a large probability that two anions may approach each other quite closely, either on the same molecular chain or on adjacent chains. The nature of the interaction between the anions is an open question, but a repulsive coulomb interaction is expected[54] to dominate because of the charges on the anions. Since the lattice constraint forces are weak in the rubbery state, the anions will be able to move away from each other to a more energetically favorable configuration by rotating about the chain, etc. The resulting large separation between anions will act as a large molecular barrier, characterized by an activation energy, $E_\mu(RS)$ for the conduction processes. Therefore, in the rubbery state, the carrier mobility will be low and characterized by a large activation energy. Also, the activated charge density is characterized by an activation energy, E_n, of similar magnitude, such that at the high temperatures accompanying the rubbery state, enough carriers are available to result in a low resistivity.

The glassy state (GS) may be considered to be a tight lattice in which the lattice constraint forces are much stronger than those in the rubbery state, and the molecular chains are frozen into a fairly rigid configuration. In this case, it is also likely that there will be many anions fixed into a configuration in which they are relatively close to one another. However, the anions cannot move away from each other since the lattice constraint forces holding the anions rigidly in place are stronger than the repulsive forces acting between anions. Consequently, there may be considerable overlap between anions which will result in a low molecular barrier of energy $E'_\mu(GS)$. Therefore, the carrier mobility in the glassy state will be larger than in the rubbery state. On the other hand, the low temperatures accompanying the glassy state will result in a low charge density, thereby yielding high resistivity.

The pertinent relations governing the conduction processes are:

$$\left.\begin{aligned}
&E_\mu(RS) > E'_\mu(GS) \\
&E_n(RS) = E_n(GS) > E'_\mu(GS) \\
&\mu(RS) < \mu(GS) \\
&E_a(RS) = E_n(RS) + E_\mu(RS) > E_a(GS) = E_n(GS) + E'_\mu(GS)
\end{aligned}\right\} \quad (16.7)$$

There is some evidence[54] that the electrical response to the onset of molecular, or of segmental, motion is slower, and delayed, relative to the results obtained from thermal methods such as differential thermal analysis because the carrier mobility is determined by the lowest potential barriers intervening. A recent study, however, claims[31] that T_g is definitely not correlated to electrical resistivity. The matter remains to be clarified.

Recent results[2] on the electrical resistance of thin metallic wires appear to have a considerable bearing on charge transfer in linear polymers: wires with diameters of the order of a few 100 Å should, theoretically,[1] exhibit a limiting resistance (not resistivity) of the order of 10,000 ohm because of the effect of impurities in the metal. The electrons then are forced to remain in localized states rather than being mobile in a conduction band and charge transport then occurs by thermally activated hopping between adjacent localized states. The resistance of the wire then increases exponentially with its length; the system is readily seen to be a model for the study of hopping transfers in more complex solids. The effect is observable only at cryoscopic temperatures because of phonon scattering of electrons from one localized state to another.

PAQR polymers have been extensively studied by Pohl and coworkers, see Table 8.7A. Some typical, additional, results are given in Tables 8.7D and 8.7E.

Doped polymers are discussed separately in Section 16.8, while polymeric charge transfer complexes are dealt with in Section 17.10. Pressure effects are discussed in Section 19.8.

Highly Conducting Polymers

Semimetallic, and one-dimensional conducting systems are treated in Chapter 21. Here, discussion will be confined to a few supplementary remarks. "Highly Conducting" will here be arbitrarily defined as exhibiting a resistivity of 10 ohm cm or less at room temperature and pressure. The inorganic polysulfurnitride,[42] which is indeed highly conducting, falls outside the scope of this book.

Some relevant, highly conducting, polymer data will be found in Table 8.7C. At the time of writing, one of the most promising systems appears to be the polypyrroles prepared by Kanazawa, Diaz and coworkers.[43] These workers have prepared shiny, blue-black films which, like acceptor-doped $(CH)_n$[44] exhibit metallic behavior even without doping with corrosive dopants such as $AgClO_4$ or $AgBF_4$ used in the case of the polyacetylenes (see below). The polypyrrole films, at room temperature, exhibit resistivities down to about 0.01 ohm cm; they are stable in air and allow heating up to 250° C without degradation. A typical composition such as $C_4N_{0.87}H_{3.5}$ $(BF)_{0.25}$ indicates that the pyrrole rings remain intact consistent with a polymer formed by linkage of the pyrrole rings *via* the α-carbons.[45] The positive charge on the pyrrole units is balanced by the negative BF_4^- in a fashion similar to that in doped polyacetylene films.[46] The conductivity is weakly temperature activated. The simplest unsaturated organic polymer is polyacetylene, $(CH)_n$, consisting of a linear carbon chain with alternating C−C bonds and an H atom attached to each C. Of the 4 valency electrons of each C atom, 3 are in σ-bonds linked to other C or H atoms, while the fourth valency electron, in a π-orbital, can delocalize along the polymer chain resulting in the appearance of a conduction band.

Shirakawa and Ikeda[47] obtained films of polyacetylene by the catalytic polymerization of acetylene; these are mechanically flexible and produced in thicknesses of 10^{-5} to 0.5 cm, and of a silvery appearance. The films consist of a felt-like random assembly of fibrils each of a few 100 Å in diameter, a structure which appears to be inherent in polyacetylene films. The unoriented film exhibits a resistivity as low as 0.001 ohm cm at room temperature, comparable to that of Hg. It may be acceptor doped at a level of about 1 to 2 mole% in which case it exhibits a semiconductor-metal transition (see Chapter 21), where its conductivity ceases to increase further upon increasing the dopant concentration. Below this critical concentration of dopant, the conductivity rises exponentially with temperature while above the concentration, the conductivity remains virtually temperature independent, exhibiting merely a very weak increase with rising temperature. If the film is mechanically stretched and heat treated,[48] the fibrils become at least partially aligned resulting in an about 3-fold reduction of the resistivity to about 3.3×10^{-4} ohm cm at room temperature, measured in the longitudinal, stretch direction, while higher by about a factor of 10 perpendicularly to it.

Details of preparation appear to play a crucial role, Goldberg et al.[52] report that the ESR signals observed in undoped $(CH)_x$ are due to a highly mobile defect arising from a neutral radical induced in the preparation of the polymer.

(Trans-) polyacetylene in its pure as well as in its, e.g., I_2-doped form shows p-type conductivity;[69] charge transfer in the undoped polymer is said[69] to be by hopping via localized states.

The highly conductive, doped polyacetylenes are discussed in Section 17.10 and more fully in Chapter 21.

Other highly conducting polymers are poly(arylacetylene)[50] and polyparaphenylene;[49] the latter is specially interesting because it appears that at higher concentrations, the dopant becomes intercalated between the $(CH)_n$ layers.

Polydiacetylenes have been extensively studied[201] because they are highly crystalline, and can even be obtained in single crystal form. They are thus, in a way, halfway between "classical" polymers and molecular crystals. Single crystals are highly anisotropic;[262] they conduct well along the polymeric chain but poorly in directions perpendicular to the chain. The electron and hole mobilities are about equal 2.8 $cm^2/Vsec$.[263] The conductivity probably involves exciton bands though this is by no means certain.[264]

Pyropolymers generally, have high conductivities (see Section 8.6). The recently developed polyimide pyrolysates[241] *e.g.*, exhibit room temperature resistivities of as low as 0.001 ohm cm.

16.5 Nomadic Polarization Effects

Nomadic polarization[73] refers to a polarization in which the motion of charges produced by the application of an external field is relatively extensive—over distances corresponding to one or more molecular lengths or several lattice sites. By way of contrast, the charge displacements in the course of simple electronic, atomic, or orientational polarization are relatively small. Nomadic polarization can be either electronic or protonic in character. Nomadic polarization which is due to the response of highly delocalized electrons moving on long molecular domains has become known as hyperelectronic polarization.[74] That for protons is referred to as hyperprotonic polarization.

Nomadic polarization has been found in materials such as selenium,[75] certain hydrated salts,[76] or in polymers.

There are two necessary conditions for the presence of appreciable nomadic polarization: first, there must be appreciable concentrations of (thermally or field-generated) roving charges. Second, there must be suitable geometrically long domains for them to rove in. Such conditions are provided by certain macromolecular or crystalline solids.

Roving charges usually arise as thermally produced (possibly field-aided) charge pairs, usually as excitons. They may be of two types: intramolecular, Frenkel excitons; and intermolecular Mott excitons. The intermolecular excitons are the more effective: the individual charges lie on separate molecules or crystallite regions which form quasi one-dimensional channels for the charges. Long range delocalization of the orbitals, such as occurs in large aromatic polymers offers a suitable domain for the required charge response to an external field. Such domains for the electronic shifts can typically extend 100 to 10,000 Å in carefully prepared molecular solids such as selenium, or in selected aromatic hydrocarbon derivative polymers. Possible formation of small polarons in this system will further enhance the polarizability since the energies of the delocalized states are then brought closer together.[73]

Numerical values of effective permittivity for a range of typical solids exhibiting nomadic polarization are tabulated in Table 10.4A. The present picture of charge motion[74] in solids exhibiting nomadic polarization is that of easy motion along the chain or channel, and a rather more difficult one across the molecular gap separating the chains. The intrachain motion may be of the wave packet drifting type; or thermally activated hopping. A model, described in detail in Ref. 77, is that intrachain transport occurs by thermally activated hopping between small polaron states, while inter-chain transfers, assumed to be much more difficult, are also of the hopping type. The existence of these highly polarizable macropolar charges is transitory by reason of their thermal birth and recombination processes.[73]

In the case of long and extensively conjugated polymers, the most effective charge carriers will be Mott excitons, *i.e.*, ion pair states in which the charges of opposite polarity will be on different molecules.[73] Frenkel excitons in which the charge pair resides on the same molecule, can be expected to contribute nothing to the dc conductivity and only little to the ac conductivity or polarizability. One may view the process in general as represented by:

$$2P \rightleftharpoons P_.^+ + P_.^- \qquad (16.8)$$

with a dissociation energy, $W = I - A$, the difference between the ionization energy and the electron affinity per local domain (molecule) in the solid.

Pohl and Pollak[73] introduce the following simplifying assumptions: (1) that the polymer molecules (*i.e.*, local domains for the roving charges) are all of the same size, and (2) that the dissociation energy, W, is the same for all the polymer molecules, and for all ion pairs formed. For this simple two energy level process, the Fermi level, E_F, must lie at a point equal to ½ the dissociation energy, *i.e.*,

$$E_F = W/2. \qquad (16.9)$$

Since the dissociation to create a Mott exciton is essentially that of the promotion of an electron to an energy W, the probability density follows Fermi statistics:

$$N(W) = \frac{n_0}{e^{-(E_F - W)kT} + 1} = \frac{n_0}{e^{W/2kT} + 1} \qquad (16.10)$$

where n_0 is the number of electronic states at the energy of excitation. Assuming that the proper wave functions are small polaron functions there are n such degenerate functions per polymer molecule, *i.e.*, one per monomer unit in a chain of n-mers, and n_0 is the number of monomers per unit volume.

Let L represent the length of the domain along which the free carriers may move freely. In a polymeric crystalline solid such as selenium, L corresponds to some average chain length uninterrupted by defects such as chain breaks or impurity atoms. In a crystalline solid such as the hydrated lithium salts in which the mobile carriers are protons, L probably represents the average length between defects in the crystal channels along which the protons move easily. In amorphous or polycrystalline organic polymers of strongly conjugated structure, L represents an average molecular length of the strongly delocalized associated π-orbitals. The problem of determining L is different for each type of solid. To date, only rather approximate means have been developed for the determination of L, and mainly for the amorphous polymers.

For a randomly oriented set of stiff, highly conjugated chains the conductivity, σ, can be expected to vary with the applied field strength, F as[74]

$$\sigma/\sigma_0 = (1/x)(e^x - 1) \qquad (16.11)$$

where σ_0 is the conductivity at zero field intensity. For small x

$$(\sigma/\sigma_0) - 1 \cong x/2 \qquad (16.12)$$

Accordingly, L is approximately given by

$$L = [(\sigma/\sigma_0) - 1] 4kT/(eF) \qquad (16.13)$$

The domain length estimates indicate a rather long region available for the travel of the nomadic charges, distances of 10^2 to 10^4 Å for typical conductive polymers.[73]

From the relation

$$P = \chi \epsilon_0 F = (\epsilon_r - 1)\epsilon_0 F \qquad (16.14)$$

where P is the magnitude of the polarization, and χ the electric susceptibility, at saturation, $F = F_{max}$,

$$P = Nel/2 \qquad (16.15)$$

hence

$$(\epsilon_r - 1) = \frac{Nel}{2\epsilon_0 F_{max}} \qquad (16.16)$$

where the concentration of polarizable centers is $N = n_0 \exp(-W/2kT)$.

Numerically,[73] for a typical conjugated polymer, taking n_0 as about $\times 10^{21}$ units/cm^3, $W/2$ as about 0.2 eV yielding for N about 8×10^{17} polarizable units ("macropoles")/cm^3, and taking L as 300 Å and F_{max} as about 200 Vcm^{-1}, the zero frequency relative permittivity ϵ_0 then becomes about 10,000. Typical experimental values for ϵ_0 are about 5000 at 300 Hz[74] though very much higher values have been encountered.[74]

Over appreciable frequency ranges, ϵ and the ac conductivity σ_{ac} obey power laws of the form

$$\epsilon \simeq BW^{-p}$$
$$\sigma_{ac} \simeq AW^{1-p} \qquad (16.17)$$

where p is a constant. Pohl and Pollack[73] have shown that p can be expressed in terms of the real and imaginary components of the complex conductivity σ^* and of the complex permittivity ϵ^* as

or
$$1 - p = \frac{2}{\pi} \arctan \frac{\sigma''}{\sigma'}$$
$$1 - p = \frac{2}{\pi} \arctan \frac{\epsilon'}{\epsilon''}$$
(16.18)

This has been experimentally confirmed as illustrated in Fig. 16.6.

Pohl and coworkers[74] find that the relative permittivity depends on the applied electric field following the empirical relation

$$\frac{\epsilon_{r0}}{\epsilon_r} - 1 = \frac{F}{F_{1/2}} \qquad (16.19)$$

where ϵ_{r0} is the value of ϵ_r, the relative dielectric constant, at zero field and $F_{1/2}$ is the value of F, the field strength when ϵ_r is equal to ½ the value of ϵ_{r0}. A typical plot of such data in this form is shown in Fig. 16.7.

Pohl and Pollak[73] conclude that the roving charge experiences a type of square-well potential within a conjugated chain molecule.

There are a number of solids, mostly organic so far, with several hydrated lithium salts as exceptions, which are protonic conductors. If the protonic carriers suffer blockage of their paths in responding to the external electric

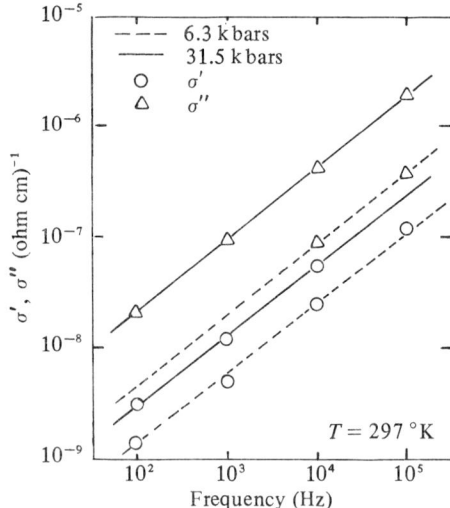

Fig. 16.6 The real σ' and imaginary σ'' parts of the complex conductivity for the acridone–PMA polymer as a function of frequency. After Pohl and Pollack.[73]

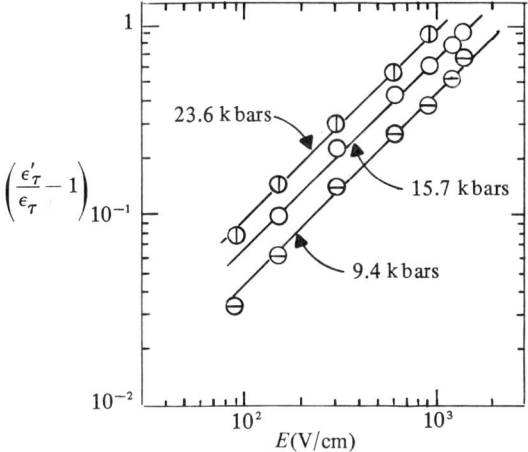

Fig. 16.7 A plot of typical experimental data for the variation of the dielectric constant with the applied electric field. The several curves shown are those observed at differing hydrostatic pressures (kbar) on the 9-thioxanthane-pyromellitic dianhydride (PAQR) polymer. After Wyhof.[77]

field, this will give rise to what is seen by the experimental observer as a polarization response, as contrasted to one which is deemed to be conduction.[73] Hyperprotonic polarization can cause some solids to display enormous dielectric constants, even in the case of single crystals.

The lithium salts listed in Table 8.22A are examples of hyperprotonic polarization. Other solids in which the phenomenon has been reported, are also listed in Table 8.22A.

It should be pointed out that appearance of extraordinarily high permittivity values, while certainly a necessary result of hyperelectronic polarization, is not, *per se,* sufficient evidence for its occurrence.

Interfacial polarizations, *i.e.,* those involving a Maxwell-Wagner effect in inhomogeneous dielectric, electrolytic capacitances at interfaces, or ferroelectric effects must always be suspected. A *prima facie* case for hyperelectronic polarization exists when the apparent high permittivity persists with different contact materials, if the system obeys Ohm's law and is independent of the sample thickness and geometry. Identification of extreme polarization as nomadic is supported in hyperelectronic cases where that polarization (1) increases strongly with pressure, (2) increases strongly with temperature, to show (3) saturation at modest field strengths, and (4) decreases with frequency.

An effect which may well be closely associated with nomadic polarization is that of shock-induced polarization:[78] An anomalous electrical phenomenon has been observed in polymers subjected to high-pressure shock-waves. This occurs in samples on which electrodes are placed and to which no electrical potential difference is applied. Then, currents are observed in external circuits connecting the electrodes when the samples are shock loaded at pressures greater than some critical stress. The form of the current pulse indicates that it is produced by a bulk polarization with a relaxation time of typically 10^{-5} to 10^{-6} sec. Such a phenomenon has not been successfully described in terms of equilibrium thermodynamic properties, and is called shock-induced polarization.

Poly(pyromellitimide), commercial Vespel SP-1, exhibits the effect[78] as well as the related one of a substantial rise in its conductivity. The process appears to be irreversible or at least to show considerable hysteresis. The effect is said[78] to be confined to polymers with complex monomers such as polymethylmethacrylate.

Solids showing nomadic polarization contain appreciable numbers of unpaired spins (giving a strong esr signal) and freed carriers of the electronic or hole type, making them interesting candidate materials for catalysts, photocopying, electrodes, dielectrics, and semiconductors. The ekaconjugated polymers are generally very stable against thermal degradation.

16.6 Dyes

Many aspects of the semiconducting activity of dyestuffs are discussed when dealing with doping in Section 16.8; there are also several reviews and books on the topic.[79]

Additional data on the solid state properties of dyes are collected in Table 8.8A; they are as incomplete as those earlier data listed in Table 8.8. Further values for the energy gap E_g for cyanine dyes will be found in Table 15.12. According to Nelson,[99] the most important and indeed characteristic property of organic dyes is the ability of an excited dye molecule to act as electron donor in a photoactivated charge transfer. In the case of sensitization, the sensitized substrate is the acceptor, while in photoconductivity, donor and acceptor are two like molecules in the bulk dye. Sensitization in order to be effective requires that the transfer of electrons from dye to substrate be exergonic; the photoconductivity of a pure dye is inherently a weak phenomenon because the transfer is endergonic.[100]

Nelson[99] points out that in most dyes, in which charge transport is predominantly by hopping, an electron, once it appears as a charge carrier, has a very high rate of changing sites, since the barrier is rather permeable, and no thermal energy is required. The rate is 10^{12}/sec, so that in the lifetime of the conductive excited state, $\sim 10^{-5}$ sec in pinacyanol, the electron may make $\sim 10^7$ jumps, carrying it ~ 1 micron in a random walk. This property is given experimental plausibility in Nelson's work on sensitization of photoconductivity in CdS by very thick films of pinacyanol under circumstances in which it seems improbable that the energy can be carried by resonance processes.[99] This rapid hopping probably accounts for the short relaxation time for photoconductivity in this dye.

Since in an applied field, the change in energy of the conduction electron per hop is very small, mobilities are likewise still being estimated to be $\sim 10^{-3}$ cm^2/Vsec for the mean mobility in pinacyanol.

In the light of later studies, Nelson's classification of cationic dyes as set out in Table 8.10, can no longer be sustained:[99] essentially all cationic dyes require a thermal energy contribution in addition to optical excitation to result in free carrier generation[99] leading to the auto-ionization model discussed in Section 14.4 and 15.1, also applicable to dyes. There is evidence[103] that the values for the energy gap, E_g, are close to those for the position of the J-band in the spectrum of a film of the cyanine dyes, but also are in agreement with the positions of the M-bands, to within an error of about 0.1 to 0.2 eV. The excited state reached by light absorption is thus close in energy to that of the separated charge state, and so corresponds to E_G.[103]

For the non-cyanine dyes E_a derived from the thermal activation of dark conductivity also agrees for the most part with E_G, which, in turn, is close to the energy of the first excited singlet level in the dye.[103] Thus, the auto-ionization model (see Section 15.1) provides the link between the state with charge separation and the excited singlet state in a film, which is closely similar in many respects to the excited singlet state of an isolated molecule.[103]

A further old observation[104] of Vartanyan probably can also be clarified by the present model. For a number of dyes Vartanyan found that the activation energy of photoconductivity lessened as the energy of the incident photons increased. A similar result was found for pentacene by Silinsh[105] and for anthracene by Chance and Braun[106] and by Lyons and Milne.[106] As the photon energy increases the distance between the charges in the ion-pair increases[106] and so the energy to dissociate the pair decreases.

Since a great deal, if not most, of current interest centers on phenomena involving excited states in dyes, Fig. 16.8 supplementing Fig. 8.28 (Part A, p. 488) shows an energy level diagram of a typical dye.[80] The singlet ground state S_0 at room temperatures is unlikely to be excited to the higher vibronic levels S_0^v; rotational levels which are very hard to resolve spectroscopically, are indicated by closely spaced short lines. Light absorption excites the dye within about 10^{-14} sec or less to the excited singlet levels $S_1, S_2 \ldots$; at the same time activating higher vibronic and rotational levels. These relax in radiationless transitions, within times of 10^{-11} sec or less, *via* phonon interactions, to S_1 or S_2. The lifetime of the S_2 level and higher singlet states is generally shorter than 10^{-11} sec; that of S_1 10^{-9} sec or less. The

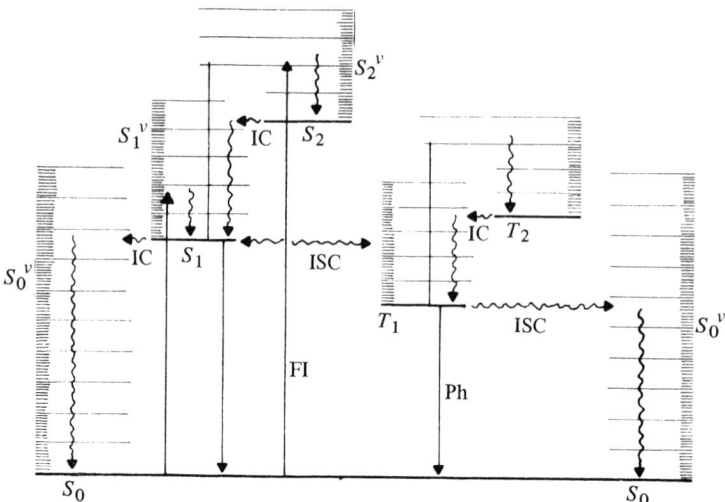

Fig. 16.8 Energy level diagram for a typical organic dye. For discussion, see text. After Leupold et al.[80]

short-lived higher singlet states S_2 ... usually decay by internal conversion, shown as IC, in radiationless transition, into an S_1^v state which in turn and in like fashion relaxes to the 1st excited singlet level $S_1^v o$.

In dyes, two types of singlet excitation may be distinguished: a very intense $\pi \to \pi^*$ which governs the color of the dye, and in those dyes which contain n orbital electrons, (e.g., in an azo group) a much less intense $n \to n^*$ transition. These latter involve much longer lifetimes than $\pi \to \pi^*$, thus providing competing processes such as those leading to triplet states. The most important of these competitors involved in $S_1^v \to S_0$ transitions in dyes are:

(1) fluorescence emission
(2) further "IC" transition
(3) normally forbidden transition to a triplet state, via a spin reversal, i.e., by intersystem crossing, "ISC."

The long-lived triplet state reverts to the ground state via

(a) exciton transfer
(b) photochemical reactions
(c) phosphorescence emission
(d) further "ISC" to S_0^v relaxing into S_0
(e) triplet-triplet annihilation sometimes with delayed fluorescence.

Higher $S_1 \to S_2$ and $T_1 \to T_2$ transitions may also occur upon absorption of further quanta of radiation.

Organic dyes are of considerable interest in the design of lasers.[81]

Of special interest is the behavior of adsorbed dye monolayers;[82] the excitation energy of a dye molecule has been shown[83] to be specifically transferable from a singlet or triplet level to an appropriate acceptor—say a thin gold layer—at an adjustable distance of the order of 100 Å. An adsorbed dye molecule, upon photoexcitation, may inject an electron into the substrate;[102] this effect is of considerable importance in catalysis (see Section 12.11).

Anomalous fluorescence, i.e., at a shorter wavelength than the exciting radiation, has been observed[84] in dyes because in the excited state, conjugation over the entire extent of the molecule is lost and the effect reflects transitions between two levels of this now decoupled system.[84] Metal-multilayer dye-metal assemblies are discussed by Polymeropoulos et al.[85]

Mixed monolayers having one of their components irreversibly adsorbed are produced[95] through chemical bonding of an active silane compound-like n-octadecyl-trichlorosilane (OTS) to the surface of the substrate. Such chemisorbed monolayers are of remarkable stability when exposed to different solvents, only the reversibly adsorbed component being removed by this treatment. Molecular holes can be produced in OTS + dye monolayers by removal of the reversibly adsorbed dye. The skeleton monolayers obtained by this method may find interesting applications as a technique for handling molecules and constructing organized structures.[95]

Experiments with mixed OTS + dye monolayers have shown[96] that, when the dye is removed from the mixed monolayer, the molecule holes thus produced in the monolayer cause a lowering of the OTS barrier height from 3.1 to 2.8 eV. This means that a new channel is opened through which electrons may tunnel, thus making the effective barrier height of monolayers with holes somewhat lower than for pure OTS monolayers. Since some molecular defects may exist in fatty acid monolayer films we may assume that the barrier height measured for fatty acid monolayers may be somewhat lower than the true value characteristic for organic monolayer.[96]

An excited fluorescent dye molecule in a dielectric but near a metal surface can decay into a surface plasmon on the metal-dielectric interface.[97] This decay mechanism is nonradiative since the radiation is trapped at the interface. If the metal film is thin enough and the refractive index of the dielectric on the opposite side of the film high enough, then the surface plasmons can radiate into the high index dielectric. This effect has been observed by Weber and Eagen.[97]

The ionization energies I_{ad} of several dyes adsorbed on a series of substrates have been measured.[87] A simple relationship appears to hold between

$I_{\alpha d}$ and the permittivity of the substrate; the effect is essentially an electrostatic one, but related to the dye structure: the governing factor is said[87] to be the distance of the conjugated chain of the dye molecule from the substrate. The magnitude of the effect, however, is small, amounting to about 0.2 eV for all cases studied.[87]

The monolayer assembly technique[95,96] discussed has been applied to studies of electro-chromism, viz., changes in the absorbance of a dye caused by the application of an externally applied electric field. The technique allows the controlled introduction of fields up to a few M V/cm. By this method thin dye capacitors are prepared by transferring monomolecular layers of dyes and fatty acids onto a glass plate covered with a semitransparent aluminum electrode, followed by the evaporation of a second aluminum film. In such multilayer systems the orientation of the dye molecules remains unchanged in the electric field, and an electric field induced linear and quadratic change in the light transmittance of the dye-capacitors is observed (electrochromic spectra).

Heesemann[98] has thus obtained highly dichroic electrochromic spectra of 4-nitro-4'amino-azobenzene and stilbene derivatives. These chemically modified dyes form monolayers on a water surface which can be transferred to a glass support. The obtained polarized absorbance spectra are assigned to H-aggregates in which the long axis of the chromophores is oriented perpendicularly to the surface plane. From the electrochromic spectra the differences of the dipole moments ($\Delta\mu$) and polarizabilities ($\Delta\alpha$) between the ground and excited states were obtained:

4-nitro-4'-amino-azobenzene: $\Delta\mu \sim 2.0$ D; $\Delta\alpha \sim 220$ Å

4-nitro-4'-amino-stilbene: $\Delta\mu \sim 2.0$ D; $\Delta\alpha \sim 210$ Å

Combination of two different dyes in controlled molecular alignment has been studied by Sugi et al.[251] A cyanine dye is arranged perpendicularly oriented to an azo dye; electrons are then transferred from the exciton cyanine dye to the azo compound in a fast reaction. The former then reverts to its ground state by accepting an electron from the contact, while the electron hops from the azo dye to the anode in a thermally activated process, the activation energy being reported[251] as 0.25 eV. A greatly improved conversion efficiency is claimed relative to the use of one single dye.

Their non-linear optical behavior has been used to devise a tunable laser based upon a single dyestuff.[88]

Merocyanine dyes,[86] of interest in electrophotography, as well as trimethinecyanines[89] have been studied in detail; in the former, the photocurrent decreases as the chain length is increased.[86] The merocyanines may be characterized[86] by this quantity which affects not only their electronic structure but also the ionization potential and electron affinity.[86]

Some dyes exhibit the phenomenon of metachromasia: in aqueous solution especially of cationic dyes, a characteristic color change is observed[43] upon changes in concentration and/or temperature; the effect also occurs if the dye molecules are bound to polyelectrolyte groups of anionic character.[93] It has been reported e.g., in acridine orange, acridine blue, proflavine, etc. Crystal violet, e.g., exhibits a well-defined metachromasia band at 506 nm in a solution buffered to pH 6.5.[93] There is a definite break in the conductance titration at a 1:1 mole ratio, if crystal violet is titrated against Na-polyphosphate. Spectroscopically, a new absorption band or a prominent shoulder appears.[93]

The effect is related to the solvatochromism exhibited by e.g., merocyanine dyes,[94] viz., color changes determined by the identity of the solvent. It has been proposed that the electronic ground state and the excited state of the dye molecules are described by two resonant structures, one polar and one non-polar. If parts of the molecule exhibit donor character D and acceptor character A, the process may be written[94]

$$D-(CH=CH)_n-CH=A \rightleftharpoons D^+ =(CH-CH)_n=CH-A^- \qquad (16.19)$$

Aniline black is an interesting dye: above 295.5 K its resistivity is said[90] to rise rapidly from the remarkably low value of about 500 ohm cm; it is also strongly field dependent, dropping above 500 V/cm by several orders of magnitude.[90] These effects await confirmation and further study.

Studies by Nelson on several dyes such as crystal violet, malachite green and brilliant green indicate that conduction is a thermally activated, trap-untrap process.[91] For planar dyes, it is reported[91] that the conductivity is dominated by the probability that a tunnelling electron gains sufficient energy by thermal activation to allow separation of the electron-hole pair.

The photochromic spiropyranes, of great interest in electrophotography, are reported[92] to assemble themselves along the lines of force of an applied electric field in form of quasi-crystalline globules of 0.1 to 0.4 μm diameter, joined together like straight strings of beads, each string being uniform within itself.[92] The implications of this effect remain to be investigated.

16.7 Doping and Sensitization

Due to the relevance of this topic to catalysis and, *a fortiori*, to electrophotography, there exists now an extensive literature on these matters, much of it in form of patents. The very bulk of it puts it beyond the scope of this book, apart from a few typical cases and, again, of direct bearing on semiconducting effects.

Doping,[107] *i.e.*, the controlled introduction of small concentrations of impurities into a matrix, may affect phase transitions in organic solids, and also drastically raise their conductivity. The topic has been reviewed by several authors.[107]

Doping of polymers is discussed in Section 16.8 while theoretical aspects of charge transfer in doped systems are treated in Chapter 15, especially in Section 15.9, structural aspects are discussed in Section 19.6.

Consider a dopant D in a host matrix A: the binding energy of an electron trapped at an acceptor site may then be approximated by[158]

$$E_t \cong A_c^D - A_c^A \qquad (16.20)$$

with a corresponding equation for holes, replacing the electron affinities A_c by the ionization energies I_c. Experimentally,[158] the method of determining trap depths for electrons $\epsilon_{(n)}$ and for holes $\epsilon_{(p)}$ is based on the temperature dependence of the effective trap-controlled mobilities (μ), where N_t is the density of trap states (generally approximated by the concentration of dopant molecules)

$$\frac{\mu_0(n,p)}{\mu(n,p)} - 1 = \frac{N_t}{N_b} \exp\left(\frac{\epsilon_{(n,p)}}{kT}\right) \qquad (16.21)$$

and N_b is the density of states in the respective band.

Experimental data[159] for tetracene doped anthracene single crystals are displayed in Fig. 16.9 and trapping levels for several dopants,[160] also in

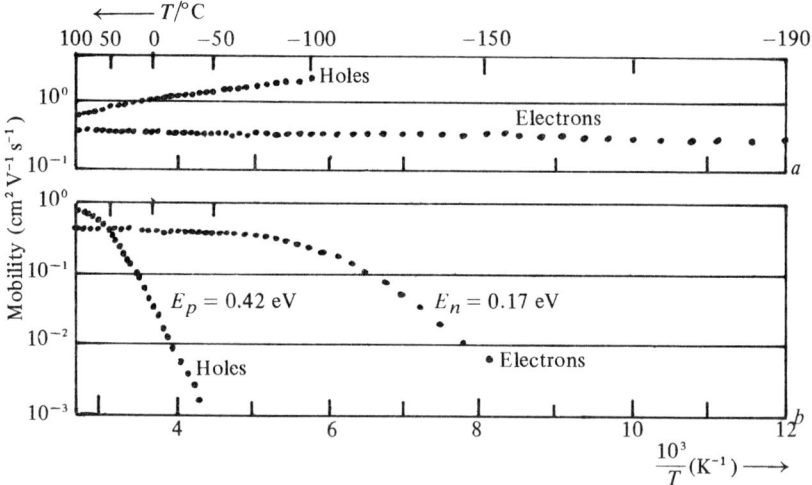

Fig. 16.9 Temperature dependences of charge-carrier mobilities in (a) pure and (b) a tetracene doped (ca. 10^{-7} mol per mol) anthracene crystal in the C' direction. After Probst and Karl.[159]

an anthracene matrix, are shown in Fig. 16.10. The agreement is seen to be reasonably good, considering the uncertainties involved in the value for the electron affinities and also, to a lesser extent, in the ionization energies. It is also seen that *e.g.*, tetracene may act as an electron trap as well as a hole trap, the trap depth being 0.17 and 0.42 eV, respectively.

Anthracene doped with tetracene and phenazine has been studied[164] with regard to its trap structure (*cf.*, Section 15.10); this apparently is considerably affected by the nature of the dopant.

Basically, doping may produce either an increase in the carrier concentration or in the carrier mobility, or even both. However, flash photoconductivity experiments[161] indicate that it is the former effect which predominates in organic solids; the mobility is reported[161] to be independent of dopant levels over a 100-fold concentration range. This has been confirmed by measurements on metal-free phthalocyanine doped with acceptors such as TCNE, TCNQ or chloranil.[162]

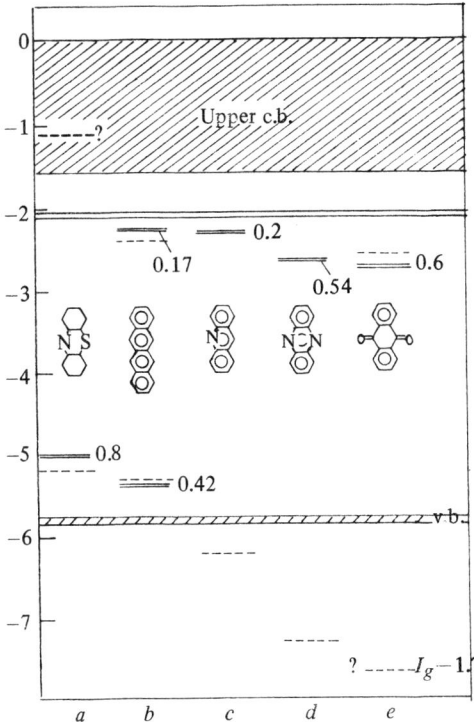

Fig. 16.10 A comparison of the measured (———) and calculated (- - - - -) trapping levels in anthracene formed by impurities. After Karl.[160]

Dopants diffuse readily in a host matrix because the dopant finds itself in a constricted, *i.e.*, higher energy, state.[163]

When excited states are created in host molecules of a doped solid, energy may be transferred to the activator by[116]

(1) one-step quantum mechanical resonance, *i.e.*, *via* a virtual photon OR
(2) *via* mobile excited states, *i.e.*, *via* excitons;

whether the transfer involves activated hopping or motion in an energy band, depends on the exciton-phonon interaction. At room temperature, it is likely that hopping would prevail; on a random walk model[107] (see Section 15.5) greater than nearest neighbor steps would be involved.

Acridine-doped single crystals of anthracene have been studied by several authors;[117] anthracene doped films by tetracene are reported on by Sakurai.[118]

The fluorescence spectra of ultra-pure anthracene have been studied by Lyons and Warren[165] who report that impurity bands, apparently arising from locally disorientated host molecules, appear on doping with most of the following, the effects increasing in the order given: carbazole, biphenyl, *p*-terphenyl, anthraquinone, naphthalene, naphthacene, phenanthrene, pyrene, chrysene. The host vibronic bands were virtually unaffected by dopants, except for small line shifts in the presence of 2-methylanthracene.

From the quenching of the fluorescence of anthracene, produced by interaction with a dye solution, Donati and Williams[166] found that the diffusion length of singlet excitons depends greatly on the quality of the crystal surface; in a direction perpendicular to the *ab* (001) cleavage plane, this quantity in acridine-doped single crystal anthracene is reported as 322 ± 80 Å comparing to 598 ± 51 Å in undoped crystals.

Drastic structural effects of doping (see also Section 19.6) are observed in TTF-TCNQ complexes which, undoped, exhibit 3 phase transitions, *viz.*, at 52.6, 49 and 38 K.[108] Doping, whether with a donor or an acceptor, wipes out the two lower transitions completely.[109] Here, we shall be more concerned with the effects of doping on electrical conductivity: Consider doping of α–TTF single crystals with bromine. The TTF molecules form stacks[110] with a stacking distance of 4.023 Å corresponding to a bandwidth of 0.3 eV. Doping can introduce holes into this band. Fig. 16.11*a* shows the effect of bromination on the conductivity of such a TTF crystal.[169] The relative conductivity is shown[109] as a function of the exposure time to bromine at constant vapor pressure. The dramatic increase of the conductivity up to values of 10 (ohm cm)$^{-1}$ is accompanied by a color change of the crystal from yellow to black, originating in the absorption spectrum of TTF-ions which are formed as a consequence of doping. Varying the vapor pressure and/or exposure time, allows one to vary the concentration of charge carriers continuously. This, in a way, is an extreme case involving the formation of

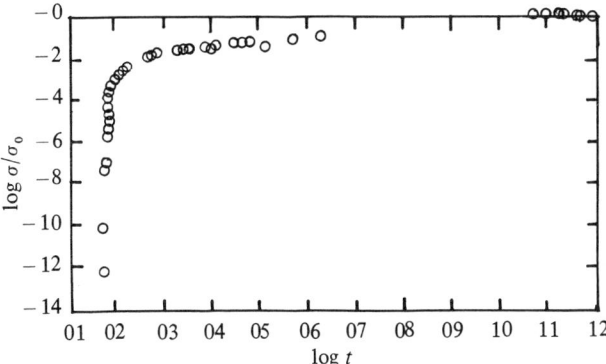

Fig. 16.11 The relative conductivity of TTF crystal as a function of exposure time to bromine gas. After Tomkiewicz et al.[109]

a definite charge transfer complex: TTF is an excellent donor and halogens are good acceptors.

The sign of the majority carrier can be altered by suitable impurity doping: thus, p-type tetrathiotetracene becomes n-type upon doping with ortho-chloranil,[121] probably by the formation of localized charge transfer complexes. Likewise, polyacetylenes may become p-type if doped with e.g., AsF_5, or n-type e.g., with Na; a heterojunction may be produced by combining them.[122]

Spectral Sensitization[123]

The mechanisms for the sensitization effect of dyes on semiconductor surfaces may be divided[130] into energy transfer models and electron (charge) transfer models. In the former the light excites the dye and the excitation energy is by resonance effects transferred to the semiconductor, where it excites an electron that ends up in the conduction band.[130] In electron-transfer models, a photo-excited electron is transferred from the dye into the conduction band of the semiconductor.[125] Such transfer is proposed to be promoted by p-n junctions,[126] Schottky barriers[127] or different kinds of surface traps.[128] Kokado et al.[129] suggest that in sensitized photoinjection, the electrons derive from surface states, dependent on the population of these levels. They introduce a critical level for sensitization in order to account for time and temperature effects.

Nelson[146] has shown that every sensitizing dye molecule acts as an independent center at which carriers may be injected into the substrates; sensitization efficiency is said to drop with decreasing temperature.[146]

The addition of a photo-sensitizer, e.g., biacetyl to azopropane[134] intro-

Table 16.1 Efficient Sensitizers for TiO$_2$. After Clark and Sutin.[133]

(a) in the blue:

thioflavin chloride; (3-ethylbenzothiazol) (1'-ethyl quinolyl) 2,4'-monomethine cyanine iodide; (3-ethyl 6-nitro benzothiazol) (1'-ethyl quinolyl) 2,4'-monomethine cyanine iodide; 3,3'-diethyl 2,2'-thiacyanine iodide.

(b) in the green:

1,1'-diethyl 2,2'-cyanine bromide; acrydin orange; 1,1'-dimethyl 2,2'-cyanine chloride; 3,3'-diethyl 2,2'-thiacarbocyanine iodide.

(c) in the red:

1,1'-diethyl 4,4'-carbocyanine chloride (cryptocyanine); 1,1'-diethyl 4,4'-dicarbocyanine iodide; 1,1'-diethyl 4,4'-tricarbocyanine iodide; (3-ethyl 6-nitro benzothiazol) (1'-ethyl quinolyl) 2,4'-dicarbocyanine chloride; 3,3'-diethyl 2,2'-thiadicarbocyanine iodide.

duces an excited state below that accessible by direct UV radiation. The efficiency of the energy transfer from ^{3}biacetyl* to azoalkanes is reported[134] to be very high.

Methylene blue, *e.g.*, reacts with O$_2$ (below pH 7) to form the semiquinone semi-methylene blue.[135]

A converse, recombination, reaction is involved in *e.g.*, electrogenerated chemiluminescence[136] in which electrogenerated free radical ions recombine *via* the formation of a very short lived CTC, as is the case *e.g.*, with ferrocene[137] or fumaronitrile.[138]

Many photochemical reactions are not initiated by direct excitation of the reacting molecule but are the result of a transfer of energy from a molecule which absorbs the light to another one which becomes excited and hence chemically reactive. Such sensitized reactions have found numerous practical applications in preparative photochemistry. On the other hand, transfer of energy results in an inhibition of chemical reactions originating from the initially excited molecule. This latter aspect of the problem is specially relevant to the stabilization of polymers against photodegradation.

As typical examples, sensitization of polyacenaphthalene and of polyvinylcarbazole will be briefly discussed, following Hayashi et al.[130] Table 9.2A displays their quantitative results for the former and Table 9.2B for the latter polymer as function of a large range of dopants. From Table 9.2A crystal violet, methyl violet, malachite green of the triphenylmethane derivatives, rhodamine 6G and rhodamine B of the xanthene derivatives, as well as the thiazine derivative methylene blue, all basic, or cationic, dyes exhibiting a structure involving $=N^+<$, are efficient sensitizers for polyacenaphthalene.

The anionic, p-type, acidic dyes such as the triphenylmethanes acid violet 6B or naphthalene green or the xanthene relations acid rhodamine G, eosine A, eosine S, fluorescein, rose bengal or phloxene are all poor sensitizers for either of the two polymers.

Effective sensitizers for PVK are cationic, basic, n-type dyes. Oil soluble non-ionic dyes are ineffective.

Photosynthetic sensitizers such as chlorophyll, pheophytin, etc., all sensitize a one-electron transfer[167] producing a ternary complex consisting of a (ground state) porphyrin analogue, alcohol cation radical and a semiquinone radical as the primary photoproduct at low temperatures. The vitally important topic of photosynthesis, for which an enormous body of literature is available, falls beyond the scope of this book.

Likewise, an extensive literature exists on the sensitization of inorganic solids, especially ZnO and TiO_2, with organic dyes; here only a very few examples can be cited: thus, ZnO sensitized with rose bengal plus rhodamine B is said[131] to allow the design of photovoltaic cells with a conversion efficiency of up to 1.5 %. ZnO sensitization with eosin is reported to be an activated process involving two kinds of surface traps.[132]

Sensitization of the n-type semiconductors TiO_2 by complexes such as polypyridine:ruthenium has been reported.[133] In view of the practical importance of sensitized TiO_2 for a variety of fields including electrophotography and solar energy conversion devices, a list of cyanine dyes capable of sensitizing TiO_2 is given in Table 16.1.

Photoconductivity in monolayer assemblies with functional units of sensitizing and conducting molecular components has recently attracted a great deal of interest.[168,171] The photoconductivity appears to be[169] in a transition region between coherent and hopping transfer (see Section 15.3) where the current tends to rise with increasing temperature because of the increase in phonons in the vibrational modes but tends to drop because of the simultaneous decrease in the electron-phonon interaction.[169] The dye adsorbed onto the metallic substrate is said[169] to result in a charge transfer interaction in the excited state, upon absorption of a photon, resulting in an increase in the photocurrent by several multiples.[170]

A monolayer of arachidic acid and dye I in a molar mixing ratio of 10:1 is sandwiched between monolayers of arachidic acid and metal electrodes as shown[170] in Fig. 16.12. A voltage is applied and the system is illuminated with light of wavelength 550 nm which is absorbed by dye I. The charge transfer process is illustrated in Fig. 16.13.

REVIEW OF PUBLISHED DATA

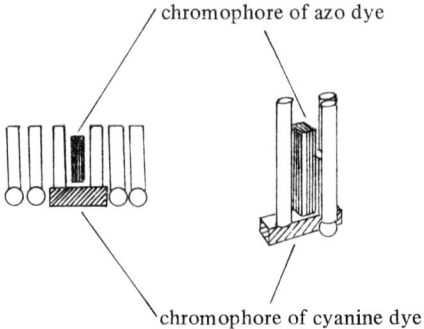

Fig. 16.12 Coupling of *I* (cyanide dye) with *II* (azo dye) in a mixed monolayer. The chromophores are represented by the shaded areas, the hydrocarbon chains by cylinders, and the carboxyl groups by spheres. After Polymeropoulos et al.[170]

The cyanine dye is excited by light and the excited electrons move through the π-electron system which acts as the conducting element.[170] In the low temperature mode the logarithm of the photocurrent decreases linearly with $(1/\text{temperature})^{1/2}$, while in the high temperature mode it decreases linearly with (1/temperature); in both modes the photocurrent is proportional to the light intensity and its logarithm increases linearly with the bias voltage. The excited electron is transferred from the cyanine dye to the π-electron system by tunnelling or by thermal activation over a barrier of 0.25 eV;

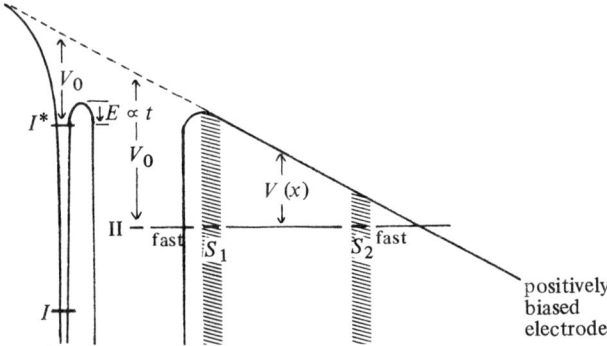

Fig. 16.13 Monolayer assembly with the cyanine dye and the azo dye as functional unit. Electron conduction model based on the existence of two-dimensionally arranged traps at the interface between neighboring monolayers. The electron in the excited state I^* of the cyanine dye jumps over the barrier or tunnels through the barrier to the azo dye, from where it proceeds to an interface state S_1, and then tunnels to some state S_2 at the next interface. Subsequently it reaches the positively biased electrode. After Möbius.[171]

from there it tunnels through the next fatty acid layer to an interface state, and then hops to the positively biased electrode.[170]

One thus has, effectively, an electron source spatially separated from an electron sink:[170] in other words, a light-driven molecular electron pump. Thus, sensitized photocurrents in monolayer assemblies appear to offer considerable scope also from the point of view of devices; for details the literature should be consulted.[171]

The term "supersensitization"[139] has been coined to express the improved injection efficiency of electrons from a solid into an electrode *via* a light-activated charge transfer complex between the excited dye and *e.g.,* halide ions[139] (see also Section 17.5). Several patents referring to electrophotographic devices and methods employ this and related systems.[172] Russian work on photochromic polymers has been reviewed by Ermakova et al.[180]

In an ingenious color electrophotographic process[140] selective color formation takes place on sensitized grains of TiO_2 by means of the reduction of a metal salt. Absorption of photons by the TiO_2 produces electron-hole pairs which migrate to the grain surface. The electrons are trapped by surface defects, thus forming reducing centers while the holes are neutralized by OH^- of a suitable adsorbate. An ionized metal salt and a complexing agent are also adsorbed on the titanium surface. The trapped electron then migrates to the metal ion M^+ reducing it to neutral metal while an OH^- migrates into the vicinity so as to maintain the charge distribution. The metal atom and a molecule of the complexing agent remain localized near a grain surface defect. Eventually the metal ion, produced by heating of the system, reacts with the complexing agent to form a colored metal complex; the process can be sensitized by means of adsorbed pigments such merocyanines or spiropyrene polymers to allow operation in the visible spectral region.

An electrophotographic method said to exhibit good panchromatic sensitivity[150] uses a metal-(*e.g.,* Cu-)phthalocyanine/ZnO system. An electrophotographic film claimed to have high sensitivity due to doping with a trinitrofluorenone:polymer matrix charge transfer complex has been developed;[151] aromatic di- or poly-nitro compounds may also serve as acceptors.[151] An electrophotographic device using a charge transfer complex between trinitrofluorenone and polyvinylcarbazole, Se-doped, has also been developed.[152] An interesting electrophotographic system based upon the development of photo-crosslinkable polymers, is said to have very good sensitivity.[153]

IBM Corporation is now producing an electrophotographic copying machine which[144] uses polyvinylcarbazole (PVK) sensitized with 2,4,7-trinitro-9-fluorenone (TNF) as the photoconductor and Matsushita Electric Corporation of Japan is producing an automatic 35 mm slide processing camera which uses sensitized PVK film.[145]

A unique and interesting method to generate an electrophotographic image, due to Robillard,[147] does so in terms of local variations of refractive index which are then explored holographically.

Catalysis[148]

Again, a very considerable body of literature is available[148] on this topic so that discussion here has to be confined to only a very few selected examples (see also discussion in Section 12.11).

Since heterogeneous catalysis is essentially a surface effect, a method has been devised to measure the dopant concentration in the surface by means of electro-reflectance,[149] using the excess carrier density generated by the dopant in its interaction with the electric field vector of an incident light wave. An alternating field is applied to the sample so that the reflected light beam is modulated at its frequency, thus greatly enhancing the signal to noise ratio in its detector.[149]

Photoexcitation of adsorbed dyes is associated[154] with catalytic activity: the excited dye molecule injects an electron into the conduction band of the substrate. A qualitative interpretation[155] of sensitized radiolytic catalysis processes, involving percolation of excitons, is attempted by Bednar.

Ortho-para hydrogen conversion catalyzed by alkali metal-doped charge transfer complexes, ion radical salts and polymers has been studied by Inokuchi and his associates.[156,157] Organic (usually dye-catalyzed) fuel cell electrodes, especially anodes, have been devised.[173]

16.8 Doped Polymers

The sensitization of polymers by dyes is discussed in Section 16.7; data are given in Tables 9.2A and 9.2B. Theoretical aspects of charge transfer in doped polymers are discussed in Section 15.10 and 15.5. Polymeric charge transfer complexes are dealt with in Section 17.10.

The photoelectric properties of the solid charge transfer complex formed between PVK and dinitrobenzene (DNB) doped with Brønsted acids, *i.e.*, with electron acceptor dopants, have been studied by Pfister et al.[174] Doping sensitizes the photoconductivity of the complex, reducing carrier recombination by several orders of magnitude. The photoconductivity is p-type:

(a) The incident light F is absorbed in the charge transfer band and excites the complex (PVK as donor D, DNB as acceptor A).

$$(DA) \xrightarrow{F} (DA)^* \qquad (16.22)$$

(b) In the absence of the Brønsted acid, under steady illumination and zero bias conditions, the excited state rapidly establishes thermal equilibrium with the ground state (DA) and the ionized donor-acceptor pair D^{+o}, A^{-o}. If the light is turned off, (DA)* decays to the ground state within the order of a fraction of a second (lifetime of the excited state). However, if the acid TCAA is present, the excited state (DA)* can react with the proton furnished by the acid to form the relatively stable radical species HA^o, viz.,

$$(DA)^* + H^+ \xrightarrow{K_1} HA^o + D^{+o} \qquad (16.23)$$

It is not necessary[174] to identify the excited state (DA)* as singlet state $(DA)^1$ or triplet state $(DA)^3$. Because the lifetimes of these states are orders of magnitude shorter [nanoseconds for $(DA)^1$ and microseconds for $(DA)^3$] than the times involved in experiments (minutes) it is reasonable to assume[174] that the concentrations $[(DA)^1]$ and $[(DA)^3]$ remain at their thermal equilibrium values, and that reaction (16.23) does not deplete the concentration of (DA)*.

(c) The radical species HA^o in turn decays according to

$$HA^o + D^+ \xrightarrow{K_2} (DA) + H^+ \qquad (16.24)$$

Because of the reaction (16.23), irradiation with light results in a separated electron-hole pair even under zero bias conditions. The electron is localized at the protonated acceptor site HA^o and the hole is localized at the donor site D^{+o}. The buildup of the photocurrent under steady illumination with light absorbed in the charge transfer band and its subsequent decay in the dark, exhibit the same time and light intensity dependence as the ESR signal which can be photogenerated in these compounds.

Relating[174] the photocurrent I to the contribution of the electrons and the holes to the photocurrent with the concentration of the species HA^o and D^{+o}, respectively, viz.,

$$I = eE(\mu_p \theta_p [D^{+o}] + \mu_n \theta_n [HA^o]) \qquad (16.25)$$

μ_p and μ_n are the hole and electron mobility, respectively, e the unit electric charge, and E the applied field. The ratio of the free to trapped charge is expressed by the factors θ_n (electrons) and θ_p (holes). Under light illumination the three species D^{-o}, A^{-o}, and HA^o coexist. After the illumination ceases, the concentration of the ionized acceptor species A^{-o} rapidly decreases so that the contribution of A^{-o} can be neglected in the decay of the photo-induced current. On the time scales of the experiments $[A^{-o}] \ll [HA^o]$ during illumination.

In many molecularly doped polymers, charge transport is governed[175] by trap-controlled hopping (see Section 15.10). In such systems, a polymeric matrix is doped with dopant molecules known to activate transport. The trap-controlled hopping process can be verified by proper choice of dopant molecules—for hole transport the difference of ionization potential can be identified with the trap depth—and by proper variation of the concentration of the hopping and trapping sites.

In the typical polymer PVK (polyvinylcarbazole) suitable dopants, or sensitizers, are *e.g.*, pyrilium, triarylmethane or cyanide dyes.[143] Transport in fluorenone doped PVK is treated in Section 15.5.

Energy transfer from an excited triplet state is spin-permitted by exchange interactions only, which requires relatively close contact between donor and acceptor molecules (see Section 15.5 and 15.6).

David et al.[176] have carried out a quantitative study of triplet energy transfer in polyvinylbenzophenone films containing naphthalene. At 77K, the emission spectrum of those films irradiated at 365 nm consists of the superimposed phosphorescence of both pure components although naphthalene is not directly excited at the wavelength. Short lived emission of the polymer ($\tau = 1.2 \; 10^{-2}$ sec) can be eliminated by the use of a phosphoroscope and pure phosphorescence of naphthalene ($\tau = 2.2$ sec) is then recorded. The intensity of this latter emission increases with naphthalene concentration in the polymer film giving evidence for an increase of the energy transfer efficiency.[177]

A critical transfer distance of 36 Å is found, which is considerably larger than that expected as a consequence of exchange interaction. This demonstrates that migration of the triplet excitation occurs prior to energy transfer. Similar measurements have been performed by the same authors[177] with polyphenylvinylketone and polymethylvinylketone films containing naphthalene. The experimental data obtained are summarized in Table 16.2.

These results show that migration of energy prior to the transfer occurs in polyvinylbenzophenone and polyphenylvinylketone, but not in polymethylvinylketone since the critical transfer distance determined experimentally in the latter case has the expected values for transfer by exchange interaction. It may be suggested[177] that migration proceeds through the aromatic chromophores in such a way that the energy becomes delocalized in regularly ordered regions of the polymer. Indeed, the critical transfer distance in

Table 16.2 Triplet-Triplet Energy Transfer to Naphthalene. After Geusekens and David[177]

Polymer (film at 77 K)	*R_0 experimental (Å)
Polyvinylbenzophenone	36
Polyphenylvinylketone	26
Polymethylvinylketone	11

polyvinylbenzophenone is found to be larger than in a glassy solution of benzophenone and naphthalene.[177] but smaller than in benzophenone crystals doped with naphthalene.[177] The behavior of the doped polymer is thus intermediate between that of an amorphous medium and that of an ordered lattice.

Excimer fluorescence in doped polymer has been studied by Frank and Gashgari,[178] using poly (2-vinylnaphthalene) (P2VN) dispersed as a low concentration guest (0.2 % by weight) in a series of fourteen polymer host matrices, twelve of which were poly(alkylmethacrylate) derivatives. The solubility parameters for these host matrices bracketed the value for P2VN, allowing the tendency toward molecular compatibility to be varied. The observed excimer fluorescence was found to be a minimum when the guest and host polymers had the same solubility parameter.

While pure polyvinylacetate exhibits a log $I/V^{1/2}$ relation between current I and voltage V, indicative of a Schottky or Poole-Frenkel effect governed conductivity (see Section 15.10 and 15.12), the eosin-doped polymer exhibits a much faster rise in the current,[179] suggesting that other transfer modes also contribute.

Polyacetylenes $(CH)_x$ may change their conductivity by a factor of 10^{12} upon chemical doping with either a donor or an acceptor impurity;[111] the former produces n-type and the latter p-type material. Similarly, undoped polyacetylene has a room temperature resistivity of about 10^7 ohm cm which, upon doping with iodine, drops to about 1 ohm cm in the bulk and about 25 ohm cm in a deposited film;[112] again, the iodine is said[112] to be present as complex involving I_3^-, I_5^- and I_2^- ions. Iodine-doped polyvinylacetate films have been studied by Mehendru.[113]

Polyesters doped with trinitrofluorenone have been investigated by Gill.[114] Polycarbonate doped with N-isopropyl carbazole was studied in detail by Mort et al.[115]

Polymers doped with the free radical salt methylacridinium:TCNQ are reported[119] to be remarkably good dielectrics: their low frequency permittivity is said to range from 40 to 106 while their resistivity is about 10^{10} ohm cm. The system is said to consist of finely dispersed crystals of the salt, separated from each other by thin polymeric layers.[119]

The dielectric strength of polymeric binders is said[142] to increase considerably by doping the binder with dyes containing 2 quinoid rings linked *via* 1 or 2 ethene-di-ylidene groups; this is probably a case of carrier scavenging.

16.9 Liquids and Glasses

Occupying a position between liquid metals, molten salts and molecular liquids in their properties, liquid semiconductors offer a specially interesting research area, also with a view to possible technological use. They relate, *inter alia*, to non-crystalline and amorphous solids, condensed matter, and polymers. The topic is physically unique because the liquid state could be defined as that state in which the mutual interactions of molecular translation and reorientation are a maximum, though in polar liquids electrodynamic interactions also affect especially the dielectric behavior.*

An extensive body of literature is available on this topic.[181] Much of what is discussed when dealing with Amorphous Systems, (*cf.*, Section 19.4), also applies to liquids; again the band structure model has to be modified so as to include band, tail and localized states.[182]

---

*The following brief note is based on a paper by Levitt and Hsieh:[222]

The "significant structure theory" (SST) of Eyring and his collaborators[223] approaches the liquid from the solid-liquid boundary. The model is based on the melting process of the solid, and suggest that the liquid inherits, more or less, most of the solid lattice structure. This works well in predicting the thermodynamic properties of a considerable number of liquids over an extensive range of temperature and pressure.[224]

The basic idea of SST is that those structures which make the major contribution to the thermodynamic properties of the liquid system are singled out and any others are ignored. Three significant structures are considered: (a) the solid-like degrees of freedom: possessed by molecules having only other molecules as nearest neighbors (Their positions are restrained in a manner similar to their situation in the solid lattice); (b) the gas-like degrees of freedom: possessed by molecules having one or more vacancies as nearest neighbors. (They will have three-dimensional translation degrees of freedom by virtue of their ability to move into the neighboring hole(s): (c) positional degeneracy of solid-like molecules: because of the existence of molecular size holes, a solid-like molecule will have a positional degeneracy other than its most stable equilibrium lattice position. This positional degeneracy is proportional to the number of the neighboring holes which exist, and inversely proportional to the energy required to preempt the neighboring hole from the competing neighboring molecules.

At the melting point, a certain large number of molecules will have enough kinetic energy to overcome the local potential energy. The result is that they vibrate too vigorously and their positions deviate too far from the lattice site, thus causing a large number of "empty sites," Eyring and his collaborators called this phenomenon "liftoff," and these lifted-off molecules are considered as still remaining in the dense phase. Because of the existence of these molecules and the large number of empty sites (which are considered as holes of molecular size), the original solid lattice equilibrium is broken. The holes, the lifted-off molecules, and the remaining solid-like molecules seek a new equilibrium, which is the liquid lattice having the same coordination sphere (the same unit cell with the same cell volume, and the same number of lattice sites per cell), since the mean intermolecular spacing is approximately the same in both states. The lattice sites are then occupied by molecules and molecular-size holes. This structure explains the short-range order found in a liquid. The introduction of molecular-size holes into the lattice explains the disappearance of the long-range order and the fluidity of the liquid, as well as the volume expansion. The net volume expansion upon melting is explained by the additional empty sites (the molecular-size holes) introduced into the the system.

A theory for localized electrons similar to the Fermi-liquid theory of Landau,[225] and applicable to glasses, has been devised by Freedman and Hertz.[226] For low temperatures and excitation energies below the Fermi energy, it is shown that the electrons form a gas of localized quasi-particles.

Drift mobility values for liquids appear to depend greatly whether intrinsic (?) carriers or injected, excess, carriers are considered; in the first case, barely measurable low values are reported while excess carriers appear to have quite high mobilities, more than 50 cm^2/Vsec for ethane, (see Tables 4.7E which lists experimental mobilities and their activation energies, see also Table 8.12).

A table of mobilities compiled by Berlin et al.[197] is reproduced in Table 4.7D: this refers to effective, or observable (macroscopic) mobilities compared to microscopic mobilities (see eq. (15.61) and discussion in Section 15.10 and 15.4). The data refer to liquids and to injected excess carriers; again very high mobility values are reported.

Some of the very low mobility values reported undoubtedly refer to ionic rather than to electron carriers; for a critical review of ionic mobilities see a discussion by Evans et al.[198]

Huang and Freedman[227] have shown that cation mobilities in liquid hydrocarbons do not obey Walden's rule ($\mu_+ \propto \eta^{-1.0}$). Their relationship with the solvent viscosity η is $\mu_+ \propto \eta^{-1.2 \pm 0.1}$, as found earlier for ions in ethers. Ion migration in liquids made up of nonspherelike molecules is more closely linked to solvent molecular rotation than to shear viscosity. Dipole (permanent or induced) rotation is required during ionic displacement to minimize changes in polarization energy. Diffusion coefficients D of ions in liquids n-hexane are twofold lower than the self-diffusion coefficient of the hexane molecules over most of the liquid range. The conductivity of liquid electrolyte, non-aqueous, solutions is discussed in a review by Barthel et al.[199] The important topic of liquid crystals is considered to be beyond the scope of this book, for a review see a synopsis by Kovshev et al.[220]

Theoretical aspects of charge transfer in liquids are discussed in Chapter 15; especially in Section 15.6 for the example of methyltetrahydrofuran glasses.

In the much-studied hydrocarbons, conductivity is said[184] to be mainly ionic plus a component of electronic transport due to charge injection from the contacts; at high values of applied field, conductivity appears to be trap-limited.[191]

The trap depth for electrons in benzene has been reported[229] as 0.5 eV, while in alkanes values of 0.26 eV have been found.[230] In benzene, conduction is reported to be mainly unipolar from carrier injection from the contacts.[195] In another study of the conductivity of benzene,[196] this is confirmed to be due to ionic and electronic hopping transfer with the electron free time severely limited due to trapping at impurity centers. The often observed reverse emf after removal of an applied direct field is shown to be a space charge effect.[196] Relaxation effects of dark conductivity in benzene are discussed by Bobyl et al.[189] The diffusion coefficient of electrons in liquid benzene has been redetermined[217] by first applying a field of 50 V/cm and measuring the current decay following its sudden removal; the value of

1.4×10^{-5} cm^2/sec is reported.[217] Liquid alcohols have been studied by Licea.[187] The electronic conductivity of insulating liquids is reported to rise considerably upon increasing concentrations of polar additives.[188]

Polyproline II containing less than 8 % of water, or, in other words, less than 0.5 % per imide group, is said[190] to conduct mainly electronically but becomes an ionic conductor above these limits. Liquid alkanes appear to behave in a similar fashion.[216] The non-ohmic behavior so often found in dielectric liquids has been studied,[218] *i.e.,* by Kazatskaya et al. who find that it is due to carrier injection from the contact into a slightly conducting liquid surface layer on the electrode; pinacoline, methylketone and cyclohexane were studied.[218] The electronic contribution to the entropy of fusion is discussed by Glazov et al.[193] In considering carrier transport in liquids and glasses, attention is focussed on questions of mobility; the subject has recently been reviewed.[194] Numerical mobility data are collected in Tables 4.7D and 4.7E. A fruitful avenue explored in such studies involves the study of excess electrons, produced in the liquid or glass by an external agency such as by Photoionization, gamma-irradiation or electron injection.

Excess electrons localized in organic liquids or glasses are termed trapped, or presolvated, if the surrounding molecules near the trap have not yet equilibrated; the electron is referred to as a solvated electron if the molecules in its vicinity have attained an equilibrium configuration.[196] The energetic characteristics,[197] including the energy levels[199] as well as reactions involving localized electrons[198] especially in ethanol glasses, have been investigated in detail.

For the electron to be localized, the energy of the ground state of the electron must be below the lowest energy of the conducting state.[228] However, an important question is the mode of transition from a delocalized to a localized state for excess electrons in organic matrices.[242] To clarify this, Hase and Higashimura[196] have measured the optical absorption spectra of solvated electrons e^-_{sol}, and of trapped electrons e^-_t, produced in an ethanol glass in the presence of an externally applied electric field. They find that both spectral absorptions were lowered in intensity upon application of the field, which effect is said[196] to be due to a decreased transition rate from the delocalized to the localized electron state.

A large number of experimental studies of long range electron transfer (see Chapter 15) between localized states in doped glasses have been reported.[209] The results may be summarized as follows. The time and temperature dependences observed are all rather similar:

(a) the decay of the population of trapped electrons in the presence of scavengers in the medium.[212]

(b) Geminate recombination of electrons with parent ions followed by recombination luminescence.[214]

(c) Electron transfer reactions of the type $A^- + B \to A + B^-$ between guest molecules in the glass.[213]

The number of initial species does not decay exponentially with time, but rather follows either an exponential or linear function of log t over several orders of magnitudes of time. The temperature dependence of most of these ET processes generally is weak in the region 4-80 $\sim$ 100 K while a much stronger temperature dependence appears at higher temperatures.

An extremely wide range of reaction rates may be obtained with different electron accepting dopants in the same medium.[211] The theory for such reactions at low temperatures assumes direct electron tunnelling between two bound states, from donor to acceptor sites.[210] By assuming that the acceptors are randomly dispersed around each donor, the exponential decay of the individual reaction rates with the donor acceptor distance can account for the logarithmic time decay of the occupied donor states.[210] In any direct long-range electron transfer mechanism the rate should depend exponentially on the donor-acceptor separation. Experimental evidence suggests[210] that many of these electron transfer reactions are exothermic. Beitz and Miller[215] have employed acceptors with electron affinities ranging from 6.1 eV for Te^+ to organic molecules with an affinity of 0.3 eV resulting in reaction rates extending over 5 orders of magnitude though no definite correlation between electron affinity and reaction rate has so far been demonstrated. For a complete discussion, see Ref. 210.

The observed mobility μ_{0b} of the excess electrons may be written[200]

$$\mu_{0b} = \mu_L P + \mu_F(1-P) \qquad (16.25a)$$

where μ_L is the electron mobility for the carrier in a localized state, P the probability of carrier localization, and μ_F the electron mobility for the carrier in a quasi-free state, meaning the electron moving in a conduction band. Experiments with liquid neopentane, tetramethylsilane, and methane[201] indicate that μ_{0b} is highest at that resistivity value for which the energy of the conduction band exhibits a minimum, and at a temperature at which the number density of the system is about 3.5×10^{21} to 10^{22} molecules cm^{-3}: even though the liquid is disordered, it appears[201] that, overall, the electron "sees" a fairly smooth potential. Quantitatively, excess electrons produced in saturated hydrocarbons by photoionization are reported[202] to show values of $\mu_{0b} = 0.04$ cm^2/Vsec at room temperatures, while μ_F is said[202] to exceed 100 cm^2/Vsec. The energy of the localized states involved is reported[202] to be 0.4 to 0.6 eV. High values of hole mobility in cyclohexane and methyl-cyclohexane are reported to be due to a resonant charge transfer between a positive ion and a neighboring neutral molecule.[203] The values quoted above for μ_F are in at least semi-quantitative agreement with values obtained for μ_F for ethane[208] and for liquid hydrocarbons by Berlin,[204] viz., 34 to 400 cm^2/Vsec. Mobility data for excess electrons taken from Berlin[197] are listed in Table 4.7D. This author[197,204] considers the electrons as being associated with a plane wave scattered by density fluctuations in the liquid. The interaction of slow electrons with benzene and its derivatives has been reviewed.[25]

The mobility appears to be thermally activated; thus, liquid olefins exhibit activation energies of the order of 0.1 to 0.2 eV;[26] 3-methylpentane is reported[27] to have a value of 0.2 eV and, as seen from Tables 4.7D and 4.7E, values in that range appear to be characteristic for electronic conduction in many liquids, though iso-octane is reported[195] to have the low value of 0.06 eV compared to 0.5 eV for the thermal activation energy in the solid state.[195] Even lower values of activation energies, close to zero, are reported for neopentane (see Table 4.7D). Upon solidification the product $\mu\tau$, τ being the carrier free time, is said[195] to rise sharply as does the radiation yield of free electrons. This effect should be confirmed and further studied.

Excitonic transfer in glasses is discussed by Kaminski and Kawski;[183] the role of biexcitons in glasses is treated by Licciardello[232] who suggests that these excited states are localized at typical sites which are not local defect structures. An intrinsic electric dipole at tunnelling sites is predicted.[232] Excitonic transfer is important in glasses of dyes such as pararosaniline, crystal violet, rhodamine B, or malachite green.[233]

Fused anthracene exhibits[185] nearly equivalent bipolar carrier injection from the contacts; fused naphthacene behaves in a qualitatively similar manner. Most molten pyridinium salts are nearly completely dissociated; and, thus, conduct ionically[192] though some form what appears to be autocomplexes resulting in a "handing-on-of-charge;" and, thus, have an electronic component.

A new form of organization of organic liquids has been reported:[219] quasi-crystalline globules of 0.1 to 0.4 μm diameter join together like straight strings of beads along the lines of an electric field; the string is uniform within the chain.[219] The effect has been studied, e.g., in photochromic spiropyrans in non-polar solvents under a constant external electric field.[219]

Injected excess carriers in silicone oils are said[231] to move along filamentary paths of less than 0.1 mm diameter; though such an inhomogeneous transport is probably an impurity effect, it might perhaps at least partially be associated with the above-mentioned effect.

16.10 Seebeck Coefficients

Further experimental values of this quantity are tabulated in Table 8.14A, supplementing Table 8.14. In addition, thermoelectric powers, or Seebeck

SEEBECK COEFFICIENTS

coefficients, for several PAQR polymers are listed in Table 8.14B. These are all, with the exception of the dibenzpyrene-PMA polymer, p-type with the majority having values of about 60 μV/K.

TCNQ adducts are as likely to be p-type as n-type while there is a distinct tendency for hole conduction to be observed in all other compounds.

Seebeck coefficients for a range of ionene:TCNQ polymeric adducts are listed in Table 8.6D; of the 30 values listed 23 are positive and 6 negative. one is zero. There is no preference for any value like 60 μV/K.

Table 8.14A shows 111 entries; it is remarkable that 17 of these again exhibit Seebeck coefficients in the vicinity of +60 μV/K.

From eq. (16.28) it follows that dS/dT should have the opposite sign to that of S itself. This appears to hold as seen from Table 8.14B. Also, in a general fashion, the higher S values are found for higher resistivity materials where E_g is large.

Fritzsche[234] writes for the thermoelectric power (Seebeck coefficient) S the following equation, which should hold generally as long as inter-electronic interaction can be neglected:

$$S = -\frac{k}{|e|} \int \frac{E - E_F}{kT} \cdot \frac{\sigma(E)}{\sigma} dE \qquad (16.26)$$

The integral has to be evaluated over all energy states E of a one electron system described by a Fermi level E_F and a conductivity which, in turn, is given by

$$\sigma = \int \sigma(E) dE = e \int g(E) \cdot \mu(E) \cdot f(E) \cdot [1 - f(E)] dE \qquad (16.27)$$

where $g(E)$, $f(E)$, and $\mu(E)$ are the density of states, the Fermi distribution function, and the mobility of the (majority) carriers. k is Boltzmann's constant and e the electronic charge. For an intrinsic semiconductor, the well-known eq. (16.28) results (see also eq. (2.33) and (2.34), Section 2.6).

$$S = \frac{k}{|e|} \frac{\mu^+ - \mu^-}{\mu^+ + \mu^-} \left[\frac{E_g}{2kT} + A \right] \qquad (16.28)$$

If one assumes that the mobilities $\mu \neq f(T)$, S should vary with $1/T$. As seen from Fig. 16.13A, this holds reasonably well at higher temperatures, but not at all in the lower temperature region. The difficulties in measuring the values of the scattering parameter A are discussed in Section 2.6; that discussion is still pertinent, since experimental values for A are incompatible with theory.[234]

Chaikin[235] has discussed the theory of the thermoelectric effect in the higher temperature region for the case of localized carriers with electron

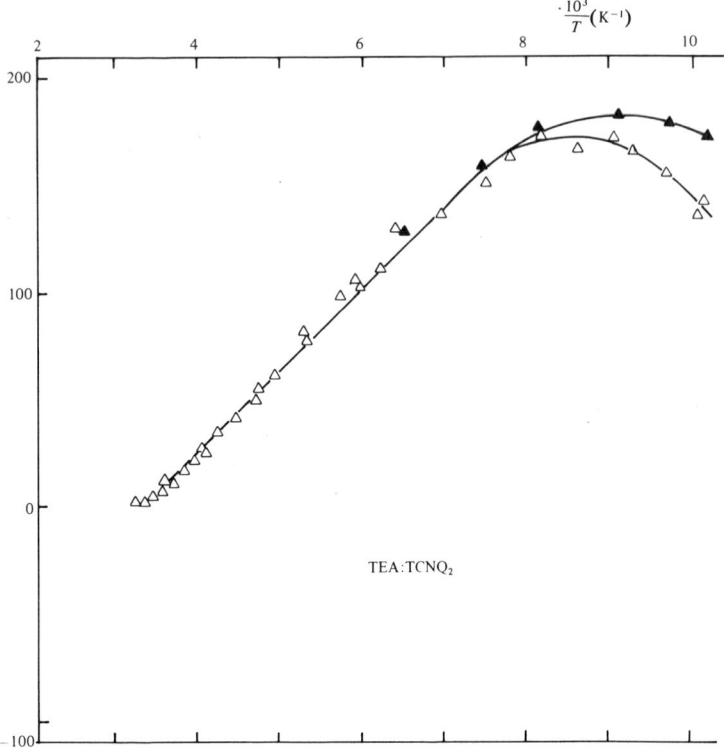

Fig. 16.13A Plot of the Seebeck coefficient for TEA:TCNQ$_2$ single crystals in the y-crystallographic direction; data are shown for two samples. S appears to vary with $1/T$ over a limited temperature interval in the high temperature region only. After Brau.[240]

interactions U, the coulombic repulsion energy with 2 carriers on one site, and with V, that for the case of both carriers occupying different sites. Neglecting the kinetic energy of the carriers, he finds that S is constant at higher temperatures and that its sign and magnitude are wholly determined by the electron density of the system and by the occupation probabilities of the carrier accepting sites. The Seebeck coefficient S then results as

$$S_{T\to\infty} = G/eT \qquad (16.29)$$

where G is the chemical potential given by

$$G = kT\frac{\partial \ln W}{\partial N} \qquad (16.30)$$

W refers to the occupation probability of any one site (see Section 6.9, eq. (6.35), also for entropy effects); it depends on the distribution of the N_A accepting sites.

If one assumes that the intra-site interactions govern the carrier localization processes, then two cases may be distinguished depending on the magnitude of the energy U relative to the thermal energy kT:

(1) $kT \gg U$ (the high temperature regime):

Writing N for the sum of electrons having upward and downward spin directions, and N_A for the density of acceptor sites,

$$S_{T\to\infty} = -\frac{k}{e}\ln\frac{2N_A - N}{N} \qquad (16.31)$$

(2) $kT \ll U$ (the low temperature regime):

In this case, double occupancy of the same site by electrons with parallel or antiparallel spins is forbidden. W then results as

$$W = \frac{N_A!\,2^N}{N!(N_A - N)!} \qquad (16.32)$$

whence

$$S_{T\to\infty} = -\frac{k}{e}\ln\frac{2|N_A - N|}{N} \qquad (16.33)$$

Beni et al.[236] have applied Chaikin's theory[235] to 1:2 complexes. In such, if the number of neutral sites equals the number of charged sites, $N_A/N = 2$. Then, numerically for $kT \gg U$

$$S_{T\to\infty} \cong -95\ \mu\text{V/K} \qquad (16.34\text{A})$$

and for $kT \ll U$

$$S_{T\to\infty} \cong -60\ \mu\text{V/K}. \qquad (16.34\text{B})$$

Experimental results, summarized in Fig. 16.14 for $TCNQ_2$:TEA along the high conductivity direction at 300 K, quantitatively agree with

$$S = -\frac{k}{e}\ln 2 = -60\ \mu\text{VK}^{-1}, \qquad (16.34\text{C})$$

i.e., for $kT \ll U$. Thus, Beni et al.[236] conclude that in these solids it is the intra-site coulomb repulsion energy U which, even at high temperatures, is one of the dominant factors in the charge localization processes.

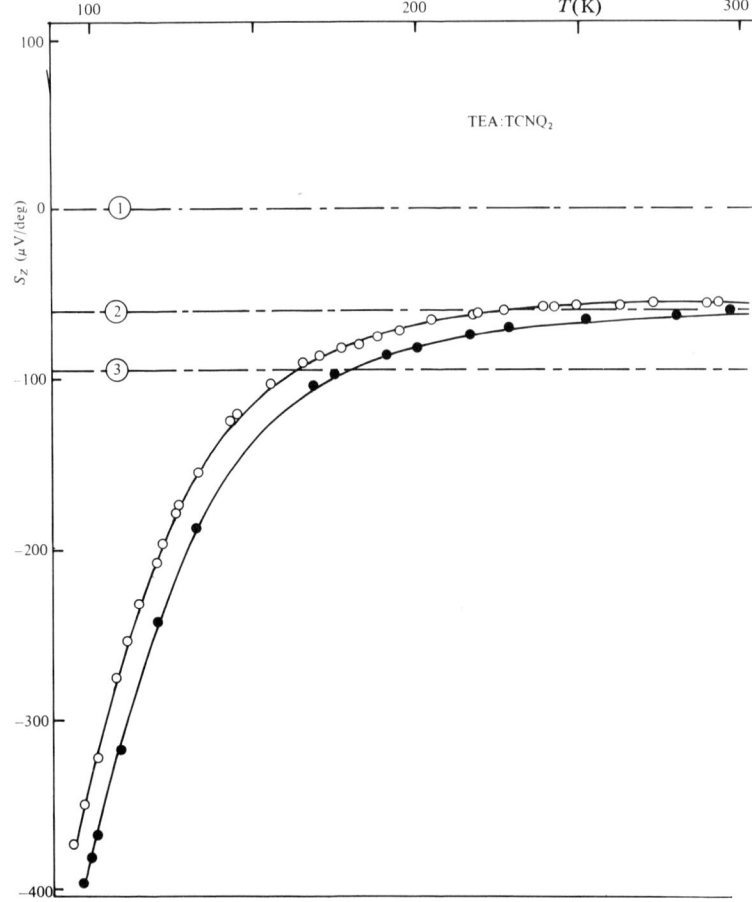

Fig. 16.14 Plot of the Seebeck coefficient, S, for TEA:TCNQ$_2$ single crystals in the high conductivity, z, crystallographic direction, *vs.* temperature T, in K. The limiting values (1) are for fermions without spin, (2) for fermions of spin ½ assuming the Coulombic repulsion energy $U \gg kT$, while (3) applies to fermions of spin ½ for $U \ll kT$. After Brau.[240]

Kwak has pointed out[237] that this agreement is largely due to an excess of neutral molecules along the columns of TCNQ if TCNQ^0/TCNQ$^-$ > 1.

Similar agreement has also been observed[239] in quinolinium:TCNQ$_2$, acridinium:TCNQ$_2$; TMPD:TCNQ$_2$, all in the high conductivity direction. Inspection of Table 8.14A shows that nearly all the TCNQ$_2$ adducts do exhibit thermoelectric powers close to 60 μV/K. However, at lower temperatures, all these compounds tend to behave differently in sign as well as magnitude of S. This behavior is due to the nature of the intra-site and inter-site interactions, to the values of the transfer integrals involved, to the percentage $\rho = N/N_A$ of the charge transferred in complexation, and to the electron phonon coupling factor g in eq. (16.27). This is illustrated in Fig. 16.15 where S is plotted for a hypothetical system with $V = 0$ and $U \to \infty$, i.e., for a system with strong electronic interactions, for various values of ρ.

Several other theories of the thermoelectric effect have been devised,[238] but they all fail to explain the behavior of S at low—say liquid nitrogen—temperatures. Measurements at liquid He temperatures should clarify the matter by indicating whether the $1/T$ regime for S goes over into a T^3 regime as expected for strong electron-phonon coupling.

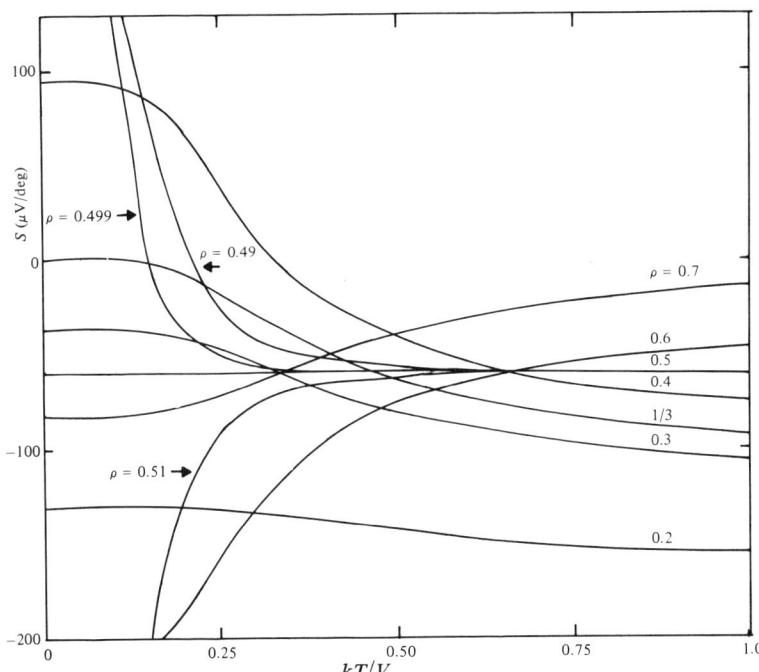

Fig. 16.15 Theoretical plot of the Seebeck coefficient vs. kT/V in arbitrary units; V is the inter-site Coulombic repulsion energy; the parameter $\rho = N/N_A$, the portion of total charge transferred in the complex formation, i.e., the electron density, for an infinitely long molecular (Hubbard) chain assuming the transfer integral to be negligibly small and presence of strong bonding. After Brau.[240]

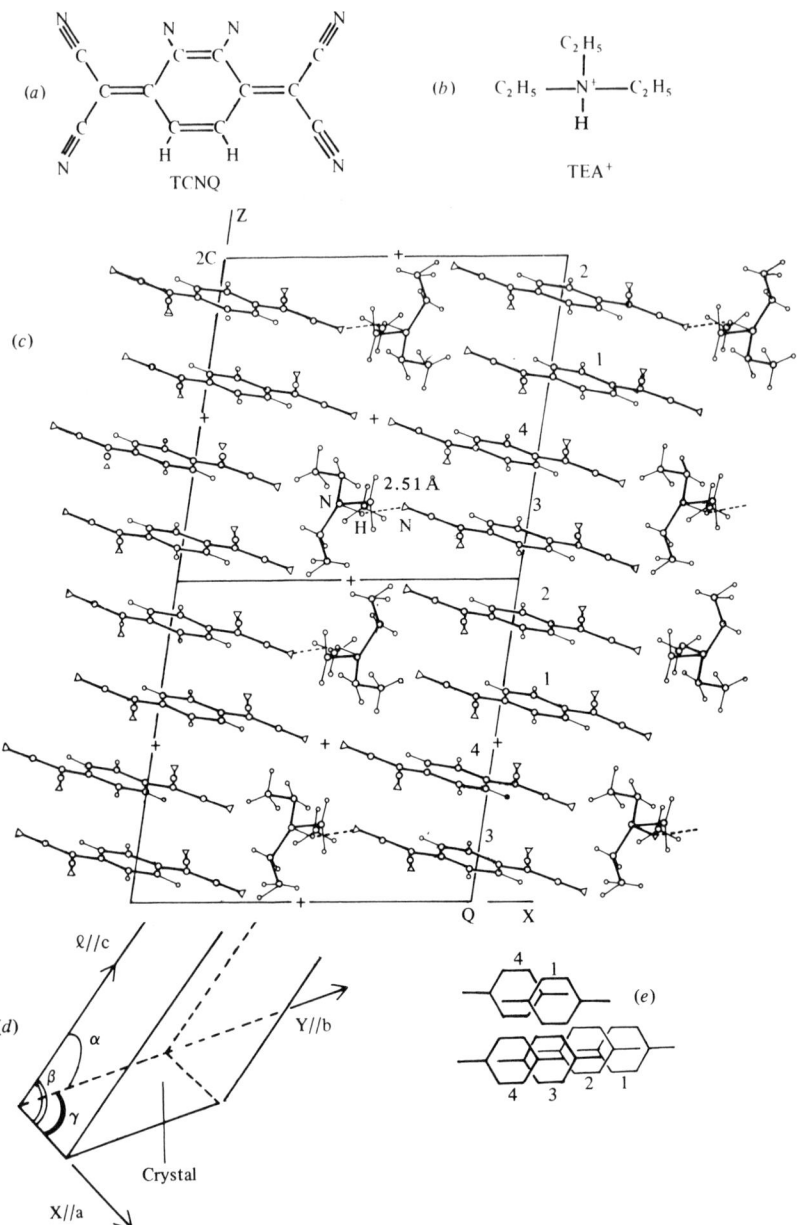

Fig. 16.16 Molecular and crystal structure of TEA:TCNQ$_2$. After Jaud et al.[243] Diagrams (a) and (b) show the chemical structures of the donor ion TEA$^+$ and of the acceptor TCNQ. The molecular structure is illustrated in diagram (c) which also indicates the alignment with the crystallographic directions shown in diagram (d). Numbering TCNQ molecules 1 to 4, diagram (e) further illustrates their spatial alignment with reference to diagram (c).

16.11 Semiconducting TCNQ Free Radical Salts

These adducts have attracted considerable interest during the last decade or so and a great deal of pertinent data have been published. Some of the most representative ones are collected in Table 8.6A which lists 121 entries. Room temperature resistivities range from 10^8 to 0.002 ohm cm. Most of these compounds are anisotropic; some exhibit anisotropy coefficients of the order of 1000. Energy gap values range from 1 to 0.6 eV. The few-only 10-mobility values available range from 0.04 to a metallic 10^4 cm^2/Vsec. Of 58 entries stating the sign of the majority carrier, 23 are positive and 35 negative; most are obtained from thermo-emf measurements. Other Seebeck coefficients are listed in Tables 8.14A and 8.7F; the latter table displays data for a range of polymeric ionene-TCNQ adducts discussed in Section 17.10. For a discussion of thermo-emf results obtained in TCNQ adducts, see Section 16.10. Ternary TCNQ inclusion compounds such as benzidine:TCNQ are dealt with in Section 17.7. Polymeric TCNQ complexes are discussed in Section 17.10 (see also Table 8.7F). Tables 8.6B and 8.6C which show resistivity values as well as several activation energy values, for a range of adducts not of TCNQ but of two other quinone analogues, $viz.$, tetrachloro-p-diphenoquinone and tetrabromo-p-diphenoquinone. Only a few of these adducts exhibit low resistivities, of the order of a few 10 ohm cm.

TEA adducts are perhaps those TCNQ compounds which have been studied in greatest detail. The crystal structure of this adduct is shown[243] in Fig. 16.16; it is seen that, essentially, TEA (TCNQ)$_2$ consists of linear chains of TCNQ dimers with slightly alternating interdimer spacings, as shown in Fig. 16.17, the crystals are triclinic.

In each unit cell, there are 4 TCNQ molecules; there is one electron transferred for every two TCNQ molecules. The temperature dependence of the conductivity σ of the adduct is illustrated in Fig. 16.18 for the 3 principal components of the conductivity tensor.[244] The slopes $-d(\log \sigma)/d(T^{-1})$, evaluated between about 80 and 350 K, exhibit a pronounced peak between 200 and 220 K; the anisotropy ratios are seen to rise with increasing temperature.[244]

The ac conductivity of the adduct is plotted[244] as a function of temperature in Fig. 16.19; included in the plot are results obtained by Shchegolev,[245] for comparison.

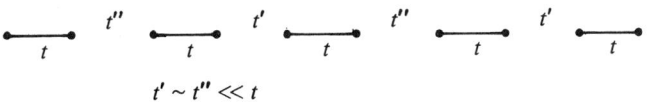

Fig. 16.17 TEA:TCNQ$_2$ may be considered as an assembly of linear chains of TCNQ dimers with slightly different and alternating interdimer spacings.

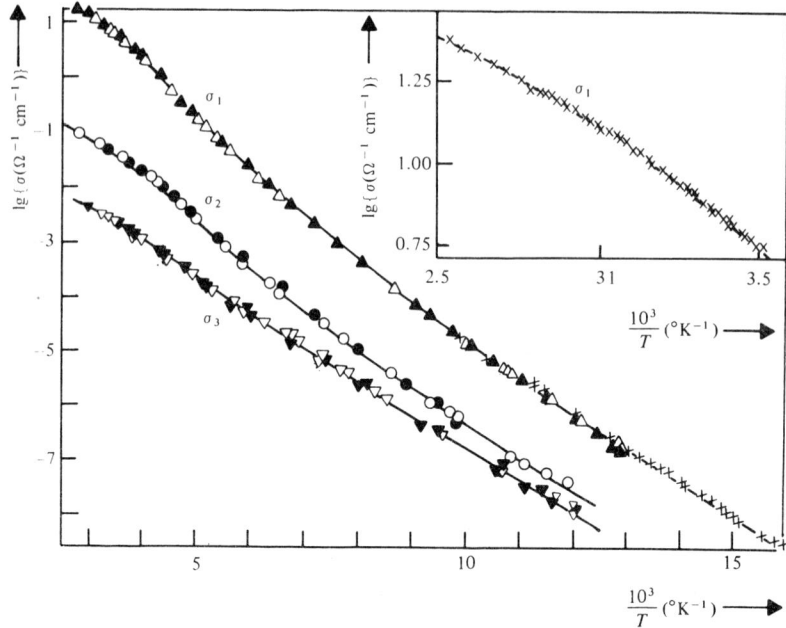

Fig. 16.18 Temperature dependence of the ac conductivity of TEA:(TCNQ)$_2$ adducts. Results are shown for ten different specimens. The subscripts 1,2,3 refer to the crystallographic axes: 1 is parallel to the Z axis, 2 is perpendicular to the 1 and 3 axes; and 2 is perpendicular to the Y and Z crystallographic axes; see also diagram in Fig. 16.16D. Direction 1 is parallel to the TCNQ chains and directions 2 and 3 perpendicular to them. Direction 2 is within and direction 3 is perpendicular to the alternating TCNQ-TEA molecular layers. After Farges, Brau et al.[244]

The permittivity at 10^8 Hz at 77 K was found to be 70 ± 20 in the 1 direction and 6.2 ± 0.7 in the 3 direction.

Polarized, near-normal-incidence, reflectance spectra (see also Section 20.5) of this adduct are displayed in Fig. 16.20; the subscripts 1, 2 and 3 again refer to the three mutually orthogonal directions of the conductivity tensor (see Fig. 16.18). Thus, in R_1, the incident beam is almost parallel to the 3 direction and polarized parallel to the 1 direction.

The most satisfactory theoretical model[247] agreeing with the experimental results given above, is based upon the theory of Rice et al.[246] From the D_{2h} symmetry of the TCNQ molecules, the b_{2g} orbital of the conduction electrons is linearly coupled *only* to the intramolecular a_g modes. These modes are *basically non-active* optically.

Rice et al.[246] have shown that, as a result of the electron-phonon coupling, these modes are *renormalized as collective modes* (or phase phonons) thus

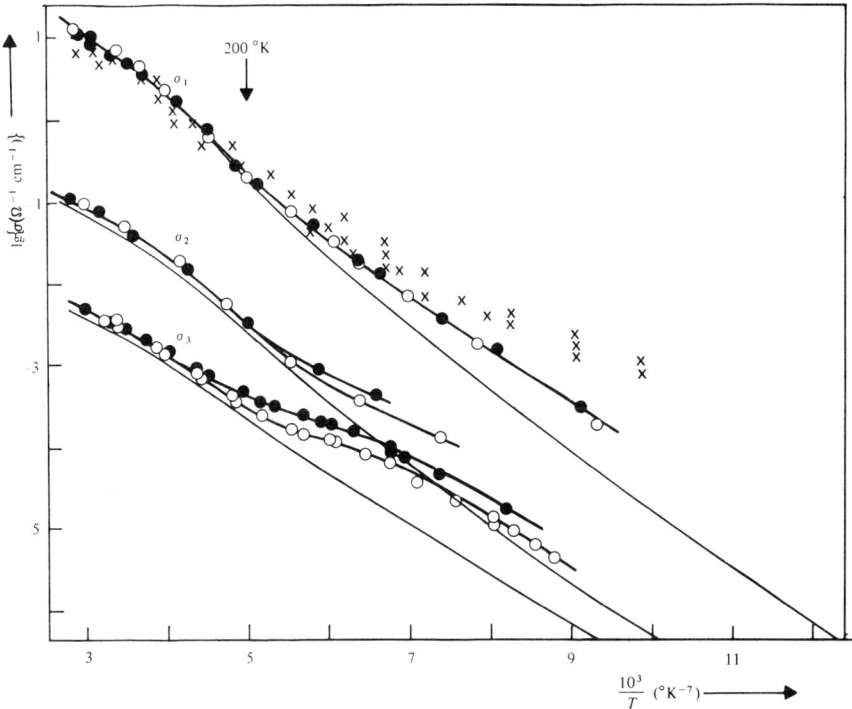

Fig. 16.19 Temperature dependence of the alternating current conductivity σ TEA:(TCNQ)$_2$ adducts. The subscripts 1,2 and 3 refer to the three principal components of the conductivity tensor, see caption to Fig. 16.18. Circles ○ refer to measurements at 100 MHz and full circles ● refer to data obtained at 200 MHz. The crosses refer to results obtained by Shchegolev[245] at 10^{10} Hz, while the continuous thin line indicates results at zero frequency, see Fig. 16.18. Farges, Brau et al.[244]

acquiring a quite anomalous optical activity, but only for a polarization parallel to the conductive TCNQ chains. This provides an explanation of the reflectivity spectra, as evident from Fig. 16.21.

From the experimental results of the Nice Group on TEA (TCNQ)$_2$, here summarized, Rice et al. have calculated[246] from their theory the coupling constant g_n for each mode n and have found a reasonable agreement with standard calculations of quantum chemistry. Also, they have found that if V is the structural gap in the absence of coupling ($g_n = 0$), and E_g the renormalized gap (as evaluated from dc conductivity), then:

$$V/E_g \simeq 0.13 \qquad (16.35)$$

This indicates that the semiconducting gap is nearly entirely due to *intra-*

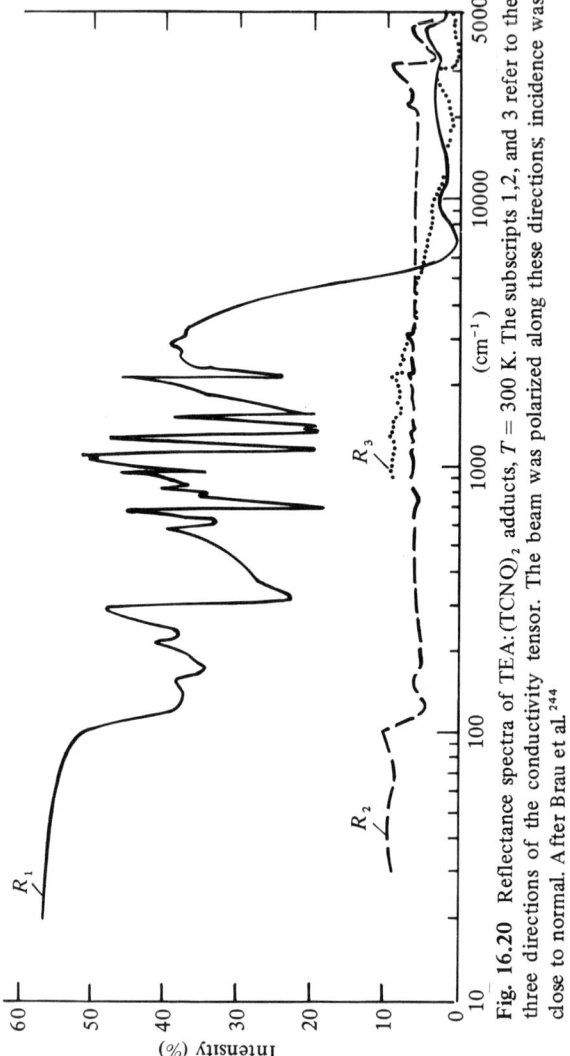

Fig. 16.20 Reflectance spectra of TEA:(TCNQ)$_2$ adducts, $T = 300$ K. The subscripts 1, 2, and 3 refer to the three directions of the conductivity tensor. The beam was polarized along these directions; incidence was close to normal. After Brau et al.[244]

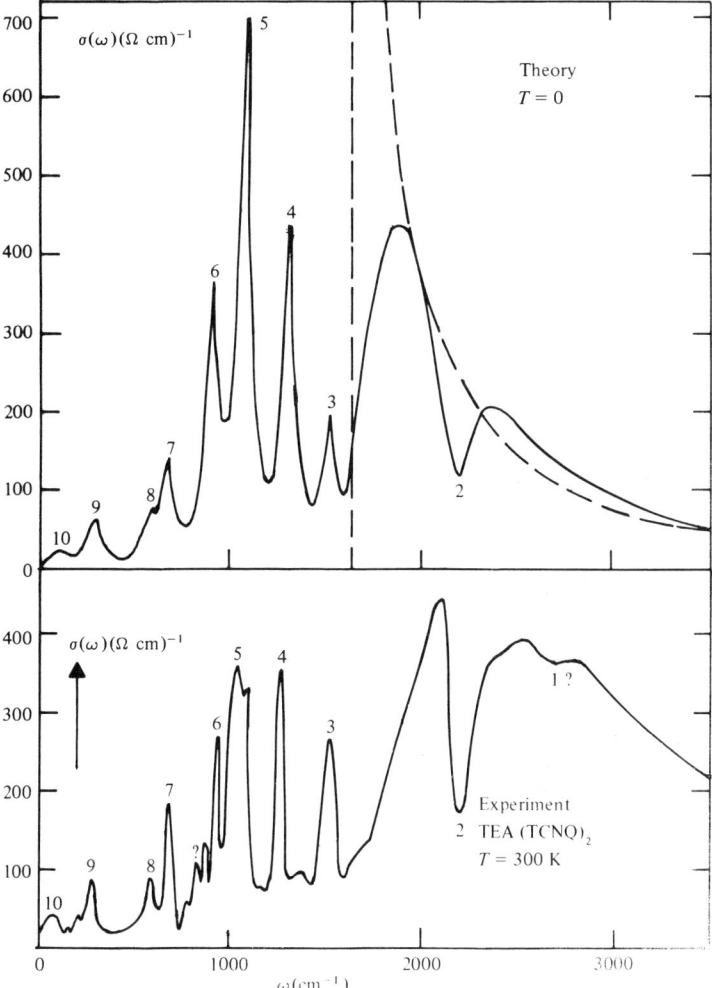

Fig. 16.21 Comparison of experimental reflection spectra of TEA:(TCNQ)$_2^?$ single crystals, see Fig. 16.20, with theoretical predictions by Rice et al.[246]

molecular distortion rather than to *intermolecular* modulation, even in the case of strong electron correlations.

The frequency dispersion of the ac conductivity at temperatures below 200 K supports charge transfer by hopping;[244] this probably involves tails of localized states in the energy gap. Such dispersions, though at lower frequencies, have also been reported by Vlasova et al. for K(TCNQ).[248] At higher temperatures, there is no evidence for a dispersion up to 10^{10} Hz, though it does become evident for σ above 10^{11} Hz.[244]

The thermo-electric behavior of TEA:(TCNQ)$_2$ is discussed in Section 16.10 (see Fig. 16.13 and 16.14).

The electrical properties of TCNQ salts have been correlated with the donor properties in experiments employing a series of donors;[250] charge transfer is said[250] to be by variable range hopping.

References

1. A. R. Blythe, *Electrical Properties of Polymers*, Cambridge U. Press, 1979; G. Geuskens and C. David, "Energy Transfer and Migration in Polymers," "Proc. 23rd Internatl. Congr. Pure and Appl. Chem.," Boston, Mass. (1971), Butterworth, London, 8, 19 (1971); D. A. Seanor, "Elect. Prop. of Polymers," *Polymer Science*, A. D. Jenkins, ed., Nth. Holland Publ., Amsterdam, 1972; A. Rembaum, "Semicond. Polymers," *J. Polymer Sci. C.*, No. 29, 157 (1970); R. G. Kepler, "Electronic Properties of Polymers" in *Treatise on Mat. Sci. and Technol.*, 10, Part B, *Properties of Solid Polymer Mat.*, J. M. Schultz, ed., Academic Press, N.Y., 1977; H. A. Pohl, *J. Biol. Phys.*, 2, 113 (1974); V. A. Ovchinnikov et al., *Soviet Phys.-USP.*, 15, 575 (1973); G. Kossmehl, *Ber. Bunsen Ges. Phys. Chem.*, 83, 417 (1979); P. Fischer, *J. Electrostat.*, 4, 149 (1978); Y. Wada, *Dielect. Relax. Mol. Processes*, 3, 143 (1977); J. O. Williams, *Adv. Phys. Org. Chem.*, 16, 159 (1978); V. Adamec and J. H. Calderwood, *J. Phys.*, D-11, 781 (1978); C. Bremmer et al., *Faserforsch. Textiltech.*, 28, 575 (1977); A. R. Blythe, *Elect. Properties of Polymers*, Cambridge Univ. Press., 1980; I. Misurkin and A. A. Ovchinnikov, *Russ. Chem. Revs.*, 46, 967 (1977); E. P. Goodings, *Quart. Rev. Chem. Soc.*, 5, 95 (1976).

2. R. G. Kepler, in Ref. 1.

3. J. M. Schultz, *Polymer Materials Science*, Prentice Hall, Englewood Cliffs, N.Y., 1974.

4. K. M. Ormitoto and Y. Murakai, *App. Opt., Suppl. No. 3*, 50 (1969); M. Lardon et al., *Molec. Cryst.*, 2, 241 (1967); P. J. Regensburger, *Photochem. Photobiol.*, 8, 429 (1968); A. I. Lakatose and J. Mort, *J. Non-Cryst. Solids*, 4, 117 (1970); D. M. Pai, *J. Chem. Phys.*, 52, 2285 (1970); A. Szymanski and M. M. Labes, *J. Chem. Phys.*, 50, 3568 (1969); H. Bauser and W. Klopffer, *Chem. Phys. Lett.*, 7, 137 (1970); H. Sato, and M. Skeda, *J. Appl. Phys.*, 43, 4108 (1972); G. Weiser, *J. Appl. Phys.*, 43, 5028 (1972); H. Seki and W. D. Gill, "Proc. Inst. Conf. Conduct. Low Mobility Mater. 2nd.," Taylor and Francis, London, 1971; p. 409; W. D. Gill, *J. Appl. Phys.*, 43, 5033 (1972); R. C. Hughes, *J. Chem. Phys.*, 55, 5442 (1971); *Chem. Phys. Lett.*, 8, 403 (1971); *J. Chem. Phys.*, 58, 2212 (1973); in "Electrophotography," D. R. White, ed., Soc. Photog. Sci. and Eng., D.C., 1979, p. 1479; G. Pfister et al., *Phys. Rev. Lett.*, 40, 659 (1978).

5. D. A. Seanor, in Ref. 1.

6. F. Gutmann, *Japan. J. Appl. Phys.*, 8, 1417 (1969).

7. P. J. Regensburger, *Photochem. Photobiol.*, 8, 429 (1968).

8. H. A. Pohl, in ref. 1.

9. M. Vodenicharova and P. Amov, *Bulg. J. Phys.*, 1, 1 (1974).

10. W. Bucker and A. Herspring, *Priv. Comm.*, (1978), cf., also W. Bücker, *Dipl. Ing. Dissert.*, T. H. Aachen, W. Germany (1972).

REFERENCES

11. F. Fuhrmann et al., *Polymer.*, **19**, 131 (1978).
12. J. C. Phillips, *J. Solid State Chem.*, **19**, 309 (1976); *Proc. Natl. Acad. Sci. USA.*, **43**, 3820 (1976); *Phys. Stat. Solidi*, **B-78**, 371 (1976).
13. J. Hirsch and E. M. Martin, Kodak Ltd. Harrow, England, Internal Report, 1971.
14. R. D. Hartmann and H. A. Pohl, *J. Polymer Sci.*, **A-1, 6**, 1135 (1967); J. Labbé and E. C. van Reuth, *Phys. Rev. Lett.*, **24**, 1232 (1970).
15. M. J. Rice and J. Bernasconi, *J. Phys.*, **C-3**, 55 (1973); J. Bernasconi, et al., *J. Phys., C. Solid State Phys.*, **5**, L127 (1972).
16. I. Schopov and M. Vodenicharova, *Macromol. Chem.*, **179**, 63 (1978).
17. F. Nordhage and G. Backstrom, *J. Electrostat.*, **2**, 317 (1973); *cf.*, also Ref. 16.
18. D. J. Thouless, *Phys. Rev. Lett.*, **39**, 1167 (1977); *cf.* also J. C. Garland et al., *Bull. Amer. Phys. Soc.*, **24**, 280 (1973).
19. J. H. Ranicar et al., *Aust. J. Phys.*, **24**, 325 (1971); Ph. D. Thesis, "Electrical Conduct. in PVC.," Monash Univ. Melbourne, Australia (1972).
20. N. Giardino et al., *Phys. Rev. Lett.*, **43**, 725 (1979); G. J. Dolan and D. D. Osterhoff, *ibid.*, p. 721; P. W. Anderson et al., *ibid.*, p. 718.
21. R. Hirohashi et al., *Kogyo Kagaku Zasshi*, **73**, 1042 (1920).
22. G. Kossmehl and M. Rohde, *Makromolec. Chem.*, **178**, 715 (1977).
23. V. S. Boishchev et al., *Vysokomol. Soedin, Ser. B*, **20**, 259 (1978).
24. H. Graovac et al., *Colloid. Polymer Sci.*, **255**, 480 (1970).
25. M. M. Khvorov et al., *Vysokomol. Soedin, B*, **20**, 27 (1978).
26. H. Maeda et al., *Appl. Phys. Lett.*, **32**, 278 (1978).
27. A. Rembaum, *J. Polymer Sci., Part C*, No. 29, 157 (1970); A. Rembaum et al., *J. Macromolec. Sci. Chem.*, **A-4**, 715 (1970); *cf.*, also A. M. Hermann and L. G. Whilhite., *Proc. Louisana Acad. Sci.*, **32**, 126 (1969).
28. H. Shirakawa et al., *Polymer J.*, **4**, 460 (1973); H. Shirakawa and S. Ikeda, *ibid.*, **2**, 231 (1971); T. Ito et al., *J. Polymer Sci. Polymer Chem.*, **12**, 11 (1974); **13**, 1943 (1975).
29. N. R. Byrd et al., *J. Polymer Sci.*, **10-A2**, 957 (1972).
30. J. O. Williams, *Adv. Phys. Org. Chem.*, **16**, 159 (1978).
31. L. Nicodemo et al., *Polymer J.*, **19**, 230 (1978).
32. Sung-il Lee, *Chosen Shogakkoi Gakujutsu Rombunshu*, **6**, 137 (1976).
33. S. D. Phadke, *Thin Solid Films*, **55**, 391 (1978).
34. L. A. Berkovich and A. Fomin, *SB. Aspir. RAB.-Kasan. Gos. Univ. Tochn. Nauki*, **3**, 12 (1976).
35. J. Tylzkowski et al., *Thin Solid Films*, **55**, 253 (1978).
36. T. Borisova, *Plast. Massy*, **1978**, 28.
37. A. Abu-Bakr et al., *Europ. Polymer J.*, **13**, 799 (1977).
38. E. T. Samulski and D. B. DuPré, "Polymeric Liquid Crystals," in *Advances in Liquid Crystals*, G. H. Brown, ed., Academic Press, N.Y., 1979; see also ref. 220.
39. C. M. Vodenicharov and A. Stanche, *Thin Solid Films*, **37**, 157 (1976); *cf.*, also ref. 17; C. M. Vodenicharov and S. G. Christov, *Phys. Stat. Solidi*, **A-25**, 387 (1974).
40. G. Pfister, *Phys. Rev.*, **B-16**, 3676 (1977); F. Gutmann et al., *Nature* (Lond.), **221**, 1237 (1969); C. B. Duke, *Tunnelling in Biological Systems*, B. Chance, ed., Academic Press, N.Y., (1979).
41. M. Audenaert et al., *Solid State Comm.*, **30**, 797, 1979; *cf.*, also Refs. 1 and 4.
42. M. M. Labes et al., *Chem. Rev.*, **79**, 1 (1979); G. B. Street and W. D. Gill, *Proc. Nato ASI, Molecular Metals*, Les Arcs, Franch, 1978, Plenum Press, N.Y., (1980).
43. K. Kanazawa and A. F. Diaz et al., IBM, San Jose, Calif., Private Comm. (1979).
44. T. C. Clarke et al., *J. Chem. Soc. Chem. Comm.*, **1978**, 489.

45. G. P. Gardini, in *Adv. in Electrocycl. Chem.*, A. R. Katritzky and A. J. Boulton, eds., Academic Press, N.Y., **15**, 67 (1973).

46. T. C. Clarke et al., *J. Chem. Soc. Chem. Comm.*, **1978**, 489.

47. H. Shirakawa and S. Ikeda, *Polymer J.*, **2**, 231 (1971); H. Shramm et al., *J. Chem. Soc. Chem. Comm.*, **1978**, 578; *Makromolekulare Chemie, 179*, 1565 (1978).

49. D. M. Ivory et al., *J. Chem. Phys.*, **71**, 1506 (1979).

48. Editorial, *Physics Today*, **32**, 19 (1979).

50. V. Percel et al., *Plaste Kautsch.*, **25**, 222 (1978).

51. C. K. Chiang et al., *Ber. Bunsen Ges. Phys. Chem.*, **83**, 407 (1979); *J. Chem. Phys.*, **69**, 5098 (1978); *J. Am. Chem. Soc.*, **100**, 1013 (1978); *Phys. Rev. Lett.*, **39**, 1098 (1978); C. R. Fincker et al., *Solid State Comm.*, **27**, 489 (1978).

52. I. B. Goldberg et al., *J. Chem. Phys.*, **70**, 1132 (1979).

53. R. W. Warfield and M. C. Petree, *Makromol. Chem.*, **58**, 139 (1962); S. Saito et al., *J. Polymer Sci.*, **A-2 6**, 1297 (1968).

54. R. Somoano, S. P. S. Yen and A. Rembaum, *Polymer Lett.*, **8**, 467 (1970).

55. J. R. Hanscomb and Y. Kaahwa, *J. Phys.*, **D-12**, 579, 587 (1979).

56. K. Hattori and Y. Wada, *Polymer Phys.*, **B**, 1963 (1975).

57. J. Bernasconi, *Phys. Revs.*, **B-7**, 2252 (1973).

58. S. Alexander et al., *Phys. Rev.*, **B-7**, 4311 (1978).

59. J. H. Ranicar and R. J. Fleming, *J. Polymer Sci.*, **A-2**, 1321 (1972).

60. K. Yoshino et al., *Am. Rep. Conf. Electr. Insul. Dielectr. Phenom.*, **1976**, Publ., 1978, p. 46.

61. H. St. Onges and A. C. Le Gloan, *Ann. Rep. Conf. Elect. Insul. Dielectr. Phenom.*, p. 56, **1976**, publ., 1978.

62. H. Hetzler and E. Kay, *J. Appl. Phys.*, **49**, 5617 (1978).

63. N. Islam et al., *Bull. Chem. Soc. Japan*, **51**, 2712 (1978); C. A. Angell, *J. Phys. Chem.*, **69**, 2137 (1965).

64. P. K. Watson, *Ann. Rept. Elect. Insul. Dielect. Phenomena*, **1976**, p. 10, publ., 1978.

65. e.g., G. P. Johari, *Ann. N.Y. Acad. Sci.*, **279**, 117 (1976); Yu. S. Lipatov, *Russ. Chem. Revs.*, **47**, 186 (1978); E. N. Dalal and P. Phillips, *Bull. Amer. Phys. Soc.*, **24**, 481 (1979).

66. M. Jarrigeon et al., *Bull. Revs. Phys. Soc.*, **24**, 286 (1979).

67. J. Hong and J. O. Brittain, *Bull. Amer. Phys. Soc.*, **24**, 481 (1979).

68. e.g., T. J. Lewis, *Ann. Rep. Conf. Electr. Insul. Dielectr. Phenomena*, **1976**, p. 533, publ. 1978; V. P. Lebev, *Russ. Chem. Revs.*, **47**, 69 (1978); see also Ref. 65; F. Gutmann and H. Keyzer, *Electrochim. Acta*, **13**, 693 (1968).

69. Y. W. Park et al., *Bull. Amer. Phys. Soc.*, **24**, 328 (1979).

70. K. Saha, J. C. Abbi and H. A. Pohl, *J. Non-Cryst. Solids*, **22**, 291 (1976); A. R. Blythe, *Elect. Prop. of Polymers,* Cambridge Univ. Press, 1979; M. Kryszewski, *Semiconducting Polymers,* Wiley, New York, 1978; V. Adamec and J. H. Calderwood, *IEE Conf. Publ.*, **1979**, 177, 263.

71. R. Pethig, Sci. Papers Inst. Org. Phys. Chem. Wroclaw Techn. Univ. No. T, Ser. Conf. No. 1 (1974).

72. H. A. Pohl, *J. Biol. Phys.*, **2**, 113 (1974).

73. H. A. Pohl and M. Pollak, *J. Chem. Phys.*, **66**, 4031 (1977), and reference cited therein; see also ref. 74.

74. R. Rosen and H. A. Pohl, *J. Polym. Sci.*, **A-4**, 1135 (1966); H. A. Pohl, *J. Polym. Sci.*, Part C-17, 13 (1967); R. D. Hartmann and H. A. Pohl, *J. Polym. Sci.*, Part **A-6**, 1135 (2968); J. R. Wyhof and H. A. Pohl, *J. Polym. Sci.*, Part **A-8**, 1741

REFERENCES

(1970); W. S. Chan and A. K. Jonscher, *Phys. Status Solidi,* **32**, 749 (1969); H. A. Pohl and J. R. Wyhof, *J. Polym. Sci.,* Part **A-10**, 387 (1972); J. R. Wyhof and H. A. Pohl, *J. Non-Cryst. Solids,* **11**, 137 (1972); H. A. Pohl, *J. Biol. Phys.,* **2**, 113 (1974).

75. J. Heleskivi, T. Salo and T. Stubb, "The Conduction Mechanism in Monocrystalline Selenium," *Valtion Teknillinen Tutkimulaitos,* No. 147, Helsinki, 1969.

76. L. K. H. van Beek, *Physica* (Utrecht), **29**, 1323 (1963); V. H. Schmidt, J. E. Drumbell and F. L. Howell, *Phys. Rev.,* **B-4**, 4582 (1971).

77. J. R. Wyhof, "The Nature of Hyperelectronic Polarization," Oklahoma Univ., Ph. D. Thesis (1970).

78. R. A. Graham, *J. Phys. Chem.,* **83**, 3048 (1979), and references cited therein.

79. J. Griffiths, *Color and Constitution of Organic Molecules,* Academic Press, N.Y., 1977; F. P. Schafer, *Angew. Chem.,* **82**, 25 (1970); J. Kosar, *Light Sensitive Systems,* Wiley, N.Y., 1965; E. Gurr, *Synthetic Dyes,* Academic Press, Lond., 1971; E. N Abrahart, *Dyes and Their Intermediates,* Arnold and Co., Lond., 1977.

80. D. Leupold et al., *J. Chem.,* **10**, 409 (1970).

81. J. I. Steinfeld, *Molecules and Radiation,* Harper and Row, N.Y., 1974; A. E. Siegman, *An Introduction to Lasers and Masers,* McGraw Hill, N.Y., 1971; V. V. Gruzinski and S. V. Davydov, *Zh. Prikl. Specktroskop.,* **28**, 224 (1978).

82. J. Sagiv and E. E. Polymeropoulos, *Ber. Bunsen Ges. Phys. Chem.,* **82**, 882 (1978).

83. O. Inacker, *Chem. Phys. Lett.,* **27**, 317 (1974).

84. R. König, *J. Lumin.,* **9**, 113 (1974).

85. E. E. Polymeropoulos et al., *J. Chem. Phys.,* **68**, 3918 (1978).

86. *Signalaufzeichnungsmaterialien,* **5**, 39 (1977).

87. R. C. Nelson, *J. Molec. Spectroscop.,* **23**, 213 (1967).

88. G. F. Nutt and S. C. Haydon, Molec. Rate Processes Symp., (1975).

89. F. Dietz et al., *Signalaufzeichnungsmaterialien,* **5**, 347 (1977).

90. J. Langer, *Pr. Naulk Inst. Chem. Org. Fiz. Politech.* Wroclaw (Poland), **16**, 221 (1978).

91. N. Petruzella, S. Takeda and R. C. Nelson, *J. Chem. Phys.,* **47**, 4247 (1967).

92. V. A. Krongauz et al., *Nature* (London), **271**, 43 (1978); *J. Phys. Chem.,* **82**, 2469 (1978); V. D. Ermakova et al., *Russ. Chem. Revs.,* **46**, 145 (1977).

93. Y. Yamaoka et al., *J. Sci.,* Hiroshima Univ., **A-34**, 1 (1970); *Bull. Chem. Soc. Japan,* **47**, 611 (1974).

94. Z. B. Pawelka and L. Sobczyk, *J. Chem. Soc. Faraday I,* **76**, 43 (1980).

95. D. Möbius, *Ber. Bunsen Ges. Phys. Chem.,* **82**, 848 (1978); H. Kuhn et al., *Phys. Methods of Chemistry,* A. Weissberger and B. W. Rossiter, eds., Vol. 1, and IIIB. Wiley, N.Y., 1972.

96. J. Sagiv and E. E. Polymeropoulos, *Ber. Bunsen Ges. Phys. Chem.,* **82**, 883 (1978); E. E. Polymeropoulos, et al., *J. Chem. Phys.,* **68**, 3918 (1978).

97. W. H. Weber and C. R. Eagen, *Bull. Amer. Phys. Soc.,* **24**, 441 (1978).

98. J. Heesemann, *Ber. Bunsen Ges. Phys. Chem.,* **82**, 868 (1978).

99. R. C. Nelson, *J. Chem. Phys.,* **47**, 4451 (1967).

100. R. C. Nelson, *J. Phys. Chem.,* **71**, 2517 (1967).

101. L. C. Isett and E. H. Perez-Albuerne, *Res. Discl.,* **158**, 59 (1977); G. D. Stucky et al., *Ann. Revs. Mater. Sci.,* **7**, 301 (1977); J. S. Miller and A. F. Epstein, *Prog. Inorg. Chem.,* **20**, 1 (1976).

102. T. Takizawa et al., *J. Phys. Chem.,* **81**, 1845 (1977); **82**, 1391 (1978); H. G. and F. Willig, *Top. Curr. Chem.,* **61**, 31 (1976).

103. L. E. Lyons, Priv. Comm. (1980).

104. A. T. Vartanyan, *Dokl. Akad. Nauk. SSSR,* **94**, 829 (1954).

105. E. Silinsh, *Electronic States of Organic Molecular Crystals*, Zinat, Riga, SSSR (1978).
106. R. R. Chance and C. L. Braun, *J. Chem. Phys.*, **59**, 2269 (1973); **64**, 3573 (1976); L. E. Lyons and K. A. Milne, *ibid.*, **65**, 1474 (1976).
107. H. J. Uth and D. Woehrle, *Bremer Briefe Chem.*, **2**, 5 (1978); R. C. Powell and Z. G. Soos, *Phys. Rev.*, **B-5**, 1547 (1972); Z. G. Soos, and R. C. Powell, *ibid.*, **B-6**, 4035 (1972); W. E. Spear, *Adv. Phys.*, **26**, 811 (1977); S. Ramdall, *Chem. Phys. Lett.*, **60**, 320 (1979); J. Weigl, *Photochem. Photobiol.*, **16**, 291 (1972).
108. S. Kagoshima et al., *J. Phys. Soc. Japan*, **39**, 1143 (1975); S. K. Khanna et al., *Phys. Rev.*, **B-16**, 1468 (1977); P. Bak and J. Emergy, *Phys. Rev. Lett.*, **36**, 751 (1976); T. D. Schulz and S. Etemad, *Phys. Rev.*, **B-13**, 4928 (1976).
109. *e.g.*, Y. Tomkiewicz et al., *Solid State Comm.*, **23**, 471 (1977); *Phys. Rev. Lett.*, **43**, 1532 (1979); NATO Conf. Ser. (Ser. 6) 1978, Publ. 1979, p. 43.
110. W. F. Cooper et al., *Chem. Commun.*, **1971** 889; *Cryst. Struct. Commun.*, **3**, 23 (1974).
111. C. R. Fincher et al., *Solid. Stat. Commun.*, **27**, 489 (1978); T. C. Clarke and G. B. Street, IBM Symp. Struct. and Prop. of Highly Condctg. Polymers and Graphite, San Jose, Calif., 1979, p. 2.
112. S. L. Hsu, *J. Chem. Phys.*, **69**, 106 (1978).
113. P. C. Mehendru, *J. Chem. Phys.*, **69**, 31 (1978).
114. W. D. Gill, "Proc. 5th Int. Conf. Am. Ph. Liq. Semicond.," Garmisch Partenkirchen, Taylor and Francis, London, 1974, p. 901.
115. J. Mort et al., *Solid State Commun.*, **18**, 693 (1976).
116. R. C. Powell, *J. Chem. Phys.*, **58**, 20 (1973).
117. D. Donati and J. O. Williams, *Mol. Cryst. Liq. Cryst.*, **44**, 23 (1978).
118. T. Sakurai, *Japan J. Appl. Phys.*, **13**, 901 (1974).
119. T. Tamura et al., *Nippon Kagaku Kaishi*, **1978**, 1209.
120. G. Pfister et al., *J. Chem. Phys.*, **57**, 2979 (1972).
121. E. Krikorian and R. J. Sneed, *J. Appl. Phys.*, **40**, 2306 (1969).
122. A. G. MacDiarmid and A. J. Heeger, IBM Symp. Struct. and Prop. Highly Conduct. Polymers and Graphite, IBM, San Jose, Calif., (1978).
123. H. Meier, *Spectral Sensitization*, Focal Press, London, 1968; K. Hauffe, *Reactionen in u. an Festen Stoffen*, Springer, Berlin, 1966; "Proc. Symp. Dye Sensitization," Bressanone, (1969); Focal Press, London, 1970; K. Rajeshwar et al., *Electrochim. Acta*, **23**, 1117 (1978); K. Rajeshwar et al., *Electrochim. Acta*, **23**, 117 (1978); W. F. Berg et al., *Dye Sensitization Symposium–Bressanone*, Focal Press, New York, 1970; H. Tsubomura et al., *Elektrokhimiya*, **13** (11), 1689 (1977); J. C. Kurracose et al., *Indian J. Chem.*, **A-16**, 568 (1978); M. V. Alfimov and O. B. Yakusheva, *Usp. Khim.*, **48**, 585 (1978); *Russ. Chem. Rev.*, **48**, 317 (1979); *Proc. Symp. Non-Silver Photog. Processes*, R. J. Cot, ed., Academic, New York, London, 1975.
124. I. A. Akimov et al., *Phys. Stat. Solidi*, **20**, 771 (1967); H. Kokado et al., *J. Phys. Chem. Solids*, **32**, 2785 (1971).
125. R. C. Nelson, *J. Phys. Chem.*, **71**, 2517 (1967); L. I. Grossweiner, *Photochem. Photobiol.*, **8**, 411 (1968).
126. S. J. Dudkowski et al., *J. Phys. Chem. Solids*, **28**, 485 (1967).
127. G. Heiland et al., *Photochem. Photobiol.*, **16**, 315 (1972).
128. L. I. Grossweiner, *Photochem. Photobiol.*, **8**, 411 (1968).
129. H. Kokado et al., *J. Phys. Chem. Solids*, **34**, 1 (1973).
130. Y. Hayashi et al., *Bull. Chem. Soc. Japan*, **39**, 1664 (1966).
131. M. Matsumura et al., *Bull. Chem. Soc. Japan*, **50**, 2533 (1977).

132. M. Vodenicharova, *J. Phys. Chem. Solids*, **36**, 1241 (1975).
133. W. D. K. Clark and N. Sutin, *J. Amer. Chem. Soc.*, **99**, 4676 (1977); C. T. Lin et al., *ibid.*, **98**, 6536 (1976).
134. S. Yamashita, *Bull. Chem. Soc. Japan*, **46**, 2744 (1973).
135. E. Ivove et al., *ibid.*, **47**, 3181 (1974).
136. C. P. Kesztheli and A. J. Bard, *Phys. Lett.*, **24**, 300 (1974).
137. M. Kikuchi et al., *Bull. Chem. Soc. Japan*, **47**, 1331 (1974).
138. Y. Shirota et al., *ibid.*, p. 991; *cf.*, also ref. 31.
139. K. Hauffe et al., *J. Electrochem. Soc.*, **117**, 993 (1970); A. D. Krishina and A. V. Vannikov, *Russ. Chem. Rev.*, **48**, 746 (1978).
140. J. J. Robillard, French Pat. 73-41-591 (1973); C. R. Acad. Sci. (Paris), **274**, 396 (1972); "Non-Silver Photogr. Processes Symp. 1973," Publ. 1975, p. 113, R. J. Cox, ed., Academic Press, London, 1975.
141. "Proc. Symp. on Non-Silver Photogr. Processes," Committee Royal Soc. London, Photogr. Soc., Oxford, U.K., 1973; Applied Optics, Suppl. No. 3, Electrophotogr. (1969); M. E. Trukhan, *Zh. Nauchn. Prikl. Fotogr. Kinematog.*, **22**, 49 (1977); J. J. Robillard and J. P. Gilbert, "Techniques Philips," No. 3, 2 (1970).
142. W. A. Huffman et al., German Pat., 2609149, 9th Sept., 1976.
143. M. Ideka, *Photo. Sci. Eng.*, **19**, 60 (1975); Sh. Marugima et al., Japan. Pat., 7145063, 23 June, 1971; German Pat., 223003, Dec. 23, 1972.
144. R. M. Schaffert, *IBM J. Res. Develop.*, **15**, 75 (1971).
145. K. Morimoto and Y. Murakami, *Appl. Optics Suppl.*, **3**, 50 (1969).
146. R. C. Nelson, *J. Photogr. Sci.*, **22**, 286 (1974).
147. J.-M. Brot and J. J. Robillard, *C. R. Acad. Sci.* (Paris), **277B**, 167 (1973).
148. H. G. Willig and F. Willig, *Top. Curr. Chem.*, **61**, 31 (1976); H. Inokucki and Y. Maruyama, in *Photoconductivity and Related Phenomena*, J. Mort and D. H. Pai, eds., Elsevier, Amsterdam, (1976); p. 155; H. Inokuchi, *Disc. Faraday Soc.*, **51**, 183 (1971); H. Kawazura et al., *Bull. Chem. Soc. Japan*, **47**, 1829 (1974); F. Gutmann and J.-P. Farges, in *Modern Aspects of Electro-Chemistry*, J. O'M. Bockris and B. E. Conway, eds., Plenum, N.Y., 1980; S. Roy Morrison, *Chem. Tech.*, **7**, 570 (1977).
149. Y. Yokomoto et al., *Japan J. Appl. Phys.*, **15**, 2137 (1976); C. E. Okeke, Yeshiva Univ. Bronx, N.Y. Diss. (1977), Univ. Microfilms, No. 7732535.
150. M. Sawoda et al., *Jpn. Kokai*, 77,141,229, Nov. 25, 1977.
151. H. Hoerhogd et al., East German Pat., 127,983, Oct. 26, 1977.
152. P. O. Sliva et al., Belg. Pat., 856,316, Oct. 17, 1977.
153. M. Satomura, *Japan Kokai*, 77,121,085, Oct. 12, 1977.
154. T. Takizawa et al., *J. Phys. Chem.*, **82**, 1391 (1978); T. Watanabe et al., *ibid.*, **81**, 1845 (1977).
155. J. Bednar, *Radiochem. Radioanal. Lett.*, **32**, 233 (1978).
156. H. Kawazura et al., *Bull. Chem. Soc. Japan*, **47**, 829 (1974), and reference cited therein.
157. K. Kimura and H. Inokuchi, *J. Catalysis*, **29**, 49 (1973), and reference cited therein.
158. J. O. Williams, *Adv. Phys. Org. Chem.*, **16**, 159 (1978).
159. K. H. Probst and N. Karl, *Phys. Stat. Solidi*, **27A**, 499 (1975).
160. N. Karl, *Adv. Solid State Phys.*, **14**, 261 (1974).
161. H. Meier and W. Albrecht, *Z. Naturforsch.*, **24**, 257 (1969).
162. H. J. Uth and D. Wohrle, *Bremer Briefe Chem.*, **1**, 23 (1977).
163. V. M. Lenchenko, *Fiz. Khim. Mikroelectron.*, **21** (1976).

164. A. Samoc et al., *Pr. Nauk. Inst. Chem. Org. Fiz. Politech.*, Wroclaw (Poland), 16, 267 (1978).
165. L. E. Lyons and L. J. Warren, *Aust. J. Chem.*, 25, 1427 (1972).
166. D. Donati and J. O. Williams, *Molec. Cryst. Liq. Cryst.*, 44, 23 (1978).
167. J. R. Harbour and G. Tollin, *Photochem. Photobiol.*, 19, 147 (1974).
168. Several Papers in: *Faraday Disc. Chem. Soc.*, 58, (1974).
169. H. Gerischer, Ref. 168, p. 219.
170. E. E. Polymeropoulos et al., *J. Chem. Phys.*, 68, 3918 (1978).
171. D. Möbius, *Ber. Bunsen Ges. Phys. Chem.*, 82, 848 (1978); W. Arden and P. Fromherz, *ibid.*, p. 868; J. Sagin and E. E. Polymeropoulos, *ibid.*, p. 883.
172. *e.g.*, N. G. Rule, German Pat., 2,557,398, (July 1, 1979); A. M. Horgan, U.S. Pat., 4,047949, (Sept. 13, 1977); M. Miyagawa, *Jpn. Kokai*, 77,64,934, (May 28, 1977); *cf.*, also refs. 140 and 143; J. W. Weigl, *Angew Chem. Inst.* (Ed. Engl.), 16, 374 (1977); L. Corrsin, Conf. Rec.-IAS Ann. Meeting, IEEE, 1975, p. 289; "Electrophotography," D. R. White, ed., *Soc. Photog. Sci. Eng.*, Washington, D.C., 1974.
173. *e.g.*, D. M. Schold, *J. Photochem.*, 8, 39 (1978); M. Olson, *Science*, 168, 438 (1970); L. Y. Johansson et al., *Electrochim. Acta*, 18, 255 (1973); F. Beck, *Ber. Bunsen Ges. Phys. Chem.*, 77, 353 (1973); Y. Umezawa and T. Yamamura, *J. Electrochem. Soc.*, 126, 705 (1979).
174. G. Pfister et al., *J. Chem. Phys.*, 57, 2979 (1972); E. P. Goodings, *J. Chem. Soc. Rev.*, 5, 95 (1976); D. J. Williams et al., *J. Am. Chem. Soc.*, 94, 7970 (1972).
175. G. Pfister et al., *Phys. Rev. Lett.*, 37, 1360 (1976).
176. C. David et al., *Europ. Polymer. J.*, 6, 537 (1970).
177. G. Geusekens and C. David, "23rd Internatl. Cong. Proc. Prop. and Appl. Chem.," Boston, USA, 1971, Butterworth, London, 1971, p. 19.
178. C. W. Frank and M. A. Gashgari, *Bull. Amer. Phys. Soc.*, 24, 257 (1979).
179. P. C. Mehendru et al., *Proc. Nucl. Phys. Solid State Symp.*, 20C, 142 (1977).
180. V. D. Ermakova et al., *Russ. Chem. Revs.*, 46, 145 (1977); *cf.*, also, *Russ. Chem. Rev.*, 48, 596 (1979).
181. A. D. Buckingham et al., *Organic Liquids*, Wiley, N.Y., 1978; Several papers in "Proc. 8th Internatl. Conf. Amorph. and Liquid Semiconductors," Cambridge, Mass., 1979, Nth Holland Publ. Co., Amsterdam, 1979; several papers in "Proc. 6th Int. Conf. Conduct. Breakdown of Dielect. Liquids," Editions Front, Dreux, Fr., 1978; N. F. Mott, *Recherche*, 9, 846 (1978); M. Cutler, *Liquid Semiconductors*, Academic Press, N.Y.,1977; "Proc. 5th Internatl. Conf. Amorph. and Liquid Semicond.," Garmisch-Partenkirchen, Taylor and Francis, London, 1974; "Conduction and Breakdown in Dielect. Liquids Proc. Int. Conf. 5th," 1975, J. M. Goldschwartz et al., eds., Delft Univ. Press, 1975; T. J. Gallagher, *Simple Dielectric Liquids*, Clarendon, Oxford 1975; "Proc. 6th Internatl. Conf. Amorph. and Liquid Semi-Cond.," B. T. Kolomeits, ed., Nauka, Leningrad, 1976; D. Chandler, "Structure of Molecular Liquids," *Ann. Rev. Phys. Chem.*, 29, 441 (1978); J. G. Powles, "Structure of Molec. Liquids by Neutron Scattering," *Adv. Phys.*, 22, 1 (1973); N. F. Mott, *Rev. Mod. Phys.*, 50, 203 (1978).
182. M. R. Belmont, in *Conduct and Breakdown of Diel. Liqu. Proc.* 5th Int. Conf., J. M. Goldshwartz et al., eds., Delft Univ. Press, 1975.
183. J. Kaminski and A. Kawski, *Z. Naturforsch.*, A-32, 1339 (1977).
184. I. M. Williams et al., "Proc. 6th Int. Conf. Breakdown Diel. Liquid.," 1978, p. 48, Editions Front, Dreux, France.
185. J. Gelsdorf and H. Krausse, ref. 184, p. 345.
186. R. G. Romanets et al., *Elektron Obrab. Mater.*, 1976, 44.
187. I. Licea et al., *An. Univ. Bucuresti. Fiz.*, 26, 53 (1977).

188. N. N. Krasnikov, *Elektron Obrab. Mater.*, **1977**, 61.
189. K. G. Bobyl et al., *Elektron Obrab. Mater.*, **1976**, 41.
190. J. Guillet, *Angew. Makromolek. Chemie*, 68, 163 (1978).
191. K. C. Kao, "Proc. 6th Int. Conf. Conduct. Breakdown Dielectr. Liquid.," Edition Front, Dreux, France, p. 65, (1978); Y. Inushi, *ibid.*, p. 351.
192. D. S. Newman et al., "Proc. Int. Symp. Molten Salts," p. 168, 1976, J. P. Pemsler, J. Braunstein and K. Nobe, eds., Electrochem. Soc., Princeton, N.Y., 1976.
193. V. M. Glazov et al., *Zh. Fiz. Khim.*, 51, 2202 (1977).
194. A. Mozumder, *IEEE Trans. Nuclear Sci.*, NS26, (1, Pt. 1) 129 (1979).
195. A. A. Balakin et al., *Khim. Vys. Energ.*, 12, 210 (1978).
196. H. Hase and T. Kigashimura, *J. Phys. Chem.*, 83, 822 (1979).
197. Yu. A. Berlin et al., *J. Chem. Phys.*, 69, 2401 (1978).
198. D. F. Evans et al., *J. Phys. Chem.*, 83, 2669 (1979).
199. J. Barthel et al., *Modern Aspects of Electrochemistry*, 13, 1 (1979); J. O'M. Bockris and B. E. Conway, eds., Plenum Press, N.Y., 1979.
200. H. T. Davis and R. G. Brown, "Low Energy Electrons in Non-Polar Fluids," in *Adv. in Chem. Phys.*, p. 32., I. Prigogine and S. A. Rice, eds., Wiley, N.Y., 1975.
201. R. A. Holroyd and N. E. Cipollini, *J. Chem. Phys.*, 69, 501 (1978).
202. L. Nyikos et al., "Proc. Tihany Symp. Radiat. Chem., 1976," Publ. 1977, 4, 179 (1977).
203. M. P. De Haas in "Conduct. and Breakdown in Dielect. Liquids, Proc. 5th Int. Conf., 1975," J. M. Goldshwartz, ed., Delft Univ. Press, 1975.
204. Yu. A. Berlin et al., Hungar. Acad. Sci. Cent. Res. Inst. Phys., KFKI-1978-16.
205. L. G. Christophorou et al., *Adv. Chem. Res.*, 36, 413 (1977).
206. J. P. Dodelet et al., *J. Chem. Phys.*, 59, 1293 (1973).
207. I. Klinowski et al., "Conduct and Breakdown of Dielect. Liquid., Proc. 5th Int. Conf. 1975," J. M. Goldschwartz, ed., Delft Univ. Press, p. 11, 1975.
208. W. Doldissen et al., "Proc. 6th Int. Conf. Conduct. Breakdown Dielect. Liquid., 1978," p. 197, Edit. Front, Dreux, France, 1978.
209. K. I. Zamaraev and R. F. Khairutdinov, *Chem. Phys.*, 4, 181 (1974); W. M. Bartzack, J. Korb, E. Romanowsky and C. Z. Stradowski, *Curr. Top. Radiat. Res. Q.*, 11, 307 (1977); E. J. Marshall, M. J. Pilling, and S. A. Rice, *J. Chem. Soc. Faraday Trans.*, 2, 71 1555 (1975); F. S. Dainton, N. J. Pilling and S. A. Rice, *J. Chem. Soc. Faraday Trans.*, 2, 71, 1311 (1975); B. G. Ershov, and F. Kieffer, *Nature* (London), 252, 118 (1974); B. G. Ershov, V. M. Byakov and N. L. Sukhov, *Dokl. Akad. Nauk. SSSR*, (1976) 1097; *cf.*, also refs. 211 and 213.
210. I. Webman and N. R. Kestner, *J. Phys. Chem.*, 83, 451 (1979).
211. J. R. Miller, *J. Phys. Chem.*, 79, 1970 (1975); J. V. Beltz and J. R. Miller, "Tunnelling Distances and Exothermic Rate Restrictions in Electron Transfer Reactions." "Proceedings of the Conference on Tunnelling in Biological Systems," Philadelphia, Pa., Fall, 1977.
212. J. R. Miller, *Chem. Phys. Lett.*, 22, 180 (1973); K. I. Zamaraev and R. F. Khairutdinov, *Chem. Phys.*, 1, 181 (1974); E. J. Marshall et al., *J. Chem. Soc. Faraday Trans. II.*, 71, 1555 (1975); F. S. Dainton et al., *ibid.*, 1311; B. G. Ershov et al., *Dokl. Akad. Nauk SSSR*, (1970) 1097; *cf.*, also ref. 31.
213. J. R. Miller, *Science*, 130, 221 (1975); R. F. Khairutdinov and K. I. Zamaraev, *Dokl. Akad. Nauk, SSSR*, 222, 654 (1975).
214. W. M. Bartzak et al., *Curr. Top Radiat. Res. Q.*, 11, 307 (1977); B. G. Ershov et al., *Dokl. Akad. Nauk. SSSR*, (1976) 1097.
215. J. V. Beitz and J. R. Miller, "Tunnelling Distances and Exothermic Rate

Restrictions in Electron Transfer Reactions," Proc. Conf. "Tunnelling in Biological Systems," Philadelphia, Pa., 1977.

216. J. Maher, "Annual Rep. Conf. Elect. Dielect. Phenom.," p. 25, 1977, Publ., 1979.

217. B. M. Dikarev, *Zh. Fiz. Khim.*, **50**, 1300 (1976).

218. L. S. Kazatskaya et al., *Elektron. Obrab. Mater.*, (1978) 64.

219. V. A. Kronganz et al., *Nature* (London), **241**, 43 (1978); *Z. Phys. Chem.*, **82**, 2469 (1978).

220. E. I. Kovshev et al., *Russ. Chem. Revs.*, **46**, 395 (1977), see also ref. 38.

221. M. Abkowitz, *Bull. Amer. Phys. Soc.*, **24**, 353 (1979).

222. L. S. Levitt and E. T. Hsieh, *J. Am. Chem. Soc.*, **101**, 4664 (1979).

223. M. S. Jhon and H. Eyring, J. Chem. Phys., 44 1465 (1966); Proc. Natl. Acad. Sci. USA, 54, 1419 (1965); H. Eyring, in Physical Chemistry, An Advanced Treatise, Vol. o6 VIII-A, Ch. 5, Academic Press, N.Y., 1971.

224. R. S. Ree, T. Ree and H. Eyring, *Proc. Natl. Acad. Sci. USA.*, **48**, 501 (1962); K. Liang, H. Eyring and R. P. Marchi, *ibid.*, **52**, 1107 (1964).

225. C. D. Landau, *Zh. Eksp. Teor. Fiz.*, **30**, 1058 (1956); **32**, 59 (1957); **35**, 97 (1958).

226. R. Freedman and J. A. Hertz, *Phys. Rev.*, **B-15**, 2384 (1979).

227. S. S-S Huang and G. R. Freedman, *J. Chem. Phys.*, **70**, 1538 (1979).

228. L. Kevan, *J. Phys. Chem.*, **82**, 1144 (1978).

229. Y. Nakato et al., *J. Phys. Chem.*, **76**, 2105 (1972).

230. S. Noda et al., *J. Phys. Chem.*, **79**, 2866 (1975).

231. T. Sato, "Dielectr. Liqu. Conf., 1978 Proc.," p. 211.

232. D. C. Licciardello. Les Houches Proc. 1978, *Bull. Amer. Phys. Soc.*, **24**, 493 (1979).

233. I. H. Leaver, *Photochem. Photobiol.*, **19**, 309 (1974).

234. H. Fritzsche, *Solid State Comm.*, **9**, 1813 (1971).

235. P. M. Chaikin and G. Beni, *Phys. Rev.*, **B-13**, 2, 647 (1976).

236. G. Beni, J. F. Kwak and P. M. Chaikin, *Solid State Commun.*, **17**, 1549 (1975).

237. J. F. Kwak and G. Beni, *Phys. Rev.*, **B-13**, 652 (1976).

238. R. A. Bari, *Phys. Rev.*, **B-9**, 4329 (1974); D. Emin, *Phys. Rev. Lett.*, **35**, 882 (1975).

239. E. Muller et al., *Phys. Stat. Solidi*, **33**, K55 (1969); D. Zosel et al., *ibid.*, **38**, 177 (1970).

240. A. Brau, D. Sc. Thesis, University of Nice (1976).

241. H. B. Brom et al., *Bull. Amer. Phys. Soc.*, **24**, 326 (1979).

242. C. B. Duke et al., *Phys. Rev.*, **B-18**, 5717 (1978); C. B. Duke, *Surf. Sci.*, **70**, 674 (1978).

243. J. Jaud et al., *C. R. Acad. Sci.* (Paris), C278, 769 (1974).

244. A. Brau and J.-P. Farges, *Phys. Stat. Solidi*, B61, 257 (1974); J.-P. Farges et al., *J. Phys. Chem. Solids*, **33**, 1723 (1972); A. Brau et al., *Phys. Stat. Solidi*, A-51, K45 (1979); J.-P. Farges and A. Brau, *ibid.*, B64, 269 (1974); B61, 669 (1974); *cf.*, also refs. 240, 247.

245. I. F. Shchegolev, *Phys. Stat. Solidi*, A21, 9 (1972).

246. M. J. Rice et al., *Solid State Commun.*, **21**, 757 (1977).

247. J.-P. Farges, D. Sc. Thesis, Université de Nice, France (1974).

248. R. M. Vlasova et al., *Sov. Phys. Solid State*, **12**, 2979 (1971); *cf.*, also S. K. Khanna et al., *Phys. Rev.*, **B-10**, 2139 (1974).

249. J. B. Torrance, "Conf. on Org. Conduct. and Semicond., Siofok (Hungary), 1976, Proc.," Publ. House Hungar., Acad. Sci., Budapest, 1977.
250. K. Holczer et al., Hungarian Acad. Sci. (Budapest), *Cont. Res. Inst. Phys.,* KFKI 1977, 58; see also K. Kral, *Czech. J. Phys.,* **B-27**, 777 (1977).
251. M. Sugi et al., *Thin Solid Films,* **27**, 205 (1975).
252. N. Ueno et al., *J. Phys. Soc., Japan,* **48**, 1254 (1980).

17

Charge Transfer Complexes

17.1 Introduction

This subject has attracted a great deal of interest and several surveys are available.[1] This chapter is intended to supplement and to extend the discussion in Section 8.4 and 9.12. Numerical data for a range of typical charge transfer complexes (CTC) are tabulated in Table 8.5A, while Table 8.5B deals with ternary CTC, Table 8.5C with protonic (*cf.*, *infra* and Section 17.9) CTC. Structural aspects of CTC are discussed in Section 19.7.

It will be convenient to separate CTCs arbitrarily into two main groups, *viz.*, weak complexes involving but little charge transfer and strong complexes involving considerable transfer of charge from donor to acceptor. Discussion of contact CTC in which interaction is but transitory, acting only while the molecules are in contact during collisions, will be omitted.

While Slifkin[2] distinguishes a third class of medium-strength complexes in which the ground state is dative to a certain degree though not to the extent encountered in the strong complexes, adducts mentioned will be somewhat arbitrarily assigned to either the weak or the strong category, though this classification, in many cases, may be debatable. The lower the ionization energy of the donor and the higher the electron affinity of the acceptor, the easier is complexation. However, the "strength" of a donor is not determined solely by its ionization potential; especially in solution it is more meaningful to express the electron-donating tendency of a donor in terms of its "donicity"[3] and, likewise to discuss the electron acceptance in terms of an "electron acceptance number."[4] Both these topics are discussed in Section 17.2. While this classification of electron donors and acceptors often proves useful, these terms are only relative; under appropriate conditions, such as when the highest occupied molecular orbital is located between the orbitals of the potential donor and acceptor, any molecule can exhibit both electron-donor and electron-acceptor properties.[356] Thus, dimethylalloxazine,[5] for example, act as a donor to the strong acceptor TCNE (tetracyanoethylene) and as an acceptor to the strong donor pyrene. This dual character applies particularly to π-bonded molecules and is especially true for large biological molecules[6] such as the purines[7] or cyclophanes.[8] Even the classic electron acceptor, oxygen, has recently been shown[9] to be capable of donating an electron to the magnesium phthalocyanine molecule, which has a high electron affinity of 14 eV, thus matching the ionization potential of O_2, *viz.*, 13 eV. The resulting CTC is of biological interest. The donor, having donated may act as an acceptor and *vice versa*; reverse charge-transfer complexes[6] are known in which, for example, the π-electron system of an aromatic compound acts as an electron acceptor rather than as a donor such as the ferrocenes[10] or benzoporphyrazines.[11]

Even the classical π-donor anthracene can accept electrons from alkali metals and form a variety of adducts.[12] Auto-complexes, ternary, surface and/or micellar complexes and adducts in which electron transfer is accompanied by proton transfer, thus forming proton complexes, are discussed in subsequent sections in this chapter. Charge-transfer complexes have also been obtained in the gas phase[13] as well as in melts.[14] The former, often weak to very weak contact CTC, are outside the scope of this book. Semiconducting free radical salts based on TCNQ are discussed in Section 16.11, while Chapter 21 deals with organic semi-metals, many of which are also TCNQ adducts though their conductivity is metallic.

Apart from the general distinction between weak and strong CTC, these complexes may be classified according to the molecular orbitals involved in the electron-donating and -accepting processes.

Donor and acceptors may be classified into three groups of (1) π-donors and acceptors, in which the reacting electron comes from or goes into the π-orbital of a conjugated system, so that charge-transfer bonding is delocalized over the π-system of the molecule: (2) σ-donors and -acceptors, in which the electron comes from or goes into a σ-orbital and (3) n-donors, in which the charge comes from a more or less localized electron: here, the charge-transfer bonding will be localized on the donor atom (*e.g.*, nitrogen) in the molecule. We may thus define six donor-acceptor combinations (the donor is named first): $\pi - \pi, \pi - \sigma, \sigma - \pi, \sigma - \sigma, n - \pi$, and $n - \sigma$.

These groupings, however, are not to be taken as mutually exclusive: thus, *e.g.*, crown ethers complex with TCNE yielding adducts having $\pi - \pi$ as well as $n - \pi$ character.[27] The magnitude of the interaction forces have been shown for many types of adducts to be of comparable magnitude,[15] though, in some cases at least it has been shown that the potential energy curve of a CTC exhibits two local minima corresponding to 2 energy states differing in their ionic character.[16] Complex formation is facilitated by a polar environment, though strong CTC in solvents of sufficiently high permittivity may dissociate into the component ions. The ground state of the CTC is destabilized as the solvent polarity is increased.[30] The dipole moment of the adduct has been related to the enthalpy of its formation,[31] which is usually[17] of the order of a few kilocalories per mole, in contradistinction to the tens of kilocalories per mole involved in conventional chemical reactions. Complex formation usually involves a lowering of the entropy of the system, because of the increased order, though in solution this may be masked by clustering, association and solvation effects. Entropy contributions are discussed in Section 6.9.

For high stability, charge transfer should be incomplete,[19] complete transfer, of course, would result in an ionic compound and not in a CTC. Thus, the entropy change involved in the ion-molecule equilibrium affects the complexation:[18] writing P for the probability that a reaction between donor and acceptor does lead to charge transfer, and employing subscripts F for the forward and R for the reverse reaction, Z being the rate constant of the reaction, the entropy change ΔS can be written[18] as

$$\Delta S = R\left[\ln\frac{Z_F}{Z_R} + \ln\frac{P_F}{P_R}\right] \tag{17.1}$$

The formation of a CTC is sometimes followed by a true chemical reaction forming a new chemical compound; this occurs, for example[20] in the reaction between aminoacids and iodine or between indole and chloranil[21] or between chlorpromazine and 6-hydroxydopamine;[22] it occurs frequently[23] in σ-bonded anthracene-alkali metal adducts. CTC complexation may become important in RedOx reactions; thus the reduction of methylene blue by various dihydroaromatic hydrocarbons has been found to be accelerated by a factor of 50 by the formation of CTC complexes with aromatic hydrocarbons.[24]

If the donor is a metal chelate, such as e.g., a metal-phthalocyanine, then it appears that the complexation with e.g., a quinone depends critically on the identity of the central, chelated, metal atom.[25] This would indicate that conductivity in such adducts is of the ligand-mediated mode (see Section 15.9 and 19.5). In extreme cases, such as in the Ni-phthalocyanine:I_2 CTC, the conductivity in the stacking direction is said[26] to approach the metallic mode. Often, the electronic conductivity is governed by the acceptors while hole conduction depends primarily on the donors component (see Section 19.7).

17.2 Donicity and Electron Acceptance Numbers

One would be tempted to equate the "donor strength" of a donor with its ionization potential. However, several donors of very nearly equal ionization potential appear to have quite differing "donor strengths;" this is so because the ionization potential measures the energy required to remove the electron from the gaseous molecule in its ground state to infinity while in the formation of a CTC the electron only has to be moved to the often contiguous acceptor molecule, the process being carried out not *in vacuo* but in a material medium such as a solvent.

Especially for the study of solutions, it has been found convenient to introduce the "donicity" DN, of donor D, defined[3] arbitrarily as the negative of the total enthalphy change measured in complexing with antimony-pentachloride:

$$DN \equiv -\Delta H_{D \cdot SbCl_5} \qquad (17.2)$$

As a reference, the solvent is 1,2-dichloroethane and the concentrations are assumed to be low. Some donicities are listed in Tables 8.21. The donicity may be used only as an approximate expression of the donor strength of a molecule towards a given substrate, though it has been found to serve as a most useful guide for the interpretation or prediction of a number of interactions,[3] such as:

(a) formation of donor-acceptor complexes,
(b) ionization of covalent compounds by means of donor molecules,
(c) autocomplex formation,
(d) rate of solvent substitution,
(e) solvation of metal ions and
(f) RedOx-equilibria in non-aqueous media.

The physical significance of the donicity concept is underlined by the following correlations:[4]

1. donicity values are proportional to the free energy changes in the respective interactions $S \rightarrow SbCl_5$
2. the donicities are linearly related to the enthalpy changes ΔH of the solvents relative to a given acceptor species
3. the donicities are linearly related to the NMR chemical shifts in different donor solvents due to adduct formation, *e.g.*, with trifluoro-iodomethane. The effect is said[4] to be independent of the nature of the functional group of the D. Chemical shifts in other systems are reported to behave likewise.[29] This matter is illustrated in Fig. 17.1.

The electron affinity, likewise, should be expected to enter the case, or otherwise, of the complexation reaction in a more subtle way than one would expect; in fact it has been shown[4] to be related to the effective dipole moment of the acceptor. It is a measure of the electrophilic behavior of a compound, in general, and of a solvent more specifically.

Electron acceptance numbers for several compounds are tabulated in Tables 8.21A and C. They represent dimensionless numbers expressing the acceptor properties of a given solvent relative to those of $SbCl_5$, which is also the reference substance for assessing the donicities.

An estimation of the coordinate bond energies is possible by making use of the formula:

$$\Delta H = \frac{DN \cdot AN}{100} + \Delta\Delta H_{Donor} + \Delta\Delta H_{Acceptor} \qquad (17.3)$$

If the latter terms are small, the coordinate bond energy can be estimated from the donor- and acceptor-numbers of the compound.

One of the most remarkable results is the extremely low electrophilic character of diethylether, while the non-polar solvents benzene and carbon tetrachloride have stronger electrophilic properties, which are not drastically different from those of most other aprotic solvents. Considerably greater acceptor numbers are found for solvents containing acidic C-H hydrogen atoms such as CH_2Cl_2, $CHCl_3$ or formamide. As expected, higher values are found for the alcohols and for water and to a greater extent for protonic acids.

For a more detailed discussion of solute-solvent effects in terms of donor/acceptor, the reader is referred to the literature.[28]

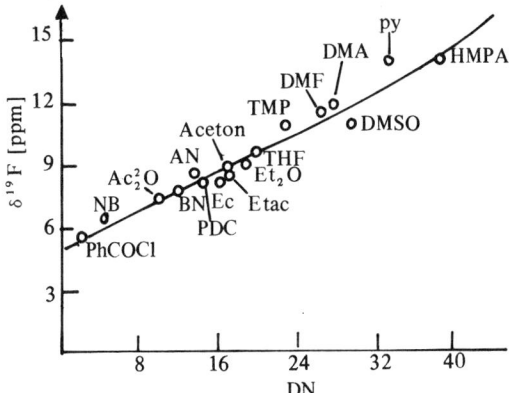

Fig. 17.1 Relationship between ^{19}F NMR chemical shifts CF_3Cl in different solvents and solvent donicity. After V. Gutmann.[4]

17.3 Weak Charge Transfer Complexes

The mainly nonionic, nondative ground state characteristic of these complexes may be associated with a high ionization potential of the donor, a low electron affinity of the acceptor, or both these effects. The stability of the complex results from van der Waals forces. The intermolecular distances are only slightly less than the sum of the van der Waals radii, and the complexes do not appear as chemical entities, but possess chemical properties close to the sum of the components. There is little charge-transfer stabilization of the ground-state. An almost complete electron transfer results from the transition to the first excited state, which may be due to the absorption of light, giving the complex the highly colored appearance often observed. Thus, for many so-called CTC^2, the charge transfer is significant only in the excited state, and it would be more appropriate to speak of "complexes showing charge-transfer absorption." It is, however, sometimes difficult to demonstrate the presence of the supposed absorption experimentally. Such radiation-(usually light)-activated adducts will be discussed in Section 17.5.

In weak charge transfer complexes, the photoemission yield curve (photo current *vs.* $h\nu$ of incident radiation) is just about identical with that of the pure donor; in weak complexes, thus, it appears that both donor and acceptor components maintain their individuality. Since the ionization potential of the acceptor is likely to be considerably higher than that of the donor, it is only the photoemission from the donor which is observed. This is exemplified in Fig. 17.2 for the anthracene-pyromellitic dianhydride (PMA) complex:[32] the electron band (1) of the complex corresponds well to the band of anthracene, while PMA does not show a photoemission band in the same energy region. Therefore, it is probable that the highest occupied electronic level of this complex consists mainly of an anthracene level as expected by general considerations of weak charge transfer complexes.

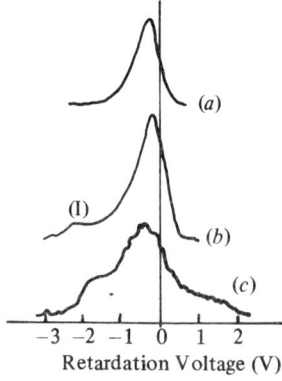

Fig. 17.2 Energy distribution curves of photo-electrons at an excitation energy of 9.18 eV. (*a*) PMA, (*b*) anthracene-PMA and (*c*) anthracene. After Ishii et al.[32]

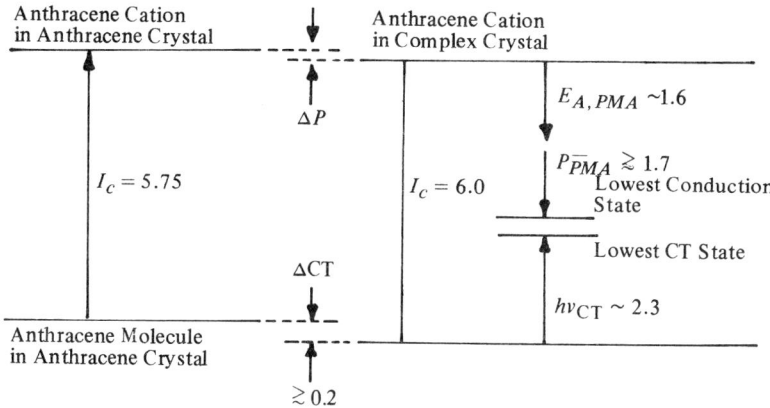

Fig. 17.3 Energy level diagram of photoemission processes in anthracene and anthracene-PMA. The energies are shown in eV. For discussion, see text. After Ishii et al.[32]

An energy level diagram for this adduct, a fairly typical example of a weak CTC, is shown in Fig. 17.3. The left hand side of the diagram[32] shows the photo-ionization process in anthracene, for comparison. The ground state represents a neutral molecule in an anthracene crystal. Through the photo-ionization process, an anthracene cation is produced in the anthracene crystal. The threshold energy of the ionization is 5.75 eV.[33] In the complex crystal shown on the right hand side of the figure, an anthracene molecule is surrounded by other anthracene and PMA molecules. The energy of an anthracene molecule under these circumstances is variable, probably lowered by the large contribution of the charge-transfer interaction, ΔCT. The photo-ionization process including the highest occupied level in the complex is considered to produce a cation state almost completely localized on the anthracene molecule. The threshold energy of the ionization is 0.6 eV. The resultant anthracene cation polarizes the surrounding molecules. This polarization energy is also different from that in an anthracene crystal. Since the dielectric constant of the complex is likely to exceed that of the donor anthracene[35] (the density of the latter is 1.25 g/cm^3 while that of the adduct is 1.49 g/cm^3), the energy of the anthracene cation state in the complex may be lowered by a small amount, ΔP. Therefore, the stabilization energy of the charge-transfer interaction, ΔCT, is not smaller than 0.2 eV. This value agrees qualitatively with an estimation by a molecular orbital calculation of the complex.[34]

The electrical conduction state contributing most to the photoconduction in the complex is considered to be an anthracene cation and PMA anion pair sufficiently separated in the crystal. The energy separation between the ground and lowest conduction states in the complex is estimated to be 2.7 eV or smaller. This value is to be compared with the charge-transfer excitation energy; the lowest band is located at 2.3 eV at room temperature.[36] Consequently, the energy separation between the charge-transfer and conduction states is inferred to be very small, and this conclusion agrees well with the recent observation of the activation energy of the photoconduction by Karl and Ziegler:[37] The activation energy of photoconduction is very small in the charge-transfer band region, and most of it is attributable to the activation energy of the mobility.

Other examples of interesting weak complexes are those of pyridine derivatives with TNB[39] or adducts of diphenyldiamine derivatives with suitable acceptors.[40]

It has been shown[38] that in substituted benzenes, oxygen-R[†] groups and sulfur-R groups, which in their usual planar configuration are π-donors, can also act as π-acceptors in their orthogonal configuration, as long as there is enough molecular orbital interaction between substrate and substituent.

The electron acceptor capability of these substituents is favored in negatively charged species since the energy gap between the occupied MO of the adjacent moiety and the vacant σ^*_{XR} bond orbital is lower than in neutral compounds. In these molecular species oxygen is a better π-donor than sulfur, while the sulfur-R group is a better π-acceptor than the oxygen R group.

Organometallic pseudohalides[††] may form CTC in which the pseudohalide acts either as donor or as acceptor; it may even form autocomplexes. This highly interesting field has been reviewed in detail.[42]

In a solvent, complexation is required by theory[41] to occur in two steps: a competitive solvation of the acceptor by the donor and solvent in the nearest-neighbor-cage where there is free molecular rotation, followed by a true D-A CTC interaction where the D and A maintain their relative and fixed positions.[41]

Calculations of the co-solubility of donors and acceptors in solid solutions leading to adduct formation are given by Glazov et al.[43] The solubility of halogens, specially of iodine, in water is said[44] to be associated with the formation of a CTC.

17.4 Strong Charge Transfer Complexes

The formation of such complexes involves both strong donors and strong acceptors: the ionization potential of the donor must be low and the electron

---

[†] R represents an organic residue.

[††] Pseudohalides are polyatomic groups resembling halides in their chemical properties; iso-cyanide, iso-cyanate, iso-thiocyanate, iso-selenocyanate, azide and fulminate. A typical organo*metallic* pseudohalide would be cacodyl cyanide, $(CH_3)_2SCN$.

affinity of the acceptor must be high.* The resulting molecular assocations exhibit a marked ionic character in their ground state. In this case, the charge-transfer forces, which contribute to the molecular interaction, are strong enough to produce an appreciable stabilization of the ground state of the complex; other contributions arise from electrostatic, induction and dispersion forces. The perturbation theory of Mulliken is less applicable under these circumstances, and other methods such as the variation method should be used.

In solution, solvation and association effects are likely; sometimes, these effects can be detected *via* NMR chemical shifts. The association constants of CTCs have thus been shown[45] to vary exponentially with the energy of the charge-transfer transition.

While no ESR absorption occurs in the case of weak complexes, broad bands are often observed with strong CTCs; the g value is usually below 2.0023 indicating that the electrons are not completely free.

Strong CTC in which the ground state is predominantly ionic exhibit very high dipole moments and therefore high permittivies; *e.g.*, the dipole moment of the triethylamine-iodine complex has been experimentally determined as 11.3 Debye (D) units,[46] though other workers report only 5.23-6.55 D, depending on the solvent.[47] The pyridine-iodine complex, for example, has a moment of 4.44 D.[48] However, their generally low resistivity renders the measurements difficult and imprecise. Weaker complexes, especially of the π-π type, formed between nonpolar donors and acceptors show high dipole moments because of the resulting mutual polarization of the constituents:[49] the mean dipole moment of pyrene-naphthalene, for example, is 2.0 ± 0.3.[50] In CTC formed between one polar and one nonpolar constituent, these same induced polarizations result in a diminution of the net polarization; *e.g.*, nitrobenzene has a dipole moment of $4.22 - 4.28$ D while the resulting CTC exhibits a value of[51] 3.78 D.

Attempts have been made to correlate the permittivity of CTC with other physical properties,[52] though with rather indifferent success.

It is unlikely[53] that CT stabilization energies are as large as those of London dispersion, charge-dipole, dipole-dipole, dipole-induced dipole, or hydrogen-bonding interactions. CT interactions are then only likely to have an effect on the mode of DA overlap when these larger forces show little sensitivity or orientation. The presence of a charged atom, strong dipoles, large polariza-

---

*Quantitatively, it has been shown by Jain et al.[60] that in 9-nitroanthracene-donor adducts, the increase in conductivity due to the introduction of a gaseous electron acceptor increases exponentially with the acceptor vapor pressure and is inversely proportional to the ionization potential of the acceptor. These results should be corroborated and further clarified.

tion atoms, or groups able to form hydrogen bonds will all negate this condition. DA orientation in the metal complexes may be dominated by charge-dipole interactions; and in the 8-hydroquinoline-TNB, chloranil and the hydroxybenzene-benzoquinone complexes by hydrogen bonding and dipole-dipole interactions.[53]

In the absence of strong intermolecular interactions such as hydrogen bonding or ion-dipole interactions, the resulting donor-accepting orientation of the adduct is that which maximizes the charge transfer stabilization energy of the complex.[53]

For a given donor, the stability of a CTC depends on the acceptor, and, more specifically, on its electron affinity; thus Kondratov[54] has shown that the following anhydrides can be arranged in order of decreasing stability of the adducts in which they participate: dichloropyromellitic anhydride, pyromellitic dianhydride, dinitrophthalic anhydride, nitrophthalic anhydride, naphthalic anhydride, phthalic anhydride.

In the important complexes based on I_2 as the acceptor, it has been proposed[55] that 2 types of adducts are formed, viz., $D:I_2$ and $D:I^-:I^-$. Similar configurations had been suggested for phenothiazine:I_2 CTC[56] much earlier and, in a more general way, by Kommandeur.[57] The role of various oxidation states of I_2 in the formation of highly conducting CTC such as perylene: iodine (1:1.5) or pyrene:iodine (1:2) has been emphasized by Teitelbaum et al.;[58] transport in such solids occurs along molecular stacks, or columns, via partially oxidized iodine,[58] and probably by ligand mediated transfer (cf., Section 9.12 and 15.9). Doi et al.[59] suggest that the low resistivity of e.g., N-methylphenothiazine:I_2 adducts, viz., 2 ohm cm (see Table 8.5A) is due to carriers generated by extra iodine introduced into the complex cation radical. Matsunaga and Shono[354] showed that charge interaction of I_2 with phenothiazinebromide (PBr) proceeded in three different stages;[2] the first stage in the mole ratio 1 PBr:$0.25I_2$ with a resistivity of 4 to 5 ohm cm, the second occurred at 1 PBr:$0.5I_2$ and the last at 1 PBr:$1I_2$, both the latter complexes having a considerably higher resistivity. These workers[354] observed no such pronounced effects on the physical properties of the p-phenylenediaminebromide:I_2 complex. For strong CTC the molar equilibrium constant K_c and the molar (spectroscopic) extinction coefficient ϵ_c may be obtained from absorption spectroscopy using the relation[61]

$$\frac{D_0 A_0 b}{\alpha'} = \left[D_0 + A_0 - \frac{\alpha'}{b \epsilon_c'} \right] \frac{1}{\epsilon_c'} + \frac{1}{K_c \epsilon_c'} \qquad (17.4)$$

where D_0 and A_0 are the initial molar concentrations of D and A, respectively b the path length of the spectrophotometer cell and

$$\alpha' \equiv \alpha - b \epsilon_A A_0; \qquad \epsilon_c' \equiv \epsilon_c - \epsilon_A.$$

In the ultraviolet $\alpha' = \alpha$ and $\epsilon'_c = \epsilon_c$. α is the total absorbance observed and ϵ_A the molar extinction coefficient of pure A. Numerical computations involve an iterative solution usually by a computer.

The correlation between spectroscopic data and thermodynamic parameters of CTC is discussed by Lux et al.[62]

Nearly complete charge transfer resulting in a predominantly ionic ground state is claimed for the adducts[63] between 1,4 diazabicyclo-(2,2,2) octane and chloranil or bromanil, though the thermal activation energies as evaluated from log (conductivity) vs. $1/T$ plots are quite high, of the order of 1.1–1.2 eV;[63] likewise for the bis(o-phenylenediamido)Ni complexes with acceptors such as chloranil or TCNE.[64]

The coupling between neutral and ionized states in crystalline strong CTC is said[129] to be mediated by a Frenkel exciton interaction.

Some highly interesting perylene:metal dithiolate complexes are reported by Alcacer and Maki;[337] it appears that it is the perylene system which is responsible for the semiconducting properties of these adducts; their conductivity is said[337] to be intrinsic and reaches values of 50.6 (ohm cm)$^{-1}$. Their structure bears a resemblance to linearly stacked TCNQ adducts (see Section 16.11).

17.5 Activated Charge Transfer Complexes

If the electron is taken not to the bottom of the conduction band but into an excited state energy level within the energy gap, below the bottom of the conduction band, the E_A is increased and the probability of complexation is correspondingly enhanced. Likewise, if the electron is raised not from the Fermi level of the donor but from a higher, excited state energy level, I_c is reduced, having the same effect. Both effects may occur simultaneously; both require an additional, external source of energy equal to the excitation potential involved. Frequently, at least experimentally, this energy comes from radiation, usually with light,[357] resulting in what is loosely termed as "light activated charge transfer complex." Such adducts are of considerable importance in biophysics and also in medicine. However, light irradiation may be a sufficient, but is by no means a necessary condition for the generation of such complexes. Since the energy differences involved in the excitation processes are, generally, quite small and may even be of the order of just a few times kT, such "excitation activated complexes" can be produced via a variety of energy supplying mechanisms e.g., by electrochemical means.

Complexes involving an excited state are termed exciplexes.[70] They are generally rather weak adducts produced in a one-quantum process involving a virtual photon. Examples of such activated complexes are those formed between phenothiazines and melanine[65] or indole exciplexes.[66] Even within a single molecular species, self-complexing, or auto-complexation, may result due to an intramolecular charge transfer *e.g.*, from a non-conjugated donor substituent having a low ionization potential to another substituent exhibiting a high electron affinity; auto-complexes are discussed in Section 17.6.

Since in an n-π^* transition an electron moves into a highly delocalized orbital the transition probability is low but the lifetime of this highly polar state is relatively long. Thus, in activated complexes it is these states which are likely to play a major role rather than the far more probable π-π^* transition states; the n-π^* state is a very reactive donor or acceptor, thus facilitating electron transfer in what amounts to a RedOx reaction.

Complexation may also take place *via* excited dimers, *i.e.*, excimers (see Section 13.7). However, excimers, *per se*, are unlikely to enter into any further CTC formation. The intermolecular spacing for the frequently resulting sandwich configuration is less than that for both molecules in their ground states.[67] Excitation leads to a considerable change in the magnitude and direction of the dipole moment.[68]

In hydrocarbon solvents, more or less free exciplexes may arise;[69] these usually involve short-lived ion pairs with strong coulombic interaction between the counter ions.[73] At low temperatures, the odd electrons of the two radicals may pair off and yield sufficient resonance between two charge transfer states, both featuring positively charged ion radicals, may lead to complex formation.[71] Perhaps the clearest single feature of an exciplex is the reversibility of its equilibrium with the isolated components.[73]

It is a characteristic of activated complexes that they are stable only in their excited state, and only as long as the excitation is maintained. Usually, exciplex formation is associated with quenching of the fluorescence exhibited by its components and/or by a new, red-shifted, absorption band. Transient absorption spectroscopy is a powerful tool for their study; both the singlet and the excited states of the activated complex show absorption bands of the cation and of the anion radicals of the component donor and acceptor.[91] Generally, such adducts tend to be rather strong complexes rather than weak adducts.

The stability of such adducts is often due to a dipole-dipole interaction though such tends to give rise to rather weak adducts; those between benzene and olefins are cases in point.[130] However, the exciplex formed in the interaction of phenazine with dimethylaniline[131] appears to be a reasonably "strong" CTC. The role of the solvent in exciplex formation is discussed by O'Connor and Ware.[132]

Ion radicals form even in moderately polar solvents,[72] say $\epsilon \geqslant 6$. Since the free energy difference between a free and solvated ion pair $D^+ - A^-$ usually exceeds kT it appears that dissociation occurs *via* a non-relaxed state of the ion pair.[82] However, some dissociation may also result from the fully relaxed exciplex in view of the positive entropy contribution arising from de-solvation.

The first step in the photo-dissociation of charge transfer complexes often[74] (*e.g.*, in 2-methyl tetrahydrofuran-pyromellitic anhydride complexes) goes *via* the excited singlet state $^1CT^*$ which generally means $D^*A + DA^*$.

This mechanism holds especially for organic charge transfer salts where the excited CTC* consists of these two neutral radical species. The photoconductivity in such salts is considered to be due[75] to intrinsic carrier generation; it is recombination-limited because the photocurrent is proportional to (light-intensity)$^{1/2}$. It peaks on the long wave side of the CT band and is unaffected by the presence of O_2. Typically, the thermal activation energies of dark conductivity and of photoconductivity for these salts are equal. The energy level of photoconductivity is very close to that of the CT* so that no further energy—in addition to $h\nu$— is required for carrier generation, which appears to arise[75] from trapping levels about 0.4 eV below the edge of the conduction band. The thermal activation energy refers only to carrier migration.

The singlet-triplet separation energy, *e.g.*, in ionic radical salts of 2,3-dichloro-5,6-dicyanobenzoquinone, which form linear chains of dimers, is 0.19 eV for the $(CH_3)_4N^+$ salt and 0.30 eV for the $(CH_2)_4N^+$ salt.[76] For chloranil-adenine, it is 0.42 eV.[77]

Exciplexes, or excimers—depending on the partner—also form from aromatic ring pairs with 3 methylene groups, becuase of the strong π-overlap. Their structure is plane parallel.[78]

In solution, excitation to $D^*A + DA^*$ is often followed[79] by the formation of solvent shared ion pairs:

$$(A^-_{\text{solvated}} + D^+_{\text{solvated}}) \rightleftharpoons (^3A^* \underbrace{+ D) + (A +}_{\text{solvent}} {}^3D^*) \tag{17.5}$$

Many dyes, such as methylene blue, rose bengal and eosin, initiate photochemical reactions *via* triplet states which, in solution, arise in this fashion.

Other exciplexes are, *e.g.*, the activated complexes between the strong donor chlorpromazine, a phenothiazine derivative, and the acceptor melanin;[80] between indoles and polar solvents[66] (many cases of sometimes puzzling solvent interactions involve exciplexes) or between nucleosides and nucleotides,[66] or between amines and hydrocarbons.[81] However, the $^1CTC^*$ may also directly dissociate into ion pairs.[82] (See also Section 6.11).

The exciplex may decay, after the supply of excitation energy has ceased, by a variety of radiative and/or non-radiative processes which are discussed in Chapter 13, especially Section 13.3. Non-radiative decay of CTC* is said,[83] *e.g.*, in TCNE: aromatic hydrocarbon complexes, to take place *via* internal conversion from a singlet to a ground state, while fluorescence in an exciplex is said to be[82] from a $D - A^*$ state.

In stoichiometric CTC, excess donor or acceptor act as dopants, resulting in the formation of mobile or trapped triplet excitons. The role of traps is profound.[75]

In some cases, an excited CTC forms as a precursor to a chemical reaction: 1,2-dichloroethane is produced by photolysis of the (weak) ethylene:Cl_2 complex.[84] This is a relatively fast reaction; that between 6-hydroxydopamine and chlorpromazine is a rather slow one.[85] The latter is a well-known and powerful electron donor, it complexes with 6-hydroxydopamine giving rise to a charge transfer band with a peak which decreases upon rising temperature, characteristic of CTC.[85] The complexation is a fast process; it is, however, followed by a proper chemical reaction yielding a hitherto unidentified reaction product. This is a slow reaction taking several hours to approach completion.[85] Significant color changes accompany the entire reaction sequence. The formation of photo-oxides also may involve the formation of a $^3CTC^*$ exciton as an initial step.[86] This is of importance in catalysis because there always exist electron transfer sites on a metal oxide surface,[87] such as are also exhibited by TCNQ salt surfaces.

The dependence of the spectra of the 2,3-dithia-(4,5)-spirodecane: bromine complex on cooling rate is most likely due[88] to a chemical reaction following the CTC interaction.

In some CTC, such as benzophene:aromatic amines, excitation is followed by proton transfer, involving the lowest excited triplet state of the acceptor[85] (see Section 17.9).

It has been shown that the injection efficiency of electrons from a solid into a contact may be improved[90] by forming an activated CTC between an excited dye and a halide ion; a case of supersensitization (see Section 16.7).

The properties of 51 activated complexes have been reported mainly in studies of the fluorescence of the component arenes and hetero-arenes.[134]

The isotopic exchange between anthracene and aliphatic amines is said to occur *via* the formation of an exciplex.[133] Some aspects of activated complexes are discussed in Section 16.7, when dealing with sensitization effects; many of these do involve the formation of excited CTC.

17.6 Intramolecular (Auto-) Complexes

Self-complexing may occur,[92] with different regions of the same molecular species acting as donor and acceptor. The resulting entities are referred to as intramolecular complexes.[93] These effects are important in solution, where they may lead to association. Donor and acceptor properties can be modified by chemical substitution, and are sensitive to steric hindrances.

In a way, intramolecular charge transfer reproduces the effect of delocalized π-electrons,[94] it may often be diagnosed from a linear variation of the absorption of the charge transfer band with concentration; the experiment must be repeated at several different wavelengths.[95] The method has been applied to intramolecular adducts in polymethylmethacrylates, where charge transfer occurs between the pendant anthracene and trinitro-fluorenyl groups. Polynuclear aromatic groups tend to act as donor and the monomer as acceptor.[95] (See also Section 17.11 and Fig. 17.5).

Both intra- and inter-molecular charge transfer may occur simultaneously as e.g., in tricyanovinyl aromatics.[96]

In lamellar compounds, such as 2,2-paracyclophane quinhydrones, intramolecular charge transfer is said[97] to be mediated *via* the π-electron system interposed between donor and acceptor groups; thus the interaction does not appear to require direct spatial contact between the donor and the acceptor.

The stability of the molecular ions formed in several auto-complexes is reported to increase with increasing "donor strength" of the substituents within a series of indole derivatives;[98] the stability of these molecular ions may be ascertained by mass spectroscopy.[136] The complexation sometimes results in a distortion of the molecular arrangements which assume a twisted configuration.[99]

Intramolecular adducts often arise in large molecules such as in flavine-adenine dinucleotides between the flavine and adenine moieties,[66] or in the intermolecular interactions reported[102] to occur between phenol and indole chromophores in compounds where they are joined by methylene and amide groups. Even in much smaller molecules, such as dichloro-anthracene, self-complexing involving an electron transfer of about 5 Å along the a-axis, has been reported.[104] Nitrodiphenylmethanes auto-complex in solution,[103] showing a distinct charge transfer band, though self-complexing in solution is often hindered. In the solid state, however, amphoteric compounds such as tricyanovinyl derivatives may exhibit up to 100% self-complexing.[100] These auto-complexes sometimes show relatively large photo-chromic, photo-voltaic and photoconductive effects; e.g., 5×10^{-3} M tetramethyl-p-phenylene-diamine in acetonitrile,[101] in a sandwich cell, discolors to blue upon irradiation, due to the formation of TMPD$^+$ and yields a photovoltage of 40 mV.[101] This voltage rises with increasing work function values of the electrode material;[101] the TMPD$^+$ cations pull electrons out of the contact.

Electrochromism has been observed in auto-complexes such as trinitrophenyl-amines.[105]

An auto-complex by itself may further act as electron donor (or acceptor) as exemplified[103] by the p-methylphenolate ion pairs which still are strongly donating.

Electronic excited states may arise in auto-complexes; thus exciplex formation has been reported[106] at temperatures above 200 K. while excimers are said[107] to result from exciton migration to suitable trapping sites. The generation of intramolecular excimers is discussed by Goldenberg et al.[108] Activated auto-complexes have been reported by several authors.[109]

n-π^* type activated auto-complexes have been observed[110] spectroscopically, e.g., by Friemans.[111] The spectroscopy of auto-complexes has been discussed for several compounds: e.g., in cyclophanes,[112] in N-purinoylaminopyrimidine,[113] in 2-N-arylamino-6-naphthalene sulfonates[114] and its derivatives,[155] in crown ethers,[116] and in substituted diphenylmethanes.[117]

Compounds having a π-electron system built up from phenyl- and keto-groups, and containing a sulfur atom, exhibit π^*-π transitions observable spectroscopically below 50 K; all the higher intensity transitions observed are said[118] to be due to auto-complexation. The sulfur atom acts as a donor while the other groups act as acceptors, as does the CO double bond in keto- or aldehyde groups.

Spectroscopic absorption due to auto-complexation in mono-, di- and tri-substituted benzenes was studied[119] by Lutskii; di-substituted derivatives containing strong donor as well as acceptor groups show 2 bands in the 3–6 eV region, while tri-substituted derivatives exhibit an additional band. The maximum energy E of the transition is said to obey

$$E = I_g - A_g - E^*_{\text{coul}} \qquad (17.6)$$

where E^*_{coul} stands for the Coulomb interaction energy between the electron-donating group having an ionization energy I_g and the accepting group having an electron affinity A_g; E^*_{coul} refers to the excited states of the substituents.

The UV spectra of benzyl-phenyl sulfones and sulfoxides containing a phenyl ring of high electron affinity and a benzyl ring of low ionization potential indicate[120] the formation of an auto-complex in a folded form; these compounds are ESR active; a folded configuration has also been suggested[121] for the auto-complexes formed in mixed ligand nucleotides.

Associations formed between two or more excitons and involving a definite binding energy, can be considered a type of auto-complex because only a single molecular species is involved. While the free complex is unstable, it becomes stabilized if associated with a fixed impurity center.[122] Each exciton retains its individuality.[122]

17.7 Ternary and Inclusion Adducts

Such adducts arise in either solid or liquid solutions.[121] The semiconducting properties of several typical adducts of this class are tabulated in Table 8.5B. In the solid state, they often involve guest molecules in a host matrix, where the host forms a charge transfer complex.

The host lattice has cavities which are occupied by guest (solvent) molecules. Intermolecular forces of the order of 0.2 to 0.5 eV bind guest to host.[123] Thus, e.g., the p-tricyanovinyl-N, N-dimethylaniline auto-complex is deep blue. In chloroform, it forms a ternary complex which is deep red.[124] The resulting electronic properties are quite different from those exhibited by the solvent-free CTC:[123] the resistivity typically drops from, say, 10^9 to $10^3 - 10^5$ ohm cm, the activation energy of the conductivity drops from, say, 0.3 eV to 0.1-0.2 eV and free radicals become evident from ESR as well as from spectroscopic studies. The new ternary complex thus formed has the structure $D{:}A{:}S{:}D{:}A$ or, perhaps more likely, $D{:}S{:}A{:}S{:}D$ where S stands for a solvent molecule.

Ternary CTC produced by alkali metal doping of an anthracene: organic acceptor adduct—such as dibenzanthracene, tetracene or pentacene—have been studied.[125] Considerable electron delocalization is reported,[125] amounting to 6 to 10 anthracene *per* dibenzanthracene molecule and 4×10^4 anthracenes *per* pentacene molecule in K-anthracene-pentacene.

Energy and charge transport in such doped CTC (see Section 16.7) occurs, at room temperatures, mainly by percolation, or variable range hopping (as distinct from nearest neighbor hopping) of excitons[126] (*cf.*, Section 13.3 and 15.5). Often these are triplet excitons exhibiting a thermally activated mobility.[127] Excimers may be formed as *e.g.*, in pyrene in a naphthalene host, as evident from fluorescent spectra.[128] The ternary complex then is $D\text{-}D^*\text{-}A$. Exciton trapping is important in these processes; the size of the trapping region depends critically on lattice distortion.[128] The energy transfer distance is about 30 Å for pyrene in naphthalene, 23 Å for anthracene in naphthalene and only 12 Å for tetracene in naphthalene. Larger values, up to 130 Å, have been suggested.[128]

A typical example of ternary adducts formed by doping, are charge transfer complexes between tetracyanoethylene and aromatic donors,[125] doped with guest donors: the guest complexes yield triplet traps in the host complex crystal and are ESR active. The electron trapped is then de-localized over one donor-acceptor pair.[125] Doping effects are discussed in Section 16.7 and 16.8

Charge transfer complexes may enter into a competitive interaction with free radicals such as the piperidinoxyl radical[149] forming what is effectively a ternary adduct.

A particularly interesting class of, effectively, ternary complexes are electrolytically formed intercalation complexes such as the deep blue graphitic $C_{24}^+(HSO_4)^-2H_2SO_4$.[137] These lamellar complexes are very good catalysts for the formation of *e.g.*, formates and acetates. They are highly ESR active indicating a high concentration of free radicals and delocalized electrons.

Upon contact with an acid, *i.e.*, with an electron acceptor, considerable electron localization takes place. The graphitic compound as such, has, apparently, a great many electron donating centers; bromine,[137] $SbCl_5$,[138] chromic anhydride,[139] Al_2Cl_6,[139] as well as other strong acids,[137] such as $HClO_4$ or H_3PO_4, may also be intercalated. These materials should have interesting conduction properties. Graphite also forms intercalation ternary adducts stabilized by metal halides. Some of these are superconducting at very low temperatures,[140] though TaS_2 intercalation compounds exhibit relatively high transition temperatures.[141] The ternary adduct TaS_2 (pyridine)$_{1/2}$ is claimed[142] to exhibit two-dimensional superconductivity.

Graphite, the prototype layered solid, can form intercalation compounds[145] with various chemical substances. The compounds of alkali metals, of the form C_8M and $C_{12n}M$ ($n \geq 2$), where M represents either potassium, rubidium, or cesium, are regarded[144] as synthetic metals. When potassium, for example, is intercalated between the hexagon layers of carbon atoms, the distance between the layers has been reported to expand from 3.35 to 5.40 Å, and the distance between the potassium atoms is 4.91 Å even for the potassium-richest compound, C_8K.[143] This means that many vacancies, which are supposed to be situated under some special electric field, are newly formed between the layers through the intercalation of potassium atoms.

Electron transfer from the alkali metal to the graphite leads to an increased charge carrier density in the carbon network of graphite. An increase in resistivity upon the admission of hydrogen onto the potassium-intercalated compounds can be explained in terms of the electron transfer from the compounds to the hydrogen, which causes a decrease of charge carrier density in the carbon network.[146]

A reversible change in resistivity observed below 200 K indicates the occurrence of a weak charge transfer inter-action between the compounds and hydrogen molecules. A similar result has been reported[147] for the perylene-cesium-hydrogen system.

The charge distribution and transport phenomena in graphite-acceptor lamellar adducts has been reviewed.[148]

Several intercalation complexes behave as organic semi-metals, (see Chapter 2). The conductivity in a direction perpendicular to the layer planes may be considered along the lines of a one-dimensional model.[351] Structural aspects of intercalation ternaries are dealt with in Section 19.9.

For a discussion of ternary metal-bridged CTC,[121] especially as far as charge transfer is concerned, see Section 15.9 and 19.7). The stability of these ternaries depends greatly on aromatic stacking interactions.[121] (See Section 9.12.) Such "indirect charge transfer" interaction in multilayered paracyclophane-quinhydrones is discussed by Staab.[97] Similar arrangements have been found[150] to exist in metallocenes of the general formula $M(C_nH_n)_2$ where M stands for a metal valency of 2,3 or higher. These sandwich complexes are said[150] to form a "skyscraper-like" arrangement.

Ternary surface complexes (*cf.*, Section 17.8) are reported[153] to be formed by adsorption on, *e.g.*, a SiO_2 surface. Such adducts may well be of far more frequent occurrence than hitherto suspected and may play a significant role in biological phenomena. Very little work has been done on this interesting topic.

A binary complex may become a ternary complex when the concentration of one of its components becomes very high.[154] Ternary complexes are reported[158] to enter the formation of coal.

If a CTC is recrystallized from a solvent, then the nature of that solvent may considerably affect the properties of the (solid) CTC; thus, *e.g.*, the 1,6-diaminopyrene:chloranil complex is dark blue when recrystallized from benzene.[156] These are two completely different chemical entities with different physical properties evident in their spectra and resistivity. This may be due[156] to the formation of a ternary inclusion compound between the CTC and the solvent; the host lattice, here that of the (solid) CTC, has cavities which are occupied by the guest molecules, *viz.*, those of the solvent. Such CTC inclusion compounds have been proposed[157] for several benzidine and TCNQ adducts. Thus, while the benzidine:TCNQ complex is[159] diamagnetic, the ternary B:TCNQ:S [$S = e.g.$, dichloromethane (CH_2Cl_2) or chloroform ($CHCl_3$)] is strongly ESR active and the charge transfer band is shifted to longer wavelengths with a new shoulder appearing in the spectrum. The spectral changes depend on the relative donor/acceptor properties of the solvent.[160] The solvent dipoles try to align themselves with the host dipole moment; the ground state of the complex is less polar than the excited state.[160] The benzidine and the TCNQ molecule are alternatively stacked on each other, forming molecular columns with intervening channels accommodating solvent molecules.[161]

Most aprotic solvents are good electron donors,[172] though, *e.g.*, ethanol or acetone are relatively weak. In such cases, the solvent may behave amphoterically as does water.[176] The frequently used solvents, dimethylsulfoxide, DMSO, and dimethylformanide, DMF, are known to form CTC with *e.g.*, chloranil in CCl_4; the latter can act as an acceptor.[174]

Many solvents enter into ternary complexes; thus, much of the work on complexes in acetonitrile as solvent is suspect because this solvent has been shown[162] to form CTC with halogens which are frequently employed as acceptors. Many reported (see Section 17.11) irregularities in conductivity titrations using CH_3CN as solvent are probably due to the formation of ternary adducts. The red-colored serotonine-picrate monohydrate,[163] which forms alternating stacks of continuous columns much like TCNQ salts, is probably another case in point.

Activated ternary complexes are reported[135] to be formed in solvents such as benzene, heptane and xylene, from pyrene: dimethylamine or anthracene: dimethylamine CTC dissolved therein; the aromatic moiety is said[135] to be primarily involved in the linkage.

Of particular interest are the inclusion complexes[164] of mononuclear aromatic hosts with α-cyclodextrin: These form weakly coupled hydrophobic cavities capable of binding hydrophobic residues of compatible size. This small, water soluble, "active site" catalyzes specific reactions with kinetics similar to those of enzymes. It thus acts as a model compound for specific enzymatic reactions. The electron donor:acceptor interaction, though, is reported to be minor.[164]

The entropy contributions to the formation, and stability, of ternary complexes must be quite complicated and do not appear to have been studied to any extent; the stability of Cu^{2+} and Zn^{2+} ternary complexes is discussed in Section 19.7. CTC formation, generally, involves a desolvation and thus an increase of entropy because of the decrease of order due to the at least partial destruction of the solvation shell. In many cases, complexation in solution occurs mainly because it involves a drop in the free energy of the system, though the enthalpy change may be endothermic. Ternary systems would thus be favored from the point of view of entropy production:[165] there is a decrease of order because the bonds active in solvation are now engaged in interactions with the occluded solvent molecules.

Solvation[166]

What could be termed "pseudo-ternary complexes" may arise in solution from solvation effects.

Solvation, generally, lowers the free enthalpy of complex formation as well as the entropy of the reaction. Especially in polymeric CTC, ion pairs and radical ions are solvated[167] to a higher degree than the neutral complex. Being therefore energetically favored, this results in increased dissociation into solvated ion pairs and solvated radical ions.[167]

However, part of the solvent effect, notably anisotropies in the distribution of solvent molecules about one another, are due to[165] interactions which disturb the intrinsic polarizabilities of the molecules.

The requirement that free ions result from an electron transfer from a donor D, to an acceptor A in a solvent can be written[169] as

$$|E_s + \beta E_c| > I - |E_A| \qquad (17.7)$$

where E_s is the sum of the solvation energies of the individual ions and

TERNARY AND INCLUSION ADDUCTS

of any ion pairs, or higher aggregates, E_c the coulombic energy to form an ion pair present in the solution in a concentration βM, M being the molarity of the complex, I is the ionization energy of the donor and E_A the electron affinity of the acceptor. Since in an organic CTC I is likely to be low and E_A high, this condition is frequently fulfilled. Thus, e.g., in acetronitrile, anions are quite stable in the absence of water and of oxidants, though the cations have lifetimes of only a few milliseconds until they react with the solvent.[170] However, competition by the solvent should also be allowed for,[171] i.e., the formation of a solvate complex following

$$A + nS \rightleftharpoons AS_n^s \tag{17.8}$$

Assuming that the dissolved complex exists in solution exclusively in the form of such complexes,[171] the exchange of the, say, acceptor A between the competing molecules of the donor D and the solvent S proceeds as

$$AS_n + D = AD + nS \tag{17.9}$$

and

$$K_e = \frac{[AD][S]^n}{[AS_n][D]} \tag{17.10}$$

The equilibrium $A + D \rightleftharpoons AD$ is governed by a constant K_S which should vary linearly with $[S]^{-n}$ as long as the donor is present in excess, while K_e should remain independent of both $[D]$ and $[S]$. So far, these relations have been tested only for the benzene/I_2 in CCl_4 system where they appear to hold to about 5%.

The effect of anion solvation on complex formation is said[175] to follow the series

$$H_2O > ROH \gg DMSO \cong ES > CH_3CN \cong TMS \cong NM \cong DMF > DMA \tag{17.11}$$

where ROH stands for methylalcohol, DMSO for dimethylsulfoxide, ES for ethylensulfite, TMS for tetramethylsulfone, NM for nitromethane, DMF for dimethylformanide and DMA for dimethylacetamide.

The effect of 32 solvents on the acenaphthene-tetrachlorophthalic anhydride CTC has been studied;[177] from these effects, solvents may be divided into protic, dipolar aprotic, aromatic with or without functional groups, and n-donors. Their data on solvent effects on the charge transfer absorption band of the acenaphthene:tetrachlorophthalic anhydride complex are listed in Table 5.4A. Ionic solvation in non-aqueous solvents has also been reviewed.[178]

CTC formation may be followed, or may be accompanied, by protonation.[179,180,181] For further discussion of proton complexes see Section 17.9.

17.8 Surface and Micellar Complexes

Organic donors, if exposed to an electron accepting atmosphere, form surface charge transfer complexes which may be identified, *e.g.*, by reflectance spectroscopy or by conductivity determinations. The converse holds for acceptors. This has been utilized to detect minute traces of contaminants of ambient gases.

These experimental techniques are discussed in Chapters 2, 3 and 13. For a brief review of some aposite topics in surface science, see Section 12.17.

An adsorbate behaves as an electron donor if the energy of its lowest unoccupied level evaluated in its adsorbed state is situated at or above the Fermi level; the adsorbate, conversely, acts as an acceptor if the highest occupied level of the system substrate plus adsorbate is at or below the Fermi level. Surface complexes thus are often dimer-like[182] with their energy states well localized within the energy gap, and resulting in a depletion of states in the vicinity of the Fermi levels; any excess charge tends to accumulate on the surface complex,[182] raising the carrier hopping probability. Some surface complexes consist of an adatom plus its nearest neighbor on the surface, again a dimer-like structure.[183] The first stage of an adsorption process involves a charge transfer, donor-acceptor, interaction between the adsorbent and the adsorbate[184] resulting in an increased effective ionization potential, or electron affinity, of the adsorbed donor or acceptor, and a raise in the local energy levels of the substrate. This stage is followed[184] by electron, or hole, transfer into stabilized vacant energy levels of the adsorbate. Surface excitons and polaritons, such as arise on *e.g.*, anthracene surfaces,[185] are discussed in Section 13.8; their binding energy has been reported to be of the order of 0.1 eV.[192]

The topic of surface CTC is basic to the theories of catalysis and is usually treated in that context;[186] thus it is beyond the scope of this work. However, see discussion in Section 2.7 and 12.11.

Many gases form surface CTC with organic semiconductors, thereby altering their surface properties to a great extent (*cf.*, Section 3.8 and 12.18). Examples are oxide surfaces, such as MgO exposed *e.g.*, to I_2.[187] Adsorbed Mg-phthalocyanine, without additional ligands, is reported[188] to have an electron affinity of 13-14 eV.

Dinitrogen-metal complexes are examples[189] of surface adsorption complexes such as are involved in the technical catalytic production of ammonia as in many enzymatic reactions.

Surface CTC appear to play a role in the observed[190] changes in the electrode potentials of merocyanine dyes in aqueous solutions of anionic, cationic and nonionic surfactants; there are also changes in permittivity, as well as in the optical spectra.

The surface complexes between donors and oxygen are of considerable importance in the design and operation of fuel cells. Thus, Asmolov and Krylov[196] have studied the donor:acceptor interactions of α-alumina surfaces with adsorbed hydrocarbons. Substrate intermolecular CTC's are of importance[191] in some enzymatic reactions, *e.g.*, in those of the substrate-porphyrin ionized complex, Subst$^+$:Porph·$^-$ they also involve excited states: other examples of CTC are between cytochromes *b* and *c*.[191]

Surface CTC of chlorpromazine have been shown to be involved[193] in hypothermia and sedation of mice; *in vitro* surface adducts between chlorpromazine and collagen have been demonstrated, as well as with lignocain and with phenytoin.[194] The surface CTC between polyethylene and SF_6 formed on the inner surfaces of the voids which are unavoidably present in the extruded polymeric insulation for high voltage cables, improves the dielectric strength of the cable by about 50%.[195]

Micellar and Colloidal Charge Transfer Complexes

Complexes are known in which at least one component exists in colloidal form: thus, chlorpromazine as well as other phenothiazine derivatives are colloidal[197] and act as powerful electron donors:[198] melanin is colloidal as well as electron accepting[199] and many biologically active molecules, such as acetylcholine,[200] are capable of forming colloidal CTC. These appear to form above the critical micelle concentration between solubilized TCNQ and the surfactant and assume a palisade type of molecular arrangement.[195]

A micellar structure results in a considerable contribution towards the free energy of the system, as can be shown by a reasoning analogous to that employed in the discussion of entropy effects (Section 6.3): Consider n electrons distributed over N molecular sites $n \ll N$. The number of ways in which similar systems may be assembled is given by

$$W = N!/(N-n)!n! \quad (17.12)$$

Let us now add ONE electron. The free energy contribution due to the entropy change ΔS involved is

$$-T\Delta S = -kT\ln N \quad (17.13)$$

where

$$S = k\ln W \quad (17.14)$$

At 300 K, this free energy *gain* is about $0.055 \log_{10} N$ eV. In a monolayer of 10^{15} molecules, the energy gain is therefore about 0.8 eV.

This reasoning is supported by the experimental finding[201] that the energy required for the photoionization of pyrene, phenothiazine and of tetramethyl benzidine from an aqueous micellar system is considerably below that required for the gas phase reaction. Several similar results have been reported,[202] e.g., for charge transfer reactions involving exciplexes in pyrene:dimethylamine micellar adducts.[206]

Excitation of an arene such as pyrene in the presence of an amine such as dimethylaniline leads to the formation of the complex excited state of dimethylaniline-pyrene,[207] an exciplex. The nature of the exciplex depends critically on the medium in which it is formed. In cyclohexane the excited triplet state of pyrene is prominent and there is evidence of fluorescence from the exciplex. In more polar solvents such as methanol, the exciplex tends to disappear and the formation of ions becomes preponderant *via* electron transfer from dimethylaniline to pyrene, leading to pyrene anions and dimethylaniline cations.[207] The pyrene negative ion is observed with λ_{max} at 493 nm and dimethylaniline cation with λ_{max} at 465 nm. Comparison of the subsequent chemistry of these systems in various micelles, shows dramatically how the micelle charge affects the photochemistry. For example, in cetyltrimethyl-ammonium-bromide, a positively charged micelle, the yield of the pyrene anion is greatly enhanced over that in any other system. The ion yield is smaller in sodium lauryl sulfate and quite small in e.g., igepal.[207]

Excitation of the pyrene molecules results in electron transfer from the dimethylamine to the pyrene held in the micelle.[207] The states and dynamics of at least some CTC, especially in their excited state, thus are profoundly affected if the system is micellar.[203] Intramolecular excimers in aqueous micelles have been reported by Emert,[204] and in detergents by Truro.[205]

Addition of surfactants stabilizes free radical ions in e.g., anthraquinone: anthrahydroquinone complexes; lifetimes of several weeks have been reported.[209] The resulting (ternary?) adduct appears to have a micellar structure; if its concentration exceeds the critical micelle concentration, a spectral blue shift results.[209] The complex is said[210] to raise the solublizing power of the surfactant. Addition of a dye yields a ternary adduct.[210] The ionic dyes fluorescein and eosine are attracted[208] to cationic micelles (e.g., cetyltrimethylammonium-bromide), as can be seen from the red shift in the absorption and emission spectra compared to solutions below the critical micelle concentration,[208] Since the dyes are not incorporated into the core of a micelle, they have no influence on its size, shape, etc. The spectral properties of these dyes are ideal for dipole-dipole energy transfer (Förster mechanism): the maximum of the band of fluorescein coincides almost exactly with the maximum of the absorption band of eosine. Thus, energy transfer from fluorescein (donor) to eosine (acceptor) competes effectively with spontaneous emission at separation distances comparable with micellar dimensions.[208]

Electron transfer in micellar solutions has been studied by Almgren et al.[211]

who suggests that it is the difference in the RedOx potentials between donor and acceptor which provides the driving force for the electron transfer; it is greatly affected by the association processes taking place on or in the micelle.[211] Highly efficient energy transfer from a photo-excited singlet state of an acridine orange-type dye dissolved in a micelle, to methylene-blue adsorbed on its surface, has been reported;[212] in effect many of these systems are ternaries consisting of an interacting system donor-acceptor-micelle.

If an acceptor is incorporated into a cationic micelle, then the rate of reactions involving hydrated electrons is increased by a factor of about 60 times that prevailing at outside negatively charged anionic micelles.[216]

Since counterions interact with, and are attracted by the, say, hydrophilic heads of the colloidal particles, micellar assemblages of such particles should be catalytically active. This, in general, has been found[213] to be the case, and should be a rewarding field for the study of micellar CTC as well as for electro-catalysis. Formation of micellar CTC is also reported[214] to result in considerably shifted ion exchange equilibria.

Suspensions of polymers such as PVC may be coated with a donor or an acceptor and then irradiated.[215] A charge transfer reaction follows in which charge is transferred from or to the additive; the system then exhibits[215] a greatly enhanced conductivity.

17.9 Proton Transfer Complexes

The electron transfer involved in the formation of a charge transfer complex may in suitable systems be coupled to a proton transfer, resulting in proton transfer complexes.[217] This holds, especially for surface reactions, *e.g.*, at electrodes, and *a fortiori* in non-aqueous media, because the proton affinity* of water is so very high, *viz.*, about 8.9 eV.[218]

Hydroxy-dinitro pyridines, *e.g.*, act as electron acceptors and/or as proton donors to *e.g.*, napthalene derivatives.[219]

The field of protonic CTC has been opened up by Matsunaga and his school,[217] and has been reviewed by Morokuma.[220] This author points to

---

*The proton affinity A_p is defined as the enthalpy of the reaction
$$X + H^+ \to XH^+ \qquad (17.15)$$
In the gas phase, where there are energy contributions from solvation or polarization, A_p may be measured by means of ion cyclotron spectroscopy.[222] Several proton affinity values are listed in Table 8.5D; see also a report by Beauchamp[222] and some other relevant publications.[224] Proton affinities are often given in kJ/mole; 1 kJ/mole equals 0.0104 eV. Other authors use kcal/mole; 1 kcal/mole equals 0.0436 eV. Exchange repulsion energies do not enter into proton affinities.[220] The "relative protonicity" of solvents, akin to the donicity concept for electrons (see Section 17.2) is discussed by Bayless et al.[223]

the close linkage existing in such adducts between proton and concurrent electron transfer; one does not occur without the other. The proton complex is said[220] to involve an electron transfer from e.g., an amine to the proton. There is evidence that the electron transfer in some cases may be catalyzed by the presence of a proton and thus enhanced above its thermally controlled rate.[221]

Proton transfer may, sometimes, be really nothing more than a case of conventional hydrogen bonding, but in many cases the simultaneous transfer of an electron and a proton produce a new different type of adduct; then the difference between mere hydrogen bonding, and complex formation may indeed be dramatic, as pointed out by Arnett and Mitchell;[225] there is no correlation between the heats of protonation and of hydrogen bonding.[225]

In organic solvents, proton transfer appears to take place[240] via an $AH \ldots B$ complex formation.

Ion pair and hydrogen bonded complexes[226] are beyond the scope of this book. Here, only a few representative examples of protonic complexes can be discussed. As one extreme, one might consider the 1:1 complex between diphenylcyclopropenone and substituted acetic acids such as phenylacetic acid: the complexation appears[227] to involve a hydrogen bond between the oxygen of the polar carbonyl group of the propenone and the acidic hydrogen of the acetic acid. The infrared spectrum exhibits a red shift.[227]

At the other extreme, there are a number of complexes in which the CTC formation is due to the simultaneous transfer of electronic charge and a proton: the already mentioned orange colored α-napthylamine-picric acid 2:1 complex[228] or the orange/red 2:1 (or higher) complexes of tetranitrobiphenyl-4,4'-diol with aromatic monoamines are examples of such adducts.[228]

Thus, one part of a diamine molecule acts as an electron donor while another part acts as proton acceptor. Conversely, in 1:2 picric acid complexes e.g., one picric acid functions as the proton donor and the other as the electron acceptor.[228]

The 1:1 and 1:2 complexes between oxalic acid and α-aminoacids also involve a proton transfer from the carboxyl group of the oxalic acid to the carboxyl ion of the amino acid; there is a similar interaction in the 1:1 complex between malonic acid and glycine.[229] These adducts may well involve a great deal of proton delocalization resulting in the formation of proton energy bands rather than energy levels. Proton tunnelling is a well established fact[230] and even halogen ions have been shown to be able to tunnel between two suitable energy levels. Ionic charge transfer complexes between benzyl radicals and halide ions have also been reported.[231]

Ions are thermally excited from localized ionic states to free-ion-like states capable of migration with a velocity v determined by their energy $E = (mv^2)/2$. Interaction with the rest of the solid then produces a finite lifetime corresponding to excitation of the free-ion-like state. This mechanism has been proposed[232] for MAg_4I_5, where M stands for NH_4 or an alkali metal, or for $Ag_4HgSe_2I_2$ adducts.

The average time of residence of the proton at a temporary equilibrium site is of the order of one vibrational period of the OH group and of the average lifetime of H_3O^+, viz., about 2.5×10^{-13} sec; since the dielectric relaxation time of water is about 10^{-11} sec, it appears that dielectric relaxation, and frictional, processes are not involved in proton transfer.[233] However, in what appears to be protonic charge transfer complexes between e.g., aliphatic amines and phenols, a considerable increase in viscosity is reported.[234]

The orange-colored α-naphthylamine-picric acid 2:1 complex is really a ternary adduct consisting of picrate ions, protonated amine and molecular α-naphthylamine. Several other complexes of that type have been reported;[217] e.g., picric acid and benzidine, or adducts between donors such as phenothiazine and the strong acceptors tetranitrobiphenyl 4,4'-diol (TNB). Other protonated complexes have been reported[235] with the strong acceptor $SbCl_5$ and solvents such as 1,2-dichloroethane. Adducts of the form of $(DH)^+ SbCl_6^-$ have also been suggested.[236] At least part of the stability of some ternary, columnar, solvent inclusion complexes is due[237] to a proton transition. The important solvent dimethylsulfoxide (DMSO) is an efficient proton donor.[238] Ultrasonic experiments indicate[239] that proton transfer plays a major role in nucleotides in aqueous solution even at pH 5.

Intramolecular proton transfer in electronically excited molecules has been reviewed;[241] in salicylic acid esters it is said to lead to de-excitation of the excited electron;[242] the free energy change involved in that proton transfer is reported[242] as about 0.13 to 0.22 eV.

The effect has been shown to accompany at least some cases of electronic excitation[30] in organic molecules; such as aromatic compounds. These effects have hardly been touched upon as far as they affect biologically important transitions, as they are bound to do. This is an important and rather new field which should be explored.

Electron excitation causes only a very small change in the enthalpy of hydrogen bonding; a value of about 0.0044 eV is reported.[243]

In photoreduction processes, electron transfer often precedes proton transfer;[244] the stability of the protonic bond is at least partially due to an $n - \sigma^*$ (*excited state) interaction.[245] From experiments with the proton complex formed between electron acceptors and e.g., N-methylimidazole or pyridine,[233] it appears that the strength of the protonic interaction is proportional to the ionization potential of the donor; it is sensitive to solvent polarity.[246] The commercially available "proton sponges,"[277] viz., 1,8-bis(dimethylamino)naphthalene compounds[248] should be capable of forming new and interesting proton complexes. Formation of a protonic complex between aliphatic amines and phenols is reported[234] to result in a considerable increase in viscosity.

Ternary, protonic, surface complexes arise[249] from the doping of a poly(N-vinyl carbazole):nitro-aromatic (*e.g.*, *o*-dinitrobenzene) CTC with strong organic acids such as trichloroacetic acid. The resulting films are highly photoconductive and exhibit a photo-induced ESR signal which might be used as a photoelectric memory.

Numerical data for several protonic charge transfer complexes are listed in Table 8.5C.

17.10 Polymeric Charge Transfer Complexes[250]

In general, polymeric charge-transfer complexes are characterized by a lower conductivity than their monomeric analog complexes.

When aromatic hydrocarbons are linked to a macromolecular chain, polymeric complexes are formed on reaction with suitable acceptors, generally in a suitable solvent and at room temperature. Thus aromatic polymers such as poly-vinyl-naphthalene, polystyrene, or polyacenaphthylene form colored complexes with tetracyanoethylene, tetrachloroquinone (chloranil), or dichlorodicyanoquinone.

These are true CTC because the tendency to form such adducts, as evident from the height of the (absorption) charge transfer band, drops with increasing temperature. This behavior is the opposite to that of a chemical reaction (see Section 17.11). Typically, the conductivity rises upon complexation, *e.g.*, by exposure to I_2 vapor,[251] by several orders of magnitude (see Section 8.6).

The electrical conductivity of the TCNQ complex of a type of ionene polymer has been studied.[252] Ionene polymers have the general structure (I) and (II); shown below, R' in structure (I) is an alkyl group, typically methyl, R'' is $(CH_2)n$, where n

$$\begin{array}{c} R' \\ | \\ -N^{\oplus}- \\ | \\ R' \end{array} \overset{Br^{\ominus}}{} \begin{array}{c} R' \\ | \\ R''-N^{\oplus}- \\ | \\ R' \end{array} \overset{Br^{\ominus}}{} \begin{array}{c} R' \\ | \\ R'''-N^{\oplus}- \\ | \\ R' \end{array} \overset{Br^{\ominus}}{} \begin{array}{c} R' \\ | \\ R''-N^{\oplus}- \\ | \\ R' \end{array} \overset{Br^{\ominus}}{} R'''- \quad (I)$$

$$\overset{Br^{\ominus}}{-N^{\oplus}\!\!\left\langle\bigcirc\right\rangle\!\!-R'-}\overset{}{\left\langle\bigcirc\right\rangle\!\!-R''-}\overset{Br^{\ominus}}{N^{\oplus}\!\!\left\langle\bigcirc\right\rangle\!\!-R'-}\overset{Br^{\ominus}}{\left\langle\bigcirc\right\rangle\!\!N^{\oplus}-} \quad (II)$$

can be 3 or higher, and R''' can be the same as R'' or have a different value of n. Aromatic rings may be incorporated into ionenes to give structure (II), where R' and R'' may be the same as R'' in structure (I) or represent an aromatic or unsaturated unit. The resistivity of aliphatic saturated as well as aromatic ionenes is greatly dependent on the presence of neutral TCNQ. The ionene-TCNQ complexes containing aromatic rings have lower resistivities than TCNQ complexes of completely saturated polymers, which are also less stable (see Table 8.7F).

The lack of stability of aliphatic polymer-TCNQ complexes is also exhibited by[253] polymers of structure (III),

$$\left[\begin{array}{c}-\text{OCH}_2\text{CH}-\\ |\\ \text{CH}_2\\ \text{TCNQ}^{\div}\;|\\ N^{\oplus}\\ \bigcirc\end{array}\right]_n \quad (\text{TCNQ})_m$$

The electrical properties of polymeric TCNQ complexes seem to depend mainly on the ratio of paramagnetic TCNQ $\div$ to neutral TCNQ and not to be influenced significantly by the nature of the backbone, which has, however, an important bearing on their stability (see Table 8.7F).

The electrical properties of a range of ionene-TCNQ polymeric adducts are summarized[252] in Table 8.7F. The interesting feature of these results is the rather pronounced minimum resistivity for about six CH_2 groups, associated, as to be expected, with a minimum in the energy gap value, but also with negative, and low, values of the Seebeck coefficient. The highest, positive, values of this quantity was always associated with the highest resistivities; again in agreement with theory (see eq. (16.28)). PVK and similar compounds have been studied by several authors in charge transfer complexation with several acceptors, of which the iodine complex has received the most attention.[254] They suggest that conduction involves phonon-assisted hopping of localized electrons in unpaired spin states, i.e., triplet states. Quite generally, electron transfer in molecularly doped polymers forming CTC appear to be by phonon-assisted hopping between dopant molecules.[267] PVK-chloranil, bromanil and poly(4-vinylpyridine) complexes are said[255] to exhibit supramolecular structures in 200–600 Å thick films, visible in the electron microscope; they are attributed to an increase in the rigidity of the polymeric chain upon complexation. PVK charge transfer complexes useful in electrophotography are reported;[256] their resistivity is said[256] to be about 0.1 ohm cm.

The electrical and spectroscopic properties of PVK charge transfer complexes are considerably affected by the structural symmetry of the acceptor molecule employed;[266] if this is asymmetric, the absorption spectra are reported to show two maxima.

Intra-chain CTC formation in solvents such as ethanol is reported[268] in polysarcosine chains having donor and acceptor groups at the chain terminals; this is an interesting development which should be followed up.

Aromatic polyimides complexed with acceptors such as TCNQ or chloranil indicate[257] that complexation lowered the ionization potential of the donor by about 0.3 eV. Copolymers of styrol or p-dimethylamino-styrol as donors and e.g., chloranil as acceptor, were studied[258] in detail: the segmental mobility of the macromolecules is reduced so that the entropy likewise is lowered causing a drop in the enthalpy of formation. In dissociation processes, the system tends to dissociate[258] into solvated ion pairs and separate free radicals. The repeat distance along the polymeric chain of polymeric charge transfer complexes is reported[259] as 6.4 Å.

Doped polyacetylenes[265] exhibit very high conductivities indeed, values in excess of 2000 $(ohm\ cm)^{-1}$ have been attained.[265] These organic (semi-) metals are discussed in Chapter 21. Carriers in these substances also exhibit a mobility considerably in excess of values usually obtained in "pure" polymers. Absorption data indicate a direct energy gap of about 1.4 eV while dielectric measurements support an "interrupted strand" charge transfer mechanism[269] (see Section 15.5). The Hall mobility in 10 atom-% AsF_5-doped polyacetylene films is reported to be 0.02 $cm^2/Vsec$ which indicates hole conduction.[270]

17.11 Identification of Charge Transfer Complexes

Spectroscopy and Other Optical Methods

There is a very large body of literature on this subject.[264] Several spectroscopic methods can be employed to determine the presence or absence of a charge-transfer interaction: absorption spectroscopy in the visible, IR, and UV regions, fluorescence and phosphorescence spectroscopy, Raman Spectroscopy[353] and flash photolysis. Polarized single-crystal spectroscopic studies, which may permit the simultaneous examination of charge-transfer and "molecular" transitions as well as the resolution of the distinction between these transitions through their different polarizations, hold the potential of providing important information in regard to this problem.[265]

Measurement of the refractive index, n_∞, (its value extrapolated to infinitely high frequencies; the optical value is usually employed) can be used to detect complexation[267] and to obtain a measure of the strength of the interaction.[267]

A change in electron cloud density in the neutral atom or molecule will lead to a change in the polarizability. Since the complex is generally more polar than the components, the electronic polarizability or the refractive index increases and the deviation depends upon the extent of interaction between the donor and acceptor. Therefore, the stronger the complex, the larger the deviation in the refractive index values.[267]

A linear plot of Δn, the difference in refractive indices of the calculated and observed values, against the calculated concentration of the complex, indicates that there is no interaction between the complex and the individual species.

A plot of n_∞^2 against the concentration of the donor yields a maximum at the complex stoichiometry, if complexation does take place. An increase in the density of the adduct as compared to its components also supports the assumption that a CTC has been formed.[268] The method has been applied, *i.e.*, to the well-known benzene:TCNQ complex; the resulting plot is shown in Fig. 17.4.

For a solid complex, reflectance spectra[265] can give a great deal of information; the diffuse reflectance has been shown to be related[84] to the excitation coefficient of the crystal as measured by absorption spectroscopy. In some of the more highly conducting CTC a nearly metallic reflectance is sometimes noted.[271] This type of reflectance would indicate the presence of an electron plasma in the surface (see Section 19.7 and 20.5).

Many of the poorly conducting solid CTC are photoconducting—the effect

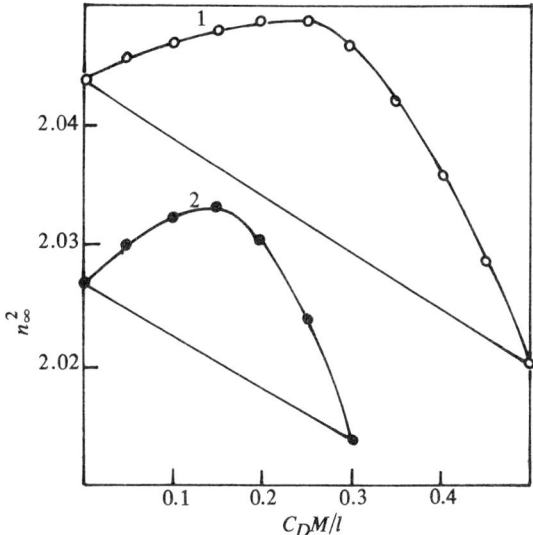

Fig. 17.4 Plot of n_∞^2 against the concentration of benzene: (1) 0.5 mol l^{-1} total concentration; (2) 0.3 mol l^{-1} total concentration. After Singh and Bhat.[267]

is masked by the high dark conductance in the case of the better-conducting complexes—allowing the photocurrent spectra to be studied.[272] In general, these spectra resemble the absorption spectra (*cf.*, Section 1.9 and 6.11).

If at least one component, either the donor or the acceptor, of a CTC, is optically active, a Cotton effect, *i.e.*, induced circular dichroism[273] (see also Section 12.10) arises especially in weaker complexes, and becomes observable in the region of the intermolecular charge transfer band, because the charge transfer transition induces the optical activity.[273] Generally, CTC exhibit exceptionally large second order optical effects in the liquid[274] as well as in the solid[275] state. These may be employed as evidence towards the occurrence of a CTC reaction though, again, by themselves they are not sufficient evidence.

Such methods have been employed to identify, *e.g.*, CTC between optically active alcohols and TCNQ;[314] they are applicable also to intermolecular adducts. In CTC exhibiting a charge transfer band, its peak is regularly about 0.2 to 0.3 eV below the intrinsic photoconduction threshold;[279] this is ascribed to the formation of ion pairs. A second and smaller threshold at lower energies (*e.g.*, 0.7 eV) than the first may be attributable to defects.

Multiple charge transfer bands[276] sometimes complicate the evaluation of the spectra, as do spectral shifts observed as a function of *e.g.*, solvation[277] (solvatochromism, see Section 16.6) or of temperature[278] (thermochromism). An extra band may also arise from triplet emission processes. Multiple charge transfer absorptions of methiodides complexes with diazines, quinaldene and acridine have been observed:[278] separation of the charge transfer bands was explained on the basis of the interaction of $2P_{1/2}$ and $2P_{3/2}$ states of the I atom with the organic portions in the excited charge transfer states. A charge-transfer band should diminish reversibly upon heating because[280] the equilibrium between the complex and its constitutent ions then shifts toward dissociation so that the band tends to be replaced by the spectra of the constituent ions. A chemical reaction, conversely, will cause the peak amplitude to rise with increasing temperature because the equilibrium then shifts towards a higher concentration of the reaction product. Only ions contribute to the peak obtained in conductivity titrations; thus the conductivity peak due to a charge transfer complex increases upon heating.

Alternatively, electronic absorption spectra of suspected charge transfer complexes may be repeated for several different donor:acceptor stoichiometries: if now absorbance is plotted *vs.* wavelength, the absorption curves, for the case of a charge transfer complex, will intersect[282] at one or more isobestic points. The method is also applicable to ternary complexes,[282] and to autocomplexes, (see Section 17.6). The method is illustrated in Fig. 17.5 for tri-octylphosphine: iodine complexes dissolved in *n*-hexane.

Isobestic points may also be obtained by keeping the concentration

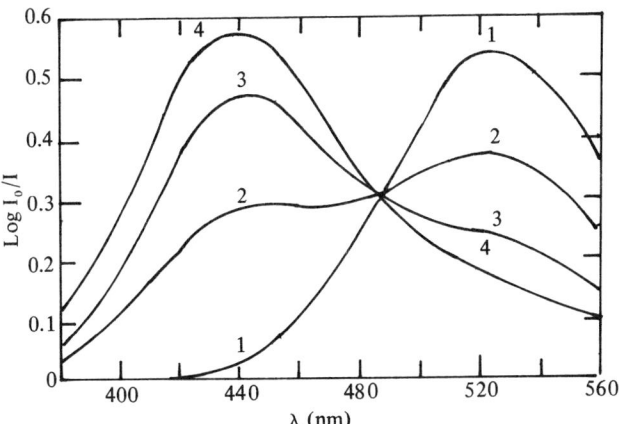

Fig. 17.5 Visible absorption spectra of iodine-trioctylphosphine oxide solutions; solvent n-heptane, temperature 20° C, cell 1 cm; curve 1, 6.0×10^{-4} M iodine, curve 2, 6.0×10^{-4} M iodine plus 1.0×10^{-3} M trioctylphosphine oxide; curve 3, 6.0×10^{-4} M iodine plus 3.2×10^{-3} M trioctylphosphine oxide; curve 4, 6.0×10^{-4} M iodine plus 5.2×10^{-3} M trioctylphosphine oxide. After Lang.[282]

constant but varying the length of the optical absorption path by varying the cell dimension.[281]

Raman spectroscopy has also been employed to identify charge transfer interactions,[266] e.g., in the benzene:Cl_2 and benzene:Br_2 adducts.

Conductivity Titrations[283]

The probability of charge transfer in solution increases[284] with increasing permittivity ϵ of the medium. If ϵ is sufficiently high, the complex may dissociate into ions, giving rise to appreciable ionic conductivity:

$$D + A \rightleftharpoons DA$$
$$DA \rightleftharpoons D^+ + A^- \quad (17.16)$$

The formation of CTC can be followed by measuring changes in the electrical conductivity of a solution of, say, the donor in an inert solvent of sufficiently high permittivity, consequent on additions of a solution of the acceptor to the same solvent, or *vice versa*. In effect, this amounts to a conductimetric titration; the donor may be titrated with the acceptor, or *vice versa*.

In the absence of interaction, chemical or other, the addition of a solution of conductivity σ_A to another solution of conductivity σ_D yields a conductivity σ given by

$$\sigma = \sigma_D + \sigma_A$$
$$= Ze[C_D(\mu_D^+ + \mu_D^-) + C_A(\mu_A^- + \mu_A^-)] \tag{17.17}$$

where C_D and C_A stand for the concentrations of the solutes A and D respectively and the μ's refer to the mobilities of the carriers contributed by D and by A. Assume that $Z = 1$ for all carriers. Concentrations will be used instead of activities and the conductivity of the solvent, or medium, itself will be neglected.

In a titration, where the volume changes on addition of titrant, donor and acceptor are present in concentrations C_D^0 and C_A^0 respectively, so that

$$C_D^0 = V_0^D C_0^D V_0^D + V_0^A$$
$$C_D^0 = V_0^A C_0^A V_0^A + V_0^D \tag{17.18}$$

where V_0^D and V_0^A refer to the volumes of stock solutions of D and A with concentrations C_0^D and C_0^A respectively.

In the absence of interaction, C_D^0 and $C_A^0 = C_A^0$. Since these conductivities are additive, it follows that σ, in the absence of interaction, should be linearly related to the concentration of titrant.

When charge transfer occurs, following eq. (17.16), it gives rise to an increase in conductivity. Let the concentration of the complex DA be given by C_{DA} and the concentrations of D^+ and A^- by C_D^+ and C_A^- respectively. C_D and C_A stand for the concentrations of unreacted donor and acceptor as before. We then can write the following equations for the equilibrium state of the system:

Complex formation: $D + A \rightleftharpoons DA$

$$\frac{C_{DA}}{C_D C_A} = K_1 = f(\epsilon, T \ldots) \tag{17.19}$$

Complex dissociation: $DA \rightleftharpoons D^+ + A^-$

$$\frac{C_D^+ C_A^-}{C_{DA}} = K_2 = f'(\epsilon, T \ldots) \tag{17.20}$$

Direct transfer and recombination: $D^+ + A^- \rightleftharpoons D + A$

$$\frac{C_D C_A}{C_D^+ C_A^-} = K_3 = f''(\epsilon, T \ldots) \tag{17.21}$$

The solution must be electrically neutral, so that

$$C_D^+ = C_A^- = n \tag{17.22}$$

which is the carrier concentration required. It follows that

$$[C_D^0 - (n + n^2/K_2)][C_A^0 - (n + n^2/K_2)] = n^2 K_3 = n^2 (K_1 K_2)^{-1} \tag{17.23}$$

The conductivity σ, which is proportional to n, is a maximum if $C_D^0 = C_A^0$ for a 1:1 complex;

$$-C_A^0 + n + n^2/K_2 = 0 \tag{17.24}$$

Thus, a σ peak will appear if

$$C_D^0 - C_D = C_A^0 \tag{17.25}$$

so that

$$n + C_{DA} = C_A^0 \tag{17.26}$$

or in terms of the titrating solutions,

$$V_0^D C_0^D = V_0^A C_0^A$$

$$V_0^D / V_0^A = C_0^A / C_0^D \tag{17.27}$$

holds for the conductivity peak. The stoichiometry of the complex may thus be determined.

In practice, conductivity peaks occur slightly off the stoichiometric ratio, thus indicating non-ideal behavior. A typical example[285] of such a conductivity titration plot is shown in Fig. 17.6. The data are consistent with two stoichiometries, viz., 1:2 and 2:3 which the method in its present instrumentation is unable to resolve.

Conversely, Fig. 17.7 illustrates the linear relationship obtained[283] if there is no complex formation or the complex fails to dissociate appreciably, generally due to the permittivity of the solvent being too low. A well-developed conductivity maximum is not always obtained. A steeply sloping base line caused by serious mismatch of the conductivities of the pure donor and acceptor solutions tends to yield ill-defined or obscured conductivity peaks. This is illustrated in Fig. 17.8 which refers to the iodine-naphthalene complex formed in acetonitrile. The decidedly nonlinear plot indicates complexing. The stoichiometry of this complex can be determined by correcting for the background conductivity obtained by linear interpolation, i.e., by plotting σ-σ_0 vs. relative concentration. This plot is shown in Fig. 17.9. A peak becomes evident at a stoichiometric ratio of 0.57 and indicates a 1:1 complex.[283]

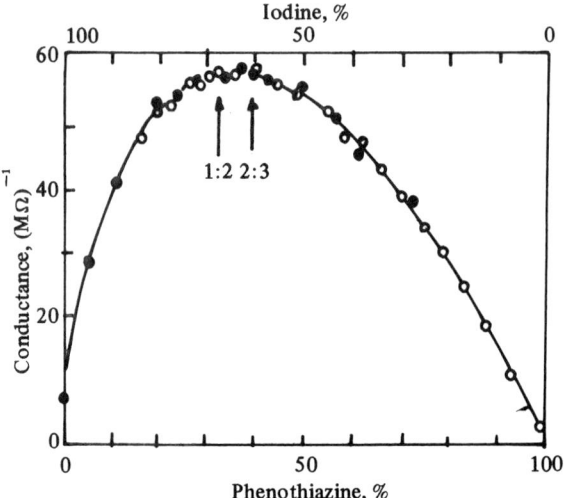

Fig. 17.6 Conductimetric titration of the phenothiazine-iodine complex formed in acetonitrile as solvent. Two stoichiometries are evident. After Gutmann and Keyzer.[285]

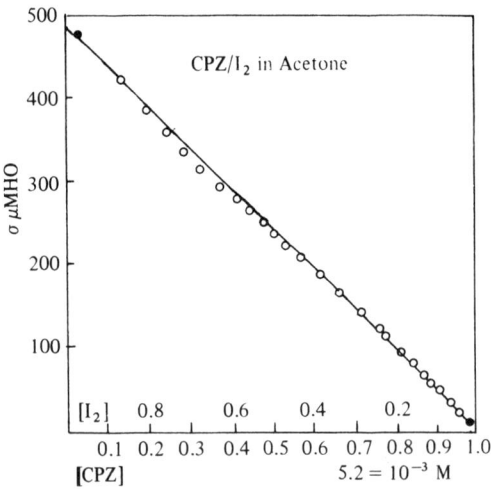

Fig. 17.7 Conductimetric titration of the chlorpromazine (CPZ) iodine complex formed in acetone as solvent. This is a well-known complex that, however, fails to yield a conductivity peak in acetone probably because the permittivity of the medium is too low to dissociate the complex. After Gutmann and Keyzer.[283]

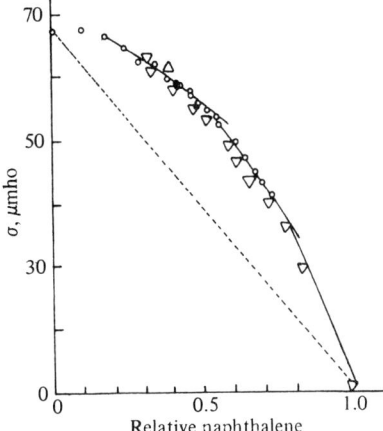

Fig. 17.8 Conductimetric titration of the naphthalene-iodine CTC, formed in acetonitrile as solvent. While complexing is evident from the departures above the base line, the steep slope makes the determination of the stoichiometry of the complex uncertain. After Gutmann and Keyzer.[283]

Solvent interactions may cause difficulties, such as ionic association effects in acetonitrile and in formamide;[291] acetonitrile is known to complex with cations.[293] Solvent donor:acceptor interactions have been studied by Fialkov and Borobikov.[292] Donor:iodine complexes in hexene are reported[294] to appear in two types due to a solvent effect; $D\ldots I_2$ and $D\ldots I^-\ldots I^-$.

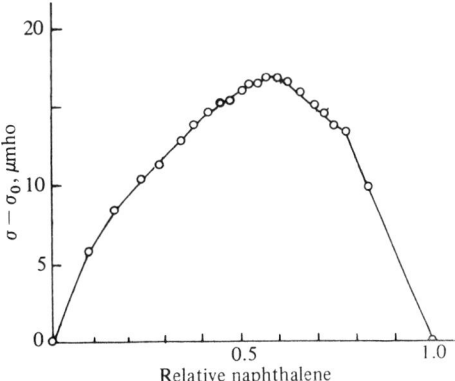

Fig. 17.9 Replot of the data displayed in Fig. 17.8. The excess conductances above the base line, i.e., $\sigma - \sigma_\infty$ are used in this plot in place of the conductances σ. A clear conductance peak results indicating a 1:1 complex stoichiometry. After Gutmann and Keyzer.[283]

o-Nitrophenol:amine 1:1 complexes in benzene are reported[295] to form ionic pairs in which the anion takes a quinone anion form. Hysteresis effects,[286] manifested in time-dependent conductivities, and in failure of the two branches of the titration curve to meet and to coincide, are sometimes evident and may cause difficulty. They have been reported for nucleic acids where the effect has been ascribed to the formation of several adducts which do not remain separate[286] but become linked to form polymeric complexes. The hysteresis effect has also been ascribed to irreversible protonation,[287] as e.g., in DMSO[288] or in chloroform.[289]

Assuming that the deviations from the linear base line indicated in Fig. 17.8 are entirely due to CTC formation, the equilibrium constant, K, can be calculated by following the procedure of Yoshida and Osawa.[290]

For an equimolar mixture of D and A (with total concentration C), the maximum deviation, Δy in any of the physical properties, (here σ) from the additivities can be written as

$$\Delta y = \alpha x \qquad (17.28)$$

where α is a proportionality constant and x is the concentration of the complex. When the total concentration is changed from C to C', then

$$\Delta y' = \alpha x' \qquad (17.29)$$

therefore

$$\frac{\Delta y}{\Delta y'} = \frac{\alpha x}{\alpha x'} = \frac{x}{x'} = k \qquad (17.30)$$

The equilibrium constant, K, is given by,

$$K = \frac{x}{(C/2 - x)^2} = \frac{x'}{(C'/2 - x')^2} \qquad (17.31)$$

and

$$K = \frac{2k^{1/2} \{k^{1/2}(C + C') - (C + kC')}{(C - kC')^2} \qquad (17.32)$$

The same reasoning may be applied to the transformation kinetics of inner-outer complex transitions.[352]

Conductivity titrations have been employed to identify a large variety of CTC e.g., chlorpromazine complexes,[297] amaranth-,[298] pyridine-,[299] neomycin-,[300] acetylcholine-,[301] 6-hydroxydopamine-,[301] serotonine-,[301] and other adducts, especially those suspected to play a role in drug interactions;[304] also in protein-[302] and in protonic complexes.[288] The original measurements[283] have more recently been repeated and confirmed by Ramamurthy and Quayum.[303]

The method supplements spectroscopic investigations because it is at its best for highly ionized complexes that do not yield a charge transfer band and are difficult to identify by spectroscopy only; the ground state of CTC is destabilized in high permittivity solvents.[296]

Dipole Moments

The formation of a CTC is generally accompanied by an increase in the dipole moment[305] though complexes between a polar and a non-polar component result in a reduction of the molecular polarization. Thus, measurement of changes in the molecular polarization following complexation provides an indication, though by no means an infallible proof, of the formation of CTC. The changes[306] are small differences between comparable quantities if both donor and acceptor are polar, and are also small if both are apolar. It is important that the changes in the molecular dipole moment be shown to be temperature-reversible in order to establish a charge-transfer interaction.

However, the polarizations being nonadditive indicates, though it falls short of proving, a charge-transfer interaction: e.g., the m-cresol-pyridine and -quinoline complexes were initially studied in this way, the interaction then being confirmed by means of conductimetric titrations.[299] For strong complexes, the measured permittivity is linearly related to the weight fraction of the dissolved complex using the same solvent;[308] the specific polarization of the complex solution has been shown to follow[309] an equation derived from the Clausius-Mosotti equation:

$$\frac{\epsilon-1}{\epsilon+2} \cdot \frac{1}{d} = P \quad (17.33)$$

having the form[309]

$$P_3 = P_{12}\left(1 - \frac{B}{P_{12}} + \frac{3A}{(\epsilon_{12}+2)(\epsilon_{12}-1)}\right) - (P_{12}-P_1)(P_2-P_1)$$

$$\times \left(w_2 + \frac{M_2}{M_3}(1-w_3)\right). \quad (17.34)$$

Here, d stands for the density, ϵ for the permittivity, M for the molecular weight, P for the specific polarization and w for the weight fraction; the subscript 1 refers to the solvent, 2 to the donor and 3 to the complex. A and B are proportionality constants derived from the linear relationship.[309]

$$\epsilon_{123} = \epsilon_{12} + Aw_3 \qquad d_{123} = d_{12} + Bw_3 \qquad (17.35)$$

Excitation leads to a significant change in the dipole moments.[310] It is often accompanied by intramolecular charge transfer.

For the π-electron components of the dipole moments in the ground (μ^0), fluorescent (μ^V) and photophorescent (μ^T) states in all compounds, the following inequality is valid:

$$|\mu^V - \mu^0| = |\Delta\mu^V| > |\Delta\mu^T| = |\mu^T - \mu^0| \qquad (17.36)$$

That is, the change of the dipole moment in the phosphorescent state is smaller than its change in the fluorescent state.

In almost all compounds, the following inequality is fulfilled:

$$\mu^V > \mu^T \qquad (17.37)$$

Excitation to higher singlet and triplet states is accompanied by intramolecular charge transfer and changes in dipole moments, though there is no relationship between the extent of this transfer and the energy of the transitions.[307]

Permittivity Titrations

The conductivity titrations discussed involve in practice a series of conductance measurements using one of the well-known bridge methods. Balancing an alternating current bridge requires a quadrature balance, *i.e.*, a measurement of the equivalent capacitance of the system. In many systems involving the formation of known CTC such as chlorpromazine-iodine in acetonitrile, there occur large capacitance changes that show a capacitance maximum at the same stoichiometry, which indicates complex formation from the conductivity changes, as well as from other evidence. Similar observations have been made for several other adducts.[310]

No capacitance change was observed[310] using DMSO as a solvent; the effect appears to be confined to relatively inert solvents of not too high a permittivity. The value of the capacitance peak depends on the direction of the titration; *e.g.*, in the titration of acetylcholine chloride *vs.* chlorpromazine-HCl in water the capacitance rise was roughly 10-fold when adding acetylcholine which the converse yielded only roughly a 2-fold rise. Both peaks occurred at the same, correct, stoichiometry, which indicated complexation.[310]

Capacitance peaks have been observed in the course of several conductivity titrations of charge transfer complexes. They cannot possibly be caused by any reasonable increase in the bulk permittivity of the solutions which were only of the order of 10^{-3} M, and which is determined by that of the solvent. These changes must then be due to processes occurring within an "active space" or reaction zone associated with the double layer.[313] During the titration, the electrochemical concentrations differ greatly from the bulk concentrations. Thus, e.g., acetylcholine[301] adsorbs on the electrode to a lesser degree than the more surface-active chlorpromazine (CPZ) and fails to displace the CPZ from the double layer in proportion to its bulk concentration. During the titrations, the capacitance values exhibit considerable changes, peaking in both directions at the stoichiometry indicated by the conductance values, viz., at 1:2 CPZ-HCl:acetylcholine. The peak capacitance values were 12.8×10^{-4} μF when adding CPZ and 34.6×10^{-4} μF for the converse, i.e., when adding acetylcholine.[301] In DMSO as the solvent, there is little evidence for such preferential adsorption and both branches of the titration curve coincide quite well. Similar results have been obtained for several other adduct interactions.[301,304] Capacitance titrations have also been applied to the evaluation of the stability constants of 1:1 CTC.[315]

The measured electron-transfer rate is markedly affected by the material of which the supposedly inert electrode is made. This is due to the extent of orbital overlap between electroactive species and electrode. The resulting differences in electron-transfer rate for the same couple are evident from the data for the iron-oxalate complexes, where the rate constant has been found[312] to vary over three orders of magnitude depending on the nature of the electrode.

Other Dielectric Methods

Other dielectric properties have been employed to aid identification of CTC. Their formation in a dielectric solvent causes the appearance of space charges which may set up an electric field, causing changes[305] in the refractive properties of the solution *via* the "electro-optic" or Kerr effect;[301] *e.g.*, chlorobiphenyl complexes have been studied by this method.[305]

Dielectric relaxation effects, though they may facilitate the identification of CTC[316] of thiazine derivatives have been identified by this method.[317] Contact CTC may give rise to dielectric absorption peaks[316] because of collision-induced dipole moments, so do ion pairs.[318] Dielectric relaxation measurements identify[319] stable CTC in contradistinction to contact CTC. The stable complex exhibits a relaxation time characteristic of a rigid molecule, while the relaxation time(s) obtained from a contact CTC depends on its lifetime and will be much shorter than for a rigid molecule,[318] *viz.*, a few times 10^{-11} to 10^{-12} sec. If the lifetime of the complex is shorter than the relaxation time of the entity itself, then the faster process will determine the experimental result.

Other Electrical and Magnetic Methods

In this section, some results of ESR (Electronspin Resonance), NMR (Nuclear Magnetic Resonance), NQR (Nuclear Quadrupole Resonance) and related methods will be briefly discussed.

Electron Spin Resonance

For a discussion of ESR methodology and applications the reader is referred to the voluminous literature[320] on this topic. Of the very many complexes studied by ESR, only a very few representative examples can be mentioned: aminoacid-chloranil,[321] melanin,[322] phenothiazines[324] and pyridine complexes with iodine[323] and other CTC have thus been studied; also organic semi-metals such as TTF:TCNQ.[349]

Nuclear Magnetic Resonance

The method is based on observing the interaction of the magnetic moment of the proton with externally applied magnetic fields. Of interest here is the shift in the proton resonance frequency as a function of its chemical environment, *i.e.*, the chemical shift.[325]

To identify a CTC, one looks for the chemical shift associated with an unpaired electron; however, this evidence is by no means decisive by itself. The method is best for short-lived, weak complexes, the observed chemical shift δ is given by[326]

$$\delta = \frac{[DA]}{[A]}(\delta_{complex} - \delta_{free}) + \delta_{free} \qquad (17.38)$$

where one considers the formation of a CTC from a donor D and an acceptor A with, say, the donor present in excess to ensure that all molecules of the acceptor do interact. Then a molecule of A interacts with a certain number of molecules of D within the characteristic time (period) of the method. The problem consists in finding the portions of this time that the molecule spends free and as part of the complex: the δ values above refer to these two situations.[326] The solvent should be inert, though this is difficult to achieve. Equilibrium constants may be thus evaluated.[327]

Relatively short lived CTC of several organic donors with oxygen as acceptor have thus been identified;[328] Auch has used NMR for studies of the anthracene: TNB adducts.[329]

For relations between NMR chemical shift and donicity, see Section 17.2 and Fig. 17.1.

Nuclear Quadrupole Resonance

This effect[326] involves an interaction between the quadrupole moment, if any, of a nucleus with the gradient of the electric field that it experiences: the resonance frequency is affected by the character of the bonds, and sometimes allows identification of a CTC as such. Only asymmetrical nuclei have a quadrupole moment—typically, Cl and other halogen nuclei. The method has been applied to a few complexes such as those containing the acceptor chloranil[330] and those between chloroform and several donors.[331] The triiodide ion, I_3^-, has thus been identified[332] in amine complexes as well in complexes with the donors picoline, pyridine, acridine, phenazine and hexamethylene-tetramine.[333] The quadrupole interaction is related to the ionization[334] potential of the donor[334] as deduced from a study of 26 CTC.[334] The method has been employed to establish the presence of an ionic (ground) state with virtually complete electron transfer.[350]

Magnetic Susceptibilities

The uncoupling of one or more electrons involved in the formation of a CTC causes a lowering of a diamagnetic susceptibility χ_d generally exhibited by most organic compounds. A minimum in χ_d thus results at the complex stoichiometry, though this is rather flat; such a minimum has been found, for example for the violanthrene-iodine 1:2 entity.

While in CTC with a nondative ground state the paramagnetic contribution from the unpaired electron(s) causes only a reduction in χ_d in strong complexes in which the ground state is mainly or largely ionic, the resulting complex becomes paramagnetic characterized[335] by a paramagnetic susceptibility χ_p. A paramagnetic complex may be studied by means of ESR, a method that, of course, does not apply to diamagnetic systems.

The magnetic susceptibilities of several other CTCs have been studied.[336] Susceptibility measurements are especially valuable when combined with ESR studies; in *e.g.*, perylene: metal dithiolate complexes[337] such studies have suggested the existence of two distinct sources of the paramagnetism of the adduct: one arises from the metal chelate system while the other is a thermally activated contribution of perylene$^+$.

Electrochemical Methods. Polarography and Voltammetry

Consider a solution of acceptor A in the presence of a suitable, inert, supporting electrolyte.[338] Polarography then will yield a half-wave potential $E_{\frac{1}{2}a}$ due to the reduction of A to A^- at the dropping mercury cathode. If now donor D is added, in general several complexes of different stoichiometries form, having the composition D_jA, where j is an integer assuming values from unity to i; for a 1:1 complex $i = 1$. Each of the i stoichiometries is governed by an equilibrium constant K_{ij} corresponding to a complex D_jA. If now the activity of the donor remains constant throughout the solution because the concentration of $D \gg A$, i.e., if there is a considerable donor excess, and introducing the Ilkovic equation, another half-wave potential $E_{\frac{1}{2}c}$ will be obtained which will be shifted by $\Delta E_{\frac{1}{2}}$ relative to $E_{\frac{1}{2}a}$. It can then be shown that[338]

$$\Delta E_{\frac{1}{2}} = \frac{RT}{F} \ln j_A \frac{I_c}{I_a} \sum_{j=0}^{j=1} \frac{K_j[D]^j}{\sum_0^i j D_j A} \qquad (17.39\text{A})$$

j_A is the activity coefficient of the acceptor, I_c and I_a the polarographic diffusion currents of the complex and of the acceptor, respectively, and $[D]^j$ stands for the activity of the donor D in the j'th complex. For a 1:1 complex, eq. (17.39A) reduces to

$$\Delta E_{\frac{1}{2}} = \frac{RT}{F} \ln j_A \frac{I_c}{I_a} \frac{K[D]}{j_{DA}} \qquad (17.39\text{B})$$

The values of the equilibrium constants K are similar to those obtained spectroscopically.

It is assumed that no RedOx reactions occur of the form

$$\begin{align} D + A &\rightleftharpoons D^+ + A^- & (a) \\ A^- + D &\rightleftharpoons D^- + A & (b) \\ D + e^- &\rightleftharpoons D^- & (c) \end{align} \qquad (17.40)$$

Reactions like (a) are likely[338] with strong donors and acceptors, i.e., with strong complexes, and in a medium of relatively high permittivity. Otherwise, the resulting shifts in the half-wave potential are small relative to $\Delta E_{\frac{1}{2}}$ in eq. (17.38b).

If the dissociation of the complex is slow, then[338] $E_{\frac{1}{2}}$ shifts in a direction opposite to that derived from the free energy of complexation, eqs. (17.39); 1 mV shift corresponding to a drop in I_c by about 4%. The same reasoning holds for added acceptor to an excess of donor.

The relation between the oxidation potential of several CTC and their equilibrium constant is illustrated in Fig. 17.10.

The half-wave potentials e.g., of phenothiazines shift, if oxygen is not expelled from the solution; this has been interpreted[341] as evidence for the formation of a CTC.

An inverse relation between the diffusion current and the molecular weight

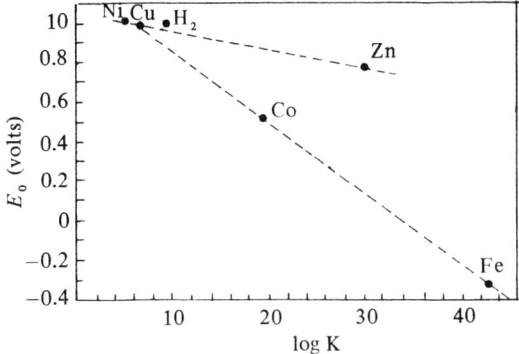

Fig. 17.10 Relation between the oxidation potentials and the logarithm of the equilibrium constants of charge transfer complex formation for several metal-tetraphenylporphyrins. After Williams.[340]

of the acceptor has been reported.[342]

Polarography has been applied to determine trace amounts of iodine by means of complexation with suitable donors and following the process by means of ac polarography.[343] Several complexes such as pyrene:TCNQ have been studied.[344]

The polarography of CTC in non-aqueous solvents such as acetonitrile is discussed by Ramaley and Gaul[339] (see also Ref. 345). Energy gap values may also be deduced from electrochemical, and especially from polarographic data (see Section 15.13).

The technique of cyclic voltammetry is also applicable to the study of charge transfer complexes; new peaks are observed *e.g.*, in the well-known CPZ:I_2 interaction which are absent in the components.[346] Likewise, the CPZ:dilantin interaction shows a new peak; these shift as a function of scan rate.[346] These matters are at present being further investigated.

The formation of a CTC can also be followed by a potentiometric titration;[346] the electrode potential of, say, a Pt electrode measured against a reference electrode exhibits a maximum, or minimum, at the stoichiometry of complexation.

Potentiometry has also been employed to confirm what appears to be the formation of surface CTC on membranes.[347]

The pyridine:iodine complex has been studied[348] in a concentration cell Pt|[Pyridine-2I_2]$_{c_1}$ + CH_3CN|[Pyridine-2I_2]$_{c_2}$ + CH_3CN|Pt and by potentiometry; it appears that the 1:2 complex tends to decompose upon dilution with acetonitrile.

References

1. B. Pullman, ed., *Molecular Associations in Biology,* Academic Press, N.Y., 1968; R. S. Mulliken and W. B. Person, *Molecular Complexes,* Wiley, N.Y., 1969; R. Foster, *Organic Charge Transfer Complex,* Academic Press, London, 1969; G. Briegleb, *Elektronen Donator-Akzeptor Komplexe,* Springer-Verlag, Berlin, 1961; M. A. Slifkin, *Charge Transfer Interactions of Biomolecules,* Academic Press, London, 1971; R. Foster, ed., *Molecular Association,* Vol. 2, Academic Press, N.Y., 1979; R. S. Mulliken, *Am. Rev. Phys. Chem.,* **29**, 1 (1978); J.-P. Farges and F. Gutmann, in *Modern Aspects of Electrochemistry,* J. O'M. Bockris and B. E. Conway, eds., Plenum Press, N.Y., **12**, 267 (1977); **13**, 361 (1979); B. P. Bespalov and V. V. Titov, *Phys. Chem. Revs.,* **44**, 1091 (1975); S. I. Peredereeva et al., *ibid.,* **295**; I. F. Shchegolev, *Phys. State. Solidi,* **A12**, 9 (1972); H. R. Zeller, *Adv. Solid State Phys.,* **13**, 31 (1973); D. M. Hanson, *Crit. Rev. Solid State Sci.,* **3**, 243 (1973); H. Sato and T. Yasuniwa, *Bull. Chem. Soc. Japan,* **47**, 368 (1974).Mo
2. M. A. Slifkin, *Chem. Phys. Lett.,* **7**, 195 (1970).
3. V. Gutmann, *Fortschr. Chem. Forsch.,* **27**, 59 (1972); *Chimia,* **23**, 285 (1969); *Chem. Brit.,* **7**, 102 (1971); V. Gutmann and U. Mayer, *Monatshefte f. Chem.,* **99**, 1383 (1968); D. Duschek and V. Gutmann, *ibid.,* **104**, 990 (1973); see also ref. 4.
4. V. Gutmann, *Electrochim. Acta,* **21**, 661 (1976); *Chimia,* **31**, 1 (1977).
5. Y. Matsunaga, *Nature* (London), **211**, 182 (1966).
6. R. S. Mulliken and W. B. Person in Ref. 1; M. A. Slifkin and R. H. Walmsley, *Photochem. Photobiol.,* **13**, 57 (1971); M. A. Slifkin, private communication, (1977).
7. T. Montenay-Garestier and C. Hélène, *Nature* (London), **217**, 844 (1968).
8. I. Schroff et al., *Tetrahedron Lett.,* 1649 (1973).
9. V. F. Gachkovskii, *Russ. J. Phys. Chem.,* **47**, 208 (1973).
10. D. W. Slocum et al., *Tetrahedron Lett.,* **46**, 4429 (1971).
11. Z. Witkiewicz, *Biul-Wojsk. Akad. Tech.,* **20**, 69 (1971).
12. J. P. V. Gracey and A. R. Ubbelohde, *J. Chem. Soc.,* **1955**, 4089; G. C. Martin, N. D. Parkyns and A. R. Ubbelohde, *ibid.,* **1961**, 4958; P. C. Li, J. P. Devlin and H. A. Pohl, *J. Phys. Chem.,* **76**, 1026 (1972).
13. F. T. Lang and R. L. Strong, *J. Am. Chem. Soc.,* **87**, 2345 (1965); J. Prochrow and A. Tramer, *J. Chem. Phys.,* **44**, 4545 (1966); M. Tamres and J. M. Goodenow, *J. Phys. Chem.,* **71**, 1982 (1967); **43**, 3393 (1965); M. Kroll and M. L. Ginter, *J. Phys. Chem.,* **69**, 3671 (1965); W. K. Duerksen and M. Tamres, *J. Am. Chem. Soc.,* **90**, 1379 (1967); M. Tamres et al., *J. Phys. Chem.,* **72**, 966 (1972); M. Tamres and J. Grundnes, *J. Am. Chem. Soc.,* **93**, 801 (1971); J. Aikara et al., *Bull. Chem. Soc. Japan,* **40**, 2460 (1967); **42**, 1842, (1969).
14. J. Aikara et al., *Bull. Chem. Soc. Japan,* **43**, 3323 (1970); F. Pellizza et al, *Thermochim. Acta,* **4**, 135 (1972).
15. P. Kollman, *J. Am. Chem. Soc.,* **99**, 4875 (1977).
16. E. I. Balabanov et al., *Dokl. Akad. Nauk. SSSR,* **238**, 1365 (1978).
17. L. Abate and G. Siracusa, *Thermochim. Acta,* **29**, 157 (1979); I. G. Orlov et al., in "Chemistry of Acetylene," "Proc. of the 3rd All-Union Conf., 1968," A. A. Petrov, ed., Nauk Publ. House, Moscow, 1972.
18. S. G. Lias and P. Ausloos, *J. Am. Chem. Soc.,* **99**, 4831 (1977).
19. B. D. Silverman, *Phys. Rev.,* **B-16**, 5153 (1977).
20. M. A. Slifkin, *Spectrochim. Acta,* **20**, 1543 (1964).
21. R. Foster and P. Hanson, *Tetrahedron,* **21**, 755 (1965).
22. F. Gutmann et al., *Adv. Biochem. Physchopharmacol.,* **9**, 15 (1974).

23. V. Chen-Teh-Bien, Univ. of Queensland, St. Lucia, Qld., Australia, Ph. D. Thesis December (1975).
24. Y. Iwasawa et al., *Bull. Chem. Soc. Japan,* **43,** 2656 (1970).
25. Y. Iida, *Bull. Chem. Soc. Japan,* **44,** 2564 (1971); S. Kiozumi and Y. IIda, *ibid.,* **44,** 1436 (1971).
26. C. J. Schramm et al., *Science,* **200,** 47 (1978).
27. Y. Jayathirta and V. Krishnan, *Natl. Acad. Sci. Lett.* (India), **1,** 365 (1978).
28. T. M. Krygowski and W. R. Fawcett, *J. Am. Chem. Soc.,* **97,** 2143 (1975); see also Ref. 4.
29. N. K. Wilson, *J. Phys. Chem.,* **83,** 2649 (1979); V. Gutmann, *Electrochim. Acta,* **21,** 661 (1976).
30. I. Deperasinska and J. Prochorov, *Adv. Molec. Relax. Interact. Process,* **11,** 51, (1977).
31. J. Giera et al. "3rd Internatl. Symp. Specific Interact. Molec. Ions Proc. 1976," **1,** 154; Inst. of Chem. Univ. Wroclaw, Poland.
32. K. Ishii et al., *Chem. Phys. Lett.,* **41,** 154 (1976).
33. M. Fujihara et al., *Chem. Phys. Lett.,* **19,** 584 (1973).
34. T. Kunii, Ph. D. Thesis, Tokyo Univ., (1966).
35. K. Ishii et al., in *Energy and Charge Transfer,* K. Masuda and M. Silver eds., Plenum Press, N.Y., 1974, p. 183.
36. D. Haarer and N. Karl, *Chem. Phys. Lett.,* **21,** 49 (1973); D. Haarer, *Chem. Phys. Lett.,* **27,** 91 (1974); **31,** 192 (1975).
37. N. Karl and J. Ziegler, *Chem. Phys. Lett.,* **32,** 438 (1975).
38. F. Bernardi et al., *J. Phys. Chem.,* **83,** 640 (1978); *cf.,* also A. Mangini, *Mem. Acad. Lincei,* **14,** Sez. II (1977).
39. H. Poradowska et al., *Rosz. Chem.,* **51,** 551 (1977).
40. V. V. Kopylov et al., *Zh. Fiz. Khim.,* **52,** 533 (1978).
41. M. H. Litt and J. Wellinghoff, *J. Phys. Chem.,* **81,** 2644 (1977).
42. J. S. Thayer and R. West, in "Adv. in Organometallic Chem.," **5,** 195 (1967); F. G. Stone and R. West, eds, Academic Press, N.Y., 1967.
43. V. M. Glazov et al., *Zh. Fiz. Khim.,* **51,** 2788 (1977).
44. A. Treinin and E. Hayon, *J. Am. Chem. Soc.,* **97,** 1716 (1975).
45. H. J. M. Andriessen et al., *J. Chem. Soc.,* Perkin Trans., **2,** 861 (1972).
46. S. Kobinata and S. Nagakura, *J. Am. Chem. Soc.,* **88,** 3905 (1966); H. Tsubomura and S. Nagakura, *J. Chem. Phys.,* **27,** 819 (1957).
47. A. Funatsu and K. Toyoda, *Bull. Chem. Soc. Japan,* **43,** 279 (1970).
48. W. E. Vaughan, *Dig. Lit. Dielect.,* **35,** 75 (1971).
49. M. J. Mantione, *Theor. Chim. Acta* (Berlin), **11,** 119 (1968); D. Bauer et al., *J. Phys. Chem.,* **74,** 4594 (1970).
50. H. Kuroda et al., *J. Am. Chem. Soc.,* **89,** 6056 (1967).
51. S. P. McGlynn, *Chem. Rev.,* **58,** 1130 (1958).
52. W. Focke and C. E. Weikowitsch, *Z. Phys. Chem.,* (Wiesbaden), **105,** 251 (1977).
53. B. Mayoh and C. K. Prout, *J. Chem. Soc.,* Faraday II, **68,** 1072 (1972).
54. V. K. Kondratov and G. M. Karpin, *Zh. Fiz. Khim.,* **53,** 211 (1979).
55. V. A. Gorodyskii and A. A. Morachevskii, *Zh. Fiz. Khim.,* **52,** 2248 (1978).
56. F. Gutmann and H. Keyzer, *J. Chem. Phys.,* **46,** 1969 (1967); A. Brau, J. P. Farges and F. Gutmann, *Electrochim. Acta.,* **17,** 1803 (1972).
57. D. Bargeman and J. Kommandeur, *J. Chem. Phys.,* **49,** 4069 (1968); J. Ludwig and J. Kommandeur, *ibid.,* **52,** 2302 (1970).
58. R. C. Teitelbaum et al., *Bull. Amer. Phys. Soc.,* **24,** 355 (1979).

59. S. Doi et al., *Bull. Chem. Soc. Japan,* **50**, 837 (1977); *cf.*, also K. Kan and Y. Matsunaga, *ibid,* **45**, 2096 (1972).
60. K. Jain et al., *Indian J. Phys.,* **52A**, 543 (1978).
61. R. P. Lang, *J. Am. Chem. Soc.,* **93**, 5047 (1971); *J. Phys. Chem.,* **78**, 1657 (1974); M. Tamres, *ibid.,* **65**, 654 (1961).
62. F. Lux et al., *Z. Chemie,* **18**, 107 (1978).
63. F. Sersen et al., *Collect. Czechoslov. Chem. Comm.,* **42**, 2173 (1974).
64. F. Babler and A. Von Zelewski, *Helv. Chim. Acta,* **60**, 2723 (1977).
65. F. Gutmann and H. Keyzer, *Electrochim. Acta,* **13**, 693 (1968); see also Ref. 80.
66. M. A. Slifkin, *Charge Transfer Interactions of Biomolecules,* Academic Press, London, 1971, p. 251.
67. E. A. Chandross and J. Ferguson, *J. Chem. Phys.,* **47**, 2557 (1967); H. A. Staab et al., *Angew. Chemie,* Internatl. Ed. (Engl.), **16**, 801 (1977).
68. W. Robertson et al., *J. Chem. Phys.,* **35**, 464 (1961); N. Tyutyuckov et al., *Theor. Chim. Acta* (Berlin), **20**, 385 (1971).
69. J. K. Roy and D. G. Whitten, *J. Am. Chem. Soc.,* **94**, 7162 (1972); see also Ref. 132.
70. M. Gordon and W. R. Ware, *The Exciplex,* Academic Press, N.Y., 1975; *Molecular Association,* R. Foster, ed., Academic Press, N.Y., 1979, p. 2; P. Fröhlich and E. L. Wehry, "The Study of Excited State Complexes (Exciplexes)," in *Mod. Fluorescence Spectroscopy,* E. L. Wehry, ed., Plenum Press, N.Y., **2**, 319 (1976).
71. K. Takemoto et al., *Bull. Chem. Soc. Japan,* **41**, 1974 (1968).
72. Y. Taniguchi and N. Mataga, *Chem. Phys. Lett.,* **13**, 596 (1972); Y. Taniguchi et al., *Bull. Chem. Soc. Japan,* **45**, 764 (1972); G. Bökestein and M. Buck, *Rec. Rev. Chim., Pays-Bas.,* **92**, 1095 (1973).
73. H. Knibbe et al., *Ber. Bunsen Ges. Phys. Chem.,* **73**, 839 (1969); S. Yomosa, *J. Phys. Soc. Japan,* **36**, 1655 (1974).
74. M. Shimada, *Bull. Chem. Soc. Japan,* **46**, 1903 (1973); *cf.*, also Ref. 75.
75. T. Tamamura, *Bull. Chem. Soc. Japan,* **47**, 448 (1974).
76. D. Gordon and M. J. Hove, *J. Chem. Phys.,* **59**, 3419 (1973).
77. V. S. Grechishkin, *Optiks y Spektroskop.,* **20**, 300 (1966).
78. H. Hirayama, *J. Chem. Phys.,* **42**, 3163 (1965); E. A. Chandros and C. J. Dempster, *J. Amer. Chem. Soc.,* **92**, 3586 (1970).
79. N. Ohrbach et al., *J. Phys. Chem.,* **77**, 2831 (1973).
80. I. Forrest et al., *Rev. d'Agressologie* (Paris), **7**, 147 (1966).
81. N. Mataga, *Bussei Kenkyu,* **18**, A37 (1973).
82. M. Itoh et al., *Bull. Chem. Soc. Japan,* **47**, 1078 (1974).
83. J. Prochorow, *J. Luminesc.,* **18-19**, 105 (1979).
84. L. Fredin and B. Nelander, *J. Mol. Structure,* **16**, 205 (1973); B. Nelander, *Theor. Chim. Acta,* (Berlin), **25**, 382 (1972).
85. F. Gutmann et al., *Adv. Biochem. Psychopharmacol.,* **9**, 15 (1974).
86. R. I. Kukhtim et al., *Opt. y. Spektrosk.,* **35**, 845 (1973); V. V. Slobodyanik et al., *Dopov. Akad. Nauk. Ukr.,* RSR, Ser. A, **35**, 1033 (1973).
87. A. Terenin, *Advan. Catalysis,* **14**, 372 (1963).
88. B. Nelander, *Spectrochim. Acta,* **29A**, 859 (1973).
89. S. Arimatsu and H. Tsubomura, *Bull. Chem. Soc. Japan,* **45**, 1357 (1972).
90. K. Hauffe et al., *J. Electrochem. Soc.,* **117**, 993 (1970).
91. M. Ottolenghi, *Acc. Chem. Res.,* **6**, 153 (1973); M. Itoh et al., *Bull. Chem. Soc. Japan,* **50**, 2509 (1977).
92. J. Ferrari et al., *J. Am. Chem. Soc.,* **95**, 948 (1973); F. Gutmann and H. Keyzer,

J. Chem. Phys., **46**, 1969 (1967); T. Tamaki, *Bull. Chem. Soc. Japan*, **46**, 2527 (1973); *J. Am. Chem. Soc.*, **97**, 3209 (1975); S. C. Abbi and D. M. Hanson, *J. Chem. Phys.*, **60**, 319 (1974); *cf.*, also M. A. Slifkin, ref. 66; A. K. Prokofev, *Russ. Chem. Revs.*, **45**, 519 (1976); L. G. Schroff and A. J. A. van der Weerdt, *Tetrahedron Lett.*, **18**, 1649 (1973); I. V. Turovskii et al., *Dokl. Akad. Nauk, SSSR, Phys. Chem.*, **223**, 746 (1975).

93. R. S. Mulliken and W. B. Person, *Molecular Complexes*, Wiley, N.Y., 1969; M. A. Slifkin and R. H. Walmsey, *Photochem. Photobiol.*, **13**, 57 (1971).

94. K. Mutai et al., *Chem. Lett.*, **1977**, 1047; M. L. Olson and K. R. Sundberg, *J. Chem. Phys.*, **69**, 5400 (1978); J. R. C. Applequist and K-K. Fung, *J. Am. Chem. Soc.*, **94**, 2952 (1972); *Acc. Chem. Res.*, **10**, 79 (1977); S. L. Mair, *Acta Cryst.*, **A34**, 656 (1978).

95. S. R. Turner and M. Stolka, *Macromolecules*, **11**, 835 (1978).

96. A. Sasaki and Y. Matsunaga, *Bull. Chem. Soc. Japan*, **47**, 2926 (1974).

97. H. A. Staab et al., *Angew. Chemie Int. Ed.*, (Engl.), **16**, 801 (1977).

98. M. A. Yurovskaya et al., *Tetrahedron*, **34**, 2931 (1978); *cf.*, also Ref. 136.

99. Z. R. Grabowski, *J. Luminesc.*, **18-91**, 420 (1979); *Acta Phys. Polon.*, **A54**, 769 (1978).

100. Y. Matsunaga et al., *Bull. Chem. Soc. Japan*, **47**, 2926 (1974).

101. I. Imura, *Bull. Chem. Soc. Japan*, **46**, 2075 (1973).

102. J. W. Longworth, *Photochem. Photobiol.*, **7**, 587 (1968); T. Tamaki, *Bull. Chem. Soc. Japan*, **46**, 2527 (1973).

103. H. Inoue et al., *Bull. Chem. Soc. Japan*, **46**, 3614 (1973).

104. S. C. Abbi and D. M. Hanson, *J. Chem. Phys.*, **60**, 319 (1974).

105. L. M. Blinov et al., *Zhur. Struct. Khim.*, **14**, 662 (1973).

106. M. Itoh, *Bull. Chem. Soc. Japan*, **47**, 1078 (1974).

107. A. P. Pivovarov, *Vysokomol. Soedin*, Ser. B., **20**, 54 (1978).

108. M. Goldenberg et al., *J. Am. Chem. Soc.*, **100**, 7171 (1978).

109. T. N. Kopylova et al., *Zh. Fiz. Khim.*, **51**, 1601 (1977); K. Itaya and Sh. Toshima, *Chem. Phys. Lett.*, **51**, 447 (1977); F. Pragst et al., *J. Luminesc.*, **17**, 425 (1978).

110. D. Simov and S. Stoyanov, *God. Sofii* (Bulgaria) Univ. Khim. Fakul., 1971-72 (Publ. 1975) **66**, 323.

111. J. Freimans et al., *Adv. Molecul. Relax. Processes*, **5**, 33 (1973); *Zh. Org. Khim.*, **11**, 2106 (1975); *Zh. Obshch. Khim.*, **45**, 2484, (1975).

112. I. Schroff et al., *Tetrahedron Lett.*, 1649 (1973).

113. S. Cohen et al., *Chem. Phys. Lett.*, **42**, 15 (1976).

114. E. M. Kosower et al., *J. Am. Chem. Soc.*, **97**, 2167 (1975).

115. E. M. Kosower et al., *J. Phys. Chem.*, **82**, 2012 (1978).

116. M. R. Johnson and I. O. Sutherland, *J. Chem. Soc. Chem. Commun.*, **1979**, 306, 309.

117. H. Inoue et al., *Bull. Chem. Soc. Japan*, **46**, 3614 (1973).

118. K. H. Giovanelli et al., *Ber. Bunsen Ges. Phys. Chem.*, **75**, 864 (1971).

119. A. E. Lutskii, *Opt. Spektroskop.*, **34**, 1076 (1973).

120. R. Van Est-Stammer and J. B. F. N. Engberts, *Canad. J. Chem.*, **51**, 1187 (1973).

121. H. Siegel and C. F. Naumann, *J. Am. Chem. Soc.*, **98**, 730 (1976); H. Siegel and P. E. Amsler, *ibid.*, **98**, 7390 (1976); H. Siegel et al., *Inorg. Chem.*, **16**, 790 (1977); E. C. Johnson et al., *Canad. J. Chem.*, **56**, 1381 (1978); P. R. Mitchell and H. Siegel, *J. Amer. Chem. Soc.*, **100**, 1564 (1978); *Helv. Chim. Acta*, **62**, 1723 (1979); C. R. Naumann and H. Siegel, *FEBS Lett.*, **47**, 122 (1974); P. Chaudhuri and H. Siegel, *J. Amer. Chem. Soc.*, **97**, 3209 (1975); **99**, 3142 (1977); H. Siegel et al., *Europ. J. Biochem.*,

41, 209 (1974); G. Cilento, *Quart. Revs. Biophys.*, **6**, 488 (1973); Y. Fukuda, P. R. Mitchell and H. Siegel, *Helv. Chim. Acta*, **61**, 638 (1978); H. Siegel et al., *Biochim. Biophys. Acta*, **148**, 655 (1967); H. Siegel and C. F. Naumann, *J. Am. Chem. Soc.*, **98**, 730 (1976); H. Siegel and P. E. Amsler, *ibid.*, **98**, 7390 (1976); A. M. Golub, *Russ. Chem. Revs.*, **45**, 479 (1976); R. Clement et al., *J. Chem. Soc. Chem. Commun.*, 654 (1974); A. N. Nesmeyanov et al., *Dokl. Nauk, SSSR*, **221**, 229 (1975); P. Pasman et al., *Chem. Phys. Lett.*, **59**, 398 (1978); L. M. Titvinenko et al., *Russ. Chem. Revs.*, **44**, 718 (1975).

122. Z. A. Insepov and G. E. Norman, *Zh. Eksper. Teor. Fiz.*, **73**, 1517 (1977).

123. H. Bretschneider, *Wiss. Z. Tech. Hochschule Karl Marxtadt*, **17**, 281 (1975); F. Cramer, *Einschlussverbindungen*, Springer, Heidelberg (1974).

124. Y. Matsunaga et al., *Bull. Chem. Soc. Japan*, **47**, 2826 (1974).

125. M. Möhwald, *Chem. Phys. Lett.*, **26**, 509 (1974).

126. *e.g.,* H. Scher and M. Lax, *Phys. Rev.*, **B-7**, 4491 (1973); G. G. Roberts and J. I. Polanco, *Phys. Stat. Solidi*, **(a)1**, 409 (1970); A. J. Grant and E. E. Davis, *Solid State Commun.*, **15**, 563 (1974).

127. *e.g.,* Ref. 126 and *cf.,* also F. Gutmann et al., *Nature* (London), **221**, 1237 (1969); A. M. Hermann, "Elect. Prop. Polymers," K. C. Frisch, ed., 1972, p. 103.

128. R. C. Powell, *J. Chem. Phys.*, **58**, 920 (1973); L. E. Lyons, *Search*, **7**, 339 (1976).

129. P. Petelenz, *Proc. Nauk Inst. Chem. Org. Fiz. Politech. Wroclaw* (Poland), **16**, 257 (1978).

130. H. Leismann et al., *J. Photochem.*, **9**, 338 (1978).

131. V. V. Osipov et al., *Teor. Eksp. Khim.*, **14**, 703 (1978).

132. D. V. O'Connor and W. R. Ware, *J. Am. Chem. Soc.*, **101**, 121 (1979).

133. J. Gebicki et al., *Chem. Phys. Lett.*, **59**, 197 (1978).

134. I. Bendiz et al., *Teor. Eksp. Khim.*, **14**, 629 (1978).

135. S. Basu, *J. Photochem.*, **9**, 539 (1978).

136. M. A. Yurovskaya et al., *Tetrahedron*, **34**, 2931 (1978).

137. J. Bertin et al., *J. Am. Chem. Soc.*, **96**, 8113 (1974).

138. J. Bertin et al., *Tetrahedron Lett.*, 763 (1974).

139. J. M. Lalancette et al., *Canad. J. Chem.*, **52**, 589 (1974); *ibid.*, **50**, 3058 (1972).

140. M. E. Volpin and Yu. N. Novikov, *Topics in Non-Benzenoid Aromatic Chemistry*, **1**, 269 (1973), T. Nozoe et al., eds., Wiley, N.Y.

141. F. R. Gamble et al., *Science*, **168**, 568 (1970); **174**, 493 (1971); M. M. Labes, *Nature* (London), *Phys. Sci.*, **246**, 122 (1973).

142. A. H. Thompson, *Solid State Commun.*, **13**, 1911 (1973).

143. D. E. Nixon and G. S. Parry, *J. Phys.*, **D-1**, 291 (1968); I. Inoshita et al., *J. Phys. Soc. Japan*, **43**, 1237 (1977); C. Underhill et al., *ibid.*, 374.

144. R. C. Croft, *Quart. Rev.*, **14**, 1 (1960); J. J. Murray and A. R. Ubbelohde, *Proc. Roy. Soc. London*, Ser. A, **312**, 371 (1969); I. T. McGovern et al., *Bull. Amer. Phys. Soc.*, **24**, 410 (1979).

145. J. E. Fischer and T. E. Thompson, *Physics Today*, **31**, 36 (1978); M. S. Dresselhaus et al., *Bull. Amer. Phys. Soc.*, **24**, 410 (1979).

146. M. Sano and H. Inokuchi, *Chem. Lett.*, **1979**, 405.

147. H. Inokuchi et al., *J. Catalysis*, **8**, 383 (1967).

148. F. L. Vogel, *Bull. Amer. Phys. Soc.*, **24**, 418 (1979); F. L. Vogel et al., "Proc. 5th London Internatl. Conf. on Carbon and Graphite," 708 (1978); Conf. Series-Inst. of Phys. (London), **43**, (Semiconduct), Publ. 1979.

149. S. Dyankov and V. Iliev, *Dokl. Akad. Nauk. Bulgar.*, **30**, 1423 (1977).

REFERENCES

150. H. Werner, *Angew. Chem.*, **89**, 1 (1977); *Angew. Chem. Int. Ed.* (Engl), **16**, (1977).
151. P. R. Mitchell and H. Siegel, *Angew. Chem.*, **88**, 585 (1876).
152. O. Yamaguchi et al., *Bull. Chem. Soc. Japan*, **48**, 2572 (1975).
153. C. M. A. Bourg and P. W. Schindler, *Chimia*, **32**, 166 (1978).
154. R. Foster and N. Kulevsky, *J. Chem. Soc. Faraday Trans.*, **69**, 1427 (1973).
155. S. Koisumi and Y. Matsunaga, *Bull. Chem. Soc. Japan*, **45**, 423 (1972).
156. L. Mandelkorn, *Chem. Revs.*, **59**, 827 (1959); J. H. v.d. Waals and J. C. Platteeum, *Adv. Chem. Phys.*, **2**, 1 (1959).
157. M. Ohmasa et al., *Bull. Chem. Soc. Japan*, **44**, 391 (1971).
158. W. Zeichmann and M. Wakill, *Erdöl, Kohle, Petrochem.*, **29**, 297 (1976).
159. M. Ohmasa et al., *Bull. Chem. Soc. Japan*, **44**, 391 (1971).
160. L. Fredin and B. Nelander, *J. Molec. Struct.*, **16**, 205 (1973); B. Nelander, *Theoret. Chem. Acta* (Berlin), **25**, 382 (1972).
161. M. Mandelkorn, *Chem. Revs.*, **59**, 827 (1959); J. B. Torrance and B. D. Silverman, *Phys. Rev.*, **B-15**, 788 (1977); K. Deguchi and K. Mcguro, *J. Colloid. Interface Sci.*, **38**, 596 (1972).
162. H. Negita et al., *Bull. Chem. Soc. Japan*, **46**, 2662 (1973).
163. U. Thewalt and C. E. Brugg, *Acta Cryst.*, **B-28**, 82 (1972).
164. F. Cramer, *Einschluss-Verbindungen*, Springer, Heidelberg, 1974; J. P. Behr and J. M. Lehn, *J. Amer. Chem. Soc.*, **98**, 1743 (1976).
165. *cf.*, Y. Iida, *Bull. Chem. Soc. Japan*, **44**, 3344 (1971).
166. *e.g.*, A. K. Covington and K. E. Newman, in *Modern Aspect of Electrochemistry*, J. O'M Bockris and B. E. Conway, eds., **12**, 41 (1977), Plenum Press, N.Y.; E. V. Goldammer, *ibid.*, **10**, 1 (1975); F. Feakins et al., *J. Chem. Soc. Chem. Commun.*, 1978, 218; T. Erdey-Gruz, "Transport Phenomena in Aqueous Solution," A. Hilger, London, 1974, p. 468; see also Ref. 174.
167. D. Braun and H. J. Sterzel, *Ber. Bunsen Ges. Phys. Chemie*, **76**, 551 (1972).
168. A. D. Buckingham et al., *Trans. Faraday. Soc.*, **67**, 577 (1971); L. Jansen and P. Mazur, *Physica*, **21**, 193–208 (1955).
169. G. Briegleb, *Elektronen-Donator-Akzeptor Komplexe*, Springer, Berlin, 1961, p. 182; D. Braun and J. H. Sterzel, *Ber. Bunsen Ges. f. Physik. Chemie.*, **76**, 551 (1972).
170. B. Case et al., *J. Electroanal. Chem.*, **10**, 360 (1965).
171. G. B. Sergeev et al., *Russ. J. Chem.*, **47**, 396 (1973).
172. M. A. Slifkin, *Charge Transfer Interactions of Biomolecules*, Academic Press, London, 1971, p. 47.
173. M. A. Slifkin, *Spectrochim. Acta*, **25A**, 1037 (1969); R. Kh. Ibragimova et al., *Izv. Akad. Nauk Uzb. SSR Ser. Fiz. Mat. Nauk.*, **17**, 55 (1973).
174. O. B. Nagy, J. B. Nagy and A. Bruylants, *J. Chem. Soc. Perkin Trans. II*, **1972**, 968; V. Gutmann and U. Mayer, *Monatshefte f. Chemie*, **99**, 1383 (1968); M. A. Slifkin, *Charge Transfer Interactions in Biomolecules*, Academic Press, London, N.Y., 1971; *cf.*, also *Non-aqueous Solutions*, V. Gutmann, ed., Pergamon Press, Oxford, 1976; S. Ahrland, *Solvation and Complex Formation in Protic and Aprotic Solvents*, J. J. Lagowski, ed., Academic Press, N.Y., 1978.
175. U. Mayer and V. Gutmann, *Monatshefte J. Chemie*, **101**, 912 (1970).
176. *Water*, F. Frank, ed., Plenum Press, N.Y., 1973; T. Kagiya, *Bull. Chem. Soc. Japan*, **41**, 767 (1968); see also refs. 3, 4, 166, 175.
177. O. Nagy et al., *J. Chem. Soc. Perkin Trans.*, (**1972**) 968.
178. J. I. Padova, *Modern Aspects of Electrochemistry*, No. 7, J. O'M. Bockris and B. E. Conway, eds., Plenum Press, N.Y., 1972, p. 1.

179. A. Kofler, *Z. Elektrochem.*, **50**, 200 (1944); G. Saito and Y. Matsunaga, *Bull. Chem. Soc. Japan*, **46**, 1609 (1973).
180. P. Stilbs and G. Olofsson, *Acta Chem. Scand.*, **A28**, 647 (1974); G. Olofsson and I. Olofsson, *Tetrahedron*, **29**, 1711 (1973).
181. Q. Appleton et al., *Tetrahedron*, **27**, 5921 (1971).
182. T. L. Einstein, *Phys. Rev.*, **B-12**, 1282 (1975).
183. T. B. Grimley, *J. Vac. Sci. and Technol.*, **8**, 31 (1971); J. R. Schriefer et al., *ibid.*, **9**, 561 (1972).
184. S. G. Gagarin and Yu. A. Kolbanovskii, *Kinet. Katal.*, **19**, 1463 (1978).
185. J. M. Turlet et al., *J. Luminesc.*, **18–19**, (Pt. 1) 47 (1979); see also R. T. Holm and E. D. Palik, *Phys. Rev.*, **B-17**, 2173 (1978).
186. O. Peshev, *J. Res. Inst. Catalysis*, Hokkaido Univ., **19**, 181 (1972); O. Johnson *ibid.*, **22**, 157 (1974); A. Nakamura, *ibid.*, **26**, 145 (1978); J. Horiuti, "Catalyzed Reactions by Solid Surfaces and Adsorption," 2nd Intern. Conf. Solid Surf. Proc., 1974; K. Kano and T. Yatsuo, *Bull. Chem. Soc. Japan*, **47**, 2836 (1974); A. J. Frank et al., *Ber. Bunsen Ges. Phys. Chem.*, **80**, 547 (1976); M. Grätzel, *ibid.*, **79**, 475 (1975); S. A. Alkatis et al., *J. Am. Chem. Soc.*, **97**, 5723 (1975); M. Chew et al., *ibid.*, 2052.
187. F. Tokiwa and K. Tsuji, *Bull. Chem. Soc. Japan*, **46**, 2684 (1973); H. Lange, *Kolloid Z.-Z. Polymere*, **243**, 101 (1971); M. L. Fishman and F. R. Eirich, *J. Phys. Chem.*, **75**, 3135 (1971).
188. J. Robillard, Priv. Commun. (1976); *cf.*, also L. I. Ahmed, *J. Phys. Chem. Solids*, **29**, 1653 (1973).
189. B. Folkesson, *Acta Chem. Scand.*, **27**, 287 (1973).
190. F. Gutmann, unpublished results.
191. *e.g.*, J. P. Williams, 5th Keilin Mem. Lect., *Biochem. Soc. Trans.*, **1**, 1 (1973); G. M. Schwab, *Fortschr. Chem. Forschg.*, **25**, 105 (1972); O. Johnson, Ref. 186.
192. R. Del Sole and E. Tossatti, *Solid State Commun.*, **22**, 307 (1970).
193. H. Keyzer et al., "4th Internatl. Symp. on Phenothiazine and Related Drugs," H. Eckert, I. S. Forrest and E. Usdin, eds., Elsevier, Amsterdam, 1980.
194. J.-P. Farges and F. Gutmann, unpublished results.
195. K. Deguchi and K. Meguro, *J. Colloid and Interface Sci.*, **38**, 596 (1972).
196. G. N. Asmolov and O. V. Krylov, *Kinet. Katal.*, **19**, 1208, 1213 (1978).
197. J. F. J. Kibblewhite and A. J. Tench, *J. Chem. Soc. Faraday Trans. I*, **70**, 72 (1974).
198. G. Heublein and S. Spange, *Z. Chem.*, **14**, 22 (1974).
199. N. G. Gaylord, *Polymer Prepr. Amer. Chem. Soc. Div. Polymer Chem.*, **13**, 505 (1972).
200. F. Gutmann et al., *Adv. Biochem. Psychopharmacol.*, **9**, 15 (1974).
201. Y. M. Thomas and P. Piciulo, *J. Amer. Chem. Soc.*, **100**, 3239 (1978).
202. S. C. Wallace et al., *Chem. Phys. Lett.*, **23**, 359 (1973); M. Grätzel and J. K. Thomas, *J. Phys. Chem.*, **78**, 2248 (1974); S. A. Alkatis et al., *J. Am. Chem. Soc.*, **97**, 5723 (1975); *Ber. Bunsen Ges. Phys. Chem.*, **79**, 54 (1975); S. A. Alkatis and M. Grätzel, *J. Am. Chem. Soc.*, **98**, 3549 (1976).
203. E. A. G. Aniansson, *Ber. Bunsen. Ges. Phys. Chem.*, **82**, 981 (1978); H. Hoffmann, *ibid.*, 988; B. Katusin-Razem et al., *J. Am. Chem. Soc.*, **100**, 1679 (1978); J. H. Fendler and Li-Jen Liv; **97**, 999 (1975).
204. J. Emert et al., *J. Am. Chem. Soc.*, **101**, 771 (1979).
205. N. Truro et al., *J. Am. Chem. Soc.*, **101**, 772 (1979).
206. B. Katusin-Razem et al., Ref. 203.
207. J. K. Thomas et al., *Ber. Bunsen Ges. Phys. Chem.*, **82**, 941 (1978).

REFERENCES

208. N. Rössler and G. V. Bünen, *Ber. Bunsen Ges. Phys. Chem.*, **82**, 949 (1978).
209. R. Memming and F. Mollers, Faraday Soc. Symp., No. 4, 1140 (1970); R. Memming and G. Kurstein, *Ber. Bunsen Ges. Phys. Chem.*, **75**, 1140 (1971); see also ref. 186 and 202.
210. J. F. J. Kibblewhite and A. J. Tench, *J. Chem. Soc. Faraday I*, **70**, 72 (1974).
211. M. Almgren et al., *J. Phys. Chem.*, **83**, 3232 (1979).
212. Y. Usui and A. Gotov, *Photochem. Photobiol.*, **29**, 165 (1979).
213. C. A. Bunton, *Progr. Solid State Chem.*, **8**, 167 (1973).
214. V. T. Gorshkov et al., *Zh. Fiz. Khim.*, **51**, 2680 (1977).
215. H. Veda and D. Yashiro, *Bull. Chem. Soc. Japan,* **44**, 391 (1971).
216. H. J. Frank et al., *Ber. Bunsen Ges. Phys. Chem.*, **80**, 547 (1976); M. Grätzel et al., *ibid.,* **79**, 475 (1975).
217. S. G. Christov, *J. Res. Inst. Catal.,* Hokkaido Univ., **24**, 27 (1976); G. Saito and Y. Matsunaga, *Bull. Chem. Soc. Japan,* **46**, 1609 (1973); **45**, 963 (1972); **47**, 2873 (1974); **47**, 1020 (1974); **46**, 714 (1973); Y. Matsunaga, *ibid.,* **48**, 37 (1975); Y. Matsunaga and R. Osawa, *ibid.,* **47**, 1589 (1974); J. M. Dumai et al., *J. Chem. Phys., Phys-Chem. Biol.,* **72**, 1185 (1975); H. Ratajcak et al., *Chem. Phys.,* **17**, 197 (1976); A. Kofler, *A. Elektrochem.,* **50**, 200 (1974); G. I. Krishtalik et al., *J. Res. Inst. Catal.,* Hokkaido Univ., **22**, 101 (1974); R. R. Dogonadze and A. M. Kuznetsov, *ibid.,* **26**, 15 (1978); I. Yu. Martynov et al., *Russ. Chem. Revs.,* **46**, 1 (1977); W. Klopffer, *Adv. Photochem.,* **10**, 311 (1977).
218. M. J. Rice and W. L. Roth, *J. Solid State Chem.,* **4**, 294 (1972); L. J. Gagliardi, *J. Chem. Phys.,* **57**, 2193 (1973).
219. J. Koziol and P. Tomasik, *Bull. Acad. Pol. Sci. Ser. Sci. Chim.,* **25**, 689 (1977).
220. K. Morokuma, *Acc. Chem. Res.,* **10**, 294 (1977).
221. K. Kalnins et al., *Dokl. Akad. Nauk. SSSR,* **244**, 400 (1979).
222. J. L. Beauchamp, *Ann. Rev. Phys. Chem.,* **22**, 527 (1971); also in *Interactions between Ions and Molecules,* P. Ausloos, ed., Plenum Press, N.Y., 1975, p. 413; N. Hartmann et al., *Top. Curr. Chem.,* **43**, 57 (1973).
223. J. H. Bayless et al., *J. Am. Chem. Soc.,* **90**, 531 (1968).
224. P. Kollmann and S. Rothenberg, *J. Am. Chem. Soc.,* **99**, 1333 (1977); J. E. Huheey, *Inorg. Chem.,* 2nd ed. Harper and Row, N.Y., 1978; D. E. Smith "The Proton Affinity of Gaseous Molecules," Xerox Univ. Microfilms, Ann. Arbor, Mich. No. 76-24, 2454 (1976); K. W. Hartmann et al., "An Annotated Bibliography of Proton Affinities," U.S. NITIS Pb. Rep. 256328 (1976); Govt. Announced Index (US) 76, (2) 142 (1976).
225. E. M. Arnett and E. J. Mitchell, *J. Am. Chem. Soc.,* **93**, 4052 (1971).
226. T. Erdey-Gruz and S. Lengyel, in *Mod. Aspects of Electrochem.,* J. O'M. Bockris and B. E. Conway, eds., Plenum Press, N.Y. **12**, 1 (1977); J. E. Crooks and B. H. Robinson, in "Proton Transfer," Faraday Symp. Chem. Soc. London, No. 10 (1975) p. 29; C. E. Bannister et al., *ibid.,* p. 78.
227. I. Agranat and S. Cohen, *Bull. Chem. Soc. Japan,* **47**, 723 (1974); *cf.,* also N. Inoue and Y. Matsunaga, *ibid.,* **45**, 3478 (1972); G. N. Felix and P. L. Huyskens, *J. Phys. Chem.,* **79**, 2316 (1975).
228. G. Saito and Y. Matsunaga, *Bull. Chem. Soc. Japan,* **46**, 1609 (1973).
229. J. Nishijo, *Bull. Chem. Soc. Japan,* **47**, 1539 (1974).
230. P. W. Anderson et al., *Phil. Mag.,* **25**, 1 (1972); J. Tauc, *Physics Today,* Oct. **1976**, 27; S. G. Christov, *Contemp. Phys.,* **13**, 199 (1972); *Phys. Stat. Solidi,* **7**, 371 (1971); *Croatica Chim. Acta.,* **44**, 67 (1972); G. Gusman and R. Deltowi, *Solid State Commun.,* **15**, 1587 (1974); R. R. Dogonadze and A. M. Kuznetsov, *J. Res. Inst.*

Catalysis, Hokkaido Univ., **22**, 93 (1974); J. H. Bush and J. R. De la Vega, *J. Am. Chem. Soc.,* **90**, 2397 (1977).

231. T. Izumida et al., *J. Phys. Chem.,* **83**, 373 (1979).

232. M. J. Rice and W. L. Roth, *J. Solid State Chem.,* **4**, 294 (1972); *cf.,* also L. F. Gagliardi, *J. Chem. Phys.,* **58**, 2193 (1973).

233. L. F. Gagliardi, ref. 232.

234. G. N. Felix and P. C. Huyskens, *J. Phys. Chem.,* **79**, 316 (1975).

235. P. Stilbs and G. Olofsson, *Acta. Chem. Scand.,* **A28**, 647 (1974); G. Olofsson and I. Olofsson, *Tetrahedron,* **29**, 1711 (1973).

236. Q. Appleton et al., *Tetrahedron,* **27**, 5921 (1971).

237. M. Mandelkorn, *Chem. Revs.,* **59**, 827 (1959); M. Ohmasa et al., *Bull. Chem. Soc. Japan,* **44**, 391, 395 (1971); K. Deguchi and K. Meguro, *J. Colloid and Interface Sci.,* **38**, 596 (1972); J. B. Torrance and B. D. Silverman, *Phys. Rev.,* **B-15**, 788 (1977).

238. W. B. Nixon et al., *Internatl. J. Mass Spectrom. Ion. Phys.,* **26**, 115 (1978).

239. L. M. Rhodes and P. R. Schimmel, *J. Am. Chem. Soc.,* **96**, 2609 (1974).

240. F. Strohbusch et al., *J. Phys. Chem.,* **82**, 2447 (1978).

241. W. Klöpffer, *Adv. Photochem.,* **10**, 311 (1977); I. Yu. Martynov et al., *Russ. Chem. Revs.,* **46**, 1 (1977).

242. Yu. I. Martinov et al., *Khim Vys. Energ.,* **11**, 443 (1977).

243. T. G. Meister and V. P. Klindukhov, *Adv. Molec. Relax. Interact. Processes,* **13**, 107 (1978).

244. G. G. Wubbles et al., *J. Am. Chem. Soc.,* **95**, 1281 (1973); A. Cu and A. C. Testa, *ibid.,* **96**, 1963 (1974).

245. P. Kollman, *J. Am. Chem. Soc.,* **99**, 4875 (1977).

246. T. Yahabe, *J. Chem. Soc. Faraday Trans. I.,* **73**, 1860 (1977).

247. Trade Name reg. to Aldrich Chem. Corp. Milwaukee, Wis.

248. R. W. Bowman et al., *Chem. Comm.,* **1968**, 723.

249. K. Morimoto and Y. Murakami, *Appl. Optics Suppl. No. 3, Electrophotog.,* **1969**, 50; G. Pfister et al., *J. Chem. Phys.,* **57**, 2979 (1972).

250. J. M. Pearson et al., "The Nature and Applic. of Charge Transfer Phenomena in Polymers," in *Molecular Association,* R. Foster, ed., Academic Press, N.Y., 1979, p.2.

251. J. Castonguay, Commun. Table Ronde Int. Trait. Surf. Polym. Plasma, **1977**, RTA, P. Fauchais, ed., Univ. of Limoges, France.

252. A. Rembaum, *Encyclopedia of Polymer Sci. and Technol.,* **11**, 318 (1969) Wiley, N.Y., 1969; V. Hadek et al., *Makromolecules,* **4**, 494 (1971).

253. J. M. Bruce and J. R. Henson, *Polymer,* **8**, 619 (1967).

254. A. M. Hermann and A. Rembaum, *J. Polymer Sci.,* **C-19**, 107 (1967).

255. S. I. Peredereeva et al., *Dokl. Akad. Nauk SSSR,* **210**, 154 (1973).

256. H. Willersinn et al., German Pat., 1,953,898, 6 May (1971).

257. T. A. Gordina et al., *Vysokomol. Soedin,* Ser. B, **15**, 378 (1973).

258. D. Braun and H. J. Sterzel, *Ber. Bunsen Ges. Phys. Chem.,* **76**, 551 (1972).

259. M. H. Litt et al., *J. Polymer Sci. Polymer Chem.,* **11**, 1339 (1973).

260. R. B. Seymour and D. P. Garner, *Rev. Plast. Med.,* **268**, 500 (1978).

261. G. Wegner, *Makromolec. Chem.,* **154**, 35 (1972); **134**, 219 (1970); Z. Iqubal et al., *J. Chem. Phys.,* **66**, 5520 (1977); R. H. Baughman et al., *J. Chem. Phys.,* **60**, 4755 (1976).

262. W. Schermann and G. Wegner, *Makromolek. Chem.,* **175**, 667 (1974); R. R. Chance et al., *Chem. Phys.,* **13**, 181 (1976); R. R. Chance and R. H. Baugman, *J. Chem. Phys.,* **64**, 3889 (1976).

263. B. Reimer and H. Bässler, *Phys. Stat. Solidi,* **32**, 435 (1975); K. Lochner et al.,

Chem. Phys. Lett., **41**, 388 (1976); *Phys. Stat. Solidi*, **676**, 533 (1976); B. Reimer et al., *Phys. Stat. Solidi*, **73**, 709 (1976).

264. e.g., B. P. Bespalov and V. V. Titov, *Russ. Chem. Revs.*, **44**, 1091 (1975); S. I. Peredereeva et al., *ibid.*, 295; Y. Matsunaga, 7th Molec. Cryst. Symp. Nikko, Japan; M. A. Slifkin, *Charge Transfer Complexes of Biomolecules*, Academic Press, London, 1971; R. Foster, *Organic Charge Transfer Complexes*, Academic Press, London, 1969; J. Rose, *Molecular Complexes*, Pergamon Press, Oxford, 1967; J. G. Heathcote et al., *Spectrochim. Acta*, **27A**, 1391 (1971); J. B. Birks, *Photo Physics*, Wiley-Interscience, N.Y., 1970; R. S. Mulliken and W. B. Person, *Molecular Complexes*, Wiley-Interscience, N.Y., 1970; M. R. Philpott, *Chem. Phys. Lett.*, **50**, 18 (1977); S. I. Peredereeva et al., *Russ. Chem. Revs.*, **44**, 295 (1975).

265. L. J. Parkhurst, Ph. D. Thesis, Yale Univ., 1965; B. J. Anex and E. B. Hill, *J. Am. Chem. Soc.*, **88**, 3648 (1966); Y. Ihaya et al., *Bull. Chem. Soc. Japan*, **45**, 1004 (1972); C. K. Chiang et al., *Phys. Rev. Lett.*, **39**, 1098 (1977); A. G. MacDiarmid, *Bull. Amer. Phys. Soc.*, **24**, 418 (1979); C. R. Fincher et al., *Phys. Rev.*, **B-20**, 1589 (1979); J. F. Kwak et al., *Solid State Commun.*, **31**, 355 (1979); A. Brilliante et al., *Chem. Phys. Lett.*, **56**, 218 (1978).

266. C. I. Simionescu and V. Percel, *Rev. Roumaine Chim.*, **24**, 171 (1979); J. W. Anthonsen and C. K. Moller, *Spectrochim. Acta*, **A-33**, 987 (1977).

267. R. A. Singh and S. N. Bhat, *J. Phys. Chem.*, **82**, 2333 (1978); J. Mort et al., *Solid State Comm.*, **18**, 693 (1976); H. Block et al., *Polymer*, **18**, 781 (1977); J. Mort et al., *Bull. Amer. Phys. Soc.*, **24**, 326 (1979).

268. K. Ishii et al., in *Energy and Charge Transfer*, K. Masuda and M. Silver eds., Plenum Press, N.Y., 1974, p. 183; J. Prakash, *J. Chim. Physique*, **73**, 696 (1976); H. Takagi et al., *Bull. Chem. Soc. Japan*, **50**, 1807 (1977).

269. B. G. Anex and W. T. Simpson, *Rev. Mod. Phys.*, **32**, 446 (1960); J. J. Eckhard and J. Merski, *Surface Sci.*, **37**, 937 (1973); Y. Matsunaga and N. Miyajima, *Bull. Chem. Soc. Japan*, **44**, 361 (1971); A. P. Kushelevsky and M. A. Slifkin, *Radiat. Res.*, **50**, 56 (1972); A. A. Bright et al., *Phys. Rev. Lett.*, **34**, 206 (1975); C. R. Fincher et al., *Bull. Amer. Phys. Soc.*, **24**, 327 (1979).

270. F. Forster and D. M. L. Goodgame, *Inorg. Chem.*, **4**, 823 (1965); M. Hashimoto et al., *Bull. Chem. Soc. Japan*, **44**, 2322 (1971); W. D. Gill et al., *Bull. Amer. Phys. Soc.*, **24**, 327 (1979).

271. R. B. Somoano et al., *Phys. Rev.*, **B-17**, 2853 (1978); Y. Matsunaga, *Bull. Chem. Soc. Japan*, **45**, 770 (1972); W. Gabes and D. F. Stufkens, *Spectrochim. Acta*, **A-30**, 1835 (1974).

272. M. Ohmasa et al., *Bull. Chem. Soc. Japan*, **41**, 1998 (1978); M. Sano et al., *ibid.*, **42**, 2204, 548 (1969); **43**, 2370 (1970); L. Alcazer and A. H. Maki, *J. Phys. Chem.*, **78**, 215 (1974).

273. Y. Ihaya and T. Yunaki, *Bull. Chem. Soc. Japan*, **45**, 3065 (1972); G. Briegleb et al., *Ber. Bunsen. Ges. Phys. Chem.*, **76**, 101 (1972); J. P. Carrion et al., *Helv. Chim. Acta*, **51**, 459 (1967).

274. B. F. Levine and C. G. Bethea, J. Chem. Phys., **65**, 2429, 2439 (1976); **66**, 1040 (1977).

275. B. L. Davydov et al., *JETP Lett.*, **12**, 16 (1970), *Opt. Spektrosk.*, **30**, 274 (1971); *Soviet J. Quantum Electron.*, **7**, 129 (1977); J. G. Bergmann and G. R. Crane, *J. Chem. Phys.*, **66**, 3863 (1977).

276. e.g., S. Bagchi and M. Chowdhury, *J. Phys. Chem.*, **83**, 629 (1979); E. M. Kosower, *An Introduction to Physical Chemistry*, Wiley, N.Y., 1968.

277. A. Hantzsch, *Ber. Dtsch. Chem. Ges.*, **52**, 1535, 1544 (1979); E. M. Kosower,

J. Am. Chem. Soc., 3253, 3261 3267 (1958); J. S. Brinen et al., *J. Phys. Chem.,* **69,** 3761 (1965).
278. S. Bagchi and M. Chowdhury, *J. Phys. Chem.,* **80,** 2111 (1976).
279. H. Akamatu and H. Kuroda, *J. Chem. Phys.,* **39,** 3364 (1963); S. Yomosa, *J. Phys. Soc. Japan,* **36,** 1655 (1964); see also ref. 272.
280. M. A. Slifkin, ref. 264; M. Ohmasa et al., *Bull. Chem. Soc. Japan,* **42,** 2402 (1969); S. Walker and H. Straw, *Spectroscopy,* Chapman and Hall, London, 1966, Vol. 1.
281. L. Skulski et al., 3rd Int. Symp. Specific Interact. Molec. Ions. Proc., 1976, **2,** 501; Inst. Chem. Wroclaw Univ. Poland (1976); V. Gutmann, *ibid.,* **1,** 182.
282. L. Skulski et al., *Bull. Acad. Pol. Sci. Ser. Sci. Chem.,* **21,** 369 (1973); R. P. Lang, *J. Phys. Chem.,* **78,** 1657 (1974).
283. F. Gutmann, *J. Sci. Indust. Res.,* Sect. **B-26,** 19 (1967); F. Gutmann and H. Keyzer, *Electrochim. Acta,* **11,** 555, 1163 (1966); N. E. Wisdom and E. O. Forster, *J. Polymer Sci.,* **C-17,** 125 (1967); L. Libera and H. Bretschneider, *Z. Chem.,* **14,** 68 (1974); V. I. Kazandzan, *Russ. J. Phys. Chem.,* **46,** 1084 (1972); M. V. Ramamurthy and A. Quayum, *J. Electrochem. Soc. India,* **27,** 33 (1978).
284. L. Ward, *J. Chem. Phys.,* **39,** 852 (1963); T. N. Misra and B. Rosenberg, *J. Chem. Phys.,* **48,** 2096 (1968); J. W. Eastman et al., *J. Am. Chem. Soc.,* **84,** 1339 (1962); H. Kainer and A. Überle, *Chem. Ber.,* **88,** 1147 (1955); L. V. Keldysh, *Usp. Fiz. Nauk.,* **86,** 327 (1967).
285. F. Gutmann and H. Keyzer, Ref. 283.
286. D. Thiele et al., *Nucleic Acids Res.,* **5,** 1997 (1978).
287. Ch. Mark et al., *Nucleic Acids Res.,* **5,** 1979 (1978); see also Ref. 288.
288. W. B. Nixon et al., *Inst. J. Mass Spectrom., Ion. Phys.,* **26,** 115 (1978).
289. J. Blasczkiewicz and Z. Pajak, *Adv. Molec. Relax. Interact. Proc.,* **13,** 83 (1978).
290. Z. Yoshida and E. Osawa, *Bull. Chem. Soc. Japan,* **38,** 140 (1965); *cf.,* also P. Job., *C. R. Acad. Sci.* (Paris), **180,** 928 (1925).
291. P. C. Carmin, *J. Solution Chem.,* **7,** 845 (1978).
292. Yu. Ya. Fialkov and A. Ya. Borovikov, *Zh. Fiz. Khim.,* **53,** 258 (1979).
293. I. Yukhnovskii and I. Dimitrova, *Dokl. Bulg. Acad. Nauk,* **31,** 547 (1978).
294. V. A. Gorodyskii and A. A. Morachevskii, *Zh. Fiz. Khim.,* **52,** 2248 (1978).
295. D. Jannakovdakis and J. Moumtzis, *Z. Naturforsch.,* **B-23,** 1303, (1968).
296. I. Deperasinska and J. P. Prochorow, *Adv. Molec. Relax. Interact. Proc.,* **11,** 51 (1977); *cf.,* also F. E. Hancock and J. N. Murrell, *J. Chem. Soc. Faraday Soc. II,* **69,** 115 (1973).
297. F. Gutmann and H. Keyzer, *J. Chem. Phys.,* **50,** 550 (1969); *Electrochim. Acta,* **13,** 693 (1968); A. Brau et al., *Electrochim. Acta,* **17,** 1803 (1972).
298. D. H. Rodgers, *J. Pharm. Sci.,* **54,** 459 (1965); W. A. Harris, *Austral. J. Pharm.,* **49,** 587 (1968).
299. Y. Y. Borovikov, *Izv. Vyssh. Uchebni Zaved Khim. Khim. Technol.,* **11,** 20 (1968).
300. B. C. Gilbert in *Essays in Chemistry,* J. N. Bradley et al., eds., Academic Press, N.Y., 1972, p. 4.
301. F. Gutmann et al., *Adv. Biochem. Psychopharmacol.,* **9,** 15 (1974).
302. M. A. Slifkin, Ref. 264, p. 73.
303. M. V. Ramamurthy and A. Quayum, Ref. 283.
304. G. Eckert et al., *J. Biol. Phys.,* **6,** 161 (1978); *J. Electroanal. Chem. Interfacial Electrochem.,* **62,** 267 (1975); D. N. Gillbanks, M. Sc. Thesis, Victoria Univ. of Wellington, N.Z., Dec. 13 (1973).
305. P. Durand and R. Fournie, Dielect. Materials, Measure and Applic. Conf. IEE

REFERENCES

Conf. Publ. No. 67, IEE London, 1970, p. 142; see also ref. 268.

306. J. M. Bonnier and R. Arnard, *J. Chim. Phys.*, **68**, 423 (1971); W. Wacklawek, *Bull. Acad. Sci. Ser. Sci. Math. Astron. Phys.*, **21**, 189 (1973); *cf.*, also M. A. Slifkin, Slifkin, ref. 264 p. 22*ff*.

307. N. Tyutyulkov et al., *Theoret. Chim. Acta* (Berlin), **20**, 385 (1971); H. Beens et al., *J. Chem. Phys.*, **47**, 1183 (1967).

308. J. Caldorford, *Complex Permittivity*, Engl. Univ. Press, London, (1971); D. Bauer et al., *J. Phys. Chem.*, **74**, 4504 (1970); W. E. Vaughan, in *Dig. Lit. Dielect.*, **35**, 1973 (1971).

309. S. Kobinata and S. Nagakura, *J. Am. Chem. Soc.*, **88**, 3905 (1966).

310. F. Gutmann and H. Keyzer, Unpublished work.

311. Z. Croitorou, *Progr. Dielectric.*, **6**, 103 (1965).

312. J. E. Randles and U. W. Somerton, *Trans. Faraday Soc.*, **48**, 937 (1952).

313. B. Breyer and H. H. Bauer, *AC Polarography and Tensammetry*, Interscience, N.Y., 1963; B. Breyer and F. Gutmann, *Austral. J. Sci.*, **8**, 163 (1946); B. Breyer and S. Hacobian, *Austral. J. Chem.*, **7**, 225 (1954); D. Smith, *Crit. Revs. Anal. Chem.*, **2**, 247 (1971).

314. A. I. Saitto and A. D. Wrixon, *Chem. Comm.*, **1969**, 1184; J. Bolard, *J. Chem. Phys. Phys-Chim. Biol.*, **66**, 221 (1969); T. B. Lakdar et al., *C. R. Acad. Sci.* (Paris), **271**, 1201 (1971).

315. W. Waclawec, *Bull. Acad. Pol. Sci. Ser. Sci. Math. Astron. Phys.*, **21**, 189 (1973).

316. R. Pottel, *Ber. Bunsenges. Phys. Chem.*, **75**, 286 (1971).

317. F. Gutmann and H. Keyzer, *Electrochim. Acta.*, **13**, 693 (1968); H. Keyzer Ph. D. Thesis, University of New South Wales, 1966; Edit. Note in *Z. Phys. Chem.* (Leipzig), **259**, 177 (1978).

318. J. E. Anderson, *Ber. Bunsenges. Phys. Chem.*, **75**, 294 (1971); G. Williams, *Adv. Mol. Relaxation Processes*, **1**, 409 (1970); G. Schwartz, *J. Phys. Chem.*, **74**, 654 (1970); *cf.*, also ref. 316.

319. R. A. Crump and A. H. Price, *Trans. Faraday Soc.*, **66**, 92 (1971); R. Pethig and V. Soni, *J. Chem. Soc. Faraday I*, **71**, 1534 (1975).

320. F. Gerson, *High Resolution ESR Spectroscopy*, Wiley, N.Y., 1970; E. E. Budzinski et al., *J. Chem. Phys.*, **59**, 2899 (1973); D. J. Ingram, *Free Radicals as Studied by ESR*, Butterworths, London, 1958; Varian Associates, *NMR and EPR Spectroscopy*, Pergamon Press, Oxford, 1960; P. G. Lykos, in *Advances in Quantum Chemistry*, P. O. Lowdin, ed., "Electron Spin Relaxation in Liquids," L. Muus and P. W. Atkins, eds., Plenum Press, N.Y., 1972; G. I. Subbotin and R. V. Grechischkins, in *Magnetic Resonance Phenomena Proceedings of the 16th Ampere Congress, (1970)*, ed. by I. Ursu, North Holland Publ., Amsterdam, 1971; Y. Tomkiewicz et al., *Phys. Rev. Lett.*, **32**, 1363 (1974).

321. E. M. Gause et al., *Biochim. Biophys. Acta*, **141**, 217 (1967).

322. F. Gutmann et al., *Rev. Aggressol.* (Paris), **7**, 147 (1966).

323. J. B. Jones et al., *Nature* (London), **211**, 309 (1966); K. Tsuji et al., *J. Chem. Phys.*, **46**, 2808 (1967).

324. H. Keyzer, ref. 317; D. N. Gillbanks, ref. 304.

325. J. A. Pople, W. G. Schneider and H. J. Bernstein, *High Resolution Nuclear Magnetic Resonance*, McGraw-Hill, N.Y., 1959; Varian Associates, *NMR and EPR Spectroscopy*, Pergamon Press, Oxford, 1966; K. Higasi, H. Baba, and A. Rembaum, *Quantum Organic Chemistry*, Interscience, N.Y., 1955; R. Blanc, ed., *Magnetic Resonance and Relaxation*, North-Holland Publishing Co., Amsterdam, 1967; N. M. Atherton and A. J. Blackhurst, *J. Chem. Soc. Faraday Trans. II*, **68**, 470 (1972); E. F. Mooney, ed.,

Annual Reports on NMR Spectroscopy, Academic Press, N.Y., 1970; R. J. Abraham, *Analysis of High Resolution NMR Spectra*, Elsevier, Amsterdam, 1970; R. Foster, *Ind. Chem. Belg.*, **37**, 547 (1972); J. L. Marklet, *Accts. Chem. Res.*, **8**, 70 (1975); A. K. Covington and K. E. Newman in *Modern Aspects of Electrochemistry*, J. O'M. Bockris and B. E. Conway, eds., Plenum Press, N.Y., **12**, 41 (1977); E. von Goldammer, *ibid.*, **10**, 1 (1975).

326. J. Homer et al., *J. Chem. Soc. Trans. Faraday II*, **268**, 68 (1972); for review see *Forschr. Chem. Forsch.*, **30** (1972); W. Caspar et al., *Biochem.*, **12**, 2649 (1973); J. L. Kaplan and G. Fraenkel, *NMR of Chemically Exchanging Systems*, Academic Press, N.Y., 1980.

327. M. W. Hanna and A. L. Ashbaugh, *J. Phys. Chem.*, **68**, 811 (1964); R. Foster and C. A. Fyfe, *Trans. Faraday Soc.*, **61**, 1626 (1965); R. Foster and D. R. Twiselton, *Rec. Trav. Chim. Pay-Bas*, **89**, 325 (1970).

328. A. Yu. Karmilov et al., *Zh. Strukt. Khim.*, **18**, 817 (1977).

329. W. Auch et al., *Phys. Stat. Solidi*, **A-49**, 563 (1978).

330. H. Shihara and N. Nakamura, *Bull. Chem. Soc. Japan*, **44**, 2676 (1971); Y. Matsunaga and Y. Suzuki, *ibid.*, **45**, 3375 (1972).

331. R. A. Bennett and H. O. Hooper, *J. Chem. Phys.*, **47**, 4855 (1967).

332. G. A. Bowmaker and S. Hacobian, *Aust. J. Chem.*, **21**, 551 (1968).

333. R. Brüggemann et al., *Z. Naturforschg.*, **A27**, 1524 (1972).

334. M. F. Shostokoskii et al., *Izv. Akad. Nauk. SSSR Ser. Khim.*, 15 (1973).

335. M. Kinoshita et al., *Bull. Chem. Soc. Japan*, **44**, 2267 (1971).

336. M. Sano et al., *Bull. Chem. Soc. Japan*, **41**, 2204 (1968); **42**, 548 (1968); **43**, 2370 (1970); L. Alcazer and A. H. Maki, *J. Phys. Chem.*, **78**, 215 (1974).

337. L. Alcazer and A. H. Maki, *J. Phys. Chem.*, **80**, 1912 (1976).

338. Yu. A. Karbainon et al., *Nov. Polyarogr. Tezisy Dokl. Vses Soveshla Polyarogr.*, *6th*, 1975, J. Stradins, ed., **49**; B. Kastening, *Progr. in Polarography*, **3**, 195 (1973); M. E. Peover, *Trans. Faraday Soc.*, **60**, 417, 479 (1964).

339. L. R. Ramaley and S. Gaul, *Canad. J. Chem.*, **56**, 2381 (1978).

340. J. P. Williams, *Diss. Abst. Int.*, **36**, (8) 3950 (B-1976).

341. M. A. Slifkin, *Charge Transfer Interactions of Biomolecules*, Academic Press, London, 1971, p. 186.

342. H. Bartlet and H. Skilandat, *J. Electroanal. Chem. Interfacial Electro-Chem.*, **24**, 207 (1970).

343. L. V. Mirovich, *Ostsillogr. Peremenotok Polyarogr.*, **1971**, 126; K. G. Boto and F. G. Thomas, *Austral. J. Chem.*, **26**, 1669 (1973); V. F. Toropova et al., *Zh. Obsch. Khim.*, **43**, 711 (1973).

344. M. Barigandi et al., *Bull. Soc. Chim. Belges*, **79**, 625 (1970); A. Delsaut, Memoire de License en Sci. Centre Univ. Mons. (Belgium), (1969); see also Ref. 343.

345. J. Wolfe et al., *Melliand Textil Ber.*, Int. Ed., **54**, 61 (1973); N. Nagy et al., *J. Chem. Soc.*, Perkins Trans., 2; (1972) 1048; E. J. Rudd and B. E. Conway, *Trans. Faraday Soc.*, **67**, 440 (1971); E. G. Chikayzova, *Usp. Perspekt. Razv. Polyarogr. Metoda*, **1972**, 164; R. N. Adams, *Accounts of Chem. Res.*, **2**, 175 (1969); G. Dayhurst and P. J. Elving, *Talanta*, **16**, 885 (1969).

346. G. Eckert and F. Gutmann, *J. Electroanal. Interfacial Electrochem.*, **62**, 267 (1975); G. Eckert et al., *J. Biol. Phys.*, **6**, 161 (1978).

347. M. E. Starzak, *J. Biol. Phys.*, **2**, 57 (1974); D. V. Lamsweerde-Gallez and A. Meessen, *ibid.*, p. 75; R. A. Llenado, *Anal. Chem.*, **47**, 2243 (1975); *cf.*, also P. Groll and F. Grass, *Electrochim. Acta*, **16**, 31 (1971).

348. A. Ya. Gorenbein and A. K. Trofinchuk, *Zh. Obshch. Khim.*, **37**, 1422 (1967).

349. Y. Tomkiewicz et al., *Phys. Rev.*, **B-15**, 1017 (1977).
350. J. Murgich and S. Pissanetzky, *Chem. Phys. Lett.*, 18, 420 (1973).
351. K. Maschke et al., "Semicond. Proc. 13th Int. Conf. (1976)," F. G. Funi, ed., Nth. Holland Publ., Amsterdam, 1976, p. 411.
352. V. Singh, *Indian Inst. of Technol. Kanpur. India*, Priv. Commun., (1980).
353. J. Yarwood and R. Arndi, in *Molecular Association*, 2, (1979), R. Foster, ed., Academic Press, N.Y., 1979.
354. Y. Matsunaga and K. Shono, *Bull. Chem. Soc. Japan*, 43, 2007 (1970).
355. S. Bagchi and M. Chowdhury, *J. Phys. Chem.*, 83, 629 (1979).
356. O. Kh. Poleshuk and Yu. K. Maksyutin, *Russ. Chem. Revs.*, 45, 1077 (1976).
357. M. Masumar et al., *Chem. Phys. Lett.*, 59, 188 (1978); W. Lachish et al., *ibid.*, 65, 574 (1979); V. Lachish and D. J. Williams, *ibid.*, 72, 225 (1980).

18

Biological Materials

18.1 Introduction

Regular and periodic organization of molecules is essential for the living state, and fundamental to its processes are structured instabilities which are reflected in the solid state by, *e.g.*, charge transfer complexes. Many *in vivo* properties are difficult to explain by classical chemical processes in solution but appear to fit solid state physical processes in cells. Solid state events involving conduction are evident in animate aggregations and may well be an essential characteristic of life, which may be an electromagnetic phenomenon. A growing body of reviews and texts is available to support these views.[11] Chelate metal complexes of purines, pyrimidines, and their nucleosides have been reviewed[127] in detail.

One noticable and intriguing example of support is the observation that cancer throughout the world appears to be a periodic function of the earth's magnetic field for very low fields.[12]

18.2 Conduction Data

Tables 8.11A and B list data for 57 systems additional to those tabulated in Table 8.11. The resistivity values, again, cover about 15 orders of magnitude; several remarkably low and thus highly interesting values are included; these also include mobility results derived from microwave Hall measurements and thus indicative of electronic and not ionic conduction. The mobility values themselves are about what is usually observed in other organic

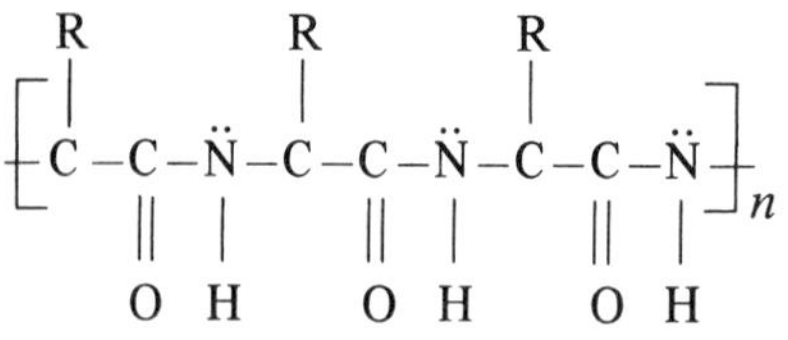

$$\left[\begin{array}{c} R \\ | \\ C - C - \ddot{N} - C - C - \ddot{N} - C - C - \ddot{N} \\ \| \ \ | \ \ \ \ \ \ \| \ \ | \ \ \ \ \ \ \| \ \ | \\ O \ \ H \ \ \ \ \ O \ \ H \ \ \ \ \ O \ \ H \end{array} \right]_n$$

R   R   R

Fig. 18.1 General structure of proteins.

semiconductors (cf., Tables 4.7C and 4.8A). The significance of these results is discussed in the following sections, especially that of the very high, viz., 100 cm^2/Vsec, mobility found[13] for collagen: methylglyoxal adducts.

18.3 Conduction and Biological Structure

Much of which follows is based on discussions by Pethig and coworkers.[13]

Proteins have the general structure of polypeptide chains with $-C-C-N-$ as a repeat unit, see Fig. 18.1.

Each peptide unit lies in a plane because it consists of a delocalized system of π-electrons associated with π-orbitals of the C− and O− atoms together with the lone electron-pair orbital on the N− atom. Such a π-electron resonance structure is sufficient to produce significant diamagnetic anistropy in the protein.[14] Some two dozen amino-acid residues make up polypeptides but only glycine and proline have a first atom in the side chain R which is not a carbon atom. Thus, many features of regularity are present which may cooperate in forming energy bands, particularly in the extended arrays of α-helices and β-pleated sheets. In fact, spectra of DNA and RNA are not basically different from those of their constituent groups because randomization of the constituents washes out all features of the electronic spectrum. Tong[15] suggests that this is due to the presence of geometry-independent lines and allowed energy bands. Altieri and Krizan,[16] using a self-consistent method show term energies of reasonable magnitude to exist if DNA and related models are characterized as intrinsic semiconductors with a band gap energy of approximately 2 eV. Many proteins have a semiconduction activation energy of about this value.[17]

Whether charge carriers can migrate freely through proteins depends greatly on these energy bands. Coherent electronic motion can occur if broad bands of electronic states are available for the ground states (valency bands) and the excited states (conduction bands). If the bands of the extended states are less than about $2kT$, or if the energy states are localized, charge transport takes place by activated hopping or by tunnelling processes[13] (see Chapter 15).

It appears that all proteins in the natural "pure" state are insulators. The valence band of the extended states is completely occupied by electrons, and the band gap is so large that at physiological temperatures the possibility of promoting electrons across the gap is negligible. Then, calculations[13] indicate that the energy band widths are greater for atomic interactions along the polypeptide backbone chains (–C–C–N–C–C–N–) than those arising from the hydrogen-bonded network interactions which stabilize the protein's tertiary structure.

The majority of semiconduction experiments have involved soluble proteins, probably because these are easily purified and crystallized. These are practically all insulators. However, the structural proteins, those which support the main functional elements, are usually discarded in purification processes, as residues. Experimental evidence has existed for some time for electronic conduction in sea-animal integuments, e.g., crab shells.[18] In order to function, such structural proteins will generally be complexed with other molecules in a manner analogous to dopants (see Section 16.7) and thereby facilitate charge transport.

It is difficult to work with structural proteins because they are not readily purified like their soluble counterparts. Therefore much of the work on proteins, and their complexes, continues to be done on the soluble proteins with the results extrapolated to the structural proteins. In order to do so one must establish two conditions: (1) that lack of hydration in purified samples is not a serious drawback in experimentation, and, (2) that conduction must be non-ionic. These two conditions will be discussed in the following sections.

18.4 Hydration of Proteins

In their natural state many proteins are bound into hydrophobic lipid matrices. The energetics of charge separation and mobility are governed largely by the dielectric constant of the surrounding medium. Calculations[13] have shown that the internal structures have an effective high frequency relative permittivity of about 2.6, which value does not increase much upon hydration. Even in the presence of a fully extended hydrogen-bonded network, as has been previously stated, conduction along the backbone is more likely, but the particular properties of a protein will depend on its precise conformation, environment and interactions with other molecules.

Considerable differences in the resistivity between that of ordinary proteins and that of electron transfer proteins have been reported[19,20] and also between ferrocytochrome-c and ferricytochrome-c. This is attributed to the electronic state of the central metal atom.[20] The anhydrous cytochromes exhibit resistivities many orders of magnitude higher; thus anhydrous ferricytochrome c_3 has a room temperature of 4.1×10^{12} ohm cm and the ferrocytochrome c_3 a value of 1.6×10^{10}; these data are even more remarkable because the ferri compound is said[125] not to follow an Arrhenius type relationship in its temperature dependence while the ferro compound is reported to yield a negative value for its thermal activation energy.

Cytochrome-c_3 is specially interesting because this globular hemoprotein (molecular weight about 12,500) is a defined chemical entity and known[21] to be the electron carrier in the respiratory chain of *Desulfovibrio vulgaris*. The electron transfer rate in its reversible electrochemical electrode[22] RedOx reaction has the enormous value of 0.1 cm sec^{-1}. Its resistivity in the ferro- as well as the ferri-form is shown in Fig. 18.2. The conductivity was ohmic and appears to be electronic in nature;[19] values are said[19] to be well reproducible. The low temperature branch of ferrocytochrome-c_3 exhibits the usual Arrhenius form but the high temperature region shows a most unusual behavior, as seen from the Figure. The resistivities of these two forms are reported[19] to differ by 11 orders of magnitude (see Table 8.11A) with the reduced form having the extraordinarily low value of 57 ohm cm at the transition temperature T_M; a value comparable to that of germanium.

The polypeptide chain within the cytochrome molecule has virtually no conductivity at all, while the complete hemoprotein exhibits a conductivity of the order of 10^{-9} to 10^{-11} (ohm cm)$^{-1}$, suggesting a considerable contribution of the heme units to the overall electron transfer. Since the porphyrin sub-unit is about 13 Å diameter, sufficient delocalization of π-electrons should be expected to impart conductivity.[19] It has been hypo-

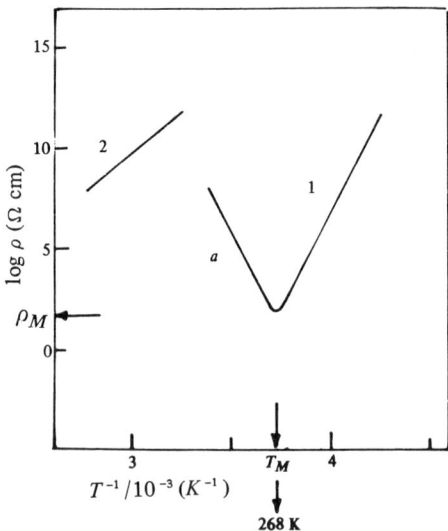

Fig. 18.2 Temperature dependence of the resistivity of (1) ferrocytochrome-c_3, and (2) ferricytochrome-c_3. After Nakahara et al.[19]

thesized that some part of the hemes in cytochrome-c_3 are exposed to the protein surface so that electrons may tunnel through a peptide residue.[19]

The nature of the high temperature conductivity of ferrocytochrome-c_3 as well as the incredibly high value of its activation energy, (see Table 8.11A) reported to be 3.85 eV in E/kT, remains to be further studied and confirmed.

Water in proteins has been determined from hydration isotherms obtained gravimetrically and the more recent quartz crystal resonator techniques.[23] The steady state conductivity of protein samples increases rapidly with water absorbed. The weight percentage, m, relates to the conductivity according to the Spivey equation.[24]

$$\lambda = \lambda_D \exp(\alpha m) \qquad (18.1)$$

where λ_D is the dry-state conductivity and α a constant. For m of about 5 wt%, a change occurs such that $\alpha = 1.3$ for $m < 5$ wt% and $\alpha = 0.9$ for $m > 5$ wt%. This may be explained by assuming that at 5 wt% nearly all the protein primary sorption sites are occupied by water molecules and the population of secondary hydration sites begins. The frequency f_m of maximum loss of dielectric dispersion is also found[13] to increase with increasing hydration following the relationship

$$\tau = \tau_0 \exp(-\beta m) \qquad (18.2)$$

where τ is the relaxation time and β a constant. Again at 5 wt% the value of the constant changes.[13] Although f_m changes, the area under the dielectric loss factors ϵ'' vs. $\log(f)$ curve does not change with hydration, suggesting[13] that the observed dielectric dispersions are not directly related to the relaxation of water dipoles.

18.5 Non-ionic Conduction in Proteins

Ionic contributions to conduction in the measurement of proteins can be determined[25] with a Ag/AgI (or $RbAg_4I_5$) cell as illustrated in Fig. 18.3.

If a constant current is passed from Ag to sample S, silver will be deposited at the interface of the sample and an ionic conductor only if electrons are mobile in S. The quantity of Ag deposited will be proportional to the electron

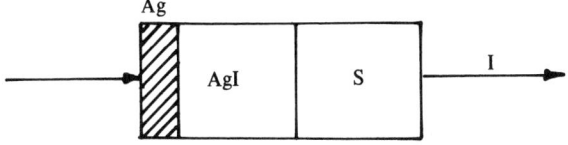

Fig. 18.3 Cell for the determination of ionic conduction. After Pethig and Szent-Györgyi.[25]

transfer number of the sample. The technique requires first the passage of a constant deposition current with S at a negative potential. The sample ionic-conductor interface is then separated and the sample surface scraped to remove the deposited silver. The scrapings are mixed with graphite and compressed into a disc which is pressed against the ionic conductor material. An oxidizing current is now passed so that the initial ionic current requires a sacrificial electrode action of the silver scrapings in the graphite disc. A rapid rise of voltage indicates that oxidation has been completed since it represents the increased potential required to dissociate other ions to maintain the oxidizing current. The electron transfer number is given by the ratio of total oxidizing charge over the original deposition charge in coulombs.

Measurements on dry casein-methylglyoxal complexes,[25] using this method, yielded values of 0.75 to 0.89 while with dry collagen values of 0.8 to 0.9 resulted.[25] Bovine serum albumen (BSA) complexes gave a value of 0.8 to 0.9 and α-conglutin complexes yielded[21] a value of 0.87; the partner always being methylglyoxal.

Since this technique underestimates rather than overestimates the electron transfer number, these high values suggest that the conductivity of the protein-methylglyoxal samples is not dominated by ionic impurities, but is mainly electronic. This technique is superior to conventional, exhaustive constant current (dc) measurements such as used by *e.g.*, Rosenberg[26] because in these it has to be assumed that there is no way in which water or at least protons may have been supplied to the sample so as to maintain the water content in spite of electrolysis, (*cf.*, Section 8.8 and Fig. 8.34). The Ag/Ag/S method also has the virtue that the results are quantitative. The theses of non-ionic conduction by Cardew and Eley[27] and by Rosenberg,[26] (see Section 8.8) in proteins and related biological compounds are thereby confirmed.

18.6 Dielectric Measurements on Protein-Complexes

Over a wide range of frequency, *viz.*, 10^{-5} Hz to 33 GHz, untreated proteins such as BSA, casein and conglutins exhibit a frequency dependent ac conductivity σ_ω of the form

$$\sigma_\omega = A\omega^n \qquad (18.3)$$

where A and n are constants with n close to unity. Since n is virtually unity this could be considered[13] to indicate the presence of dipoles and activated hopping charge carriers with a wide distribution of relaxation times. A model based on such a mechanism, *i.e.*, one in which activated charge carriers hop over potential barriers, (see also Section 15.4 and 15.5), has been developed by Eden, Gascoyne and Pethig.[28] In this model long-range hopping dominates the steady state conductivity while localized hops govern dielectric dispersion effects. Dielectric dispersion arising from a hopping carrier transfer has been confirmed[13] from methylglyoxal complexes of casein, collagen, bovine serum albumin (BSA) and lysozyme. Since the uncomplexed proteins do not exhibit this dielectric dispersion one may assume that they arise from the methylglyoxal molecules in the protein structure. Detailed analyses[29,30] of the dispersions support the conclusion that the dispersions do not arise from relaxation of the methylglyoxal molecular dipole moments, or from conventional molecular dipole moments, but that they can only be due to the relaxation of delocalized charge carriers hopping over potential energy barriers. These delocalized charges are thought to arise from charge transfer interactions between peptide donor units and the acceptor Schiff bases which link methylglyoxal to the lysine side-chain.

This appears to be a case of a ligand-mediated charge transfer as discussed in Section 15.9.

Dielectric dispersion effects in perylene:chloranil complexes have also been ascribed[31] to a carrier hopping mechanism. A theoretical analysis by Lewis[32] also agrees with the model here discussed.

18.7 Cytochrome Oxidase

Although many proteins are morphologically similar, exhibiting wide band gaps in the pure state, some proteins have unusual structures specially adapted for high electronic conductivities. For example, Cope[33] showed that cytochrome oxidase had kinetic properties consistent with rate limitations imposed by solid state conduction. Straub[17,34] and Cope[34] found the band gap energy of this compound to be 0.3 eV. Microwave Hall effect techniques (see Section 12.5, 12.13 and 15.11) showed electron conductivities within individual biological particles to be much higher than conductivities of compressed pellets of such particles. Cytochrome oxidase activity was positively correlated with the microwave Hall mobility of electrons.[35]

18.8 Methylglyoxal Complexes

The oxygen molecule is preeminent in living systems because it is a universal electron acceptor. Oxygen, however, tends to accept electrons in pairs which renders it unsuitable as a direct electron acceptor for proteins, because these would then be degraded too rapidly. Charge transfer interaction usually involves single electron transfer which may lead to a biradical complex with an unpaired electron on both the donor and the acceptor molecule. The majority of charge transfer interactions are weaker, generally transferring electronic charge only fractionally. A charge transfer complex is formed in which no new chemical bonds arise but in which the electron shuttles between the donor and the acceptor preferring the neighborhood of the donor parent. This behavior may convert a "dead" molecule to a "living" molecule. The carbonyl molecule is an alternative to the oxygen molecule as an electron acceptor even though it has poor acceptance characteristics. However, in methylglyoxal this problem is overcome by joining two C=O groups to form glyoxal and adding a methyl group to form methyglyoxal (CH_3COCHO). This molecule has an aldehyde group quite reactive to proteins, and a ketonic group with a low-lying energy level ideal for accepting electronic charge.

Methylglyoxal has been found[36] to be bound to structural proteins of beef liver. A very active and widespread enzyme, glyoxalase[37] transforms methylglyoxal efficiently to D-lactic acid. If this were the sole purpose of glyoxalase, a puzzle would exist because methylglyoxal and D-lactic acid together lie on no presently known metabolic pathway. However, the problem acquires a solution when solid state concepts of charge transfer and electronic desaturation are applied to proteins interacting with methylglyoxal. Proteins suspended in water in the dark with methylglyoxal assume a stable brown color, and the dc conductivity increases by two orders of magnitude.[13] Large ESR signals are observed.[13] Methylglyoxal binds to form colored complexes with bovine serum albumin (BSA), casein, lysozyme, chymotrypsin, chymotrypsinogen, cytochrome-c and fibrinogen.[13,38,39] Lorand showed[40] that all the arginine and 85% of the lysine groups in these proteins are involved in the interaction with methylglyoxal. It has further been shown by Pethig and colleagues[13] that the color and electron spin resonance signal intensities are directly related to the number of unblocked lysine groups: it is likely that the first step in the reaction is the formation of a Schiff base linkage to the ε-amino groups. Otto et al. have adduced evidence[41] that such a Schiff base is comparable to methylglyoxal as an electron acceptor, and that no change of the main polypeptide chain configuration is required to allow the lysine group to "bend back" to a neighboring peptide unit for the donation of electronic charge to the Schiff base. A similar mechanism operates in the case of retinal linked[42] by a Schiff base to a lysine residue in the rhodopsin and bacteriorhodopsin isolated from *Halobacterium halobium*.[13] Sulfhydryl groups[13] may also be involved in producing the color of the protein-methylglyoxal complexes, but this is not the case in complexes with bacterio-rhodopsin which contain no sulfhydryl groups.[43] The electron spin resonance signals observed in methylglyoxal-protein complexes, 3 orders of magnitude higher than for proteins alone,[44] probably arise directly from the polypeptide backbone. Thus, the steady state conductivity is associated with the long-range mobility of one or both unpaired charges, while the dielectric dispersion is due to localized motion. The observation that the conductivity increase mirrors that of the ESR signal increase in the complexes (see Section 18.6) is evidence for the first contention (see Section 18.3). Since the free radical concentration and as well as the dielectric dispersion are found to be independent of temperature,[44,13] it appears that also the second contention (*cf.*, Section 18.3) is complied with. Mobilities of charge carriers for the BSA, casein and lysozyme-methylglyoxal complexes were determined[13] to be 0.01 $cm^2/Vsec$, and for the collagen complex slightly more than 100 $cm^2/Vsec$. From a hopping charge carrier model used by Bone and Pethig[45] and Lewis[46] the average hopping distances were estimated to be 60 to 80 Å for the former complexes, and about 750 Å for the latter complex.

The borderline between charge carrier mobilities at one extreme in which coherent wave-like motion occurs through a broad band of extended energy states, and the other extreme in which the charge carriers diffuse slowly by a process of thermally activated hopping *via.* localized states (see Chapter 15), is about 1 $cm^2/Vsec$. Thus, the value found for the collagen complex, about 100 $cm^2/Vsec$, provides strong evidence for broad bands of extended energy states in its molecular structure. The charges in the collagen complex are said[13] to be holes, delocalized in the protein valence band.

18.9 Methylglyoxal-Ascorbic Acid-Protein Interaction

Ascorbic acid, or one of its metabolic derivatives, is thought[47] to play a role in charge transfer processes with proteins by assisting the electron accepting action of aldehydes such as methylglyoxal and so aiding electron transfer to the final acceptor, oxygen. Quantum mechanical calculations have recently indicated[48] that molecular conformations are possible in which electronic charge can be transferred from the nitrogen atom of a neighboring peptide group to the Schiff base formed between the ϵ-amino group of a lysine side chain and the ascorbic acid methylglyoxal acetal. Thus, charge transfer reactions should be stronger for an ascorbic acid-Schiff base than for the Schiff base alone.[48]

18.10 Integrated Biological Systems. Photobiology

Three major response differences exist between inorganic semiconductors and organic molecular solids. In an inorganic solid, such as germanium, photon absorption yields two separated charge carriers, an electron in the conduction band and a positive hole in the valency band. In an anthracene crystal, with much weaker bonding between the molecules, the absorption produces a crystal state close to the first excited singlet state of an isolated molecule. Secondly, while drift mobilities in *e.g.*, germanium are about 1000 $cm^2/Vsec$, in anthracene or phthalocyanine they are about 1 $cm^2/Vsec$ (see Chapter 15 and Table 4.7C). The third difference is in the mobility type as discussed in Chapter 15. Thus, in germanium a pair of carriers is formed which separate easily, whereas in organic solids carrier pair production is usually followed by geminate recombination.

When a molecule is excited by a sufficiently energetic photon, an electron can be ejected as a photoemission event. The resulting molecular cation polarizes electrons from its immediate environment in a time of about 10^{-16} sec, and this polarization energy stabilizes the final state of the system (see Chapter 13).

When two opposite charges in a solid of relative permittivity ϵ_r approach one another, the coulombic energy of attraction between them equals kT at a critical distance R_c so that[49]

$$R_c = e^2/4\pi\epsilon_r\epsilon_0 kT$$

or (18.4)

$$R_c = 55.9/\epsilon_r \text{ nm}.$$

A limit therefore exists[49] for the quantity of charge of one kind which can be imposed on a molecular crystal without resulting in recombination. This limit amounts to about 10^{17} electronic charges/cm^3, while there are about 10^{21} molecules/cm^3. Unlike in inorganic crystals, electric neutrality need not be preserved in organic solids. In an organic layer of about 5 nm the total of such uncompensated space charge is approximately 10^{-10} coulomb/cm^2, which, if it were uniformly distributed, would yield a rather small electric field of about 10^3 Vcm^{-1}. Potential differences of about 5 mV across 5 nm, *i.e.*, 10^4 Vcm^{-1}, do not represent a serious problem for space charge limitation of current flow through biological systems typically several nm thick. It also follows that in organic solids only one molecule in several thousand can be charged if compensating charges are absent. Thus, modelling a biological system, a localized representation based on individual charged species is preferable to a representation of a charge density distributed over all the molecules.[49]

If the distance between M^+ and M^- is less than R_c, they exist as an ion pair, *i.e.*, in a charge transfer state. If M^+ and M^- are locked into original lattice sites then the distance can have only certain values, corresponding to the charge transfer states of which the lowest ones is given the symbol E_{CT}^1. From the energies for linear polyacenes, determined accurately by Silinsh,[50] it appears that, as the number of rings increased, charge separation in the charge transfer state was facilitated. This could be significant[49] for biological systems containing highly conjugated molecules such as chlorophyll.

Solutions of chlorophyll have been shown to be photo-conductive even in the absence of extrinsic acceptors.[51] In acetonitrile as solvent, ions were formed from two molecules in the triplet state. In petroleumether, red light yielded ions from dimers if the concentration exceeded 10^{-4} M, the dimerization point.[52] In chlorophyll solutions containing ascorbic acid (here a donor) or acceptors such as quinones, triplets were involved to produce positive or negative[53-57] ions. Bromberg et al.[58] concluded that in films of chlorophyll-*a* carrier generation occurred *via* a 1-photon process involving a singlet state. They excluded triplet states because the rise time of photoconduction was less than 1 sec.

The subject of photosynthesis,[59] briefly alluded to in Section 8.8-3, has now become so large, involving an enormous wealth of published work, that it is impossible to do it justice within the necessarily limited confines of this book; attention thus will be confined to a very few topics bearing directly on the main theme of this volume.

Watanabe et al.[60] report quantum efficiencies of about 15% in chlorophyll-*a* films monolayered with 50% stearic acid and immersed in an electrolyte, similar to efficiencies found by Meilanov et al.[61] Photoconduction similar for both polarities was found[62] in chlorophyll-*a*; H_2O adducts at 820 nm, *i.e.*, at 1.5 eV. Lyons[49] suggests that this may be the energy needed to form a *CT* state but further experimental evidence is needed to support this.

The energy gap for the process in a solution, *i.e.*,

$$2 \text{ chlorophyll} \rightarrow \text{chlorophyll}^+ + \text{chlorophyll}^-$$

can be obtained from $E_{ox} - E_{red}$. In moderately polar solvents

$$E_{ox} - E_{red} \approx 1.6 \text{ eV}.$$

If $\epsilon_r \approx 4$ for the system, the corresponding energy gap ≈ 2.1 eV. *CT* states will lie below this value. Nelson[63] indicates that the energy gap for ethylchlorophyllides-*a* and -*b* is 2 eV.

Chlorophylls in condensed states therefore can produce separated charges, sometimes with high efficiencies. In chloroplasts energy transfer and charge generation at reaction centers involving more than one electron acceptor are likely to be the key processes. Lyons[49] has deduced from his experiments with model compounds (phthalocyanines) that the primary process in photosynthesis involves a carrier generation rate which is associated with acceptors and exceeds the generation of carriers without the acceptors.

Bacteriorhodopsin is a pigment found as a single protein component of the purple membrane of *Halobacterium halobium*[64,65,66] and similar extreme halophiles. The purple membrane converts light energy by translocating protons across the membrane to generate an electrochemical potential. A review[67] of this subject has recently been published. The naturally recurring crystalline structure has been determined with electron microscopy at 7 Å resolutions. The protein is folded into seven α-helical chains, all of which span the hydrophobic core of the membrane.[67] The chromophore has been located in a lysine residue in sequence of the second helical chain from the amino terminal group,[68] cf., previous discussion of methyl-glyoxal-protein complexes. The chromophore's double bond chain makes an angle of about 20° to the plane of the membrane,[69] with its ionone ring close to the center of the membrane.[70,71] The position of the Schiff base lies about 10 Å below the surface, and is probably located on the intracellular half of the membrane.[68]

A unique feature of the structure of higher plant chloroplasts is the differentiation of the internal membranes of the chloroplasts (thylakoids) into stacked and unstacked regions. Boardman et al.[59] show that the degree of thylakoid stacking depends on electrostatic screening rather than ion binding. The light absorbing pigments, chlorophyll and carotenoids, are associated with proteins and organized into energy-transferring units within the thylakoid membrane. Quanta absorbed by approximately 200 light harvesting molecules are transferred to one special molecule of chlorophyll-a where primary conversion of light into chemical free energy takes place. This reaction center chlorophyll-a is in close association with an electron donor and an electron acceptor in the thylakoid membrane. Electron transfer takes place in the membrane of the chloroplasts in a predominantly non-aqueous phase; it commences vectorially at each chlorophyll so that one component of transfer is perpendicular to the membrane. This is achieved by anisotropically arranged molecules with the electron donor at the inner surface and the electron acceptor at the outer surface.[72] Gochev and Christov[73] have given a quantum mechanical treatment of the primary event in photochemical cycles and of the primary reaction in the photochemical cycle of *H. halobium,* the electron transfer between primary and secondary acceptors in bacterial photosynthesis and the photoinduced oxidation of cytochrome-c by bacterial chlorophyll. This transfer is said[74,75] to be extremely fast, thus avoiding a back reaction; an electron transfers from the photoexcited state of a chlorophyll dimer to a pheophytin in about 10^{-11} sec. The resulting hole is filled by electron transfer from cytochrome-c taking about 20 μsec. This transfer, to the chlorophyll dimer, is reported to occur by adiabatic tunnelling.

Slater[76] has shown that energy is available for oxidation-reduction reactions in chromatophores of photosynthetic bacteria. In this context it is interesting to note that Murano et al.,[126] have shown that methyl-viologen is a free radical with a resistivity of the order of only 10^5 to 10^6 ohm cm at room temperature; its activation energy in $E/2kT$ is reported to be about 0.4 to 0.5 eV, and its ionization energy has the remarkably low value of 3.6 eV. The compound is a very good RedOx indicator, independent of solution pH. There is said to be about 1 eV difference in the photoemission threshold values between the oxidized and the reduced state.

Electron spin resonance can be used to observe free radicals or unpaired electrons in illuminated particles such as the chromatophores of bacteria[77] and melanin granules of the eye.[78] The decay of these free radicals in the dark is slow; for several seconds a barrier exists, preventing the return of electrons to their ground state. Considering these biological systems as solid particles suspended in an aqueous medium, Cope[79] suggested that the laws of solid state physics apply within the particle, while at the particle solid-liquid interface electronic reactions occur. On the simple hypothesis[79] that the slow free radical decay is due to an activation energy barrier at the particle surface, such as exist at solid interfaces,[80,81] the Tafel equation should be applicable:

$$V = b - \frac{RT}{Fa}\ln(i) \qquad (18.5)$$

V is the voltage and i the current across the interface, a and b are constants, R the gas constant, F the Faraday constant and T the absolute temperature. While a direct test is impossible due to the small particle size involved, the Tafel equation predicts the shape of the decay curve of the unpaired electrons versus time. Cope[82] therefore derives the Rojinski-Zeldovich or Elovich equation on the basis of solid state and electrode physics,

$$\frac{-dx_r}{dt} = me^{nx_r} \qquad (18.6)$$

where x_r is the concentration of the decaying species and m and n are new constants. This equation has been shown to describe well the decay of free radicals in eye melanin,[83] photosynthetic particles,[83] the decay of molecules involved with delayed light emission from green leaves[84] and the decay of photoconductivity in nerve.[79] Further evidence for solid state electronic conduction in eye melanin has been supplied through the microwave Hall effect.[42,85] The applicability of the Elovich equation to the surface physics of the inorganic semiconductors Ge and Si has been demonstrated.[86]

8.11 Nerve Conduction

Cope[86] showed not only that the Elovich equation described the decay of free radicals in eye melanin but also the decay of electronic photoconductivity at about 500 mμ across the membrane of the *Aplysia* photoneuron.[86,87] Nerve fibers consist of a solid matrix of lipid and protein with interstitial water containing i.e., Na^+, K^+ and Cl^-. Evidence[79,86] is being accumulated for a number of different solid state physical processes in the solid matrix of the nerve: superconduction, piezoelectricity and pyroelectricity.

The presence of superconductive processes in axons is suggested by the temperature dependence of neural processes which is characteristic of single electron superconductive tunnelling, while other neural processes exhibit a negative temperature dependence analogous to two-electron superconductive tunnelling.[88] Numerous organisms are effected by magnetic fields less than 2 gauss for which no mechanism is presently known other than the superconductive Josephson effect,[88,89] though magnetic interactions with ionic, or even perhaps electronic, microcurrents cannot be entirely ruled out. Cholesterol has been predicted to superconduct at physiological temperature[90] and solid state phenomena which suggest superconduction, have been observed experimentally at high field.[91] Nerve tissues involve thin lipid layers which contain cholesterol, across which high electric fields exist.[79] The role of superconductivity in nerve[92] needs a great deal of further study, before it can be fully accepted.

Nerve tissue has been observed to show pyroelectric response.[93] Since the pyroelectric voltage is proportional to the time derivative of temperature, dT/dt, two signals of opposite polarity should be observed upon raising, then upon lowering of the temperature. Indeed, in the thermo-receptors of certain fishes, such a response has been observed[94] upon application of a square thermal pulse to the system.

Piezoelectricity is thought[79] to play a role in neural response. Muscle stretch receptors also exhibit a linear neural response to the logarithm of an applied mechanical force,[94] while the membrane potential of an invertebrate nerve cell at rest exhibits a temperature dependence for which the activation energy is a linear function of temperature as required by piezoelectric theory.[95]

The transmission of nerve signals has been extensively reviewed by Moore[96] in terms of electrochemistry. The Hodgkin-Huxley[97] theory of nerve impulse conduction is based on free ions in liquid water, and the hypothesis of biological cation pumps. These tenets have been shown to be thermodynamically impossible.[98,99] The concepts which find increasing acceptance in the treatment of many biological phenomena are Ling's association-induction hypothesis,[100] the ion exchange resin, cytotonous theory of Damadian[101] and the solid state physical theory of Cope.[102] A full treatment of these theories is beyond the scope of this work and the reader is referred to a review by Cope.[103]

For a kinetic analysis of nerve excitation Cope has suggested to replace parts of the Hodgkin-Huxley theory[104] with phase transition theories based on a statistical mechanical approach[100,105] and the metallurgical growth of nuclei.[79] A mathematical treatment of phase transition kinetics has been developed by Avrami[106] which should also have application[107] to biological systems.[108]

Takenaka[109] noted that the activity of enzymes which catalyse the breakdown of protein molecules in the giant axon of squids increased upon the excitation of an action potential. While it is unlikely that signal processes in nerves can be explained entirely by either solid state theory or by a modified solution electrochemistry, the former indeed does occupy a legitimate and perhaps a central place in neural research.

Intermolecular, overlapping π orbitals can provide an electron conducting pathway as in graphite. A similar kind of a conducting pathway in neurons has been proposed by Virtanen and Kinnunen.[128] The nerve cell membrane should contain a linearly arranged pathway for electrons, *i.e.*, a continuous chain of double bonds with overlapping π orbitals. The basic *unit* consists of three cis-monosaturated fatty acid side-chains of the actual membrane lipids: two oleic acids (cis-9-18:1), and one nervonic acid (cis-15-24:1). These acyl chains are abundant in the lipids of the nervous system.[129] When a number of these units are arranged in the membrane to produce a "pseudopolymer," a continuous *semiconjugated* system of overlapping π orbitals is formed, extending from one end of the neuron to the other.

The cis-9 double bonds of the oleic acids are brought into close proximity in order to provide good overlap of the π orbitals. The long nervonic acid chain is twisted so that the cis-15 double bond lies parallel to the plane of the membrane. Thus, a strong π orbital overlap is formed between the cis-9 double bond of oleic acid and the cis-15 double bond of the nervonic acid. It should be noted that an essentially identical arrangement of the spiraloconjugated bonds is found[128] in helical dolichol.

Such an arrangement of ethylenic double bonds may provide the excitable cell with a continuous chain of overlapping π orbitals, capable of transferring electrons.

White matter and myelin contain, in addition to oleic acid and nervonic acid, appreciable amounts of saturated fatty acids and cholesterol. The latter lipids are presumably involved in the stabilization of the membrane structures to maintain the semiconjugated system. Some may contain conjugated or homoconjugated double bonds. An example of the first class of lipids is parinaric acid which has been used in biochemistry for its fluorescence properties. A homoconjugated system is formed when double bonds are separated by methylene groups. In this case conjugation is not mediated by the methylenes. Photoelectron spectroscopy shows[130] that the double bonds of 1,4,7-cyclononatriene are directly conjugated; the degree of conjugation was estimated to be 0.3 (for benzene the corresponding value is 1).

Anomalous UV-spectra of some naturally occurring unsaturated fatty acids have been observed. These show shifts to longer wavelengths with an increase in the number of double bonds, suggesting a conjugation which is almost as strong as those in truly conjugated polyenes. Virtanen and Kinnunen[128] propose *conjugation* to occur separated by ethylene groups as in farnesol. By analogy to homoconjugated system they classify this type of spatial arrangement of double bonds as a spiraloconjugated system.

Docosahexaenoate (DHE) is abundant[131] in synaptosomes, synaptic vesicles and mitochondria. It has also been found[132] in high concentrations in phospholipids associated with acetylcholine receptors. Model building studies[128] and UV-spectra suggest that it is homoconjugated. Therefore, phospholipids containing this acyl chain are capable of vertical membrane transport of electrons *via* the π orbitals of DHE.

Dolichol is linked to the glycosylation of proteins.[133] It is also found in mitochondria and lysosomes,[134] suggesting that dolichol may well be involved in the acidification of the lysosomal interior by *pumping* of electrons into the mitochondrial respiratory chain. The dimensions of dolichol allow this lipid to span through two bilayer phospholipid membranes. It should be pointed out that the arrangement of the spiraloconjugated ethylenic double bonds of the dolichol is essentially identical with the *semiconjugated* system formed by two oleic acids and one nervonic acid in a nerve cell membrane.

One more conjugated-type arrangement can exist[128] in biological membranes due to the presence of well-organized ethylenic double bonds. This type of conjugation is referred[128] to as a semiconjugated system: a semiconjugated double bond system is obtained with properly organized oleic acid sidechains containing phospholipids in membranes. Accordingly, monolayers of dioleylphosphatidicholine might be conductors. Another kind of semiconjugated system can be constructed from two oleic acids and one nervonic acid (three double bond units) arranged in repeating sequences of an axon membrane. This kind of semiconjugated system may form a semiconjugated lateral spiral of double bonds in membranes.[128]

Conjugated carbon systems consist of sp^2-hybridized carbon atoms connected by σ bonds. Conjugation is thus an intramolecular phenomenon. If molecular association is accompanied by intermolecular conjugation, such a complex then forms[128] a semiconjugated system. Two ethylenic double bonds can form different semiconjugated systems. The three most interesting types are illustrated in Fig. 18.4, with type *b* being the most likely.

To obtain some qualitative idea of the behavior of type *b* semiconjugation Virtanen and Kinnunen[128] make three approximations: I. Type *b* semiconjugation is approximated by type *a*. II. Strong type *a* semiconjugation is approximated by polyenes. III. Hückel molecular orbital calculations of the polyene type can be used; in particular, that two strongly associated ethylenic double bonds can be treated like butadiene. By analogy, three strongly associated double bonds compare to 1,3,5-hexatriene, and so on.

1,3,5,7,9-decapentene is considered as the strong association limit of five ethylenic double bonds. If all electrons are in bonding orbitals the delocalization energy DE would be 2.05β, β being the resonance integral.

The orbital ψ_5 is antibonding between ethylenic double bonds but ψ_6 is bonding. Therefore it is assumed that in systems of five double bonds the orbitals ψ_1, ψ_2, ψ_3, ψ_4, and ψ_6 are always populated. The ψ_5 to ψ_6 shift promotes association and semiconjugation with net release of energy, $\Delta E = 0.92\beta$. The ψ_5 to ψ_6 shift tends to stabilize the association of five ethylenic double bonds. When the number of double bonds m is increased, the ψ_m to ψ_{m+1} stabilization is stronger in the middle of the chain and weaker at the ends.

Once formed, the system would easily accommodate an extra pair of electrons either on ψ_m or possibly on ψ_{m+2}, because of increased association. The charge of the electrons is compensated by cations or positive groups.

Five ethylenic double bonds form roughly an optimal unit. In this model,[128] the units are arranged linearly and interact to form a conduction band occupied by electron pairs, which cannot relax without breaking the semiconjugation and increasing the energy of whole system. The situation resembles superconductivity. Auxiliary polarizing groups, *e.g.*, the guanidyl group of arginine, may improve the effect.

The formation of a semiconjugated system with extra pairs of electrons requires an electron generator, *e.g.*, mitochondria, which can translocate electrons into the membrane. Mitochondria are attached to the neurofibrils during nerve growth.[135]

In polyenes, the excitation of an electron from HOMO (ψ_m) to LUMO (ψ_{m+1}) requires a considerable amount of energy because the σ bonds keep the distances between the atoms nearly constant, and conjugation is not increased in lower molecular orbitals. In a semiconjugated system the ψ_m to ψ_{m+1} shift increases the association and conjugation. This results in a decrease of energy levels which more than compensates the energy needed for the ψ_m to ψ_{m+1} shift.

There is direct evidence for the presence of free, paired, electrons in the axon:[136] "One unexpected result from the experiments was the discovery of reductants that were bound to the nerve. Radicals that came in direct

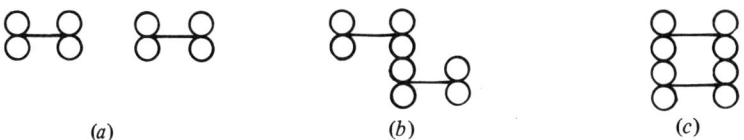

Fig. 18.4 Three types of semiconjugated ethylenic double bond pairs.

contact with the nerve were found to be gradually reduced ... the reductant could not be extracted from the nerve with seawater. Although this reduction of the radical interfered with our experiments, the presence of the nerve-bound reducing agent is itself of interest." No increase was reported to occur in the rate of reduction of the spin label upon firing of the axon. The fact that nerve impulses produced no signals in the ESR spectrum means that the electrons are paired.

Magnetic fields of the action potential of an isolated frog sciatic nerve have been recorded.[137] A field of 120 pT at a distance of 1.3 mm from the nerve could be detected. The two magnetic field vectors of the action potential have opposite signs: the first, initial field, resembles that of electrons moving opposite to the direction of the impulse; the second, larger field, agrees with this model,[128] *i.e.,* electrons moving in the direction of signal propagation. The dual nature of the magnetic field of the nerve signal can be explained by charge transfer taking place *via* the orbitals of the semiconjugated double bonds.[138] Formation of charge transfer complexes has been proposed[139] for the analysis of the electric fields generated by the propagating nerve impulses.

Silinsh[140] estimated the energy (E_{CT}) of the charge transfer state of pentacene to be about one order of magnitude smaller than that of napthalene. As the number of conjugated rings increases it becomes easier to separate the charges in a CT state; the photo-induced M^+M^- thermalized CT state of pentacene has, under some conditions, charge separation distances as large as 12.5 nm.[141] The semiconjugated molecular orbital system model of nerve cells proposed[128] should have some remarkable physical and electrochemical properties.

18.12 Bone

It has been observed that the bending of bones generates piezoelectricity,[110] dc potentials applied to the ends of broken bones stimulate growth.[111] Becker, Basset and Bachman[112] proposed that these two phenomena might act as a closed-loop feedback system by which the body controls the growth of bones in response to an applied stress. Implantation of piezoelectric dusts which stimulate tissue growth has been successfully employed by Evans and Zeit.[113] Bone heating assisted by pulsating currents has also been successfully attempted.[114] Since the mere application of a direct current modifies the chemical composition[115] of the extracellular medium, osteogenesis is difficult to ascribe directly to such current effects. The use of pulsating current, first with contacting electrodes,[114] followed by capacitive[116] and inductive coupling, has also been employed. Capacitive coupling gave ambiguous[117] results but inductive coupling was shown to be effective in clinical trials.[117]

Inductive couplings allows the use of selective waveforms to influence specific cellular processes *in vivo*. Specific interactions and charged species at a cell's surface and junctions are considered[115,118] to play a central role in growth regulation.

18.13 Cooperative Phenomena

It has been suggested[120] that the concepts of long-range order might apply to biological systems. A simple model has been developed[120] describing long-range correlation and cooperative phenomena based on the Einstein condensation of a Bose gas. In this the bosons condense into a single quantum state at low temperature. Electrical vibrations with frequencies of about $10^{11} - 10^{12}$ Hz may be excited coherently in active biological materials by chemical (metabolic) processes.[120] Such selective long-range interactions appear to apply to the control of biological growth. Membranes sustain enormous polarizations impressed upon them by a field of about 10^7 Vm^{-1} (100 mV/100Å). Pohl has related these concepts to cell reproduction:[121] Webb and colleagues[122] observed resonances at $10^{10} - 10^{12}$ Hz in laser-Raman and microwave studies of human sarcoma cells and baby hamster kidney cells, infected with sarcoma viruses. A remarkable splitting of peaks was observed when cancer was present. More notably, strong irradiation of infected cells produced a marked lowering of their infectivity with very little loss of vitality, a phenomenon suggested to be linked to an unstable (pseudo)-ferroelectric mode.[121] Then Rodan et al.[123] observed that externally applied fields (1200 Vcm^{-1}, 5 Hz) stimulated DNA synthesis in the proliferative layer of the embryomic chick epiphysis, which effect may be related to a oscillatory dipole mode. The reversion of cancer cells suggests[124] that the collective cooperative dipole modes affect electron transport across RedOx systems, as in *e.g.*, the cytochrome-*c* system.[124]

References

1. A. Szent-Györgyi, *Introduction of a Submolecular Biology,* Academic Press, N.Y., 1969.
2. A. Szent-Györgyi, *The Living State,* Marcel Dekker Publ., N.Y., 1972.
3. A. Szent-Györgyi, *Electronic Biology and Cancer,* Marcel Dekker Publ., N.Y., 1976.
4. A. Szent-Györgyi, *The Living State and Cancer,* Marcel Dekker Publ., N.Y., 1978.
5. R. Pethig, *Dielectric and Electronics Properties of Biological Materials,* Wiley, Chichester, 1979.
6. F. W. Cope, *J. Biol. Phys.,* **3**, 1 (1975).
7. H. Ti. Tien, *Solid State Phys.,* **2**, 847 (1974).
8. W. Sedlak, Bioplasma, Mater. 1st Konf. Bioplasmie, 1973, Publ. 1976.
9. M. V. Volkenstein, *Molecular Biophysics,* Plenum Press, N.Y., 1977; R. Gabler, *Electrical Interactions in Molecular Biophysics,* Plenum Press, N.Y., 1978.

10. M. Cignitti, p. 31, P. G. Kestyuki, p. 129, C. Simionescu et al., p. 151, *Topics Bioelectrochem. Bioenerg.*, **2**, (1978).
11. *Advances in Biol. and Medical Physics*, J. W. Lawrence, J. W. Gofman and T. L. Hayes, eds., Academic Press, N.Y., (1978).
12. J. P. Marton, *Physiol. Chem. and Physics*, **5**, 259 (1973).
13. R. Pethig and A. Szent-Györgyi, in *Bioelectrochemistry, Proceedings of the U.S.-Australia Joint Seminar on Bioelectrochemistry*, (1979), H. Keyzer and F. Gutmann, eds., Plenum Press, N.Y., 1980.
14. D. L. Worcester, *Proc. Natl. Acad. Sci. U.S.A.*, **75**, 5475 (1978).
15. B. Y. Tong, *J. Non-Cryst. Solids*, **4**, 455 (1978).
16. J. Altieri and J. E. Kirzan, *J. Biol. Phys.*, **3**, 103 (1975).
17. K. D. Straub, Ph.D. Thesis, Duke U. (1968).
18. P. S. B. Digby, *Proc. Rev. Soc. London*, **B-161**, 502 (1965); *Proc. Linn. Soc. Lond.*, **178**, 129 (1967); *Symp. Zool. Soc. London*, **19**, 159 (1967).
19. Y. Nakahara et al., *Chem. Lett.*, **1979**, 877.
20. Y. Nakahara et al., *Chem. Phys. Lett.*, **47**, 251 (1977).
21. C. M. Dobson et al., *Nature* (London), **249**, 425 (1974); K. Ono et al., *J. Chem. Phys.*, **63**, 1640 (1975); K. Kimura et al., *Biochim. Biophys. Acta*, **567**, 96 (1979).
22. K. Niki et al., *J. Electrochem. Soc.*, **124**, 1889 (1977).
23. P. R. C. Gascoyne and R. Pethig, *J. Chem. Soc. Faraday I*, **73**, 171 (1977).
24. D. Spivey, *Discuss. Faraday Soc.*, **27**, 239 (1959).
25. R. Pethig and A. Szent-Györgyi, *Proc. Natl. Acad. Aci. U.S.A.*, **74**, 226 (1977).
26. B. Rosenberg, *Nature* (London), **193**, 364 (1962).
27. M. H. Cardew and D. D. Eley, *Disc. Faraday Soc.*, **27**, 115 (1959).
28. J. Eden, P. R. C. Gascoyne and R. Pethig, *J. Chem. Soc. Faraday I*, **75**, (1979), in print.
29. R. Pethig, *Int. J. Quantum Chem. Quantum Biol. Symp.*, **5**, 159 (1978).
30. S. Bone and R. Pethig, *Submolecular Biology and Cancer, CIBA Foundation Symposium 67* (new series), Elsevier, Amsterdam, pp. 83-105 (1979).
31. P. Camochan and R. Pethig, *J. Chem. Soc. Faraday I.*, **72**, 2355 (1976).
32. T. J. Lewis, *Inst. J. Quantum Chem. Quantum Biol. Symp.*, **5**, 149 (1978).
33. F. W. Cope, *J. Biol. Physics*, **3**, 1 (1975).
34. F. W. Cope and K. D. Straub, *Bull. Math. Biophys.*, **31**, 761 (1970).
35. D. D. Eley, R. J. Meyer and R. Pethig, *J. Bioenergetics*, **3**, 271 (1972); **4**, 389 (1973).
36. G. Foder, R. Mujunder and A. Szent-Györgyi, *Proc. Natl. Acad. Sci. USA.*, **75**, 4317 (1978).
37. H. D. Dakin and H. W. Dudley, *J. Biol. Chem.*, **14**, 155 (1913); C. Neuberg, *Biochem. Z.*, **49**, 202 (1913).
38. S. Bone et al., *Proc. Natl. Acad. Sci. USA.*, **75**, (1), 315 (1978).
39. A. Bonsignore et al., *Ital. J. Biochem.*, **26**, 162 (1977).
40. L. Lorand, in R. Pethig, Ref. 29.
41. P. Otto, J. Ladik and A. Szent-Györgyi, *Proc. Natl. Acad. Sci. USA*, **75**, 3548 (1978).
42. C. Sybesma, *An Introduction to Biophysics*, Academic Press, N.Y., 1979, p. 212.
43. R. A. Bogomolni, in *Bioelectrochemistry, Proceedings U.S.-Australia Joint Seminar on Bioelectrochemistry*, Pasadena (1979), H. Keyzer and F. Gutmann, eds., Plenum Press, N.Y., 1980, p. 152.
44. R. Pethig and A. Szent-Györgyi, *Proc. Natl. Acad. Aci. USA*, **74**, 226 (1977).

REFERENCES

45. S. Bone and R. Pethig, *Submolecular Biology and Cancer, Ciba Foundation Symposium 67* (new series), Elsevier, Amsterdam, 1979.
46. T. J. Lewis, *Int. J. Quantum Revs. Quantum Biol. Symp.*, **5**, 149 (1978).
47. P. R. C. Gascoyne, J. McLaughlin and R. Pethig, in Ref. 4, pp. 46–59.
48. P. Otto, J. Ladik and A. Szent-Györgyi, *Proc. Natl. Acad. Sci. USA*, **76**, 3849 (1979).
49. L. E. Lyons, in *Bioelectrochemistry. Proceedings U.S.-Australia Joint Seminar on Bioelectrochemistry*, Pasadena (1979), H. Keyzer and F. Gutmann, eds., Plenum Press, N.Y., 1980.
50. E. Silinsh, *Phys. Stat. Solidi* (a), **25**, 339 (1974).
51. T. Imura, T. Furutsuka and K. Kawabe, *Photochem. Photobiol.*, **22**, 129 (1975).
52. V. B. Estigneev et al., *Doklady Akad. Nauk. SSSR*, **230**, 726 (1976).
53. A. Chibisov, *Doklady Akad. Nauk. SSSR*, **175**, 230 (1967); *Photochem. Photobiol.* **10**, 331 (1969).
54. K. Seifert and H. T. Witt, *Naturwiss.*, **55**, 222 (1968).
55. A. R. Kelly and G. Porter, *Proc. Roy. Soc. London*, **A319**, 319 (1970).
56. R. Livingston and P. J. McCartin, *J. Amer. Chem. Soc.*, **80**, 4826 (1963).
57. V. B. Estigneev et al., *Doklady Akad. Nauk. SSSR*, **203**, 1346 (1972).
58. A. Bromberg, C. W. Tang and A. C. Albrecht, *J. Chem. Phys.*, **60**, 4058 (1974).
59. N. K. Boardman, W. S. Chow, J. T. Duniec and S. W. Thorne in *Bioelectrochemistry, Proceedings U.S.-Australia Joint Seminar on Bioelectrochemistry*, Pasadena, (1979), H. Keyzer and F. Gutmann, eds., Plenum Press, N.Y., 1980; G. A. Seely, *Photochem. Photobiol.*, **27**, 639 (1978).
60. T. Watanabe, T. Mujasaka, A. Fujishima and K. Honda, *Chem. Lett.*, **4**, 443 (1978).
61. Y. S. Meilanov, V. A. Benderskii and L. A. Blyumenfeld, *Biophys.*, **15**, 851 (1970).
62. J. Nakata, T. Imura and K. Kawabe, *J. Phys. Soc. Japan*, **42**, 146 (1977).
63. R. C. Nelson, *Photochem. Photobiol.*, **8**, 448 (1968).
64. P. Oesterhelt and W. Stoeckenius, *Nature* (New Biol.), **233**, 149 (1971).
65. A. E. Blaurock and W. Stoeckenius, *Nature* (New Biol.), **233**, 152 (1971).
66. S. C. Kushawa, M. Kates and W. Stoeckenius, *Biochem. Biophys. Acta*, **426**, 703 (1976).
67. W. Stoeckenius, R. H. Lozier and R. A. Bogomolni, *Biochem. Biophys. Acta*, **505**, 215 (1979); P. N. T. Unwin and R. Henderson, *J. Mol. Biol.*, **94**, 425 (1975); R. Henderson and P. N. T. Unwin, *Nature*, **94**, 957 (1975).
68. Yu. Orchinnokov, N. G. Abdulaev, M. Yu. Feigina, A. V. Kiselev, N. A. Lobanov and I. V. Nasimov, *Bioorganiyica Chimica*, **4**, 1593 (1978).
69. M. P. Heyn, R. J. Cherry and N. Muller, *J. Mol. Biol.*, **117**, 607 (1977); R. A. Bogomolni, S. B. Hwang, Y. W. Iseng and W. Stoeckenius, *Biophys. J.*, **17**, 98a (1977).
70. G. I. King, R. A. Bogomolni, S. G. Hwang, W. Stoeckenius and B. P. Schoenborn, *Biophys. J.*, **17**, 97a (1977).
71. G. I. King, W. Stoeckenius, H. Cresp and B. P. Schoenborn, *Biophys. J.*, **130**, 395 (1979).
72. H. T. Witt, *Quart. Revs. Biophys.*, **4**, (4), 365 (1971).
73. A. D. Gochev and S. G. Christov, *Dokl. Bulg. Akad. Nauk.*, **32** (3), 321 (1979).
74. M. Redi and J. J. Hopfield, *Bull. Amer. Phys. Soc.*, **24**, 346 (1979).
75. M. J. Potasek and K. W. Beeson, *Bull. Amer. Phys. Soc.*, **24**, 344 (1979); K. W. Beeson and M. J. Potasek, *ibid.*, 344.

76. E. C. Slater, *Quart. Rev. J. Biophys.*, **4**, (1) 36 (1971).
77. R. H. Ruby, I. D. Kuntz and M. Calvin, *Proc. Nat. Acad. Sci. USA*, **51**, 515 (1964).
78. F. W. Cope, R. J. Sever and B. D. Polis, *Arch. Biochem. Biophys.*, **100**, 171 (1963).
79. F. W. Cope, *J. Biol. Phys.*, **3**, 1 (1975).
80. G. Kortüm and J. O'M. Bockris, *Textbook of Electrochemistry*, Elsevier, Amsterdam, 1951.
81. D. R. Turner, in *The Electrochemistry of Semiconductors*, P. J. Holmes, ed., Academic Press, London, 1962.
82. F. W. Cope, *J. Chem. Phys.*, **40**, 2653 (1964).
83. F. W. Cope, *Proc. Nat. Acad. Sci. USA*, **51**, 809 (1964).
84. F. W. Cope, *Bull. Math. Biol.*, **37**, 79 (1975).
85. E. M. Trukhan, N. F. Perewoschikof and M. A. Ostrowski, *Biofizika*, **15**, 1052 (1970).
86. F. W. Cope, *Proc. Nat. Acad. Sci. USA*, **61**, 905 (1968).
87. N. Chalozonites, *Photochem. Photobiol.*, **3**, 539 (1960).
88. F. W. Cope, *Physiol. Chem. and Physics*, **3**, 403 (1971).
89. F. W. Cope, *Physiol. Chem. and Physics*, **5**, 173 (1973).
90. D. E. Bercher, *Ann. N.Y. Acad. Sci.*, **188**, 324 (1971).
91. F. W. Cope, *Physiol. Chem. and Physics*, **6**, 405 (1974).
92. S. Goldfein, *Physiol. Chem. and Physics*, **6**, 261 (1974).
93. H. Athenstaedt, *Z. Zellforsch.*, **98**, 300 (1969).
94. F. W. Cope, *Bull. Math. Biol.*, **35**, 31 (1973).
95. F. W. Cope, *Ann. N.Y. Acad. Sci.*, **204**, 416 (1973).
96. W. J. Moore, *Special Topics in Electrochemistry*, P. A. Rock, ed., Elsevier, Amsterdam, 1977.
97. A. L. Hodgkin and A. L. Huxley, *J. Physiol.*, **117**, 500 (1952).
98. G. N. Ling, *Amer. J. Phys. Med.*, **34**, 89 (1955).
99. R. Damadian, *Ann. N.Y. Acad. Sci.*, **204**, 249 (1973).
100. G. N. Ling, *Internatl. Rev. Cytol.*, **26**, 1 (1961); *A Physical Theory of the Living State*, Blaisdell Press, N.Y., 1960.
101. L. Minkoff and R. Damadian, *Physiol. Chem. and Physics*, **8**, 349 (1976).
102. F. W. Cope, *Bull. Math. Biophysics*, **27**, 99 (1965); **29**, 691 (1967).
103. F. W. Cope in *Bioelectrochemistry, Proceedings of U.S.-Australia Joint Seminar in Bioelectrochemistry*, Pasadena, (1979), H. Keyzer and F. Gutmann, eds., Plenum Press, N.Y., 1980.
104. F. W. Cope, *Physiol. Chem. and Physics*, **9**, 155 (1977).
105. G. Karreman, *Bull. Math. Biophys.*, **33**, 483 (1971).
106. M. Avrami, *J. Chem. Phys.*, **7**, 1103 (1939); **8**, 212 (1940); **9**, 177 (1941).
107. J. W. Christov, *The Theory of Transformations in Metals and Alloys*, Pergamon Press, London, 1975; V. Raghavan and M. Cohen, in *Treatise on Solid State Chemistry*, Vol. 5, N. B. Hannay, ed., Plenum Press, N.Y., 1975; P. Duhaj, D. Barancock and A. Ondreijka, *J. Non-Cryst. Solids*, **21**, 411 (1976).
108. F. W. Cope, *Physiol. Chem. and Physics*, **9**, 383 (1977).
109. I. Takenaka, in L. Bau and W. J. Moore, *Brain. Rev.*, **33**, 451 (1971).
110. E. Fukada and I. Yasuda, *J. Physical Soc. Japan*, **12**, 1158 (1957); C. A. L. Bassett and R. O. Becker, *Science*, **137**, 1063 (1962); M. H. Shamos, L. S. Lavine and M. I. Shamos, *Nature* (London), **197**, 81 (1963).

REFERENCES

111. *e.g.,* C. A. L. Bennet, R. J. Pawluk and R. O. Becker, *Nature* (London), **204**, 652 (1964); S. D. Smith, *Anat. Record.,* **158**, 89 (1967).
112. R. O. Becker, C. A. L. Bennet and C. H. Bachman, in *Bone Biodynamics,* H. M. Frost, ed., Little Brown and Co., Boston, 1964.
113. S. M. Evans, *J. Indust. Hygiene and Tox.,* **30**, 353 (1948); S. M. Evans, and W. Zeit, *J. Lab. Clin. Med.,* **34**, 592, 610 (1949).
114. T. E. Jorgenson, *Acta Orthop. Scandin.,* **43**, 421 (1972); W. Krans and F. Lechner, *Munch. Med. Wschr.,* **114**, 814 (1972); C. A. L. Basset, A. A. Pilla and R. J. Pawluk, *Clin. Orthop.,* **124**, 117 (1977).
115. *e.g.,* A. A. Pilla, in *Electrochemical Bioscience and Bioengineering,* I. Miller, A. Salkind and H. Silverman, eds., Electrochem. Soc. Inc., Princeton, N.Y., 1973; R. O. Becker and A. A. Pilla, in *Modern Aspects of Electrochemistry,* J. O'M. Bockris, ed., Plenum Press, N.Y., Vol. 10, 1975.
116. A. A. Pilla, in *Bioelectrochemistry, Proceedings U.S.-Australia Joint Seminar on Bioelectrochemistry,* Pasadena, (1979), H. Keyzer and F. Gutmann, eds., Plenum Press, N.Y., 1980.
117. C. A. L. Bassett, A. A. Pilla and R. J. Pawluk, *Clin. Orthop.,* **124**, 117 (1977).
118. A. A. Pilla, in *Surface Chemistry in Biological and Medical Bioelectrochemistry,* M. Blank, ed., *Amer. Chem. Soc.,* Washington, 1979.
119. R. Harrison and B. G. Hunt, *Biological Membranes, Their Structure and Function,* Wiley, N.Y., 1975.
120. H. A. Pohl, in *Bioelectrochemistry, Proceedings U.S.-Australia Joint Seminar on Bioelectrochemistry,* Pasadena, (1979), H. Keyzer and F. Gutmann, eds., Plenum Press, N.Y., 1980.
121. H. A. Pohl, Research Note 90, June 1979, Oklahoma State Univ., Stillwater, Oklahoma.
122. S. J. Webb, R. Lee and M. E. Stoneham, *Int. J. Quant. Chem. Quant. Biol. Symp.,* **4**, 277 (1977).
123. G. A. Rodan, L. A. Bourret and L. A. Norton, *Science,* **199**, 690 (1978).
124. A. Szent-Györgyi, *Inst. J. Quant. Chem. Quant. Biol. Symp.,* **3**, 45 (1976).
125. K. Kimura et al., *J. Chem. Phys.,* **70**, 3317 (1979).
126. K. Murano et al., *Bull. Chem. Soc. Japan,* **49**, 2407 (1976).
127. J. Th. Kistenmacher and C. G. Margilli, *Jerusalem Symp. Quant. Chem. Biochem.,* **9**, pt 1, 7 (1977).
128. J. A. Virtanen and P. K. S. Kinnunen, Dept. Medical Chem., Univ. of Helsinki, Finland, Private Commun., 1981.
129. J. S. O'Brien and G. Rouser, *J. Lipid Res.,* **5**, 339 (1964); A. Montfoort, L. M. G. van Golde and L. L. M. van Deenen, *Biochim. Biophys. Acta,* **231**, 335, (1975); F. A. Manzoli, S. Stefoni, L. Manzoli-Guidotti and M. Barbieri, *FEBS Lett.,* **10**, 317 (1970); J. S. O'Brien, E. L. Sampson and M. B. Stern, *J. Neurochem.,* **14**, 357, (1967).
130. P. Bischof et al., *Helv. Chim. Acta,* **53**, 1425 (1970).
131. W. C. Breckenbridge, G. Gombos and I. G. Morgan, *Biochim. Biophys. Acta,* **266**, 695 (1972); C. Cotman, M. L. Blank, A. Moehl and F. Snyder, *Biochemistry,* **8**, 4606 (1969); J. W. Deutsch and R. G. Kelly, *Biochemistry,* **20**, 378 (1981); T. Richardson, A. L. Tappel, L. M. Smith and C. R. Houle, *J. Lipid Res.,* **3**, 344 (1962); A. Emilsson and A. Gudbjarnason, *Biochim. Biophys. Acta,* **664**, 82 (1981).
132. J.-L. Popot et al., *Eur. J. Biochem.,* **85**, 27 (1978).
133. A. J. Parodi and L. F. Leloir, *Biochim. Biophys. Acta,* **559**, 1 (1979); C. J. Waechter and W. J. Lennarz, *Ann. Rev. Biochem.,* **45**, 95 (1976).

134. G. S. Adrian and R. W. Keenan, *Biochim. Biophys. Acta*, **663**, 637 (1981).
135. R. S. Geiger, *Nature* (London), **182**, 1674 (1958).
136. M. Calvin et al., *Proc. Natl. Acad. Sci., USA*, **63**, 1 (1969).
137. J. P. Wikswo Jr., J. P. Barach and J. A. Freeman, *Science*, **208**, 53 (1980); S. J. Williamson and L. Kaufman, *J. Magnetism Magnetic Materials*, **22**, (2), 129 (1981).
138. J. M. Pearson, S. R. Turner and A. Ledwith, in *Molecular Association*, R. Foster, ed., Academic, 1979, Vol. 2, p. 79.
139. A. C. Scott, *Rev. Mod. Phys.*, **47**, 487 (1975).
140. E. Silinsh, *Phys. Stat. Sol.*, **(a) 25**, 339 (1974).
141. E. Silinsh, *Electronic States of Organic Molecular Crystals*, Zinat., Academy of Science, Latvian SSSR, Riga, 1978.

19

Structure

19.1 Introduction

A serious problem facing research into the properties of well-conducting organic semiconductors is presented by the fact that so many are refractory, black, amorphous, insoluble in all solvents and virtually chemically inert. This statement applies particularly to polymers. As a consequence their structure, and physico-chemical properties, are almost inaccessible by conventional investigation techniques. Some of these electronically conducting organic and organo-metallic compounds exhibit giant polarization and enormous dielectric constants, as high as 300,000.[1] Correlation of such effects, possibly arising from nomadic polarization with chemical properties as described by Pohl and coworkers[2,3] and discussed in Section 16.5, mandates structure determinations. Several avenues towards solution of these problems have been attempted. One may introduce functional groups which could improve solubility in ordinary solvents, or make the compound's structure amenable to further chemical reaction or physio-chemical techniques such as nuclear magnetic resonance. Complexation with paramagnetic ions, for instance, yields proton relaxation effects resulting in selective broadening of signals from protons close to binding sites.[4] Information from model compounds can be very helpful. X-ray studies of single crystals of model TCNQ-polymer compounds showed that the electrical properties depended mainly on the crystal geometry and not on the length of the chains in the polymer.[5] Inevitably one must make assumptions as to structure-electrical property relationships from the analogous characteristics observed. X-ray diffraction patterns can yield information on van der Waals spacing in some cases such as have been determined for conjugated conductive polymers of the polyacenequincne type.[6] Electron diffraction and dark-field imaging has also been used in this respect with some success.[7] To overcome problems of insolubility some workers[8] have used organic glasses to investigate structure-electrical property dependence as in the case of charge transfer complexes of anthracene with tetrachlorophthalic anhyride and trinitrobenzene, (see Chapter 17). Where the conventional methods such as X-ray and light-scattering measurements of polymeric structure determination fail, in specific cases other methods have been invoked. For instance, the charge transfer complexes of poly-(1-naphthylmethyl)-L-glutamate with organic electron acceptors were subjected to circular dichroism and optical rotatory dispersion measurements[9] to show that the complexes had α-helix structures. The preparation of films of poly-(N-vinyl-carbazole) and its complexes in the presence of electric fields which induced preferential alignment of dipoles (Kerr effect), and magnetic fields aligning magnetic dipoles allowed the conclusion[10] to be drawn that the carbazole rings were planar with corresponding anisotropic ring currents which could orient monomer units. The determination of glass transition temperatures (T_G) (see Section 16.9) of elastomeric ionenes[11] complexed with LiTCNQ *via* direct current resistivity temperature measurements has proved useful in lattice constraint studies; the glassy state is considered a tight lattice in which the constraint forces are much stronger than those of the non-glassy state. In the latter case TCNQ anions can move away and the electrical resistivity can be considerably decreased giving rise to an activated hopping model.

19.2 Anisotropy

Anisotropy often accounts for unusual conductivity effects in organic semiconductors. Phonon-assisted mobilities (see Chapter 15) caused by rotational vibrations, for example, are important[12] in anthracene. This increases electronic mobility in the c' axial direction of the single crystal by an order of magnitude. Takahashi and coworkers[13] showed that the conductivity parallel to the cleavage plane of ureanitrate single crystals was 1.5×10^{-8} (ohm cm)$^{-1}$, three orders of magnitude higher than that perpendicular to the cleavage plane. A highly anisotropic crystal may even be *p*-type in one direction and *n*-type in another as demonstrated[14] for auramine which is *p*-type in the *c*-direction and *n*-type in the *a*-direction. In one case,[15] such as the tetraselenafulvalene:TCNQ complex (see Chapter 21), the different resistivities along the *a*- and *b*-axes above the critical 3-dimensional ordering temperature is probably dominated by an anomaly in electron-phonon scattering with the onset of 3-dimensional Peierls distortion.

Studies[16] of TCNQ salts of asymmetric and symmetric bipyridine donors (see Section 16.11) showed that the latter led to small band gap semiconductors with large anisotropy in conductivities. For these, susceptibility measurements indicated strong electron correlation, the complexes with the asymmetric donors exhibiting electron localization due to disorder. A specific heat anomaly at 72°C in phenanthrene has been shown to give rise to a change in the anisotropic permittivity of single crystals.[17]

Lochner and co-workers[18] determined that carrier mobilities parallel to the chain axis on the (100) plane of polydiacetylene semiconductors were 800 times larger than those perpendicular to the chain axis. They also showed that the quantum yield for photoelectric carrier generation was highest for fields parallel to the chain.

19.3 Crystalline Transitions

The theory of reversible crystallization in polyvinylidene-fluoride, PVF, has been investigated by Dvey-Akharon and co-workers[19] in relation to their large piezoelectric and pyroelectric coefficients; they suggest that these may be due to stress-induction at the amorphous-crystallite boundary. Wang[20] showed that piezoelectricity was associated with quasi single crystal orientation in β-phase PVF. While Wang induced crystal orientation in PVF by mechanical means, Herchenroeder, Naegele and Yoon[21] oriented crystalline and amorphous regions electrically, showing that the polar β-form crystallites were easily affected and contributed strongly to the observed piezo- and pyroelectric phenomena.

Phase transitions have been observed[22] in TCNQ complex salts for the cation (methyl-1-N-ethylbenzimidazolium)$^+$. During cooling the conductivity in different needles of this compound showed drops by factors of 3 to 300 at a discontinuity temperature of 250 K. This was never observed in large specimens of the material after recrystallization; here instead, the discontinuity was replaced by a steep but continuous conductivity anomaly (cf., Section 16.11).

Phase transitions in solid camphor compounds were studied conductimetrically and confirmed colorimetrically by Swiatkiewicz and Pigon;[23] these substances, though, are ionic conductors.

19.4 Amorphous Systems

During the last 15 years or so, the physics of amorphous semiconductors has progressed by leaps and bounds and several excellent reviews on the subject are available.[24] We shall confine ourselves very briefly to some of the basics and a review of some of the apposite work on organics. For a general discussion of charge transfer theories, see Chapter 15.

The stochastic transport features are common to all amorphous solids in which hopping occurs whether they are characterized by covalent or molecular bonding (*cf.,* Chapter 15). However, a characteristic distinction exists between the disordered molecular state and disordered covalently bonded solids.[25] In the latter case the effect of disorder is to cause, in varying degrees, the "tailing" of states (to be discussed in detail shortly) into a forbidden gap while preserving the existence of bands or extended states. For molecular solids, on the other hand, the situation is different and gives this class of disordered solids some distinctive properties. In molecular single crystals, the weak intermolecular forces give rise at most to extremely narrow transport bands with typical carrier mobilities of ~ 1 cm^2/Vsec and usually weak temperature dependence.[26] In such a situation the transition to the disordered state, from the point of view of charge transport, must be quite different from the case of covalent solids; (see Chapter 15).

White and Ngai[27] have studied the effects of weaker/stronger bonds on the electronic structure of homopolar disordered "tetrahedrally" bonded semiconductors. This study employed a parameterized tight-binding Hamiltonian which included all possible interactions between s and p states associated with nearest neighboring atoms. Model calculations were made by treating exactly the local environment of the strained bond while replacing the remainder of the system by an effective medium. Interest was focussed on how such strained bonds which do not represent defects *per se* could produce states in the gap with energetic positions that were shifted appreciably by changes in the strained bond length.

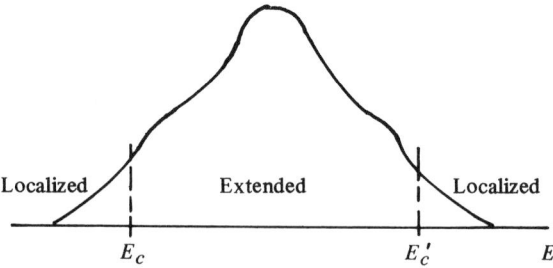

Fig. 19.1 The simplest model for the density of states of a single isolated band of extended states inside the energies E_c and E_c' with tails of localized states outside. These are expected to be universal features of the electronic structures of disordered materials. After Cohen.[28]

Electron Transfer in Amorphous Solids

In very disordered systems such as amorphous solids[28] the density of states within an isolated energy band can be depicted as in Fig. 19.1, 19.2, and 19.3. Instead of sharp band edges, we find tails of localized states caused by the potential energy fluctuations due to the structural disorder. If there is, in addition, disorder of composition, *i.e.*, if the chemical nature of the system varies from region to region, these tails may overlap. The carrier mobility of the band edges, *i.e.*, in the tails, drops from a high value characteristic of motion within an energy band, to the low value associated with thermally activated hopping (see Fig. 19.4). Within the tails, however, where the energy states change from an extended to a localized regime, there prevails a transport mechanism that is akin to Brownian motion; *i.e.*, it is a Markovian process with a mobility intermediate between that of the propagation of an electron wave within an energy band and thermally assisted hopping (see Fig. 19.5). This mode of charge transfer is sometimes referred to as percolation. In poorly conducting charge transfer complexes, *e.g.*, this is likely to be the predominant mode of conduction.[28] (See Chapter 15.)

In a summary of the dependence of electronic properties on amorphous semiconductors Cohen[24] considers a model in which the energy gaps, typical of crystalline materials, are replaced by pseudo-gaps, in which the density of states is low but non-zero. The wavefunctions of the states in the pseudo-gap extend only over limited distances, and so the states are "localized." However, even in amorphous solids the existence of extended states is expected,[30] as typified in crystals, because an electron is but slightly scattered by fluctuations of potentials sufficiently small compared to the electron energy. Thus,

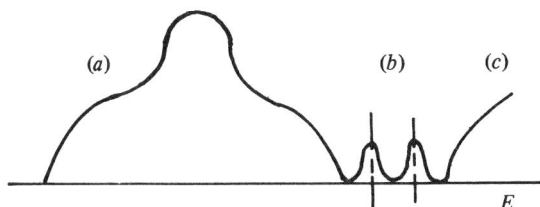

Fig. 19.2 Density of states $n(E)$ as a function of energy E for (*a*) a perfect crystal (*b*) a crystal containing only one localized imperfection and (*c*) a crystal containing a low concentration of localized imperfections. In all three cases there are continuous bands of energy levels separated by gaps with square root singularities at the band edges. The square root singularities associated with saddle points within the band in (*a*) and (*b*) are eliminated by scattering in (*c*). If the potential change ΔV introduced by the imperfection is strong enough in (*b*), localized levels split off from the bands (off the bottom for attractive ΔV or the top for repulsive), which broadens into bands in (*c*). After Cohen.[28]

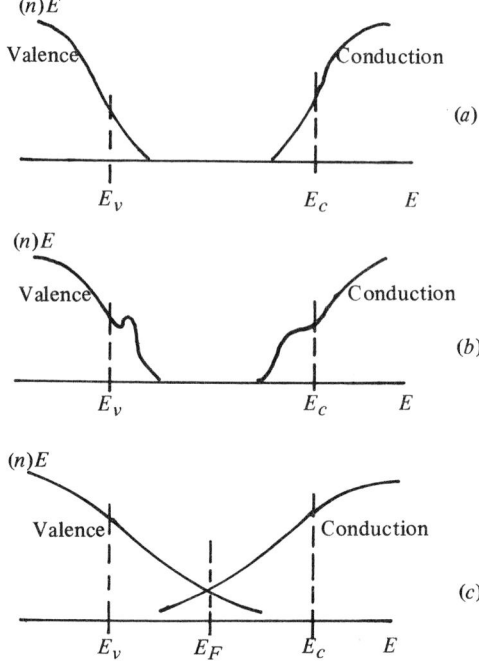

Fig. 19.3 Band models for amorphous semiconductors. In (a) the simplest possible case, there are valence and conduction bands with extended states for $E < E_c$ and $E > E_c$ and tails of localized states on each band. For elements and compounds (b) the well-defined local order leads to the presence of well-defined structure defects and non-monotonicity in $n(E)$ in or near the tails. For alloys (c) there is, in addition, compositional disorder and translational disorder enhanced by the requirement that valences be locally satisfied. The tails of the conducting and valence bands therefore overlap, which leads to a finite density of states at the Fermi energy $n(E_F)$ and a finite concentrate of localized charged states near E_F. After Cohen.[28]

in a homogeneous system the separation of localized and extended states, must be sharp, the so-called "mobility edge," because these two kinds of states cannot exist at the same energy.

Experimental evidence[30] suggests that the "intrinsic" states in the gap do not form from broad distributions deep within the gap but are restricted to energy intervals very close to the band edges.

Griffith[31] found that the mobility gap in non-polar amorphous semiconductors decreased with increasing temperature. Thus, as has been pointed out by Scher and Lax,[32] charge conduction entirely by "free carrier" band transport is the biggest casualty in going from an ordered to a disordered solid, though in most crystals, such as azulene, enough disorder occurs even

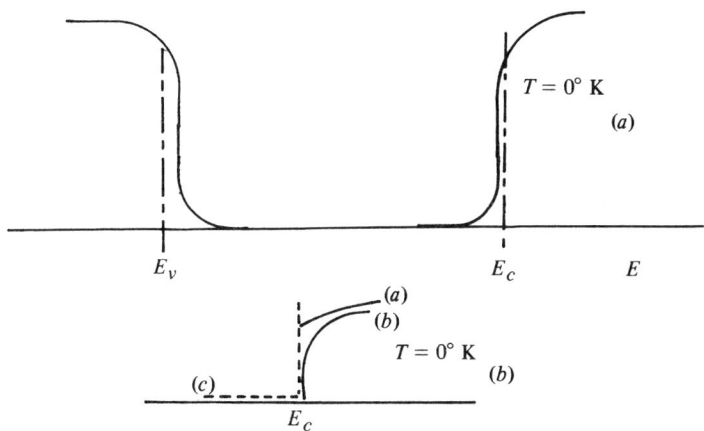

Fig. 19.4 Simple models for the energy dependence of the mobility μ: $T = 0$ K; (b) $T = 0$ K. In (a) E_v and E_c play the role of mobility edges because the mobility drops here from values characteristic of band conduction to values characteristic of phonon-listed hopping conducting within the mobility gap between E_v and E_c. This drop is termed a mobility shoulder. In (b), various possible energy dependencies of μ near E_c are shown for $T = 0$ K that are consistent with: case (a) (solid line) a continuous drop from a finite value to zero, case (b) (solid line) an abrupt but continuous drop toward 0. and case (c) (dotted line) a finite though small value of μ within the mobility gap. After Cohen.[28]

in single crystals that it has been suggested that even these are part of the way already to an amorphous state.[33] The lack of long-range structural order is the central essence of the problem; this lack of order causes a random potential to exist in the solid. The random potential can give rise to a distribution of localized electronic states[34] as well as the extended bandlike states.[35] Therefore, the nature of electronic transport can either[36] be the more familiar band type (free-carrier motion with occasional scattering from potential fluctuations), with the localized states acting as traps, or there can be transport among the localized states[37] (this is further discussed in

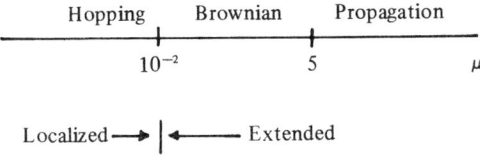

Fig. 19.5 Mobility scale for semiconductors at room temperature in units of cm^2/Vsec. The various regimes proposed in the text are indicated above the scale and the character of the wave function below the scale. After Cohen.[28]

Chapter 15). Thus, we are faced with three possible modes of electronic conduction in amorphous semiconductors:

1. Classical wave propagation in a conduction band, above a mobility edge (see Section 15.2).
2. Hopping of electrons between localized states below a mobility edge; percolation or variable range hopping (see Section 15.4).
3. Hopping conduction in localized states at the Fermi level; this is "classical," known as nearest neighbor hopping (see Section 15.5).

The major difference between 2 and 3 is that in percolation both the mobility and the carrier concentration are activated while in classical, nearest neighbor hopping the statistics are degenerate. Silver[38] has used Monte Carlo simulation techniques to examine diffusion and drift in disordered systems where transport takes place only by hopping. $9 \times 9 \times 9$ and $20 \times 20 \times 20$ cubic lattices were used as backbones in which various fractions of the available sites were filled at random. Hopping was allowed only between filled sites. All possible jumps were included up to second-nearest neighbors with jump frequencies dependent exponentially upon the jump length. A fraction of the particles equal to the percolation fraction for nearest neighbor jumps diffused quickly with a rate proportional to the $(C - C_{th})^{1.8}$ where C is the concentration of filled sites.[38] The remainder diffused slowly. Including all jumps, the rate at which particles left the volume approached a power law. These results show dispersive decay even though the jump frequencies are the same for the same kind of jump. Comparison between these predictions and experiments[40,41] showed excellent agreement. It is not possible to give here more than a highly condensed version of the theory of variable range hopping (see also Chapter 15 and Ref. 42).

The localized energy states within the forbidden zone of an amorphous solid are inherently improbable[43] and may be described in terms of an "abnormal bonding model,"[43] being associated with local distortions. They constitute traps controlling the hopping probability;[44] in carbazole they yield p-type conductivity.

Gagara et al.[45] have pointed out that a real solid is likely to be a heterogeneous matrix of amorphous and crystalline regions, so that the current becomes primarily governed by "interfacial potential wells for the carriers"— in other words, traps localized at the interfaces, though the resistivity of the crystalline regions may be several orders of magnitude below that of amorphous regions;[45] part of the total carrier concentration becomes localized in potential wells arising from the medium range disorder predominant in amorphous solids.[46]

Which of the two types of conductivity predominates, depends on the position of the Fermi level relative to the edge of the conduction band:

At higher temperatures, electrons are excited to the edge of the mobility gap so that

$$\sigma = \sigma_0 e^{-(E_{CB} - E_{FL})/kT} \tag{19.1}$$

while at lower temperatures electrons hop from one localized energy state to another, activated *via* phonon collisions so that, for a three-dimensional solid,

$$\sigma = \sigma e^{-BT^{\frac{1}{4}}}$$

This has been experimentally verified in several cases.[47] For a two-dimensional system, the $T^{1/4}$ term has to be replaced by $T^{1/3}$ (see Chapter 21). The Hubbard intraatomic energy U is defined as the average of e^2/r_{12} when two electrons are on one of the centers. If the centers are far apart, the energy necessary to take an electron from one center and put it on another is just this quantity U; it is the difference between the ionization energy and the electron affinity.

An extra electron placed on one of the centers can move to the next and so on through the lattice, with a Bloch wave function of the type

$$\psi_k = e^{ikx} U_k(x,y,z) \tag{19.3}$$

and bandwidth B_1; this band of energies is called the upper Hubbard band. Similarly the "hole" from which an electron has been removed can move, the width of its "lower Hubbard band" being denoted by B_2.

Mott[42] has shown that interaction between electrons on localized states produces a band gap for single-particle excitations, but not for those of the many-electron system. Variable-range hopping with σ proportional to $\exp(-B/T^{1/4})$ occurs in the limit of low T, but interaction may lead to deviations at higher T, and decreases the preexponential factor. Let E denote the energies of the one-electron states, defined without the inclusion of U. These will be occupied by at least one electron up to the Fermi energy E_F. States with one-electron energies less than $E_F - U$ will be doubly occupied; if the localization radius is large, then the energy required to remove one electron from a pair in a state at an energy $E_i (E_i \leq E_F - U)$ and put it into an unoccupied state just above E_F is $(E_F - E_i) - U$. Thus, hops with vanishingly small energy are possible for electrons both in singly occupied states at E_F and for electrons in doubly occupied states at an energy $E_F - U$.

If U is larger than the range of energies E_i, only singly occupied states exist in the lowest state of the whole system. If there is one electron per state, $N(E_F)$ vanishes and no hopping is possible, the two *Hubbard bands* are separated by a gap. Hopping is possible only if either the number of electrons is less than one per atom (as in impurity conduction in a compensated semiconductor) or if the two Hubbard bands overlap, which means that U is not larger than the range of E_i.

Mott[42] now points out that the random field acting on a given electron depends on the positions of all the other electrons, and that if a given state i is occupied, this raises the level of the surrounding states so that some of them will be empty, while, in the system's ground state, they would be full if the level i were empty. An *electronic polaron* is therefore formed.

Considering an electron at the Fermi level, the question is whether there are empty states at about the same energy to which it can hop *without taking its polaron with it*. Since a new polaron must be formed at the final site, the electron is looking for sites which, before the hop, are lower in energy than the initial site. In equilibrium such sites will lie below the Fermi level and are therefore occupied. Thus no empty states exist at the required energy.

The term $e^2/\kappa R$ therefore opens up a gap in single particle excitations so that variable-range hopping is then impossible: In the limit of low T, the hopping electron *carries its polaron with it*. However, consider an electron which hops a large distance R. Then electrons in the neighborhood of the initial and final sites, say in sites j, suddenly see a change of potential $e^2/\kappa r_{ij}$. Thus, defining their wave functions as f_i before the electron jumps, g_j after, then the functions f_j, g_k for two different sites are not orthogonal. The chance that, simultaneously with the main hopping process over a distance R, another electron jumps a small distance from site j to k, is

$$P_j k = |\smallint f_j g_k d^3 x|^2 \tag{19.4}$$

The hopping probability must be multiplied by a product of such terms having a mean value of p.

Mott[42] now asks how many simultaneous short-range hops are necessary if the electron is to find a state at energy $\sim 1/N(E)R^3$ above the Fermi level, as assumed in variable-range hopping. Let this number be n. Choose the centers j, k in such a way as to increase or decrease the polaron energy, so that 2^n new levels will be produced. They should lie in a range of order

$$\text{const}(e^2/\kappa a)n^{-1/3} \tag{19.5}$$

where a is the intersite distance, because they lie at a distance $\sim a n^{1/3}$ from the center under consideration. Thus variable range hopping over a distance R, with an energy change $1/N(E)R^3$, is possible if n is so large that, setting the constant equal to unity,

$$1/N(E)R^3 \geqslant (e^2/\kappa a)/2^n n^{1/3}. \tag{19.6}$$

This gives

$$n \geqslant 3\log_2(R/R_1), \tag{19.7}$$

with

$$R_1^3 = 1/\{N(E)n^{-1/3}(e^2/\kappa a)\}. \tag{19.8}$$

Mott's[42] argument is that the pre-exponential factor v_0 in the normal hopping probability, namely $v_0 \exp(-2\alpha R - \Delta E/kT)$ where $\Delta E = 1/N(E)R^3$, must be multiplied by p^n. This can be written, [cf., eq. (19.22)], taking the equality sign in (19.7):

$$p^n = \exp\{-3\ln(1/p)\log_2(R/R_1)\}. \tag{19.9}$$

The optimum hopping distance is thus obtained by finding the maximum value of

$$-3\ln(1/p)\log_2(R/R_1) - 2\alpha R - 1/N(E)R^3 kT. \tag{19.10}$$

This occurs when

$$-3\ln(1/p)/(R\ln 2) - 2\alpha + 3/R^4 N(E)kT = 0. \tag{19.11}$$

Mott[42] now discusses the effect of the first term in eq. (19.11) in the limit of low temperature. Since R tends to infinity, the term is small compared to 2α. It has the effect, then, of increasing α by $\delta\alpha$ where

$$\delta\alpha/\alpha = 3\ln(1/p)/2\alpha R \ln 2. \tag{19.12}$$

Although p may be small, the $T^{1/4}$ behavior, which is immediately derivable from eq. (19.9), is thus preserved at low temperatures. However, deviations at higher temperatures, not predicted by the one-electron theory, are expected, as also is a smaller pre-exponential factor.[42]

So far, transport has predominatly been envisioned as a one-electron excitation, electron-electron interactions and correlations being neglected. However, in systems characterized by electrons in localized states owing to the local nature of the states, there are strong Coulomb interactions between electrons.[48] Such interactions lead to the formation of a Coulomb gap in the density-of-states for one-electron excitations.[49] This gap may be either an absolute gap, *i.e.*, the density-of-states may be identically equal to zero, or the density-of-states may be statistically insignificant in the gap since there is not a significant contribution to the conductivity by excitations through these states. The width of the gap E_G is of the order of the inter-electron Coulomb interaction E_e, *i.e.*, $E_G = \theta E_e = \theta e^2/\kappa R$, R is approximately equal to the average distance between carriers, κ is the background dielectric constant, and θ is of the order of unity.[48]

This holds when the disorder energy $\Delta \sim E_e$, as in impurity conduction with compensation $K \sim \frac{1}{2}$. When Δ is appreciably larger than E_e the gap is smaller because fewer electrons can arrange their position to minimize the Coulomb repulsion in the ground state. Srinivasan finds[49] that

$$E_G \simeq E_e \sqrt{(E_c/\Delta)}.$$

In such a system, electrons must surmount the gap in order to transport charge; transport is by hopping and typically the hopping rate proportional to (i) a term $|M|^2$, which arises from initial and final state overlap and (ii) a term of the form $\exp(-E_g/\kappa T)$ which arises from the occupation statistics of carriers and/or phonons. $|M|^2$ is usually given as $\propto \exp(-\alpha R)$ where $\alpha = 2/a$, and a is the *radius* of the localized state. Thus the conductivity should obey (*cf.*, eq. 19.22 and 19.26).

$$\sigma \propto \exp[-\alpha R - (\beta/R)] \qquad (19.13)$$

where

$$\beta = \theta e^2/\kappa kT \qquad (19.14)$$

In Fig. 19.6 the two terms in the exponential of equation (19.13) are plotted against R. It is seen that at high carrier densities, small values of R, the conductivity is energy limited while, conversely, at low carrier density, when R is large, conduction is limited by $|M|^2$. The plot is divided into

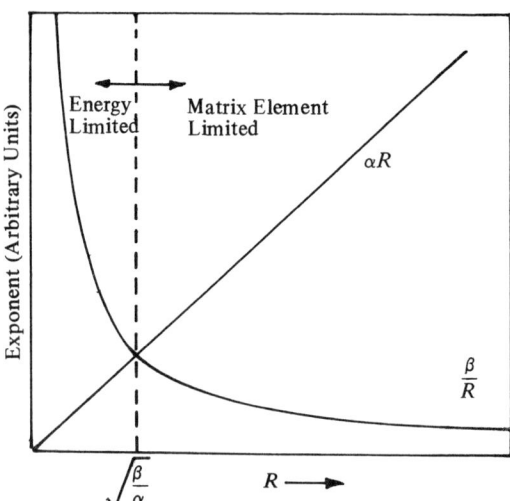

Fig. 19.6 The two terms in the exponent of the impedance (see eq. 19.13 in text) are plotted separately *vs.* R, the average separation between carriers. At high densities, the energy term, β/R, is dominant while at lower densities, the term αR, arising from the matrix element between initial and final states, is dominant. At high densities, processes which lower the excitation energy will begin to contribute. After Srinivasan.[49]

two parts, one termed matrix-element-limited and the other energy-limited. In the matrix-element-limited case the rate is not limited by the necessity to surmount the Coulomb gap but rather conditions are such that the system will maximize $|M|^2$ (or minimize αR). In such a case activated behavior results with a characteristic activation energy E_G.

In the energy-limited region, there may be a net gain in the transition rate if there exists a mechanism whereby transitions, with lowered values of $|M|^2$, can be made to states with an energy less than E_G. In the correlated transition of more than one electron the matrix element contains a product of factors $\exp(-\alpha R)$ for each electron.[50] The energy gained in such a transition over the one-electron transition is of the order of the electron-electron Coulomb energy, $e^2/\kappa R$. Thus the exponent in eq. (19.13) is diminished by an amount, Δ, which is proportional to, and of the order of, $(-\alpha R + \beta/R)$. Such rates will dominate when $\Delta > 0$. The condition for this to hold is that the system under study be in the energy-limited regime in the figure ($R < R_0 = \sqrt{(\beta/\alpha)}$ where the quantity β/R is the larger of the two. This requires that the site density, N, be greater than a critical density, N_c given by

$$N_c = \frac{1}{\frac{4}{3}\pi R_c^3} = \frac{3}{4\pi}\left(\frac{\alpha}{\beta}\right)^{3/2} \tag{19.15}$$

It is seen that, as R is further reduced or T is lowered, the number n of electrons involved in correlated transitions important to transport will increase. This will manifest itself in a continual lowering of the energy involved in the transitions as the temperature is lowered. Thus a temperature dependence not unlike Mott's $T^{-1/4}$ dependence,[42] with a continually decreasing slope on a $1/T$ plot may be expected when this mechanism is operative. Therefore:

1. At high temperatures such that $kT > E_G$ there should be no vestige of the Coulomb gap in the temperature dependence.
2. At moderate temperatures and low site densities such that $kT < E_G$ and $N < N_c$ (resulting in $kT > E_G a/2R$), we expect an activated behavior with a characteristic energy E_G.

Efros and Shklovskii find that the gap is not complete and the density-of-states varies as $(E - E_F)^2$ in the vicinity of the Fermi level.[48] In such a case one would not see an activated, T^{-1}, behavior, but a $T^{-1/2}$ behavior at very low temperatures. However, at such temperatures the correlated multi-electron hops will be important, so that the one-electron density-of-states in the vicinity of the Fermi energy is not relevant.

3. At moderate temperatures and moderate densities such that $kT \simeq E_G a/2R$,

one expects to see an activation energy which continually decreases as the density of localized states increases above the value N_c, but equalling E_G for $N < N_c$.

4. At very low temperatures such that N is appreciably greater than N_c, or $kT < E_G a/2R$, the conductivity should have a temperature dependence in $\alpha \propto T^{-x}$ where $x < 1$. Mott argues[42] that, in fact, the mechanism can lead to a $T^{-1/4}$ temperature dependence.

Schreiber and John[51] have devised one, at least didactically, very valuable model to explain these processes: As the electronic energy drops, the electronic "ocean"—the Fermi "sea"—dries up and decomposes into a series of disconnected "lakes," or electronic "droplets" beginning at a certain energy value, viz., the percolation level. The electrons then are localized within these "droplets." The conductivity is of the Arrhenius form, with an activation energy equal to the difference between the Fermi level and the percolation level. If this difference is appreciable, charge transport occurs by tunnelling between "droplets." At very low temperatures, there occur few hops but over large distances because the "droplets" have large separation distances. At high temperatures, the hops are frequent but cover small distances only, because the "droplets" are closely spaced. If the Fermi level is above the percolation level, the "lakes" or "droplets" may be connected by "channels."

Variable range hopping involves more distant, usually impurity, centers. The transition between the two hopping regimes—nearest neighbor or variable range—is difficult to observe experimentally; for the relatively more straight-forward case of compensated impurity semiconductors containing both electron donor and acceptor impurities, Shklovskii[52] obtains for the transition temperature T_{trans}:

$$\text{For } \kappa \ll 1: \quad T_{trans} \simeq \frac{e^2 N_A^{2/3}}{\epsilon \kappa^{2/3} \alpha} \tag{19.16}$$

$$\text{For } \kappa > 0.5: \quad T_{trans} \simeq \frac{e^2 N_A^{2/3}}{\epsilon \kappa^{2/3} \alpha} \tag{19.17}$$

where $\kappa = N_D/N_A$, a measure of the degree of compensation, i.e., of the extent to which the n-type character of the solid, due to domain impurities, is compensated by acceptor impurities. N_D and N_A represent the impurity concentrations, ϵ the background permittivity of the solid and the delocalization parameter α is the reciprocal of the localization length characterizing the spatial extension of the wave function; a quantity similar to the Debye length $1/\kappa$ in double layer theory. Qualitatively, the outcome of all these theories is that the variable range hopping energy is generally less than that indicated by Arrhenius plots of the mobility[53] and that the hopping lengths are generally quite appreciable.[53] This is illustrated in Figs. 19.7A, B where the quantity n is defined *via* eq. (19.18) for the density of electron states $N(E)$ near the bottom of the (conduction) band.

$$N(E) = \text{constant} \cdot (E - E_A)^n \qquad (19.18)$$

E is the energy and E_A the energy at the band bottom, as illustrated in Fig. 19.7A. The Fermi energy E_F is sufficiently removed from E_A that Boltzmann statistics can be used to describe the occupancy of states. The number of electrons in the band is then given by

$$\int_{E_A}^{\infty} N(E) \exp\{-E - E_F\}/kT\} \, dE \qquad (19.19)$$

The maximum of the integrand occurs at an energy

$$E_m = E_A + nkT \qquad (19.20)$$

The density of states at E_m is then

$$N(E_m) = \frac{N(E_c)}{(\Delta E)^n} (nkT)^n \qquad (19.21)$$

where $N(E_c)$ is the density of states at E_c.

Following Mott,[42] at constant density of states for the degenerate case, the hopping probability around E_m (under the assumption that the energy range of states involved is small) may be written

$$\nu = \nu_{ph} \exp(-2\alpha r) \exp(-w/kT) \qquad (19.22)$$

where r and w are the variable (with T) averaging hopping length and average hopping energy, respectively, α is a tunnelling factor and ν_{ph} a quantity related to the phonon frequency.

The hopping energy, w, is now written[53] as $3/4\pi r^3 N(E_m)$ with $N(E_m)$ given by equation 19.21. Most probable hops then occur when

$$2\alpha = 9(\Delta E)^n/4\pi r^4 N(E_c) n^n (kT)^{n+1} \qquad (19.23)$$

or

$$r = \left(\frac{9(\Delta E)^n}{8\pi \alpha N(E_c) n^n (kT)^{n+1}} \right)^{1/4} \qquad (19.24)$$

The hopping energy is then

$$w = \frac{3}{4} \left(\frac{8}{9} \right)^{1/4} \left(\frac{(kT)^{3-n}(\Delta E)^n \alpha^3}{\pi n^n N(E_c)} \right)^{1/4} \qquad (19.25)$$

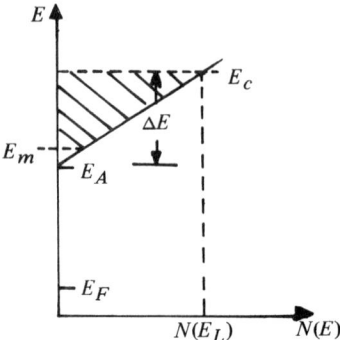

Fig. 19.7A Density of states and energy levels leading to Fig. 19.7B. Localized states are shaded. Conduction occurs in a small energy range near E_m. After Grant and Davis.[53]

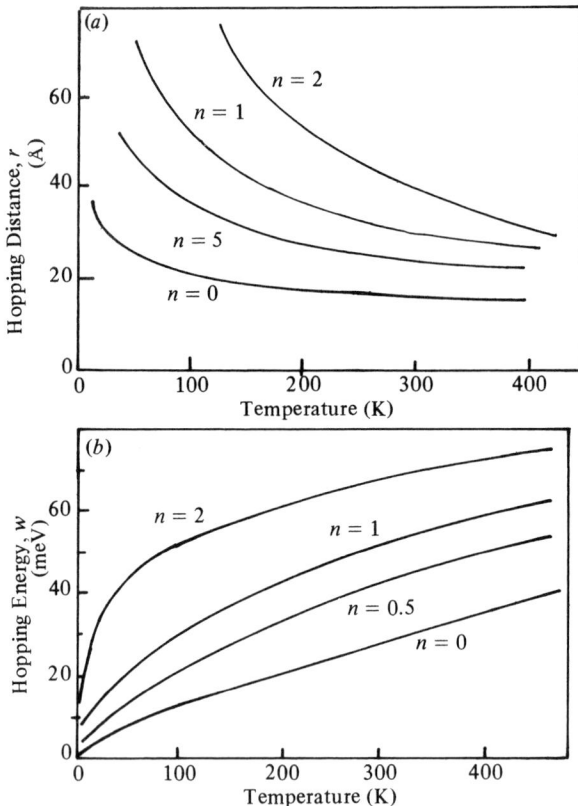

Fig. 19.7B Temperature variation of (a) the hopping length r and (b) the hopping energy w for various values of n $|N(E) \sim E^n|$. After Grant and Davis.[53]

AMORPHOUS SYSTEMS

Assuming the mobility is proportional to p, it will vary with T as

$$\ln(\mu) = A - BT^{-(n+1)/4} \qquad (19.26)$$

where

$$B = 3\left(\frac{8}{9}\right)^{1/4}\left(\frac{(\Delta E)^n \alpha^3}{\pi N(E_c) n^n k^{n+1}}\right)^{1/4} \qquad (19.27)$$

will contain a term $\exp\{-(E_m - E_F)/kT\}$ which represents the carrier concentration activation energy and this temperature dependence will generally dominate over that occurring in the mobility[52] (cf., also eq. 19.13).

For a linear density of states tail, $n = 1$, the above expressions reduce to

$$r = \left(\frac{9\Delta E}{8\pi\alpha N(E_c)(kT)^2}\right)^{1/4} = 0.7736\left(\frac{\Delta E \alpha^{-1}}{N(E_c)(kT)^2}\right)^{1/4} \qquad (19.28)$$

$$w = \frac{3}{4}\left(\frac{8}{9}\right)^{3/4}\left(\frac{\Delta E \alpha^3 (kT)^2}{\pi N(E_c)}\right)^{1/4}$$

$$= 0.5157\left(\frac{\Delta E(kT)^2}{N(E_c)(\alpha^{-1})^3}\right)^{1/4} \qquad (19.29)$$

where

$$B = 2.0628\left(\frac{\Delta E}{N(E_c)(\alpha^{-1})^3 k^2}\right)^{1/4} \qquad (19.30)$$

Thus, using reasonable values of the parameters to estimate the average hopping length r and the average hopping energy w, $\Delta E = 0.3$ eV, $\alpha^{-1} = 10^{-7}$ cm, $N(E_c) = 2 \times 10^{21}$ cm^{-3} eV^{-1}, at $T = 300°$ K

$$r = 20.5 \text{ Å}, \qquad w = 0.05 \text{ eV}$$

The calculated values, particularly that of r, are rather insensitive to the parameters chosen due to the $\frac{1}{4}$ exponent.[53]

The hopping energy seems to be quite small (justifying the assumption that the energy range of states involved around E_m is small) and, in this example, is about 1/6th of the total range of localized states ΔE. It decreases with decreasing temperature. On the other hand the hopping length is quite large and increases with decreasing temperature.[53]

The magnetic and electrical properties flowing from the application of the Hubbard model to a free lattice are discussed in detail by Grensing et al.[54]

The problem of how structural disorder affects electronic properties may be treated[55] thus: Assume that the atoms, or molecules, within a microscopic region can rearrange themselves without changing the energy of the system. The energy then may assume two alternative values, both about equal to the ground state. If the potential barrier separating these two energy minima is not too high, a carrier may tunnel through in a local rearrangement mode, which is called a "softon" because it is associated with a "soft," easily deformable, local chemical structure. It is a function of some local, atomic or molecular, coordinate. A softon may capture a hole or an electron, giving rise to a softaron state, it may even bind two carriers in its vicinity, forming a bisoftaron. While this has been proposed mainly for chalcogenide glasses, it is applicable also to amorphous systems such as organics.

Carrier combination in these circumstances has been studied by Street[55] while Malikov[56] suggests that it is the entropy factor which really dominates the carrier transport in these systems.

In many organic films, such as low density polyethylene, adsorbed water may act as a bridge across localized potential barriers and thus determine carrier transport.[57] Percolation of carriers in mixed insulator-semiconductors systems has been discussed in a review article by Coutts;[58] while Fisch and Harris[59] treat percolation in terms of a random network of discrete resistors.[59] The one-dimensional hopping problem in a linear chain (see also Chapter 21) is discussed by Fedder.[60] The frequency dependence of hopping electron transfer is discussed by Boettger and Brysuin[61] (see also Section 15.5 and 15.7).

The interaction between tunnelling modes of electrons and holes in amorphous systems may produce localized states which in turn lead to an effective electron-electron interaction giving rise to localized carrier pairs and resonance states with a strong effective mass increase[62] (such cooperative excited states are discussed in Section 13.10). Charge transfer theories are discussed at length in Chapter 15, while conductivity in glasses is dealt with in Section 16.9.

Amorphous materials are finding imaginative applications in semiconductor technology as threshold and memory switching devices.[63] Detailed mechanisms of these effects in organochalcogenides are not clear but they appear to be related to the non-rigidity of the network structures and the particular structure of the compound. When light creates a hole in a lone-pair localized state, it produces a singly occupied orbital on the chalcogen. This state has high reactivity which may be satisfied by an atomic rearrangement.[64] Since the structural changes are accompanied by shifts in the absorption edge and in refractive index, these effects have photographic applications. Organic materials used for this purpose are organotellurides of the form R_2TeCl_2 in which the organic portion R, is a ketone such as acetophenone or its derivatives.[65] In keeping with configurational rearrangements of slightly differing energies it has been suggested that the barriers between them are thin enough to allow atoms to tunnel from one local minimum to another.[66]

Such a process has been suggested to[67] explain an extra contribution to the specific heat in glasses at very low temperatures.

19.5 Intramolecular Changes

It has been pointed out in Section 19.1 that the electrical properties of model TCNQ-polymer compounds depend mostly on crystal geometry and less on chain length[68] (see also Section 17.10 and 16.11). Summers and Litt,[68] investigating charge transfer complexes of the electron donors

10-methylphenothiazine,
4-(methylthio)anisole,
poly-[N-[4-10-methyl-3-phenothiazinyl)butyryl]iminoethylene] and
poly[N-[4[4-(methylthio)phenoxy]butyryl]iminoethylene]

with the acceptors dichlorodicyanoquinone, TCNQ, tetranitrofluorenone and tetracyanoethylene, found that the electronic conductivity was higher in the polymers than in the model donor complexes and increased with increasing crystallinity. Thermal activation energies for the complexes ranged from the 0.5-1.8 eV and were roughly the same for the model and the polymer complexes.[68] Electronic spin concentrations appeared to be independent of temperature.[68] Much experimentation continues to involve metallo-organic compounds. An interesting feature concerns the probability of charge transfer complex formation of 8-hydroxyquinoline which has been shown to depend on the species of metal ion utilized.[69] In the case of the metallophthalocyanines (vanadium oxide, Th, Pb, Cu, Zn) derivatives under static high pressure in excess of 200 kbar, the lowest resistivity of 0.1 ohm cm is observed for the vanadyl-centered phthalocyanine, then increasing in order of the above series. The metal-free phthalocyanine has the highest resistivity (see Table 19.1). The electronic performance of metal-free-, Zn- and Ni-phthalocyanine semiconducting electrodes correlates with conductivity which is influenced by ambient factors controlling this parameter.[71] These p-type materials have a high density of intermediate states.[71]

Metallo-organics can accept or donate electrons without affecting the coordination number of the metal as shown by Freeman[72] for poplar plastocyanin, which has a Cu center. This center is embedded deep in the protein and has no contact with the external environment.

As intimated in Section 19.1, semiconductive properties of polymers depend mainly on crystal geometry and not greatly on chain length and must be considered as a substitution phenomenon. $\alpha\omega$-substituted polyenes with $(2n + 2)\pi$ electrons and homologous polyendiones had energy gaps about the same as those for the unsubstituted compounds.[72] However, it must be pointed out that for the 3 homologous series investigated by Zalukaev and

Table 19.1 Minimum Resistivities of Phthalocyanines under Pressure. After Onodera, Kawai and Kobayashi.[70]

Phthalocyanine	Minimum resistivity Ω-cm	Pressure kbar	Deviation Å
VOPc	0.1	290^a	$<2.6^{125}$
ThPc$_2$	0.2	330^a	1.45^{126}
SnPc$_2$	2×10	360^a	1.35^{127}
PbPc	2	490^a	0.4^{128}
CuPc	4	620^b	
ZnPc	1×10	640^b	
H$_2$Pc	4×10	570^b	
CuPcCl$_{16}$	8×10^5	520^a	

Pc: Phthalocyanine, VO: vanadyl. The deviation values refer to the departures of the central metal atom from the central plane of the molecule. a: pressure for resistivity minimum, b: highest pressure applied.

colleagues[74] the energy gap for conductivity increased depending on the number of CH$_2$ groups present.

It was found[75] that TCNQ complexes of saturated linear quaternary ammonium polymers, known as ionenes, containing an odd number of methylene groups, generally had lower resistivities (80-3000 ohm cm) than those with an even number (about 10^4-10^8 ohm cm) see Section 17.10 and Table 8.7C).

In spite of the importance of metallo-organic complexes in biological and synthetic chemical processes scant attention has been paid to the measurement of their electrical properties. Yet results of fundamental interest have been obtained. Thus, in complexes of valerolactones with transition metals, for example, Malikov[76] and coworkers have shown that an alternation of activation energy values exists in the compounds ML$_2$ add M'L$_3$ where ML = 2-acetyl-α-chloro-α-valerolactone, M = VO$_2$, Mn, Fe, Co, Ni, Cu, Zn and M' = Ti and Cr; elements with odd numbers of electrons leading to a higher activation energy than for adjacent elements. The one-dimensional Pt complex,[77] K$_{1.81}$Pt(C$_2$O$_4$)$_2$·2H$_2$O which exhibits a super-structure, is discussed in Chapter 21.

Another interesting result has been observed for Al-polyvinylcarbazole copper.[78] At low field, below 1.05 V, the electrical conductivity was found to be ionic with an activation energy of 0.8 eV whereas at higher temperatures an electronic conductivity was observed with an activation energy of 0.4 eV.

For macromolecules of poly-Schiff bases obtained from *p*- or *m*-phenylenedialdehydes with *p*- or *m*-phenylenediamines only the *p-p* isomer is photoconductive.[79]

In the crystalline polyacetylene films[80] of cis-trans composition, the resistivity and activation energy dropped up to a trans content of 80%. Beyond 80% the resistivity increased while the activation energy remained constant. The electrical properties of polymers are discussed in more detail in Section 16.4 and 16.8. See also Chapter 23.

19.6 Doping and Structure

Baughman and coworkers[81] who investigated cis-polyacetylene doped with I_2 concluded that the complexes consisted of linear polyhalide ion arrays with I_3^{-m} probably substituting in positions of displaced acetylene chains; at large iodine concentrations alternate close-packed layers of polyacetylenes were separated by layers of such polyhalide arrays. The mean limiting intramolecular charge transfer amounted to 0.11 electrons per carbon atom in the polyacetylene chain. Sometimes, as *e.g.*, in the non-stoichiometric complex tetrathiotetracene $I_{3+\delta}$, impurities tend to act as dopants.[82] They destroy correlations between electrons on different chains to suppress 3-dimensional phase transitions, *i.e.*, conversion of the 3-dimensional organic chain conductor to a 1-dimensional conductor, thus enhancing superconductivity fluctuations of the single chain.[82]

Wyhof and Pohl have found hyperelectronic polarization in Ca- and Ni-doped pyropolymers,[83] the latter exhibited a resistivity of 7.7×10^{-6} ohm cm (see Section 16.8 and 16.5). Other aspects of doping are dealt with in Section 16.7 and 16.8.

19.7 Charge Transfer Complexes

The structures of charge transfer complexes relative to their electrical properties have received considerable attention. Tamamura and Yamane[84] were able to show that charge transfer interactions between cations and anions affected crystal structure to a lesser degree than hydrogen bonding or electrostatic attraction. Charge transfer stabilization energies of π-π^* complexes are unlikely to be as large as those of London dispersion, charge-dipole, dipole-dipole, dipole-induced dipole, hydrogen bonding interactions, so that charge transfer interactions should have an effect on donor-acceptor overlap only when these larger forces show little sensitivity to orientation.

Some exceptions should be expected where such stabilization would be maximized by a center-on-center orientation. It is of interest to note that the most frequently found[85] interplanar distance for solid charge transfer complexes is 3.3 Å. In charge transfer complexes which have segregated stacks of organic donor and acceptor molecules the charges are partially localized and may vary from site to site down the stack, forming a modulated charge density, or a charge density wave (see also Chapter 13). In such a crystal, e.g., tetrathiafulvalene-TCNQ[86] attractive interactions exist in some directions and repulsive interactions in other directions, (see Chapter 21). The use of acceptors containing iodine continues to be popular in the characterization of charge transfer complexes. The well-known contact charge transfer complex of benzene with iodine has recently been shown[87] to posses an axial structure. Hadek[88] studied the structure of the iodine complex of N,N'-diphenyl-p-phenylenediamine which has a resistivity of 3.4×10^{-1} ohm cm, and found the donor molecules stacked in layers spaced 3.77 Å apart while their distance in the (010) plane was 3.3 Å apart. The distribution of distances of the iodine atoms in chains perpendicular to the (010) plane followed the sequence...*BAABAA*..., where *A* was 2.99 Å, a little larger than the triiodide ion, and *B* equaled 3.44 Å, almost 1 Å less than the van der Waals distance for I_2. The polyacceptor iodine chains are coordinated to the NH groups from opposite sides of the donor leading to the donor molecules interacting with two acceptor units in two different parts of the skeleton. See Fig. 19.8.

Hadek and Ulbert[90] showed that for charge transfer complexes the electron conduction was governed by the acceptor component, and the hole conductance by the donor component. Hadek[88] concludes that the electron conductivity in the N,N'-diphenyl-p-phenylenediamine:I_2 complex at a higher temperature was dominated by the polyiodide chains, while the low temperature hole conductivity is governed by the relatively closely spaced donor centers.

Purely organic acceptors such as TCNE, substituted benzenes, naphthalene and pyrene form complexes with inert polymer films of polyethylene, poly-

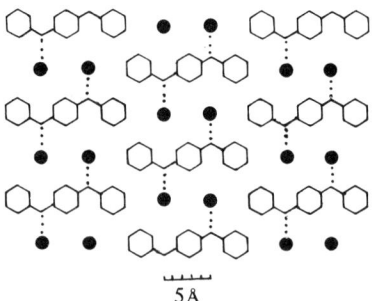

Fig. 19.8 Partial structure of the iodine complex of N,N'-diphenyl-p-phenylenediamine in the (010) plane as determined by Huml.[89] The NH groups between every two benzene rings of the donor molecular are not illustrated: each of the iodine chains is indicated by a full point and donor-acceptor interactions by dotted lines. After Hadek.[88]

methylmethacrylates and polystyrene, apparently forming donor-acceptor sandwiches.[91] Hexamethylbenzene form a 2:1 complex with TCNE exhibiting a stacked configuration in the crystal.[92] TCNQ complexes are discussed in Section 16.11. The relation between crystal structure and metallic conductivity of linear organics is discussed[93] *via* a microdomain model by Philips and explained in terms of strong lateral interactions between the chains (see also Chapter 16). Several reviews of the structure of charge transfer complexes are available;[94] the ionicity and structure of solids have been examined by Huebner and Barkenov.[95]

Several charge transfer complexes of organics with metal compounds have been synthesized. Chlorpromazine hydrochloride was found[96] to complex 1:1 with $CuCl_2$ producing brown, monoclinic single crystals with 4 complex products ($H_2CPZ.CuCl_4$) per unit cell. The $CuCl_4$ ion formed a slightly flattened tetrahedron.[96] Charge transfer complexes involving Cu(II) with thiocarbazones of benzaldehyde, salicaldehyde, biacetyl, benzil, acetylacetone, acetonylacetone, gluteraldehyde, ω-acetylvalderaldehyde, 1,3-indandione and 1,3-nitroindandione in dimethylformamide are reported. The structure and bonding of these complexes are relatively insensitive to the 3d coordination sphere but strongly dependent on the Cu-N bondlength.[97] Metal-ion-bridged CTC too have been reported,[98] *e.g.*, for Cu^{2+} adducts such as Cu-bipyridyl $(ATP)^{--}$. This has a plane-parallel, lamellar structure with the metal forming a bridge between the two aromatic ring systems. Such charge transfer interactions are of considerable importance in biological systems;[99] thus metal-ion-bridged ternary purine-indole adducts[100] as well as some nucleic acid-protein interactions *via* an ionic bridge, have been reported.[101] The stability of the complex may be determined[102] by potentiometric titration of the ternary and of the binary adducts, *i.e.*, with and without *e.g.*, tryptophane; this refers to the system (metal)-(adenosine triphosphate)-(tryptophane). These complexes exist predominantly in a "folded" form.[102] The most likely structure[100] of the adduct is shown in Fig. 19.9.

The stability constants of ternary Cu^{2+} and Zn^{2+} complexes, each of which contains 2,2'-bipyridyl and a carboxymethylaryl sulfide, were determined in 50% aqueous dioxane.[98] A comparison of the stability of these ternary complexes with those formed with simple carboxylates demonstrates an enhanced stability of the carboxymethylaryl sulfide containing mixed ligand complexes. This enhanced stability is due to an intramolecular aromatic stacking interaction between the aryl moiety of the carboxymethylaryl sulfide and 2,2'-bipyridyl.[98]

Depending on the kind of the ternary metal ion complex, the stability enhancement, due to the intramolecular stacking between the aromatic parts of the coordinated ligands, is between about 0.2 to 0.5 log unit.[98] A similarly enhanced stability due to stacking has also been observed, *e.g.*, in ternary

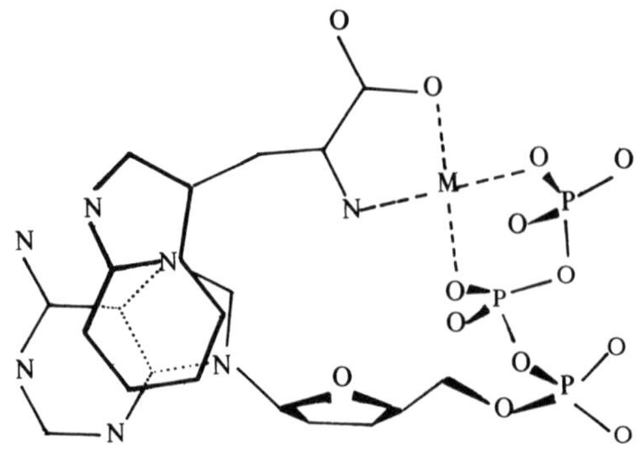

Fig. 19.9 Proposed structure of the metal-adenosine-triphosphate-tryptophane adduct. After Siegel and Naumann.[100]

metal ion-adenosinetriphosphate-phenanthroline complexes.[124] Other atoms may also act as bridging ligands; *e.g.*, iodine may link the ether oxygens in different dioxane molecules forming crystalline iodine-dioxane complexes.[103] For a discussion of ternary adducts, see Section 17.7.

19.8 Pressure

Mechanical stress is a major cause of changes in conductivity in semiconducting organics. Hermann and coworkers[104] stretched polymers of the tetramethylamino-poly-(propyleneglycol) type complexed with TCNQ and report increased conductivities suggestive of stress-induced orientations. The effect of pressure up to 6000 atm on π-π charge transfer complexes of the neutral donors hexamethylbenzene (HMB), benzene, naphthalene, anthracene, and pyrene with the neutral acceptors tetracyanoethylene (TCNE), s-trinitrobenzene, 2,4,6-trinitrochlorobenzene and chloranil caused the stability constants to be increased in every case,[105] accompanied by a red shift of the charge transfer band except for the TCNE-HMB complex for which the shift changes sign above 1000 atm. The observed shift is a balance between the red shift due to decreased separation in the ground state, and a blue shift due to the lowering of the ground state energy by increased resonance between the no-bond and the dative-bond structure, as suggested by Offen and Kadhim.[106] Charge transfer complexes of the cation acceptor 4-methoxy-N-methylpyridinium-iodide with anion donors show a decrease in stability constants and a blue shift of the charge transfer band with increasing pressure,[107] while between ions and molecules (NaI and trinitrobenzene, tropyllium tetrafluoroborate and hexamethylbenzene, potassium pentamethoxycarbonylcyclopentadienylide and trinitrobenzene) little effect on stability constants and charge transfer maxima is observed. This is probably due to changes in solvation of the components with changing pressure. The resistivity of iodanil,[108] 10^{12} ohm cm at 1 kbar drops to 5×10^{-2} ohm cm at 400 kbar; and that for hexaiodobenzene to 5 ohm-cm at 500 kbar. Similar behavior is observed for tetraselenonaphthacene[108] (10^{-1} ohm cm), and tetrathionaphthacene,[110] bromanil and chloranil.[109] Resistivity minima were observed in these cases which were attributed to the presence of heavy atoms, such minima being absent for naphtacene.[111] Moreover, the pressure at which the resistance minima were observed is roughly proportional to the reciprocal of the atomic radius of the heavy atoms.

Resistivity minima were also observed[70] for Co, Fe, Pb and metal-free phthalocyanines under applied pressure over 100 kbar, while none were found for the Cu and Zn phthalocyanines under pressure (see also Section 19.5).

Cu phthalocyanine-$Cl_{1.6}$ is extremely insensitive to pressure, probably due to mutual repulsion of the chloride ions obstructing overlap between π-electrons of neighboring molecules.[70]

Pohl[112,113] and coworkers, working with the semiconducting, p-type, polyacene quinone radical (PAQR) polymers, showed that the conductivity depended on pressure P according to the relationship

$$\ln(\sigma/\sigma_0) = (b^*/k)P^{1/2} \qquad (19.31)$$

in which b^* is an inverse function of the number of fused rings in the monomer.

The conductivity of organic compounds generally increases[129] with increasing pressure. In the case of ferrocene,[136] the conductivity rises with pressure up to 5000 atm, where it reaches a maximum, and then decreases with increasing pressure. The logarithm of the conductivity is proportional to the volume change, ΔV on compression,[130] up to 300 kbars.

Several authors,[131,133,135] have pointed out that the increase can be accounted for by a decrease in the activation energy for charge carrier formation. Batley and Lyons[129] propose a mechanism for this change in activation energy. This mechanism would apply to the initial rise in conductivity in ferrocene but at higher pressures, increasing crystal field strength arising from shortening of the distance between the iron and the cyclopentadienyl rings, could decrease the conductivity, as proposed by Okamato et al.[137] A detailed account of this work is available.[129,138]

The observation[113,136] of Pohl et al., that the logarithm of the conductivity of polymers is proportional to the square root of the pressure in the 0-40 kbar range, is consistent with the abovementioned findings of Samara and Drickamer[130] on molecular crystals, since a plot of Bridgeman's compressibility data[139] against the square root of the pressure is linear in this pressure range, although above 60 kbars the plot begins to deviate significantly from linearity in the case of molecular crystals. Pohl[136] did not calculate changes in the energy gap.

If charge carriers are generated in the bulk of the material, the energy term E in the expression for the current i, namely

$$i = i_0 \exp(-E/2kT) \qquad (19.32)$$

can be evaluated as[129]

$$E = I - A - 2P \qquad (19.33)$$

If conduction is intrinsic then I and A are the ionization energy and electron affinity of a molecule in the gas phase; P is the energy of polarization of a crystal by a single charge on one molecule.[140] Equations (19.32) and (19.33) are still valid if carrier generation in the bulk is extrinsic, provided that carriers of one species are free to move independently of those of the other type. I and A then require different but analogous interpretations.[129]

The decrease in volume of the crystal at high pressures will produce an increase in the polarization energy P and hence a decrease in E. The magnitude of the change in P can be estimated for various simple models.[129]

In the simplest model the ion is treated as a unit charge in a spherical cavity, radius r, in a uniform dielectric of relative permittivity κ. The polarization energy is then given by

$$P = e^2(1 - 1/\kappa)/2r \qquad (19.34)$$

In changing from pressure a to pressure b,

$$\Delta P = P_b - P_a = P_a \left[\frac{r_a}{r_b} \cdot \frac{\kappa_a - \kappa_a/\kappa_b}{\kappa_a - 1} - 1 \right] \qquad (19.35)$$

Using the Mossotti-Clausius relation between the dielectric constant and the density of the solid, it is possible, from the compressibility data of Bridgeman,[139] to determine ΔP in terms of P_a for a given κ_a. While the Mossotti-Clausius relation holds approximately for non-polar liquids, there appear to be no experimental data for non-polar molecular crystals.[141] Camphor[142] and diamond,[143] for which experimental data are available, are different crystal systems from those under consideration in this section. Taking a as atmospheric pressure one obtains P_a as the difference between the gas phase and crystal ionization potentials. P values[144] calculated for $a = 3.5$ are given in Tables 19.2 and 19.3.

Table 19.2 contains results for substances with activation energies measured at high pressures. E for these materials is thus directly determined from experiment. E for the materials in Table 19.3 is evaluated from the change

Table 19.2 Change in the Polarization Energy and the Observed Activation Energy for Conductivity with Pressure. After Batley and Lyons.[129]

Substance	Pressure Range (kbars)	P(exptl.) (eV)	$2\Delta P$(eV)				ΔE(obs)[a] (eV)	Ref.
			$\kappa_a = 3 \cdot 5$	Ion-Dipole only	Ion-Dipole and Dipole-Dipole			
Pentacene	0–100	1·6	0·9	1·7	0·80[d]		0·75	132
	0–160		1·1	2·1	0·90[d]		1·36	134
	0–200	b	1·2	2·2	0·94[d]		1·4	132
Quaterrylene	0–160	1·3[c]	0·9	1·6	1·3		0·45	134
Violanthrone	0–160	1·6[c]	1·1	2·1	1·6		0·58	134
3(p-Phenylenediamine):2(p-chloranil)	0–50	unknown					0·45	132
	0–300[a]						0·49	132
Metal free phthalocyanine	0–50	1·8[c]	0·8	1·3	1·0		0·37	135
Copper phthalocyanine	0–50	1·8[c]	0·8	1·3	1·0		0·13	135

[a] At 100 K
[b] Calc. 1·44
[c] Gas-phase ionization energy estimated from spectroscopic data
[d] ΔP was evaluated by use of the calculated polarization energy

Table 19.3 Change with Pressure in the Polarization Energy and the Activation Energy for Conductivity Obtained from the Observed Change in Conductivity Assuming i_0 Constant. After Batley and Lyons.[129]

Substance	Pressure Range (kbars)	P(exptl.) (eV)	$2\Delta P$(eV)			ΔE^a (eV)	Ref.
			$\kappa_a = 3.5$	Ion-Dipole Only	Ion-Dipole and Dipole-Dipole		
Naphthacene	0–100	1·63	1·0	1·76	0·78[e]	—	130
	100–200		0·23	0·48	0·18[e]	0·33	130
	200–300	[b]	0·03	0·04	0·02[e]	0·06	130
Hexacene	0–100	1·6[c]	1·0	1·7	1·3	0·55	131
Isoviolanthrone	100–300	1·6[c]	0·26	0·54	0·38	0·50	130
DPPH	0–100	unknown				0·72	134
Perylene:tetracyanoethylene complex	0–100	1·50[d]	0·9	1·6	1·2	0·64	132
	100–200		0·2	0·4	0·30	0·10	132
$NNN'N'$-Tetramethyl-p-phenylenediamine:p-chloranil complex	0–280	1·9[d]	1·5	2·7	2·0	0·24	132
$NNN'N'$-Tetramethyl-p-phenylenediamine:p-bromanil complex	0–280	1·9[d]	1·5	2·7	2·0	0·24	132
Perylene:p-chloranil complex	0–12·5	1·5[d]	0·29	0·34	0·32	0·11	145
1,6-Diaminopyrene p-chloranil complex	0–12·5	2·1[d]	0·40	0·47	0·46	0·13	145
	0–100		1·3	2·2	1·8	0·28	132

[a] Assuming constant i_0
[b] Calc. 1·15
[c] Gas-phase ionization energies estimated from spectroscopic data
[d] Assuming that the photoionization threshold of the solid reflects a property of the neutral rather than the charged molecule; Batley, M., unpublished data.
[e] ΔP was evaluated by use of the calculated polarization energy. The other results were calculated by use of the experimental value of 1·63 eV.

in conductivity with pressure. For more detailed discussions of the data in these Tables consult Refs. 129 and 140.

As one might expect of organic solids with pressure dependent electrical properties, upon lowering of pressure a hysteresis of conductivity is often observed. The effect may be due to a permanent change in phase, pressure-induced chemical modification or molecular reorientation, or both as observed[114] for the piezoconduction of the coordination polymers N,N'-dimethyldithiooxamido-Cu(II) and N,N'-dicyclohexyldithiooxamido-Cu(II). Conductivity hysteresis has been noted by Onodera and coworkers[70] for tetraselenonaphthacene and tetrathionaphthacene. Low molecular weight organic substances in high pressure experiments generally exhibit hysteresis in conductivity but not in permittivity upon pressure relaxation.

For a review of the effect of pressure on the electronic properties of organic semiconductors the reader is referred to the work of Jerome and Weger.[115] The methodology of pressure studies is discussed in Section 12.21. The pressure dependence of mobility is treated in Section 15.4.

19.9 Intercalation Compounds

Fischer and Thompson[116] have published a review of graphite intercalation compounds. Very high conductivities can be achieved for these compounds which often utilize strongly acidic acceptors, which presents some difficulties with experimental technique. Vogel[117] has overcome this by swaging e.g., graphite powder intercalated with SbF_5 in a copper tube. After subtracting the contributions to the total electronic conductivity of the graphite and copper this worker[117] found for this intercalation compound a conductivity of 10^6 (ohm cm)$^{-1}$, 40 times that of graphite and more than 1.5 times that of pure copper. Generally, he showed that the conductivity of graphite intercalated with strongly acidic molecules increased in a direction perpendicular to the crystalline c-axis. A large class of new semiconductors has been introduced[118] based on the formation of intercalation complexes of Lewis bases and layered sulfides. For further discussion, see Section 17.7 and Chapter 21.

19.10 Layered Materials and Thin Films

Maschke and Overhof report[119] that stacking disorder is present in many layered materials which leads to the localization of electronic states along the layer normal. The effect of layering on the charge carrier transport has been reviewed by Fivac and Schmid.[120] In another review, Besson treats[121] the effect of high pressure on layered semiconductors. Shirotani et al. have investigated[146] the effect of pressure on the absorption spectra of oriented and amorphous organic films. Kamura et al.[147] discovered that the electronic spectra of perylene and coronene evaporated films were functions of crystallinity in their compounds.

Thin films present themselves as natural candidates for switching phenomena. Sadaoka and Sakai[122] have reported such behavior in naphthacene thin films. Switching effects have been observed in which field-induced bistable structures alter molecular dimensions. Sadaoka and coworkers[123] have demonstrated the application of metal (*e.g.*, Cu) phthalocyanine thin films as sensors for gases such as NO_2, NO, SO_2, O_2, N_2, and CO_2 (see Section 12.18). The topic of thin films is further discussed in Section 12.22.

References

1. H. A. Pohl, C. J. Norrell, M. Thomas and K. D. Berlin, "Anthrone Polymers with Semiconducting Properties," Research Note 77, Oklahoma State University, Stillwater, Sept., 1978.
2. M. Pollak and H. A. Pohl, *J. Chem. Phys.*, **63**, 2980 (1975).
3. H. A. Pohl and M. Pollak, *J. Chem. Phys.*, **66**, 4031 (1977).
4. H. Sigel and D. B. McCormick, *J. Am. Chem. Soc.*, **93**, 2041 (1971).
5. V. Hadek, H. Noguchi and A. Rembaum, *Macromolecules*, **4**, 494 (1971).
6. J. R. Wyhof, Ph.D. Thesis, Oklahoma State Univ. (1970).
7. A. Howie, C. Lima and H. A. Pohl, "The Nature of Hyperelectronic Polarization," *J. Biol. Phys.*, **2**, 113 (1974).
8. J. Prochorow and A. Tramer, *J. Chem. Phys.*, **47**, 775 (1967).
9. M. Hatano, T. Enomoto, I. Ito and M. Yoneyama, *Bull. Chem. Soc. Japan*, **46**, (12); 3698 (1973).
10. M. Nagao and A. M. Herman, *Polymer Letters*, **12**, 69 (1974).
11. R. Somoano, S. P. S. Yen and A. Rembaum, *Polymer Letters*, **8**, 467 (1970).
12. I. Vilfau, *Lect. Notes Phys.*, **65**, 629 (1977).
13. T. Takahashi, S. Tanase and O. Yamamoto, *Solid State Commun.*, **26**, (11), 705 (1978).
14. Y. Aikawa et al., *Mol. Cryst. Liq. Cryst.*, **36**, 235 (1976).
15. D. Guidotti, P. M. Horn and E. M. Langler, *Lect. Notes Phys.*, **65**, 469 (1977).
16. G. Mihaly et al., *Lect. Notes Phys.*, **65**, 553 (1977).
17. D. H. Spielberg et al., *J. Chem. Phys.*, **54**, 2597 (1971).
18. K. Lochner et al., *Chem. Phys. Lett.*, **41** (2), 388 (1976).
19. H. Dvey-Akharon, T. J. Sluckin, A. J. Hopfinger and P. L. Taylor, *Bull. Amer. Phys. Soc.*, **24**, 415 (1979).
20. T. T. Wang, *Bull. Amer. Phys. Soc.*, **24**, 415 (1979).
21. P. Herchenroeder, D. Naegele and D. Y. Yoon, *Bull. Amer. Phys. Soc.*, **24**, 415 (1979).
22. J.-P. Farges, H. Grassi, A. Brau and P. Dupuis, *Phys. Stat. Solidi*, (a), **55**, 89 (1979).
23. J. Swiatkiewicz and K. Pigon, *Acta. Phys. Polon.*, A-**53**, (2) 165 (1978).
24. J. Tauc, *Physics Today*, p. 23, Oct. 1976; M. H. Cohen, *J. Non-Cryst. Solids*, **4**, 395 (1970); *Electronic and Structural Properties of Amorphous Semiconductors*, P. G. LeComber and J. Marty, eds., Academic Press, London, 1973; P. W. Anderson,

Nobel Address, *Rev. Mod. Phys.*, **50**, 191 (1978); N. F. Mott, Nobel Address, *ibid.*, p. 203; E. A. Davis, *Endeavour*, **30**, 55 (1971); M. Saitoh, *Progr. Theoret. Phys.*, **45**, 746 (1971); M. H. Cohen, *Canad. J. Chem.*, **55**, (11), 1906 (1977); G. Pfister, *Phys. Semiconductor Proc., 13th. Int. Conf., (1976)* p. 537, J. G. Fuoni, ed., Nth. Holland, Amsterdam, 1976; Several Papers in "Amorphous and Liquid Semiconductor. Proc. 7th. Conf., 1979, W. E. Spear, ed., U. of Edinburgh 1977; M. H. Cohen, *Physics Today*, May (1971) p. 26; *Amorphous and Liquid Semiconductors*, J. Tauc, ed., Plenum Press, N.Y., 1974; J. P. Suchet, *Electrical Conductivity in Solid Materials*, Pergamon, Oxford, 1975; *Amorphous and Liquid Semiconductors*, J. Stuke and W. Brenig, eds., Taylor and Francis, London, 1974; *Proc. 12th Internatl. Conf. Phys. Semicond.*, Teubner, Stuttgart, W. Germany, 1974; J. Hoshino and M. Watanabe, *J. Phys.*, **C-11**, (7) 1381 (1978); A. E. Owen and W. E. Spear, *Phys. Chem. Glasses*, **17**, 174 (1976); J. Klafter and J. Jortner, *Chem. Phys.*, **26**, 421 (1977); R. C. Albers and J. E. Gubernatis, *Phys. Rev.*, **B-17**, 448 (1978).

25. G. Mort et al., *Solid State Commun.*, **18**, (6) 693 (1976).

26. N. Karl, *Advances in Solid State Physics*, **14**, 261 (1974); *e.g.*, coronene at room temperature has practically no disorder (less than 2%), and can be regarded as an almost perfect organic semiconductor–G. S. Pawley and T. Rayment, *J. Phys. Chem. Solids*, **40**, 715 (1979).

27. C. T. White and K. L. Ngai, *Bull. Amer. Phys. Soc.*, **24**, 307 (1979).

28. M. H. Cohen, *J. Non-Crystalline Solids*, **4**, 396 (1970).

29. M. H. Cohen, *Physics Today*, May 1971, p. 26.

30. J. Tauc, *Physics Today*, October 1976, p. 23.

31. R. W. Griffith, *J. Non-Cryst. Solids*, **24**, (3) 413 (1977).

32. H. Scher and M. Lax, *Phys. Rev.*, **B-7**, 4491 (1974).

33. N. F. Mott, *USP. Fiz. Nauk*, **127** (1), 41 (1979).

34. N. F. Mott, *Phil. Mag.*, **17**, 1259 (1968); **22**, 1 (1970).

35. B. I. Halperin and M. Lax, *Phys. Rev.*, **148**, 722 (1966); M. H. Cohen, *Physics Today*, **24**, 26 (1971); R. Zallen and H. Scher, *Phys. Rev.*, **B4**, 4471 (1971).

36. E. A. Davis in *Conduction in Low-Mobility Materials*, N. Kein, D. S. Tannhauser, and M. Pollak, eds., Taylor and Francis, London, 1971.

37. P. G. LeComber and W. E. Spear, *Phys. Rev. Lett.*, **25**, 509 (1970).

38. M. Silver, *Bull. Amer. Phys. Soc.*, **24**, 308 (1979).

39. H. Baessler, *J. Chem. Phys.*, **49**, 5198 (1968).

40. G. Pfister, *Phys. Rev.*, **B16**, 3676 (1977).

41. J. P. Straley, *J. Phys.*, **C-10**, (11) 1903 (1977); R. M. Hill, *Thin Solid Films*, **51** (2), 133 (1978).

42. N. F. Mott, *Phil. Mag.*, **19**, 835 (1969); **34**, 643 (1976); *Physics Today*, Nov. (1977); A. J. Grant and E. A. Davis, *Solid State Commun.*, **15**, 563 (1974); B. I. Shklovskii, *Soviet Phys. Semicond.*, **6**, 1053 (1973); A. L. Epros and B. I. Schklovskii, *J. Phys.*, **C-9**, 149 (1975); M. Pollak, *J. Non-Cryst. Solids*, **8-10**, 486 (1972); **11**, 1 (1972); B. I. Shklovskii et al., *Phys. Stat. Solidi*, **50**, 451 (1972); H. Schreiber and W. John, *Phys. Stat. Solidi*, **B-78**, (1) 991 (1976).

43. K. L. Ngai and P. C. Taylor, *Phil. Mag.*, **B-37**, (2), 175 (1978).

44. G. Pfister, in *Physics Semicond. Proc. 13th Int. Conf., 1976*, J. G. Fumi, ed., Nth. Holland, Amsterdam, (1976), p. 537.

45. L. S. Gagara et al., *Depos. Doc.*, **1975**, Viniti 1529.

46. B. Pistoulet et al., *J. Phys. (Paris) Colloqu.*, **1977** (7) 207.

47. *e.g.*, N. F. Mott et al., *Proc. Royal Soc.*, **A-345**, 169 (1975); P. Thorua and D. Wuertz, *Phys. Stat. Solidi*, **B-86**, (2), 541 (1978).

48. A. L. Efros and B. I. Shklovskii, *J. Phys.*, **C-8**, 149 (1975).
49. M. Pollak, *Disc. Faraday Soc.*, **50**, 13 (1970); *Proc. Royal Soc.*, **A-325**, 383 (1970); M. L. Knotek and M. Pollak, *J. Non-Cryst. Solids*, **8-10**, 505 (1972); *Phys. Rev.*, **B-9**, 664 (1974); G. Srinivasan, *Phys. Rev.*, **B-4**, 2581, (1971); also ref. 47.
50. M. L. Knotek and M. Pollak, *Phys. Mag.*, **35**, 113 (1977).
51. J. Schreiber and W. John, *Phys. Stat. Solidi*, **B-78**, 9, 199 (1976).
52. B. I. Shklovskii, *Solid Phys. Semicond.*, **6**, 1053 (1973).
53. A. J. Grant and E. A. Davis, *Solid State Commun.*, **15**, 563 (1979); N. F. Mott, *Phil. Mag.*, **19**, 835 (1969); M. Pollak, *J. Non-Cryst. Solids*, **11**, 1 (1972); M. Pollak et al., *Phys. Rev. Lett.*, **30**, 853, 856 (1972); J. J. Hauser and A. Staudinger, *Phys. Rev.*, **B-8**, 607 (1973).
54. D. Grensing et al., *Phys. Rev.*, **B-17**, 2221 (1978).
55. R. A. Street, *Phys. Rev.*, **B-17**, (10) 3984 (1978).
56. B. F. Malikov et al., *Dokl. Akad. Nauk. SSSR*, **241**, (4) 856 (1978).
57. R. P. Dupey and N. K. Chaudhuri, "Proc. Nucl. Phys. Solid State Phys. Symp.," 1976, 19C, p. 14.
58. T. J. Coutts, *Thin Solid Films*, **38**, 313 (1976).
59. K. Fisch and A. B. Harris, *Phys. Rev.*, **B-18**, 416 (1978).
60. P. A. Fedder, *Phys. Rev.*, **B-18**, 45 (1978).
61. H. Boettger and V. V. Bryskin, in *Amorphous Semiconductors*, "Proc. Internatl. Conf." 1976, Publ. 1977, p. 71, I. K. Somogyi, ed., Akad. Kiado, Budapest (1977).
62. E. N. Economou, *Amorph. Liquid Semicond. Proc. 7th. Int. Conf. 1977*, W. E. Spear, ed., Cent. Indust. Consultancy Liaison, Edin., Scotland 1978, p. 296.
63. S. R. Ovshinsky and H. Fritzsche, *IEEE Trans.*, **ED-20**, 91 (1973).
64. S. R. Ovshinsky and K. Sapru, in *Amorphous and Liquid Semiconductors*, J. Stuke and W. Brenig, eds., Taylor and Francis, London, 1974.
65. H. Fritzsche, *J. Japan Soc. Appl. Phys.*, **43** (Suppl.), 32 (1974).
66. P. W. Anderson, B. I. Halperin and C. M. Varma, *Phil. Mag.*, **25**, 1 (1972); W. A. Phillips, *J. Low Temp. Phys.*, **7**, 351 (1972).
67. R. C. Zeller and R. D. Pohl, *Phys. Rev.*, **B4**, 2029 (1971).
68. J. W. Summers and M. H. Litt, *J. Polym. Sci., Polym. Chem. Ed.*, **11**, (6), 1379 (1973).
69. S. Koizumi and Y. Iida, *Bull. Chem. Soc. Japan*, **44**, 1436 (1973).
70. A. Onodera, N. Kawai and T. Kobayashi, *Solid State Commun.*, **17**, 775 (1975).
71. F. Ren Fan and L. R. Faulkner, *J. Am. Chem. Soc.*, **101** (17), 4779 (1979).
72. H. C. Freeman, *J. Proc. Royal Soc. N.S.W.*, **112**, 60 (1979).
73. N. Tyutyulkov, J. Petkov, O. E. Polansky and J. Fabian, *Theoret. Chim. Acta*, Berlin, **38**, 1 (1975).
74. P. Zalukaev, *Dokl. Akad. Nauk. SSSR*, **233** (1), 117 (1977).
75. A. Rembaum, W. Baumgartner and A. Eisenberg, *Polymer. Lett.*, **6**, 159 (1968).
76. B. F. Malikov et al., *Dokl. Akad. Nauk, SSSR*, **241** (4), 856 (1978).
77. A. Kobayashi, Y. Saraki, I. Shirotani and H. Kobayashi, *Solid State Commun.*, **26**, (10) 653 (1978).
78. Sung-Il Lee, *Chosen Shogakkai Gakujutsu Rembunshu*, **6**, 137 (1976). (C. A. **88**, 43979u (1978).
79. N. A. Vasilensko, R. N. Nurmukhametov, V. M. Vozzhennikov and A. N. Pravednikov, *Vysokomol Soedin, Ser. A.*, **19**, (9), 1982 (1977).
80. H. Shirakawa et al., *Makromolec. Chem.*, **179**, (6) 1565 (1978).
81. R. H. Baughman et al., *J. Chem. Phys.*, **65**, (12) 5405 (1978).
82. E. Abrahmas, L. P. Gor'kov and G. A. Kharadze, *J. Low Temp. Phys.*, **32**, 637 (1978).

REFERENCES

83. J. R. Wyhof and H. A. Pohl, *J. Polymer. Sci.*, A-2 **8**, 1741 (1970).
84. J. Tamamura and T. Yamane, *Bull. Chem. Soc. Japan*, **47**, (4), 832 (1974).
85. J. C. A. Boeyens and I. H. Herbstein, *J. Phys. Chem.*, **69**, 2160 (1965).
86. J. H. Torrance and B. D. Silverman, *Phys. Rev. B*, **15**, (2) 788 (1977).
87. L. Fredin and B. Nelander, *J. Am. Chem. Soc.*, **196**, 1672 (1974).
88. V. Hadek, *J. Chem. Phys.*, **49**, 5202 (1968).
89. J. Huml, *Acta Cryst.*, **22**, 29 (1967).
90. V. Hadek and K. Ulbert, *Rev. Sci. Instr.*, **38**, 991 (1967).
91. M. J. Mobley, K. I. Rieckhoff and E. M. Voigt, *J. Phys. Chem.*, **82**, (18) 2005 (1978).
92. B. Hall and J. P. Devlin, *J. Phys. Chem.*, **71**, 465 (1976).
93. J. C. Philips, *J. Solid State Chem.*, **19**, (3) 309 (1976); *Proc. Natl. Acad. Sci. USA*, **73**, (11) 3820; *Phys. Stat. Solidi*, **B-78**, (1) 371 (1976).
94. e.g., *Proc. Conf. Org. Conductors and Semiconductors*, Siofok, Hungary, Springer, Berlin, 1977; Z. G. Soos, *J. Chem. Ed.*, **55**, (9), 546 (1978); Z. Soos and D. J. Klein, in *Molecular Association*, **1**, R. Forster, ed., Academic Press, N.Y., 1975; F. H. Herbstein, *Perspectives in Struct. Chem.*, **4**, D. Dunitz and J. Z. Ibers, eds., Wiley, N.Y., 1971; *Lecture Notes Physics*, **65**, (1975).
95. K. Huebner and V. K. Barkenov, *Phys. Stat. Solidi*, **B-77**, (2), 473 (1976).
96. J. Harris and S. Gabay, 2nd Phenothiazine Conference, Paper 118, Cham. Dept., Arizona State University.
97. G. K. Budniko et al., Sb. Nekot, Probl. Org. Khim. Mater. Nauk. Sess. Inst. Org. Fiz. Khim. Akad. Nauk. SSSR (1972), p. 229.
98. H. Siegel et al., *Europ. J. Biochem.*, **41**, 209 (1979).
99. *Metal Ions in Biological Systems*, H. Siegel, ed., Marcel Dekker, N.Y., 1974.
100. C. F. Naumann and H. Siegel, *J. Am. Chem. Soc.*, **96**, 2750 (1974); H. Siegel and C. F. Naumann, **98**, 730 (1976); C. F. Naumann et al., *Europ. J. Biochem.*, **41**, 209 (1974).
101. H. Siegel, *J. Am. Chem. Soc.*, **97**, 3209 (1975).
102. P. Chaudhori and H. Siegel, *J. Am. Chem. Soc.*, **98**, 3142 (1977).
103. D. Hassel and C. Romming, *Quarterly Rev.* (London), **16**, 1 (1962).
104. A. M. Hermann, S. P. S. Yen, A. Rembaum and R. F. Landel, *Polymer Letters*, **9**, 627 (1971).
105. A. H. Ewald, *Trans. Faraday Soc.*, **64**, (3), 733 (1968).
106. H. W. Offen and A. H. Kadhim, *J. Chem. Phys.*, **45**, 269 (1966).
107. A. H. Ewald and J. A. Scudder, *H. Phys. Chem.*, **76**, (2) 249 (1972).
108. A. Onodera, I. Shirotani, H. Inokuchi and N. Kawaki, *Chem. Phys. Lett.*, **25**, (2), 296 (1974).
109. I. Shirotani, A. Onodera, Y. Kamura, H. Inokuchi and N. Kawai, *J. Solid State Chem.*, **18**, 235 (1976).
110. I. Shirotani, H. Inokuchi and S. Minomura, *Bull. Chem. Soc. Japan*, **39**, 386 (1966); I. Shirotani, K. Kawada and H. Inokuchi, **43**, 2381 (1970).
111. G. A. Samara and H. G. Drickamer, *J. Chem. Phys.*, **37**, 474 (1962).
112. J. H. T. Kho and H. A. Pohl, *J. Polymer Sci. Part A-1*, **7**, 139 (1969).
113. K. Saha, S. C. Abbi and H. A. Pohl, *J. Non-Crystalline Solids*, **21**, 117 (1976).
114. R. D. Hartman, S. Kanda and H. A. Pohl, *Proc. Okla. Acad. Sci.*, **47**, 246 (1968).
115. D. Jerome and M. Weger, *NATO Adv. Study Inst. Ser. B*, **B25**, 341 (1976, publ. 1977).
116. J. E. Fischer and T. E. Thompson, *Physics Today*, **31**, (7) 36 (1978).
117. F. L. Vogel, *J. Mater. Sci.*, **12** (5) 982 (1977).

118. F. R. Gamble et al., *Science,* **174**, 493 (1971).
119. K. Maschke and H. Overhof, *Phys. Rev.,* **B-15**, (4) 2058 (1977).
120. R. C. Fivac and P. E. Schmid, *Phys. Chem. Mater. Layered Structure,* **4**, 343 (1976).
121. J. M. Besson, *Nuovo Cimento Soc. Ital. Fiz.,* **38-B** (2) 478 (1977).
122. Y. Sadaoka and Y. Sakai, *Bull. Chem. Soc. Japan,* **51**, (1) (1978).
123. Y. Sadaoka et al., *Denki Kagaku Oyobi Kogyo Butsori Kagaku,* **46**, 597 (1978).
124. H. Siegel and C. F. Naumann, *J. Am. Chem. Soc.,* **98**, 737 (1976).
125. J. M. Assour, J. Goldmacher and S. E. Harrison, *J. Chem. Phys.,* **43**, 159 (1965).
126. T. Kobayashi and N. Uyeda, "Abst. Proc. Sympos. Struct. of Molecules," Sendai, Japan **119**, (1972).
127. W. E. Bennet, J. E. Broberg and N. C. Baenziger, *Inorg. Chem.,* **12**, 930 (1973).
128. K. Ukei, *Acta Cryst.,* **B29**, 2290 (1973).
129. M. Batley and L. E. Lyons, *Aust. J. Chem.,* **19**, 345 (1966).
130. G. A. Samara, and H. G. Drickamer, *J. Chem. Phys.,* **37**, 474 (1962).
131. R. G. Aust, W. H. Bentley and H. G. Drickamer, *J. Chem. Phys.,* **41**, 1856 (1964).
132. W. H. Bentley and H. G. Drickamer, *J. Chem. Phys.,* **42**, 1573 (1965).
133. Y. Harada, Y. Maruyama, I. Shirotani, and H. Inokuchi, *Bull. Chem. Soc. Japan,* **37**, 1378 (1964).
134. H. Inokuchi, I. Shirotani, and S. Minomura, *Bull. Chem. Soc. Japan,* **37**, 1234 (1964).
135. R. S. Bradley, J. D. Grace and D. C. Munro, *Trans. Faraday Soc.,* **58**, 776 (1962).
136. H. A. Pohl, A. Rembaum and A. Henry, *J. Am. Chem. Soc.,* **84**, 2699 (1962).
137. Y. Okamato, J. Y. Chang and M. A. Rantor, *J. Chem. Phys.,* **41**, 4010, (1964).
138. L. E. Lyons, "Intrinsic and Extrinsic Conductivity of Organic Solids," Third International Organic Crystal Symposium, Univ. Chicago, May, 1965.
139. P. W. Bridgman, *Proc. Am. Acad. Arts. Sci.,* **76**, 9 71 (1945).
140. L. E. Lyons and J. C. Mackie, *Proc. Chem. Soc.,* **71**, (1962).
141. S. D. Mamann, *Physico-Chemical Effects of Pressure,* Butterworth, London, 1957.
142. D. R. Wheeler and J. R. Green, *Bull. Am. Phys. Soc.,* **9**, 742 (1964).
143. D. F. Gibbs and G. J. Hill, *Phil. Mag.,* **9**, 367 (1964).
144. H. Kronberger and J. J. Weiss, *J. Chem. Soc.,* 464 (1944).
145. M. Schwarz, H. W. Davies and B. J. Dobrianski, *J. Chem. Phys.,* **40**, 3257 (1964).
146. Shirotani et al., *Mol. Cryst. Liquid Cryst.,* **28**, 345 (1975).
147. Y. Kamura et al., *Bull. Chem. Soc. Japan,* **49** (2), 418 (1976).

20

Space Charge Effects

20.1 Space Charge Limited Currents

Experiments involving space charge limitation of currents in insulators often fail to obey the scaling law:

$$d_{\text{eff}} = [1 - (j_0/j)^{1/(2l+1)}]d. \tag{20.1}$$

where j is the current density, d is the insulator thickness and V is voltage. Dresner[1] found $j \propto V^5/d^7$ at low fields, and $j \propto V^3, V^2$ or V^4 at high fields for anthracene. Williams and coworkers[2] found $j \propto V$ at low fields; and $j \propto V^3/d^4$ at high fields, with intermediate $j \propto V^n$ ($6 < n < 12$) in some cases. To reconcile experimental data with theory, several workers[3] calculated transient photocurrents in insulators with trapped space charges, while Meaudre and Mesnard[4] considered transient currents with two types of carriers and two blocking electrodes.

If diffusion is also considered, the theory becomes mathematically difficult due to the non-linearity of the differential equations for the electric field as a function of distances for the contact; the resulting integrals are often divergent.[5] Thus, while the "classical" treatment of space charge limited currents (SCLC), (see Section 10.2), neglects diffusion effects, several authors have examined the effects of spatial variation of trapping parameters: Nicolet et al.[6] consider an insulator with a shallow trap distribution uniformly distributed in a narrow region near the injecting electrode, and with a trap-free interior. Delannoy, Schott and Berrehar[7] discuss a similar model, but with a finite trap density especially in tetracene. Sworakowski[8] gives a general formulation for an arbitrary spatial variation of trap-density for shallow and exponential traps. He obtains results for the case where the trap density decreases away from the surface as $\exp(-x/r)$ or as $[1 + \exp\{(x-a)/r\}]^{-1}$ where r and a are parameters. The latter case resembles the step-function model employed by Nicolet et al.[6] and by Delannoy, Schott and Berrehar.[7]

Hwang and Kao give formulations[9] identical with Sworakowski's[8] for shallow and exponential traps, and also consider an energetically uniform distribution, as well as the case of a trap density decreasing exponentially with distance from one or both surfaces. They apply a similar treatment to a Gaussian (in energy) trap distribution[10] for the special cases[11] where it behaves as a shallow or exponential trap distribution, and also enact a Gaussian spatial dependence of trap density.

The main results of all these treatments which neglect diffusion are:[12]

1. A high trap density near the back electrode (*i.e.*, the negative electrode for hole injection) has a negligible effect on the current.
2. A high trap density near the injecting electrode lowers the current at a given voltage, but does not alter the shape of the current-voltage curve.
3. The reduction in current can be described in terms of an increased effective thickness of the insulator. This is sometimes convenient but, as Nicolet[6] has shown, it is more appropriate to consider the current as SCLC through the region of high trap density near the injecting electrode, driven by an effective voltage which is lower than the applied voltage by the ratio of the thickness of the high trap-density region to the total thickness.

When the trapping parameters are spatially uniform, the diffusion-free SCLC theory only applies at rather high voltages[13] (above about 5 V for shallow traps and somewhat higher for an exponential distribution).

Rozental and Paritskii[14] have included diffusion in a study of shallow traps with an exponential spatial distribution. Their study was entirely numerical—done with an analog computer—and generalization from their work is not possible.

The following discussion will be based upon a detailed treatment of the problem by Bonham,[12,15] and Bonham and Jarvis.[13,16] The concentrations of (electron) donors, acceptors and of any intrinsically present carriers are taken as negligible relative to the concentration of injected holes. The positive contact is taken as ideal while the other contact is assumed to be blocking, *i.e.*, its charge density is finite.

An exponential trap distribution is defined by

SPACE CHARGE LIMITED CURRENTS

$$h(E_t) = (H/kT_c)\exp(-E_t/kT_c); \qquad T_c/T = l > 1 \qquad (20.2)$$

where $h(E_t)$ is the trap density per unit energy range at a depth E_t from the band, and H, T_c are parameters. The free charge density ρ_f is related to the trapped charge density $\rho(\rho \gg \rho_f)$ by[12,15]

$$\rho_f = eN_0 [\sin(\pi/l)/eH(\pi/l)]^l \rho^l \qquad (20.3)$$

where e is the electronic charge and N_0 the effective density of states in the band. It is convenient to define the following set of dimensionless variables:

$$i = (e/kT_c)^{l+1} \frac{d^{2l+1}}{\mu e N_0} \left[\frac{eH\pi/l}{\epsilon \sin(\pi/l)}\right]^l j, \qquad \gamma = (ed^2/\epsilon kT_c)\rho,$$

$$E = (ed/kT_c)F, \qquad u = (e/kT_c)U, \qquad s = x/d \qquad (20.5)$$

where d is the sample thickness, μ the charge carrier mobility, ϵ the insulator's permittivity, j the current density, F the electric field, and U is the applied voltage. The transport and Poisson equations then become

$$i = E\gamma^l - \gamma^{l-1}d\gamma/ds,$$
$$dE/ds = \gamma \qquad (20.5)$$

which can be combined to give

$$dy/df = f - y^{-l} \qquad (20.6)$$

where $y = i^{-2/(2l+1)}\gamma$ and $f = i^{-1/(2l+1)}E$.

For a particular solution to eq. (20.6) a dimensionless position variable is obtained:

$$z = i^{1/(2l+1)}s = \int_{-\infty}^{f} y^{-1} df \qquad (20.7)$$

whence

$$i^{1/(2l+1)} = \int_{-\infty}^{f_2} y^{-1} df \qquad (20.8)$$

where $f = f_2$ at $s = 1$.

For a finite $\gamma(0)$ it is convenient to measure the dimensionless potential, ψ, relative to a reference $\psi_1 = \ln(y_1/2\pi^2)$ where y_1 corresponds to $\gamma(0)$. Then, taking the limit as $y_1 \to \infty$ and using eq. (20.5) to substitute for E:

$$\psi - \psi_1 = -\int_0^s E\, ds - \ln(y_1/2\pi)^2 = -\int_{-\infty}^{f} y^{-(l+1)} df - \ln(y/2\pi^2) \qquad (20.9)$$

The dimensionless Fermi level, ψ_f is defined by $i = \gamma^l \, d\psi_f/ds$ whence

$$\gamma \propto \exp[-(\psi - \psi_f)]$$

and

$$u = \int_0^1 \psi_f \, ds = \int_{-\infty}^{f_2} y^{-(l+1)} \, df \tag{20.10}$$

Integration of eq. (20.6) gives

$$y - \frac{1}{2}f^2 - c = \int_{-\infty}^{f} y^{-l} \, df \tag{20.11}$$

from which c can be determined numerically. Some values are listed in Table 20.1. Applying the boundary condition $E(0) = -\infty$ for the ohmic electrode, gives the charge distribution near the electrode as

$$y = c \sec^2[(c/2)^{1/2} z - \pi/2] \tag{20.12}$$

Expressions for E and ψ can readily be obtained, for finite as well as infinite y_1. Kalinowski and Godlewski[17] applied to experimental results a zero-current equation which is equivalent, for an ohmic contact, to $y = f^2/2$, yielding

$$y = 2z^{-2} \tag{20.13}$$

Neglecting the diffusion term dy/df yields

$$y = f^{-1/l} \tag{20.14}$$

indicating that this theory reduces to the conventional SCLC treatment for $f \geqslant 3$. Only in this case can a closed solution be obtained for the integrals in eqs. (20.8) and (20.10), also requiring the additional restriction that $y_2 < f_2^{-1/l}$. The following current vs. voltage relations then are obtained:[12,15]

Table 20.1 Values of Constants. After Bonham.[15]

l	c	$i_0^{1/(2l+1)}$	u_0
1	2.87	2.75	3.83
1.5	2.17	2.34	2.43
2	1.88	2.19	1.80
4	1.48	2.05	0.91
7	1.30	2.05	0.54

$$\Delta\psi_F = \ln(f_2^{-1/l}/y_2)$$

$$i^{1/(2l+1)} = i_0^{1/(2l+1)} + [l/(l+1)]f_2^{(l+1)/l} \quad (20.15)$$

$$u = u_0 + \left[\frac{l}{2l+1}\right]f_2^{(2l+1)/l} + \Delta\psi_F$$

In terms of dimensioned equivalents j_0, U_0 and E_F for the dimensionless quantities, i_0, u_0 and $\Delta\psi_F$ there follows:

$$\left[j^{\frac{1}{2l+1}} - j_0^{\frac{1}{2l+1}}\right]^{(2l+1)} = \mu e N_0 \left[\frac{l e \sin(\pi/l)}{eH\pi/l}\right]^l \frac{l(2l+1)(V - V_0 - \Delta E_F)}{[(l+1)d]^{2l+1}}$$

Having allowed for diffusion results in correction terms appearing in the current and voltage terms of eq. (10.12), however, the current correction may also be incorporated in d by defining an effective thickness

$$d_{\text{eff}} = [1 - (j_0/j)^{1/(2l+1)}]d \quad (20.17)$$

The main source of error in neglecting diffusion is the incorrect placement of the potential maximum at $s = 0$. In Fig. 10.2 it can be seen that the Fermi level drops only slightly before the potential maximum (i.e., most of the voltage drop occurs across the remainder). Thereafter, ψ_F remains almost parallel to ψ so that the charge density gradient is slight and diffusion is weak, except very close to $s = 1$ where the sudden drop $\Delta\psi_F$ occurs.

The general shape of the curve for ψ_F in Fig. 20.1 closely resembles that for the mean local potential in the well known Stern model of the electrolytic double layer,[18] exhibiting a charge density minimum. In calculating the current-voltage relation one first specifies $\gamma(l)$ and then for various values of f_2 calculates i, $\Delta\psi_F$ and u in turn. Fig. 20.2 shows some results for $l = 4$

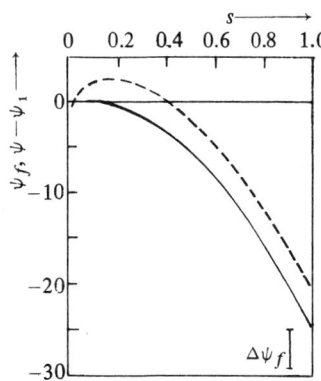

Fig. 20.1 Dimensionless Fermi level, ψ_F (solid line), and potential, $\psi - \psi_1$ (dashed line), as functions of dimensionless position, $s = xd$, for the case $l = 2$, $f_2 = 5$ and $y_2 = 5 \times 10^{-5}$. After Bonham.[15]

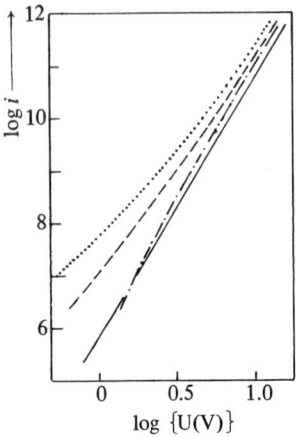

Fig. 20.2 Dimensionless current, i, as a function of voltage, U, for $l = 4$, with $\gamma(l) = \infty$ (dotted line); $\gamma(l) = 20$ (dashed line) and $\gamma(l) = 0.02$ (dash-dotted line). The solid line is predicted by the Mark-Helfrich eq. (10.19). After Bonham.[15]

to indicate the magnitude of the diffusion effect and also the effect of varying $\gamma(l)$.

Bonham[15] also develops a phenomenological theory of double injection, considering injected plasmas in an insulator containing traps, acting as recombination centers. In the case of discrete hole and electron traps as well as an exponential trap distribution, a large range of current vs. voltage relationships become possible. Recombination radiation may cause detrapping, resulting in the scaling law to be violated, as has been often observed.

Summarizing for voltages of tenths of a volt the current density is given by

$$j \propto U/d^{2l+1} \tag{20.18}$$

while at high voltages the usual equation is valid, viz.,

$$j \propto U^{l+1}/2^{2l+1} \tag{20.19}$$

where U is the applied voltage, and l is related to the trapped charge density by eq. (20.20). The free charge density, ρ_f, is related to ρ by

$$\rho_f = \theta \rho^l \tag{20.20}$$

where θ is a parameter inversely related to the charge density ($\theta = 1$ for the trap-free case; otherwise it is assumed that $\rho_f \ll \rho$); $l = 1$ for the trap-free and shallow-trap cases, and $l > 1$ for exponential traps.

For shallow traps of density N_t, lying at an energy E_t below the valence band,

$$\theta = (N_0/N_t)\exp(-E_t/kT), \quad \text{and} \quad l = 1 \tag{20.21}$$

where N_0 is the density of states in the valence band. Bonham and Jarvis[16] considered the theory of space charge limited currents using one blocking electrode. Modifying the constant field approximation gives a good description of the current-voltage curve when the charge density at the blocking electrode is small, and the current lies well below the purely space charge limited current which flows if both electrodes were ohmic. Bonham[12] also developed a theory for space charge limited currents for insulators with traps varying in position which included the diffusion aspects. After showing that the current may be bulk or surface limited, or follows a transition from one to the other as the voltage in the symmetric case is raised, he shows that this yields a considerable structure in the current-voltage curves which was not predicted by theories neglecting diffusion. In the asymmetric case, distribution may exhibit rectification at low voltages. Comparison of open-circuit profiles suggests that the system is insensitive at low voltage to the nature of spatial distribution of traps. Only the trap density near the surface is important. Jaros claims[19] that all defects dominated by short range forces are deep traps.

General reviews on phenomena in insulators and dielectrics are available.[20] Stiles[21] has published an extensive discussion of charge carrier transport mechanisms in 2-dimensional space charge layers.

Although anthracene has continued to be the target of much space charge limited current experimentation,[22] data has been obtained by Caillon, Reboul and Toureille[23] for various parameters of caoutchouc and polyvinyl chloride filled with carbon black. The number of carriers (n_0) and mobilities (μ_n) were shown to be 10^{15} and 10^{-7} for caoutchouc, respectively, and 4×10^{15} and 10^{-5} for polyvinyl chloride, respectively. For films of poly-α-naphthol, Vidadi and coworkers[24] found the trapping level depths to be approximately 0.8 eV with a density of about 10^{15} cm^{-3}. At room temperature the carrier density was 2×10^9 cm^{-3} and the mobility 10^{-3} cm^2/Vsec. Local level cross-section was 8×10^{-19} cm^2, and the conductivity rose sharply with applied electric field and persisted after removal of the field.

Kasica and Swiatek[25] evaluated trap parameters of vacuum evaporated tetracene films and found them to be shallow. Photo-traps were observed in tetrathiotetracene and explained in terms of conduction mechanisms and refined energy structure. Sadaoka and Sakai[27] determined the effect of the trap density in naphthacene thin films provided with an Au anode and Al cathode. They report that the threshold voltage for crystalline samples was proportional to the (thickness)^2. For the amorphous material it was irreproducible. An explanation was advanced in terms of changes in trap density. El-Wahaidy,[28] measuring the capture coefficient of electrons by positively charged traps, found that it increased with increasing applied electric field, but also found the process to be insufficiently rapid to lead to negative differential conductivity. Zablowska and Waclawek[29] found that traps in the energy gap of Cu phthalocyanine depended strongly on the nature of gas ambients.

Bobyl and coworkers[30] studied the relaxation of dark current in liquid benzene; they found that the reverse current was a space charge effect. In weak fields the reverse EMF was 95% of the applied voltage. They[30] suggested that the conductivity was due to charge injected from the anode where potential barriers increased with time. Sheng and Hanson[109] have used Stark spectroscopy to measure space charge distributions in organic semiconductors which showed correlations of spectral curve shifts or splittings with changing electric field.

20.2 Glow Curves[110]

Thomas and colleagues[33] carried out thermally stimulated current discharge studies on anthracene and report that the glow curve peak at 0°C was an intrinsic property based on structural faults. The trapping levels were about 0.7 eV above the valence band for hole carriers. The magnitude of the peak decreased six-fold after a hundred-fold increase in the content of dislocations which slide in the (010) direction on the (100) and (201) planes. Structural imperfections are discrete traps for positive charge carriers. Only one peak is obtained with high purity anthracene.

From transient photoconductivity and glow curve current measurements on anthracene, Samoc and Zboinski[34] deduce that mechanical roughening introduced trapping levels with depths greater than 0.9 eV for electrons and 0.46 eV for holes. Samoc and coworkers[35] experimenting with anthracene doped with phenazine observed electron and hole currents suggesting that carriers were released from deep traps with a thermal energy of about kT at the glow peak of 245 K. They[35] measured the hole trap in tetracene-doped anthracene to be 0.42 eV deep. Swiatek[36] found that aging of thin polycrystalline p-terphenyl layers caused the disappearance of shallow traps which were replaced by deeper traps.

Van Turnhout[37] used thermally stimulated discharge methods to determine low frequency molecular motions, such as glass transition temperatures and other kinds of relaxation phenomena in polymers. These experiments were done on thirteen polymers and copolymers which included polymethylmethacrylate, polyethyleneterephthalate, polycarbonate, polyvinylchloride, polystyrene and polytetrafluorethylene. Like van Turnhout, Ranicar and Fleming,[38] working with the latter four compounds plus polyimide, obtained glow curve peaks and thermoluminescent peaks occurring at temperatures similar to those assigned to the onset of specific molecular motions, indicating that these peaks were due to the liberation of electrons from traps. These might be cavities or entanglements in amorphous regions, or mainchain branching points in the polymers. Peaks not correlated with molecular motion were assigned[38] to impurity traps. Glow curve peaks for polyvinylchloride were not appreciably affected by applied electric fields of high strengths during the heating process which suggested that the electron traps were neutral when empty, and electrically charged when filled. The carrier mobility in this polymer was dominated by motions of the molecular chain, i.e., an interchain transport or hopping process of conductivity[39] (cf., Section 15.10 and 15.11). Trap depths were estimated[39] to be about 0.44 eV. Charging currents in polyethylene are also said to be not only space charge limited but also affected by hopping transitions over multiple morphological barriers.

Trap studies by thermal stimulation emf methods have also been conducted by Makarov-Mironov and coworkers.[40] Glow curve experiments are reported on copper-doped polyvinylacetate,[41] in which ESR measurements indicated the dopant to be present as Cu^{++}. As the dopant concentration (cf., also Section 16.7) increased, the low temperature peaks decreased and new glow peaks appeared at higher temperatures. A negative peak evident at higher copper concentration disappeared at high polarizing temperatures. Hampe[42] irradiated one half of a sample of low density polyethylene film and found that the other half became polarized. The glow curve obtained from the material so treated exhibited[42] thermal detrapping maxima at $-120, -80, +20°C$. A maximum at $-20°C$ was due to dipolar rearrangements.

Samoc and colleagues[43] have attempted an interpretation of thermally stimulated and isothermal decay currents. Other interpretations of thermally stimulated conductivity and thermoluminescence experiments have been given by Kivits.[44] Maxia[45] has investigated nonequilibrium thermodynamics of thermoluminescent phenomena. Glow curves for polymethylmethacrylate electrets (see Section 19.6 and 20.4) are also reported[46] and explained in terms of dipolar volume polarization involving cationic space charges.[46] Vakser has presented[111] a theory of thermoelectric effects.

20.3 Ferro-, Pyro- and Piezo-electric Effects

The subject of ferro-electricity has been the object of very many studies, so that, within the limits of this book, it is not possible to do it justice. Thus, the reader is referred to the specialist journal dealing with this topic, viz., "Ferroelectrics," as well as to the several reviews and bibliographies which deal with it.[47]

Reviews of piezo- and pyro-electric phenomena are also available.[48,49] A number of workers,[50,51] suggest that about 16% of the observed pyroelectric voltages in polyvinylidene fluoride are a primary effect, *i.e.*, not due to any piezo-electric effect also present.

Broadhurst et al.[50] suggest that the polymer exists in two forms, a nonpolar one which can be converted into a polar, polarizable modification by poling it in a field in excess of 1 MV/cm at room temperature. At least part of the residual polarization is said[50] to reside in the crystalline phase of the polymer. He also reports that charges at the contact electrodes of approximately 10^{-3} C/m^2 are induced upon applying strain in the thickness direction. Kepler and Anderson[52] attempted to confirm the presence of primary pyroelectricity in this compound and concluded that it was present below 15% if it existed at all. Later, more refined experiments[53] showed that the upper limit of the magnitude of the primary pyroelectricity set by these experiments should be determined before it could be said to exist at all. However, a copolymer of vinylidene fluoride with tetrafluoroethylene is reported[105] to be ferro-electric as well as piezoelectric and also exhibits electret behavior.[106]

Niguchi and Suzumura and others[54] found that polymer films stretched unidirectionally while cooled at 50°C below the softening point exhibited polarized dark currents. Extrusion of polystyrene gives rise to the injection of charges a hundred times in excess of those expected purely from friction.[55] The piezoelectric activity of unoriented and biaxially oriented polyvinylidene fluoride polarized at very high fields, up to 4.0×10^6 V/cm, was investigated[56] with respect to changes in molecular orientation, morphology and crystal structure. It is a function of polarizing voltage, time and temperature. These studies[56] showed the existence of an endothermic peak intermediate between phase I and phase II peaks for films polarized with fields greater than 1.6×10^6 V/cm. The relative magnitude of the endothermic peaks varied with polarizing conditions.[56] Of great interest is the discovery of an electrocalorific effect in drawn and poled polyvinylidene fluoride.[57] This effect, an inverse piezoelectric effect, was the first of its kind to be discovered;[57] The measured temperature change *vs.* the applied field, at constant stress and entropy, $(dT/dE) = 2.4 \times 10^{-9}$ K. mV^{-1}, allowed computation of the electrocalorific coefficient at constant stress and temperature

$$p^\sigma = (dS/dE)_{\sigma T} = 2.2 \times 10^{-5} \text{ J m}^{-2} \text{ K}^{-1} \text{ V}^{-1}.$$

The measurement of the direct pyroelectric effect (polarization change $p^\sigma = (dP/dT)_{\sigma E} = -2.4 \times 10^{-5}$ C m^{-2}K^{-1}) confirmed the equality of the two effects, as predicted by the thermodynamic model of homogeneous pyroelectrics.[57] Dreyfus and Lewiner[58] showed that two different effects may produce results similar to piezoelectricity in thin films of polypropylene; the electrodes could be set in vibrational motion by the applied electric fields, the static field produced by either an external bias voltage or by polarization of the field itself (electret behavior). Interestingly, Litt and collaborators[59] showed that volume polarization contributed to pyroelectricity in nylon-11 unless the sample was thoroughly silvered.

Both pyro- and piezoelectricity in this compound was found[59] to be quite high, probably due to dipole orientation in crystalline regions. Charge injection or volume polarization also contributed to pyroelectricity.[59] Hysteresis effects in piezo-electric metalloorganic polymers are reported by Hartman et al.[60]

Pyroelectricity has been demonstrated for saccharose,[104] the coefficient is said[104] to be 2×10^{-4} $\mu C/cm^2 K$ in single crystals at cryogenic temperatures, peaking at 14 K. Triglycine sulfate doped with α-alanine is said[107] to exhibit an increased piezoelectric and ferroelectric effect; doping extends the temperature interval of the ferroelectric phase.

Piezoelectric and related electrical effects in bioorganic compounds may have an important bearing on the living state. For instance, α-keratins resemble artificial electrets (see Section 20.4) and ferroelectrics in having a high degree of dipole ordering which may be significant for biological membranes.[61] Chepel and Laurentlev[62] demonstrated piezoelectricity in stretched collagen, collagen-containing structures and bone. Vasilescu and coworkers[63] noted piezoelectric behavior in a large number of aminoacids which depended on their crystal structure. Piezoelectricity and pyroelectricity as a basis for force and temperature determinations by nerve receptors have been suggested by Cope.[64] This topic is further treated in Section 18.11.

Piezoelectric amino acids are listed in Table 10.2A; the piezoelectric coefficients of nuclear acids are tabulated in Table 10.2B. However, Mascarenhas[102] maintains that, e.g., DNA is not ferroelectric at all and that results which have been interpreted in terms of ferroelectric behavior are really caused by hydration changes. DNA, though, does exhibit electret effects.[102] This important matter remains to be clarified.

Temperature coefficients of the piezoelectric voltages observed in various solids are collected in Table 10.2C.

Biological application of piezo- and pyroelectricity have been reviewed by Cope.[65] Photovoltaic effects in ferroelectrics are discussed by Kraut and Von Baltz.[103]

20.4 Electrets[112]

Interest in electrets has burgeoned with the introduction of the first foil electret in 1962 by Sessler and West[66] and its subsequent commercial use in condenser microphones.[67] These workers determined time constants for polymers as high as 100 years: polyimides 0.1 year, polyesters 1 year, fluorocarbons and polycarbonates in excess of 100 years. They also separately measured the homocharges and heterocharges in electrets.[66,67] The former are due to space charges located close to the surface of the dielectric, whereas the latter consist of a volume polarization caused by dipole orientation or space charge polarization induced by migration of ions over distances comparable to sample thickness. van Turnhout[37] recommended thermally stimulated discharge measurements as a powerful method for finding, analyzing and developing electret materials. This worker[68] also was able to make dielectric measurements at ultralow frequencies by preheating of the electrets.

Roos[69] showed that electrets provided with evaporated electrodes do not have to be stored under short-circuit conditions to remain effective. He also indicated that the arc and spark transfer across the electret-electrode gap of thick electrets with detachable electrodes disturbes all charge measurements. Using thermograms, Ong and van Turnhout[70] determined that thermoelectrets produce distorted, distributed polarizations when forming temperatures are either too low or storage times too long. Model calculations for specific distribution functions in relaxation times suggest[70] that the deformations are due to the incomplete filling of subpolarizations. Polypropylene homoelectrets were obtained at low temperatures, but heteroelectrets resulted from the utilization of high temperatures.[71] The homoelectret in this case had a longer life-time than the heteroelectret, which exhibited a glow curve peak agreeing with a temperature dispersion of internal friction.

Several workers[72,73,101] have examined the electret behavior of polymethylmethacrylate and polyvinyl chloride. The temperature dependence of the steady state conductivity, for polyvinylchloride an inherent property of the material independent from charge carriers injected by electrodes, showed a reproducibly convex relationship in the temperature range of 75-100°C,[73] which cannot be approximated by two straight lines without introducing serious errors. This property can be determined only at high temperature. Measurement of the transit time of the charge carriers between electrodes after the application of a voltage pulse allowed the mobility of charge carriers for polyvinylchloride[74] to be determined, $viz.$, 8×10^{-4} cm^2/Vsec, at room temperature.

Formaldehyde-phenol copolymer electrets have recently been formed.[75] An electret requiring only the remarkably low polarizing field of 85 V/cm, and that at room temperature, has been reported by Teodorescu,[76] using deposited films of the n-type 4-(dimethyloctylammonia)stilbene.

Using single crystals of naphthalene to study the formation of thermoelectrets, Campos et al.[77] obtained what is claimed to be evidence for the existence of Schottky barriers at the electrodes. Jonscher[78] developed a theory of ac loss to cover the case of nonlinear behavior under high amplitudes of exciting electric fields, and the loss at high field related to low-field loss and nonlinear dc current field characteristics. The theory offers a rationalization of high-field time-dependent mechanisms in siliconmonoxide and stearic acid films. A study by Hartmann and O'Brien[79] involving reflectance infrared spectroscopy has provided the first direct evidence for polar molecule alignment in electrets. This work also showed which components were responsible for charge retention in carnauba wax, the earliest electret ever studied.

Electret depolarization currents are observed when e.g., polymers are heated after electric polarization at elevated temperature.[80] The peaks occur at temperatures where the polymeric solid undergoes a structural transition;[81] their intensity, or height, depends on the thermal history of the sample. The effect has been interpreted[81] as due to changes of the texture, or structure, of the solid. The depolarization currents in nonpolar polymers, e.g., in polyethylene or in polystyrene, however, are too high to be explained entirely due to dipolar chain effects. They may be caused, at least partially, to depolarization of dipoles arising from the Maxwell-Wagner interfacial polarization process[82] arising at internal interfaces between structural, or textural, subregions of the polymer. Polyethylene, e.g., is known[83] to exhibit a lamellar structure associated with interfacial polarization effects. Polyvinylchloride also exhibits an aggregate structure even above the glass transition temperature T_G. The thermal history affecting these structures is reflected in the depolarization current spectra.

Assuming that the texture, or micro-structure, of the solid polymer can in a first approximation be described in terms of only one parameter, viz., the concentration of structural defects, N_d, then the interfacial polarization P is described by

$$P = kN_d \qquad (20.22)$$

where k is a constant of proportionality. At a structural transition, temperature T_0, when N_d changes abruptly, a sudden change in P will occur, causing an observable current peak:

$$I = k\left(\frac{dp}{dt}\right)_{T_0} = k\left(\frac{dN_d}{dT}\right)_{T_0} \qquad (20.23)$$

Upon rapid cooling, a high concentration of structural defects will become frozen in. Consequently, upon subsequent heating the change in N_d will be less than that occurring upon slow cooling. Therefore, the depolarization current peak may be considered to be a measure of the physical, structural, state of the polymer at the temperature of the commencement of the heating process.

A new method[84] for forming electrets is beginning to receive widespread attention; with the use of magnetic fields, Grawaz and Bhatnagar[85] have developed a magnetically polarized naphthalene electret. Bhatnagar and Qureshi[86] report a change in the value of the dielectric constant of polymethylmethacrylate upon its conversion into a magneto-electret. Photodepolarization of magneto-electret charges in photoconductors, results in a decrease of the capacitance in a parallel plate condenser configuration.[84]

Work on photo-electrets is reviewed by Kovalskii and Schneider.[87] Ametov et al.[88] report that α-irradiation of electrically polarized electron-ion pairs in organic solids produced electrons trapped near the positive ions in 3-methylpentane.

Phenanthrene single crystals become photoelectrets[89] if illuminated in a field applied above 72°C, followed by cooling. A photoconductive long wavelength limit was found to be 4000 Å, while the photon flux was 4×10^{14} cm^{-2} sec^{-1} above 72°C. The photocurrent was 5×10^{-3} A through an area of 0.077 cm^2. Energy gaps for this compound, measured from room temperature to 82°C, were 1.5 eV below 72°C and 1.1 eV above 72°C. Volume polarization charges were observed[89] if the phenanthrene single crystals were cooled through the transition temperature.

The resistivity at 25°C for this compound with electrolyte contacts was 9.2×10^{14} ohm cm, indicative of charge injection,[89] while its resistivity with SnO$_2$/Cu contacts was 10^{17} ohm cm. Human, bovine, canine and rodent bone samples exhibit[90] electret behavior with a saturation polarization of about 10^{-8} C/cm^2, the effect being due to collagen components which involve ionic or protonic space charge polarization. In fact, polarization storage appears to be a property of collagens in general.[91]

Bornzin and Miller[92] have demonstrated electret behavior of synthetic polyelectrolyte membranes which thus might be used as models for biological systems. Gross[93] has reviewed charge transport and storage in solid dielectrics, including irradiated polymers. Polymer electrets are reviewed by Kepler.[94] A general review of electrets is also available.[95]

Table 10.3A lists charge densities of electrets produced from several polymers, including commercial plastics.

20.5 Plasmas

Optical reflectance spectroscopy (see Section 12.10 and 16.11) has become a valuable tool in the study of semiconductivity and semimetallic organics. Such studies are specially informative if carried out over a range of applied pressure.[96] Thus, Welber et al.[97] have shown that the Drude edge of the plasma frequency in the near infrared shifts to higher values with increasing pressure. The "optical resistivity" appears to be the sum of the dc or low frequency resistivity plus a pressure independent residual resistivity component.[97] Examples of optical reflectance spectra are discussed in Section 16.11, Figs. 16.20 and 16.21.

Another important and typical example refers to the room temperature polarized reflectivity[98] of single crystals of the adduct tetrathio-tetracene $(TTT)_2$: iodine, illustrated in Fig. 20.3. The anistropy of this system is seen to be quite pronounced. The sharp rise in reflectivity near 0.6 eV is indicative of a plasma edge, while the prominent structure centered around 1.85 eV is due to[98] interband transitions. The minimum near 0.6 eV also suggests contributions from nearby interband transitions.

Somoano et al.[98] have analyzed these data on the basis of a Drude-Lorentz model: the permittivity is given by

$$\bar{\epsilon}_{DL} = \bar{\epsilon}_f + \bar{\epsilon}_b \tag{20.24}$$

where

$$\bar{\epsilon}_f = -\omega_p^2/(\omega^2 - i\omega/\tau) \tag{20.25}$$

and

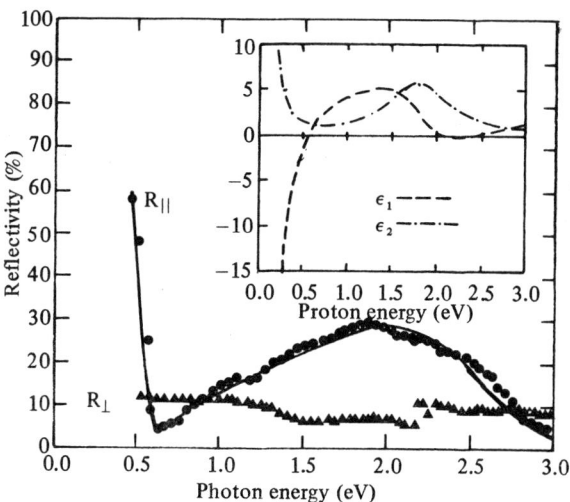

Fig. 20.3 Optical reflectivity of a single crystal of $(TTT)_2I_3$. $R_\perp$ and $R_{||}$ are the reflectivities for light polarized parallel and perpendicular to the conducting needle b axis, respectively. The solid line represents a fit of eq. (20.24) to the $R_{||}$ data. The inset shows the real and imaginary parts of the total dielectric constant ϵ_1 and ϵ_2, respectively, as computed from eq. (20.24) where $\bar{\epsilon}_{DL} = \epsilon_1 + i\epsilon_2$. After Somoano et al.[98]

$$\bar{\epsilon}_b = \epsilon_c + \sum_1 \frac{\omega_{pi}^2}{(\omega_i^2 - \omega^2) - i\omega/\tau} \tag{20.26}$$

$\bar{\epsilon}_f$ and $\bar{\epsilon}_b$ are the free and bound electron contributions to the dielectric constant, respectively. $\omega_p^2 = 4\pi N_h e^2/m^*$ is the free-electron plasma frequency, N_h is the density of holes, m^* the optical effective mass, $\tau \equiv \Gamma^{-1}$ the relaxation time, ϵ_c the contribution to the dielectric constant from the core electrons, and ω_1 is the frequency of the interband transition. A fit of this model to the data using one Lorentzian oscillator yields the following parameters, $\epsilon_c = 2.61 \pm 0.27$, $\hbar\omega_p = 1.33 \pm 0.15$ eV, $\hbar\Gamma = 0.076 \pm 0.056$ eV. $\hbar\omega = 1.87 \pm 0.18$ eV, $\hbar\Gamma = 0.81 \pm 0.04$ eV, and $\hbar\omega_p = 294 \pm 0.42$ eV. The minimum in the reflectivity occurs approximately at $\hbar[\omega_p^2/\epsilon_1^b(\omega_{min})]^{1/2}$ since the free-electron plasma frequency is screened by the nearby interband transitions. Fig. 20.3 shows the real and imaginary parts of the total dielectric constant, ϵ_1 and ϵ_2, respectively, as computed from Eq. (20.24) where $\bar{\epsilon}_{DL} = \epsilon_1 + i\epsilon_2$. From the plasma energy 1.33 eV and the density of carriers, one can estimate the optical effective mass. Using the X-ray data and assuming complete charge transfer, $N_h \sim 1.2 \times 10^{21}$ holes/cm^3, one finds $m^* \sim 0.93$ m_0. For a three quarter-filled electron band, using the tight-binding approximation, this gives a crude estimate of the bandwidth of ~ 1.5 eV.

Some features of plasmas in solids may be discussed in terms of plasmons, *i.e.*, of collective excitation states such as arise when dipoles are embedded in a nonpolar dielectric;[99] this occurs, *e.g.*, in liquid interfacial multilayers (see also Section 13.10). Such dispersive plasmas may be probed by infrared methods: let $1/\alpha_0$ equal the extension of the electric field away from the interface which has a thickness d. For $\alpha_0 d \ll 1$ the frequency ω_p^2 of the bulk dipolar plasmon is then given by

$$\omega_p^2 = \frac{8\pi\mu_0^2}{3\epsilon_0 I^*}\frac{\rho}{d} \tag{20.27}$$

where I^* is the effective moment of inertia of a polar molecule of dipole moment μ_0; ϵ_0 is the permittivity of the medium and ρ/d the volume density of (free) dipoles. Quite generally,

$$\frac{2\epsilon_0}{\alpha_0} = \frac{c\Omega_p}{\omega^2} \tag{20.28}$$

where c is the velocity of light *in vacuo*, and Ω_p the characteristic dipolar frequency of the dipoles.

Nonlinear vibrational excitation waves in a molecular plasma containing negative ions are reported[108] to be associated with instabilities which are relieved by these excitation waves until stability is attained.

References

1. J. Dresner, *RCA Rev.*, **30**, 322 (1969).
2. W. G. Williams, G. Spong and D. J. Gibbons, *J. Phys. Chem. Solids*, **33**, 1879 (1972).
3. A. M. Hermann, *J. Appl. Phys.*, **44**, 926, 5200 (1973); *cf.*, also M. A. Lampert and P. Mark, *Current Injection in Solids*, Academic Press, N.Y., 1970; S. Nespurek and E. A. Silinsh, *Phys. Stat. Solidi*, **A34**, 747 (1976).
4. R. Meaudre and G. Mesnard, *Rev. de Phys. Appl.* **7**, 213 (1972).
5. R. de Levie et al., *J. Membr. Biol.*, **9**, 241 (1972); **10**, 171 (1972); G. Ormancey and G. Godfray, *J. Phys.* (Paris), **35**, 135 (1974); T. J. O'Reilly and J. de Lucia, *Solid State Electronics*, **18**, 965 (1975); A. I. Rosental and A. Sapar, *J. Appl. Phys.*, **45**, 2785 (1974); A. I. Rosental and L. G. Paritskii, *Soviet Phys. Semicond.*, **5**, 2100 (1972); S. Nespurek and E. A. Silinsh, *Phys. Stat. Solidi*, **A-34**, 747 (1976); W. G. Williams et al., *J. Phys. Chem. Solids*, **33**, 1879 (1972).
6. M. A. Nicolet, *J. Appl. Phys.*, **37**, 4224 (1966); M. A. Nicolet et al., *Surf. Sci.*, **10**, 146 (1968).
7. P. Delannoy et al., *Phys. Stat. Solidi*, **A32**, 577 (1975).
8. J. Sworakowski, *J. Appl. Phys.*, **41**, 292 (1970).
9. W. Hwang and K. C. Kao, *Solid-State Electron.*, **15**, 523 (1972).
10. W. Hwang and K. C. Kao, *Solid-State Electron.*, **19**, 1045 (1976).
11. S. Nespurek and E. A. Silinsh, *Phys. Stat. Solidi A*, **34**, 747 (1976).
12. J. S. Bonham, *Aust. J. Chem.*, **31**, 2291, 2117 (1978).
13. J. S. Bonham and D. H. Jarvis, *Aust. J. Chem.*, **30**, 705 (1977).
14. A. I. Rozental and L. G. Paritskii, *Sov. Phys. Semicond.*, **5**, 2100 (1972).
15. J. S. Bonham, *Phys. Stat. Solidi*, **55A**, 61 (1979); *Aust. J. Chem.*, **31**, 2103 (1978).
16. J. S. Bonham and D. H. Jarvis, *Aust. J. Chem.*, **31**, 2103 (1978).
17. J. Kalinowski and J. Godlewski, *J. Chem. Phys.*, **32**, 201 (1978).
18. B. Breyer and F. Gutmann, *J. Chem. Phys.*, **21**, 1323 (1953); J. Ross MacDonald and M. K. Brachman, *ibid.*, **32**, 1314 (1954); J. Ross MacDonald, *ibid.*, 1317.
19. M. Jaros, *Proc. Nauk. Inst. Chem. Org. Fiz. Polytech.*, Wroclaw, **16**, 281 (1978).
20. Ann. Rep. Conf. Elect. Insul. Diel. Phenomena, see issues of.
21. P. J. Stiles, *Conf. Ser. Inst. Phys.*, **43**, 41 (1978, publ. 1979).
22. W. Nehl and W. Buchner, *Z. Phys. Chem.*, (N.F.) **47**, 76 (1965); J. Dresner, *RCA Rev.*, **30**, 322 (1969); F. Lohmann and W. Mehl, *J. Chem. Phys.*, **50**, 500 (1969); I. F. Williams and M. Schadt, *J. Chem. Phys.*, **53**, 3480 (1970); H. P. Schwob, J. Fünfschilling and I. Zschokke-Granacher, *Mol. Cryst. Liquid Cryst.*, **10**, 39 (1970); W. G. Williams, P. L. Spong and D. J. Gibbons, *J. Phys. Chem. Solids*, **33**, 1879 (1972); M. A. Lampert and P. Mark, *Current Injection in Solids*, Academic Press, N.Y., 1970; see also ref. 31 and 32.
23. P. Caillon, J-P. Reboul and A. Toureille, *C. R. Acad. Sci. Paris*, **B-272**, 1074 (1971).
24. Y. A. Vidadi et al., *Dokl. Akad. Nauk. Az. SSSR*, **33** (3), 26 (1977).
25. H. Kasica and Y. Swiatek, *Proc. Nauk, Inst. Chem. Org. Fiz. Polytech.*, Wroclaw, **16**, 191 (1978).
26. D. Balode et al., *Zinat. Akad. Vestis, Fiz. Teh. Zinat. Ser.*, **1978**, (1) 35.
27. Y. Sadaoka and Y. Sakai, *Bull. Chem. Soc. Japan*, **51**, (1) 45 (1978).
28. E. F. El-Wahaidy, *Indian J. Phys.*, **51A**, 278 (1977).

29. M. Zablowska and W. Waclawek, *Proc. Nauk. Inst. Chem. Org. Fiz. Polytech.*, Wroclaw, **16**, 329 (1978).
30. V. C. Bobyl et al., *Elektron*, OBRAB mater (1976) (3) 41.
31. J. Dresner in Ref. 22.
32. W. G. Williams, *ibid.*
33. J. M. Thomas, J. O. Williams and G. A. Cox, *Trans. Faraday Soc.*, **64**, 2496 (1968).
34. A. Samoc and Z. Zboinski, *Phys. Stat. Solidi*, **A-46**, (1) 251 (1978).
35. A. Samoc et al., *Pr. Nauk. Inst. Chem. Org. Fiz. Politech.*, Wroclaw, **16**, 267 (1978).
36. J. Swiatek, *Pr. Nauk. Inst. Chem. Org. Fiz. Politech.*, Wroclaw, **16**, 299 (1978).
37. J. van Turnhout, *Polymer J.*, **2**, 173 (1971).
38. J. H. Ranicar and R. J. Fleming, *J. Polymer Sci.*, **10**, 1979 (1972).
39. I. M. Talwar and D. L. Sharma, *J. Electrochem. Soc.*, **125**, 434 (1978).
40. A. M. Makarov-Mironov et al., "Vopr. Fiz. Elektrolyumin," in *Naukova Domka*, M. V. Fok, ed., Kiev, USSR, 1975.
41. V. J. Jain et al., *Thin Solid Films*, **48**, 175 (1978).
42. A. Hampe, *Prog. Colloid Polymer Sci.*, **62**, 154 (1977).
43. A. Samoc et al., *Phys. Stat. Solidi*, **A-36**, (2) 735 (1976).
44. P. Kivits, *J. Luminescence*, **16** (2) 119 (1978).
45. V. Maxia, *Phys. Rev.*, **B-17**, 3262 (1970); General references in Proc. Inst. Conf. Luminescence, Budapest, (1966).
46. J. Vanderschueren and A. Linkens, *J. Polymer Sci. Phys. Ed.*, **16**, 223 (1978).
47. K. Toyoda, *Ferroelectrics*, **14**, 849 (1976); W. Heywang et al., "Ferroelectrics," in *Ullmanns Encyclop. d. Tech. u. Chem.*, 4th ed., 1976, Verlag Chemie, Weinheim, W. Germany; V. M. Fridkin and B. M. Popov, *Usp. Fiz. Nauk.*, **126**, 657 (1978); G. A. Smolenskii et al., "Status of Physics of Ferroelectrics," Izv. Vyssh. Uchebn. Zayed Fiz., **22**, 5 (1979); F. Micheron, Rev. Tech. Thompson-CSF, **10**, 445 (1978); "Bibliography of Ferroelectrics" by K. Toyoda, *Ferroelectrics*, **19**, 61, 211, 235 (1978); J. Ravez and F. Micheron, *Actual. Chim., 1979*, 9; G. Chanussot, Proc. 4th Int. Meeting. Ferroelect. Leningrad, USSR, in *Ferroelectrics*, **20**, 37 (1978); I. S. Zheludev, *Dielektr. Poluprovodn.* **13**, 3 (1978); see also references 49 and 100.
48. S. Lang, *Ferroelectrics*, **14**, (3-4), 807 (1976); **193**, 175 (1978); S. C. Abrahams, *Mater. Res. Bull.*, **13**, 1253 (1978).
49. R. G. Kepler, *Ann. Revs. Phys. Chem.*, **29**, 497 (1978); V. M. Rudyak, ed., "Ferroelectrics and Piezoelectrics," Kalininskii Univ., Kalinin, USSR, 1978; "Internatl. Symp. Electrets Dielectr. Proc. 1975," publ., 1977, Soares de Campos, ed., Acad. Brasileira Cienc. Rio de Janeiro; Y. Wada and R. Hayakawa, *Japan J. Appl. Phys.*, **15**, 2041 (1976); M. Oshiki and E. Fukuda, *J. Mater. Sci., 1975*, 10.
50. M. G. Broadhurst, G. T. Davis and J. E. McKinney, *J. Appl. Phys.*, **49**, 4998 (1978).
51. J. Cohen, S. Edelman and C. F. Vezzetti, "Electrets," M. M. Perlman ed., The Electrochem. Soc. Princeton, N.Y., 1973, p. 505.
52. R. G. Kepler and R. A. Anderson, *J. Appl. Phys.*, **49**, 4918 (1978).
53. R. G. Kepler and R. A. Anderson, *Bull. Amer. Phys. Soc.*, **24**, 287 (1979).
54. H. Niguchi and M. Suzumura, *Japan Kokai*, 7, 754, 999, 4th May, 1977; see also I. Sumita, 7, 787, 497, *Japan Kokai*, July 21, 1977.
55. J. Kedzia, *J. Electrostat.*, **4** (3), 291 (1978).
56. C. H. Yoda, J. I. Scheinbein, B. A. Newman and K. D. Pae, *Bull. Amer. Phys. Soc.*, **29**, 288 (1979).

57. T. Brossat, G. Bichou, C. Lemonon, M. Royer and F. Micheron, *Bull. Amer. Phys. Soc.*, **24**, 288 (1979).
58. G. Dreyfus and J. Lewiner, *Phys. Rev.*, **B-14**, 5451 (1976).
59. M. H. Litt et al., *J. Appl. Phys.*, **48**, (6), 2208 (1977).
60. R. D. Hartman, S. Kanda and H. A. Pohl, *Proc. Okl. Acad. Soc.*, **47**, 246 (1968).
61. E. Menefee, "Electrets," M. M. Perlman, ed., The Electrochem. Soc. Princeton, N.Y., 1973, p. 661; E. Fukada et al., *Biochem. Biophys. Res. Commun.*, **62**, 415 (1975).
62. V. F. Chepel and V. V. Laurentlev, *Mekh. Polim.*, 1978, (4) 702.
63. D. Vasilescu, R. Cornillon and G. Mallet, *Nature* (London), **225**, 635 (1970).
64. F. W. Cope, *Bull. Math. Biol.*, **35**, 31 (1973).
65. F. W. Cope, *J. Biol.*, **3**, 1 (1975).
66. G. M. Sessler and J. E. West, *J. Acoust. Soc. Am.*, **34**, 1787 (1962).
67. G. M. Sessler and J. E. West, *J. Electrochem. Soc.*, **115**, (8), 836 (1968).
68. J. van Turnhout, "Inst. Symp. Electrets, Dielectric Proc.," M. Soares de Campos, ed., Acad. Brasil. Rio de Janeiro, 1975, publ. 1977, p. 97.
69. J. Roos, *J. Appl. Phys.*, **40**, (8), 3135 (1969).
70. P. H. Ong and J. van Turnhout, "Electrets," M. M. Perlman, ed., The Electrochem. Soc. Princeton, N.Y., 1973, p. 213.
71. M. Matsui and N. Murasaki, *ibid.*, p. 141.
72. M. Latour, *C. R. Acad. Sc. Paris*, **272**, 1062 (1971).
73. V. Adamec, *Kolloid-Z. u. Z. Polymere*, **249**, 1085 (1971); **237**, 219 (1970).
74. M. Kryszeweski et al., *J. Polymer Sci. Part C*, **16**, 3915 (1968).
75. M. Goel et al., *Polymer*, **19**, (8), 905 (1978).
76. G. Teodorescu, Bull. Inst. Politech. "Gheorghe Gheorghi-Dej," Bucuresti, Ser. Chim-Metal., **40**, (1) 3 (1978).
77. M. Campos, S. Mascarenhas and G. Leal Ferreira, in "Internatl. Symp. Electrets Dielectr. Proc., 1978," M. Soares de Campos, ed., Acad. Brasil. Cienc., Rio de Janeiro, 1978.
78. A. K. Jonscher, "Electrets," M. M. Perlman, ed., The Electrochem. Soc., Princeton, N.Y., p. 269.
79. A. K. Hartmann and R. M. O'Brien, *ibid.*, p. 444.
80. J. van Turnhout, *Polymer J.* (Japan), **1**, 1011 (1970).
81. P. Hedvig, "Proc. Conf. on Polarization and Conduction in Polymers," Bratislava, (Czechoslavakia) 1972, p. 91 and p. 99, Müanyag Es Gumi, 7, 283 (1970).
82. C. Maxwell, "Lehrbuch der Elektrizität und des Magnetismus," Berlin, 1883; K. W. Wagner, *Archiv f. Elektrotechnik.*, **2**, 371 (1914); J. S. Dryden and R. J. Meakins, *Revs. Pure and Appl. Chem.*, **7**, 15 (1957).
83. S. Kavesh and J. M. Schultz, *J. Polymer Sci.*, **A-28**, 243 (1970); **A-29**, 85 (1971).
84. C. S. Bhatnagar and M. L. Khare, *Indian J. Pure Appl. Phys.*, **8**, 700 (1976).
85. B. M. A. Grawaz and C. S. Bhatnagar, *Indian J. Pure Appl. Phys.*, **16**, 50 (1978).
86. C. S. Bhatnagar and M. S. Qureshi, *Indian J. Pure Appl. Phys.*, **17** (9), 575 (1979).
87. P. N. Kovalskii and A. D. Shneider, "Photoelectret Effect in Semiconductors," Vishcha Shkola, Kiev, SSSR, 1977.
88. K. K. Ametov et al., *Radiat. Phys. Chem.*, **10**, (1) 43 (1977).
89. R. A. Arndt and A. C. Damask, *J. Chem. Phys.*, **45**, 4627 (1966).
90. S. Mascarenhas, in "Electrets," M. M. Perlman, ed., The Electrochem. Soc. Princeton, N.Y., 1973, p. 650.
91. S. Mascarenhas, *ibid.*, p. 657.
92. G. A. Bornzin and I. F. Miller, *J. Electrochem. Soc.*, **125**, 409 (1978).
93. B. Gross, Annual Rep. Conf. Electr. Insul. Dielectr. Phenom., 1978, 55.

94. R. G. Kepler, in *Treatise on Materials Science and Technology,* Vol. 10, J. M. Schultz, ed., Academic Press, N.Y., 1977, p. 669.

95. "Electrets and Related Electrostatic Phenomena," L. M. Baxt and M. M. Perlmann, eds., Symp. Ser. Electrochem. Soc., N.Y., 1968; G. M. Sessler and J. E. West, *J. Acoust. Soc. Amer.,* 53, 1589 (1973).

96. *Optical Reflectance Spectroscopy at High Pressures,* F. G. Fumi, ed., Nth. Holland, Amsterdam, 1976.

97. B. Welber et al., ref. 61, p. 349; G. Weber, *Colloid Polymer Sci.,* 256, (9) 923 (1978).

98. R. B. Somoano et al., *Phys. Rev.,* B-17, 2853 (1978).

99. B. Banville, *J. Chem. Phys.,* 66, 3664 (1977).

100. R. G. Kepler, in *Treatise on Material Science and Technology,* Vol. 10B, p. 670, J. M. Schultz, ed., Academic Press, N.Y., 1977.

101. K. Mazur, *J. Appl. Phys.,* 17, 265 (1978).

102. S. Mascarenhas, in "Internatl. Symp. Electrets Dielectr. Proc. 1978," M. Soares de Campos, ed., Acad. Brasileira Cienc., Rio de Janeiro, 1978, p. 71.

103. W. Kraut and R. von Baltz, *Phys. Rev.,* B-19, 1548 (1979).

104. J. Mangin and A. Hadni, *Phys. Rev.,* B-18, 7139 (1978).

105. J. C. Hicks et al., *J. Appl. Phys.,* 49, 6092 (1978); M. G. Broadhurst et al., Annual Report Conf. Electr. Insul. Dielectr. Phenomena, 1976, publ., 1978, p. 38.

106. S. M. Yata et al., *Japan Kokai,* 78, 111, 500, 29 Sept. 1978.

107. G. Galstyan and L. G. Lomova, *IZV. Akad. Nauk Armen., SSR, Fiz.,* 13, 384 (1978); G. Galstyan and A. A. Filimonov, *ibid.,* p. 305.

108. V. G. Dresvyannikov and O. I. Fisum, *Zh. Eksp. Teor. Fiz.,* 75, 2141 (1978).

109. S. J. Sheng and D. M. Hanson, *J. Appl. Phys.,* 45, 4954 (1974).

110. *e.g.,* I. Diaconu and S. Dimitrescu, *Europ. Polym. J.,* 14, 971 (1978).

111. A. I. Vakser, *Sov. Phys. Semicond.,* 12, 847 (1978).

112. G. M. Sessler, *Electrets,* Springer, Berlin, Heidelberg, New York, 1980; J. Mayoral and I. F. Miller, *J. Electrochem. Soc.,* 126, 1708 (1979).

21

Quasi-One-Dimensional Organic Conductors

Robert Somoano

Jet Propulsion Laboratory

California Institute of Technology

Pasadena, California 91103

The field of organic conductors has advanced considerably since the last edition of this book. This chapter is meant to provide a brief, and general, description of the field of organic conductors for those scientists not active in this area of research. There are now numerous excellent review articles[1-14] available which cover all aspects of this field and can provide the interested reader with additional information. The emphasis of this article will be on the electrical properties of these materials. Organometallic compounds have also been thoroughly reviewed elsewhere[4,15] and will not be considered in this Chapter.

The materials of concern are quasi-one dimensional organic compounds in which planar organic molecules stack face-to-face in a pancake-like fashion to form stacks, or chains. In an actual crystal, there are weak interactions between chains, thereby giving rise to the highly anisotropic "quasi"-one-dimensional nature of the system. The stacks may contain anion radicals, resulting in electronic conductivity along the chains, or cation radicals, leading to hole conductivity. It is also common for the structure to consist of both types of stacks such that conductivity is due to electrons and holes. Fig. 21.1 is a schematic of the structure of (tetrathiafulvalene) (tetracyanoquinodimethane), (TTF), (TCNQ), in which both types of chains, or stacks, transport charge. TTF is an electron donor (cation radical) and TCNQ is an electron acceptor (anion radical).

The term "organic metal" is commonly used to denote an organic material, with a quasi-one-dimensional type of crystal structure, in which the temperature coefficient of the electrical conductivity, σ, parallel to the stacking axis is negative ($\partial\sigma/\partial T < 0$) over a given temperature range. The magnitude of the conductivity at room temperature, σ_{RT}, is usually in the range of $\sim 10^2 - 10^3$ (ohm cm)$^{-1}$. These criteria are obviously ambiguous and whether or not an organic conductor is truly "metallic" is still a difficult question to answer. Certainly one needs to supplement electrical studies with other measurements, such as magnetic, optical or structural studies, in order to gain additional insight into the nature of the microscopic charge transport processes. Nevertheless, the above criteria are commonly used to describe metal-like behavior of organic conductors. In a typical organic metal, the temperature range of metallic behavior usually extends from room temperature down to a temperature, T_{PK}, where the conductivity reaches its highest or peak value, σ_{PK}. Below T_{PK}, the conductivity decreases with decreasing temperature in a non-metallic fashion ($\partial\sigma/\partial T > 0$). This change from metallic to nonmetallic behavior in the conductivity reflects the occurrence of a

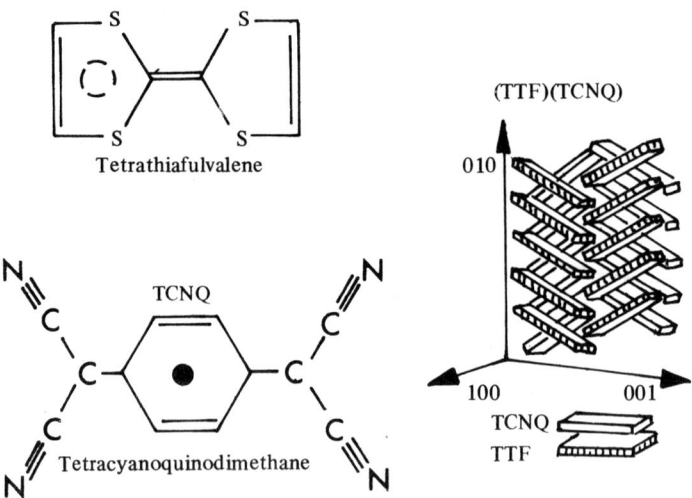

Fig. 21.1 Schematic view of the structure of (TTF) (TCNQ).

metal-to-insulator phase transition. The temperature of this transition, T_{MI}, is less than T_{PK} and is usually determined as the temperature at which $\partial(\ln \sigma)/\partial(T^{-1})$ is a maximum. At temperatures below T_{MI} the conductivity may vary as $\sigma = \sigma_0 e^{-\Delta/T}$, but the low value of the conductivity of the sample together with a limited range of measurement makes it difficult to unambiguously confirm this behavior in many cases. The materials are metallic above room temperature but usually decompose near <400° K. Fig. 21.2 shows a hypothetical situation in which the relevant parameters are shown.

The basic underlying physics responsible for the above situation is based on the Peierls instability. Peierls[16] has shown that a one dimensional metal consisting of a chain of equally spaced atoms or molecules is unstable relative to the formation of a 1-D semiconducting state. Associated with this transition from a metallic to a semiconducting state is the occurrence of a periodic lattice distortion along the chain axis. Because of electron-phonon coupling, the electronic distribution also undergoes a periodic distortion to form what is known as an electronic charge density wave (CDW). Thus, the lattice distortion leads to the formation of an electronic energy gap, 2Δ, at the Fermi energy and semiconductivity of the form

$$\sigma(T) = \sigma_0 e^{-\Delta/T} \qquad (21.1)$$

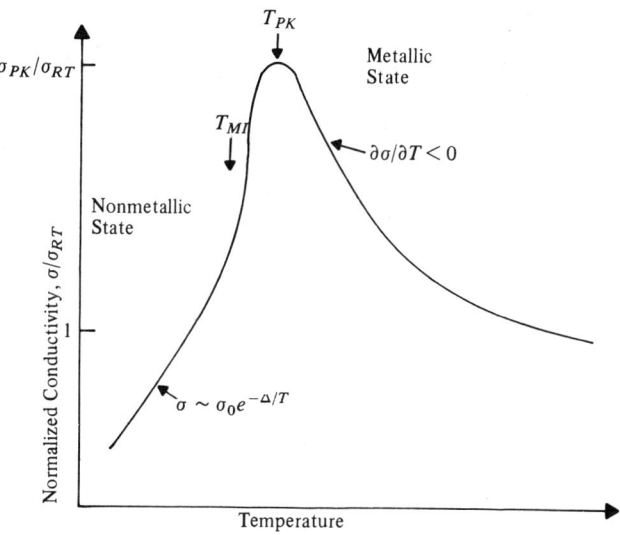

Fig. 21.2 The electrical conductivity, normalized to the room temperature value, for a hypothetical organic conductor.

In Mean Field (MF) theory (which neglects critical fluctuations) the Peierls transition temperature is given by[17]

$$T_P = \alpha T_F e^{-1/\lambda} \quad (21.2)$$

where T_F is the Fermi temperature, λ is the dimensionless electron-phonon coupling constant, and α is a constant which depends on the nature of the approximation used, e.g., tight-binding or free-electron model. The electron-phonon coupling, λ, is

$$\lambda = \frac{g^2 N(\epsilon_F)}{\hbar\omega(2k_F)} \quad (21.3)$$

where g is the electron-phonon coupling constant, $N(\epsilon_F)$ is the density of states at the Fermi energy, ϵ_F, and $\hbar\omega(2k_F)$ is the energy of a phonon of momentum $2k_F$. The metal-to-nonmetal transition is of the soft phonon type in which a particular phonon of frequency $\omega(2k_F)$ is driven soft by the divergent response function of the 1-D electronic system at $q = 2k_F$.[9] The energy gap follows a BCS-type[18] of temperature dependence below T_F with T_F being the characteristic temperature. At $T = 0°$ K, the gap is given by

$$\Delta_{(T=0)} = 4 T_F e^{-1/\lambda} \quad (21.4)$$

The MF Peierls transition temperature, T_P, is taken to be the temperature at which the energy gap becomes negative, and thus, is equal to T_{PK} described above.

It is well known that a phase transition cannot occur in a 1-D system at a finite temperature since fluctuations will drive the transition temperature to zero.[19] Thus, the Peierls instability should occur at $T = 0°$ K. However, organic conductors do not have purely 1-D structures, but rather have a weak three dimensionality as provided by the electronic interchain coupling, η, between the chains. It is this transverse interaction which permits the Peierls transition to occur at finite temperature (and which is implicitly assumed to be present in MF theory).

Thus, in realistic "quasi"-one-dimensional conductors, the Peierls instability leads to a well-defined transition from a metallic to a nonmetallic state at a temperature T_{MI} (also known as T_{3D}).[20] An example of interchain coupling is the Coulomb interaction between CDW's on adjacent chains. In this case, as one approaches T_{MI}, 1-D fluctuations allow the coherence length of a CDW on a given chain to grow in size. Eventually, the CDW's on adjacent chains will "sense" each other via the Coulomb interaction. This transverse coupling suppresses fluctuations and allows the Peierls transition to occur at temperature T_{MI}. The variation of T_{MI} with η has been considered theoretically and is summarized schematically in Fig. 21.3. At low values of η, T_{MI} varies as η^2 while at high values of η, the one-dimensionality and the Peierls instability loose relevance and T_{MI} decreases sharply.

As mentioned above, the true nature of the ground state in many of these materials is still unresolved and controversial. However, there are also many other fundamental questions concerning these materials which are in a similar state of controversy. Examples of some of these questions are: (a) the relative importance of Coulomb interactions versus electronic bandwidth effects; (b) the exact nature of the driving forces responsible for the metal-insulator instability and the associated structural phase transition; (c) the existence of collective modes, such as the incommensurate Fröhlich sliding mode, and their contribution to transport properties; (d) the existence and contribution of solitons,[23] a charged, elementary excitation of the pinned Fröhlich condensate; and (e) the possible occurrence of superconductivity *via* either the excitonic[24] or BCS-type mechanism.

The challenges posed by these questions, together with a desire to achieve a better understanding of the 1-D metallic state, in general, has sparked a tremendous activity in the study of the physics and chemistry of these materials. Numerous new compounds have been discovered, purified, and grown as single crystals for investigation. Many of these compounds are shown in Table 21.2 together with the values of some of the relevant parameters discussed above. The compounds have been segregated according to whether the conductivity is due to two types of conducting chains (*i.e.*, cation radicals and anion radicals) or to one type only. Further segregation of the single

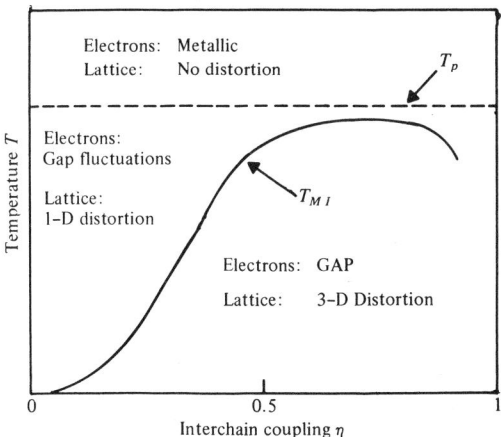

Fig. 21.3 Phase diagram for a quasi-one-dimensional conductor with a Peierls instability. Reprinted with permission.

Table 21.1 Summary of Electrical Conductivity

Substance		How Purified	Form	Contact
Electron Donor	Electron Acceptor			
A. Materials Containing Two Types of Charge Carriers				
1. Materials containing both TTF and TCNQ, or their derivatives				
TTF	TCNQ	R, GS	SC	Au, Ag, C
		R, GS	SC	Ag
		R, GS	SC	Ag
			SC	μ-Wave
		S	CF	Ag
			CP	Ag
TTF	Methyl-TCNQ	R, GS	SC	Ag
	$(TCNQ)_{0.97}(MTCNQ)_{0.03}$		SC	Ag
TTF	Diethyl-TCNQ		SC	
Dimethyl-TTF	TCNQ		CP	μ-Wave
Tetramethyl-TTF	TCNQ		—	
	TCNQ		SC	μ-Wave
$(Tetramethyl)_{1.66}$	$TCNQ_2$		SC	μ-Wave
(Tetramethyl)	TCNQ	R, GS	SC	Ag
			CP	Ag
Tetramethyl-TTF	Methyl-TCNQ	R, GS	CP	Ag
Hexamethylene-TTF	TCNQ			
			SC	Ag
Tetra-n-propyl-TTF	TCNQ		SC	
TSF	TCNQ			
			SC	Ag
	(TCNQ)(MTCNQ)$_X$ $X = 10$ MOLE%		SC	
	$X = 3\%$		SC	
	$X = 10\%$		SC	
	$X = 12\%$			

Data of Quasi-One-Dimensional Organic Conductors

Temp. Range of Metallic Behavior (K)	Room Temp. Conductivity, σ_{RT} (ohm cm)$^{-1}$	σ_{PK}/σ_{RT}	Temp. of σ_{PK}, T_{PK} (K)	Metal Insulator Transition Temp., T_{MI} (K)	Sign of Majority Carrier at Room Temp., Estimated From	Ref.
4 – 300	200 – 920 ~200 – 600 (AVG)	3 – 150 ~10 – 50 (AVG)	~58 – 72			1
4 – 300	200 – 920	10 – 150	~58	53		2
					n:TEP($\|$b)	3
					p:TEP($\|$a)	4
					p:Hall effect	5
40 – 300	30	1.2	125			6
77 – 300	15 – 40	1.4	~140			7a, 7b
4 – 300	200 – 500	1.2	210			8
~4 – 300					n:TEP	
	200			111		10
15 – 300	50	25	~50			11, 12
	1000	1	–	–		13
4 – 300	600	7	65			14
4 – 300	50	2	150			14
4 – 300		11	60			7b
4 – 300	–	2	60			7b
4 – 300	4.5					8
10 – 300	500	4	80		p:TEP	11, 15
	400 ± 50	3.5	80	50; 43	p:TEP	16
180 – 354	6.2 × 10^{-3}					17
	800	12	40			11, 18
4 – 300	800 ± 100	12	38 – 42	29	p:TEP	19, 20, 21
12 – 300	~500	~3	75	–		22
4 – 300					p:TEP	9a
4 – 300					p:TEP	9a
4 – 300					n:TEP	9a

Table 21.1

Substance		How Purified	Form	Contact
Electron Donor	Electron Acceptor			
Dithia-diselena fulvalene	TCNQ			
			SC	
			SC	
Tetramethyl-TSF	TCNQ		SC	
	Methyl TCNQ			
	Dimethyl TCNQ	R, GS	SC	
Hexamethylene-TSF	TCNQ		SC	
	Dimethyl TCNQ		SC	
$(TTF_{1-X})(TSF)_X$	TCNQ	R, GS	SC	Ag
$X = 0.001$				
0.01				
0.02				
0.03				
0.18				
0.42				
0.68				
0.95				
0.98				
[structure with M_eS, S, SM_e groups]	TCNQ		SC	Ag
[structure with $(CH_2)_n$ groups]	TCNQ			
$n = 2$			CP	Ag
$n = 3$			CP	Ag

(Continued)

Temp. Range of Metallic Behavior (K)	Room Temp. Conductivity, σ_{RT} (ohm cm)$^{-1}$	σ_{PK}/σ_{RT}	Temp. of σ_{PK}, T_{PK} (K)	Metal-Insulator Transition Temp. T_{MI} (K)	Sign of Majority Carrier at Room Temp., Estimated From	Ref.
	500	7	64		11	11
4 – 300	350 ± 100	7	60 – 64	45		19
4 – 300	550				n:TEP	21
	1000 – 1200	6	65 – 71			11, 23 24
	2 × 10^{-5}					8
4 – 300	300	10	50			25
1.1 – 300	2000	3.4	45 – 75	No transition;	p:TEP	11, 26
	< 10^{-8}					25
4 – 300						
		52.8	60 – 62			20
	280	51.5	62 – 63		n:TEP	20, 27
		51	60 – 65			20
	340	50.5	62 – 63		n:TEP	20, 27, 9a
	270	47	65 – 68		n:TEP	20, 27
	260	42	62 – 67		n:TEP	20, 27
	350	41 ± 1	60 – 65		n:TEP	20, 27
	510	39	50 – 55		n:TEP	20, 27
	640	36	45 – 50		n:TEP	
	10^{-5}					28
	50					28
	80					28

Table 21.1

Substance		How Purified	Form	Contact
Electron Donor	Electron Acceptor			
A. 2. Other Two Carrier Materials				
TTT	(TCNQ)$_2$	R, S	SC	μ-Wave
TTF	TNAP		SC	Ag
Hexamethylene-TSF	TNAP		SC	
B. Single Carrier Materials				
1. Materials Containing Cation Radicals Based on TTF or its Derivatives				
TTF	Cl$_X$ X = 0.67 0.70	R, GS	SC	C
	0.68 0.77			
	1.0		SC	
	2.0		SC	
	Br$_X$ X = 0.71 – 0.76		SC	
	0.7		SC	Ag
	1.0			
	2.0		SC	
	I$_X$ X = 0.71	R, GS	SC	Au
	0.71		SC	Ag
	0.71		SC	Au
	2.0		SC	Au
	SCN$_X$ X = 7/12 X = 6/11	R, GS	SC SC	Au, C Au
	SeCN$_X$ X = 7/12 X = 6/11	R, GS	SC SC	Au, C Au

(Continued)

Temp. Range of Metallic Behavior (K)	Room Temp. Conductivity, σ_{RT} (ohm cm)$^{-1}$	σ_{PK}/σ_{RT}	Temp. of σ_{PK}, T_{PK} (K)	Metal-Insulator Transition Temp. T_{MI} (K)	Sign of Majority Carrier at Room Temp., Estimated From	Ref.
4 – 300	20 – 160	2 – 35	90			29
88 – 370	40			185		20
1.5 – 300	2400 ± 600	6	47			31
100 – 360	150			250	p:TEP	32
	100 – 500					33
	Insulating					33
	Insulating					33
	200 – 500			170		34
~70 – 300	300 – 550			180		35
	< 10^{-6}					33
	Insulating					33
140 – 330	100 – 400			230 ± 15	p:TEP	36
125 – 300	100 – 450			200 – 270		
	360 ± 50					37
	10^{-3}					35
60 – 360	550 ± 250	1.1 – 1.3	225 ± 5	169 ± 1	p:TEP	38
60 – 300	250 ± 50					
60 – 360	750 ± 150	1.1 – 1.3	230 ± 10	173 ± 5	p:TEP	38
60 – 300	13 ± 5					37

Table 21.1

Substance		How Purified	Form	Contact
Electron Donor	Electron Acceptor			
TTF	(quinone structure with X, X, X, N substituents)			
	X = Cl		CP	
	Br		CP	
	F		CP	
Tetramethyl-TTF	(quinone structure with four X substituents)			
	X = Cl		CP	
	Br		CP	
	F		CP	
(Tetramethyl-TSF)$_2$	X			
	X = PF$_6$		SC	Au
	AsF$_6$		SC	Au
	SbF$_6$		SC	Au
	BF$_4$		SC	Au
	NO$_3$		SC	Au
B. 2. Materials Containing Other Cation Radicals				
TTT	X			
	X = Cl	R	CP	Ag
	Br	R	CP	Ag
	I	R	CP	Ag
	SCN	R	CP	Ag
TTT$_2$	I$_3$	R, GS	SC	C
TTT$_2$	I$_3$	R, GS	SC	Pt.
TTT$_2$	I$_3$	R, GS	SC	C
	High disorder			
	Medium disorder			
	Least disorder			

(Continued)

Temp. Range of Metallic Behavior (K)	Room Temp. Conductivity, σ_{RT} (ohm cm)$^{-1}$	σ_{PK}/σ_{RT}	Temp. of σ_{PK}, T_{PK} (K)	Metal-Insulator Transition Temp. T_{MI} (K)	Sign of Majority Carrier at Room Temp. Estimated From	Ref.
	8×10^{-4}					39
	6×10^{-4}					39
	10					39
	20					39
	1					39
	0.6					39
4 – 300	540	~100	19	15	p:TEP	40
4 – 300	530	~15	~20	15	p:TEP	40
4 – 300	500	–		17	p:TEP	40
4 – 300	540	~10	70	39	p:TEP	40
4 – 300	780	~400	~15	12	p:TEP	40
	4.3×10^{-4}				n:TEP	41
	0.2				p:TEP	41
	1.4				p:TEP	41
	0.23				p:TEP	41
3.3 – 300	1000 ± 200	3.3	~70	21	p:TEP	42
4 – 300	$10^3 - 10^4$	8	35			43
4 – 3000	1000 ± 250	1.66	82	24 ± 1	p:TEP	44
	850 ± 250	2.1	92	~30	p:TEP	44
	500 ± 250	2.2	100	35 ± 1	p:TEP	44

Table 21.1

Substance		How Purified	Form	Contact
Electron Donor	Electron Acceptor			
TTT	$I_{1.6}$	R	SC	Au, Ag
TTT	I_X		SC	C
	$1.5 \leqslant X \leqslant 1.55$. For initial mixing conc. of I_2/TTT; $I_2/TTT = 0.5$			
	0.85			
	1.1			
	1.25; 1.50			
TTT	$I_{1.1}Br_{0.5}$		SC	
	$I_{0.8}Br_{0.7}$		SC	
$TSeT_2$	Cl		SC	
$TSeT_2$	Br		SC	
	CP			

B. 3. Materials Containing TCNQ Anion Radicals

NMP	TCNQ	R, GS	SC	Ag
			Sc	Ag
$NMP_X(Phen)_{1-X}$	TCNQ		SC	
X = 0.63				
X = 0.81				
X = 0.94				
X = 1.00				
Qn	$TCNQ_2$		SC	Ag
			CP	
			SC	Ag
Ad	$TCNQ_2$		SC	Ag
			SC	
N-methyl Ad	$TCNQ_2$		SC	Ag
2,2′Bipyridinium	$TCNQ_2$		SC	Ag

(Continued)

Temp. Range of Metallic Behavior (K)	Room Temp. Conductivity, σ_{RT} (ohm cm)$^{-1}$	σ_{PK}/σ_{RT}	Temp. of σ_{PK}, T_{PK} (K)	Metal-Insulator Transition Temp. T_{MI} (K)	Sign of Majority Carrier at Room Temp., Estimated From	Ref.
10 – 100	700 – 1500		100		p:TEP	45
4 – 300	1000 ± 100					
		~3	90	45		46
		~4	60	38		46
		~7	45	30		46
		~9	35	30		46
4 – 300	500	150 – 180				47
4 – 300	500	150 – 180				47
	2000		26			48a
	~10^3					48a
	20					48b
4 – 300	120 – 380	1.0 – 1.25	300 – 225			49, 50
~50 – 300					n:TEP	51
4 – 360						
	70	1.26	175		n:TEP	52
	100	1.85	155			52
	100	1.27	205		n:TEP	52
	200	1.17	220		n:TEP	52
10 – 300	70	1.07	240			50,53,54,55
					n:TEP	53
					n:TEP	51
10 – 300	70	1.5	140		n:TEP	50, 53
10 – 300	80	~1.5	150			54, 55
10 – 300	110	~1.5	180			55
10 – 300	40	~1.3	160			55

carrier materials has been made based on the nature of the conducting species, *e.g.,* TTF or TCNQ. The great variety of compounds shown reflects the desire, and capability, of the scientist to selectively modulate, *via* chemical synthesis, etc., various molecular parameters such as structural disorder, interchain coupling, the intrastack stacking configuration, the number of types of conducting species, and the concentration of impurities and defects. Electrical, magnetic, optical, thermal, and structural data exist for many of these compounds and the interested reader is referred to the references and review articles for further information.

The use of pressure to modulate the transport properties of quasi-one-dimensional conductors has proven to be one of the most fruitful techniques available.[9,25-28] Pressure studies have provided valuable insight into the nature of collective modes and incommensurate-commensurate transitions in (TTF)(TCNQ) and its derivatives.[29] Similarly, the metallic state in some organic conductors has been stabilized by pressure to very low temperatures. Even more dramatic, superconductivity has been observed, for the first time in an organic conductor, in $(TMTSeF)_2(PF_6)$ at 0.9 K and 12 kbar.[30]

The great interest and activity in these materials has naturally spilled over into the inorganic field. The quasi-one-dimensional polymer, polysulfur nitride, $(SN)_x$, has been found to be metallic down to 0.25 K, below which it becomes superconducting.[31] Similar properties have been observed in the brominated form.[14,32] However, recent studies indicate that $(SN)_x$ is more of an anisotropic three-dimensional metal than a quasi-one-dimensional conductor. Studies of another very interesting inorganic "chain" conductor, $NbSe_3$, have revealed the existence of large anomalies in the electrical resistivity which appear to be associated with the pinned Fröhlich mode.[33] However, $NbSe_3$, like $(SN)_x$, appears to be a highly anisotropic three-dimensional metal in spite of the one-dimensional features of its crystal structure. Finally, the search for novel one-dimensional organic conductors has led to the discovery, and study, of highly conducting polymers, *e.g.,* polyacetylene doped with halogens.[34] These materials provide a versatile and manipulatable environment to study many of the fundamental questions mentioned above,[35] and may also represent the most promising opportunity to realize important device applications of quasi-one-dimensional conductors.

The author wishes to acknowledge support from NASA, Contract NAS7-100.

Abbreviations Used in Table 21.1

σ_{PK}	peak conductivity at temperature, T_{PK}
R	recrystallized
GS	gradient sublimed
SC	single crystal
CP	compacted pellet
CF	cast film
C	aquadag
μ-wave	microwave
TEP	thermoelectric power
TTF	tetrathiafulvalene
TCNQ	tetracyanoquinodimethane
TSF	tetraselenafulvalene
MTCNQ	methyl TCNQ
TTT	tetrathiatetracene
TNAP	tetracyanonapthoquinodimethane
SCN	thiocyanate
SeC_N	selenocyanate
TSeT	tetraselenatetracene
NMP	N-methyl phenazine
Phen	Phenazine
Qn	quinolinium
Ad	acridinium
AVG	average
Me	methyl

References to Table 21.1

1. G. A. Thomas, D. E. Schafer, F. Wudl, P. M. Horn, D. Rimai, J. W. Cook, D. A. Glocker, M. J. Skove, C. W. Chu, R. P. Groff, J. L. Gillson, R. C. Wheland, L. R. Melby, M. B. Salamon, R. A. Craven, G. DePasquali, A. N. Bloch, D. O. Cowan, V. V. Walatka, R. E. Pyle, R. Gemmer, T. O. Poehler, G. R. Johnson, M. G. Miles, J. D. Wilson, J. P. Ferraris, T. F. Finnegan, R. J. Warmack, V. F. Raaen, and D. Jerome, *Phys. Rev.*, **B-13**, 5105 (1976), and references therein.

2. Marshall J. Cohen, L. B. Coleman, A. F. Garito, and A. J. Heeger, *Phys. Rev.*, **B-13**, 5111 (1976).

3. P. M. Chaikin, J. F. Kwak, T. E. Jones, A. F. Garito, and A. J. Heeger, *Phys. Rev. Lett.*, **31**, 601 (1973).
4. J. F. Kwak, P. M. Chaikin, A. A. Russell, A. F. Garito, and A. J. Heeger, *Solid State Commun.*, **16**, 729 (1975).
5. Nai-Phuan Ong, *Phys. Rev.*, **B-15**, 1782 (1977).
6. P. Chaudhari, B. A. Scott, R. B. Laibowitz, Tomkiewicz, and J. B. Torrance, *Appl. Phys. Lett.*, **24**, 439 (1974).
7a. M. J. Cohen, L. B. Coleman, A. F. Garito, and A. J. Heeger, *Phys. Rev.*, **B-10**, 1298 (1974).
7b. L. B. Coleman, *Review Sci. Instrum.*, **49**, 58 (1978); Ph.D. dissertation, University of Pennsylvania, 1975.
8. C. S. Jacobsen, J. R. Andersen, K. Bechgaard, and C. Berg, *Solid State Commun.*, **19**, 1209 (1976).
9a. P. M. Chaikin, R. L. Greene, and E. M. Engler, *Solid State Commun.*, **25**, 1009 (1978).
9b. Y. Tomkiewicz, R. A. Craven, T. D. Schultz, E. M. Engler, and A. R. Taranko, *Phys. Rev.*, **B-15**, 3643 (1977).
10. A. J. Schultz, G. D. Stucky, R. Craven, M. J. Schaffman, and M. B. Salamon, *J. Am. Chem. Soc.*, **98**, 519 (1976); A. J. Schultz and G. D. Stucky, *J. Phys. Chem. Solids*, **38**, 269 (1977).
11. D. O. Cowan, A. Bloch, T. Poehler, T. Kistenmacher, J. Ferraris, K. Bechgaard, R. Gemmer, C. Hu, P. Shu, W. Krug, R. Pyle, V. Walatka, T. Carruthers, T. Phillips, and R. Banks, *Mol. Cryst. Liq. Cryst.*, **32**, 223 (1976).
12. A. N. Bloch, J. P. Ferraris, D. O. Cowan, and T. O. Poehler, *Solid State Commun.*, **13**, 753 (1973).
13. J. P. Ferraris, T. O. Poehler, A. N. Bloch, and D. O. Cowan, *Tet. Lett.*, **27**, 2553 (1973).
14. E. Ehrenfreund, S. K. Khanna, A. F. Garito, and A. J. Heeger, *Solid State Commun.*, **22**, 139 (1977).
15. R. Schumaker, M. Ebenhahn, G. Castro, and R. L. Greene, *Bull. Am. Phys. Soc.*, **20**, 495 (1975).
16. R. L. Greene, J. J. Mayerle, R. Schumaker, G. Castro, P. M. Chaikin, S. Etemad, and S. J. LaPlaca, *Solid State Commun.*, **20**, 943 (1976).
17. L. B. Coleman, F. G. Yamagishi, A. F. Garito, A. J. Heeger, D. J. Dahm, M. G. Miles, and J. D. Wilson, *Phys. Lett.*, **51A**, 412 (1975).
18. E. Engler and V. V. Patel, *J. Am. Chem. Soc.*, **96**, 7376 (1975).
19. S. Etemad, T. Penney, E. M. Engler, B. A. Scott and P. M. Seiden, *Phys. Rev. Lett.*, **34**, 741 (1975); S. Etemad, *Phys. Rev.*, **B-13**, 2254 (1976).
20. S. Etemad, E. M. Engler, T. D. Schultz, T. Penney, and B. A. Scott, *Phys. Rev.*, **B-17**, 513 (1978).
21. P. M. Chaikin, R. L. Greene, S. Etemad, and E. Engler, *Phys. Rev.*, **B-13**, 1627 (1976).
22. Edward M. Engler, Robert A. Craven, Yaffa Tomkiewicz, Bruce A. Scott, Klaus Bechgaard, and Jan R. Andersen, *J. Chem. Soc. Chem. Comm.*, 337 (1976).
23. K. Bechgaard, D. O. Cowan, and A. N. Bloch, *J. Chem. Soc. Chem. Comm.*, 937 (1974).
24. R. E. Pyle, A. N. Bloch, D. O. Cowan, K. Bechgaard, T. O. Poehler, T. J. Kistenmacher, V. V. Walatka, R. Banks, W. Krug, and T. E. Phillips, *Bull. Amer. Phys. Soc.*, **20**, 415 (1975).

25. J. R. Anderson, C. S. Jacobsen, G. Rindorf, H. Soling, and K. Bechgaard, *J. Chem. Soc. Chem. Comm.*, 883 (1975).
26. A. N. Bloch, D. O. Cowan, K. Bechgaard, and T. O. Poehler, *Phys. Rev. Lett.*, **25**, 1561 (1975).
27. P. M. Chaikin, J. F. Kwak, R. L. Greene, S. Etemad, and E. M. Engler, *Solid State Commun.*, **19**, 1201 (1976).
28. Masao Mizuno, Anthony F. Garito, and Michael P. Cava, *J. Chem. Soc. Chem. Commun.*, **18**, (1978).
29. L. I. Buravov, O. N. Eremenko, R. B. Lyubovskii, L. P. Rozenberg, M. L. Khidekel, R. P. Shibaeva, I. F. Shchegolev, and E. B. Yagubskii, *JETP Lett.*, **20**, 208 (1974).
30. P. A. Berger, D. J. Dahm, G. R. Johnson, M. G. Miles, and J. D. Wilson, *Phys. Rev.*, **B-12**, 4085 (1975).
31. K. Bechgaard, C. S. Jacobsen, and N. Hessel Andersen, *Solid State Commun.*, **25**, 875 (1978).
32. R. Williams, C. Lowe Ma, S. Samson, S. K. Khanna, and R. B. Somoano, *J. Chem. Phys.*, (1980).
33. B. A. Scott, S. J. LaPlaca, J. B. Torrance, B. D. Silverman, and B. Welber, *J. Am. Chem. Soc.*, **99**, 6631 (1977).
34. S. J. LaPlaca, P. W. R. Corfield, R. Thomas, and B. A. Scott, *Solid State Commun.*, **17**, 635 (1975).
35. R. J. Warmack, T. A. Callcott, and C. R. Watson, *Phys. Rev.*, **B-12**, 3336 (1975).
36. R. B. Somoano, A. Gupta, V. Hadek, T. Datta, M. Jones, R. Deck, and A. M. Hermann, *J. Chem. Phys.*, **63**, 4970 (1975).
37. F. Wudl, D. E. Schafer, W. M. Walsh, Jr., L. W. Rupp, F. W. DiSalvo, J. V. Waszczak, M. L. Kaplan, and G. A. Thomas, *J. Chem. Phys.*, **66**, 377 (1977).
38. R. B. Somoano, A. Gupta, V. Hadek, M. Novotny, M. Jones, T. Datta, R. Deck, and A. M. Hermann, *Phys. Rev.*, **B-15**, 595 (1977).
39. J. B. Torrance, J. J. Mayerle, V. Y. Lee, and K. Bechgaard, preprint.
40. K. Bechgaard, C. S. Jacobsen, K. Mortensen, H. J. Pedersen, and N. Thorup, preprint.
41. E. A. Perez-Albuerne, H. Johnson, Jr., and D. J. Trevoy, *J. Chem. Phys.*, **55**, 1547 (1971).
42. L. C. Isett and E. A. Perez-Albuerne, *Solid State Commun.*, **21**, 433 (1977); L. C. Isett, *Phys. Rev.*, **B-18**, 439 (1978).
43. B. Hilti and C. W. Mayer, *Helv. Chim. Acta*, **61**, 50 (1978).
44. S. K. Khanna, S. P. S. Yen, R. B. Somoano, P. M. Chaikin, C. Lowe Ma, R. Williams, and S. Samson, *Phys. Rev.*, **B-19**, 655 (1979); R. B. Somoano, S. P. S. Yen, V. Hadek, S. K. Khanna, M. Novotny, T. Datta, A. M. Hermann, and J. A. Woollam, *Phys. Rev.*, **B-17**, 2853 (1978).
45. G. Mihaly, A. Janossy, and G. Gruner, *Solid State Commun.*, **22**, 771 (1977).
46. V. F. Kaminskii, M. L. Khidekel, R. B. Lyubovskii, I. F. Shchegolev, R. P. Shibaeva, E. B. Yagubskii, A. V. Zvargkina, and G. L. Zverena, *Phys. Status Solidi*, **A-44**, 77 (1977); L. I. Buravov, G. I. Zereva, V. F. Kaminskii, I. F. Shchegolev, and E. B. Yagubskii, *J. Chem. Soc. Chem. Commun.*, 720 (1976).
47. K. Kamaras, G. Mihaly, G. Gruner, and A. Janossy, *J. Chem. Soc. Chem. Comm.*, 979 (1978).
48a. S. P. Zololukhin, V. F. Kaminskii, A. I. Kotov, R. B. Lyubovskii, M. L. Khidekel, R. P. Shibaeva, I. F. Shchegolev, and E. B. Yagubskii, *Zh. Eksper. Teor. Fiz.*, Pisma, in press.

48b. M. G. Kaplunov, K. I. Pokhodnya, A. I. Kotov, E. B. Yagubskii, T. A. Kitaeva, and Yu. G. Borodko, *Phys. Status Solidi,* **A-43,** K73 (1977).

49. A. J. Epstein, E. M. Conwell, D. J. Sandman, and J. S. Miller, *Solid State Commun.,* **23,** 355 (1977); A. J. Epstein, S. Etemad, A. F. Garito, and A. J. Heeger, *Phys. Rev.,* **B-5,** 952 (1972); L. B. Coleman, J. A. Cohen, A. F. Garito, and A. J. Heeger, *Phys. Rev.,* **B-7,** 2122 (1973).

50. I. F. Shchegelov, *Phys. Status Solidi,* **A-12,** 9, (1972); L. I. Buravov, M. L. Keidekel, I. F. Shchegelov, and E. B. Yagubskii, *JETP Lett.,* **12,** 99 (1970).

51. J. F. Kwak, G. Beni, and P. M. Chaikin, *Phys. Rev.,* **B-13,** 641 (1976).

52. A. J. Epstein and J. S. Miller, *Solid State Commun.,* **27,** 325 (1978); A. J. Epstein, J. S. Miller, and P. M. Chaikin, *Phys. Rev. Lett.,* **43,** 1178 (1979).

53. L. I. Buravov, D. N. Fedutin, and I. F. Shchegelov, *Sov. Phys. JETP.,* **32,** 612 (1971).

54. G. Mihaly, K. Ritvay-Emandity, and G. Gruner, *J. Phys. C: Solid State Phys.,* **8,** L361 (1975).

55. K. Holczer, G. Mihaly, A. Janossy, G. Gruner, and M. Kertesz, *J. Phys. C: Solid State Phys.,* **11,** 4707 (1978).

References

1. I. F. Shchegelov, *Phys. Status Solidi* (a), **12,** 9 (1972).
2. Z. G. Soos, *Ann. Rev. Phys. Chem.,* **25,** 121 (1974).
3. A. F. Garito and A. J. Heeger, *Accts. of Chem. Res.,* **7,** 232 (1974).
4. Galen D. Stucky, Arthur J. Schultz, and Jack M. Williams, *Ann. Rev. Mater. Sci.,* **7,** 301 (1977).
5. A. J. Berlinsky, *Contemp. Phys.,* **17,** 331 (1976).
6. I. N. Bulaevskii, *Usp. Fiz. Nauk,* **115,** 263 (1975); *Sov. Phys.-Usp.,* **18,** 131 (1976).
7. H. J. Keller, ed., *Low Dimensional Cooperative Phenomena, The Possibility of High Temperature Superconductivity,* Plenum Press, N.Y., 1974.
8. H. J. Keller, ed., *Chemistry and Physics of One-Dimensional Metals,* Vol. 25, NATO Advanced Study Institute Sereis B-Physics, Plenum Press, N.Y., 1977.
9. R. H. Friend and D. Jerome, *J. Phys. C: Solid State Phys.,* **12,** 1441 (1979).
10. G. A. Toombs, *Phys. Rep.,* **40C,** 181 (1978).
11. L. Pal, G. Gruner, A. Janossy, and J. Solyom, eds., *Lecture Notes in Physics,* Series No. 65, Springer-Verlag, Berlin, N.Y., 1977.
12. H. G. Schuster, ed., *Lecture Notes in Physics,* Series No. 34, Springer-Verlag,
13. J. H. Perlstein, *Angew. Chem.* (Int. Ed. Engl.), **16,** 519 (1977).
14. J. S. Miller and A. J. Epstein, eds., "Proceedings of Conference on Synthesis and Properties of Low Dimensional Materials," *Annals of N.Y. Acad. Sci.,* **313,** 1978.
15. J. S. Miller and A. J. Epstein, *Progr. Inorg. Chem.,* **20,** 1 (1976).
16. R. E. Peierls, *Quantum Theory of Solids,* Clarendon, Oxford, 1955, p. 108.
17. M. J. Rice and S. Strassler, *Solid State Commun.,* **13,** 125 (1973).
18. J. Bardeen, L. N. Cooper, and J. R. Schriefer, *Phys. Rev.,* **108,** 1175 (1957).
19. L. D. Landau and E. M. Lifshitz, *Statistical Physics,* Pergamon, Oxford, 1958.
20. K. Carneiro, A. S. Petersen, A. E. Underhill, D. J. Wood, D. M. Watkins, and G. A. Mackenzie, *Phys. Rev.,* **B-19,** 6279 (1979).
21. B. Horowitz, H. Gutfreund, and M. Weger, *Phys. Rev.,* **B-12,** 3174 (1975).
22. V. Emery, Ref. 8.

REFERENCES

23. P. A. Lee, T. M. Rice, and P. W. Anderson, *Solid State Commun.*, **14**, 703 (1974); M. J. Rice, A. R. Bishop, J. A. Krumhansl, and S. E. Trullinger, *Phys. Rev. Lett.*, **36**, 432 (1976).
24. W. A. Little, *Phys. Rev.*, **A134**, 1416 (1964).
25. J. R. Cooper, D. Jerome, S. Etemad, and E. M. Engler, *Solid State Commun.*, **22**, 257 (1977); J. R. Cooper, M. Weger, D. Jerome, D. LeFur, K. Bechgaard, A. N. Bloch, and D. O. Cowan, *Solid State Commun.*, **19**, 749 (1976); A. Andrieux, P. M. Chaikin, C. Duroure, D. Jerome, C. Weyl, K. Bechgaard, and J. R. Andersen, *J. Physique*, **40**, 1199 (1979).
26. R. H. Friend, D. Jerome, J. M. Fabre, L. Giral, and K. Bechgaard, *J. Phys. C. Solid State Phys.*, **11**, 263 (1978).
27. R. H. Friend, M. Miljak, and D. Jerome, *Phys. Rev. Lett.*, **40**, 1048 (1978).
28. R. L. Greene, J. J. Mayerle, R. Schumaker, G. Castro, P. M. Chaikin, S. Etemad, and S. J. LaPlaca, *Solid State Commun.*, **20**, 943 (1976).
29. A. Andrieux, H. J. Schultz, D. Jerome, and K. Bechgaard, Preprint.
30. D. Jerome, A. Mazaud, M. Ribault, and K. Bechgaard, *J. Physique Lett.*, **41**, L95 (1980), M. Ribault, G. Benedek, D. Jerome, and K. Bechgaard, preprint.
31. H. P. Geserich and L. Pintschovius, in *Advances in Solid State Physics*, Vol. XVI, J. Treusch, ed., Vieweg, Braunschweig, Germany; C. Hsu and M. M. Labes, *J. Chem. Phys.*, **61**, 4640 (1974); A. G. MacDiarmid, C. M. Mikulski, M. S. Soran, P. J. Russo, M. J. Cohen, A. A. Bright, A. F. Garito, and A. J. Heeger, in "Advances in Chemistry Series," No. 150 (1976); R. H. Baughman, P. A. Apgar, R. R. Chance, A. G. MacDiarmid, and A. F. Garito, *J. Chem. Phys.*, **66**, 401 (1977); R. L. Greene and G. B. Street in Ref. 8.
32. C. Bernard, A. Herold, M. Lelauvain, and C. Robert, *C. R. Acad. Sci. Ser.*, **C283**, 625 (1976); W. D. Gill, W. Gludau, R. H. Geiss, P. M. Grant, R. L. Greene, J. J. Mayerle, and G. B. Street, *Phys. Rev. Lett.*, **38**, 1305 (1977).
33. P. Haen, P. Monceau, B. Tissier, G. Waysand, A. Meerschaut, P. Molinie, and J. Rouxel, *Proceedings Fourteenth International Conference on Low Temperature Physics*, Otaniami, Finland, 1975, **5**, p. 445; N. P. Ong and P. Monceau, *Phys. Rev.*, **B16**, 3443 (1977); R. M. Fleming, D. E. Moncton, and D. B. McWhan, *Phys. Rev.*, **B18**, 5560 (1978); N. P. Ong, J. W. Brill, J. C. Eckert, J. W. Savage, S. K. Khanna, and R. B. Somoano, *Phys. Rev. Lett.*, **42**, 811 (1979).
34. C. K. Chiang, Y. W. Park, A. J. Heeger, H. Shirakawa, E. J. Louis, S. C. Gau, and A. G. MacDiarmid, *Phys. Rev. Lett.*, **39**, 1098 (1977); J. F. Kwak, T. C. Clarke, R. L. Greene, and G. B. Street, *Solid State Commun.*, **31**, 355 (1979); C. R. Fincher, Jr., M. Ozaki, M. Tanaka, D. Peebles, L. Lauchlan, A. J. Heeger, and A. G. MacDiarmid, *Phys. Rev.*, **B20**, 1589 (1979).
35. M. J. Rice, *Phys. Lett.*, **71A**, 152 (1979); W. P. Su, J. R. Schriefer, and A. J. Heeger, *Phys. Rev. Lett.*, **42**, 1698 (1979).

22

Photo Effects

22.1 Introduction

Basically, all photoeffects rest upon the absorption of photons producing either an excited state or liberating electrons directly. In the latter case, photoemission may occur in that the electrons are emitted from the surface of the solid. The excited state may revert to the ground state, by any of a number of mechanisms discussed in Chapter 13, or it may result in a photochemical reaction, or else, the electron-hole pair, *i.e.*, the exciton which is the primary photoproduct, may dissociate into at least one free carrier, the other one usually remaining localized.

It is the way in which the electron-hole pair is separated which determines which photoeffect is being measured in an experiment: if the separation occurs by means of an externally applied electric field, photoconductivity is observed and a photocurrent is measured. If the separation occurs by means of an externally applied magnetic field, photomagnetic effects arise. Separation may also occur by means of differences in the mobility and diffusivity of the electrons and holes, giving rise to photovoltaic effects. All these will be briefly discussed in the following Sections which are supplementary to Sections 1.23, 2.16, 2.17, 2.18, 2.19, and 2.20 to 6.11, 8.11 and 8.12 in part A.

Since the primary photoeffect is always the same, *viz.*, the absorption of photons, the action spectra of photoelectric effects bear a close resemblance to optical absorption spectra.

Photoeffects usually are very fast events involving extremely short relaxation times. Schein et al.[1] report observations of current, time-resolved to 800 psec, in the transient photoconductivity of anthracene. These measurements indicate that the transient currents associated with photogeneration, polaron formation, and carrier motion are over within 800 psec of the light illumination. A new transient photoconductivity effect, *viz.* a current peak whose duration equals the duration of the laser illumination, is reported but it origin is not clear at all.[1]

Photoemission threshold values are listed in Tables 6.7A1 and A2 for organic solids and liquids including several charge transfer complexes, though a few entries are duplicated in that values from different sources are listed without preference. Sensitization of photoconductivity is treated in Section 22.4, and Tables 9.2A and B gives numerical values for the sensitization of polynaphthalene, polyacenaphthylene, and of polyvinylcarbazole with dyes and with Lewis acids these tables contain entries for acids, ketones, aldehydes, quinones, and others by phenylmethane, xanthene, azine, acridine, thiazine, cyanine and azo-dyes. Photoconductivity data are tabulated in Tables 8.15A1 and 8.15A2; a review by Lakin et al., is available.[99]

22.2 Contacts

Contact potentials and contact energy barriers play a decisive role in all photoeffects (see Sections 12.9, 15.12, 22.1). Schottky barriers are discussed in detail in Section 15.12, and the Frenkel-Poole effect in Section 15.10. Charge injection from contacts is dealt with in Section 12.16. The present Chapter is intended to supplement these discussions by focussing on aspects closely related to photoeffects.

Cations arising from the dissociating of charge transfer complexes may[3] pull electrons out of a metallic contact; this has been observed[3] in sandwich cells containing tetramethyl-*p*-phenylene-diamine (TMPD) in a polar solvent capable of forming a charge transfer complex with the TMPD such as *e.g.*, acetonitrile. These electrons are then trapped giving rise to a blue discoloration.[3] Thus, the work function of metal electrodes employed in such studies should be high enough to prevent similar effects.

Adsorption on an organic substrate of a suitable reducible solute influences the magnitude of the hole saturation photocurrent and hence the charge-carrier injection efficiency.[4] Efficiencies up to 0.65 have been observed[4] with organic photoinjectors, and 0.31 for a hexacyanoferrate (III) hole photoinjector in the singlet region of anthracene (λ 395 nm). About 5 ± 4 % of the holes formed in the initial photoinjection event do not reach the external circuit, presumably[6] because they react at the injecting surface.

Excited dye molecules bound to the surface of a *n*-type semiconductor electrode may inject an electron into the conduction band, if the level of the excited state is positioned at an appropriate energy. This process leads to a photocurrent through the electrolyte electrode interface to the bulk of the semiconductor and if the oxidized dye is regenerated by an electron donor in the electrolyte.[5]

This current may be modified in its spectral and chemical characteristics by coupling the excitation of the injecting dye to another dye exhibiting desired physical and chemical features: the injecting dye A is excited by energy transfer from a donor dye D. Due to the spectral shift of donor and acceptor, the electron injection from dye A is promoted by light not effective in dye A alone. Moreover, while the injecting dye A is optimized with respect to the injection process. D may be selected independently with additional chemical or physical sensitivities. If certain photophysical properties of the donor are affected, the energy transfer process mediates such a sensitivity to the acceptor dye, leading to a modulation of the injected current.[5]

In the coupled system the donor dye D need not be located at the surface of the semiconductor, and thus does not compete with A for space, and does not affect the injection yield of this dye.[5] This has been used to construct photoelectric cells by the assembly of mixed monolayers,[6] *e.g.*, of a cyanine dye with the chromophores in the layer plane and a chainlike π-electron system oriented perpendicular to the layer plane, sandwiched between fatty acid monolayers and metal electrodes.[6] For further discussion see Section 22.5 and 23.2.

Mulder et al.,[7] using dilute Na_2SO_3 solutions as electrolytic contacts, measured photocurrents in single crystals of anthracene due to light absorption in the 25000 to 30000 cm^{-1} region. The photocurrent consequent upon illumination of the positive electrode, i_{pos}, was 50-100 times larger than that resulting from illumination of the cathode, i_{neg}. They deduce that i_{pos} arises virtually exclusively at the anode-crystal interface due to injection of positive holes into the solid. i_{neg} is said[7] to be due to re-absorption of fluorescent radiation. Doping of the crystal with acridine inhibited fluorescence and i_{neg} did drop by 2 to 3 order of magnitude.[8]

Other reports indicate[9] that the photoconductivity of anthracene crystals equipped with electrolytic contacts depends on both the electrolyte and crystal parameters such as depth of penetration of the incident light, and trap-concentration and -depth. The following RedOx systems were employed as hole injecting contacts: Ce^{4+}/Ce^{3+}, Fe^{3+}/Fe^{2+}, Tl^{3+}/Tl^{2+}, $IrCl_6^{2-}/IrCl_6^{3-}$. and $PtCl_6^{2-}/PtCl_6^{3-}$. Although direct reactions of light generated excitons with the electron accepting electrolyte are not impossible, they are unlikely because of the well-substantiated[10] presence of surface traps in anthracene. It is more likely that the exciton decays in the layer adjoining the surface, but the electron does not pass immediately into solution. It first arrives at the surface level and then passes into the electrolyte. Matters are further complicated because of the uncertainty in local permittivity of the interface; at high light intensity there is also the chance, that the rate of the electrochemical electrode reaction may be insufficient to sustain the current. Concepts of tunnelling at electrolytic contacts have been critically reviewed and discussed by[11] Duke.

22.3 Carrier Generation

The generation of charge carriers in organic materials is a complex process,[100] particularly for light of photon energy lower than the first electronic transition. However, surface generation of charge carriers from singlet-excitons, produced either in the bulk or on the surface, is often the most important charge carrier generation process.[12] This has been found to be true *e.g.*, for durene,[12] for which an activation energy of $\simeq 0.1$ eV for both hole and electron generation in the range 150-300 K, has been reported. For light from a nitrogen laser ($\lambda = 3371$ Å), the charge-carrier generation process involves a direct two-photon absorption to produce singlet excitons which then dissociate at surface states giving charge carriers. A lower limit of 10^{-48} cm^4 sec/(photon molecule) was estimated for the molecular singlet-singlet excitation-rate constant.[12] Steady state one-photon intrinsic photogeneration in anthracene crystals was measured by Lyons and Milne[13] who used 1M Na_2SO_3 aqueous solutions as electrodes, as a function of photon energy and electric field (up to 6×10^7 Vm^{-1}). With aqueous electrodes, values of the absolute photoconduction yield now agree for aqueous, blocking, and metal electrodes, and for pulsed and steady state measurements. See Fig. 22.1. Interpretation of the current-field curves in terms of Onsager's theory (see Section 14.4) yield values of r_0 (the initial separation of the ion pair) as a function of photon energy and dielectric constant, assuming a delta function distribution for r_0. Two ion pair states were formed: at $h\nu = 3.9$ to 4.2 eV, $r_{0A} = 1.8$ nm (for $\epsilon_r = 3.8$); at 4.5 to 5.0 eV, $r_{0B} = 2.5$ nm. For $h\nu = 4.2$ to 4.5 eV, r_0 experienced a smooth transition. The maximum value (0.1) of ϕ_1, the one-photon ion pair production efficiency, occurred at 4.2 ± 0.15 eV, near the maximum in $\bar{\eta}$, the two-photon ion pair yield. From the energy dependence of ϕ_1 and the magnetic field effect on prompt fluorescence, it was concluded[13] that, in the process of singlet exciton fission into two triplet excitons, the lowest ion pair state was not the fissionable singlet. Only a very small [0.07($\pm$0.03)%], direction-independent, magnetic field-effect on the intrinsic photocurrent was observed at a photon energy of 4.5 eV. A plot of photon energy *vs.* photoconduction yield is given in Fig. 22.2 for anthracene.

It is satisfying to note from Fig. 22.2 that the photoconduction quantum yields as measured by two different workers do agree reasonably well with

CARRIER GENERATION 423

each other as well as with theoretical predictions.

While it is seen from Fig. 22.1 that the data could be explained by the assumption that a Schottky energy barrier (see Section 15.12) determines the photocurrent at constant illumination, field emission and tunnelling effects appear to be negligible, since the slopes do not agree with what would be expected from a Poole-Frenkel effect as the current determining factor (see Section 15.10).

However, it appears that the most satisfactory explanation of the photoconduction yield, or quantum yield, as expressed in terms of the photocurrent generated, is based on the Onsager theory of geminate recombination:[14] Extrinsic photoconduction in the case of anthracene crystals is generally caused by the reactions of singlet excitons at the crystal surface, although a biexcitonic mechanism may be operative. The characteristics[13] of these photocurrents are:

1. their spectral dependence follows the absorption coefficient,

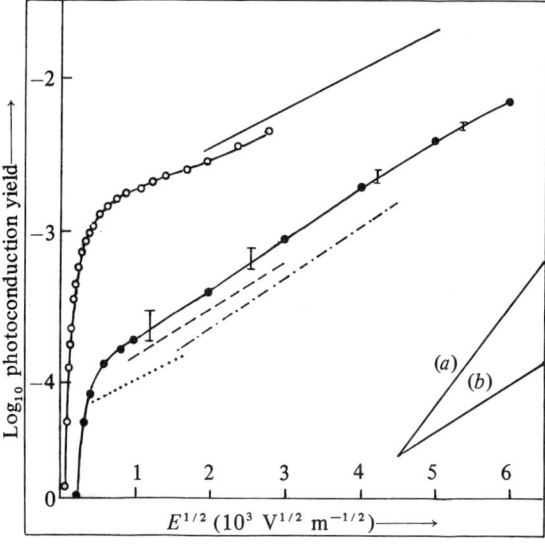

Fig. 22.1 Variation of photoconduction yield with $E^{1/2}$ for anthracene single crystals. (E, electric field strength). ——— H_2O contacts $h\nu = 4.4$ eV. – – – – Al contacts, $h\nu = 4.4$ eV. ········ blocking contacts $h\nu = 4.9$ eV. – · – · – blocking contacts $h\nu = 4.4$ eV. –o–o– present work: H_2O contacts $h\nu = 4.5$ eV. –•–•–: average of four crystals. Polyvinyl alcohol on back surface and aqueous 1.0 M Na_2SO_3 contacts, $h\nu = 4.5$ eV, (a) Poole-Frenkel slope ($\epsilon_r = 3.8$). (b) Schottky slope ($\epsilon_r = 3.8$). Error bars indicate reproducibility of the yield-field relationship. ϵ_r is relative permittivity. After Lyons and Milne.[13]

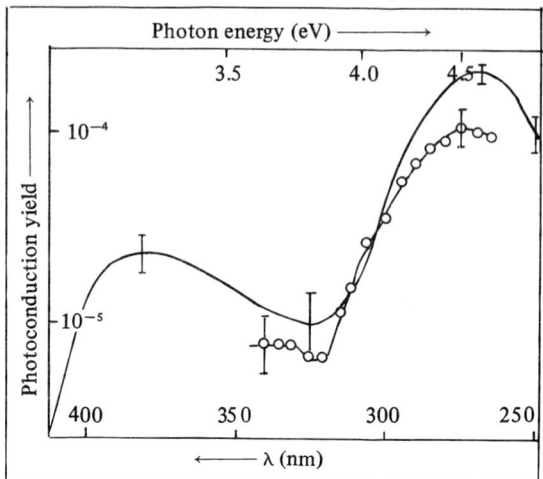

Fig. 22.2 Variation of electron photoconduction yield with photon energy, for anthracene crystals. ——— by E. Schmid in article by N. Karl, *Adv. Solid State Physics*, **14**, 261 (1974); $10^5 - 10^6$ Vm^{-1}; bare crystal, high vacuum. ο ο ο experimental results; polyvinyl alcohol film on back surface; 1.0 M Na_2SO_3 contacts; thickness, 12 μm; field 4×10^6 Vm^{-1}. After Lyons and Milne.[13]

2. they are dependent on the electrodes and the nature of the crystal surface,
3. they are decreased if fluorescence quenchers are introduced into the crystal lattice,
4. photon flux dependences are usually linear, but could be quadratic if a biexcitonic mechanism is operative,
5. hole yields are usually much greater than electron yields.

In contrast, intrinsically generated photocurrents are:

(a) independent of, or only very weakly related to the absorption coefficient (hence independent of the polarization of the light),
(b) independent of electrodes and the nature of the surface,
(c) not quenched by fluorescence quenchers incorporated in the crystal lattice,
(d) not greater than linearly dependent on photon flux,
(e) approximately equal for hole and electron currents.[7]

Figure 22.1 shows that use of polyvinylalcohol on the back surface of the crystal, plus the use of 1.0 M Na_2SO_3 contacts, significantly decrease the current. This indicates that extrinsic charge injection (*e.g.*, photoinjection

CARRIER GENERATION

from aqueous O_2) from the back surface was prevented, so that the observed current was intrinsic, supported by the following observations:

1. The absolute steady state photoconduction yields, using aqueous contacts, agree in magnitude and field dependence with those for blocking and aluminum contacts and measurements by pulse techniques (Fig. 22.1).
2. The action spectrum for photogeneration of free charge carriers at high fields is independent of the use of aqueous or blocking contacts (Fig. 22.2),
3. Photoconduction yields for holes and electrons are equal at the highest fields,
4. The photocurrent is proportional to the photon flux at either high or low fields over the range 5×10^{10} to 2×10^{12} photons cm^{-2} sec^{-1},
5. The magnitude of the photocurrent is independent of the polarization of the light.

The process of intrinsic photogeneration of free charge carriers may be divided into three steps:

1. Absorption of light to form an excited crystal state of lifetime 10^{-13} sec involving the vibronic manifold of an excited singlet,
2. Formation of an ion pair, with an efficiency ϕ_1, the distance between the positive and negative charge being r_0. The Onsager model[14] assumes an initial ion pair, and calculates the probability of the carriers escaping geminate recombination (see Section 14.4 and 22.3).
3. Escape of the carriers, to form free carriers in the conduction band.

The theory of Knights and Davis[15] assumes that the electron-hole pair is formed by thermalization of the hot electron, *via* diffusion and scattering, to a position on the potential curve, at a distance r_0, dependent on field, from the positive charge. The theory leads to

$$\log_{10}\left(\frac{1}{\phi_{es}} - 1\right) + bE^{1/2}/2.303\,kT = (eEr_0/2.303\,kT)$$

$$+ \frac{e^2}{(4\pi\epsilon r_0)2.303\,kT} - \log_{10}(\nu_e \tau_r) \quad (22.1)$$

where $b = (e^3/\pi\epsilon)^{1/2}$; $\epsilon = \epsilon_r \epsilon_0$; ϵ_r, relative dielectric constant; ϵ_0, permittivity of a vacuum (Fm^{-1}); k, Boltzmann's constant; T, absolute temperature (K); e, the magnitude of the electronic charge (C); ν_e, an attempt-to-escape frequency (sec^{-1}); τ_r, the recombination lifetime of the electron-hole pair (sec). However, the Knights and Davis model can be made to fit the experimental results only for $E > 10^7$ Vm^{-1}; *i.e.*, over a smaller range than does the Onsager theory, see below, which theory is further discussed in Section 14.4. Lyons and Milne[13] conclude that the Onsager model is adequate to explain the results. The Poole-Frenkel, Schottky, and Knights and Davis models each fails to consider the effect of three-dimensional diffusion which greatly enhances the probability of carrier separation at lower fields, especially in solid anthracene where the mean free path is of the order of a lattice spacing and the potential barrier maximum. At high fields the Knights and Davis model agrees with the Onsager model. The drop at low fields ($10^4 < E < 10^5$ Vm^{-1}), of the intrinsic photoconduction yield in anthracene below that predicted by the Onsager[14] theory was first reported by Batt, Braun and Hornig[16] for pulsed measurements. They explained this reduction by assuming bulk recombination of holes and electrons in the regeneration region: at low fields, free carriers are not drawn from the generation region quickly enough, and, although they escape geminate recombination, it is possible for them to encounter other opposite charges and recombine before escaping the generation region.

Karl and Sommer[17] studied the pulsed intrinsic photoconductivity of anthracene and also suggested an explanation of the low-field reduction, in terms of recombination between free electrons and free holes: The presence of traps of chemical or structural origin causes (in the presence of excess charge density) the concentration of free carriers to be several orders of magnitude below the concentration of trapped carriers. Thus the recombination rate of free carriers with free carriers is negligible compared with the recombination rate of free carriers with trapped carriers. Their neglect of the effect of traps resulted at low fields in a less than linear dependence of the current on photon flux, an effect not seen in experimental work.[18] Chance and Braun[19] discussed the problem further in order to explain their pulsed photoconductivity results: (a) that the individual slope/intercept ratios of the high-field ($E > 10^5$ Vm^{-1}) yield-field relationships showed a definite correlation with the absorption coefficient of the exciting light, the results for shallow penetrating light being higher than for deep penetrating light; (b) that a strict linear variation of the number of carriers with photon flux (even at the lowest fields) was observed. Chance and Braun[19] suggested that free-carrier-trapped-carrier bulk recombination was responsible for the observed low-field deviations from the predictions of the Onsager[14] theory. The excellent agreement between their experimental second-order rate constant for free-electron-trapped-hole recombination and diffusion-controlled recombination rate constant confirmed that the assumption of free-trapped recombination is sufficient to explain the reduced yields. The investigations of Batt, Braun and Hornig,[16] Karl and Sommer[17] and Chance and Braun[19] were concerned with pulsed photoconductivity in which both time and position in the crystal are variables.

In the steady-state a reduction of the photoconduction yield at low fields, similar to that for pulsed photoconductivity, has been observed.[18] The current varied linearly with photon flux at low fields (as well as at high fields[20]), which indicates that free-carrier-trapped-carrier bulk recombination may explain the observed yield-field relationship.

Free carriers produced by intrinsic photogeneration, even if they escape geminate recombination, are subject to fates other than contributing to the external current. At high fields, the only pathway for loss of carriers from the external current is that of geminate recombination. At low fields, however, bulk recombination accounts for the loss of even more free carriers. Surface ejection of both positive and negative carriers at the illuminated crystal surface also modifies the external current. For a fuller discussion of the theory, the original paper[18] should be consulted. In an earlier study, however, Silver and Jarnagin[21] concluded that geminate recombination does not dominate photoconductivity phenomena. While they base their argument mainly on the measurements by Chaiken and Kearns on the photocurrent obtained from fused anthracene,[22] the weight of the present evidence suggests that the Onsager theory[14] of geminate recombination does indeed provide the most satisfactory model at least at this juncture (see discussion in Ref. 23). Further support for the dominating role of geminate recombination in photoconductivity comes from the studies of Borsenberger[23] on 11 μm thick films of aggregates consisting of 2 phases with a ternary structure: a crystalline phase imbedded in an amorphous matrix. This has been realized[23] by employing a dye plus a polymer such as a bisphenol-polycarbonate polymer as the crystalline aggregate while the amorphous matrix consisted of the same polymer or its 60% solution. The electrodes were mercury. The dc dark conductivity ranged from 10^{-13} to 10^{-15} (ohm cm)$^{-1}$, depending on the applied field and is said[34] to be dominated by carrier injection from the contacts. Under illumination, electrons as well as holes are said[23] to be generated, exhibiting the remarkably low mobility of about 10^{-8} cm^2/Vsec. For percolation theory (see Section 13.3 and 19.4) to apply, at least 25 volume% must be assumed to be occupied by conducting spheres or equivalent, while in this case, the aggregate occupied only 5%. The field dependent quantum yield was found to be about 0.5 at 1000 kV/cm; the author concludes that in view of the photoconductivity being affected by both electron and holes Onsager's geminate recombination model does apply. The Onsager critical distance in the non-sensitized material is increased to 54 Å.

The aggregate phase is reported[23] to exhibit a filamentary structure giving rise to essentially one-dimensional conductivity.

The photogenerated carriers are said[34] to have drift ranges of the order of a few Å per unit applied field as long as this is kept above about 60 kV/cm.

This interesting system also shows[23] an electret effect (see Section 20.4); if it is charged in the dark in a field of about 1000 kV/cm, it is found that the subsequent dark discharge current rises sharply upon heating the material above the glass transition temperature of the amorphous phase, *viz.*, about 50° C.

The theory of photon-induced dc hopping conductivity in disordered systems has been treated by Keiper and Schuchard[24] who find that the photocurrent is mainly dependent on the electron mobility and, at very low light intensities, proportional to the square of the light frequency ω. At high illuminations, it is said[24] to become proportional to $\sinh^{-1/3}(\omega)$. The process appears to be highly sensitive to the spectrum of energy states within the mobility gap and to result in charge transfer by variable range hopping (see also Section 15.4, 15.5).

The corresponding ac photoconductivity has been studied by Böttger and Bryskin.[25]

For an authoritative discussion of carrier photogeneration in amorphous solids, the reader is referred to a review by Mott.[26]

Gailis et al.,[27] report that the activation energy for carrier photogeneration in pentacene in a field of 300 kV/cm approximately equals the thermal energy kT, which would indicate quite a high quantum efficiency. Tetracene is said[27] to yield similar results though requiring higher values for the applied field; in both these compounds, the activation energy dropped with increasing field. Somewhat similar results are reported by Balode[28] for thiotetracene; the carriers are said to be positive.

Geacintov et al.,[29] studied the photocurrent and photovoltage of tetracene single crystals with distilled water electrodes as a function of the excitation wavelength in the 220-560 mμ region. At wavelengths longer than 410 mμ, the photocurrent was due to injection of holes at the illuminated electrode. A bulk-generated electron photocurrent was produced with excitation energies in excess of 3 eV, *i.e.*, at wavelengths less than 410 mμ. The relative fluorescence efficiency (λ) began to decrease with decreasing wavelengths at 410 mμ. The drop in $\phi(\lambda)$ at excitation energies greater than 3 eV was said to be due to the appearance of a nonradiative competitive process. This process involved mostly the formation of nonconducting and nonfluorescent (or weakly fluorescent) charge-transfer states. Formation of separated charges with low quantum efficiency ($\phi_i < 5.10^{-3}$) also occurred which gave rise to a bulk-generated negative current. Since this current appeared only for excitation energies >3 eV, this value was associated with the smallest band gap for conductivity in tetracene.

The wavelength dependence of the bulk-generated negative photocurrent is related to varying probabilities of transition from the different excited electronic states of the crystal to ionized states.[7] The positive hole current was mostly due to the diffusion of the lowest-energy singlet excitons to the surface and their dissociating at the electrode. An exciton diffusion length of 2000 Å was calculated[7] from the experimental data.

Quite high values of quantum yield are obtainable with organic liquid systems. Thus, naphthalene-sensitized charge transfer complexes between ruthenocene and CCl_4 in a methanol or CCl_4 solvent are reported[30] to produce a quantum efficiency of 0.72 at 313 nm; this is said[30] to proceed *via* a triplet state.

Photoconductivity in a sandwich-type cell of a poly(N-vinyl carbazole) film with transparent Au electrodes was investigated[31] under high vacuum and in the visible region. In low applied fields, no difference was found between the photoconductivity with positive-electrode illumination and that with negative-electrode illumination. In moderate or high fields, however, the photoconductivity with positive-electrode illumination was significantly different from that with negative-electrode illumination. The former showed a[31] superlinear dependence on the applied field, $i_{pos} \gg i_{neg}$ and a spectral response coinciding with the absorption spectrum was observed,[31] as illustrated in Fig. 22.3.

In intermediate applied fields as large as 35000 V/cm, with negative-electrode illumination, a photoresponse curve of the B-type was observed only at 300–310 mμ. On the other hand, with positive-electrode illumination it was still observed at 270–340 mμ. In high applied fields as large as 150000 V/cm, almost every photoresponse curve, was of the A type.[31]

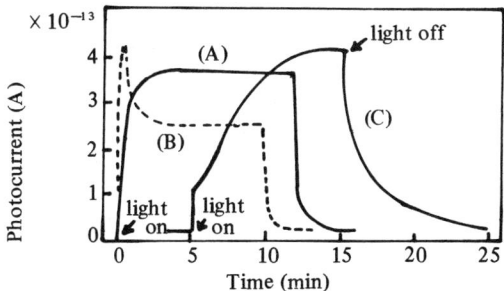

Fig. 22.3 Photoresponse curves obtained in a polyvinylcarbazole sandwich type cell. (A) low applied field; (B) applied field of about 35 kV/cm; (C) applied field of about 15 kV/cm. After Okamoto et al.[31]

The authors propose that carriers are produced by field-assisted thermal dissociation of exciplexes plus detrapping of trapped carriers by singlet excitons.[31] They assume the carriers then to migrate in an essentially coherent fashion *via* the overlap of π-electron orbitals of adjacent carbazyl rings within the same polymeric chain (see discussion in Section 15.9, 16.4 and 16.8).

22.4 Photoconductivity

Aidelis et al.,[32] studied films of poly(N-epoxypropyl-carbazole) in a sandwich cell furnished with SnO_2 electrodes. These systems are reported[32] to behave somewhat similarly to those using PVK and discussed above though the current yield was less and the mobility of the positive majority carriers was higher, ranging from 10^{-4} to 10^{-5} cm²/Vsec and depending on the value of the applied field. The activation energy, likewise, was field-dependent, between 0.24 and 0.32 eV.[32] Residual potentials were observed and attributed to detrapping from relatively deep traps.[32] The polymer can be sensitized by dyes such as rhodamine 6G.

Ieda et al.,[33] studied a number of polymers, *viz.*, low and high-density polyethylene, polypropylene, polystyrene, polyvinylchloride, polyvinylfluoride and irradiated polyethylene, in sandwich type cells with Au electrodes. Their results are summarized in Table 22.1.

These authors distinguish with respect to photoconductivity, two classes of polymers: Class I which are only UV sensitive, while those of Class II exhibit a photoresponse extending into the visible and infrared part of the spectrum. The relation between the photocurrent and the value of the applied field is illustrated in Fig. 22.4, for the case of polyethylene. It is seen that the current is ohmic for low fields, where the photocurrent produced by illumination of the positive electrode exceeds that from the negative contact. This relation inverts at high fields where the current becomes superlinear.

The activation energies for the photocurrent[33] in polyethylene were 0.09 eV below 45° C and 0.64 eV above that temperature, while that of the dark current (in E/kT) was[33] 1.37 eV.

The superlinearity of the current may be due to (1) space charge limited currents or (2) Poole-Frenkel (or Schottky) effects. The small dependence of the photocurrent on the film thickness at the same field for 20 μm and 100 μm films (see Fig. 22.4) eliminates (1). At high fields, above 5×10^4 V/cm, I_{phot} is related to E by $\log I_{phot} \propto \sqrt{E}$, which supports Poole-Frenkel or Schottky type conduction. The slopes of $\log I_{phot}$ *vs.* $\sqrt{E}$ plots are smaller than those of the dark current (see Table 22.2). Continuous radiation optically excites electrons to higher energy levels from a steady state distribution, which may be approximated by an effective temperature T_{eff}.[34] This excited distribution is due to excited carriers requiring a finite time to lose their excess

Table 22.1 Photocurrents in Several Polymers. (Applied field = 5×10^4 V/cm at room temperature) After Ieda et al.[33]

Polymer	Sample thickness (μm)	in AIR (A/cm^2)		in VACUUM (A/cm^2)	
		Dark Current	Photo Current	Dark Current	Photo Current
CLASS I					
n-PE (polethylene)					
Low density PE	20	1.8×10^{-15}	2.3×10^{-13}		7.1×10^{-13}
High density PE	20	2.0×10^{-15}	2.3×10^{-13}		5.1×10^{-13}
Polypropylene	25		7.2×10^{-14}	1.2×10^{-15}	3.5×10^{-13}
Polystyrene	28	1.6×10^{-15}	4.8×10^{-14}		1.3×10^{-13}
CLASS II					
γ-PE (10^5 rad)	33	3.5×10^{-11}	3.9×10^{-11}	2.5×10^{-11}	1.2×10^{-10}
Polyvinylchloride	30	9.4×10^{-14}	2.0×10^{-13}	1.0×10^{-13}	1.7×10^{-12}
Polyvinylfluoride	100	4.8×10^{-9}	4.9×10^{-9}	1.4×10^{-9}	1.6×10^{-8}

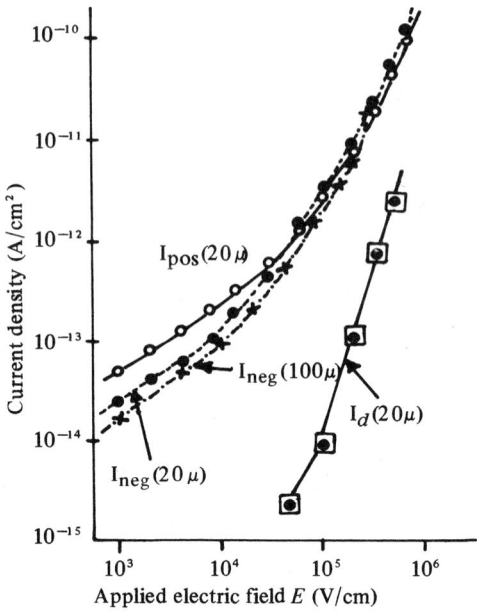

Fig. 22.4 Photocurrent vs. applied field for polyethylene films. ($d = 20$ and $100 \ \mu$m) in vacuum: I_{pos}, positive electrode illuminated; I_{pos}, negative electrode illuminated; I_d, dark current. After Ieda et al.[33]

energy. Thus the lowered slope of the log I_{phot} vs. $\sqrt{E}$ may be interpreted[35] as an increase of the effective temperature T_{eff}:

$$\text{slope} = e\beta_{PF}/2kT_{eff} = (e\beta_S/kT_{eff}) \quad (22.2)$$

Where β_{PF} and β_S are, respectively, the Poole-Frenkel and Schottky coefficients.[34] The steady photocurrent I_{ph} at 5×10^4 V/cm was not proportional to the light intensity L:

$$I_{phot} \propto L^m \quad (22.3)$$

Assuming a trap distribution[35] n_t as a function of energy E,

$$n_t(E)dE = A \exp(-E/kT_1) \quad (22.4)$$

where A is a material constant, and T_1 a "critical" temperature defining the trap distribution in terms of Eq. (22.4). Then,

$$m = T_1/(T + T_1) \quad (22.5)$$

The concentration of free, photoexcited, carriers n_p may then be written[35]

Table 22.2 Slope of log (I/E) versus E For Low Density Polyethylene. I stands for the photocurrent, and E stands for the applied electric field; d represents the thickness of the polymer film.

		Slope
theory	$e\beta_{PF}/2kT$ $(e\beta_S/kT)$	0.83×10^{-2} $(V^{-1/2} \text{ cm}^{1/2})$
experiment $(d = 20 \ \mu m)$	I_d	0.87
	I_{pos}	0.33
	I_{neg}	0.37

$$n_p = C^{T_1/(T+T_1)} \cdot N_c^{T/(T+T_1)} \tag{22.6}$$

where C is a constant and N_c is the number of accessible states at the bottom of the conduction band. Expressing the photoconductivity σ_p in terms of n_p and the carrier mobility μ, e being the electronic charge:

$$\sigma_p = en_p\mu \propto \exp(-E_{\sigma_p}/kT) \tag{22.7}$$

$$E_{\sigma_p} = (-d\ln\sigma_p)/d(1/kT) \tag{22.8}$$

and assuming an activated, hopping type of mobility governed by an activation energy E_μ so that

$$\mu = \mu_0 \exp(-E_\mu/kT) \tag{22.9}$$

one obtains

$$E_{\sigma_p} = E_\mu + (T/T+T_1)E_{fP} = E_\mu + (1-m)E_{fP}$$

where E_{fP} is the quasi-Fermi level.[16] The activation energy $E_{\sigma d}$ of the dark current is derived in the same manner:

$$E_{\sigma d} = E_\mu + E_{fd} \tag{22.11}$$

It is thus possible to derive values for the mobility activation energy E_μ and Fermi energy E_f from the experimental values of E_{σ_p} and $E_{\sigma d}$. Putting $E_{\sigma_p} = 0.09$ eV and $E_{\sigma d} = 1.37$ eV in the low temperature region (20-45° C), $E_{fd} = 1.4$ eV, $E_{fp} = 1.3$ eV and $E_\mu = 0$ eV is obtained.

The values of E_f are much less than the band gap and support the existence of donors and traps. The low value of mobility activation energy is consistent with the band model. However, the application of the band model to polymers is highly debatable, if not altogether unlikely (see discussions in

Section 15.4, 15.3, 16.4 and 16.8). The photoconductivity of high density polyethylene has also been studied Tanaka and Inuishi.[98]

The photoconductivity of polyvinylcarbazole:trinitrofluorenone charge transfer complexes has been investigated by Hughes[17] using cast films in a sandwich cell with Al electrodes. Optical excitation, derived from a Q-switched laser, in the red spectral region, is said[17] to have a high quantum yield though the electron mobility was very low, *viz.*, 3.2×10^{-8} cm^2/Vsec. The hole mobility was lower still. The field was 100 kV/cm. Recombination is reported[17] to occur within the bulk of the solid.

Polydiacetylene single crystals furnished with Ag electrodes were studied by Reimer and Bässler[33] using the photoconductive transients technique discussed in Section 12.10, 13.3 and 15.5. The free carriers are reported[33] to have a mobility of 4.8 cm^2/Vsec and a lifetime of somewhat in excess of 400 nsec, limited by trapping. While the photoconductivity activation energy was found[33] to be 0.056 eV, that for carrier injection from the contacts was only 0.01 eV. Since the mobility increased with decreasing temperature proportional to $T^{-0.5}$ this would suggest the applicability of a band model to these highly conducting systems. The authors conclude that the energy required to separate the electron hole-pairs does not depend on the primary generation mode. The photoconductivity is said[36] to be highly anisotropic and it appears that, again, the carriers preferentially move along the polymeric chains.

Binks et al.,[37] attributed the photocurrent in several polymers to the photoinjection from the electrode. It is, however, difficult to interpret the relation of $I_{\text{phot}+} > I_{\text{phot}-}$ in low fields by electron-injection from the electrode. The photocurrent may be due to positive holes produced by the ultraviolet excitation, which are not uniformly distributed across the thickness of the sample. In high field, $I_{\text{phot}-}$ exceeds $I_{\text{phot}+}$ probably because of the enhancement of the photoinjection from the photoinjection from the electrode by the lowering of the contact-barrier height by the applied field (Schottky effect).

Tamamura suggests[38] that photoconductivity in charge transfer complex salts is dominated by electrons liberated from traps 0.4 eV below the conduction band which in turn appears[38] to be situated below the excited energy level of the complex. Thus, there is no energy required for carrier generation, but only for charge transport. Defect centers in organic materials are of importance in photoconduction because this is one way by which electrons and holes may recombine. The parameters for recombination centers depend strongly on structural details and on chemical purity of the sample.

Since, quite generally, the photocurrent I_{phot} is given by

$$I_{\text{phot}} = nevg\tau\mu E/L \qquad (22.12)$$

where g stands for the number of quanta absorbed per cm^3 in the sample of volume v and electrode spacing L and subjected to an electric field of E; each photon then generates n carriers of lifetime τ and mobility μ. In different specimens of the same compound, τ may vary by many orders of magnitude[39] causing considerable differences in the observed photocurrent. The latter may be increased, again by several orders of magnitude, by suitable doping (see Section 16.7 and 16.8); thus I_{phot} of phthalocyanine can be raised by a factor of 400 – 10^4 by doping with tetracyanoethylene or o-chloranil,[39] respectively.

In a planar sandwich structure of the type $D{:}A{:}D$, the donor D being imizadole and the acceptor A a quinone, Hirohashi et al.[40] find that the dark current increases in the sequence

α-naphthoquinone > chloranil > benzoquinone,

while I_{phot} upon illumination at 230 nm increased in the order

benzoquinone > bromanil > α-naphthoquinone > chloranil.

In another study of doped phthalocyanines, Benderskii and Usov report[41] that doping with the acceptors p-chloranil, I$_2$ or tetracyanoethylene results in a drop of the activation energy of the dark current which rises accordingly. While the steady state photocurrent also increases upon doping, a new, "inertial" component of I_{phot} is said to appear; the quantum yield rises. Actual charge transfer is reported to be very weak.[41] These results are summarized in Table 22.3.

The number of impurity centers created by the acceptor is reported to be about 3×10^{13} in a sample with a 1 cm^2 area, which is close to the number of molecules in the surface layer of phthalocyanine (5×10^{13} cm^{-2}). The volt-ampere curves were found[41] to be characteristic of currents (space charge limited) in the presence of traps. Ohm's law holds up to fields of

Table 22.3 Effect of Acceptor Doping of Metal-free Phthalocyanine. After Benderskii and Usov.[41]

Acceptor	i_T/i_T^0	i_ϕ/i_ϕ^0	g/g^0
p-Chloranil	5×10^2	4×10^2	15
Iodine	2×10^6	8×10^4	2
Tetracyanoethylene	2×10^3	9×10^2	40

The superscript 0 refers to the undoped solid. i_T stands for the dark currents and i_ϕ for the photocurrent, while g is the value of the resulting quantum yield.

10^3 V/cm, after which there is a rapid increase in the current ($i \approx V^{3.5}$). Above $3.5\ 10^3$ V/cm (gap 0.03 mm), Child's law applies (see Section 10.2). The concentration of traps is said[41] to coincide with the number of centers created by the acceptor.

On the basis of their results Benderskii and Usov propose[41] the following model of the action of the acceptor: the absence of CTC bands shows that the interaction with the acceptor is limited to the surface layer. This confirms the above observation. Since the acceptor does not diffuse deeply into the photoconductor, it should not influence the rate of diffusion and mobility of the carriers. It follows that the rate of diffusion of excitons does not limit the quantum yield. The observed equality of the photocurrents in the case of frontal and rear illumination of the material tends to confirm this.[41] The time constant of increase in the photocurrent corresponds to the time constant of ionization of a dissociation center capturing an exciton and dissociating it into a trapped electron and a free hole. It thus should be characteristic of the centers of this type. The observed conservation of the time of the photoresponse after the application of an acceptor could be explained by the fact that the acceptors do not create new centers of dissociation. The increase in the quantum yield observed at the same time is due to the fact that the quantity limiting it in pure phthalocyanine is the number of centers of dissociation unoccupied by electrons. The role of the acceptor reduces to a destruction of these centers by the creation of surface levels, to which the electrons are transferred from the centers. As a result, the number of free centers of dissociation and, consequently, the quantum yield is increased. The dissociation centers for excitons are identified[41] with the donor surface centers, situated below the Fermi level. The nature of these centers remains unknown.

The inertial component mentioned above appears as a result of exchange and recombination of electrons in the acceptor levels with holes in the valence of phthalocyanine. The time constant is determined by probability of surmounting the potential barrier at the interface, and its activation energy should coincide with the activation energy of the dark current. The nonlinearity of the volt-ampere curves is also determined by this process.

Ambient gases may profoundly affect the photocurrent as mentioned in Section 12.18 (see also data in Table 22.1). Yasunaga et al.[42] studied the affect of ambient oxygen on single crystals of Pb-phthalocyanine; holes are said to be indirectly photogenerated *via* charge transfer states at (probably absorbed) oxygen centers to recombine with oxygen ions (probably also absorbed). The photoconductivity of 1,4-diamino-anthraquinone is reported to be lowered in the presence of oxygen.[43] Upon exposure of an aqueous solution of the important psychotropic drug chlorpromazine, CPZ, as the $CPZ.S^+$ to UV light (303 nm) in the presence of oxygen, a green, dimeric, photoproduct has been isolated;[44] the photocurrent dropped upon increasing oxygen concentration.

Spitler and Calvin[45] have studied the photoconductivity changes in a ZnO single crystal caused by the adsorption of rhodamine on its 001 face from an aqueous solution, employing an electrochemical method. Electrons injected from the photoexcited dye are drawn to the anode giving rise to a measurable current even when the quantum yield is quite low, as for photooxidation in the case mentioned, *viz.*, 0.027; it did not depend on surface coverage.[36] At its isoelectric point, *i.e.*, at pH 6.5, rhodamine B is reported to be unaffected by adsorption.[45] At a flux of 2.1×10^{15} photons cm^{-2} at 560 nm, the efficiency of electron transfer in a KCl solution thus was found to be 0.037. The photocurrent, of the order of 5×10^{-8} A, varied very little with dye concentration, and no major changes in the excited structure of the chromophore appeared to occur upon dye adsorption.[45] Since complete surface coverage by a dye monolayer was found to be 1.1×10^{-7} mmoles cm^{-2}, an effective area of 150 Å^2 per molecule results.[45] Problems encountered in doped ZnO response are reviewed by Hauffe.[46]

In systems involving salt-like, ionic dyes, especially cationic dyes such as basic brilliant green, it has been shown[47] that the identity of the anion as well as the state of aggregation of the dye affect the photoconductivity of the system to a great extent; even the sign of the majority carrier has been shown[47] to depend on these quantities, as do the relaxation times of the photoelectric processes.

22.5 Photovoltaic Effects

There has been considerable interest in photovoltaic cells consisting of an organic solid (*e.g.*, tetracene,[48] phthalocyanines,[49] porphyrins,[50] or similar compounds) sandwiched between semi-transparent metal electrodes. At present these devices are very inefficient, and their behavior is still not well understood.

Irrespective of the nature of the photogeneration process of carriers, whether photoinjection, including exciton-induced carrier injection, from the contact(s), or photogeneration of carriers within the bulk or near the surface, the photovoltaic current should be strongly influenced by the built-in field arising from contact potential differences or related effects. Tang and Albrecht,[51] working with chlorophyll-*a* and a variety of electrodes appeared to have observed a correlation with contact potential, in that the active surface (site of photogeneration or injection of holes) was that of the electrode with lower work function. However, their work was done in air, and their results possibly may reflect the blocking action of oxide layers on the electrodes.

The photovoltaic effect involves an internally produced electric field in order to separate the photogenerated electron-hole pairs. There are several ways to produce this field, among them: *p-n* junctions, Schottky barriers, semiconductor-liquid-electrolyte interfaces.

In a *p-n* junction a semiconductor is doped to make one side *p*-type material and the other *n*-type material (see Section 2.20). The resultant electric field is located in the junction or "depletion" region. The thickness of this region depends on the doping, usually a few mecrons.

A Schottky barrier can arise from the transfer of electrons from a *n*-type semiconductor (or holes from a *p*-type) to a metal layer deposited on it; the effect is mostly seen in junctions between *n*-type semiconductors and high work functions metals. When a metal-semiconductor sandwich contaning such a barrier is illuminated, electrons flow into the semiconductor and holes into metal, producing a photovoltage.

The same situation prevails in semiconductor-liquid-electrolyte interfaces in which electrolytes play very much the same role as metals do in Schottky barriers. The holes and electrons transferred into the electrolyte produce chemical reactions, and for appropriate combinations of semiconductors and electrolytes, these reactions can be used to produce fuels directly. Such devices are of considerable interest for solar energy conversion.[52]

Regenerative photoelectrochemical cells have been constructed utilizing single crystal *n*-GaAs in acetonitrile solutions. Solution RedOx couples (anthraquinone, *p*-benzoquinone, dimethylferrocene, ferrocene, hydroxymethylferrocene, and tetramethyl-*p*-phenylenediamine)whose standard RedOx potential varied by over 1.2 V, were photooxidized at the semiconductor electrode and reduced at a Pt counterelectrode converting light directly into electrical energy. A power conversion efficiency of 14% was observed for the *n*-GaAs electrode in a ferrocene-ferricenium-acetonitrile solution at a radiant intensity of 0.52 mW/cm^2 of 720-800 nm light. The efficiency and stability were found to be very dependent upon the residual water concentration, radiant power, and concentration of electroactive species.[53]

A detailed analysis of photoelectrochemical systems[54] by Guruswamy and Bockris suggests that the four important parameters which have to be varied are: (a) the energy gaps; (b) the flat band potential; (c) electrode conductivity; and (d) competing electron reactions, such as corrosion reactions where the key criterion is the flat band potential of the semiconductor. With the energy gap this determines the potential range in which the semiconductor may undergo competitive (instability-causing) reactions.

Detailed theoretical considerations suggest that the criteria for suitable anodes in photogenerators is: an energy gap of 0.7 to 2 eV, an electron affinity between -5 and -5.6 eV, and a flat band potential of -0.3 to -0.8 V on the hydrogen scale of potentials.

These authors emphasize[54] that a semiconductor in contact with an electrolyte usually exhibits a significant concentration of surface states so that most of the potential drop does occur across the electrodic double layer and not across the bulk of the system. Moreover, at low overpotentials it is probably interfacial charge transfer rather than bulk carrier mobility which is rate determining.

Mn(III)phthalocyanine and porphyrin complexes have been known to disproportionate into Mn(II) and Mn(IV) complexes upon illumination with visible light and with a Mn(III)porphyrin modified Pt electrode, conversion of visible light (including red) into electricity has been demonstrated.[56]

The compound used was a surface-active

Mn(III)porphyrin,
meso-tetrakis-(4-N-stearylpyridinium)porphine Mn(III)-tetraiodide-monohydroxide,
Mn(III)(OH$^-$)St$_4$ PyPI$_4$,
St=Porphine, St-Stearyl, Py=Pyridinium).

Upon illumination with visible light, a disproportionation reaction may take place in Mn(III) porphyrin layers to form Mn(II) and (IV) porphyrins, Mn(II)-P and MN(IV)-P, 2MN(III)-P → Mn(II)-P + Mn(IV)-P. The species thus formed may be subject to reaction with H$^+$, OH$^-$, and trace amounts of reducing or oxidizing agents, if any, in the electrolyte solution and leave excess charges of opposite sign in the contact. The excess charge thus accumulated may be cancelled by current flow to/from the counter electrode. The system has efficiencies of 0.005-0.01% for incident photon to electricity conversion, and has much higher efficiency for converting red light (620-650 nm) vs. blue.

Anthracene crystals easily transport holes, and when these emerge into an aqueous electrode they ordinarily are discharged, giving rise to the oxidation of the crystal surface itself. If, however, suitable solutes are present, e.g., sodium-p-anilino-benzene-sulphonate, or ferroin, the emergent holes are 100% consumed in oxidizing the solute. Reactions of the holes with water or with anthracene therefore do not occur at all.

Table 22.4 lists a number of solutes which have been found[57] to give rise to a high efficiency of injection of holes into anthracene crystals under conditions of (a) =395 nm, (b) high stirring rates, (c) absence of oxygen, (d) 10^4 Vcm^{-1}. It would appear from these results that the limiting factor in the injection efficiency is the successful motion of the exciton to within a reaction distance of the surface.[8]

From the limited number of solutes in Table 22.4 for which the solute has a RedOx potential out of reach of a triplet exciton (1.75 eV in anthracene), namely the last four, single excitons are responsible for injection efficiencies of 45%. Triplets are responsible therefore either for a further

Table 22.4 Solutes Giving Efficient Injection of Holes into Anthracene Crystals. After McGregor.[57]

Hole injector[a]	Standard reduction potential[b] (volts; NHE)	Charge carrier injection efficiency $\phi(\pm 0.03)$
1. 1,4-benzoquinone	+0.69	0.65
2. 1,2-dichlorophenol indophenol (DCP)	+0.668	0.65
3. Methylene blue (MB)	+0.53	0.66
4. Cu^{+2}-1,10-phenanthroline complex. $(CuP_2)^{+2}$	+0.174	0.66
5. N,N,N',N'-Tetramethyl-p-phenylene diamine (TMPD)	+0.14	0.63
6. β-Nicotinamide adenine dinucleotide (NAD)	−0.32 (pH = 7)	0.65
7. Rhodamine B (RB)	−0.54 (pH = 7)	0.45
8. Fluorescein (Na Salt)	<−.60	0.45
9. Eosin (Na Salt)	<−.60	0.45
10. Erythrosin	<−.60	0.45

[a]The hole photo-injector solution contained 1×10^{-4} mol ℓ^{-1} photo-injector and $[O_2] < 1 \times$ mol ℓ^{-1}. The solution was stirred at $\omega > 2000$ rpm.
[b]The standard RedOx potentials are at pH = 0 in aqueous solution unless otherwise indicated.

20% or else for zero, depending on the RedOx potential of the solute. This conclusion, however, holds only tentatively.

There is no difficulty in arranging a cell of this type to have an average electric field of at least 10^4 Vcm^{-1}. The actual field at interfaces, however, often departs significantly from the average field, and a field-dependent injection process depends on the actual local field. It is therefore not easy to predict phenomena in the absence of an applied field.

Several authors[49,50,51] have discussed organic photovoltaic systems in terms of the Schottky barrier effect (see Section 15.12), which theory, however, predicts strongly voltage-dependent capacitances because the thickness of the depletion layer arising at the contact varies with applied voltage. Hall et al., however, report[58] that the measured capacitances of such systems were unaffected by illumination and independent of the applied voltage. Thus, it appears that donors and acceptors are not present in sufficient concentrations to produce an exhaustion layer.

The following discussion will concentrate on systems of the MIM type, *i.e.*, an organic semiconductor sandwiched between two metal electrodes. Experimentally,[58] such a cell consists of a glass substrate onto which is deposited a thin film of metal, referred to as the substrate electrode, followed by another film of the compound to be studied, on which is then deposited a semi-transparent layer of metal.

The study by Hall et al.,[58] to be discussed in the following, employed metal-free phthalocyanine H_2Pc_4 films, triply sublimed and in ultra-high vacuum.

If the concentration of donors is negligible, then the current in the dark is carried by injected holes and/or electrons. The work functions of the metals used are approximately: Al 4.25 eV; Au 4.80 eV; Pb 4.00 eV. The ionization energy of the phthalocyanine is about 5 eV and the band gap is probably about 2.5 eV, though it is not known with certainty. In anthracene and tetracene the band gap is considerably greater than the singlet exciton energy[59] and the intrinsic photogeneration process can be seen clearly in the photoconduction spectral response. The corresponding process apparently has not been observed in phthalocyanine. One may thus assume that only holes are injected. This reasoning is supported[58] by the finding that on illumination of the non-substrate electrode with strongly absorbed light the H_2Pc develops a photovoltage with the illuminated electrode being negative. This is consistent with photoinjection of holes.

At low voltages, the photocurrent-voltage curves can be quantitatively explained by a space-charge-free theory of conduction. Energy barriers to hole injection, and hence the built-in field, are not determined by the work function of the metal, but in each case the higher barrier occurs on the non-substrate side of the phthalocyanine. This, together with the observed[58] variation of the built-in field with irradiance, and the failure of the space-charge free conduction model at high voltages, are explained by assuming[58] that the metals make ohmic contact with the phthalocyanine and that the effective barriers to the injection are determined by space-charge of holes trapped near the metal-insulator interface. The trap density is highest at the nonsubstrate side of the film and is approximately uniform in energy. The dark current exceeds the saturation photocurrent at high voltages, suggesting that the mechanism of photoinjection into the bulk is probably exciton dissociation at defects near the illuminated electrode rather than exciton-induced photoinjection directly from the metal. Results are summarized in Table 22.5.

In another study[60] of photovoltaic sandwich cells based on metal-free as well as metal-phthalocyanines, a quantum efficiency of 14% at 6328 Å was reported; the photo-emf on open circuit is said[60] to be 0.8 V. Since quite high rectification ratios, up to about 10^3, were observed, as well as considerable spontaneous, dark, potentials (several 100 mV) across the cells,

Table 22.5 Photovoltaic Properties. After K. J. Hall, J. S. Bonham and L. E. Lyons, Aust. J. Chem., 31, 1661 (1978).

Cell	Injection quantum efficiency[A] ($j_s/e\Pi$)	Short-circuit quantum efficiency[A]	Max. Power conversion efficiency[A]	$U_{OC}(V)$[B]	d (nm)	Transmittance of $g/M_2 \leftarrow$ at 620 nm	Temperature of measurement (K)	Π_0[C] (m^{-2} sec^{-1})	W_{calc}[D] (eV)
g/Au/H$_2$Pc/Au$\leftarrow$	0.56%	0.16%	0.009%	0.31	290 ± 10	0.52	298	2.3 × 10^{18}	0
g/Au/H$_2$Pc/Al$\leftarrow$	0.16%	0.06%	0.003%	0.29	150 ± 10	0.16	298	2.2 × 10^{18}	0.55
g/Au/H$_2$Pc/Pb$\leftarrow$	0.25%	0.08%	0.006%	0.34	290 ± 10	0.22	298	2.3 × 10^{18}	0.8
g/Al/H$_2$Pc/Au$\leftarrow$	0.58%	0.06%	0.003%	0.39	420 ± 10	0.66	283	2.8 × 10^{18}	−0.55
g/Al/H$_2$Pc/Al$\leftarrow$	0.15%	0.02%	0.0006%	0.20	430 ± 20	0.20	296	2.3 × 10^{18}	0
g/Al/H$_2$Pc/Pb$\leftarrow$	0.02%	0.0008%	0.0003%	0.23	430 ± 20	—[E]	297	2.3 × 10^{18}	0.25

[A] Based on photon flux incident on M$_2$ for λ = 620 nm.
[B] For $\Pi = \Pi_0 \cdot \Pi_0$ = photon flux
[G] Glass substrate
U_{OC} Photo-voltage (open circuit)
Π Photon Flux
d Organic film thickness
$W_{calc} = W_1 - W_2$ = difference in the work functions of the two electrodes
[C] Maximum photon flux used for $j - U$ measurements at 620 nm
[D] Calculated from differences in the work functions quoted in the text
[E] Pb oxidized rapidly on exposure to air
The arrow indicates the direction of incident light

it appears[60] that these systems do involve Schottky contact barriers, the height of which has been estimated to be about 1.3 eV using In and Au electrodes.

When light irradiates one side of a thin layer, crystal or film of, say, anthracene, the steady-state distribution of singlet and triplet excitons depends on the quenching of the excitons at the surfaces. Figure 22.5 shows some results calculated for anthracene single crystals under the assumption of no re-absorption of fluorescent light.

Both metal and electrolyte interfaces with an anthracene crystal quench excitons, and thus reduce significantly the observable photofluorescence.[8]

Exciton reactions at the interface may or may not result in the production of free charge carriers in the organic. For example, a transfer of energy from the exciton to the metal, followed by the fast thermal degradation of the energy in the metal, produces no charge injection into the organic. Nonetheless, exciton reaction with a metal often does result in about 5% of the excitons giving rise to positive holes free to move in the organic. Such phenomena are often responsible for the photovoltaic effects in, *e.g.*, MIM (metal-insulator-metal) cells containing napthacene or other aromatics.

The following is now assumed: (1) a uniform field across the solid, *i.e.*, neglecting any bending of the energy bands at the contacts. This is defensible, at least to a zeroth approximation, since theory[63] suggests that at least the open circuit voltage is largely independent of any band bending. (2) absence of surface states (see also Section 12.17 and 13.8). (3) Neglect of image forces. This may not be so serious as it may appear: carrier injection from the electrode to a plane in the crystal just beyond the range of image forces is all that need to be fed into theoretical discussion.

Thus, the concentration of carriers at the illuminated interface increases, so that the Fermi layer F moves nearer to the conduction band. The rise ΔF given by

$$\Delta F = kT \ln(n_L/n_D)$$

where (22.13)

$$n_D = N\exp[-(C - F_D)/kT]$$

n_L, n_D, concentration of free carriers in light, dark;
N, density of states in conduction band: $N = 10^{21}$ to 10^{22} cm^{-3};
C, energy of conduction band;
F_D, Fermi energy in dark and at equilibrium.

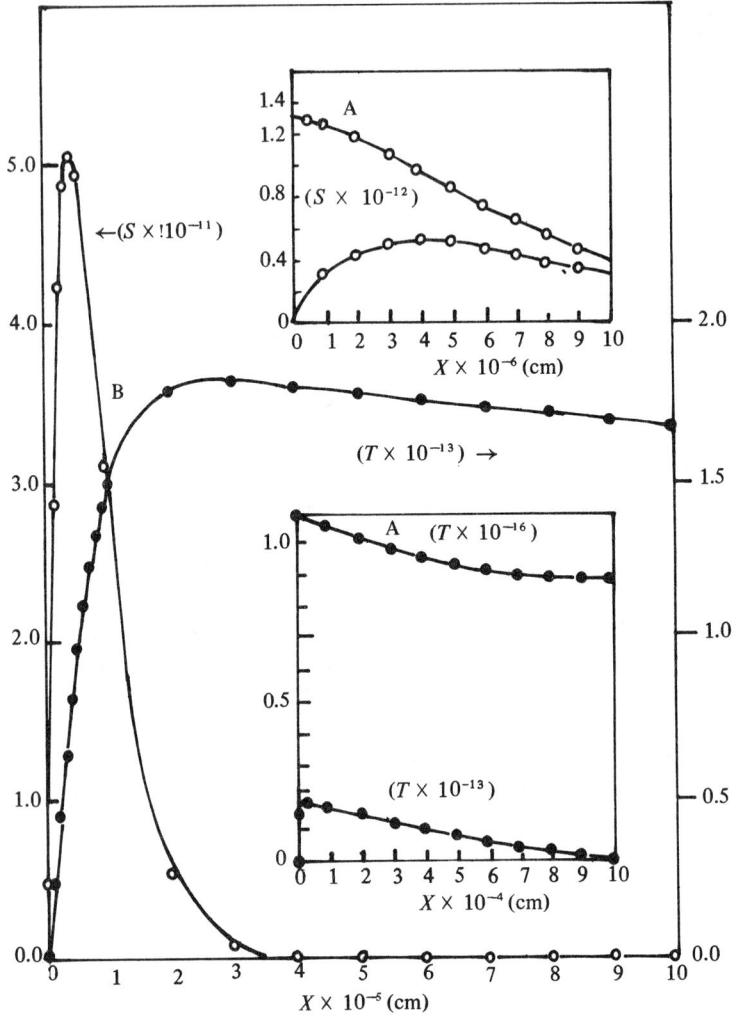

Fig. 22.5 Calculated steady-state exciton concentration as a function of distance into the crystal from the irradiated surface. S, concentration (cm^{-3}) of singlet excitons; T, concentration of triplet excitons; A, zero quenching of excitons at surface; B, fast quenching of excitons at surface, *viz.*, for singlets, rate constant = 10^4 cmsec^{-1}; for triplets = 10^2 cmsec^{-1}. After Lyons.[62]

The open-circuit photovoltage $U_{PV,OC}$ is given by

$$U_{PV,OC} = \Delta F/e, \qquad n_L/n_D = 1 + a\Pi \tag{22.14}$$

where Π is the photon flux, and $a\Pi$ is the ratio of rate of light to dark injection of carriers into the organic from the electrode. Photoinjection here denotes either light-or-exciton-induced injection. Thus

$$U_{PV,OC} = (kT/e)\ln(1 + a\Pi) \tag{22.15}$$

It is thus possible from the observed dependence of $U_{PV,OC}$ on Π to determine a; furthermore, a can be determined at different wavelengths, and information then derived about exciton-induced injection as a function of the penetration depth of the exciting light.

Bonham[63] has extended the above to allow some light to reach the back surface and there set up a photovoltage opposing that at the front surface. He has also considered injection into traps near the interface followed by excitation to the carrier band. For both surfaces active but without trap intermediates, eq. (22.15) is replaced by

$$U_{PV,OC} = (kT/e)\ln[(1 + a\Pi)(1 + b\Pi)] \tag{22.16}$$

where $b\Pi$ is the ratio of light to dark injection rates at the back surface. Equation (22.16) is equivalent to eq. (22.17), with $U_{PV,OC} \equiv U$,

$$\{\exp[(eU/kT) - 1]\}^{-1} = \Pi^{-1}/(a - b) + b/(a - b) \tag{22.17}$$

This can be tested experimentally, and a and b determined from slope and intercept as seen from Fig. 22.6.

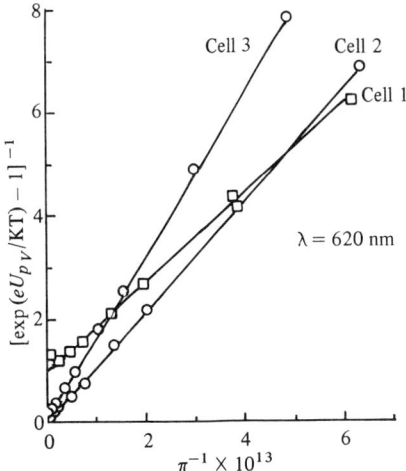

Fig. 22.6 Test of the theory of photovoltage U_{PV} dependence on incident photon flux Π, for Au-phthalocyanine Al cell in ultra-high vacuum, at $\lambda = 620$ nm. After Hall et al.[58]

Further confirmation of the general theoretical approach comes from the values of a determined as a function of wavelength. Figure 22.7 shows that the spectral dependence of a follows the absorption spectrum of the film, as expected from the physical significance assigned to a. Where the absorption coefficient is large the injection rate increases.

Figure 22.8 shows conditions under load; the photovoltaic cell is assumed to be illuminated from the left.

For attaining the theoretical maximum power conversion efficiency, which is 100% for monochromatic light and 45% for solar radiation, the conditions are[63]

(a) the exciton energy to equal the difference in work function;
(b) the exciton energy to be about 1 eV (for solar radiation); and
(c) the photovoltage to exceed the difference in work functions.

For attaining the maximum yield of charge carriers it is necessary that (a) the penetration depth of the light be less than the exciton diffusion length, which is itself less than the thickness; (b) every exciton reaching the surface injects a carrier; and (c) the ejection rate of carriers back into the electrode from which they are injected is less than the current. In turn,

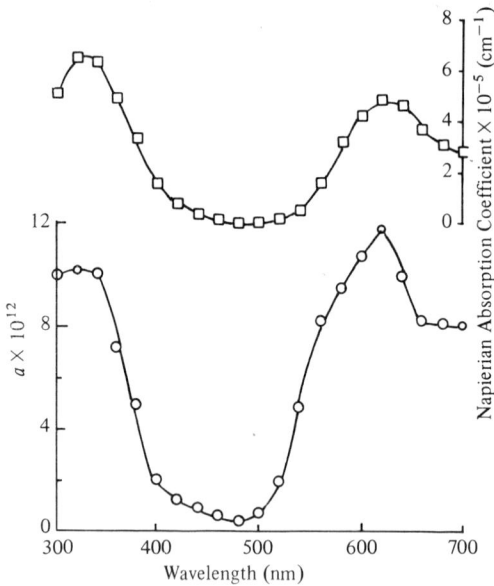

Fig. 22.7 The dependence on wavelength of the parameter a (lower curve) compared with the absorption spectrum of the phthalocyanine film in a cell with Au and Al electrodes measured in ultra-high vacuum. After Lyons.[62]

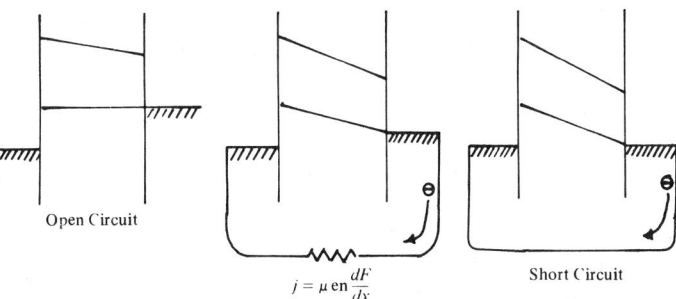

Fig. 22.8 Fermi levels of carriers and conduction band energy in the irradiated insulator of a photovoltaic cell. Left, at open circuit. Center, with a load resistor. Right, under short-circuit condition. After Lyons.[62]

(c) requires high mobility in the organic, a low space-charge density and a thin sample.

Hall et al.[58] point out that the maximum single-carrier current an undoped material can carry is the space-charge-limited current (SCLC). Assume that the insulator contains only shallow traps. For $U < 1$ V, the SCLC equation is[64]

$$j \cong 4\pi^2 \mu \epsilon \theta (kT/e)(U - U_{OC})/d^3 \qquad (22.18)$$

where θ is the ratio of free to trapped carrier densities. The power delivered is then

$$P = -jU = -4\pi^2 \mu \epsilon \theta (kT/e)(U^2 - UU_{OC})/d^3 \qquad (22.19)$$

which has a maximum value, when $U = \tfrac{1}{2} U_{OC}$, of

$$P_{max} = \pi^2 \mu \epsilon \theta (kT/e) U_{OC}^2 / d^3 \qquad (22.20)$$

In this, U is the measured photovoltage, ϵ the permittivity of the solid, d the electrode spacing, U_{OC} the open circuit value of U, e the electronic charge, k Boltzmann's constant, T the temperature, j the current density and μ the mobility.

Carrier mobilities in organic compounds are usually around 10^{-4} m^2 V^{-1} sec^{-1} or less; ϵ is unlikely to exceed about 4×10^{-11} F m^{-1}; and a reasonable expectation for U_{OC} is about 0.5 V. Inserting these values:

$$P_{max} \cong 2.6 \times 10^{-16} \theta / d^3 \qquad (22.21)$$

High values of P_{max} therefore require a thin layer of the organic insulator. On the other hand, substantial absorption of the light is unlikely if $d < 10^{-8}$

m and then only (a) if the organic has a very high absorption constant over a wide range of wavelengths, or else (b) if the cells can be constructed in such a way that light transmitted by one film is absorbed in a subsequent passage through either the same film (after reflection) or a different film (after passing through a transparent electrode). In the analysis below the cells of the ultimate thinness (10 nm) and also cells of thickness 500 nm, typical of sublimed films, are considered.[58]

If the insulator is trap-free ($\theta = 1$), then P_{max} is 2.6×10^5 kW m^{-2} for $d = 10$ nm and 2 kW m^{-2} for $d = 500$ nm. As the peak solar irradiance at earth is approximately 1 kW m^{-2}, it is clear that space-charge limitation alone does not prevent high solar conversion efficiencies, provided the insulator is not too thick.

The presence of traps, however, is probably unavoidable and causes further limitations. For $d = 10$ nm, a solar conversion of 10% is possible provided $\theta > 4 \times 10^{-7}$ which seems to be no severe restriction. However, if the thickness is increased to 500 nm, 15% efficiency requires $\theta > 5 \times 10^{-2}$. Such a high value of θ appears to be very difficult to obtain in a film, although single crystals of pyrene have been reported[65] with $\theta \gtrsim 10^{-2}$. Crystalline expitaxially-grown films of phthalocyanine have been prepared,[66] and this method may lead to much less trapping than occurs in evaporated films.

These considerations do not apply to cells in which a Schottky barrier mechanism operates.

Doped metal-phthalocyanine barrier layer photocells are said[67] to have solar energy conversion efficiencies of about 0.5%, yielding about 10 mW/cm^2.

Several patents[68] deal with organic, *e.g.,* tetracene, pentacene photovoltaic cells (solar), some involving *p-n* junctions produced by surface etching or by pyrolysis. Conversion efficiencies of up to 20% are claimed.[53]

p-n Junctions formed between *n* and *p*-type dyes, such as malachite green (*n*-type) and merocyanine (*p*-type) have been reported[69] and employed as photoelectric cells.[70] See also Table 22.6.

Dyes[71] can also be utilized in ferroelectric ceramics, which are very sensitive to light and have a response time of about 10^{-3} sec. Fridkin and coworkers[72] observed optical distortion in illuminated ferroelectric semiconductors in the intrinsic or impurity region which is associated with the formation of an anomalously high photo-emf effect in the direction of light impingement. The spontaneous polarization may be due to the Jahn-Teller effect.

The photovoltaic process is in principle closely related to the photosynthetic process. An intriguing question concerns the possibility of manmade photosynthesis with an efficiency of 10 to 30%, more than 10 times higher than that typical of natural photosynthesis.

Table 22.6 Output N_{max} of Organic pn Photoelectric Cells. After Meier et al.[61]

System		U_∞ [mV]	I_0 [A]	N_{max} [W]
n	p			
Malachite green	Phthalocyanine	100	10^{-8}	10^{-9}
CdS	Phthalocyanine	300	10^{-8}	3×10^{-9}
Triphenylmethane dyes	Phthaleins	100	0.5×10^{-9}	6×10^{-10}
Tetramethyl-p-phenylenediamine	Mg-Phthalocyanine	200		3×10^{-12}
Ion exchange membrane with thionine/ascorbic acid		15	1.6×10^{-6}	2×10^{-8}

U_∞ = open circuit voltage
I_0 = short circuit current
N_{max} = maximum power output

22.6 Photoemission

Photoemission from organic molecular crystals has been measured by several workers.[73] Schechtman and Spicer[74] used Spicer's k-independent indirect-transition model[75] (disregarding photoelectron scattering in the bulk and at the surface) and succeeded in determining the energy structures of phthalocyanine and protoporphyrin. Subject to the applicability of band theory to organic crystals, and also assuming that the state density of the conduction band is constant in molecular crystals, Spicer's model appears to be applicable.[75]

Results of photoemission studies are usually expressed in terms of electron distribution curves where quantum yield is plotted against incident photon energy, as illustrated in Fig. 22.9 for pyrene and some of its derivatives,[76] or, alternatively, the quantum yield is plotted against the value of the applied potential as shown in Fig. 22.10. It is usually found that the cube root of the quantum yield is proportional to the incident photon energy (see eq. (22.23)). The methodology is treated in Chapter 12, while the theoretical aspects of this topic are discussed in Chapter 14. The energy distribution curves exemplified in Fig. 22.8 are obtained from the experimental current vs. voltage curves such as those in Fig. 22.9 by differentiation. The most important quantity derivable from such measurements is the ionization energy, which is discussed in Section 14.3.

Solid and vapor phase photoelectron spectra of aromatic hydrocarbons which have the same structures in both phases often show a good correspondence with each other, when a small shift in energy levels is assumed.[77]

Vacuum photoemission spectroscopy enables one to determine the degree

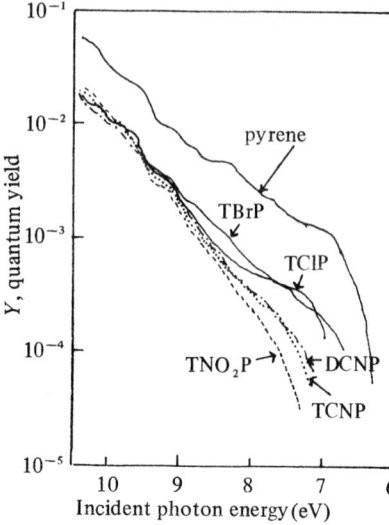

Fig. 22.9 Spectral dependences of quantum yields Y of pyrene and its derivatives, plotted as a function of incident photon energy. TBrP refers to 1,3,6,8-tetrabromopyrene, TCNP to 1,3,6,8-tetrachloropyrene and TNO_2P to 1,3,6,8-tetranitropyrene. After Hino et al.[76]

of electron transfer in many compounds, such as TCNQ derivatives. The broad peaks seen are indicative of strong coupling of the electron to the molecular modes. Because of the narrow bands and strong electron-molecule interaction, the results obtained with TCNQ salts are dominated by the electron structure of $TCNQ^-$ and not by the delocalized energy bands.[78] Several studies of the photoemission from aromatic-alkali metal salts have been made in order to clarify the mechanism of the charge-transfer of the complexes.[79] In photoemission, 3 steps are involved: (1) the formation of a quasi-free electron by photon absorption, (2) transport process to the surface, and (3) transmission through the surface. The characteristic aspects of these processes in molecular crystals are,[80] (a) not only direct ionization (DI) but also autoionization through a highly excited molecular exciton (AI) might take place; (b) on ionization, k-vector conservation is not an effective selection rule, because of the localization of the hole;[75] (c) quasi-free electrons must escape from geminate recombination in the Coulomb field formed by the hole, and (d) electrons can suffer energy loss due to the excitation of the exciton or the intramolecular vibration of other molecules, in addition to the other processes observed in inorganic solids.

The theory of photoemission for organic solids has already been discussed by several workers,[81] and is treated in Chapters 12 and 14.

It is frequently assumed[82] that pair production is the main electron-energy-loss process in electron scattering processes. However, results on the energy loss of slow electrons[83] show that exciton excitation (including triplet

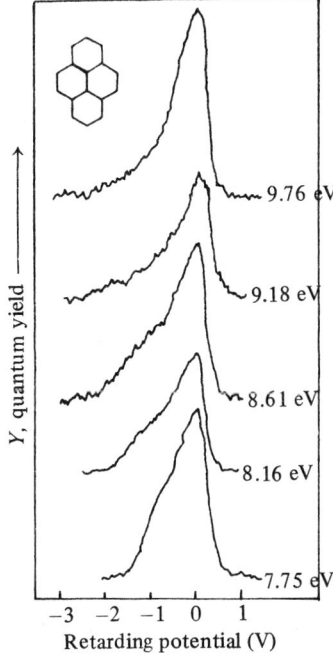

Fig. 22.10 Energy distribution curves of pyrene, plotted as a function of retarding potential V. Numerical values beside each curve are the energies of the incident photons. Y is in arbitrary units. After Hino et al.[76]

excitons) makes a greater contribution to scattering than does pair-production. It seems that the conclusions[84] that either the scattering length $L(E)$ decreases monotonically with E, or that secondary electrons tend to have only a small kinetic energy are also valid for the exciton-excitation process.[84] The main change in going from pair-production to exciton excitation will be the change in the threshold for scattering: from $2E_g$ for pair-production to $E_g + E_x$ for exciton excitation, where E_g is the band gap and E_x is the energy of the lowest exciton level. Considering the width of the exciton bands and the fact that the intermolecular interaction increases with the electron energy, the energy of the bottom of the *broad conduction band*, E_c, may be evaluated from

$$E_c = 2E_{th} - I_g \qquad (22.22)$$

to approximate the energy of the bottom of the conduction band, E_g. The quantities mentioned above are shown in Table 22.7 for both the lowest triplet and singlet excitons. The results shown in Table 22.7 demonstrate that almost all the electrons above the threshold can suffer inelastic scattering if triplet excitons are excited. However, no detailed information on the energy dependence of $L(E)$ is available at present.

Table 22.7 Energy Parameters of Aromatic Hydrocarbons (in eV). After Seki et al.[84]

	$I_g^{a)}$	$E_{th}^{b)}$	$E_c(\text{calc})^{c)}$	$E_T^{d)}$	$E_S^{e)}$	$E_T + E_c(\text{calcd})$	$E_S + E_c(\text{calcd})$
Anthracene	7.47	5.70	3.93	1.85	3.11	5.8	7.1
					3.16		
Naphthacene	7.01	5.28	3.55	1.27	2.38	4.7	7.1
					2.46		
Pentacene	6.64	4.85	3.06	0.95	1.85	4.0	4.9
Perylene	7.00	5.37	3.74	1.56	2.65	5.2	6.4
Coronene	7.34	5.52	3.70	2.37	3.44	6.1	7.1
p-Terphenyl	8.20	6.1	4.0	2.55	4.48	6.6	8.5

$a)$ First ionization potential of a free molecule
$b)$ Threshold energy of photoemission.
$c)$ Energy at the bottom of the conduction band.
$d)$ Energy of the lowest triplet state.
$e)$ Energy of the lowest singlet state.

There are other mechanisms of energy loss in organic solids: the excitation and absorption of the phonon and intramolecular vibration, and the excitation of plasmons (see Section 13.10). The former two do not affect the quantum yield to any considerable extent. Plasmons, (see Section 20.5) may play a minor role.

It is useful[85] for the identification of plasmons to compare the optical absorption with the electron energy-loss spectra. Plasma oscillation is intense in the energy-loss spectra, but it should be weaker in the absorption spectra. For p-terphenyl and naphthacene, the absorption coefficient is fairly large at the predicted photon energies, 7.13 eV and 7.56 eV.[84] For perylene, there is a peak in the energy-loss spectrum at about 7 eV, but this may be due to a one-electron excitation at 6 eV in optical absorption. Plasma oscillations at these energies thus are not certain.

The theory of photoemission is treated[86] in detail by Belkind et al., who derive the cube law of quantum yield Y vs. photon energy $h\nu$ (see Section 12.8); E_{ph} being the photoemission threshold:

$$Y = (h\nu - E_{ph})^3 \qquad (22.23)$$

by assuming that the positive holes are not fully screened and that the energy of the hole band may be approximated by the polarization energy of the solid.

The escape depth of electrons derived from organic solids is an important quantity pertinent to photoemission; in naphthacene and perylene films it was found[39] to be 8-15 nm and 30-37 nm, respectively by Hino and Inokuchi. Belkind et al., confirmed[87] this for naphthacene; finding a value of 14 nm. Several workers found a small electronic escape depth in TTF:TCNQ complexes, 1 nm or less, while Pong and Smith determined[89] a value of 1.1 nm for Cu phthalocyanine. The latter value may be somewhat small. All these compounds had activation energies a few eV above the Fermi level.[39,87,88,89,90] For further discussion of this topic, see Section 12.8 and 14.3. On this topic several reviews exist.[91]

22.7 Other Photoeffects

Luminescence

The luminescence of organic solids (see Section 2.9, 2.17, 5.13, 6.4 and 10.3, as well as Chapter 13) has become a subject truly vast in extent so that, within the limitations of this book, the reader has to be referred to the literature,[92] especially to a review by Williams.[93]

The topic deals with light emission from excited states and thus is closely related to fluorescence, and especially to delayed fluorescence phenomena.

Photomagnetic Effects

These effects, often referred to as the photomagnetoelectric effect, or PME (see Section 2.20 and 10.4), still has not been substantiated for organic materials.

The physical mechanism consists of photoinduced transitions of electrons between cations on different lattice sites, resulting in a redistribution of magnetic ions or "centers" and thus modify the magnetic properties. At low temperatures, the photoinduced changes are persistent due to the low mobility of the electrons, at higher temperatures a competition between photoinduced transitions and thermal electron motion occurs.

Thus, the effect should be looked for in cryoscopic experiments involving free radicals and, more specifically, free radical salts; the material should be paramagnetic though it need not possess ferromagnetic centers or atoms. The photo-Hall effect is discussed in Section 12.13.

Dye-Lasers

This subject is briefly discussed in Section 16.6. Here it is desired just to mention a laser which is based for its action on the solid state properties of dyes, *viz.*, the Q-switched, giant-impulse laser.[95] Light impulses of enormous power (10^{10} W) though only of extremely short duration-typically $10^{-2} - 10^{-6}$ μsec are attainable. The usual solid state laser material is doped with a dye which features a certain absorption at the emission frequency of the laser parent material. This absorption drops the Q-factor, or quality factor, of the system and thus prevents lasing until the population inversion has reached a value where it causes lasing in spite of the drop in Q. The photon density in the resonator then attains values which are so high that the absorptivity of the dye drops drastically, causing a rise in the effective Q and thus a still greater photon density. This positive feedback causes the entire stored energy to be released in one very large and brief emission impulse. The effect is thus based upon an effect which is typical for organic dyes, *viz.*, their saturable optical absorptivity. The Beer-Lambert law of constant absorptivity holds only for low levels of illumination, at least as far as organic dyes are concerned. A great many dyes show the effect, such as the phthalocyanines, polymethin dyes, and even chlorophyll.[95] It appears that an n-π^* transition is involved.

Photocatalysis

This topic is discussed in Section 12.1 and 23.2; it frequently involves a dye adsorbed on a Zn^{96} or $TiO_2{}^{97}$ crystal surface, for example. It has been shown[101] that the photoelectron stimulated desorption process breaks some adsorption bonds selectively. Photoelectric charge transfer at phthalocyanine-hydrocarbon interfaces has been studied by Hartmann and Noolandi;[102] these systems are reported to exhibit negative photoconductivity controlled by energy levels situated between the Fermi level and the conduction band of the phthalocyanine.

References

1. L. B. Schein et al., *J. Chem. Phys.*, **71**, 3189 (1979).
2. H. Scher, *Photoconductivity and Related Phenomena*, J. Mort and D. M. Pai, eds., Elsevier, Amsterdam, 1975.
3. T. Imura, *Bull. Chem. Soc. Japan*, **46**, 2075 (1973).
4. L. E. Lyons and K. G. McGregor, *Aust. J. Chem.*, **29**, 13 (1976).
5. W. Arden and P. Fromherz, *Ber. Bunsen Ges. Phys. Chem.*, **82**, 868 (1978).
6. E. E. Polymeropoulos et al., *J. Chem. Phys.*, **68**, 3918 (1978).
7. B. J. Mulder, J. de Jonge and G. Vermeulen, *Rec. de Travaux Chim. de Pays-Bas*, **85** (1), 31 (1966).
8. B. J. Mulder and J. de Jonge, *Rec. Trav. Chim.*, **84**, 1503 (1965); G. L. Mulder, Philips Res. Repts. Suppl. No. 4 (1968).
9. L. I. Boguslavsky and B. T. Lozhkin, *Surface Sci.*, **38**, 413 (1973).
10. A. V. Vannikov, B. T. Lozhkin and L. I. Boguslavsky, *Fiz. Tverd. Tela*, **12**, 557 (1970); G. Aprilesi, T. Garofano, F. Nava and M. Santangelo, *Nuovo Cimento*, **1**, 85 (1971); A. Bree and R. A. Kydd, *J. Chem. Phys.*, **40**, 1775 (1964); A. V. Vannikov, L. I. Boguslavsky and V. B. Margulis, *Fiz. Tekhn. Poluprovudnikov*, **1**, 935 (1967).

REFERENCES

11. C. B. Duke, Proc. 3rd Symp. Electrode Processes, 1979, The Electrochem. Soc. Proc., 1980, p. 15.
12. Z. Burshtein and D. F. Williams, *Phys. Rev.*, **B-15**, 5769 (1977).
13. L. E. Lyons and K. A. Milne, *J. Chem. Phys.*, **65**, (4) 1474 (1973).
14. L. Onsager, *Phys. Rev.*, **54**, 554, 1938.
15. J. C. Knights and E. A. Davis, *J. Phys. Chem. Solids*, **35**, 543 (1974).
16. R. H. Batt, C. L. Braun and J. F. Hornig, *Appl. Opt.*, Suppl., **3**, 20 (1969).
17. N. Karl and G. Sommer, *Phys. Status Solidi*, **A-6**, 231 (1971).
18. K. A. Milne and L. E. Lyons, *Aust. J. Chem.*, **31**, 699 (1978); if also ref. 276.
19. R. R. Chance and C. L. Braun, *J. Chem. Phys.*, **59**, 2269 (1973).
20. L. E. Lyons and K. A. Milne, *J. Chem. Phys.*, **65**, 1474 (1976); Proc. 7th Molec. Cryst. Symp. Nikko (Japan) 1975, **B-5**.
21. M. Silver and R. C. Jarnagin, *Molecular Crystals*, **3**, 461 (1968).
22. R. F. Chaiken and D. R. Kearn, *J. Chem. Phys.*, **45**, 3966 (1966).
23. P. M. Borsenberger et al., *J. Appl. Phys.*, **49**, 5555 (1978).
24. R. Keiper and R. Schuchard, *Phys. Stat. Solidi*, **B-85**, 415 (1976).
25. H. Böttger and V. V. Bryskin, *Phys. Stat. Solidi*, **B-78**, 415 (1976).
26. N. F. Mott, *Phil. Mag.*, **36**, 413 (1977).
27. A. Gailis et al., Latv. PSR Timet. Akad Vestis, Fiz. Tek. Zinat, Ser., **1978** (1) 28.
28. D. Balode, et al., *ibid.*, p. 35.
29. N. Geacintov, M. Pope and H. Kallman, *J. Chem. Phys.*, **45**, (7) 2639 (1966); N. Geacintov and M. Pope, *ibid.*, **47**, 1194 (1967).
30. P. Barrell and E. Henderson, *J. Chem. Soc. Dalton*, 1975 (5) 432.
31. K. Okamoto, S. Kusabayashi and H. Mikawa, *Bull. Chem. Soc. Japan*, **46**, 2324 (1973).
32. V. G. Aidelis et al., *Thin Solid Films*, **38** (1) 9 (1976).
33. M. Ieda, T. Mizutami and Y. Jakai, in *Proc. Internatl. Micro-symposium on Polarization and Conduction in Insulating Polymers*, May 1972, Bratislava, Czech.
34. A. K. Jonscher and A. A. Ansari, *Phil. Mag.*, **23**, 205 (1971).
35. H. J. Wintle, *Photochem. Photobiol.*, **6**, 683 (1967).
36. B. Reimer and M. Bässler, *Phys. Stat. Solidi*, **B-85**, 145 (1978).
37. A. E. Binks et al., *J. Polymer Sci.*, **A-28**, 529 (1970).
38. T. Tamamura, *Bull. Chem. Soc. Japan*, **47**, (2) 448 (1974).
39. S. Hino and H. Inokuchi, *J. Chem. Phys.*, **70**, 1142 (1979).
40. R. Hirohashi et al., *Nippon Shashin Gakuaishi*, **40**, 328 (1977).
41. V. A. Benderskii and N. N. Usov, *Dokl. Akad. Nauk SSSR*, **167**, (4) 848 (1966).
42. H. Yasunaga et al., *J. Phys. Soc. Japan*, **46**, 839 (1979).
43. M. R. Padhye, *Proc. Nucl. Phys. Solid State Phys. Symp.*, **19C**, 85 (1976).
44. J. Maruchin, in *Proc., 4th Internat. Symp. on Phenothiazines and Related Drugs*, I. Forrest, E. Usdin and H. Eckert, eds., Elsevier Nth.-Holland, N.Y., 1980.
45. M. Spitzler and M. Calvin, *J. Chem. Phys.*, **67**, 5193 (1973).
46. K. H. Hauffe, in *Electrochemistry—The Past Thirty Years and the Next Thirty Years*, H. Bloom and F. Gutmann, eds., Plenum Press, N.Y., 1977, p. 209.
47. A. M. Meshkov, *Dokl. Akad. Nauk SSSR*, **169**, 154 (1966).
48. L. E. Lyons and O. M. G. Newman, *Aust. J. Chem.*, **24**, 13 (1971); A. K. Ghosh and T. Feng, *J. Appl. Phys.*, **44**, 2781 (1973).
49. N. N. Usov and V. A. Benderskii, *Sov. Phys. Semiconductors*, **2**, 580 (1968); A. K. Ghosh, D. L. Morel, T. Feng, R. E. Shaw and C. A. Rowe, *J. Appl. Phys.*, **45**, 230 (1974).
50. F. J. Kampars and M. P. Gouterman, *J. Chem. Phys.*, **81**, 690 (1977).

51. C. W. Tang and A. C. Albrecht, *J. Chem. Phys.*, **62**, 2139 (1975).
52. J. O'M. Bockris and K. Uosaki, Solar Energy Conf., Charan, Saudi Arabia, 1975 Proc.; H. Ehrenreich and J. H. Martin, *Phys. Today*, **32**, 25 (1979). V. Guruswamy and J. O'M. Bockris, *Solar Energy Materials*, **1**, 141 (1979). D. Haneman, *Proc. Royal Aust. Chem. Inst.*, Feb. 1977, p. 37.
53. H. Hora, German Pat., P 24-15-339. 9-33 (19 Feb. 1976).
54. V. Guruswamy and J. O'M. Bockris, *Solar Energy Material*, **1**, 141 (1979).
55. G. Engelsma et al., *J. Phys. Chem.*, **66**, 2517 (1962); M. Calvin, *Rev. Pure Appl. Chem.*, **15**, 1 (1965); J. M. Olson, *Science*, **168**, 438 (1970).
56. Y. Umezawa and T. Yamamura, *J. Electrochem. Soc.*, **126**, 705 (1979).
57. K. McGregor, Ph. D. Thesis Univ. of Queensland; *Aust. J. Chem.*, **29**, 13 (1976); L. E. Lyons and K. McGregor, *ibid.*, 1401 (1976).
58. K. J. Hall et al., *Aust. J. Chem.*, **31**, 1661 (1978).
59. N. Geacintov and M. Pope, *J. Chem. Phys.*, **50**, 814 (1969).
60. F. R. Tan and L. R. Faulkner, *J. Chem. Phys.*, **69**, 3334 (1978).
61. H. Meier et al., *Angew. Chem.*, Internatl. Ed. (Engl.), **11**, 1051 (1972).
62. L. E. Lyons, *Search*, **7**, 339 (1976); L. E. Lyons and M. Newman, *Aust. J. Chem.*, **24**, 13 (1971).
63. J. S. Bonham, *Aust. J. Chem.*, **29**, 2123 (1976); J. S. Bonham, et al., *Aust. J. Chem.*, **31**, 1661 (1978); see also Ref. 58.
64. J. S. Bonham and D. H. Jarvis, *Aust. J. Chem.*, **30**, 705 (1977); J. S. Bonham, *Aust. J. Chem.*, **29**, 2123 (1976); **31**, (1978); J. S. Bonham, *Phys. Stat. Sol.*, to be published.
65. S. Z. Weisz and W. B. Whitten, *Chem. Phys. Lett.*, **23**, 187 (1973).
66. K. Kaneto, Y. Ido, K. Yoshino and Y. Inuishi, *Tech. Rep., Osaka Univ.*, **26**, 161 (1976).
67. V. A. Benderskii et al., *Dokl. Akad. Nauk. SSSR*, **239**, 856 (1978).
68. U.S. Pat., 35 30 007.
69. H. Meier, *J. Phys. Chem.*, **69**, 719 (1965); M. I. Fedorov et al., *Izv. Vyssh. Uchebn. Zaved. Fiz.*, **20** (3) 158 (1977).
70. H. Meier, W. Albrecht and U. Tschirwitz, *Ber. Bunsen Ges. Phys. Chem.*, **73**, 795, (1969); H. Hoegl, *J. Phys. Chem.*, **69**, 755 (1965); U. Tschirwitz, Dissertation, Universität Nürnberg (1970).
71. Y. A. Vidadi, L. D. Rozenshtein and E. A. Chistayokov, *Sov. Phys. Semicond.*, **1**, 1049 (1968).
72. V. M. Fridkin et al., *Fiz. Tekh. Polujerovodn*, **11** (1), 135 (1977).
73. *e.g.*, A. A. Zagruskii and F. I. Vilesov, *Fiz. Tverd. Tela*, **13**, 2300 (1971); T. Hiruoka et al., *Bull. Chem. Soc. Japan*, **42**, 1481 (1969); M. Kochi et al., *ibid.*, **43**, 2690 (1970); T. Hiruoka et al., *Chem. Phys. Lett.*, **18**, 390 (1973); A. A. Zagrubskii and F. I. Vilesov, "Uspechi Fotoniki," F. E. Vilesov, ed., **4**, 109 (1974) publ. Leningrad Univ.
74. B. H. Schechtman and W. E. Spicer, *Chem. Phys. Lett.*, **2**, 207 (1968).
75. W. E. Spicer, in *Optical Properties of Solids*, F. A. Abeles, ed., Nth.-Holland Publ., Amsterdam, 1972.
76. S. Hino et al., *Bull. Chem. Soc. Japan*, **48**, 1133 (1974).
77. S. Hino et al., *Chem. Phys. Lett.*, **36**, 335 (1975); K. Seki et al., *Bull. Chem. Soc. Japan*, **49**, 904 (1976).
78. P. Neilsen et al., *Solid State Commun.*, **15**, 53 (1974).
79. *e.g.*, T. Hiruoka et al., *Bull. Chem. Soc. Japan*, **42**, 1481 (1969); M. Batley and L. E. Lyons, *Molec. Cryst.*, **3**, 357 (1968); H. Kawamura and M. Inokuchi, *Bull. Chem. Soc. Japan*, **45**, 710 (1972).

REFERENCES

80. D. M. Hanson, *Crit. Revs. Solid State Sci.*, **3**, 243 (1973).
81. T. Hiruoka et al., *Chem. Phys. Lett.*, **18**, 930 (1973); cf. also Ref. 79.
82. e.g., A. I. Belkind et al., "Elect. Prop. of Org. Solids," Conf. Proc., Karpacz Poland, 1974, Publ. Polish Acad. Sci.; see also K. Seki et al., Ref. 77.
83. e.g., N. Veno et al., *Chem. Phys. Lett.*, **35**, 31 (1975); W. Pong and J. A. Smith, *J. Appl. Phys.*, **44**, 174 (1973); P. B. Merkel and W. M. Hamill, *J. Chem. Phys.*, **55**, 1409 (1971).
84. K. Seki et al., Ref. 77.
85. M. B. Robin, *Higher Excited States of Polyatomic Molecules*, Academic Press, N.Y., 1974.
86. A. I. Belkind et al., *Phys. Stat. Solidi*, **B-85**, 465 (1978).
87. A. I. Belkind et al., "Elect. Properties of Org. Solids Conf.," 1974, Karpacz, Poland.
88. S. F. Liu et al., *Phys. Rev.*, **B-12**, 418 (1975); P. Neilsen et al., *Solid State Comm.*, **17**, 1067 (1975).
89. W. Pong and J. A. Smith, *J. Appl. Phys.*, **44**, 174 (1973).
90. H. H. Kosche, in *Curr. Probl. Electrophotogr. 3d Eur. Colloq.* 1971, W. F. Berg, ed., Walter de Gruyter, Berlin, 1972, p. 301.
91. B. Feuerbacher et al., *Photoemission and the Electronic Properties of Surfaces*, Wiley, N.Y., 1978; L. Ley and M. Cardona, *Photoemission of Solids*, Springer, Berlin, 1979.
92. N. G. Basov, ed., *Exciton and Domain Luminescence of Semiconductors*, Plenum Press, N.Y., 1980; *J. Luminescence*, Nth.-Holland, Amsterdam; M. D. Lumb, "Organic Luminescence" in *Luminescence Spectroscopy*, M. D. Lumb, ed., Academic Press, N.Y., 1978.
93. J. D. Williams, Ann. Repts., A-1977, The Chemical Society, London, p. 51.
94. e.g., R. Hemming, *Faraday Disc. Chem. Soc.*, **58**, 261 (1974).
95. e.g., Z. Chemie, **10**, 409 (1970); A. E. Siegman, *An Introduction to Lasers*, McGraw-Hill, N.Y., 1971; R. K. Willardson and A. C. Beer, eds., *Semiconductors and Semimetals*, Vol. 14, "Lasers, Junctions, Transport," Academic Press, N.Y., 1979; M. Ross and J. W. Goodman, eds., *Laser Applications*, Vol. 4, Academic Press, N.Y., 1980; *Excimer Lasers*, C. K. Rhodes, ed., Springer, N.Y., 1979; N. G. Bason, *Lasers and Their Applications*, Plenum Press, N.Y., 1976; H. C. Casey and M. B. Panish, *Heterostructure Lasers*, Academic Press, N.Y., 1978.
96. K. H. Hauffe and U. Bode, *Faraday Disc. Chem. Soc.*, **58**, 261 (1974); K. H. Hauffe, in *Electrochemistry—The Past Thirty Years and the Next Thirty Years*, H. Bloom and F. Gutmann, eds., Plenum Press, N.Y., 1977, p. 209.
97. e.g., R. I. Bickley and R. K. M. Jayanty, *Faraday Soc. Disc. Chem. Soc.*, **58**, 194 (1974).
98. T. Tanaka and Y. Inuishi, *Japan J. Appl. Phys.*, **5**, 974 (1966).
99. W. D. Lakin et al., *Phys. Rev.*, **B15**, 5834 (1977).
100. Z. D. Popovic and R. Menzel, *J. Chem. Phys.*, **71**, 5090 (1979); "Thermal and Photostimulated Currents in Insulators," D. M. Smyth, ed., The Electrochem. Soc., Princeton, New York, 1975.
101. M. L. Knotek et al., *Phys. Rev. Lett.*, **43**, 300 (1979); D. P. Woodruff et al., *Phys. Rev.*, **B21**, 5642 (1980).
102. G. C. Hartmann and J. Noolandi, *J. Chem. Phys.*, **66**, 3498 (1977).

23

Retrospect, Outlook and Speculations. Summary.

23.1 Retrospect

Since the writing of the first part (A) of this book, work in the field of organic semiconductors has grown at an exponential rate, becoming so vast as to just about defy codification. Several practical applications have now entered the market and as perusual of the present volume shows, considerable and significant progress has been made in experimental sophistication as well as in theoretical interpretation. There is now a reasonable understanding of amorphous solids, developed nearly wholly during the last 15 years or so. Questions of mobility, central to the whole field of organic semiconduction, have received a great deal of attention resulting in significant advances in understanding (see Chapter 15).

The field of organic semi-metals (see Chapter 21) did not exist when the first part (A) of the volume was compiled; by now it has become one of the most important and fruitful areas of present day solid state research.

The discovery of hyperelectronic polarization (discussed in Section 16.5) occurred during the last 12 years and this topic too is attracting increasing attention.

Biological applications have now entered the clinical phase. If progress continues, even at the present rate, a review of just the biological aspects of the field will soon warrant a treatise of its own.

23.2 Outlook and Speculations

Devices

There is still no organic transistor, though such a device could readily be made.

Tanikawa et al[1] have found that the p-type poly(γ-(β-N-carbazole ethyl)-L-glutamate can be converted to n-type by doping with trinitrofluorenone in benzene or methylethylketone. A rectifying p-n junction was obtained this way. An electron tunnel-junction for tetracyanoquinodimethane compounds in contact with poly-tin, below 4.5 K, has been reported,[2] p-n heterojunctions have already been reported[3] in polyacetylenes, $(CH)_n$.

In the presence or excess catalyst, trans-polyacetylene has been prepared as a metal-like film consisting of a cross-connected mass of fibrils. The p-type polyacetylene had a resistivity of 10^5 to 10^6 ohmcm which could be increased by four orders of magnitude upon addition of electron donor dopants. Doping this material with sodium or lithium yielded an n-type semiconductor with a resistivity of 10^{-3} ohmcm. Doping with arsenic pentafluoride above a critical concentration of about 1% produced a change from semiconductivity to metallic conductivity. When the film is stretched the longitudinal conductivity, in the fibril direction, is 16 times higher than in the transverse direction. This subject has been reviewed[85] extensively. The material is presently being investigated for use as a substrate for solid-state batteries by some researchers.[86]

McAlear has reported[87] the use of deposited synthetic polylysine on glass to form conduction pathways. In this application the polylysine provides a surface for metal adherence. However, it is envisaged that it may be possible to assemble electrical components from proteins and molecules grown in genetically modified organisms.

Organic switching devices are likely to find practical employment in relatively low speed (because of the inherently low mobility of carriers in organic materials) applications; several types have already been proposed and tested. Organic switching has been observed[4] in pure tetracene thin films, roughly 3000 Å thick, sandwiched between metal electrodes. When a critical voltage between 4-8 V is exceeded, the material switches from a high (10^{10} ohm) to a low resistive state (10^5 ohm). This process is reversible although the critical voltage is not precisely reproducible. The effect is insensitive to changes in temperature and light, and is said[4] to involve unusual space charge limited currents with apparent hopping carrier motion in a trap-free amorphous matrix. However, space charge in a trap-free system is a concept not easy to accept, and a different interpretation may need to be sought.

Williams and Sworakowski[5] interpret non-reproducible switching in organic materials in terms of trapping and recombination of electrons and holes but much more work needs to be done before fully reliable explanations are available. Bistable switching between two impedance states has also been observed[6] in Cu-phthalocyanine films; the on-state is said to involve[7] partially carbonized regions formed by Joule heating along filamentary paths. In anthracene films less than 3 μm thick, threshold switching involves a transition from a shallow trap to a deep trap space charge limited regime.[6]

Switching has also been observed[8] in hydrated melanin which changes from a high to a low resistivity form when a critical value of applied field is attained. The effect is attributed to a cooperative interaction: the solid is said[8] to contain two electrically different phases, *viz.*, electron-rich domains embedded in an electron-poor matrix.[8] In effect, it forms a gel, a solid emulsion. At sufficiently high fields, rapid injection of electrons takes place, resulting in an extension of the electron rich phase until an emulsion inversion occurs and the resistance of the system becomes quite low.[8]

Organic charge transfer complexes have exceptionally large second-order, non-linear optical coefficients in liquids[9] and in solids,[10] offering a wide scope for the development of instrumental transducers.

The electret microphone is now a commercial proposition and several other transducers based upon electrets have been proposed.[11,12] Polymeric molecular solids containing eka-conjugated structures exhibiting giant polarizations should have a wide range of applications, *e.g.*, as rectifiers,[13] diodes and negative resistance devices for use in microwave generation,[14] thermoelectric devices with voltages of up to 3000 μV/°C, gas detectors,[15] pressure transducers[16] and switches.[16] Several of these applications have been reviewed by Meier et al.[17] Piezoelectric and pyroelectric organics also offer possibilities as transducers.[18]

Electron scavenging by organic molecules such as the fluorocarbons and naphthalene has been mooted[19] for the control of ionospheric phenomena. Aerosol filtration by fibrous, electrostatically charged organic materials on the electret principle may be a viable possibility.[20]

An electrical heating tape has been introduced involving[21] a conductive polymer arranged between two wire electrodes which allows the automatic adjustment of the heater to heat-loss conditions at any point of the circuit in a self-attenuating manner.

In the context of corrosion, Yamamoto and Hirano showed[22] that a current was generated in phosphate esters between two rotating cylinders. As the velocity gradient built up the potential difference increased with increasing conductivity of the ester, which may present an area of research where the properties of lubricants may be improved by the application of electrical currents.

Electro-Photography

This topic is discussed in Section 16.7. Commercial electro-photographic apparatus based upon organic photoconductors is now becoming available. Photochromism has been observed in spirochromenes in the amorphous state.[23] Robillard[24] and other workers have used[25] the photochromicity of spiropyranes adsorbed on semiconductors in photographic processes. Other photopolymers, such as acrylamide have also been used[26] to provide high efficiency recording phase holograms. Meier and Albrecht show[7] the electronic conduction mechanism in a large variety of organic compounds to be similar to that of inorganic photoconductors, after eliminating electrode contact problems with the use of the vidicon method, which replaces the electrode by a scanning electron beam. This work shows that organic photoconductors could be used in vidicon television pick-up tubes. Morimoto and Murakami[27] sensitized photoconductive organic layers with triarylcarbonium salts which differ from the well-known triphenylmethane dyestuffs possessing two radical centers, in that the former have no radicals. These novel sensitizers have a high capacity to increase the photosensitivity of organic photoconductors, and are said to be superior to ZnO paper.[27] The photoconductor-triarylcarbonium salt sensitizer applied to cellophane on plastic sheets yields electrographic material capable of use as diazo-masters.

Yokoyama and coworkers have developed[28] a multicolor diazocopying process of promise. Robillard has devised[29] a new dry photographic process involving photodissociation of metallic azides under various modes of controlled excitation with light an electric fields. High-speed electrophotographic elements producing images of good quality have been manufactured[30] in the form of charge transfer complexes of trinitrofluorenone with dibenzofuran-formaldehyde and dibenzothiophene-formaldehyde resins.

Energy Conversion and Storage

Primary cells based upon solid charge transfer complexes are now commercially available as cardiac-pace-maker energizers;[31,32] polyvinylpyridine-iodine-lithium cells with lifetimes of 10 years in actual use, have been reported.[33]

An electrochemically active reactant may be stored in the form of a charge transfer complex.[34] Since the adduct is generally rather weakly bonded, the system may assume a state of lowered potential energy by the release of, say, the acceptor component to act in an electrochemical reaction. Thus, e.g., iodine may be stored as the iodine-phenothiazine complex, facing a magnesium anode. The complex acts as cathode. The cell is completed by an inert, say carbon or platinum, counter electrode. The devise becomes practicable because complexation reduces the very high resistivity of elemental iodine by many orders of magnitude so that the cell shows a reasonably low internal resistance. Its open circuit voltage[34] is about 1.5 to 2.5 V. As current is drawn, more iodine is released from the solid complex. The resulting device may be used as a primary battery of long shelf life, lasting up to several years in low-power, intermittent operation such as in cardiac pacemakers.[35]

Solar energy conversion and storage materials undergoing photoinduced RedOx reactions are only just beginning to be developed.[36,37] Regenerative photoelectrochemical cells can be constructed[38] from single crystal n-GaAs in acetonitrile solutions utilizing such organic solutes as anthraquinone, p-benzoquinone, dimethylferrocene, ferrocene, hydroxymethylferrocene and tetramethyl-p-phenylenediamine. A power conversion efficiency of 14% was noted for the system using ferrocene-ferricenium acetonitrile solution with light of radiant intensity of 0.52 mW/cm^2 at 720-800 nm. A considerable number of workers[39] have been active in the area of photo-RedOx reaction photocells based on dye/reductant couples, such as the thionine/Fe(II) salt systems. In the context of organic pigment RedOx couple systems it should be pointed out that ultrasound has been used to produce[40] a photographic image in them similar to an effect observed by Ernst and Gutmann[41] and others.[42] Umezawa and Yamamura[43] have investigated the possibility of a wet solar cell based on Mn (III)porphyrin in a 0.01 molar air-free aqueous solution of ethyl ammonium-perchlorate. This system has a low conversion efficiency, approximately 0.01-0.1%, depending on the radiant frequency, but does represent an interesting break-through.

Alkali metal-treated polynuclear aromatic compounds, *e.g.*, naphthalene or anthracene, in which formation of electron donor-acceptor complexes take place, as a result of charge transfer between the alkali metal and the organic compound, can activate hydrogen and absorb considerable quantities to form the metal hydride.[88] Further, the rare earth intermetallic-polynuclear aromatic complex, $SmMg_3$-anthracene, instantaneously absorbs[89] high-purity hydrogen at room temperature at about 400 mm Hg pressure. The rate of hydrogen uptake was proportional to the hydrogen pressure and increased considerably with increasing anthracene:$SmMg_3$ ratio. Hydrogen is absorbed by $SmMg_3$ alone, after activation by evacuation at elevated temperatures. Recently synthesized tertiary phosphine complexes of manganese, $MnLX_2$ (L = tertiary phosphine, but not PPh_3, X = halide ion) resemble hemoglobin and myoglobin to the extent that they behave as reversible oxygen gas carriers, both in the solid state and in non-aqueous solutions.[90] The complexes are also reactive toward other small gas molecules, including hydrogen.[91] These approaches may open up new avenues for storing hydrogen gas.

Model organic light-energy converters have been tested[44] by Benderskii and collaborators in the case of InCl doped phthalocyanines which can be utilized as photocells with a rating of 10 mW/cm^2.

PhotoRedOx activity in a combination of tolusafranine-ethylene-diamine-tetra-acetic acid and a dye chelating agent occurs at 844 mV, comparable to that in green plants.[45]

p-n Heterojunctions produced in highly conductive $(CH)_n$ are reported[46] to exhibit open circuit photovoltages as high as 0.8 V.

Catalysis

Polymeric iron phthalocyanine is reported[47,48] to be a better electro-catalyst than Pt, at least under certain conditions. This is certain to be of considerable importance in the development of commercial fuel cells. Since in fuel cells using atmospheric oxygen as the electron acceptor, it is always the oxygen electrode which present difficulties and problems, it is of interest to note that Shold has shown[49] that the production of singlet atomic oxygen may be sensitized by excimers derived from aromatic hydrocarbons; this ought to be studied with a view of devising suitable electro-catalysts.

Catalysis on phthalocyanines (PC) is the subject of an extensive discussion by Hopf and Steinbach.[50] Metal chelates of PC, poly-PC and tetra-aza-annulene (H, Fe, Co and Cu form) provides examples[51] of high activity of electro-catalysts for the reduction of oxygen, and this may be extremely significant in fuel cell technology. This phenomenon has also been observed[52] by Johansson et al. who consider the question of the potential difference between air and oxygen electrodes. Hanke[53] found a correspondence of the PC type catalysis energy with the intramolecular oscillations of the substrate in the reduction of air and formic acid. The activation energy for conductivity of poly-PC-Cu was 0.21 eV, that of poly-PC-Ni double that; for PC-Ni and PC-Cu they were 1.60 and 1.77 eV, respectively. Emulsion polymerization of 2,6-xylenol catalyzed by a Cu-polyvinylpyridine complex proceeded[54] only at the H_2O-benzene interface. An extensive review[55] is available which discusses catalysis on semiconductors. The mechanism of polymerization of vinyl monomers with charge transfer complexes has also been found[56] to be a catalytic process. The photopolymerization of styrene in the presence of oxygen involves[57] the formation of a charge transfer complex. So does the solid phase photopolymerization of acrylamides[58] in the presence of dyes such as riboflavin and methylene blue.

Biological Applications

Many drug interactions of clinical interest are likely to involve the formation of charge transfer complexes[59] (see Section 17.5). This is a relatively new field which offers exciting possibilities in collaboration with pharmacologists and physicians.

Thus, interactions between the anticoagulant heparin and other drugs have been studied by Eckert et al.[60] and by Gutmann and Eckert.[61] Charge transfer interactions of psychotropic drugs as well as their interactions with biologically important compounds, have received considerable attention[61] and the results have shed light on some important problem areas such as the melanosis effects sometimes associated with long time administration of the phenothiazine derivative chlorpromazine.[62] It has now become feasible to test such charge transfer interactions *in vivo*, e.g., in mice.[63]

The electrical conduction properties or organic materials affect a wide variety of biologically significant processes.[64] The topic, especially as it refers to charge transfer complexes—see above—has been reviewed.[65] Thus, a few additional examples must suffice:

Khourkyamen et al. have shown[66] that the interaction of a protein with a polar solvent involves the transfer of an electron from a donor group into the conduction band of the protein and eventually to spatially distant receptor sites[66] (see also Chapter 18).

Osteogenesis may be influenced[67] by the application of continuous as well as pulsed currents *via* implanted electrodes and also by induced electric fields. Limb regeneration,[68] wound healing,[69] carcinosarcoma regression[70] and thrombosis[71] all are also affected by changes in the cellular electric environment.

Charge transfer complexes have been discussed in the context of catalysis and energy transfer or storage in nonliving systems. Szent-Györgyi points[72] out that sulfhydryl groups such as those in SH-glutathione have a general catalytic effect on charge transfer between methylglyoxal and bioamines. Ladik made the unexpected discovery[73] that SH may act as an electron donor and as an electron acceptor with far-reaching biological implications since thus SH has a central role in the electronic desaturation of proteins. Szent-Györgyi also suggested[72] that autocomplexation occurs in protein macromolecules in order to confer longevity on free radicals in reactive environments allowing compatibility in living systems. He further intimates that cancer cells are unable to arrest their growth because they are unable to activate their glyoxylase which destroys the ketone-aldehyde which keeps the cells in the radical state and at rest. Thus, a means of cancer prevention may be to decrease the electron donor to acceptor ratio in the cell.

Another area where semiconductive properties of living macromolecules may prove to be of profound interest involves the molecular dynamics of memory. Moore points out[74] that storage of specific information is not localized in specific regions of the brain, rather that the electric circuitry probably stores the information analogous to that of holography. Modifying the neuronal network by constant use most readily occurs at the neuronal rather than at the synaptic level. Resetting neuronal thresholds should thus involve a coupling mechanism of specific protein concentration dependence of the electrical activity of the membranes and *vice versa*. Protein degradation rates, modulated by electric field-dependent conformation changes, should be of key importance. The rates may also be affected by protein transfer across the cell membrane. Moore envisages[74] that the distribution of charged proteins across the membranes of nerve cells can be effected by electrostatically active fixative proteins.

In this context, it may be apposite to note that a photoelectric memory effect has been reported[75] in acid-doped polyvinylcarbazole-nitroaromatic ternary charge transfer complexes, involving thermally erasable ESR signals.[76] This is a very promising field with direct biological implications: since storage here occurs in the form of a localized excited state, similar effects might also be looked for in nervous tissues.

The discovery[77] that suspended living cells can be distinguished and sorted on the basis of their electrical polarization by the application of non-uniform electric fields offer intriguing possibilities. Pohl[77] has been able to follow reproduction of cells by microdielectrophoresis. Fröhlich has presented[78] a theoretical basis for expecting oscillatory and cooperative electric dipole effects in biological substances in the frequency range 10^{11}-10^{12} Hz. This has subsequently been confirmed[79] in growth rate experiments with a yeast. The growth rate was sharply dependent on an irradiation frequency of 4.18×10^{10} Hz (7.1 8 mm). Further confirmation[80] was found in accelerated bacterial colicin synthesis irradiated at 6.5 mm but depressed at 6.56 mm. Utilizing frequency resonances between 10^{10} and 10^{12} Hz, Webb et al.[81] observed peaks for human sarcoma cells and baby hamster kidney (BHK) cells infected with sarcoma virus. They noted a splitting of the peaks when cancer was present, and a new line at 7.5×10^{10} Hz. Particularly important was the observation that strong irradiation of infected BHK cells caused a marked lowering of infectivity without loss of viability of the normal cells. Cooper[82] pointed out that not only is the dividing cell in a cooperative polar mode, but that the fluidity of the cell surface is increased. Fröhlich's concepts[78] coupled with this concept and with Szent-Györgyi's prediction[72] of the lack of mobile electrons in functional proteins of sarcomas, confirmed[83] by the inhibition of respiration in oncogenous substances and by direct testing with esr measurements, promise[84] to be nothing short of revolutionary in the biological sciences.

References

1. M. Tanikawa et al., *J. Appl. Phys.*, 48, (6), 2424 (1977).
2. P. M. Chaikin et al., *Phys. Rev.*, B-17, 179 (1978).
3. Editorial, in *Physics Today*, September, 1979, p. 20.
4. A. Szymanski, D. C. Larson and M. M. Labes, *Appl. Phys. Lett.*, 14, (3), 88 (1969).

REFERENCES

5. J. O. Wiliams and J. Sworakowski, *Mater. Sci.*, **3**, (4), 99 (1977).
6. Y. Sadaoka and Y. Sakai, *Bull. Chem. Soc., Japan*, **50**, (9), 2223 (1977).
7. H. Meier and W. Albrecht, *Ber. Bunsen Ges. Physik. Chemie*, **73**, 86 (1969).
8. J. W. Cope, *Physiol. Chem. Phys.*, **9**, 543 329 (1977).
9. B. F. Levine and C. G. Bethner, *J. Chem. Phys.*, **65**, 2429, 2439, (1976); **66**, 1070 (1977).
10. B. L. Davydov et al., *J. Chem. Phys. Letters*, **12**, 16 (1970); *Opt. Spektrosc.*, **30**, 274 (1977); J. G. Bergman and G. R. Crane, *J. Chem. Phys.*, **66**, 3803 (1977).
11. G. M. Sessler and J. E. West, *J. Accoust. Soc. Amer.*, **34**, 1589 (1973).
12. G. M. Sessler and J. E. West, *J. Accoust. Soc. Amer.*, **53**, 1589 (1973); *Electrets*, M. M. Perlmann, ed., The Electrochem. Soc. Princeton, N.Y. (1973).
13. C. Hamann, *Phys. Stat. Solidi*, **4**, K97 (1964).
14. H. Meier, W. Albrecht and U. Tschirwitz, *Ber. Bunsen Ges. Physik. Chem.*, **73**, 795 (1969).
15. H. Meier, *Forschritte der Chemischen Forschung*, **61**, 85 (1976).
16. H. A. Pohl, A. Rembaum and A. Henry, *J. Amer. Chem. Soc.*, **84**, 2699 (1962).
17. H. Meier, W. Albrecht and U. Tschirwitz, *Angew. Chem.*, **11**, (12), 1051 (1972).
18. R. Yayakawa and Y. Wada, *Adv. Polymer Sci.*, **11**, 1 (1973); R. G. Kepler, *Phys. Rev.*, **B-9**, 4468 (1974).
19. P. R. Hammond, *J. Chem. Phys.*, **55**, 3468 (1971).
20. L. V. Radushkevich, *Doklady Adad. Nauk. SSSR*, **170**, 375 (1966).
21. Ed. Note, in *European Chem. News*, Dec. 14, (1973), Chemlex. Heat Ltd.
22. Y. Yamamoto and F. Hirano, *Wear.*, **50**, (2) (1974).
23. K. G. Dzhaparidze, *C.A.*, **81**, 119330q, m. 3902 (1974).
24. J-J. Robillard, *C. R. Acad. Sc. Paris*, **274**, 396 (1972).
25. J-M. Brot and J. J. Robillard, *C. R. Acad. Sc. Paris*, **277**, 167 (1973); I. B. Sidaravichyus, *Voprosyl. Radioelektroniki*, **1965**, (12), 105; N. Y. Gerulaitis and A. I. Korolev, *Zh. Vses. Khim. Obshchestro. D. I. Mendeleev*, **9**, 78 (1966); First used by E. I. Sponable, *J. Motion Pict. Tech.*, **60**, 337 (1953).
26. M. J. Jeudy and J-J. Robillard, *Optics Communications*, **13**, 25 (1975).
27. K. Morimoto and Y. Murakami, *Appl. Optics Suppl.*, **3**, 50 (1969).
28. T. Yokoyama et al., *Japan Kokei*, **73**, 38 728, June 7, (1973); see also ref. 29.
29. J. J. Robillard, "C. R. Symp. on Non-Silver Photographic Processes," Oxford, Sept. 11–14, (1973).
30. J. J. Robillard, French Pat., 73-41-591 (1973).
31. F. Gutmann, A. M. Hermann and A. Rembaum, *J. Electrochem. Soc.*, **114**, 323 (1967); **115**, 359 (1968).
32. M. Pampallona et al., *J. Appl. Electrochem.*, **6**, 269 (1976).
33. W. Greatbach et al., *IEEE Transactions on Biomed. Eng.*, **BMI-18**, 317 (1971); A. M. Hermann and E. Luksha, *J. Cardiovasc. Tech.*, **6**, 15 (1978).
34. F. Gutmann, A. M. Hermann and A. Rembaum, *J. Electrochem. Soc.*, **114**, 323 (1967); *ibid.*, **115**, 359 (1968); M. Pampallona et al., *J. Appl. Electrochem.*, **6**, 269 (1976); J. K. Louzos, US Pat., 3,653,968, April 4, 1972; F. Gutmann et al., US Pat., 3,660,164 (1972).
35. A. M. Hermann and E. Luksha, *J. Cardiovascular and Pulmonary Technol.*, **6**, (1), 15 (1978).
36. W. D. K. Clark and J. A. Eckert, *Sol. Energy*, **17**, 147 (1975).
37. Y. Umezawa and T. Yamamura, *J. Electrochem. Soc.*, **126**, 705 (1979).
38. P. A. Kohl and A. J. Bard, *J. Electrochem. Soc.*, **126**, 603 (1979).
39. K. Shigehara et al., *Electrochim. Acta*, **23**, 855 (1978); D. E. Hall et al., *J.*

Electrochem. Soc., **123**, 1705 (1976); W. J. Albery and M. D. Archer, *ibid.*, **124**, 688 (1977); W. D. K. Clark and J. A. Eckert, *Sol. Energy*, **17**, 147 (1975).
40. A. Eckardt and D. Fintelmann, *Naturwissensch.*, **20**, 555 (1955).
41. L. Ernst and F. Gutmann, *J. Soc. Leather Trades' Chemists*, **34**, 454 (1950).
42. P. Gourlay, *Rev. Techn. Ind. Cuir.*, **43**, 179 (1951).
43. Y. Umezawa and T. Yamamura, *J. Electrochem. Soc.*, **126**, 705 (1979).
44. V. A. Benderskii et al., *Dokl. Akad. Nauk. SSSR*, **239**, (4), 856 (1978).
45. M. Kaneko and A. Yamada, *J. Phys. Chem.*, **81**, 1213 (1977).
46. Editorial, in *Physics Today*, September 1979, p. 20.
47. H. Jahnke and M. Schonborn, paper presented at the 19th CITCE meeting, Detroit, Michigan (Sept. 26, 1968).
48. J. W. Johnson in *Electrochemistry—the Past Thirty Years and the Next Thirty Years*, H. Bloom and F. Gutmann, eds., Plenum Press, N.Y., 1977, p. 257.
49. M. Shold, *J. Photochem.*, **8**, 39 (1978).
50. H. Kopf and F. Steinbach, ed., *Katalyse an Phthalocyaninen*, Stuttgart, 1973.
51. F. Beck, *Ber. Bunsen Ges.*, **77**, 353 (1973).
52. I. Y. Johansson, J. Mrha and R. Larsson, *Electrochim. Acta*, **18**, 255 (1973).
53. W. Hanke, Zeitschrifft fur Anorg. u. Allgem. Chemie, **347**, 67 (1966).
54. E. Tsuchida, H. Nishide and H. Nishikawa, *Nippon Kagaku Kaishi*, **1972**, (12) 2416.
55. S. R. Morrison, *Chem. Tech.*, **7**, 570 (1977).
56. R. F. Tarvin, *Diss. Abstr. Inst.*, **B-33**, (7) 3030 (1972).
57. T. Kodaira, K. Hayashi and T. Ohnishi, *Polym. J.*, **4**, (1), 1 (1973).
58. I. M. Barkalov et al., *Dokl. Akad. Nauk. SSSR*, **169**, (5) 1111 (1966).
59. F. Gutmann, in *Bioelectrochemistry Proc. U.S.-Australia Joint Seminar on Bioelectrochemistry*, Pasadena, July 1979; H. Keyzer and F. Gutmann, eds., Plenum Press, N.Y., 1980, p. 159; G. Eckert et al., *ibid.*, p. 171; J. P. Farges and F. Gutmann, in *Modern Aspects of Electrochemistry*, J. O'M. Bockris and B. E. Conway, eds., Vol. 12, 267 (1978).
60. G. Eckert et al., *J. Biol. Phys.*, **6**, 160 (1978); G. Eckert and F. Gutmann, *Electroanal. Chem. and Interfacial Electrochem.*, **62**, 267 (1975).
61. F. Gutmann et al., in *The Phenothiazines and Structurally Related Drugs*, I. S. Forrest, C. J. Carr and E. Usdin, eds., Raven Press, N.Y., 1974, p. 15; F. Gutmann and H. Keyzer, *J. Chem. Phys.*, **50**, 550 (1969); *Electrochim. Acta*, **13**, 693 (1968); **12**, 1255 (1967); *Nature*, **205**, 1102 (1965); *Rev. Aggressol. (Paris)*, **7**, 27 (1968); F. Gutmann and A. Netschey, *Nature*, **191**, 1390 (1961); *J. Chem. Phys.*, **36**, 2355 (1962); C. M. Gooley, H. Keyzer and F. Setchell, *Nature*, **232**, 80 (1969); A. Brau, J.-P. Farges and F. Gutmann, *Electrochim. Acta*, **17**, 1803 (1972); J. R. Cann, *Biochem.*, **6**, 3435 (1967); R. Foster and P. Hanson, *Biochem. Biophys. Acta*, **112**, 482 (1966); A. Fulton and L. E. Lyons, *Aust. J. Chem.*, **21**, 873 (1968); E. Labos, *Nature*, **209**, 201 (1966); N. J. Seghatchian, et al., *Chem. Biol. Interact.*, **3**, 413 (1971); J. E. Bloor et al., *J. Med. Chem.*, **13**, 922 (1970); R. Foster and C. A. Fyfe, *Biochim. Biophys. Acta*, **112**, 490 (1966); R. Foster, *Organic Charge Transfer Complexes*, Academic Press, N.Y., 1969; M. A. Slifkin, *Charge Transfer Interactions of Biomolecules*, Academic Press, N.Y., 1971; M. G. Woronkow and A. J. Deitsch, *Ost. Chemiker Ztg.*, **67**, 1 (1966); J. S. Mahood et al., *Proc. 4th Internatl. Symp. Phenothiazines, Zurich, 1979*, I. S. Forrest et al., eds., Elsevier, Nth. Holland Publ., 1980.
62. I. Forrest et al., *Rev. Aggressologie*, **7**, 147 (1966); A. C. Greiner and K. Berry, *Canad. Med. Assoc. J.*, **90**, 363 (1964); A. C. Greiner and G. A. Nicolson, *ibid.*, **91**, 627 (1964); I. S. Forrest et al., *Rev. Aggressologie*, **4**, 259 (1963).

63. H. Keyzer et al., in *Proc. 4th International Symp. Phenothiazines and Related Drugs*, Zürich, 1979, H. Eckert, I. S. Forrest and E. Usdin, eds., Elsevier, Nth. Holland Publ., 1980.

64. F. Gutmann, in *Bioelectrochemistry Proc. US-Australia Joint Seminar on Bioelectrochemistry*, H. Keyzer and F. Gutmann, eds., Plenum Press, N.Y., 1980; J. P. Farges and F. Gutmann in Ref. 59.

65. e.g., F. Gutmann et al., *Adv. Biochem. Psychopharmacol.*, 9, 15 (1964); G. Eckert et al., *J. Biol. Phys.*, 6, 161 (1978); W. Stoekenius, R. H. Lozier and R. A. Bogomolni, *Biochim. Biophys. Acta*, 505, 215 (1979); see also references in *Bioelectrochemistry Proceedings U.S.-Australia Joint Seminar on Bioelectrochemistry*, Pasadena, 1979, H. Keyzer and F. Gutmann, eds., Plenum Press, N.Y., 1980.

66. V. N. Kharkyanen et al., *J. Theoret. Biol.*, 73, (1) 29 (1978).

67. I. Yusuda, K. Noguchi and T. Sata, *J. Bone Joint Surg.*, 37A, 1292 (1955); C. A. I. Bassett, R. J. Pawluk and R. O. Becker, *Nature (London)*, 204, 652 (1964); C. A. I. Bassett and R. J. Pawluk, Conference on Diffuse Electric Currents on Physiological Mechanisms, Milwaukee, Wisc., October (1967); C. Minkin, B. R. Poulton and W. H. Hoover, *Clin. Orthoped.*, 57, 303 (1968); B. T. O'Conner, H. M. Charlton, J. D. Currey, D. R. S. Kirby and C. Woods, *Nature (London)*, 222, 162 (1969); H. J. Hambury, J. Watson, A. Sivyer and D. J. B. Ashley, *Nature (London)*, 231, 190 (1971); Z. B. Friedenberg, E. T. Andrews, B. I. Smolenski, B. W. Pearl and C. T. Brighton, *Surg. Gynecol. Obstet.*, 131, 894 (1970); L. S. Lavine, I. Lustrin, M. H. Shamos and M. L. Moss, *Acta Orthopaed. Scand.*, 42, 305 (1971); L. S. Lavine, I. Lustrin, M. H. Shamos, R. A. Rinaldi and A. R. Liboff, *Acta Orthopead. Scand.*, 42, 305 (1971); D. D. Levy, *J. Electrochem. Soc.*, 118, 1438 (1971); J. Richez, A. Chamay and L. Bieler, *Virchows. Arch. Pathol. Anat.*, 257, 11 (1972); W. Kraus and F. Lechner, *Münch. Med. Wochschr.*, 114, 1814 (1972).

68. R. O. Becker, *Nature (London)*, 235, 109 (1973); R. O. Becker and J. A. Spadaro, *Bull. N.Y. Acad. Med.*, 48, 627 (1972); S. O. Smith, *Anat. Rec.*, 158, 89 (1967).

69. e.g., G. Colaccico and A. A. Pilla, *5th International Symposium on Bioelectrochemistry*, Weimar, East Germany, Sept. 1979; See also Refs. in A. A. Pilla, in *Bioelectrochemistry Proc. U.S.-Australia Joint Seminar on Bioelectrochemistry*, H. Keyzer and F. Gutmann, eds., Plenum, N.Y., 1980, p. 353 ff.

70. C. E. Humphrey and E. H. Seal, *Science*, 130, 388 (1959); C. A. L. Basset, R. J. Pawluk and A. A. Pilla, *Am. N.Y. Acad. Sci.*, 238, (1974); see also ref. 68.

71. L. E. Wolcott, P. C. Wheeler, A. M. Hardwicke and B. A. Rowley, *Southern Med. J.*, 795 (1969); P. N. Sawyer and B. Deutch, *Amer. J. Physiol.*, 187, 473 (1956); L. Duic, S. Srinivasan and P. N. Sawyer, *J. Electrochem. Soc.*, 120, 348 (1973).

72. A. Szent-Györgyi, *Proc. Natl. Acad. Sci.*, 74, (7), 2844 (1977).

73. J. Ladik, Private Communication (1979).

74. W. J. Moore, *Search*, 9, 313 (1978).

75. D. J. Williams et al., *J. Am. Chem. Soc.*, 94, 7970 (1972).

76. D. F. Ilten and M. Calvin, *J. Chem. Phys.*, 42, 3760 (1965); F. Stewart and M. Eisner, *Mol. Phys.*, 12, 183 (1967).

77. H. A. Pohl, *Dielectrophoresis*, Cambridge University Press, N.Y., 1978; H. A. Pohl, in *Bioelectrochemistry Proceedings of the U.S.-Australia Joint Seminar on Bioelectrochemistry*, Pasadena, July, 1979, H. Keyzer and F. Gutmann, eds., Plenum, N.Y., 1979, p. 273 ff.

78. H. Frohlich, *Int. J. Quantum Chem.*, 2, 641 (1968); cf., also "Cooperative Phenomena," H. Haken and M. Wagner, eds., Springer, N.Y., 1973, p. VII.

79. N. D. Devyatkov, *Sov. Phys. Usp.*, **16**, 568 (1974).
80. A. Z. Smolyanskaya and R. L. Vilenskaya, *Sov. Phys. Usp.*, **16**, 571 (1974).
81. S. J. Webb, R. Lee and M. E. Stoneham, *Int. J. Quant. Chem. Quant. Biol. Symp.*, **4**, 277 (1977).
82. M. S. Cooper, *Phys. Lett.*, **65**, 71 (1978); *Physiol. Chem. and Physics*, **10**, 289 (1978).
83. A. Szent-Györgyi, *Int. J. Quant. Chem. Quant. Biol. Symp.*, **3**, 45 (1976).
84. H. A. Pohl, P. Gascoyne and A. Szent-Györgyi, *Proc. Natl. Acad. Sci. U.S.A.*, **74**, 1558 (1977).
85. A. J. Heeger and A. G. McDiarmid, NATO Adv. Study Inst., Ser. C, **1980**, 353; A. G. McDiarmid and A. J. Heeger, 1980, 393; E. M. Conwell, Xerox Webster Research Center, Rochester, N.Y., at a regular meeting of the American Physics Soc., New Orleans, November 1981.
86. Private commun., eds., R. Baughman and R. Elsenbaumer, Allied Chemical Corp., New Jersey (1982).
87. J. H. McAlear, EMV Associates, Rockville, Md., in *Sci. Am.*, 246 (1), 78 (1982).
88. K. Tamaru and M. Ichikawa, "Catalysis by Electron Donor-Acceptor Complexes," Kodansha Ltd., Tokyo (1975).
89. H. Imamura and S. Tsuchiya, *J. Chem. Soc. Chem. Commun.*, 1981, 567.
90. C. A. McAuliffe et al., *J. Chem. Soc. Chem. Commun.*, 1979, 736.
91. C. A. McAuliffe et al., U.S. Pat. 4,251,452 (1981).

24

Addenda Tables

Abbreviations used in the Addenda Tables, p. 474

4.7A. Experimental Values of Mobility and its Activation Energy E_μ for Anthracene.

4.7B. Experimental Values of Mobility and its Activation Energy E_μ (eV) for Naphthalene

4.7C. Mobilities of Molecular Crystals

4.7D. Effective Mobilities.

4.7E1. Experimental Values of Mobility and its Activation Energy E_μ (eV) for Organic Liquids and Melts

4.7E2. Experimental Values of Mobility and its Activation Energy E_μ for Organic Liquids and Melts. Supplemental.

4.8A1. Experimental Values of Mobility and its Activation Energy for Organic Substances other than Anthracene and Naphthalene

4.8A2. Experimental Values of Mobility and its Activation Energy for Organic Substances other than Anthracene and Naphthalene. Su Supplemental.

4.9A. Activation Energies E_a of Organic Substances

5.4A Solvent Effect on the Charge Transfer Absorption Band of the Acenaphenetetrachlorophthalic Anhydride Complex at 20°C

6.6A1 Ionization Energies I_g and Work Function ϕ of Organic Molecules (in eV).

6.6A2	Ionization Energies I_g of Organic Molecules (in eV). Supplemental.
6.6B	Ionization Potential of Various Molecules and Free Radicals. Experimental Values Obtained by Retarding Potential Technique
6.7A1	Experimental Data for Photo-emission Threshold I_c and Molecular Ionization Energies I_g (in eV).
6.7A2	Experimental Data for Photo-emission Threshold I_c and Molecular Ionization Energies I_g (in eV). Supplemental.
6.8A	Absolute Electron Affinities A_g of Organic Materials (in eV).
6.8B	Polarization Energies P and Adiabatic Ionization Energies I of Organic Solids Determined by Ultraviolet Photoelectron Spectroscopy (in eV).
6.8C	Electron Affinities (A_g) of Organic Molecules (in kcal/mole), Measured by the Electron Attachment Method, Forming Negative Ions in a Magnetron Device.
6.8D	Group Contributions to the Electron Affinity A_g.
6.8E	Polarographic Electron Affinities A_g.
6.8F	Electron Affinities of Acceptors and CT Band Maxima ($\bar{\nu}_{max}$ 298 K) of Methiodides.
6.9A	Relative ($A_{G,rel}$) and Absolute ($A_{G,abs}$) Electron Affinities (in eV)
6.10A	Electron Affinity A of Organic Compounds (in eV)
6.17A	Electron Affinities A_g of the Elements.
6.18A	Electron Affinities A_g of Substituents
6.19A	Calculated Ionization Potentials and Electron Affinities (in eV) of Purines, Pyrimidines, Porphins, and Poly-pyrroles.
6.20A	Electron Affinities (A_g) of Acceptors, and Excited State Energies (E*) of Their Atoms.
8.2A	Metal-free Molecular Crystals
8.2B	Some Metal Complexes
8.5A	Charge Transfer Complexes
8.5B	Ternary and Inclusion Complexes
8.5C	Proton and Protonated Complexes
8.5D	Proton Affinities (in eV)
8.6A	Semiconducting Free Radicals and Radical Salts: TCNQ as Acceptor
8.6B	Electrical Properties of Semiconducting Anion-Radical Salts Derived from Tetrachloro-p-Diphenoquinone
8.6C	Anion Radical Salts Derived from Tetrabromo-p-Diphenoquinone
8.7C	Long-Chain Compounds and Polymers

8.7D	Room Temperature Resistivities of PAQR Polymers and Copolymers.
8.7E	Resistivities and Activation Energies for Ten PAQR Polymers after Precompression to 32 kbar at 450 K, Evaluated at 1.82 kbar and 300 K
8.7F	Electrical Properties of 7,7′, 8,8′-Tetracyanoquinodimethane Salts of Ionene Polymers and Their Model Compounds
8.8A	Organic Dyes
8.11A	Biological Materials
8.11B	Biological Materials Supplementary Data
8.14A	Seebeck Coefficients (Thermo-EMF) of Organic Semiconductors
8.14B	Thermoelectric Power Data for Conductive PAQR Polymers
8.15A1	Photo-Conductivity Data
8.15A2	Photo-Conductivity Supplementary Data
8.16A1	Photo-Effects.
8.16A2	Photo-EMF Supplementary Data
8.20A	Glass Transition Temperatures Evaluated from Glow Curves Compared to Literature Values
8.21A	Donicities and Electron Acceptance Numbers of Organic Compounds
8.21B	Supplementary Table of Selected Donicities DN vs. Antimonypentachloride in 1,2-Dichloroethane as Solvent
8.21C	Supplementary Table of Electron Acceptance Numbers AN vs. Hexane as the Reference Solvent
8.22A	The Dielectric Constants and Conductivities of Homogeneous Solids Showing Nomadic Polarizations
9.1A	Minimum Resistivities of Phthalocyanine under Pressure
9.2A	Sensitization of Photoconductivity of Polynaphthalene, Polyvinylcarbazole and Polyacenaphthalene
9.2B	Sensitization of the Photoconductivity of Polyvinylcarbazole
10.2A	Piezoelectric Substances
10.2B	Piezoelectric Nucleic Acids
10.2C	Piezoelectric Temperature Coefficients of Various Substances
10.3A	Reduced Heterocharge for Various Polymer Electrets
10.4A	Dielectric Constants and Conductivities of Organic Homogeneous Solids Showing Nomadic Polarization
10.5A	List of All Tables in Parts A and B

Abbreviations used in Addenda Tables

BPH	Benzophenothiazine
Chromat.	Chromatography
CLL	Chloranil
CPO	Contact potential
CTCS	Charge transfer spectra
DBP	Dibenzophenothiazine
DDB	2,5-Dichloro-3,6-dibromo-*p*-benzoquinone
DDQ	2,3-Dichloro-5,6 dicyano-*p*-benzoquinone
DEPE	1,2-di-(*N*-ethy-4-pyridinium)-ethylene
Depos.	Deposited
DI	Deionized
DMPD	N,N'-dimethyl-*p*-phenylenediamine
DP	Diaminopyrene
EC	Electron Capture
E_e	Thermal Activation Energy of PHC
EI	Electron injection
$f(\ )$	Function of
FRONT	Front electrode
FQ	Fluorescence quenching
Gen.	Generation
$h\nu$	Quantum
Intens.	Intensity
I_{Phot}	Photocurrent
I_{sh}	Short Circuit Current
MTA	Mellitic trianhydride
o/c	Open Circuit
PA	Phthalic anhydride
PAQR	Polyacenequinone
PD	Phthalic anhydride
PE	Photo-emission
PES	Photoelectron spectra
PG	Polarography
PH	Phenothiazine
PHC	Photoconductivity

PHTHC	Phthalocyanine
PI	Photoionization
PMA	Pyromellitic dianhydride
Power ζ	Power efficiency of conversion
PVK	Polyvinylcarbazole
Pϕ	*p*-phenylene diamine
Q	Quantum yields $\equiv \left(\dfrac{\text{No. of conduction electrons}}{\text{No. of photons absorbed}}\right)$
R	Recrystallized
S	Sublimed
SC	Single Crystal
Sensit.	Sensitized with
SO	Solvent extraction
Sol.	Solution
SU	Surface Cell
Sub.	Substrate
SW	Sandwich Cell
SY	Spectroscopic
TBDQ	Tetrabromo-*p*-diphenoquinone
TCDQ	Tetrachloro-*p*-diphenoquinone
TEA	Triethyl ammonium
TH	From thermoelectric data
TMPD	Tetramethyl-*p*-phenylenediamine
TMTTF	Tetramethyltetrathiafulvalene
Thresh	Threshold (Photoelectric)
Transp.	Transparent
TTF	Tetrathiofulvalene
TTT	Tetrathiatetracene
Z	Zone Refined
μ	Mobility
ζ	Efficiency

TABLE 4.7A Experimental Values of Mobility and Its Activation Energy E_μ(eV) for Anthracene (*signifies that μ decreases with increasing temperature)

Substance	T, (°C)	Direction	μ(cm²/Vsec)	Sign of Carriers	E_μ	Remark	Ref.
Anthracene	25		0.35 to 0.39	n			2202
			0.4			Z,SC	2210
	25		0.342 ± 0.034	n		Perdeuter.	2150
	25		0.378 ± 0.003	n			
	25		0.77 to 0.87	p			2160
	25		1.1 trap-free	p		Perdeuter.	
	25		1.49 to 1.43	n			
	25		1.7 trap-free	n			
			1.0	p		SC	2216
			1.0	n			
	−83 to 57		0.43 ± 0.02	n	0		2215
	25		1.1 + 0.06	p		$\mu=kT^{-2}$ (260–330K)	
	25	a	9	n		Hall	2162
		b	7	n			
		c	2	n			
			6	p		Hall	2222
	25	c	0.4	n		—	2212
	25		50	n		Hall	2213
	25		30	p		Hall	2213

25	c	0.2 to 0.3	n	Drift	2214
25	c	0.5 to 0.1	p	Drift	2214
25		~100	composite	Hall	2214
25		34–92	p	Photo Hall	2214
25		0.058 to 0.001	p	Magneto-Resist	2214
25		0.0015	n		2214
27	a	9	n	SC, Hall	
	b	7	n	No anomaly	
	c'	<2	n	}	2183
	a	1.7	n	Drift	
	b	1.0	n		
	c'	0.38	n		
27		0.82 ± 0.05	p	SC, Z, Photo ↓ with T ↑	2272
		0.37 ± 0.03	n	↑ with T ↓	
27		0.5	p	SC, injected	2294
		0.1	n	electrons	
		0.5 ± 0.1	p	Microwave	4115
		0.02			
	ac'	0.11 ± 0.04			4128
	c'a	0.2 ± 0.1			
77 to 300 K	c'	0.4	n	Photo Transient. Cleaved in ab plane	4129

477

TABLE 4.7B Experimental Values of Mobility and its Activation Energy E_μ (eV) for Naphthalene

Substance	T, °C	Direction	μ (cm²/Vsec)	Sign of Carriers	E_μ	Remark	Ref.
Naphthalene	20	a	0.51 0.88	n p			
		b	0.63 1.41	n p		Z,SC	2078
		c	0.68 0.98	n p			
	25		10^{-8}		0.6	From Charge Decay	2228
	25	b c' a b c' a	1.7 to 1.3 1.8 to 1.5 0.55 to 0.54 0.46 to 0.19 2.9 to 4.9 4.7 to 1.4	n n anomalous n p p anom. p anom.		Photo Hall, SC,S,Z	2180
	25	a b c' a b c'	0.6 0.5 0.6 0.8 1.2 0.5	n n n p p p		Drift SC,S,Z	2180

TABLE 4.7C Mobilities of Molecular Crystals. After Schein [2210]

Crystal	Sign of Carrier	Direction	μ (cm²/Vsec) ($T = 300$ K)	$\mu \propto T^n$ n	Range (K)
Anthracene	n	a	1.6	-1.0	77–300
	n	b	1.0	$-0.2, -0.6$	170–380
	n	c'	0.4	$+0.8, 0$	80–450
	p	a	1.2	-1.0	300–400
	p	b	2.0	-1.0	300–400
	p	c'	0.8	-1.0	170–450
Perdeuterated Anthracene	n	a	1.7	-1.8	280–400
	n	b	0.99	-1.4	280–400
	n	c'	0.35	0	280–400
	p	a	1.1	-1.7	280–400
	p	b	2.0	-1.4	280–400
	p	c	0.78	-1.0	280–400

TABLE 4.7C *Continued*

Crystal	Sign of Carrier	Direction	μ (cm^2/Vsec) ($T = 300$ K)	$\mu \propto T^n$ n	Range (K)
Azulene	n		0.15	0	220–360
Benzene	n	a	1.5 ($T=5°$C)	−2.0	173–278
1,4 diBromonaphthalene	n	a	0.017	≅−2	270–300
	n	b	0.013	≅−2	270–300
	n	c'	0.034	≅−2	270–300
	p	a	0.66	≅−2	270–300
	p	b	0.25	≅−2	270–300
	p	c'	0.87	≅−2	270–300
Naphthalene	n	a	0.51	−0.1	220–300
	n	b	0.63	0	220–300
	n	c'	0.68	−0.9	220–300
	p	a	0.88	−1.0	220–300
	p	b	1.41	−0.8	220–300
	p	c'	0.5, 0.99	−2.1	220–300
N-Isopropylcarbazole	n	c'	1.0	0	244–370
p-Diiodobenzene	p	a	12	−0.5	240–310
	p	b	4	0	240–310
	p	c	2	−0.8	240–310
Phenazine	n	a	0.29	0	180–360
	n	b	1.1	−0.65	230–360
	n	c	0.5	−0.1	230–360
Phenothiazine	n	cleavage plane	1.7	−3	300–350
	p	⊥ to plane	5.0	−3	300–400
β-phthalocyanine	p	c	1.1	−1.3	290–600
	n	c	1.4	−1.5	290–600
Pyrene	n	a–b	0.7	−1.5	260–350
	n	c'	0.5	−2.0	260–350
	p	a–b	0.7	−1.6	260–350
	p	c'	0.5	−1.3	260–350
S_8	n	all	10^{-4}	exp($-0.17/kT$)	220–413
	p	⊥(111)	10	−1.5	300–400
Se$_3$ (Monoclinic)	n	⊥(101)	2	−1.5	300–400
Tetracyanoquinodimethane (TCNQ)	n	⊥(001)	0.4	0	204–306
	p	⊥(001)	0.4	0	204–306

TABLE 4.7D Effective Mobilities. After Berlin et al.[2273]

μ_{obs} refers to the observable, macroscopic mobility, calculated from the equation (see Section 15.10),

$$\mu_{obs} = \mu_L P + \mu_F(1-P) \qquad (15.61)$$

where μ_L is the carrier mobility in the localized state and μ_F refers to the free, microscopic mobility of the carrier.

Compound	μ_F (cm²/Vsec)	E_μ (eV)	P	μ_{obs} (cm²/Vsec)
2,2 Dimethylbutane	440	0.074	0.972	12.3
n-Hexane	27	0.19	0.998	0.093
n-Hexane	42	0.16	0.995	0.24
Iso-octane	145	0.051	0.957	7
Neopentane	440	about 0	0.841	70
n-Pentane	43	0.15	0.997	0.15
2,2,4,4-Tetramethylpentane	630	0.061	0.962	24

TABLE 4.7E1 Experimental Values of Mobility and its Activation Energy E_μ (eV) for Organic Liquids and Melts.

Additional values of effective mobilities are tabulated in Table 4.7D.

Substance	T (°C)	μ (cm^2/Vsec)	Sign of Carriers	E_μ (eV)	Remark	Ref.
Cyclohexane	At critical temperature	0.005	cation			2276
Cyclopentane	At critical temperature	0.0043	cation			2276
n-Dodecane				0.22		2274
n-Eicosane				0.19		2274
Ethane	−36	5	n			2179
	+34	⩾50	n			2179
n-Hexane		0.0056	p	0.092	0	2118
n-Hexane	At critical temperature	0.0056	cation			2276
Iso-octane				0.06		2287
3-Methyl-pentane		0.2	n			2286
Neopentane	At critical temperature	42				2227
n-Octadecane	25	10^{-9} to 10^{-11}			SG from charge decay	2138
Phenylene-diamine		0.04 quasi-free >100	n			2275
Polyethylene	154	0.25			melt	2274
Silicone oil, polymerized		10^{-11}	n			2056

TABLE 4.7E2 Experimental Values of Mobility and its Activation Energy E_μ for Organic Liquids and Melts. Supplemental. See also, B. S. Yakovlev, *Russ. Chem. Revs.*, 48, 615 (1978).

Substance	$T(°C)$	$\mu(cm^2/Vsec)$ Proton	$\mu(cm^2/Vsec)$ Solvated electron	Sign of Carriers	E_μ (eV)	Remark	Ref.
Ammonia	−60	0.0013	0.01				4120
Dimethylethane	−90	—	4×10^{-4}				4120
Isopropanol	20	2.4×10^{-4}	4.6×10^{-4}				4120
Isopropanol	20	3.6×10^{-4}	4×10^{-4}				4120
Methanol	20	0.0014	0.0028				4120
Monobutylamine	20	—	0.0027				4120
Water	20	0.004	0.002				4120
Butene-1	293 K		0.064	n			4124
cis-Butene	293 K		2.2	n	3.7		4124
trans-Butene	293 K		0.029	n	5.4		4124
Cyclohexene	293 K		1.0	n	3.6		4124
2,2 Dimethylbutane	200		73	n		μ = continuous f (density)	4144
2,3 Dimethylbutane	293 K		0.9			Injected electrons	4126
	500 K		26				

Compound	Temperature	Value	Type	Value 2	Notes	Ref
2,3 Dimethylbutene-2	293 K	5.8	n	2.0		4124
2,3 Dimethylpentane	293 K	2.0			Injected electrons	4126
	536 K	26				
Ethane	21		n			4122
Hexane		0.21	n		X-ray injected	4123
Isobutene	293 K	1.44	n			4124
Methane	91–155 K	460 to 400	n			4121
3 Methylpentane	293 K	0.22			Injected electron	4126
	504 K	19			Injected electron	4126
Neopentane		93	n		X-ray injected	4119
Pyranthene in Benzene		2×10^{-4}			Non-linear to E	4023
2,2,4 Tetramethylpentane	293 K	29			Injected electrons	4126
	556 K	31				
Tetramethylparaphenyldiamine	25	7.6	p	0.154	10^{-5} M in hexane	4016
	25	10.1	n	–		
Tetramethylparaphenylenediamine	26	1.8	p	–	10^{-5} M	4016
	26	3.2	n	0.110		
Tetramethylsilane	175 K	130	n			4125
	299.5 K	100	n			4125
	451 K	10	n			4125
2,2,4 Trimethylpentane	293 K	4.5			Injected electrons	4126
	544 K	24				

TABLE 4.8A1 Experimental Values of Mobility and its Activation Energy for Organic Substances other than Anthracene and Naphthalene.

Substance	$T(°C)$	Direction	μ (cm^2/Vsec)	Sign of Carrier	E_μ (eV)	Remark	Ref.
Anthracene:TNB	20	a	0.52	n			
		b	0.59	n			
		c	2.89	n			
	70	a	1.23	n			
		b	1.33	n			
		c	5.95	n		SC	2299
	20	a	3.13	p			
		b	9.45	p			
		c	13.87	p			
	70	a	3.58	p			
		b	9.99	p			
		c	16.25	p			
Anthraquinone	25	c'	0.28	n			
		b	0.20	n			
		a	0.022	n		SC, PHC	2143
Azulene	25		0.19	n			
			0.2	p			2209
	20	⊥ to ab	>0.002	p		S, SC	2207
Benzophenone	−100 to +27	a	0.16	n		SC	2181
		b	0.049	n			
		c	0.055	n			
Benzophenothiazine:I$_2$ (2:3)	25		2.3				2290

Bovine serum albumin (BSA)	20		0.01		Microwave Hall	2308
Casein	20		0.01		Microwave Hall	2308
Ceres Yellow			0.12	p		2224
			0.06	n		
Chloroplasts			0.1	p	Spinach, clover	2032
			0.58 ~ 0.8	p		2031
Collagen:Methylglyoxal complex	20		100	p	Microwave Hall	2308
Cu-Phthalocyanine	25		2.5×10^{-4}	n	Transient PHC	2175
			1.5×10^{-4}	p		
Cytochrome			1.0 ± 80%	n	Dry	2033
			0.1 ~ 0.5	n	Damp	2031
Durene	25	ab	8	n	$kT^{-2.5}$	2176
		ab	5	p	Transient PHC	
		c'	0.15	p	$kT^{-2.8}$	
DNA			0.3	p	Dry	2031
			~9.5	n	Damp	2031
			0.5 ± 40%	p	Damp	2033
N-Ethylphenothiazine:I(2:3)	25			n	Seebeck, $\mu_-/\mu_+ = 5$	2290
Hemoglobin			2 ± 60%	p	Dry	2033
			2 ± 50%	n	Damp	2033
			0.1 ~ 0.5	n	Damp	2031
Iodine:Polyvinylcarbazole			0.4			2225
Lysozyme:Methylglyoxal Complex	20		0.01		Microwave Hall	2308
N-Methylphenothiazine:I_2 (2:3)	25		2.3	n	Seebeck, $\mu_-/\mu_+ = 2.08$	2290
Pb-Phthalocyanine			0.6–4	p	SCL	2223

TABLE 4.8A1 Continued

Substance	T (°C)	Direction	μ (cm²/Vsec)	Sign of Carrier	E_μ (eV)	Remark	Ref.
Perylene	27		0.17 ± 0.001	n	0.2	Photoelectric, SC	2184
Phenanthrene (non-virgin)	20	b	1.14	p		C, S, Z, SC	2076
(virgin)	20	c'	0.39	p		Trap-free	
	20		0.15 to 0.21	p		Trap-controlled	
	35–65	b	1.1	p		C, S, Z, SC	
	20		1 to 0.94	p			
(virgin)	20	a	2.1	p		Trap-free, C, S, Z, SC	
	20		0.52 to 0.63	p		Trap-controlled	
Phenolformaldehyde-(resin)-pyrolysates	25		7.84	p		Pyrolysis at 1200°C	2226
	25		0.0013	n		Pyrolysis at 600°C	2226
Phenoselenazine:I$_2$ (2:3)	25		2.6	n		Seebeck	2290
Phenothiazine:I$_2$ (1:3)	25		0.6 to 0.9	n		Seebeck	2290
Phthalocyanine:o-chloranil			1.4	p			2224
			0.3	n			
Poly-acrylonitrile			0.01	n		Hall, EF	2140
				p		Seebeck	
Poly-diacetylene	in chain direction		2.8			$\mu_+ + \mu_-$, SC	2149
Poly-ethylene (linear)	80		4.5×10^{-10}	p	0.55		2220
	95		9×10^{-10}	p			
Poly-ethylene (linear)	20		4.5×10^{-10}		0.55		2217
Poly-ethylene terephthalate	20		1.5×10^{-6}		0.2–0.3		2217

Compound						
Poly-ethylene terephthalate	20	1.5×10^{-6}	n	0.2		2220
	80	2×10^{-5}	n			2221
Poly-isoprene		2×10^{-12}	n	1.1	Hall constant = -7.5×10^{-3} C/cm^{-3} anomalous	2182
Poly-phthalocyanine			n			
Poly-styrene (atactic)	20	1×10^{-6}	p	0.2		2220
	80	5×10^{-5}	p			
Poly-styrene (atactic)	20	1.6×10^{-6}		0.2–0.75		2217
	80	5×10^{-5}				
Poly-n-vinylcarbazole	20	10^{-7}	p	0.7		2221
Poly-vinylcarbazole: trinitrofluorenone (1:1)		3.2×10^{-8}	n		Film	2144
Polyvinylchloride		10^{-6}	p	0.4–0.7		2163
		0.01	n			2056
Polyvinylchloride	27	$4.7 \pm 0.5 \times 10^{-4}$	n			2056, 2002
		$3.4 \pm 0.5 \times 10^{-3}$	n			2056, 2002
Pyrene	~72	$3.8 \pm 0.1 \times 10^{-3}$	n	0.43	SC	2211
	(60–100)	$3.3 \pm 0.4 \; 10^{-3}$	p	0.43	SC	2211
(1,2-bis-(4-Pyridyl)ethane$^+$): $\overline{\text{TCNQ}^-}$	25	0.04	p		Hall, SC	2020
Rhodamine-B		0.04	p			2224
		0.04	n			
TCNE	27	a 0.32	p			2174
		b 0.10				
		c 0.18				

TABLE 4.8A1 Continued

Substance	T(°C)	Direction	μ (cm$_2$/V sec)	Sign of Carrier	E_μ (eV)	Remark	Ref.
TCNE	<−30	a	0.32	p		Trans., PHC	2295
		b	0.10	p			
		c	0.18	p			
TCNQ	10 to 170		<0.3	n	0.48		2052
TEA(TCNQ)$_2$	27		1.3	n but p along y-axis		Uncorrected Hall	2009, 2010
	27		8.7			Corrected for anisotropy	
	27		1.75			From reflectance spectra and carrier concentration	
TEA(TCNQ)$_2$	27		2×10^{-4}	p		Hall, CP	2010
n-Terphenyl			$\mu_+ = 10^{-4}$	p		Electron bombardment	2059
			$\mu_- = 10^{-5}$	n			
Tetrabenzoperylene	27		0.8	$\mu_+ = \mu_-$		Drift	2177
TTF-TCNQ	∼−200		3000–14700	p		Semi-metal	2051

TABLE 4.8A2 Experimental Values of Mobility and its Activation Energy for Organic Substances other than Anthracene and Naphthalene Supplemental.

Substance	T (°C)	Direction	μ (cm²/Vsec)	Sign of Carrier	E_μ (eV)	Remark	Ref.
Anthracene/I_2			8–15	p, n		Along b axis, Hall	4118
Aryl-substituted Thiapyrilium Dye and Dialkyamino substituted Triphenylmethanes in poly(isopropyl- idene diphenylcarbonate)			~5 × 10⁻⁸ ~5 × 10⁻⁸	n p		2:1 Dye in Polymer, PHC	4030
Benzophenone	25	a b c	0.16 0.04 0.055		0.07	$T < 230$K, Thermal activation	4131
Biphenyl		a' c b	0.42 0.51 1.25	n n n		Pulsed PHC T^{-1} $T^{-1.25}$	4130
Bromanil			1.8 × 10⁻⁵	n			4012
Chlorophyll				p			4012
Chlorophyll a	30			p	0.4		4020, 4021
α-9,10 Dichloroanthracene	183 to 295 K		3.83 ± 0.2	p		$\propto T^{-2.4}$, Isotropic	4132

Material	Temp	Direction	Value	Type	Notes	Ref
β-9,10 Dichloroanthracene	25 25	b* ac plane	0.18 0.6		0.25	PHC, Isotropic	4074
β-9,10 Dichloroanthracene	25	ac plane ⊥ ac	0.6 0.185	p	0.04	PHC, Thermal activation ~0.25 eV	4074
Furan-quinone pigment yellow			3.3 to 3.7 × 10^{-8}	n		PHC, $\phi \propto E^2$	4046
I$_2$	20	b b	0.66 0.6	p n	0.08		4134
3-Methylpentane	4.2 to 85 K		0.02 to 0.11	n		PHC $\propto T^{-0.8}$	4038
Mg Phthalocyanine			0.1	n		Xenon flash	4018
Mg Phthalocyanine			0.1	n			4028
Polyvinylcarbazole			10^{-5}	p		At 10^6 V/cm	4127
Polyvinylcarbazole	−30 to +60		10^{-6} to 4 × 10^{-7}	p	~0.5	PHC	4116
p-Terphenyl	25		~10^{-3}	p			4133
Tetracene	27 87	⊥ab ⊥ab	10^{-2} 5 × 10^{-3}	p p			4133
Trinitrofluorenone			6 × 10^{-5}			Xerographic, $f(E)$	4117
Triphenylamine			10^{-7} to 10^{-5}	p		Se sensitized	4030

TABLE 4.9A Activation Energies E_μ of Organic Substances

See also extensive tables: V. Kampars and O. Neilands, *Russ. Chem. Revs.*, **46**, 945 (1977).

Substance	E_μ in $E/2$ kT (eV)	Ref.
Benzyl radical	0.88 ± 0.06	4135
Cyclopentadiene (Gas phase)	1.84 ± 0.03	4138
Deuterothiomethoxyl, (CD_3S) (Liquid)	1.858	4139
2,5 Dibromo-3,6 dicyano-p-benzoquinone	3	4137
Diethyl phthalate	0.54	4136
Dimethyl isophthalate	0.55	4136
Dimethyl phthalate	0.55	4136
Dimethyl terephthalate	0.64	4136
Dimethyl tetrachloro terephthalate	0.77	4136
Methylcyclopentadiene, (Gas phase)	1.67 ± 0.04	4B8
Terephthalaldehyde	0.56	4136
Thiomethoxyl, (CH_3S) (Liquid)	1.86	4139

TABLE 5.4A Solvent Effect on the Charge Transfer Absorption Band of the Acenaphthenetetrachlorophthalic Anhydride Complex at 20°C. After Nagy, Nagy and Bruylants, *J. Chem. Soc. Perkin II Trans.*, **1972**, 968.

The charge transfer absorption maxima ν_{max} of the complex acenapthene-tetrachlorophthalic anhydride in 32 different solvents belonging to the following classes (a) halogen-containing, (b) aromatic, (c) aromatic with one functional group, and (d) n-donor solvents.

Solvent	ν_{max}/kK	Solvent	ν_{max}/kK
Class (a)		Class (d)	
Carbon tetrachloride	24·6	n-Butyraldehyde	25·15
n-Butylchloride	24·9	Tetrahydrofuran	25·2
Dichloromethane	24·3	Tetrahydropyran	25·3
Chloroform	24·1	Acetonitrile	25·0
1,2-Dichloroethane	24·3	Nitromethane	24·8
1,1,2,2-Tetrachloroethane	24·15	Nitroethane	24·9
1-Chlorohexane	24·95	Dimethylformamide	25·3
Class (b)		Dimethyl sulphoxide	25·45
Benzene	24·5	Ethyl acetate	25·0
m-Xylene	24·7	Diethyl ether	25·1
p-Xylene	24·8	Acetone	25·15
Toluene	24·65	Dimethylacetamide	25·35
		1,4-Dioxan	24·65
Class (c)		1,2-Dimethoxyethane	25·0
Chlorobenzene	24·25		
Bromobenzene	24·15	Miscellaneous	
Benzonitrile	24·5	Cyclohexane	25·05
Benzaldehyde	24·55	Carbon disulphide	24·1
Pyridine	24·2		

TABLE 6.6A1 Ionization Energies I_g and Work Function ϕ of Organic Molecules (in eV).

Substance	I_g	ϕ	Method	Ref.
Acenaphthene	7.6		PE	2262
Amitryptilene	8.32		CTCS	2271
Aniline	7.00		CTCS relative to TCNQ	2250
o-Anisidine	7.57		FQ	2244
Anthracene	7.4(I_c)		PI	2265
Anthracene	7.5		PI	2245
Anthracene	7.41		SY.	2245
Anthracene	7.42		CP (graphite)	2258
Anthracene		4.71	CP (graphite)	2258
Anthracene:PMA complex		4.4	PES	2230
Azulene	8.59		Calculated	2264
Azulene	7.41		PE	2245
Benzaldehyde	9.85		EC	2189
Benzene	9.50		EC	2189
Benzene	9.77		Calculated	2264
Benzidine	6.87		FQ	2244
trans-1,3-Butadiene	9.78		Calculated	2264
β-Carotene (all trans)	5.44		PG	2263
Chlorpromazine	7.38		CTCS	2271
Chlorprothixene	7.68		CTCS	2271
Coronene	7.34		PE	2229, 2248
Coronene	5.59	4.70	PE	2202
Cypraheptadiene	7.94		CTCS	2271
Dialkylaniline	7.0–7.69		CTCS relative	2250
Dibenzophenothiazine	5.21		PE	2201
Dibenzophenothiazine: DDQ		4.63, 4.58	PE CP graphite	2257 2258
Dibenzophenothiazine:DDQ		4.62	PE	2257
Dibenzophenothiazine: DDQ(1:1)	0.37	4.63	PE	2202
Dibenzophenothiazine$_2$:DDQ	0.61	4.62	PE	2202
Dibenzophenothiazine$_2$:DDQ		4.60	CP (graphite)	2258
Dicyanopyrene	6.3		PE	2259
Diethylenetriamine	8.23		FQ	2204
N,N-dimethylbenzylamine	8.45		FQ	2244
Dimethylformamide	8.16		PES	2266
Dimethyl-isophthalate	9.84		EC	2189

TABLE 6.6A1 *Continued*

Substance	I_g	ϕ	Method	Ref.
Dimethylphthalate	9.64		EC	2189
N,N-dimethyl-p-phenylene-diamine	6.04		FQ	2244
N,N-dimethyl-β-naphthylamine	6.54		FQ	2244
N,N-diphenylethylenediamine	7.38		FQ	2244
Dimethyl-tetra-chloro-tere-phthalate	9.57		EC	2189
Dimethyl-terephthalate	9.78		EC	2189
Dimethylthioacetamide	7.86		PES	2266
DDQ	5.68	4.75	PE	2202
Durene	8.03		PE	2262
Ethylene	10.62		Calculated	2264
Ferrocytochrome-c	6.1		PES	2269
Ferricytochrome-c_3	5.4		PES	2269
Ferricytochrome-c	5.8		PES	2269
Indene	8.08		PE	2262
Isoviolanthrone		4.62	CP (graphite)	2258
Methdilazine	7.25		CTCS	2271
9-Methylanthracene	7.25		PES	2268
o-Methylphenyl(o-crescol)	8.49		CTCS relative	2250
Naphthacene	5.83		PES	2267
Naphthacene	8.14		PE	2245
Naphthacene	7.01		PE	2229, 2247
Naphthalene	8.12		PE	2262
Naphthalene	8.13		SY	2245
Naphthalene	9.21		Calculated	2264
Nortriptylene			CTCS	2271
Pentacene	6.64		PE	2246, 2229
Pertofran	7.39		CTCS	2271
Perylene	7.00		PE	2246, 2229
Perylene		4.54	PE	2256
Perylene		4.63	CP (AU)	2258
Perylene	7.10		CP (graphite)	2258
Phenanthrene	7.69		SY	2245
Phenanthrene	7.75		PE	2245

TABLE 6.6A1 *Continued*

Substance	I_g	ϕ	Method	Ref.
Phenol	8.5		CTCS relative to TCNQ	2250
Phenothiazine	6.96		CTCS	2271
Phenothiazine	7.7		PE	2245
Phenothiazine	7.62		SY	2245
m-Phenylenediamine	7.2		FQ	2244
o-Phenylenediamine	7.29		FQ	2244
p-Phenylenediamine	6.52		FQ	2244
n-Phenyl-α-naphthylamine	6.78		FQ	2244
n-Phenyl-β-naphthylamine	6.76		FQ	2244
PHTHC		4.41	CP (graphite)	2258
PHTHC (metal free)	6.35		PES	2253
Co:PHTHC	6.38		PES	2253
Cu:PHTHC	6.38		PES	2253
Cu:PHTHC		4.56	CP (graphite)	2253
Fe:PHTHC	6.36		PES	2253
Ni:PHTHC	6.38		PES	2253
Zn:PHTHC	6.37		PES	2253
Pipamazine	7.15		CTCS	2271
Polyethylene		4.7	PE	2261
Prochlorpromazine	7.25		CTCS	2271
Promazine	7.23		CTCS	2271
Promethazine	7.25		CTCS	2271
PVC (pure)		6.0	PE	2260
Pyranthrene	5.25	4.54	PE	2101
Pyranthrene		4.54	PE	2257
Pyranthrene		4.47	CP (graphite)	2258
Pyranthrene		4.47	CP(AU)	2258
Pyranthrene	6.98		CP (graphite)	2258
Pyranthrone	5.83	4.52	PE	2101
Pyrene	5.8		PE	2259
Quaterylene		4.56	PE	2016
Stelazine	7.43		CTCS	2271
TCNQ		5.0	CP (graphite)	2258
Tegretol			CTCS	2271
Terephthalate-aldehyde	10.13		EC	2189
p-Terphenyl	8.20		PES	2242

TABLE 6.6A1 *Continued*

Substance	I_g	ϕ	Method	Ref.
p-Terphenyl	8.20		PE	2237
p-Terphenyl	8.20		PE	2229, 2249
p-Terphenyl	7.6		SY	2245
Tetrabromopyrene	6.0		PE	2259
Tetracene		4.30	PE	2255
Tetracene		4.51	CP (graphite)	2258
Tetracene	6.94		CP (graphite)	2258
Tetrachloropyrene	6.3		PE	2259
Tetrachloropyrene	6.2		PE	2259
Tetrakis(dimethylamino)-ethylene	5.36		PI, adiabatic	2280
Tetrakis(dimethylamino)-ethylene	6.11		PI, vertical	2280
N,N,N,N-tetramethyl-benzidine	6.12		FQ	2244
Tetramethylthiourea	7.84		PI, vertical	2280
Tetramethyl-TTF	6.39		FQ	2244
Tetramethylurea	8.67		PES	2266
Tetranitropyrene	6.6		PE	2259
Tetrathiotetracene		4.24	PE	2256
Tetrathiotetracene		4.24	CP (graphite)	2258
Thioacetamide	8.69		PES	2266
Thioformamide	8.69		PES	2266
Thiopropazate	7.31		CTCS	2271
Thioridazine	7.20		CTCS	2271
Thiourea	8.50		PES	2266
TMPD	6.84		Calculated	2251
Tofranil	7.35		CTCS	2271
Tri-n-butylamine	7.22		FQ	2244
Trimethyl-TTF	6.31		Calculated	2252
TTF	6.81		Calculated	2252
Urea	10.33		PES	2266
Violanthrene	6.84		CP	2254
Violanthrene	6.86		CP (graphite)	2258
Violanthrene		4.50	PE	2257
Violanthrene		4.49	CP (graphite)	2258
Violanthrene		4.48	CP(Au)	2258
Violanthrone		4.60	CP (graphite)	2258
Violanthrone		4.66	CP(Au)	2258

TABLE 6.6A2 Ionization Energies I_g of Organic Molecules (in eV). Supplemental.

See also H. M. Rosenstock et al., *J. Phys. Chem.*, Ref. Data, **6**, Suppl. 1, (1979); G. G. Hall, *Faraday Disc. Chem. Soc.* (London), **54**, 7, (1972).

Substance	I_g	Method	Ref.
Adenine	9.6	PI	4111
Benzene	9.24		4104
Butadiene	4.08	PE	4103
Carbontetrafluoride	15.9	Adiabatic	4107
β-Carotene	6.4	Adiabatic	4114
Chlorophyll a	6.1	Adiabatic	4114
Difluoramine	12.38 11.53	Vertical, Adiabatic	4112
1-1 Difluoroethylene	10.69	584 A	4101
N-dimethylamino purine	9.3–9.2	PE	4111
3,3′-Dimethyl-2,2′-dibenzothiazolinylidene	5.7	Adiabatic	4114
Fluoroethylene	10.56	At 584 A	4101
Isoquinoline	8.54		4104
7-methyladenine	9.4	PI	4111
9-Methyl Adenine	9.4	PI	4111
N-methylaminopurine	9.5	PI	4111
6-Methylpurine	9.6	PI	4111
7-Methyl purine	9.4	PI	4111
9-Methyl purine	9.4	PI	4111
Naphthalene	4.9	PE	4103
Naphthalene	8.11		4104
Perfluoroethylene	1.56	584 A	4101
Phthalazine	8.68		4104

Phthalocyanines
  metal free
  Co
  Cu
  Fe
  Mg
  Ni
  Zn
	
	6.41
	6.38
	6.38
	6.36
	6.31
	6.38
	6.37
			
			4102

Purine	9.6	PI	4111
Pyrazine	9.36		4104
Pyridaline	8.90		4104
Pyridine	9.31		4104
Pyrimidine	9.42		4104

TABLE 6.6A2 *Continued*

Substance	I_g	Method	Ref.
Quinoazuline	8.99		4104
Quinoazuline	9.02		4104
Quinoline	8.62		4104
Quinoline	8.51	4104	4104
Tetra-aminoethylenes	5.4	Adiabatic	4114
1,1,4,4, Tetrakis-(Dimethyl-amino)-butadiene	5.6	Adiabatic	4114
Tetraphenylporphine & metallic derivative	6.3–6.5		4105
Trifluoroethylene	10.54	584 A	4101
Tryptophane in H_2O	4.5		4106

TABLE 6.6B Ionization Potentials of Various Molecules and Free Radicals. Experimental Values Obtained by the Retarding Potential Technique. After Williams and Hamill.[2270]

These data are in reasonable agreement with data obtained spectroscopically and from photo-ionization.

Molecule	(eV)	Radical	(eV)
CH_4	12.99	CH_3	9.87
C_2H_2	11.39	C_2H_5	8.34
	11.61	n-C_3H_7	8.13
	11.91	i-C_3H_7	7.57
C_2H_4	10.30		
	10.80	n-C_4H_9	8.01
C_2H_6	11.66	C_2H_5O	9.11
C_3H_8	11.09	n-C_3H_7O	9.20
CH_3Cl	11.28	i-C_3H_7O	9.20
CH_3Br	10.56	n-C_4H_9O	9.22
	10.87		
CH_3I	9.59		
	10.16		
i-C_3H_7Cl	10.65		
	10.81		
n-C_3H_7Br	10.24		
$(C_2H_5)_2O$	9.58		
$(i$-$C_3H_7)_2O$	9.16		
$(n$-$C_4H_9)_2O$	9.28		

TABLE 6.7A1 Experimental Data for Photo-emission Threshold I_c and Molecular Ionization Energies I_g (in eV).

Substance	I_g	I_c	Remark	Ref.
Acetone	9.72			2157
	9.694			2151
	9.71			2153
	9.69			2154
	9.709			2155
	9.70			2156
Anthracene		5.68	PE	2202
		5.88	PE	2239
		5.85	SC	2129
Anthracene-PMA Complex		6.0	PE	2230
		5.26	PE	2231
		6.0	$\phi = 4.9$ eV	2238
		5.03	PE	2231
Anthracene-TCNQ Complex		6.17	PE	2239
		5.03	PE	2231
Anthracene-1,3,5-trinitrobenzene		5.88	PE	2231
1,2-Benzanthracene-TCNQ Complex		5.00	PE	2231
Benzidine		5.10	PE	2231
Benzidine-TCNQ Complex		5.03	PE	2231
Benzonitrile-Alkali Salts				
Radical Salts:Cs		5.04	PE	2233
Na		5.01	PE	2233
Rb		5.05	PE	2233
Rb		5.05		2243
p-Bromanil		5.4	SC	2129
Carbazole		6.66		2239
Carbazole-TCNQ		6.76		2239
Chrysene-1,3,5-trinitrobenzene		5.50	PE	2231
p-CLL		5.0	SC	2129
Coronene		5.15	PE	2231
		6.2	PE	2236
		5.3	PE	2229
	7.34	6.1	PES	2229
Decamethylferrocene		5.1		2240
Decamehtylferrocene-TCNQ complex		5.3		2240
1,6-DP		4.71	PE	2231
1,6-DP-p-bromanil complex		4.65	PE	2231
1,6-DP-o-CLL complex		4.88	PE	2231

TABLE 6.7A1 *Continued*

Substance	I_g	I_c	Remark	Ref.
1,6-DP-p-CLL complex		4.39	PE	2231
1,6-DP-TCNQ complex		4.58	PE	2231
Ferrocene		6.1		2240
Indanthrone		5.17	PE	2202
Li-TCNQ		5.6		2239
Metal-free PHTHC		5.12	PE	2231
N-Methylphenathizinium: TCNQ		6.0		2239
Napthacene		5.31	PE	2231
		5.3	PE	2229
	7.01	5.9	PES	2229
		5.43	PE	2202
		5.83	PE	2235
Napthalene-1,8-disulfide	7.15	5.75		2139
Na-TCNQ		5.6		2239
Pentacene	6.64	5.6	PES	2229
		5.04	PE	2202
		5.51	PE	2236
		4.49	PE	2229
Perylene		5.2	PE	2229
	7.0	5.8	PE	2229
		5.33	PE	2202
		5.88	PE	2236
		5.40	PE	2231
		5.40	SC	2129
Perylene-o-CLL		5.35	PE	2231
Perylene-p-CLL		5.40	PE	2231
Perylene-dicyanomethylenetrinitrofluorene		5.23	PE	2231
Perylene-PMA		5.16	PE	2231
Perylene-TCNQ		5.50	PE	2231
Polyacetone, $(CH_3COCH_3)_4$	9.02			2151
Poly-Cu PHTHC		5.10	PE	2151
Pyrene		5.6–5.8	PE	2231
Pyrene-TCNQ		5.70	PE	2231
Pyrene-1,3,5-trinitrobenzene		5.60	PE	2231
Pϕ		5.2	CP	2129
Pϕ-bromanil		5.4	CP	2129

TABLE 6.7A1 *Continued*

Substance	I_g	I_c	Remark	Ref.
Pϕ-CLL		5.4	CP	2129
Quinolinium-(TCNQ)$_2$		5.35	PE	2239
Rubrene		5.10	CP	2231
TCNQ		7.88		2239
p-Terphenyl		6.2	PE	2229
	8.20	6.9	PES	2229
	8.20	6.1	PE	2228
	6.1			2242
		6.6	PE	2237
Tetracene		5.3	Flake	2129
N,N,N',N'-Tetramethyl benzidine		4.71	PE	2231
N,N,N',N'-Tetramethyl benzidine-1,3,5-trinitrobenzene		4.72	PE	2231
Tetramethylsilane	8.1			2288
Tetrathionaphthacene		4.42	PE	2202
TMPD		4.63	PE	2231
		4.7	CP	2129
TMPD-p-bromanil		4.87	PE	2231
		4.85	CP	2129
TMPD-CLL		5.0	CP	2129
		4.88	PE	2231
TMPD-TCNQ (from p-xylene)		4.86	PE	2231
TMPD-TCNQ (from glacial acetic acid)		4.72	PE	2231
1-Tryptophane	7.3			2242
TTF		5.0		2239
TTF-TCNQ		5.07		2239

TABLE 6.7A2 Experimental Data for Photo-emission Threshold I_c and Molecular Ionization Energies I_g (in eV). Supplemental.

Substance	I_g	I_c	Remark	Ref.
Cytochrome c		6.1	Ferriform	4109
Cytochrome c		5.8	Ferroform	4109
Cytochrome c	6.1		Ferriform	4113
Cytochrome c	5.8		Ferroform	4113
Cytochrome c_3		5.4	Ferriform	4109
Cytochrome c_3		4.6	Ferroform	4109
Cytochrome c_3	5.4		Ferriform	4113
Pentacene		5.07	PE	4010
PMA Complex		6.0	$\phi = 4\cdot 4$ eV	4108
p-Terphenyl	8.2		PE	2232
	7.83 adiabatic	6.6		
Tetracene		5.25		4024
l-Tryptophan	8.1, 9.7	7.2, 9.1		4110

TABLE 6.8A Absolute Electron Affinities A_g of Organic Materials (in eV).
Additional values are given in Table 6.8C. See also, V. I. Vedenoyev et al., *Bond Energies, Ionization Potentials and Electron Affinities*, E. Arnold, London (1966).

Compound	A_g	Ref.
Benzaldehyde	0.42	2189
Chloranil	1.37	2053
Cyclopentane	0.28 (Liquid)	2200
DDQ	1.95	2053
Diethylphthalate	0.54	2189
Dimethyl-iso-phthalate	0.55	2189
Dimethylphthalate	0.55	2189
Dimethyl-terephthalate	0.64	2189
Dimethyl-tetrachloro-terephthalate	0.77	2189
Fluorenone and derivatives *See* Table 6.8E		
Hexafluorobenzene	0.4	2187
n-Hexane	−0.4 (Liquid)	2200
Methylbenzoate	0.18	2189
Naphthalene	0.2	2188
Naphthalene	0.15	2187
Neopentane	0.43 (Liquid)	2200
Pentacene	2.7-2.87	4010
n-Pentane	0.01 (Liquid)	2200
Perfluorocyclobutane	0.5	2188
Pyromellitic acid	0.34	2190
Pyromellitic imide	0.46	2190
Pyromellitic phenylimide	0.39	2190
TCNE	1.80	2053
TCNQ	2.35	2186
TCNQ	2.88	2185
TCNQ	1.7	2053
TCNQ	2.42	2189
Terephthal-aldehyde	0.56	2189
Tetramethylsilane	0.62 (Liquid)	2200
TNB	0.7	2053
2,2,4-Trimethylpentane	0.18 (Liquid)	2200

TABLE 6.8B Polarization Energies P and Adiabatic Ionization Energies I of Organic Solids Determined by Ultraviolet Photoelectron Spectroscopy in eV.

N. Sato, H. Inokuchi and I. Shirotani, *Ann. Review Inst. Molec. Sci.* Okazaki (Japan), pp. 81 and 85 (1980).

Compound	I_{gas}	I_{solid}	P	Ref.
Anthracene	7.36	5.70	1.7	
Benz [a] anthracene	7.38	5.64	1.7	
Benzene	9.17	7.58	1.6	
Benzoperylene	7.12	5.4	1.7	
Chrysene	7.51	5.8	1.7	
Dibenz [a,h] anthracene	7.35	5.55	1.8	
Dibenzo-tetrathiafulvalene	6.68	4.4	2.3	
Dimethyl-tetrathiafulvalene	6.00	5.1	0.9	
Naphth [2,1-a] anthracene	7.2	5.45	1.8	
Napthacene	6.89	5.10	1.8	
Napthalene	8.12	6.4	1.7	
Pentacene	6.58	4.85	1.7	
Perylene	6.90	5.2	1.7	
Picene	7.5	5.7	1.8	
Pyrene	7.37	5.8	1.6	
Tetramethyl-tetrathiafulvalene	6.03	4.9	1.1	
Tetrathiafulvalene	6.4	5.0	1.4	
Tetrathia-napthacene	6.07	4.4	1.7	
Triphenylene	7.81	6.2	1.6	
Ferricytochrome c	6.1			4156
Ferrocytochrome c	5.8			4156
Ferrocytochrome c_3	5.4			4156

TABLE 6.8C Electron Affinities (A_g) of Organic Molecules (in kcal/mole), Measured by the Electron Attachment Method, Forming Negative Ions in a Magnetron Device. After Page and Goode, *Negative Ions and the Magnetron*, Wiley-Interscience, N.Y., 1969.
T is the temperature of the filament. The ionic species is also given. 1 kcal/mole = 0.0434 eV.

Compound	$T\,°K$	Ions Formed	A_g
Trigonal Ions			
Pentafluorobromobenzene	1220	$C_6F_5^-$	63.9 ± 0.1
Pentafluorobromobenzene	1400	$BrC_6F_4^-$	63.3
p-Dibromobenzene	1500	$BrC_6H_4^-$	59.3
Allene	1537	$C_3H_3^-$	53.8
Octafluorobut-2-ene	1600	$C_4F_7^-$	61.9 ± 1.1
Fumaronitrile	1430	$C_4HN_2^-$	40.1 ± 1.0
Hexachlorobenzene	1500	$C_6Cl_5^-$	63.6 ± 2.3
Pentafluorochlorobenzene	1390	$C_6F_5^-$	64.7
1-Chloro-2-nitrobenzene	1500	$C_6H_4NO_2^-$	54.1 ± 0.8
1-Chloro-3-nitrobenzene	1500	$C_6H_4NO_2^-$	37.1 ± 0.3
1-Chloro-3-nitrobenzene	1500	$C_6H_4NO_2^-$	38.6
1-Chloro-4-nitrobenzene	1450	$C_6H_4NO_2^-$	56.1 ± 1.9
Chlorobenzene	1410	$C_6H_5^-$	49.4
1:2-Dichlorobenzene	1490	$C_6ClH_4^-$	52.3 ± 0.7
1:2-Dichlorobenzene	1400	$C_6ClH_4^-$	53.4 ± 1.6
1:3-Dichlorobenzene	1410	$C_6ClH_4^-$	47.9 ± 1.4
1:3-Dichlorobenzene	1490	$C_6ClH_4^-$	46.9 ± 2.0
1:4-Dichlorobenzene	1410	$C_6ClH_4^-$	53.4 ± 1.8
1:4-Dichlorobenzene	1500	$C_6ClH_4^-$	54.1 ± 1.8
1:4-Dichlorobenzene	1500	$C_6Cl_2H_3^-$	47.2
1:2:3-Trichlorobenzene	1410	$C_6Cl_2H_3^-$	61.6 ± 1.4
1:3:5-Trichlorobenzene	1410	$C_6Cl_2H_3^-$	51.6 ± 1.0
1:2:3:4-Tetrachlorobenzene	1410	$C_6Cl_3H_2^-$	55.9 ± 1.3
1:2:4:5-Tetrachlorobenzene	1410	$C_6Cl_3H_2^-$	58.9 ± 2.2
Fluorobenzoquinone	1590	$C_6FH_2O_2^-$	55.1 ± 2.8
Hexafluorobenzene	1500	$C_6F_5^-$	63.3
p-Benzoquinone	1470	$C_6H_3O_2^-$	46.0
Benzene	1300	$C_6H_5^-$	56.7 ± 1.4
Benzene	1560	$C_6H_5^-$	55.1 ± 1.2
2:3-Dicyanobenzoquinone	1490	$C_6HN_2O_2^-$	41.9 ± 1.8

TABLE 6.8C *Continued*

Compound	$T\,°K$	Ions Formed	A_g
Tetrafluoronaphthalene	1510	$C_{10}F_4H_3^-$	49.8
Octafluoronaphthalene	1475	$C_{10}F_7^-$	58.2
Naphthoquinone	1405	$C_{10}H_5O_2^-$	50.7 ± 1.4
Naphthalene	1330	$C_{10}H_7^-$	51.0 ± 0.5
Mercury diphenyl	1500	$C_6H_5^-$	52.2 ± 1.8
Anthracene	1400	$C_{14}H_9^-$	48.6 ± 0.4
Mercury di-*p*-tolyl	1423	$CH_3C_6H_4^-$	58.1 ± 2.8

Tetrahedral Ions

Compound	$T\,°K$	Ions Formed	A_g
Monobromotrifluoromethane	1408	CF_3^-	41.0 ± 1.9
Bromoform	1285	CBr_3^-	41.4 ± 6.0
Carbon tetrabromide	1365	CBr_3^-	35.3 ± 2.0
Monochlorotrifluoromethane	1724	CF_3^-	40.0 ± 1.4
Chloroform	1450	CCl_3^-	28.0 ± 0.7
Carbon tetrachloride	1430	CCl_3^-	28.0 ± 0.7
Fluoroform	1342	CF_3^-	42.7 ± 1.8
Carbon tetrafluoride	1390	CF_3^-	42.7 ± 1.8
Hexachloroethane	1470	$C_2Cl_5^-$	35.8 ± 3.3
Octafluoropropane	1600	$C_3F_7^-$	45.9 ± 4.2
Toluene	1540	$C_6H_5CH_2^-$	18.9 ± 1.9
Duroquinone	1320	$C_{10}H_{11}O_2^-$	18.5 ± 1.1
Duroquinone	1600	$C_{10}H_{11}O_2^-$	18.0 ± 2.7
Thiocyanic acid	1870	SCN^-	48.4 ± 0.5
Dimethylamine	1500	$C_2H_6N^-$	24.0
Thiocyanogen	1800	SCN^-	48.4 ± 0.5
Selenocyanogen	1500	$SeCN^-$	63.7
Tetramethyltetrazene	1400	$C_2H_6N^-$	24.0
Pentafluoroaniline	1450	$C_6F_5NH^-$	38.1 ± 2.3
Aniline	1667	$C_6H_6N^-$	35.7 ± 1.5
Methylaniline	1500	$C_7H_8N^-$	30.0
Diphenylamine	1653	$C_{12}H_{10}N^-$	27.3 ± 3.5
Diphenylhydrazine	2012	$C_6H_6N^-$	35.7 ± 1.5
1:4-Dimethyl-1:4-diphenyltetrazene	1400	$C_7H_8N^-$	30.0
Tetraphenyl hydrazine	1980	$C_{12}H_{10}N^-$	27.3 ± 3.5
Tetrafluorohydrazine	1250	NF_2^-	69.0
Ammonia	1700	NH_2^-	25.7 ± 2.0
Hydrazine	1900	NH_2^-	25.7 ± 2.0

TABLE 6.8C *Continued*

Compound	$T\,^\circ K$	Ions Formed	A_g
σ-Oxygen Ions			
Methanol	1430	CH_3O^-	8.7 ± 0.4
Bis(trifluoromethyl) peroxide	1400	CF_3O^-	31.2
Ethanol	1430	$C_2H_5O^-$	13.7 ± 1.1
Iso-propanol	1430	$C_3H_7O^-$	15.5 ± 0.3
n-Butyl alcohol	1720	$C_4H_9O^-$	14.9
di-t-Butylperoxide	1330	$C_4H_9O^-$	20.5
Dibenzoyl peroxide	1400	$C_7H_5O_2^-$	44.0
Water	2000	O^-	31.0
Halogen Ions			
Carbon tetrachloride	1560	Cl^-	85.1 ± 0.6
Fluorobenzoquinone	1380	F^-	79.4 ± 0.8
σ-Sulphur Ions			
Bis(trifluoromethyl) disulphide	1350	CF_3S^-	36.1 ± 0.8
Ethyl mercaptan	1450	$C_2H_5S^-$	27.2 ± 5.8
Dimethyl disulphide	1450	CH_3S^-	30.4 ± 1.8
Dimethyl disulphide	1450	CH_3S^-	31.3 ± 3.0
Dimethyl disulphide	1600	CH_3S^-	32.1 ± 4.3
Diethyl disulphide	1440	$C_2H_5S^-$	$37.2 \simeq 5.3$
Thionyl fluoride	1700	FO_2S^-	63.8 ± 3.5
Other Ions			
Chloranil	1350		55.3 ± 5.9
Fluorobenzoquinone	1260		49.8
Hexafluorobenzene	1300	27.6	27.6 ± 1.6
1:3:5-trinitrobenzene	1370		60.5 ± 2.4
p-Benzoquinone	1450		30.9 ± 0.2
Tetracyanoethylene	1360		66.9 ± 1.2
Fluoranil	1370		52.2
o-Phthalonitrile	1520		24.3 ± 2.2
sym Tetracyanopyridine	1320		48.7 ± 1.7
sym Tetracyanobenzene	1300		49.5 ± 5.1
Hexacyanobutadiene	1370		74.7 ± 2.4
$77':88'$-Tetracyanoquinodimethane	1330		65.1 ± 4.4
Hexacyanobenzene	1390		57.2 ± 3.1
Anthraquinone	1405		26.5 ± 2.1
Fumaronitrile	1230		17.2 ± 2.8

TABLE 6.8D Group Contributions to the Electron Affinity A_g. After Page and Goode, *Negative Ions and Magnetron*, Wiley, N.Y., 1969.

Group	A_g (eV)	Group	A_g (eV)
CH	0.087	CNO_2	1.05
CF	0.2	$=C-(CN)_2$	1.43
CCN	0.74	CCl	0.165
C=O	0.87		

TABLE 6.8E Polarographic Electron Affinities A_g. After Kuder et al., *J. Electrochem. Soc.*, **125**, 1750 (1978).

Compound	A_g (eV)
Fluorenone	1.19
2,7-Dinitrofluorenone	1.81
7,9-Dicyanomethylene-fluorenone	1.83
2,4,7-Trinitro-fluorenone-9-ol	1.87
2,4,7-Trinitro-fluorenone acetate	1.93
2,4,7-Trinitro-fluorenone butyrate	1.93
3,6-Dinitronaphthalic anhydride	1.97
2,4,7-Trinitro-fluorenone	2.07
2-Carbo-ethoxy-4,5,7-trinitro-fluorenone	2.20
2-Carb butoxy-4,5,7-trinitro-fluorenone	2.22
2,5-Dinitro-dicyano-methylene-fluorenone	2.23
2,4,5,7-Tetranitro-fluorenone	2.35
2,4,7-Trinitro-dicyano-fluorenone	2.49
4,6-Dinitrobenzofurazan-1-oxide	2.55

TABLE 6.8F Electron Affinities of Acceptors and CT Band Maxima ($\bar{\nu}_{max}$ 298 K) of Methiodides. After Bagchi and Chowdury, *J. Phys. Chem.*, 83, 629 (1979)

Compound	$\bar{\nu}_{max}$ 298 K (obs) in CH_2Cl_2, (cm^{-1})	lowest π-I_P of the neutral azine,[a] (eV)	lowest A_g of the cation,[b] (eV)
Pyridine methiodide	26 600	9.73	−4.7083
Pyridazine methiodide	22 990	10.61	−5.3883
Pyrimidine methiodide	24 100	10.41	−4.7572
Pyrazine methiodide	21 050	10.18	−5.4536

[a] As obtained from the photoelectron spectra.
[b] Calculated

TABLE 6.9A Relative ($A_{G,rel}$) and Absolute ($A_{G,abs}$) Electron Affinities (in eV). After Lyons and Palmer, *Aust. J. Chem.*, 29, 1919 (1976).

$A_{g,rel}$ values are relative to that for tetracyanoethylene as 2.3 ± 0.3 eV.

Compound	$A_{G,rel}$[A]	$A_{G,abs}$[B]
Tetracyanoquinodimethane	+0.1	2.4
Tetracyanoethylene	(0)	2.3
2,3-Dicyano-*p*-benzoquinone	−0.08	2.2
Nitro-*p*-benzoquinone	−0.9	2.2
Dicyanomethylenetrinitrofluoren-9-one	−0.14	2.2
o-Bromanil	−0.1$_2$	2.2
o-Chloranil	−0.1$_7$	2.1
p-Bromanil	−0.2$_7$	2.0
p-Chloranil	−0.3$_0$	2.0
p-Iodanil	−0.3$_4$	2.0
1,4,9,10-Anthradiquinone	−0.2$_6$	2.0
2,5-Dichloro-*p*-benzoquinone	−0.3$_5$	1.9

TABLE 6.9A *Continued*

Compound	$A_{G,\text{rel}}$[A]	$A_{G,\text{abs}}$[B]
Cyano-p-benzoquinone	-0.4_2	1.9
Trichloro-p-benzoquinone	-0.4_7	1.8
2,6-Dichloro-p-benzoquinone	-0.5_4	1.8
2,6-Dibromo-p-benzoquinone	-0.5_4	1.8
Trinitrofluorene	-0.5_9	1.7
1,2-Dicyano-1,2-dicarboxyethylene	-0.6_1	1.7
Acetyl-p-benzoquinone	-0.6_7	1.6
Chloro-p-benzoquinone	-0.6_8	1.6
Methoxycarbonyl-p-benzoquinone	-0.6_9	1.6
Bromo-p-benzoquinone	-0.6_9	1.6
Iodo-p-benzoquinone	-0.7_0	1.6
1,2,4,5-Tetracyanobenzene	-0.7_3	1.6
Pyromellitic acid dianhydride	-0.7_3	1.6
Fluoro-p-benzoquinone	-0.7_9	1.5
Indanetrione	-0.8_1	1.5
p-Benzoquinone	-0.8_7	1.4
Methyl-p-benzoquinone	-1.0_2	1.3
1,3,5-Trinitrobenzene	-1.0_3	1.3
1,2-Naphthoquinone	-1.0_4	1.3
Tetrachlorophthalic anhydride	-0.1_9	1.2
2,5-Dimethyl-p-benzoquinone	-1.1_4	1.2
9,10-Phenanthraquinone	-1.1_6	1.1
Methoxy-p-benzoquinone	-1.1_7	1.1
1,4-Dinitrobenzene	-1.2_0	1.1
2,4,6-Trinitrotoluene	-1.2_0	1.2
Maleic anhydride	-1.3_6	0.9
Phthalimide	-1.4_0	0.9
1,3-Dinitrobenzene	-1.4_4	0.9
Phthalic anhydride	-1.4_5	0.8
4-Chloronitrobenzene	-1.9_6	0.3
Nitrobenzene	-2.0_7	0.2
Dinitrodurene	-2.2_8	0.0

[A] $\pm 0.1_5$ [B] ± 0.4

TABLE 6.10A Electron Affinity A of Organic Compounds (in eV).

Compound	A_g	A_c	Reference
p-Benzoquinone		1.4	2142
p-Benzoquinone	1.34 ± 0.009		2204
p-Benzoquinone	2.08		2205, 2206
p-Benzoquinone	1.89 + 0.02, −0.03		2206
p-Bromanil		2.44 ± 0.2	2141
t-Butanol	1.96 ± 0.12		2146
p-Chloranil		2.76 ± 0.2	2141
p-Difluorobenzene	−0.35		2164
Ethanol	1.83 ± 0.12		2146
Fluoranil		2.92 ± 0.2	2141
Lignocain	1.0		2203
Methanol	1.74 ± 0.12		2146
iso-Propanol	1.92 ± 0.12		2146
Pyrazine		−5.45	2165
Pyridazine		−5.39	2165
Pyrimidine		−4.76	2165
Pyridinemethiodide (cation)		−4.708	2165
Violanthrene		5.3 − 5.4	2201
Violanthrene		5.2	2202
$OH^- \to OH$	1.829 + 0.010, −0.014		2207
$NH_2^- \to NH_2$	0.779 ±		2207
$NH^- \to NH$	0.38 ± 0.03		2207
$SO_2^- \to SO_2$	1.097 ± 0.036		2207
$S_2^- \to S_2$	1.663 ± 0.04		2207
$O^- \to O$	1.462 + 0.003, −0.007		2207

TABLE 6.17A Electron Affinities A_g of the Elements. (For Purposes of Comparison). After Hotop and Bennett.[2208]

Element	A_g (eV)	Element	A_g (eV)	Element	A_g (eV)
O	1.465	Cr	0.65	Tc	0.63
OH	1.829	Ni	1.13	Ru	1.04
Cu	1.226	Cs	0.39	Rh	1.12
Ag	1.303	Ba	−0.54	Pd	0.40
K	0.47	La	0.44	Ta	−0.005
Ca	−1.62	Hf	−0.78	W	0.98
Sc	−0.80	Rb	0.42	Re	0.09
Ti	−0.11	Sr	−1.74	Os	1.10
V	0.51	Y	−0.76	Ir	1.58
Cr	0.85	Zr	0.08	Pt	2.12
Mn	−1.19	Nb	0.77	Au	2.31
Fe	0.14	Mo	0.86		

TABLE 6.18A Electron Affinities A_g of Substituents. After Kampars and Neylands, *Russ. Chem. Revs.*, **46**, 503 (1977).

Substituent	A_g (eV)	Substituent	A_g (eV)
CH_3O	−0.15	CH_3OCO	0.20
CH_3	−0.05	CH_3CO	0.25
F	0.13	CF_3	0.30
I	0.17	CN	0.43
Cl	0.18	NO_2	0.70
Br	0.19		

TABLE 6.19A Calculated Ionization Potentials and Electron Affinities (in eV) of Purines, Pyrimidines, Porphins, and Poly-Pyrroles. After Kunii and Kuroda, *Theoret. Chim. Acta* (Berlin), *11*, 97 (1968).

Compound	I_g	E_A
Pyrimidine	9.71	−0.47
Alloxane	10.77	1.66
Cytosine		
(lactam, 1-H)	8.01	−0.24
(lactam, 2-H)	8.12	−0.24
(lactim)	8.08	−1.25
Uracil		
(lactam)	9.16	0.02
(lactim)	8.42	−1.12
Purine		
(9-H)	8.89	−0.17
(7-H)	9.12	−0.10
Adenine		
(9-H)	7.99	−0.77
(7-H)	7.26	−0.63
Guanine		
(lactam, 9-H)	7.66	−0.96
(lactam, 7-H)	7.86	−1.00
(lactim, 9-H)	7.57	−1.00
(lactim, 7-H)	7.70	−0.82

TABLE 6.19A *Continued*

Compound	I_g	E_A
Xanthine		
(lactam, 9-H)	8.67	−0.66
(lactim, 9-H)	7.86	−0.90
Hypoxanthine (lactam)	8.22	−0.32
Uric acid		
(lactam)	8.45	0.06
(lactam, 9-H)	8.48	0.25
(lactim:lactam)	8.47	0.16
2-hydroxypurine	8.46	0.09
8-hydroxypurine	8.60	−0.03
2-aminopurine	7.87	−0.52
8-aminopurine	7.90	−0.60
Porphin	7.14	1.88
1,3-divinylporphin	7.11	1.92
1-vinyl-5-formylporphin	7.18	1.91
1-vinyl-5,8-diformylporphin	7.31	2.12
Dipyrrole	8.12	1.21
Tripyrrole	8.02	1.95
Tetrapyrrole	7.89	2.32
Pentapyrrole	7.84	2.62
Biliverdin		
(keto)	7.37	2.14
(enol)	7.21	1.98

TABLE 6.20A. Electron Affinities (A_g) of Acceptors, and Excited State Energies (E*) of Their Atoms. After Beitz and Miller, *Proc. Conf.*, "Tunnelling in Biol. Systems," Philadelphia, Penna., 1977.

$-E_{1/2}$ represents the polarographic half wave potential vs. saturated calomel electrode in polar aprotic solvents.

A_{gMTHF} refers to relative electron affinities measured at 77 K in a 2-methyltetrahydrofuran glass by electron survival techniques. (See text).

E* is the energy of the long wavelength threshold of absorption by the negative ion of the acceptor.

For comments on some of these data the original publication should be consulted. All data in eV.

Compound	$-E_{1/2}$	Electron Affinity gas phase	$E_{1/2}$	A_{gMHTF}	E*
Acenaphthylene	1.58		1.14	1.14	(1.4), 1.7
Acetophenone	1.97	0.33	0.35	0.55	1.0
Acridine	1.59		1.13	1.13	1.4
Azobenzene	1.31		1.31	1.36	1.7
Azulene	1.63	0.59	0.99	1.04	(0.9), 1.7
Alizarin (1,2,-di-hydroxyanthraquinone)	0.9		1.82	1.82	1.0
Benzene	...	(−1.1)		−0.9 ± 0.5	1.0
Benzonitrile	2.42	0.24	−0.1	0.1	
Benzoquinone	0.52	1.9	1.86	2.03	⩽2.4
Biphenyl	2.58		0.04	0.09	0.9
2-Chloroanthraquinone	0.84		1.88	1.88	1.1
2,5- and 2,6-Dichloro-benzoquinone	0.18		2.20	2.43	⩽2.4
2,6-Dichloro-3,5-dimethylbenzoquinone	0.35		2.03	2.2	⩽2.4
2,3-Dichloro-5,6-dicyanobenzoquinone	−0.51		3.13	3.17	<2.25
p-Dicyanobenzene	1.66		0.72	0.89	2.4
1,4-Dimethylanthraquinone	1.08		1.64	1.64	1.1, 2.1
1,5- and 2,6-Dimethyl-benzoquinone	0.67		1.71	1.88	⩽2.4
p-Dinitrobenzene	0.59		1.79	1.96	1.25
Fluoranthene	1.74		0.98	0.98	1.0
9-Fluorenone	1.3		1.5	1.5	1.1
Hexafluorobenzene	...	1.2 1.8		1.15 1.75	2.2

TABLE 6.20A *Continued*

Compound	$-E_{1/2}$	Electron Affinity			
		gas phase	$E_{1/2}$	A_{gMHTF}	E^*
Maleic anhydride	0.86	1.4	1.76	1.81	(2.8)
9-Methylanthracene	1.94		1.82	0.78	1.2
Methyl-*p*-benzoquinone	0.56		1.82	1.99	≤2.4
Methylmethacrylate	(2.38)		−0.06	0.14	
2-Methyl-1,4-naphtho-quinone	0.7		1.88	1.93	1.65
Naphthalene	2.52	0.15	0.10	0.15	0.7
Nitrobenzene	1.12	0.18, 1.1	1.20	1.40	
Phenazine	1.21		1.51	1.51	(1.7)
Phenyl-*p*-benzoquinone	0.51		1.93	2.08	≤2.0
Phthalic anhydride	1.28		1.04	1.24	0.9
Pyrazine	2.11		0.21	0.41	(1.6)
Pyrene	2.09	0.58	0.63	0.63	0.87
Pyridine	2.65		−0.33	−0.13	(2.3)
Pyrimidine	2.33		−0.01	0.19	
o-Tetrachlorobenzo-quinone	−0.17		2.76	2.83	<1.65
p-Tetrachlorobenzo-quinone	−0.03	2.4 2.8	2.65	2.69	≤2.4
p-Tetrafluorobenzo-quinone	0.04	2.3 2.9	2.58	2.62	≤2.6
Tetracyanobenzene	0.66	2.15	1.96	2.00	2.7
Tetracyanoethylene	−0.17	2.88 2.03	2.49	2.69	2.6
Tetranitromethane			(1.63)	>1.83	
Triazine	2.08	0.25	0.45		(1.7)
Trichlorobenzoquinone				2.61	≤2.4
Trimethylbenzoquinone	0.72		1.66	1.83	≤2.4
Trimethylpyridine	2.80		−0.48	−0.28	
2,3-Dimethyoxy-5,6,-dimethylbenzoquinone	(0.87)		1.51	1.68	≤2.4

TABLE 8.2A Metal-free Molecular Crystals

Substance	Purification	Form	Contacts	Temperature range (°C)
Anthracene	Z	SC	Ag	55–72
Anthracene		SC	Ag	−20 to 80
1,2-Benzanthrene				
Biphenyl				
Chrysene				
4-(Dimethyloctyl-ammonia)-stilbene				
1-Methylimine	R	SC	Electrolyt.	
Naphthalene				
Pentacene				
Perylene				
Phenanthrene				
Phenanthrene	Z, Chrom.	SC	Electrolyt. SnO_2	$t < 72$ $t > 72$ 20–82
Phenothiazine bromide	RC	CP		4 to 35
β-Phthalocyanine	S	SC	Au	<90 >90
Pyrene				
Quaterrylene	R R	SF CP	Mo Mo	27 to 27
TCNQ	Chromat.	SC	Ag	−35 to 170
Tetracene				
Tetrathiotetracene	R, S R, S R, S S S S	SF SF CP CP SC CP	Ag, SFC Ag, SAC	−60 to +40 −64 to +27
Urea nitrate		SC		
Urea nitrate		SC		

E in $E/2kT$ (eV)	Resistivity at room temp. (ohm cm)	Mobility μ (cm^2/Vsec)	Sign of Majority Carrier Estimated from	Ref.
4.37				2285
1.12	1.2×10^{10}, 20°C, $\perp$ab			4143
4.27				2285
7.01				2285
4.70				2285
	1.8×10^5			2124
43 kJ/mole	4.2×10^9 {∥ to (102)} 10^{15} {⊥ to (102)}			2159 2159
5.61				2285
2.9				2285
3.50				2285
5.33				2285
3.0	9.2×10^{14}			2160 2160
2.2	10^{17}			2160
0.6	1.5×10^5			2055
2.1				2167
0.76				2167
4.37				2285
1.2	10^6 10^{10}		p, Effect of O$_2$ p, Effect of O$_2$	2015 2015
0.96 above 10°C	5×10^8	$<10-3$ Hall, est.	n, photocond., Thermo-EMF	2052
3.55				2285
0.57	1.3×10^6			2016
0.41	4.2×10^{10}			2016
0.41	8.5×10^3			2016
	3×10^3	$<10^{-2}$	p, Thermo-EMF	2046
0.74 to 0.94	10^7			2122
0.64	10^4			2122
	6.7×10^7 (∥ to cleavage plane			2123
	8.3×10^{10} (⊥ to cleavage plane			2123

TABLE 8.2B Some Metal Complexes

Substance	Purification	Form	Contacts	Temperature range (°C)
Cu[Cupferron]$_2$		CP		0 to 150
Ni[Cupferron]$_2$		CP		
Zn[Cupferron]$_2$		CP		
Lutetium (Phthalocyanine)$_2$		S film	Au, electrolytic	
Lutetium (Phthalocyanine)$_2$ Reduced, acidic		S film	Au, electrolytic	
Mg Phthalocyanine	Chromatog., R, 3S	SW	Ag	
Mg Phthalocyanine	S	SW	Ag, Al	
Pb Phthalocyanine	S	SC	Ag, Au	$\dfrac{10^3}{T} = 3.4$ to 2
Metal Complexes				
Phthalocyanine:Cu	R	CP		20 to 300
Phthalocyanine:Ni	R	CP		20 to 300

TABLE 8.5A Charge Transfer Complexes

Electron donor	Electron acceptor	Purification	Form	Contact	T Range (°C)
Benzidine	(0.75:1) I$_2$	R, Z	CP		−80 to +50
	(1:1)	R, Z	CP		−80 to +50
	(1.25:1)	R, Z	CP		−80 to +50
5,6:11,12-Bis (epidithio)- napthacene	I$_2$ (2:3)		CP		
	(1:2,7)		CP		
Bis(TTT)$_2$	I$_3$ 0.5 to 0.52		SC		−26 to +27
BPH	I$_2$ (1:3)	RC	CP	graphite	−90 to +25
	I$_2$ (2:3)	RC	CP	graphite	−90 to +25
Chlorpromazine	I$_2$ (2:3)	S	CP	Au	25
Cs	Tetra-cyanopyrene	R	SF		
	Tetra-cyanopyrene	R	SF		

E in $E/2kT$ (eV)	Resistivity at room temp. (ohm cm)	Mobility μ (cm^2/Vsec)	Sign of majority carrier estimated from	Ref.
$\sigma = Ae^{-B/T^{1/4}}$	3×10^{14}, 25°C			4146
$\sigma = Ae^{-B/T^{1/4}}$	8×10^{14}, 25°C			4146
$\sigma = Ae^{-B/T^{1/4}}$	2×10^{13}, 25°C			4146
	0.37×10^3	4×10^{-6}		4145
	1.3×10^3			
0.746				4147
0.6			p	4148
1.08	$\sigma_a/\sigma_c \cong 20$		p, Thermo-EMF	4149
1.77	10^5			2063
1.60	10^8 at 170°C			2063

E in $E/2kT$ (eV)	Resistivity ρ (ohm cm) at room temp.	Mobility μ (cm^2/Vsec)	Sign of majority carrier est. from	Ref.
0.56	1.12×10^6		p, TH	2035
0.38	1.16×10^6		p, TH	2035
0.38	1.19×10^6		p, TH	2035
–	0.3			2102
0.26	–			2102
				2130
0.30	20	$\mu = 0.6$ to 0.9	n, Seebeck	2290
		$\mu = 0.26$	n, Seebeck	2289
	420			2017
0.7	6×10^6		p, TH	2021
0.94	2×10^5		p, TH	2021

TABLE 8.5A (*Continued*)

Electron donor	Electron acceptor	Purification	Form	Contact	T Range (°C)
Cu-PHTHC	$I_{0.6}$				
	$I_{1.0}$				
	$I_{1.7}$				
	I_2 or Br_2				
1,4-Diazabicyclo-(2,2,2)octane	CLL Bromanil				
	Bromanil				
DP	CLL non-stoich.	R	CP		25 to 100
	2:3	R	CP		25 to 100
	1:1	R	CP		25 to 100
	CLL(1:12)	R	CP		25 to 100
	(1:1.4)	R	CP		25 to 100
	(1:16)	R	CP		25 to 100
	DDB(2:3)	R, evap.	CP		25 to 100
	p-Bromanil	R	CP		25 to 100
Diphenyl-Pϕ	I_2	R	CP		−50 to +50
N-EthylPH	I_2 (2:3)	RC	CP		−90 to 25
Fe-PHTHC	$I_{1.93}$				
	$I_{2.74}$				
	$I_{3.85}$				
Metal-free PHTHC	$I_{2.2}$				
N-MethylPH	I_2 (2:3)		SC		
	I_2 (2:3)	RC	CP	graphite	−90 to +25
Ni-PHTHC	$I_{0.56}$		CP		
	$I_{1.0}$		CP		
	$I_{1.74}$		CP		
Perylene	Cu-dithiolate	R	SC	Ag	−196 to 27
	Ni-dithiolate	R	SC	Ag	
	Pd-dithiolate	R	CP	Ag	
PH	$SbCl_5$ (1:1)	R, Z	CP		−160 to +50
	$SbCl_4$ (1:1)	R, Z	CP		
PH	I_2 (1:3)	R, Z	CP		−80 to +50
	I_2 (1:1)	R, Z	CP		−80 to +50
	I_2 (3:1)	R, Z	CP		−80 to +50
PH	I_2 (2:3)	S	CP	Au	25
PH	I_2 (2:3)	RC	CP	graphite	−90 to +25
PH	I_2 (2:3)	RC	CP	graphite	−90 to +25
PH	I_2 (3:1)	R	CP		4 to 35
Phenoselenazine	I_2 (2:3)	RC	CP		−90 to +25

E in $E/2kT$ (eV)	Resistivity ρ (ohm cm) at room temp.	Mobility μ (cm^2/Vsec)	Sign of majority carrier est. from	Ref.
0.13	10			2131
0.165	17			2131
0.042	0.24			2131
<0.13			p	2293
0.11				2111
0.12				2111
200				2036
7				2036
870				2036
	13			2036
	6			2036
	4			2036
	5			2036
	3			2036
0.14	2.94		n, Seebeck	2130
0.30	74	$\mu_-/\mu_+ \approx 5$	n, Seebeck	2290
0.254	250			2131
0.14	500			2131
0.508	1000			2131
0.080	0.43			2131
0.96	0.022			2291
0.28	10	$\mu_-/\mu_+ = 2.11$	p, Seebeck	2289
0.048	1.4			2131
0.042	1.4			2131
0.042	1.25			2131
0.128	0.167			2067
0.102	0.02			2067
0.168	14.3			2067
0.9	$0.36 \times 10_{11}$			2035
0.9	0.32×10^9			2035
0.86	0.11×10^5		p, TH	2035
0.34	0.38×10^4		n, TH	2035
0.52	0.11×10^5		p, TH	2035
	350			2017
0.38	19		p, Seebeck	2289
			p, Seebeck	2289
0.2	4.5			2054
		$\mu_- = 2.6$		2290

TABLE 8.5A *Continued*

Electron donor	Electron acceptor	Purification	Form	Contacts	T Range (°C)
Phenoxazine	I₂ (2:3)	RC	CP	graphite	−90 to +25
Poly(p-Dimethyl -aminostyrene)	TCNE		CP	graphite	20 to 90
	DDQ		CP	graphite	20 to 90
	CLL		CP	graphite	20 to 90
	TCNQ		CP	graphite	20 to 90
Polyvinyl-carbizole	I₂ (1:1)	R	CP	Ag	−50 to 125
Polyvinyl-carbizole	I₂ (1:2)	R	CP	Ag	−50 to 125
Pyridine	Hexachloro-antimonates				
Pφ	I₂ (0.8:1)	R, Z	CP		−80 to 50
	I₂ (1:1)	R, Z	CP		−80 to 50
	I₂ (1.6:1)	R, Z	CP		−80 to 50
	I₂ (1:2)	R	CP	Ag	−50 to +50
	I₂ (1:2)	R	CP	Ag	< −50
TTT	I₂ (1:15)	C	CP		−120 to 20
		SC	CP		−76 to +25
Tetrahalo-p-benzoquinones	(1:1) Various (no TCNQ)		CP		
TetramethylTTF	Fluoranil		SC		27

E in $E/2kT$ (eV)	Resistivity ρ (ohm cm) at room temp.	Mobility μ (cm^2/Vsec)	Sign of majority carrier est. from	Ref.
0.28	10	$\mu_-/\mu_+ = 2.11$	p, Seebeck	2289
1.86				2053
1.6				2053
0.6				2053
19				2053
1.16	0.125×10^6	0.3 (Hall, estim.)	n, Hall, Dember; p, Seebeck	2053
0.78	10^5	0.7 (Hall, estim.)	n, Hall, Dember; p, Seebeck	2053
	10^6 to 10^9			2101
	4×10^8		p, TH	2035
	3.8×10^8		p, TH	2035
	6.3×10^8		p, TH	2035
0.14		3.2	n, TH	2042
0.42			p, TH	2042
0.26	0.3		n, Seebeck	2292
	10^{-3} (along needle axis)		p, Seebeck	2292
	various			4153
0.11	5			4153

TABLE 8.5B Ternary and Inclusion Complexes

Substance	Purification	Form	Temperature range (°C)
Benzidine:TCNQ (methylene-chloride)	R	CP	At 18
Benzidine:TCNQ (methylene-chloride)	R	SC	At 18
Benzidine:TCNQ (ethylbromide)	R	CP	At 18
Benzidine:TCNQ (acetonitrile)	R	CP	At 18
Benzidine:TCNQ (acetone)	R	CP	At 18
Benzidine:TCNQ (no solvent)	R	CP	At 18
Cyclohexamylose:BaI$_2$:I$_2$	R	SC	25
Cyclohexamylose:BaI$_2$:Fe	R	SC	25
DP:CLL (chloroform)	R	CP	25 to 100
DP:CLL (chloroform)	R	CP	25 to 100
DP:CLL (e.g., benzene)	R	CP	25 to 100
DP:CLL (e.g., benzene)	R	CP	25 to 100
DP:Bromanil (benzene)	R	CP	25 to 100
DP:DDB (chloroform)	R	CP	25 to 100
DP:DDB (benzene)	R	CP	25 to 100
Hydroquinone:SO$_2$ (clathrate)		CP	25 to 80

E in $E/2kT$ (eV)	Resistivity at room temp. (ohm cm)	Mobility μ (cm^2/Vsec)	Sign of majority carrier estimated from	Ref.
0.2	7×10^3			2023
	1.4×10^3			2023
0.24	6×10^3			2023
0.32	7×10^4			2023
0.42	4×10^5			2023
0.34	2×10^9			2023
	5×10^9			4140
	2×10^9			4140
Low T, 0.28	1.3×10^6			2036
High T, 0.40				2036
Low T, 0.22	2.7×10^5			2036
High T, 0.28	2.7×10^5			2036
0.20	2300			2036
Low T, 0.3	3.3×10^6			2036
High T, 0.38				
Low T, 0.24				2036
High T, 0.26	4000			
0.77		4 to 0.1	n (Seebeck)	2100

527

TABLE 8.5C Proton and Protonated Complexes

Donor	Acceptor	Remark	Reference
α-Naphthylamine	Picric Acid (2:1)	Protonated Amine	2068, 2069
Benzidine	Picric Acid	Protonated Amine	2068
Aromatic Amines	3,3′,5,5′-Tetranitrophenyl-4,4′-Diol	Proton transfer	2069
1,2-Dichloroethane	$SbCl_5$	Protonated	2070
Diphenylcyclopropenone	Substituted Acetic Acids	Hydrogen Bonded complex	2071
Glycine	Malonic Acid (1:1)	Proton Transfer	2072
α-Aminoacids	Oxalic Acid (1:1) & (1:2)	Proton Transfer	2072
Anilines	Dinitrophenyl	Proton Transfer	2073
o-Dianisidine	Tetramethylbenzidine	Proton Transfer	2069
o-Toluidine	Tetramethylbenzidine	Proton Transfer	2069
Aromatic Monoamines	Poly(Nitrobenzoic Acid) (3:2) (2:1)	True Benzoate also formed	2074
2,5-Dichloroaniline	Picric Acid (1:3)		2075

TABLE 8.5D Proton Affinities (in eV)

Substance	(eV)	Reference
Radicals OH⁻	17.3 to 17.48	2137
CN⁻	15.7	2137
CO	6.11	2137
Alkylamine Radicals		
NH_3	0.97	2296
NH_2CN_3	10.05	2296
$NH(CH_3)_2$	10.3	2296
Acetaldehyde	8.05	2121
Acetamide	10.04	2120
Acetic acid	8.43	2107
Acetone	8.1	2107
	9.05	2104
	8.18	2116
Acetonitrile	8.09	2116
Allyl esters of Furan-2-carboxylic acid	0.14 to 0.16	2103
Benzene	6.48 to 7.07; 7.94 ex 2110	2106, 2110
Benzenes substituted	7.97 to 9.13	2297
Borazine	9.05	2109
Butanal	9.03	2107
m-Butanol	8.79	2120
Butanoic Acid	8.7	2107
Carbon Dioxide	5.4	2107
Carbon Monoxide	6.18	2116
Diethylether	8.13	2116
	8.11	2298
Dimethylamine	9.80	2117
Dimethyl phosphine	9.53	2118
Ethylamine	9.63, 10.74	2117, 2120
Ethyl Nitrate	7.83	2122
Formaldehyde	7.31	2121
	6.9	2107

TABLE 8.5D *Continued*

Substance	(eV)	Reference
Formamide	3.0	2108
Formic Acid	8.25	2107
Isobutanoic Acid	8.7	2107
Methane	5.6	2105
	5.48	2114
	7.0	2115
Methanol	8.12	2107
	7.85	2298
	7.91	2116
Methylamine	9.51	2117
Methyl Chloride	7.13	2116
Methylethylketone	8.25	2107
Methyl Phosphine	9.01	2118
Methyl Sulfide	8.05	2116
Nitromethane	9.09	2120
Phenol	8.35	2120
Propanal	8.12	2107
Propanoic Acid	8.7	2107
Pyridine	9.79	2120
Styrene	8.03 to 7.3	2107
Toluene	8.36	2110
	8.12	2115
Trimethylamine	9.98	2117
Trimethyl Arsine	9.11	2119
Trimethyl Phosphine	9.92	2118
Water	7.98	2107
m-Xylene	8.38	2110
o-Xylene	8.38	2110
p-Xylene	8.38	2110

TABLE 8.6A Semiconducting Free Radicals and Radical Salts: TCNQ as Acceptor.

Substance	Purification	Form	Contacts	Temperature range (°C)
Acridine:TCNQ		SC	Microwave	−270 to +27
Acridine:(TCNQ)$_2$				
Acridinium · TCNQ salt	R	CP	Hg, Ag	−30 to +60
Benzidine:TCNQ	R	CP		
DEPE:(TCNQ)$_4$, monoclinic	R	SC	Ag	−170 to +30
	R	SC	Ag	−170 to +30
	R	SC	Ag	−170 to +30
DEPE:(TCNQ)$_4$, triclinic	R	SC	Ag	−170 to +30
	R	SC	Ag	−170 to +30
	R	SC	Ag	−170 to +30
Diaminodurene:TCNQ(1:1)	R	CP		
Diaminodurene:TCNQ(1:2)	R	CP		
Diethylcarbocyanine:(TCNQ$^-$) · TCNQ^0				
Diethylcyanine:(TCNQ$^-$) · TCNQ^0				
Diethylcyclohexylammonium TCNQ Salt	T	CP	Hg, Ag	−30 to +60
Diethylcycloammonium:(TCNQ)$_2$	R	CP	Hg, Ag	−30 to +60
1,1'-Diethyl-2,2'-cyanine [TCNQ]$_2$	R	SC	Ag	−269 to +39
N, N'-Diethyl-Pϕ-diamine:TCNQ	R	CP		
N, N-Diethyl-Pϕ:TCNQ	R	CP		
Dimethyl-1,2-benzimidazolium · TCNQ Salt	R	CP	Hg, Ag	−30 to +60
N, N'-Di-β-naphthyl-Pϕ:TCNQ	R	CP		
Dipyridylium:TCNQ Polymers (1:2)	R	SC	Au	
Polymers (1:2)	R	SC	Au	
Di-Se-TTF:TCNQ	R, S	SC		−170 to +30
DMPD:TCNQ	R	CP		
N-Ethylbenzothiazolium · TCNQ Salt	R	CP	Hg, Ag	−30 to +60

E in $E/2kT$ (eV)	Resistivity at room temp. (ohm cm)	Mobility μ (cm^2/Vsec)	Sign of majority carrier estimated from	Ref.
	0.01			2049
	0.013		n, TH	2026
0.052	0.3		n, TH	2003
0.16	5.1×10^3 at 12°C			2022
0.12	0.02 ∥b_0		n, TH	2093
0.12	3 ∥a_0		n, TH	2093
0.12	30 ∥c_0		n, TH	2093
∥c_0 0.46	400		n, TH	2093
∥b_0 0.38	8.75		n, TH	2093
∥a_0 0.38	2200		n, TH	2093
0.54	1.5×10^6, 15°C			2022
0.08	15, 15°C			2022
	5.6×10^4			2095
	400			2095
0.68	3.3×10^6		p, TH	2003
0.24 ($T > 27°$C)	7.1		n, TH	2003
0.36 ($T < 27°$C)	7.1		n, TH	2003
0.070 0.34	0.27, 20°C			4142
0.16	3.3×10^3 15, 15°C			2022
0.66	4.9×10^5 at 15°C			2022
0.8	1.25×10^7		p, TH	2003
1.34	6.7×10^7 at 15°C			2022
0.258	5×10^5			2020
0.206				2020
	0.002			2091
0.96	4×10^7 at 15°C			2022
0.84	1.25×10^6		p, TH	2003

TABLE 8.6A *Continued*

Substance	Purification	Form	Contacts	Temperature range (°C)
N-Ethyldimethyl-1,2-benzimidazolium. TCNQ Salt	R	CP	Hg, Ag	−30 to +60
N-Ethyl-methyl-1-benzimidazolium: TCNQ⁰ (1:0.5)	R	CP	Hg, Ag	−30 to +60
N-Ethylmethyl-1-benzimidazolium: (TCNQ)$_2$	R	SC	Hg, Ag	−30 to +60
Ethyl-methyl-2-benzothiazolium:TCNQ⁻, TCNQ⁰	R	CP	Hg, Ag	−30 to +60
N-Ethyl-o-phenanthrolinium. TCNQ salt	R	CP	Hg, Ag	−30 to +60
N-Ethyl-methyl-2-quinolinium⁺: TCNQ⁻ TCNQ⁰	R	CP	Hg, Ag	−30 to +60
Imidazolium. TCNQ Salt	R	CP	Hg, Ag	−30 to +60
Ionene: TCNQ Salts				see Table 8.6B
N-Methylacridinium: TCNQ⁻ TCNQ⁰	R	CP	Hg, Ag	−30 to +60
N-Methylbenzimidazolium. TCNQ Salt	R	CP	Hg, Ag	−30 to +60
Methyl-1-benzimidazolium⁺:TCNQ⁻ (TCNQ⁰)$_{0.5}$	R	CP	Hg, Ag	−30 to +60
2-Methyl-2-benzothiazolium⁺: TCNQ⁻ TCNQ⁰	R	CP	Hg, Ag	−30 to +60
N-Methylbenzothiazolium. TCNQ Salt	R	CP	Hg, Ag	−30 to +60
Methyldiethylcyclohexylammonium. TCNQ Salt	R	CP	Hg, Ag	−30 to +60
Methyldiethylcyclohexylammonium: (TCNQ)$_2$	R	CP	Hg, Ag	−30 to +60
N-Methyl-dimethyl-1,2-benzimidazolium: TCNQ⁰ (1:0.5)	R	CP	Hg, Ag	−30 to +60
Methyl-N-ethylbenzimidazolium. TCNQ Salt	R	CP	Hg, Ag	−30 to +60
(Methyl-1-N-ethylbenzimidazolium)⁺: (TCNQ⁻)$_2$		SC	Ag	
(Methyl-1-N-ethylbenzimidazolium)⁺: (TCNQ⁻)$_2$		SC	Ag	

E in $E/2kT$ (eV)	Resistivity at room temp. (ohm cm)	Mobility μ (cm^2/Vsec)	Sign of majority carrier estimated from	Ref.
1.28	1.1×10^{10}		p, TH	2003
0.19	2		n, TH	2003
0.19	Mean 1.8 Anisotropic 0.15, 23, 435		n, TH	2003
0.24	15		n, TH	2003
0.72	4×10^6		p, TH	2003
0.42	35.7		n, TH	2003
0.36	3.3×10^2		p, TH	2003
				2048
0.064	8.3		n, TH	2003
0.94	1.5×10^7		p, TH	2003
0.096	0.9		n, TH	2003
0.082	1.1		n, TH	2003
	3.8×10^5		p, TH	2003
0.98	5×10^7		p, TH	2003
	1.5 to 15 depending on solvent			2003
0.074	1.5		n, TH	2003
1.1	5.9×10^8		p, TH	2003
0.18	Along needle axis, 0.154 $\perp$ 23.7 $\perp$ 440			2307
	Along needle axis, 0.044			2030

TABLE 8.6A Continued

Substance	Purification	Form	Contacts	Temperature range (°C)
N-Methyl-1-benzimidazolium: TCNQ^0 (1:0.5)	R	CP	Hg, Ag	−30 to +60
N-Methyl-o-phenanthrolinium. TCNQ Salt	R	CP	Hg, Ag	−30 to +60
N-Methylphenanthrolinium$^+$: TCNQ$^-$ (TCNQ$^-$)$_{0.5}$	R	CP	Hg, Ag	−30 to +60
N-Methylphenazine: TCNQ		SC	Microwave	−270 to +27
N-Methylphenazinium: TCNQ	R	CP		
N-Methylphenazinium: TCNQ		SC		−243 to +27
N-Methylquinolium$^+$: TCNQ$^-$ (TCNQ^0)$_{0.5}$	R	CP	Hg, Ag	−30 to +60
N-Methyl-2-quinolinium$^+$: TCNQ$^-$ TCNQ^0	R	CP	Hg, Ag	−30 to +60
o-Phenanthrolinium$^+$: TCNQ$^-$ (TCNQ^0)$_{0.5}$	R	CP	Hg, Ag	−30 to +60
Phenazine: TCNQ		SC		
o-Phenylenediamine: TCNQ (2:1)	R	CP		0.9
Pyridinium. TCNQ Salt	R	CP	Hg, Ag	−30 to +60
1,2-bis(4-Pyridyl)ethane = D_S				
D_S^+TCNQ$^-$	R	CP	Au	−160 to +25
D_S^{2+}(TCNQ$^-$)$_2$	R	CP	Au	−160 to +25
D_S^{2+}(TCNQ$^-$)$_2$. ½TCNQ	R	CP	Au	−160 to +25
D_S^{2+}(TCNQ$^-$)$_2$ TCNQ	R	CP	Au	−160 to +25
D_S^{2+}(TCNQ$^-$)$_2$ polymer	R	CP	Au	−160 to +25
D_S^{2+}(TCNQ$^-$)$_2$ TCNQ polymer	R	CP	Au	−160 to +25
D_S^+ : (TCNQ$^-$)	R	CP	Au	−196 to +27
D_S^{2+} : (TCNQ$^-$)$_2$. TCNQ^0	R	CP	Au	−196 to +27
D_S^{2+} : (TCNQ$^-$)$_2$. (TCNQ)$_{½}$	R	CP	Au	−196 to +27
D_S^{2+} : (TCNQ$^-$)$_2$ polymer	R	CP	Au	−196 to +27
D_S^+ (TCNQ$^-$)	R	CP	Au	−196 to +27
1,2-bis(4-Pyridyl)ethylene = D_u				
D_u^+TCNQ$^-$ (polycrystalline)	R	CP	Au	−196 to +27
D_u^+TCNQ$^-$ (single crystal)	R	CP	Au	−160 to +25
D_u^{2+}(TCNQ$^-$)$_2$	R	CP	Au	−160 to +25
D_u^{2+}(TCNQ$^-$)$_2$ TCNQ	R	CP	Au	−160 to +25
D_u^{2+}(TCNQ$^-$)$_2$ TCNQ	R	CP	Au	−160 to +25
D_u^{2+}(TCNQ$^-$)$_2$ polymer	R	CP	Au	−160 to +25
D_u^{2+}(TCNQ$^-$)$_2$ TCNQ polymer	R	CP	Au	−160 to +25
D_u^{2+} : (TCNQ$^-$)$_2$	R	CP	Au	−196 to +27
D_u^{2+} : (TCNQ$^-$)$_2$. TCNQ^0 polymer	R	CP	Au	−196 to +27

E in $E/2kT$ (eV)	Resistivity at room temp. (ohm cm)	Mobility μ (cm^2/Vsec)	Sign of majority carrier estimated from	Ref.
0.18	1.5		n, TH	2003
0.9	1.4×10^7		p, TH	2003
0.066	1.25		n, TH	2003
	0.01			2049
0.09	2			2080
0.02 to 0.1	0.013	T dependent		2304, 2305
0.14	2.5		n, TH	2003
0.07	1.5		n, TH	2003
0.076	7.1		n, TH	2003
	58 ± 10	1 at −173°C		2064
1.4×10^7 at 15°C				2022
0.58	3.3×10^4		p, TH	2003
0.11	1.55×10^5	0.04	p, Hall, Seebeck	2306
	4×10^6		p, Seebeck	2306
0.08	4.7			2306
0.07	3.4			2306
0.3	1.5×10^6		p, Seebeck	2306
	92			2306
0.22	1.55×10^5	0.04	p, Hall	2020
0.07	3.4			2020
0.08	4.7			2020
0.3	1.5×10^6			2020
0.28				2306
0.26	1.1×10^4	0.3	p, Hall, Seebeck	2306
0.28				2306
	2.1×10^5			2306
0.08	13			2306
0.07	3.8			2306
0.3	5.2×10^5		p, Seebeck	2306
	80			2306
0.07 − 0.08	3.8 − 13			2020
	80			2020

TABLE 8.6A *Continued*

Substance	Purification	Form	Contacts	Temperature range (° C)
Pφ:TCNQ	R	CP		At 15° C
Quinoline: (TCNQ)$_2$		SC	Ag	−263 to +27
Quinolinium$^+$:TCNQ$^-$ TCNQ^0	R	CP	Hg, Ag	−30 to +60
Quinolinium: (TCNQ)$_2$	R, Z	SC		−75 to +50
Solithane: TCNQ	R	CF	Au	5 to 140
TEA-cyclohexylammonium. TCNQ Salt	R	CP	Hg, Ag	−30 to +60
TEA: (TCNQ)$_2$	RC	SC	Hg, Ag	−30 to +60
TEA: TCNQ^0 (1:2)	R	SC	Ag	−196 to +77
	R	CP	Ag	−196 to +77
TEA: TCNQ(1:2)	R	SC	Ag	−196 to +77
TEA: TCNQ(1:2)	R	SC		
	R	SC		
	R	CP		
	R	CP		
TEA: TCNQ(1:2)	R	SC	Ag	−196 to +82
	R	CP	Ag	−196 to +82
	R	CP	Ag	−196 to +82
	R	SC	Ag	−196 to +82

E in $E/2kT$ (eV)	Resistivity at room temp. (ohm cm)	Mobility μ (cm^2/Vsec)	Sign of majority carrier estimated from	Ref.
0.34	4.2×10^3	4.2×10^3		2022
	0.013		n, TH	2026
0.056	0.5		n, TH	2003
0.04 ∥ 0.08 ⊥	10^{-2} 10^2			2035
(<38°C) 1.0 (>38°C) 2.4				2082
0.92	5.3×10^6		p, TH	2003
0.36 (<27°C) 0.24 (>27°C)	Mean 6.6; 0.3, 33, 714 Anisotropic		n, TH	2003
$0.2 < E < 0.4$	0.135, 22.2, 385 Anisotropic			2009
$0.2 < E < 0.4$	0.135, 22.2, 385 Anisotropic			2009
	250, 1000, 3.4×10^4 (10^8 Hz; $-133°$C) Anisotropic	1.3, Hall un- corrected, 8.71, Hall corrected for anisotropy	n, but p along y axis. Hall and TH	2009, 2010
0.28 – 0.30	0.2, 15, 630			2011
	0.3-25-1250			2012
0.30	15			2013
0.32	0.3			2014
0.274 at −196° 0.4 at −38° 0.22 at +57° all ±0.005	0.175 ∥ to needle axis	0.25 at −196°C	SCL	2018
0.27 at −196° 0.4 at −38° 0.22 at +57° all ±0.005	15.4	0.6 at 25°C	Hall	2019
−196°C, 0.274 −38°C: 0.4 +57°C: 0.22 all ±0.005	15.4			2303
−196°C: 0.274 −38°C: 0.4 +57°C: 0.22 all ±0.005	0.175 Along needle axis	0.25 SCL Current at −196°C		2303

TABLE 8.6A *Continued*

Substance	Purification	Form	Contacts	Temperature range (° C)
TEA: (TCNQ)$_2$	R, Z	SC		−75 to +50
Tetramethyl-Pϕ:TCNQ	R	CP		
TMPD: (TCNQ)$_2$	R, S	CP		−200 to 240 −200 to −25 > −25
TMPD: TCNQ	R, S	CP		< −25 > −25
TMTTF: TCNQ	S, Z	SC	Microwave	
(TMTTF)$_{1.66}$:(TCNQ)$_2$	S	SC	Microwave	
Triphenylmethylphosphonium: TCNQ		SC		
TTF: TCNQ	R	SC		−269 to 27
TTF: TCNQ		CP		27
		SC		27
TTF: TCNQBr$_2$		CP		27
		SC		27
TTF: TCNQ.BrCH$_3$		CP		27
		SC		27
TTF: TCNQ(C$_2$H$_5$)$_2$		CP		27
		SC		27
TTF: TCNQ.ClCH$_3$		CP		27
		SC		27
TTF: TCNQ.I$_2$		CP		27
		SC		27
TTF: TCNQ.ICH$_3$		CP		27
		SC		27
TTF: TCNQ.(OCH$_3$)$_2$		CP		27
		SC		27

E in $E/2kT$ (eV)	Resistivity at room temp. (ohm cm)	Mobility μ Mobility μ (cm^2/Vsec)	Sign of majority carrier estimated from	Ref.
Along a-axis 0.3 Along b-axis 0.28 Along c-axis 0.30	Along a-axis 0.33 Along b-axis 25 Along c-axis 125		n, Seebeck	2035
0.72	6.8×10^5 at 15°C			2022
0.096 0.134 0.28	50		p, TH	2080
0.378 0.52			p, TH	2080
	0.0017 0.02			2094 2094
‖ [011] 50				2065
0.0062	ρmin = 0.65 $\times 10^{-4}$ at 66 K 0.005 to 0.0015 Below 8 K	3000 – 147000		2051 2051
	0.1 2.4			4152 4152
	0.2 3.4×10^{-2}			4152 4152
	0.1 2.3×10^{-3}			4152 4152
	0.1 3.5×10^{-3}			4152 4152
	0.2 4×10^{-2}			4152 4152
	45 12			4152 4152
	0.1 5.4			4152 4152
	0.2 5.8			4152 4152

TABLE 8.6B Electrical Properties of Semiconducting Anion-Radical Salts Derived from Tetrachloro-p-diphenoquinone. After Y. Matsunaga and Y. Narita, *Bull. Chem. Soc.*, Japan, 45, 408 (1972)

Cation	Composition	Resistivity (ohm cm)
Sodium	1:1	2×10^3
Triethylammonium	2:3	7×10^3
Tetraethylammonium	1:2	4×10^3
Di-n-propylammonium	2:3	6×10^2 [a]
Methyldi-n-propylammonium	2:3	1×10^3 [b]
Ethyldi-n-propylammonium	1:2	1×10^3 [c]
Tri-n-propylammonium	1:2	3×10^2 [d]
Trimethylphenylammonium	2:3	6×10^2 [e]
Triethylphenylammonium	1:2	4×10^3
p-Phenylenediamine	1:2	120
Quinolinium	2:3	2×10^5
N-(n-Propyl)-quinolinium	2:3	1×10^3
N-Methyl-isoquinolinium	2:3	7×10^3
N-Ethyl-isoquinolinium	2:3	6×10^3
N-(n-Propyl)-isoquinolinium	2:3	3×10^3
Lepidinium	2:3	1×10^4
N-Methyl-lepidinium	2:3	1×10^4
N-Ethyl-lepidinium	2:3	7×10^4
N-Methyl-acridinium	2:3	3×10^3
N-Ethyl-acridinium	2:3	5×10^4
Ferricinium	1:2	24 [f]
Cobalticinium	1:2	50 [g]

[a] Activation energy 0.21 eV,
[b] 0.37 eV,
[c] 0.17 eV,
[d] 0.18 eV,
[e] 0.19 eV,
[f] 0.08 eV,
[g] 0.07 eV.

These activation energies refer to E/kT.

TABLE 8.6C Anion Radical Salts Derived from Tetrabromo-*p*-Diphenoquinone. From Y. Matsunaga and Y. Narita, *Bull. Chem. Soc. Japan*, *45*, 408 (1972).

Cation	Composition	ρ(ohm-cm)
Na	1:1	2000
Triethylammonium	2:3	5000
Tetraethylammonium	1:2	600
Di-*n*-propylammonium	2:3	3000
Methyldi-*n*-propylammonium	2:3	1000
Ethyldi-*n*-propylammonium	1:2	140
Tri-*n*-propylammonium	1:2	120
Tetra-*n*-propylammonium	1:1	2×10^7
Trimethylphenylammonium	2:3	130
Triethylphenylammonium	2:3	84
p-phenylenediamine	1:2	66
N-(*n*-propyl)-quinolinium	2:3	2000
N-methyl-isoquinolinium	1:1	900
N-ethyl-isoquinolinium	1:1	2000
N-(*n*-propyl)-isoquinolinium	2:3	2000
N-methyllepidinium	1:1	2000
N-ethyl lepidinium	2:3	6×10^4
N-methyl acridinium	1:1	3000
N-methylacridinium	2:3	2000
N-ethylacridinium	2:3	4000
Ferricinium (1)	1:2	32
Cobalticinium (2)	1:2	103

E_a in E/kT for (1) 0.07 eV, for (2) 0.1 eV.

TABLE 8.7C Long-Chain Compounds and Polymers.

For extensive data on polythiophenediylvinylenes, polyarylenevinylenes, fluorene polymers, polyarylene, butadienylenes, metal free and Ni azomethines, see Ref. 4151.

Substance	Purification	Form	Contacts	Temperature range (°C)
Acrylic acid-amyl-proparylaniline copolymers				
Acrylic acid-methyl-proparylaniline copolymers				
Acrylic acid-octyl-proparylaniline copolymers				
Anthrone polymers	SO	CP		
$[CH(AsF_5)^{0.1}]_x$		Stretched film		
$[CH \cdot I_{0.22}]_x$ (undoped)		Crystalline film		
1,6-Diacetylenes (cyclopolymerized)				
Ionene elastomers	RC	CP	Au	−80 to 60
Ionene:TCNQ CTC polymers *See* Table 8.7F				
1,3,4 Oxydiazole polymers	SO	CP	Al	+20 to +140
	SO	DF, SW	Al	+20 to +140
	SO	DF, SW	Al	+20 to +140
Oxy-pyrrole polymer		Flexible film		
PAQR polymers				
Phenylformaldehyde polymeric pyrolysates; Pyrolysis Temp.				
600° C		CP		
1200° C		CP		

E in $E/2kT$ (eV)	Resistivity at room temp. (ohm cm)	Mobility μ (cm^2/Vsec)	Sign of majority carrier estimated from	Ref.
	10^9–10^{10}		10^{15}–10^{16}, electron spins/g	2096
	10^9–10^{10}		10^{15}–10^{16}, electron spins/g	2096
	10^9–10^{10}		10^{15}–10^{16}, electron spins/g	2096
0.28	$\geqslant 100$ at 1.8 kbar $\geqslant 2$ at 33 kbar			2083
	0.0005		p, Seebeck	2132
1.9	trans 10^5 cis 10^9			2128, 2136
	10^{10}–10^{14}		paramagnetic	2098
$t < $ Tg 0.36–0.47	2.7×10^7 to 2.2×10^8			2088
				2048
0.81	3×10^{12}			2169
1.32	8.5×10^{12}			2169
1.5	1.5×10^6			2169
	0.125			2133
see Table 8.7D, 8.7E				2084
	27	0.0014	n (From carrier concentration ESR)	2007
	0.0044	7.84	p (From carrier concentration ESR)	2007

TABLE 8.7C *Continued*

Substance	Purification	Form	Contacts	Temperature range (° C)
Phenylthiocyanate polymers				
Polyacetylene				
Polyacetylene (doped)				
Polyacetylene (undoped)				
Polyacetylenes (halogen doped)		Film	C, Au	−270 to 27
Polyacetylene (I_2 doped)	Washed	DF		
Polyacetylene (I_2 doped)	Washed	DF		
Polyacetylene (cis-rich) (undoped)	Washed	DF		
trans-Polyacetylene		Stretched film		
trans-Polyacetylene (I_2 doped, 0.22 mole %)		Stretched film		
Polyacrolynitrile	heat treated 700° C	EF, DF		−100 to +100
Poly-5,5′-bi-isatyl-thiophene-indophene	R	CP	Ag	20 to 140
Poly bis(amino)-phosphazenes	RC	CP	Pd	20 to 180
Poly-5,5′-diisatyl-methane-thiophene-indophenine	R	CP	Ag	20 to 140
Polyethylene		Film	Au	20 to 70
Polyethylene (low density)				above Tg
Polyethylene copolymer of methacrylic acid				21 to 114
Poly-imide		DF		
Poly-malonitrile		CF	SW	
Poly(metalphthalocyanines:Cu	Soxhlet extracted			
:Fe	Soxhlet extracted			
:Sb	Soxhlet extracted			
:Zn	Soxhlet extracted			

E in $E/2kT$ (eV)	Resistivity at room temp. (ohm cm)	Mobility μ (cm⁴/Vsec)	Sign of majority carrier estimated from	Ref.
0.5 to 0.8	10^5–10^8			2112
			p, TH	2281
	0.001			2279
	10^{10}			2279
0.0274	0.002			2127; 2166
	0.04			2145
	1			2145
	10^7			2145
			p, Seebeck	2132
			p, Seebeck	2132
		0.01	n, Hall	2140
			p, Seebeck	2140
air 0.84	air 2.6×10^9		p, TH	2037
vacuum 1.0	vacuum 3.1×10^9			2038
1.75	1.8×10^{11}			2097
2.2	8.1×10^{14}			2097
0.45	0.73×10^5		p, TH	2060, 2061
2.74				2055
0.17	4×10^9			2278
				2278
2.84				2277
1.72			n, Photo	2170
0.12	7×10^6			2173
0.15	1.1×10^6			2173
–	3.1×10^6			2173
0.12	5300			2173

TABLE 8.7C *Continued*

Substance	Purification	Form	Contacts	Temperature range (°C)
Poly-N-methylpyrrole		Flexible film		
Poly-oxypyrrole (black)		CP	Ag	−173 to +27
Polyphthalocyanine	Extracted	CP	Pt	
Polyphthalocyanines	Soxhlet extracted			
PolyPHTHC (metal free)	RC	CP	Pt	
Poly(metal)PHTHC:Cu :Fe :Sb :Zn				
PolyPHTHC:Cu :Ni	R R	CP CP		20 to 300 20 to 300
Polypyrrole		Flexible film		−193 to 250
Polypyrroline II	Containing ⩽8%H$_2$O			
Polyselenomethylene				
Polyselenomethylene		CP		20 to 120
Polytetrafluoro-ethylene		in dry air		
Poly(2-vinylpyridine):I$_2$ (1:2)		DF		−73 to 27
PVC(commercial)	−	Film	Au	T < Tg T > Tg
PVC(pure)	R	DF	Au	0 to 30
TTF-acetylacetonate polymers TTF-metal polymers				

548

E in $E/2kT$ (eV)	Resistivity at room temp. (ohm cm)	Mobility μ (cm^2/Vsec)	Sign of majority carrier estimated from	Ref.
	2×10^6			2134
0.044	1790			2171
0.01	6.8 at 19.6 kbar		p, Seebeck	2182
0.01	7 to 58			2173
0.02	6.8 at 19.6 kbar		n, Hall	2087
	5.8 at 1.8 kbar		p, TH	2087
0.56 at 1.8 kbar	5300			2086
0.38 at 1.8 kbar	1.1×10^6			2086
–	3.1×10^6			2086
0.44 at 1.8 kbar	7×10^6			2086
0.21 to 0.25	10		n, TH	2063
0.46	100		n, TH	2063
0.01			p, Seebeck	2134
0.033 at 80°C				
1.74			n, liquid transport	2172
	1490			2057
0.7 to 2.62	$> 10^{13}$			2058
	volume 1.1×10^{18}			2113
	surface 2.5×10^{17}			2113
> 7°C 0.12	1000			2168
> 7°C 0.9				2168
2.82–3.04	–	–	n	2008
1.24–1.96	–	–	n	2008
1.0 ± 0.1		4.7 ± 0.5	n. PHC, transients	2008
	1.6×10^4		p, TH	2099
	1.6×10^4		p, TH	2099

TABLE 8.7D Room Temperature Resistivities of PAQR Polymers and Copolymers. After Kho and Pohl[2084]

Resistivity of PAQR Polymers: Pyrene vs. various aromatic acids (AA) (Pyrene:AA:$ZnCl_2$ = 1:1:1). Pt electrodes.

Aromatic acid or anhydride	Room temperature resistivity at 1800 kg/cm^2 (ohm cm)
o-Iodobenzoic acid	2.7×10^2
Chloroacetic acid	5.9×10^2
o-Chlorobenzoic acid	7.3×10^2
m-Chlorobenzoic acid	1.3×10^3
p-Nitrobenzoic acid	2.7×10^3
1,12-Benzoperylene dicarboxyl anhydride	2.8×10^4
p-Fluorobenzoic acid	7.8×10^4
Gallic acid	1.3×10^7
Xanthene-10-carboxylic acid	1.5×10^7
Syringic acid	2.2×10^7
9-Fluorenecarboxylic acid	9.8×10^{11}
m-Aminobenzoic acid	4.7×10^{12}

Resistivity of PAQR Copolymers: Pyrene, PMA vs. various aromatic hydrocarbons (pyrene:SAHC:PMA:$ZnCl_2$ = 1:1:1:1.5). Pt electrodes.

Substituted aromatic hydrocarbon	Room temperature resistivity at 1800 kg/cm^2 (ohm cm)
Violanthrone	2.8×10^2
Pyrene	5.6×10^3
Alizarin	1.3×10^4
Indigo	1.5×10^5
p-Naphtholbenzein	7.5×10^5
Eosin Y	1.0×10^5
Aniline Black	1.4×10^5
Meldola Blue	6.8×10^5
Fluorescein	7.1×10^7
Amaranth	1.0×10^8
Congo Red	7.4×10^9
Benzoazurine G	2.4×10^{10}

TABLE 8.7E Resistivities and Activation Energies for Ten PAQR Polymers after Precompression to 32 kbar at 450 K, Evaluated at 1.82 kbar and 300 K.

Pt electrodes were used. After Pohl.[2085]

Acene	MTA		PMA	
	ρ (ohm cm)	E_a (eV)	ρ (ohm cm)	E_a (eV)
Acridine	3.4×10^4	0.269	4.2×10^4	0.235
Anthracene	2.2×10^7	0.215	6.9×10^4	0.263
Phenanthrene	1.7×10^6	0.211	1.1×10^5	0.225
Phenazine	1.1×10^5	0.145	3.2×10^4	0.245
Phenothiazine	3×10^2	0.15	8.4×10^2	0.130

TABLE 8.7F Electrical Properties of 7,7′,8,8′-Tetracyanoquinodimethane Salts of Ionene Polymers and Their Model Compounds. After Hadek, Noguchi and Rembaum.[204,8]

Numbers of CH$_2$ groups, x-y	One unit segment to two TCNQ$^-$			One unit segment to two TCNQ$^-$, two TCNQ$^\circ$		
	Resistivity, ρ (ohm cm)	Activation energy, E (eV)	Seebeck coefficient, S (mV/°C)	Resistivity, ρ (ohm cm)	Activation energy, E (eV)	Seebeck coefficient, S (mV/°C)
3-4	2.8×10^4	0.22	+0.12	1.5×10^4	0.20	+0.09
6-3	9.5×10^5	0.30	+0.31	3.4×10^2	0.24	+0.06
6-5	1.5×10^8	0.56	Positive[a]	7.7×10	0.140	−0.027
6-6	3.2×10^8	0.58	Positive[a]	5.2×10^7	0.46	+0.93
6-8	7.2×10^7	0.52	+0.72	1.4×10^5	0.45	+0.66
6-10	3.0×10^6	0.39	+0.39	7.3×10^4	0.29	+0.27
6-16	9.6×10^6	0.48	+0.04	4.7×10^2	0.23	−0.008

[a]The Seebeck coefficient was positive, but its exact value could not be determined because of polarization effects.

Electrical Properties of the TCNQ Complexes Derived from Poly[Methylenebis(Trimethylammonium Halides)]

Number of CH_2 groups, n	With negative TCNQ			With negative and neutral TCNQ		
	ρ (ohm cm)	E (eV)	S (mV/°C)	ρ (ohm cm)	E (eV)	S (mV/°C)
2	1.9×10^3	0.12	+0.12	9.0×10	0.10	+0.043
3	1.3×10^5	0.24	+0.59	4.2×10^2	0.16	−0.060
4	3.1×10^7	0.41	+0.52	6.5×10^2	0.26	+0.132
5	4.0×10^6	0.29	+0.41	1.2×10^2	0.22	−0.150
6	5.0×10^{10}	0.73	Positive[a]	1.0×10^2	0.21	−0.143
8	3.7×10^7	0.41	+0.72	7.0	0.090	−0.010
10	3.6×10^7	0.44	+0.69	4.1×10^5	0.28	+0.64
16	1.4×10^9	0.46	+0.75	1.5×10^2	0.16	0

[a] The Seebeck coefficient was positive, but its exact value could not be determined because of the high impedance.

These polymeric salts were prepared[2048] by the reaction of x-y ionene bromides with LiTCNQ in the presence or absence of TCNQ°. The specific resistivity and the activation energy for conduction of 6-6($x = 6$, $y = 6$) ionene polymeric salts were considerably higher than those of a numbers of similar polymeric salts with different x and y values. A homologous series of model TCNQ compounds representing unit segments of the polymers was synthesized[2048] from poly(methylenebis(trimethylammonium) halides)), and their electrical properties were compared with those of TCNQ polymeric salts. The large variations in electrical properties were found[2048] to depend mainly on the geometrical configuration and crystal packing and not on the distances between positively charged nitrogen atoms in the polyelectrolytes.

TABLE 8.8A Organic Dyes

Substance	Temperature range (°C)	E in $E/2kT$ (eV)	Resistivity at room temp. (ohm cm)	Mobility μ (cm^2/V sec)	Sign of majority carrier estimated from	Ref.
Acriflavin		2.53				2285
Capri Blue		1.82				2285
Eosin		2.37				2285
Erythrosine		2.29				2285
Fluorescein		2.49				2285
Methylene Blue		1.81				2285
Phenosafranine		2.11				2285
Pinacryptol Green		2.09				2285
Thionine		1.87				2285
Aniline Black	22		500			2283
Pinacyanole				0.001	PHC	2284

555

TABLE 8.11A Biological Materials

Substance	Purification	Form	Contacts	Temperature range (°C)
Albumen, Serum (Bovine)	D	CP	Ag, Steel	35 to 104
Albumen (Serum): Chloranil Complex	D	CP	Ag, Steel	25 to 76
Albumen (Serum): Coenzyme Q_{10}	D	CP	Ag, Steel	32 to 84
Albumen (Serum): Thyroxine	D	CP	Ag, Steel	32 to 82
Aldehyde Oxidase	D, DI	CP	Ag, Steel	27 to 73
Bovine Serum Albumen		CP		at room temp.
Casein		CP		at room temp
Chloroplasts (Spinach)			μ Wave Cavity 9 GHz	
Chloroplasts (Spinach)			μ Wave Cavity 9 GHz	
Collagen: Methylglyoxal Complex		CP		at room temp.
Cytochrome-c	DI	CP	Ag, Steel	34 to 87
Cytochrome-c, Oxid.	D	CF		50 to 80
Cytochrome-c, Red.	D	CF		50 to 80
Cytochrome-c3 Oxid.	D, Dry	DF		20 to 70
Cytochrome-c3 Red.	D, Dry	DF		−40 to −5
Cytochrome-c, Dry			μ Wave Cavity, 10 GHz	
Cytochrome-c3, Anhydrous		Surface cell		
Cytochrome-c, Damp			μ Wave Cavity, 10 GHz	
Cytochrome-c, Damp			μ Wave Cavity, 9 GHz	
Cytochrome Oxidase	D	CP	Ag, Au	17 to 65

E in $E/2kT$ (eV)	Resistivity at room temp. (ohm cm)	Mobility μ (cm^2/Vsec)	Sign of majority carrier estimated from	Ref.
2.21	10^{13}			2024
2.22 to 2.43	10^{13}			2024
1.8	5×10^{15} at 30°			2024
1.7	5×10^{15} at 30°			2024
1.71 ± 0.2	5×10^{12}			2024
		0.01	Microwave Hall Effect	2308
		0.01	Microwave Hall Effect	2308
		0.1	p, Microwave Hall Effect	2029, 2032
		0.1		
		0.5 to 0.8	p, Microwave Hall Effect	2029, 2031
		100	p, Microwave Hall Effect	2308
1.74	6×10^{13}			2024
1.2	3.1×10^{11}			2310
1.2	3.1×10^{9}			2310
3.3	2.3×10^{12}			2309
7.7	57 at $-5°$C			2309
	2700	1 ± 80%		2029, 2030
3.472	$\propto$ Hydrogen P			4140
	700	0	n, μ-Wave Hall	4140
	700	0.1–0.5	n, μ-Wave Hall	2029 2031
0.43, T < 45–50°C	10^{12}			2024
1.74, T > 45–50°C				

TABLE 8.11A *Continued*

Substance	Purification	Form	Contacts	Temperature range (°C)
Cytochrome Oxidase	D	CP	Ag, Au	28 to 80
Cytochrome Oxidase (Depleted of Cu)	D, DI	CP	Ag, Steel	25 to 75
Cytochrome-c: Phospholipid Complex	D	CP	Ag, Steel	45 to 84
DNA Damp			μ Wave Cavity 10 GHz	
DNA Dry	D		μ Wave Cavity 10 GHz	
DNA Dry	D		μ Wave Cavity 9 GHz	
DNA Damp			μ Wave Cavity 9 GHz	
DNA Damp	Theoretical Analysis			2.0
DNA:Na (Calf Thymus)	Denatured	Fibers	Pt	
DNA:Na (Calf Thymus)		Dry Fibers		
DNA:Na (Calf Thymus)		Fibers	Cu	
DNA:Na (Herring Sperm)		CP	Cu	
DNA:Na (Herring Sperm)	Dried	Oriented Film	Pt	
DNA:Na (Herring Sperm)		Oriented Film		
Ferricytochrome c	Chromat.	CF	Su	At 55
Ferrocytochrome c	Chromat.	CF	Su	At 55
Hemoglobin Damp			μ Wave Cavity 9 GHz	
Hemoglobin Dry			μ Wave Cavity 9 GHz	

E in $E/2kT$ (eV)	Resistivity at room temp. (ohm cm)	Mobility μ (cm^2/Vsec)	Sign of majority carrier estimated from	Ref.
0.6, T < 55°C 1.99, T > 55°C				2024
1.93	10^{12}			2024
3.0	10^{13} at 45°C			2024
	1000	0.5 ± 40%	p, μ-Wave Hall	2030 2029
	2 × 10^4	0		2030 2029
	6 × 10^5	0.3	p, μ-Wave Hall	2031 2029
	2500	9.5	n, μ-Wave Hall	2031 2029
2.0	10–100		n	2126
2.42				2039
2.36				2025 2026 2027
1.8				4150
1.8				4150
2.18	10^{13} at 40°C			4150
2.18				2028 2027
1.2	6.5 × 10^{10}			2310
	6.5 × 10^8			2310
	5000	0.1 to 0.55	n, μ-Wave Hall	2031 2029
	3000	2 ± 60%	p, μ-Wave Hall	2030 2029

TABLE 8.11A *Continued*

Substance	Purification	Form	Contacts	Temperature range (°C)
Hemocyanin (Crab Blood)	D	CP	Ag, Steel	25 to 75
Lysozyme	D, Dry	CF		At 30
Lysozyme: Methylglyoxal Complex				At room temp.
Melanin (from L-Dopa)	Dialyzed	CP	Ag	25
Myoglobin	D, Dry	CF		10 to 70
Pigment Epithelium (Eye)			μ-Wave Cavity Effect 10 GHz	
Ribonuclease	D, Dry	CF		At 30
Trypsin	D, Dry	CF		At 30

E in $E/2kT$ (eV)	Resistivity at room temp. (ohm cm)	Mobility μ (cm^2/Vsec)	Sign of majority carrier estimated from	Ref.
1.91, 1.85	$10^{12}-10^{14}$			2024
	$> 10^{14}$			2309
		0.01	Microwave Hall Effect	2308
2.0	10^{12}			4141
0.31	3.6×10^{10}			2309
	15		p, μ-Wave Hall	2033 2099
	$> 5 \times 10^{14}$			2309
	$> 10^{14}$			2309

TABLE 8.11B Biological Materials Supplementary Data

Substance	Form	Contacts	Temperature range (°C)	E in $E/2kT$ (eV)	Resistivity (ohm cm) at temperature indicated (°C)	Comment	Ref.
Cytochrome c oxid.	CF, SU	Au	—	2.7	6.1×10^{16} (30)	Anhydrous, Lyophilized	4155
Cytochrome c oxid.	CF, SU	Au	50 to 85	1.2	3.1×10^{11} (30)	Anhydrous, Lyophilized	4155
Cytochrome c red.	CF, SU	Au	10 to 60	1.2	3.1×10^{9} (30)	Anhydrous, Lyophilized	4155
Cytochrome c_3 oxid.	CF, SU	Au	20 to 70	3.3	2.3×10^{12} (30)	Anhydrous, Lyophilized	4155
Cytochrome c_3 red.	CF, SU	Au	−40 to −5	7.7(?)	5.7×10^{1} (−5)	Anhydrous, Lyophilized	4155
Ferricytochrome c	CF, SU	Au	—	1.2	6.5×10^{8} (55)	Anhydrous, Lyophilized	4154
Ferrocytochrome c	CF, SU	Au	—	1.2	6.5×10^{8} (55)	Anhydrous, Lyophilized	4154
Lysozyme	CF, SU	Au	—	—	$>10^{14}$ (30)	Anhydrous, Lyophilized	4155
Myoglobin	CF, SU	Au	—	0.31	3.6×10^{10} (30)	Anhydrous, Lyophilized	4155
Ribonuclease	CF, SU	Au	—	—	$>5 \times 10^{14}$ (30)	Anhydrous, Lyophilized	4155
Trypsin	CF, SU	Au	—	—	$>10^{14}$ (30)	Anhydrous, Lyophilized	4155

TABLE 8.14A Seebeck Coefficients (Thermo-EMF) of Organic Semiconductors.
The polarity indicated is that of the COLD JUNCTION and thus the sign of the majority carriers.

Substance	Form	Contacts	Temperature range (°C) From	To
Acridine-(TCNQ)$_2$	SC	Ag		
Acridinium:TCNQ	CP	Hg		
Alizarin:PMA				
Alizarin-PMA Polymer	CP	Pt		
Anthraquinone-PMA	CP	Pt		
Anthraquinone:PMA				
Benzidine-I$_2$ (0.75:1)	CP			
Benzidine-I$_2$ (1:1)	CP			
Benzidine-I$_2$ (1.25:1)	CP			
Bipyridilium:TCNQ$_2$		Au		
4,4'Bipyridilium. 2CH$_3$I:TCNQ$_2$		Au		
Bipyridilium:TCNQ$_2$:TCNQ°$_2$				
β-Cu-PHTHC	SC			
β-Cu-PHTHC	SC			
DEPE:TCNQ$_4$, Monoclinic	SC	Au		
Diammonium:TCNQ$_2$		Au		
Diammonium:TCNQ°				
Diethylcyclohexylammonium: TCNQ	CP	Hg		
Diethylcyclohexylammonium: (TCNQ)$_2$	CP	Hg		
Dimethyl-1-1,2-benzimidazolium: TCNQ	CP	Hg		
N-Ethyl-1-benzimidazolium:TCNQ	CP	Hg		
N-Ethyl-benzothiazolium:TCNQ	CP	Hg		
N-Ethyl-methyl-1-benzimidazolium:(TCNQ)$_2$	SC	Hg, Ag		

Average Temperature (°C)	Seebeck coefficient S (mVdeg^{-1})	$\Delta S/\Delta T$ (mV/°C)	Temperature differential ΔT(°C)	Reference
				2026
25	+0.2		Near room temp. only	2003
	+0.012			2001, 2002
	+0.012			2006, 2005
	+0.220			2005
	+0.220			2002
30	+1.0			2035
30	+1.2			2035
30	+0.1			2035
	−0.056 to −0.066			2090
	+0.15			2090
	−0.037 to −0.066			2090
25	∥ b-axis −1000 ⊥ b-axis −2			2300 ~~2300~~
25	−0.9			2302
Saturates at 80°C	∥ b_0-axis −0.06			2093
	+0.410 to 0.720			2090
	−0.01 to −0.15			2090
25	+1.1		Near room temp. only	2003
25	+0.060		Near room temp. only	2003
25	+0.3		Near room temp. only	2003
25	+1.1		Near room temp. only	2003
25	+0.6		Near room temp. only	2003
25	−0.05			2003

TABLE 8.14A *Continued*

Substance	Form	Contacts	Temperature range (°C) From	To
N-Ethyl-methyl-1-benzimidazolium$^+$. TCNQ$^-$:TCNQ	CP	Hg, Ag		
N-Ethyl-methyl-2-benzothiazolium$^+$. TCNQ$^-$:TCNQ$^\circ$	CP	Hg, Ag		
N-Ethyl-methyl-2-quinolinium$^+$. TCNQ$^-$:TCNQ$^\circ$	CP	Hg, Ag		
N-Ethyl-o-phenanthrolinium:TCNQ	CP	Hg		
Imidazolium:TCNQ	CP	Hg		
Ionene:TCNQ Complexes			See Table 8.6B	
Methyl-1-benzimidazolium$^+$. TCNQ$^-$:TCNQ$_{0.5}$	CP	Hg, Ag		
N-Methyl-1-benzimidazolium:TCNQ	CP	Hg		
N-Methyl-benzimidazolium:TCNQ	CP	Hg		
Methyldiethylcyclohexylammonium:TCNQ	CP	Hg		
Methyldiethylcyclohexylammonium:(TCNQ)$_2$	CP	Hg		
N-Methylphenazine:(TCNQ)	SC	Ag		
Ni-Phthalocyanine	SC			
PAQR Polymers	CP	Pt	30	see Table 8.14B
N-Methyl-o-phenantroline:TCNQ	CP	Hg		
Phenolformaldehyde Resin Polymeric Pyrolysate Pyrolysis Temp. <600°C	CP	Ni	+18	+160
>600°C	CP	Ni	+18	+160
Phenoselenazine:I$_2$ (2:3)	CP	Graphite		
Phenothiazine:I$_2$ (1:1)	CP			
Phenothiazine:I$_2$ (1:3)	CP			
Phenothiazine:I$_2$ (1:3)	CP	Graphite		
Phenothiazine:I$_2$ (2:3)	CP	Graphite		
Phenothiazine:I$_2$ (3:1)	CP			
trans-Polyacetylene	Stretched film			

Average temperature (°C)	Seebeck coefficient S (mVdeg^{-1})	$\Delta S/\Delta T$ (mV/°C)	Temperature differential ΔT(°C)	Reference
25	−0.06			2003
25	−0.07			2003
25	−0.15			2003
25	+0.7	Near room temp. only		2003
25	+0.6	Near room temp. only		2003
25	−0.02			2003
25	+1.1	Near room temp. only		2003
25	+0.6	Near room temp. only		2003
25	+1.3	Near room temp. only		2003
25	−0.020	Near room temp. only		2003
				2026
25	$\perp$ b-axis, −1			2301
	$\mu_-/\mu_+ = 0.686$ to 0.97			2085, 2084
25	+0.6	Near room temp. only		2003
40	−0.005. Near zero		2 to 10	2007
40	+0.001 to 0.0012	5×10^{-6}	2 to 10	2007
25	+0.004			2289
30	−0.150			2035
30	+0.132			2035
25	0.2			2290
25	+0.153			2289
30	+0.3			2035
	+0.85			2132

TABLE 8.14A *Continued*

Substance	Form	Contacts	Temperature range (°C) From	To
Polyacetylene (Undoped)	DF			
Polyacetylene (I$_2$ Doped)	DF			
Poly(p-dimethylaminostyrene): TCNQ	CP	C	30	15 to 65
Polymethylmethacrylate (Commercial)	Film	Cu		
Poly-PHTHC	CP	Pt		
Poly-PHTHC (Metal free)	CP	Pt		
Poly-PHTHC:Cu	CP		20	140
Poly-PHTHC:Ni	CP		20	140
Polystyrene (Commercial)	Film	Cu		
Polyvinylcarbazole:I$_2$ (1:2)	CP	Ag		
PVC (Pure)	DF	Cu	23.44	32.84
PVC (Pure)	DF	Cu	36.74	49.54
PVC (Commercial, Nylex Corp.)	Film	Cu	22.54	37.14
	Film	Cu	69.84	99.84
Pyrene:1,2-benzoperylene				
Pyrene-chloracetic acid polymer	CP	Pt		
Pyrene:Chloracetic acid				
Pyrene:m-Chlorobenzoic acid				
Pyrene-m-chlorobenzoic acid polymer	CP	Pt		
Pyrene:o-Chlorobenzoic acid				
Pyrene-o-chlorobenzoic acid polymer	CP	Pt		
Pyrene:p-Fluorobenzoic acid				
Pyrene-p-Fluorobenzoic acid polymer	CP	Pt		
Pyrene:Iodobenzoic acid				
Pyrene-iodobenzoic acid polymer	CP	Pt		
Pyrene:p-Nitrobenzoic acid				

Average temperature (°C)	Seebeck coefficient S (mVdeg^{-1})	$\Delta S/\Delta T$ (mV/°C)	Temperature differential ΔT(°C)	Reference
	+0.85			2281
	+0.03 to 0.018			2281
	−0.8			2053
30.84	+1.40			2008
	+0.009			2182
27	+0.0095			2087
	−0.02			2063
	−0.0216			2063
74.84	−1.50			2008
25	+0.22		5	2039, 2040
28.14	−1.45	Changes sign at 40°C		2008
43.14	+0.36	Changes sign at 40°C		2008
29.84	−1.52			2008
84.84	−2.14			2008
	+0.0593			2001, 2002
	+0.056			2006, 2005
	+0.056			2001, 2002
	+0.0661			2001, 2002
	+0.0661			2006, 2005
	+0.052			2001, 2002
	+0.0528			2006, 2005
	5×10^{-5}			2001, 2002
	5×10^{-5}			2006, 2005
	+0.0645			2001, 2002
	+0.0645			2006, 2005
	+0.0601			2001, 2002

TABLE 8.14A *Continued*

Substance	Form	Contacts	Temperature range (°C) From	To
Pyrene-*p*-Nitrobenzoic acid polymer	CP	Pt		
1,2-bis(4Pyridyl)ethane.$^+$ TCNQ$^-$	CP	Au		
Pyridinium:TCNQ	CP	Hg		
Pϕ-I$_2$ (0.8:1)	CP			
Pϕ-I$_2$ (1:1)	CP			
Pϕ-I$_2$ (1.6:1)	CP			
Pϕ:I$_2$ (1:2)	CP		−50	+50
			−130	+50
Quinizarin:PMA				
Quinizarin-PMA Polymer	CP	Pt		
Quinoline:(TCNQ)$_2$	CP	Ag	−173	−123
Quinolinium$^+$:TCNQ$^-$:TCNQ$^\circ$	CP	Hg, Ag		
TCNQ-Ionene Model Compounds			See Table 8.7F	
TCNQ-Ionene Polymers			See Table 8.7F	
TEA:(TCNQ)$_2$	CP	Hg		
TEA:(TCNQ)$_2$	SC	Ag		
TEA:(TCNQ)$_2$	R, Z		−75	+50
Thiazolinocarbocyanine:(TCNQ)	CP	Ag		
TMPD:TCNQ$_2$	R, S	Ag + C	> −1500	
Triethylcyclohexylammonium:TCNQ	CP	Hg		
Triethylcyclohexylammonium:(TCNQ)$_2$	CP	Hg		
TTF:(I)$_{0.71}$	SC	Ag + C		
(TTF)$_7$:(I)$_5$	SC	Au, C		
TTF:TCNQ	CP			
(TTF)$_{12}$:(SCN)$_7$	SC	Au, C		
(TTF)$_{12}$(SeCN)$_7$	SC	Au, C		
(TTF)$_{12}$:(SCN)$_7$	SC	Au, C		

Average temperature (°C)	Seebeck coefficient S (mVdeg^{-1})	$\Delta S/\Delta T$ (mV/°C)	Temperature differential ΔT (°C)	Reference
	+0.0601			2006, 2005
25	+0.6		5	2020
25	+0.5	Near room temp. only		2003
30	+0.04			2035
30	+0.05			2035
30	+0.06			2035
	−0.01 +0.130			2042
	+0.0081			2001, 2002
	+0.0081			2006, 2005
	−0.06			2026
25	−0.05			2003
				2048
				2048
25	−0.050	Near room temp. only		2003
27	Along z-axis −0.06 Along y-axis +0.0015 Along x-axis −0.069		<10	2009, 2010
	−0.25	0		2035
				2026
	+0.07			2080
25	+1.1	Near room temp. only		2003
25	−0.040	Near room temp. only		2003
25	+0.0065			2081
20	+0.065			2081
	−0.026			2102
20	+0.009			2080
20	+0.009			2080
25	+0.009			2050

TABLE 8.14A *Continued*

Substance	Form	Contacts	Temperature range (°C) From	To
TTT	CP			
TTT:CLL 10:1	CP			
TTT:CLL 2:1	CP			
TTT:CLL (3:1) + 3% Mg	CP			
+ 3% Ag	CP			
TTT:o-CLL 10:1	CP			
TTT:o-CLL 1:1	CP			
TTT:o-CLL 10:3	CP			
TTT:o-CLL 2:1	CP			
TTT:I_2 (1:1.5)	CP			
TTT$_2$:I_3	SC	C	−173	+45
bis(Tetrathiotetracene):$(I_3)_x$				
Violanthrone:PMA				
Violanthrone-PMA Polymer	CP	Pt		

Average temperature (°C)	Seebeck coefficient S (mVdeg^{-1})	$\Delta S/\Delta T$ (mV/°C)	Temperature differential ΔT (°C)	Reference
5	+1			2046
25	+0.7			
50	+0.38			
50	+0.1			2046
50	+0.01			2046
50	+0.30			2046
50	−0.08			2046
50	+0.2			2046
50	+0.1			2046
50	−0.03			2046
50	−0.08			2046
25	−0.06			2292
25	+0.00166 to +0.0022			2089
	$1.124 + 1.07 \times 10$			2147
	+0.061			2001, 2002
	+0.061			2006, 2005

TABLE 8.14B Thermoelectric Power Data for Conductive PAQR Polymers. After H. A. Pohl, *J. Biol. Phys.*, 2, 113 (1974)

Temperature differential in all cases was 30 K. Pt electrodes.
The mobility ratio μ_-/μ_+ varied from 0.68 to 0.97.

Polymer	Seebeck coefficient, S ($\mu V/°C$)	Resistivity ρ_{25} ohm cm at 1.8 Kbar	ds/dt ($\mu V/°C/°C$)
Dibenzpyrene-PMA	−19.0	9.5×10^2	+0.55
Pyrene-PMA (made at 306°C)	+69.6	7.6×10^3	−0.90
Pyrene-PMA (made at 253°C)	+21.8	1.6×10^4	−1.76
Phenanthrene-PMA	+156	1.0×10^5	−0.50
Chrysene-PMA	+123	1.6×10^5	−6.87
Anthracene-PMA	+346	8.3×10^5	−1.80
Chrysene-PA	+123	1.2×10^6	−6.85
Pyrene, *p*-fluorobenzoic acid	0.05	7.8×10^4	—
Quinizarin-PMA	+8.1	8.2×10^5	
Alizarin-PMA	+12.0	1.5×10^6	
Pyrene, *o*-chlorobenzoic acid	+52.8	7.3×10^2	
Pyrene, chloracetic acid	+56.0	5.9×10^2	
Pyrene; 1, 12-benzoperylene dicarboxyl anhydride	+59.3	2.8×10^4	
Pyrene, *p*-nitrobenzoic acid	+60.1	2.7×10^3	
Violanthrone-PMA	+61.0	5.9×10^3	
Pyrene, iodobenzoic acid	+64.5	2.7×10^2	
Pyrene, *m*-chlorobenzoic acid	+66.1	1.3×10^3	
Anthraquinone-PMA (A-P)	+220	$\sim 1 \times 10^4$	

TABLE 8.15A1 Photo-Conductivity Data

See also Text Tables 22.1 and 22.3, and Tables 9.2A and 9.2B.

Substance	Purification	Cell Type	Threshold	Other experimental conditions
Anthracene	Z	SW		SC, in a b plane, 10^{17} hv/cm^2 sec
Anthracene	Z, S	SW		Electrolytic contacts, $>4 \times 10^{25}$ hv/cm^2 sec
Anthracene	S	SW, SC		Aqu. Na_2SO_3 electrodes
Auramine	EF	SU		Pd electrodes
Azulene	S	SW	3600 Å	$\perp$ to ab, Ag electrodes
Brilliant Green	EF	SU		Pd electrodes
β-Carotene	Dried	SW, CF	~400 nm	Glass
Chlorophyll-a		SW		Cr|Chlor.|Hg
				Au|Chlor.|Hg
Congo Red	EF	SU		Pd electrodes
Crystal Violet	EF	SU		Pd electrodes
3,3'-Diethylthiacarbo-cyanine	EF	SU		Pd electrodes
3,3'-Diethylthiadicarbo-cyanine	EF	SU		Pd electrodes
3,3'-Diethylthiatricarbo-cyanine	EF	SU		Pd electrodes
Eosin	EF	SU		Pd electrodes
Erythrosin	EF	SU		Pd electrodes
N-Isopropylcarbazole doped with 5% bisphenol-A-polycarbonate	R	SW	~770 nm	Ni electrodes

Magnitude of effect	Thermal activation energy (eV) (E/kT)	Remarks	Ref.
$I_{phot} = 10^{-12}$ A		At 4K, linear, 500 V, Ag electrodes	2311
Hot electrons, $\mu = 100$ cm^2/Vsec		Laser, electron mean free path = 11 Å eject. depth = 200 to 500 Å	2312
Quantum yield $\sim 10^{-4}$ at 270 nm			2045
Linear to Illum., ohmic 10–100 V	0.26	n-type	2041
		Superlinear p-type	2077
$I_{phot} \propto$ Illum.$^{1/2}$, ohmic 10–100 V	0.40	n-type	2041
$Q = 10^{-5}$, linear at low $h\nu$		p-type, $\mu_+ = 0.0015$ cm^2/Vsec	2313
	0.38 to 0.68 Depends on direction	Rectifying	2314
	0.25 to 0.37 Depends on direction	ohmic, p-type	2314
Linear to Illum., ohmic 10–100 V	0.21	p-type	2041
$I_{phot} \propto$ Illum.$^{1/2}$, ohmic 10–100 V	0.36	n-type	2041
Linear to Illum., ohmic 10–100 V	0.31	n-type	2041
Linear to Illum., ohmic 10–100 V	0.19	n-type	2041
Linear to Illum., ohmic 10–100 V	0.12	n-type	2041
Linear to Illum., ohmic 10–100 V	0.27	p-type	2041
Linear to Illum., ohmic 10–100 V	0.32	p-type	2041
$Q = 0.05$			2315

TABLE 8.15A1 *Continued*

Substance	Purification	Cell Type	Threshold	Other experimental conditions
Ternary Aggregate Polymer and Dye and Aromatic	R, Z	SW		DF
Kryptocyanine	EF	SU		Pd electrodes
Malachite Green	EF	SU		Pd electrodes
Metal-8-Hydroxyl Quinolates				
Metal = Cu	Washed	SU		
Metal = Pd	Washed	SU		
Zn	Washed	SU		
Cd	Washed	SU		
Mg	Washed	SU		
Phenanthrene	C, S, Z	SW		SnO/Ag electrodes, SC
Phenanthrene	Z, Chrom.		4000Å	SC, $h\nu$ flux = 4×10^{14} cm^2 sec^{-1}
α-PHTHC Metal free	S	SU		at 610 mμ
Pinacyanole	EF	SU		Pd electrodes
Polydiacetylenes (Chain Substituted)		SU SU	750 60	DF, Blue Form DF, Red Form
Polydiacetylene		SW	3 eV	SC, along chain
Polydiacetylene		SW		SC, Ag electrodes

Magnitude of effect	Thermal activation energy (eV) (E/kT)	Remarks	Ref.
$I_{phot} = 3I_{dark}$	0.3	$I_{phot} \propto L^{0.7}$; $\mu = 10^{-7}$ to 10^{-5} cm^2/Vsec	2315
I_{phot} Linear, ohmic 10–100 V	0.20	n-type	2041
$I_{phot} \propto$ Illum.$^{1/2}$, ohmic 10–100 V	0.42		2041
		Photocurrent Max.	
$Q = 6 \times 10^{-6}$	0.4 low $h\nu$ 0.5 high $h\nu$	SC, Depends on electronegativity of metal atom	2316
$Q = 8 \times 10^{-7}$	0.4 low $h\nu$ 0.5 high $h\nu$	SC, Depends on electronegativity of metal atom	2316
		CP, Depends on electronegativity on metal atom	2316
		CP, Depends on electronegativity on metal atom	2316
		CP, Depends on electronegativity on metal atom	2316
$10^{-5} - 10^{-6}$ A/cm^2		Xenon Flashtube	2076
$I_{phot} = 5 \times 10^{-3}$ A–0.077 cm^2		$t > 72°C$	2158
$I_{phot} \cong L^{0.7}$ $L = 10^{17}$ Photons/sec	0.56 ± 0.03		2317
Linear to Illum., ohmic 10–100 V	0.31	n-type	2041
$Q = 0.2$ $Q = 0.2$		Up to Langmuir monolayers	2318
$\mu \pm > 0.001$ cm^2/Vsec		Sublimes at low field, linear at high field	2319
	0.056		2320

TABLE 8.15A1 *Continued*

Substance	Purification	Cell Type	Threshold	Other experimental conditions
Polyethylene		SW		Au electrodes
Polyethylene (High Density)	DF	SW		Au electrodes
Polymers		For Photoconductivity Data of Polymers, see Table 22.1 in Text.		
Polyvinylcarbazole	DF	SW		
Rose Bengal	EF	SU		Pd electrodes
Polyvinylcarbazole	R			DF
Porphyrins			Review with 100 references	
Rubrene	Z	SU		SnO, Al elect. EF
(Polymeric)TCNQ:Pyrene Complexes:		SW, SnO_2 and Au		Thin films dissolved in polystyrene, cast from CH_2Cl_2
(1:1)		Elect.	900 nm	Thin films dissolved in polystyrene, cast from CH_2Cl_2
(1:1) + TCNE		Elect.	900 nm	Thin films dissolved in polystyrene, cast from CH_2Cl_2
(1:2)		Elect.	900 nm	Thin films dissolved in polystyrene, cast from CH_2Cl_2
(1:2) + TCNE		Elect.	900 nm	Thin films dissolved in polystyrene, cast from CH_2Cl_2
p-Terphenyl		SW		near UV, SC

Magnitude of effect	Thermal activation energy (eV) (E/kT)	Remarks	Ref.
	0.09; $T < 45°C$ 0.64; $T > 45°C$	Ohmic at low field	2055
$\tau_1 = 20$ sec; $\tau_2 = 130$ sec; 10^{-12} A/cm²		p-type	2044
4×10^{-13} A/cm²			2062
Linear to Illum., 10-100 V	0.30	p-type	2041
$Q_{max} = 0.2$	0.05	p-type; I_{phot} saturates <10 V/cm	2321
			2322
$I_{phot} = 10^{-11}$ to 10^{-10} A		300V; 10^{14} $h\nu$/cm² sec	2323
Linear, $Q = 10^{-5}$, $I_{phot} \propto V^{\frac{1}{2}}$		24-60°C Response Time ~50 sec.	
Linear, $Q = 10^{-5}$ $I_{phot} \propto V^{\frac{1}{2}}$	0.79	24-60°C Response Time ~50 sec.	2024
Linear, $Q = 10^{-5}$ $I_{phot} \propto V^{\frac{1}{2}}$	0.74	24-60°C Response Time ~50 sec.	2024
Linear, $Q = 10^{-5}$ $I_{phot} \propto V^{\frac{1}{2}}$	0.61	24-60°C Response Time ~50 sec.	2024
Linear, $Q = 10^{-5}$ $I_{phot} \propto V^{\frac{1}{2}}$	0.59	24-60°C Response Time ~50 sec.	2024
$I_{phot} \cong 50 \times I_{dark}$ rises 6 to 9 times in polarized light		$I_{phot} \propto$ (Light Intensity)$^{3/2}$	2324

TABLE 8.15A1 *Continued*

Substance	Purification	Cell Type	Threshold	Other experimental conditions
Tetracene	S (4 times)	SW	$\lambda > 410\,m\mu$; $550\,m\mu$ $\lambda < 40\,m\mu$	H_2O electrodes
Pb-PHTHC		SW		SC, Ag electrodes
β-Zn PHTHC	S	SW		0–120°C, 7580 Å, Ag electrodes

TABLE 8.15A2 Photo-Conductivity Supplementary Data
See Also Text Tables 22.1 and 22.3, and Tables 9.2A and 9.2B.

Substance	Purification	Cell Type	Threshold	Other experimental conditions
Anthracene			3.75 eV	SC, In N_2, 200–300 K
Anthracene	Z	SW		SC; at 390 nm
Anthracene	S, Z	SW, SC	3.92 eV	
Benzanthrone		Thin film	390 nm	
Bromanil	R, S	SW		Ethanol electrodes; 4010 Å
p-Chloranil	R, S	SW		Ethanol electrodes; 4010 Å

Magnitude of effect	Thermal activation energy (eV) (E/kT)	Remarks	Ref.
Quantum effic. $\geqslant 4 \times 10^{-3}$ Linear, at 505 mμ = 0.075, at 230 mμ = 0.11		p-type, sc p and n-type, sc	2043
	as $f[O_2]$; 0.45–0.27 eV; *in vacuo* 0.2 eV		2325
$I_{phot} \propto L^p$; $p = 0.39$ to 1 $= f(T)$		n-type	2326

Magnitude of effect	Thermal activation energy (eV) (E/kT)	Remarks		Ref.
Max. Quant. Eff. at $h\nu = 442$ eV: 10^{-4}	at 320 nm: 0.193 at 280 nm: 0.076			4011
Quant. yield ~ 3.6×10^{-6}		Light intensity	Carrier concentration	4008
		low medium high	$\propto L^{0.5}$ $\propto L^{1.0}$ $\propto L^2$	
		$\mu_+ = 0.82$ cm^2/Vsec; $\mu_- = 0.37$ cm^2/Vsec.		
10^{-3} electrons per $h\nu$ absorbed				4037
$I_{Phot} \propto T$; peaks 250 nm				4069
Quant. yield $\cong$ 0.003		$I_{Phot} \propto E, n > 2$, linear to light		4012
$I_{Phot} \cong 120 \times I_{Dark}$		p-type; $I_{Phot} \propto E^2$, linear to light		4012

TABLE 8.15A2 *Continued*

Substance	Purification	Cell type	Threshold	Other experimental conditions
Chlorophyll *a*				10^{12}–10^{16} quanta cm^{-2}sec^{-1}
Chlorophyll (Microcrystalline)		SW, Front-Cr, Back Al		77 to 273 K. Pulsed laser 710 to 780 nm. Deposited from solution
Chlorophyll Photosystems I	Centrifuged fraction			$\cong 10^{13}$ photosystem cm^{-2}
Chlorophyll *a*: Stearic acid	Chromat.	DF SnO$_2$		1:1 Monolayer
Chloroplasts				
Cyanine dyes, polymeric		SW		CP, 102 Wm^{-2} white light, 1 V, 300K
1,4 Diaminoanthraquinone				Max. 596 nm
β-9,10 Dichloroanthracene	Chromat, Z	SW		3371 Å Laser pulse, 2800 Vcm^{-2}
β-9,10 Dichloroanthracene				25° C
Dinitrobenzene: Acid doped Polyvinylcarbazole CTC		SW, film		PVK:DNB:Trichloroacetic acid dope 24:1:1; at 3990 Å, 2×10^{14} $h\nu$ cm^{-2} sec^{-1}
Diphenanthraperylene	S	SU, EF	600 nm	Al or Au elect; at 450 nm
Furan-Quinone yellow pigment		SW, sublimed film	$\cong 580$ nm	5×10^{13} Photons cm^{-2} sec^{-1} at 480 nm

Magnitude of effect	Thermal activation energy (eV) (E/kT)	Remarks	Ref.
$I_{Phot} \neq f(T)$, 0–90°C		$I_{Phot} \propto \text{Light}^{0.9}$	4067
Quant. yield = 0.05		n-type; $\mu > 0.15$ cm^2/sec. Thermal act. $E = 0.06$ eV	4004
Max. Quant. yield $\cong 0.1$ at 700 nm			4005
Quant. yield 12–16%. I_{Phot} at $pH = 4$ is 1.5×10^{-7} A/cm^7		Linear to light intensity	4006, 4020
	$\cong 0$		4061
$I_{Phot} \leqslant 270$ I_{Dark}	0.34 to 0.57		4097
$I_{Phot} \propto \text{Light}^{0.9}$		I_{Phot} nearly flat 250 to 400 nm	4069
Quant. yield 2×10^{-7}, $I_{Phot} = 5\,\mu\text{Acm}^{-2}$	200–250 K, 0.25 · 300 K, $\cong 0.04$	Linear to light intensity and field. 2.5×10^{24} quanta cm^{-2} sec^{-1}	4074
I_{Phot} isotropic in ac plane	0.25	μ in ac plane = 0.6 cm^2/Vsec. μ in b_2* direction = 0.18 cm^2/Vsec.	4074
$I_{Phot} = 3 \times 10^{-5}$ Acm^{-2} at 10^5 Vcm^{-1}		$\mu_+ \propto E^2$ for $E \geqslant 50$ KV/cm, p-type	4043
6.5×10^{-7} electrons/Photon; $I_{Phot\,Max} \cong 5 \times 10^{-11}$ A			4053
Quant. yields 0.2, $I_{Phot} = 10^{-6}$ Acm^{-2}		Linear to light; xerographic conditions	4046, 4047

TABLE 8.15A2 Continued

Substance	Purification	Cell type	Threshold	Other experimental conditions
Merocyanine:Chloranil Sandwich		SW,EF, SC	700 nm	6 Monolayers arachdic acid atop dye; 600 nm; ≥ 1000 V
3-Methylpentane				4.2 to 85 K
Methyltetrahydrofuran Glass, γ-irradiated			780 nm	Peak at 520 nm
Napthalene		SW, SC	350 nm	3×10^{13} Photons cm^{-2} sec^{-1}, 2537 Å
Napthalene	Z, S	SW	~330 nm	SC, Ni mesh elect. $\perp$ to ab plane, other contact Al
Napthalene + 7×10^{-6} M/M Coronene				Melt, 3×10^{15} $h\nu$ cm^{-2} sec.$^{-1}$
N-Ethylcarbazole	synthesized		~450 nm	Ag elect. $\perp$ to a-axis, SC
Nitrobenzene				At 434 nm
N-Vinylcarbazole 96% + 4% Vinylnapthacene copolymer	R	SW, cast film	~600 nm	
Oxacyanine		SW	480	Monolayer on In$_2$O$_3$; 30 nm, photoelectrochem.
Pentacene	S	SW	2.2 eV	Sublimed film from SC 10^4 V/cm, 10^{13} $h\nu$/cm^2 sec.
Phenanthrene	S, Z		260 nm	
Phthalocyanines	Washed	SW		
β-Phthalocyanine, metal free		SW, film		Light pulses 360–740 nm. Au on SiO$_2$ blocking electrode
β-Phthalocyanine, metal-free	S			In vacuo

Magnitude of effect	Thermal activation energy (eV) (E/kT)	Remarks	Ref.
$I_{Phot\,max} \cong 2 \times 10^{-8}$ A		6.6×10^{-4} W/cm^2	4052
Quant. yield, 10^{-3} $\rho_{Phot} \cong 10^{11}$ Ωcm	0.01 to 0.02	μ_- at 85 K is 0.11 cm^2/Vsec, n-type	4038
$I_{Phot} \cong 10^{-12}$ Acm^{-2}, Quant. yield 0.19 ± 0.09	0.001	Ohmic	4039
$I_{Phot} \cong 8 \times 10^{-12}$ A		7×10^5 to 26×10^5 Vm^{-1}, at peak 230 nm	4049, 4057
intrinsic (?) carrier generation rate $= 2.3 \times 10^{-14}$ cm^3 sec^{-1}.		$\mu_- = 0.35$, $\mu_+ = 0.31$ cm^2/Vsec. $\rho_{Dark} = 10^{15-16}$, $I_{Phot} \propto L^2$ for $h\nu$ is 3.8 to 5 eV	4003
$I_{Phot} \cong 10^{-9}$ cm^{-2}		Linear to light	4041
	0.06 in vacuum, 0.08 in air	p-type; main inpurity is anthracene	4002
$I_{Phot} \cong 10^{-8}$ Acm^{-2}		$\mu_+ + \mu_- \cong 10^{-4}$ cm^2/Vsec.	4060
$I_{Phot} \cong 10^{-9}$ Acm^{-2} for 10^{13} $h\nu$/sec cm^2		p-type, Superlinear to field	4033
$I_{Phot} \cong 10^{-7}$ Acm^{-2}			4099
$I_{Phot}/I_{Dark} \cong 500$ to 1000, Quantum yield $\geqslant 0.3$; at lower field goes to 1		p-type	4010
		p-type, sublinear to light	4044
Carrier generation eff. 28%			4007
Quant. yields 0.001		Linear to field	4035
5×10^{19} trapping sites cm^{-3}	$T < 33$ K $= 0.38$ $T > 337 = 1.05$	In H$_2$:0.95 for $T < 337$ K 1.35 for $T > 337$ K. In O$_2$:0.82 for $T < 337$ K	4015

TABLE 8.15A2 *Continued*

Substance	Purification	Cell type	Threshold	Other experimental conditions
Phthalocyanine metal free As coating 3:1		Su		300 K
Cu-Phthalocyanine	S	SW		SC; Ag, Au electrodes
Cu-Phthalocyanines (α)		SW		EF, Ag electrodes to ab plane
Cu-Phthalocyanines (β)		SW		EF, Ag electrodes to ab plane
Mg-Phthalocyanine, O_2 doped	R, S Chromat.	SW, 2500Å film		10^{-4} W/cm^2 light, 20°C, 670 nm
Pb-Phthalocyanine, O_2- doped		SW, SC	~1450 nm	897 nm
Polydiacetylene			~860 nm	SC; 10^{14} $h\nu$ cm^{-2} sec.
Polydiacetylene (butadiene-doped)			~300 nm	
Polydiacetylene: Bis-(Toluenesulfonate)		SW, SC, also Su		In direct. of polymer chain, *viz.*, along b-axis $\perp$ to electrodes; laser 1.6×10^{26} photons cm^2/Vsec at 1.06 μ_-; $\cong$ 30 mWcm^{-2}
Polyferrocene				From glow discharge polymers.
Poly(β-N Carbazoethyl)-L-glutamel complex	SW, film		>700 nm	In_2O_3 and Au electrodes; $I_{Phot} \cong 1$ μA/W incident light
Polystyrene: Napthalene Complex		SW, SU	~1400Å	Ag electrode
Poly(Thiophene diylvinylene-phenylenes alkoxylate)		SW		CP, 1 V, 102 Wm^{-2} white light, 300 K
Polyvinylcarbazole		SW, DF		

Magnitude of effect	Thermal activation energy (eV) (E/kT)	Remarks	Ref.
Xerographic gain $\cong 0.3$	0.09		4013
0.5% increase		In magnetic field $\leqslant 8000$ gauss	4040
	$n = f(T); 0.9$	a.c. method, $\sigma \propto \omega$, $n = 0.5$ to 0.8	4001
	$n = f(T); 0.94$	a.c. method, $\sigma \propto \omega$, $n = 0.5$ to 0.8	4001
$I_{Phot}/I_{Dark} = 150$ to 200	0.25 to 0.30	p-type	4086
$I_{Phot} \sim 10^{-9}$ A/cm^2 $= f(P_{O_2})$;	0.45 to 0.27	I_{Phot} rises with time and T	4085
300 K: $I_{Phot} = 300 \times I_{Dark}$ 120 K: $I_{Phot} < 600 \times I_{Dark}$		Linear to light	4032
$I_{Phot} \cong 300\, I_{Dark}$			4051
I_{Phot} Max. $\cong 250$ Acm^{-2}; $I_{Phot} \parallel$ $I_{Phot} \perp$ Chain > 100		$\mu \cong 1$ cm^2/V sec	4054, 2149
$I_{Phot} \cong 100\, I_{Dark}$		n-type	4070
		p-n junction, PCLG is p-type, TNF is n-type	4045
Quant. yield $\cong 500$ electrons/$h\nu$		At peak 220Å, 300 V	4022
$I_{Phot} \leqslant 10^2$ Dark	0.16 to 1.84		4097
Sublinear to illum.		Superlinear to yield, p-type	4062

TABLE 8.15 A2 Continued

Substance	Purification	Cell type	Threshold	Other experimental conditions
Polyvinylcarbazole		SW		SC
Polyvinylcarbazole	R	SW	600 mμ	Film 15 μ thick
	R	Su	600 mμ	Au or Ag electrode
Polyvinylcarbazole: Methylene Blue:TCNE Ternary Complex				At max. 490 nm
Polyvinylcarbazole:Phthalocyanine (metal free) as coating, 3:1				300 K
Polyvinylcarbazole, pinacyanol doped				At 620 nm
Polyvinylcarbazole:Pinacyanol:TCNE Ternary Complex				At max. 490 nm
Polyvinylcarbazole:Trinitrofluorene (1:0.6 M)Complex		SW		$\geqslant$ 6300Å, 10^4–10^5 Vcm^{-1}, Au electrodes
Pyranthrene	Chromat. Dist.		1.6 eV	Dissolved in benzene, sat. sol. 3×10^{-5}M
Pyrene	R, Chromat., Z	SW, SC	300 nm	260 nm, 32–146°C, 300 to 600 V
p-Terphenyl	Z	SW, EF	5.8 eV	For 9 eV quanta.
Tetrabenzoperylene	S	SU, EF	550 nm	Al or Au elect.; at 550 nm
Tetracene	S, SC	SW		
Tetramethyl-paraphenylenediamine	C		3.86	10^{-5} M in hexone at 25°C, 8×10^{14} photons/cm^2 sec at 320 nm

Magnitude of effect	Thermal activation energy (eV) (E/kT)	Remarks	Ref.
	0.1–0.3 in vacuum 0.5–0.6 in air	Main impurity is anthracene, E_{Dark} (E/kT) 0.5 eV in vacuo, 0.45 in air	4002
$I_{Phot} \cong 10^{-13}$ Acm^{-2}	0.07 to 0.15	In visible, dep. on cond. and prehistory	4079
	0.06 to 1 eV	Dep. on cond. and prehistory	4079
$I_{Phot} \cong 15 \times 10^{-9}$ Acm^{-2}			4059
Xerographic gain $\cong 0.3$	0.09		4013
$I_{Phot} \cong 5 \times 10^{-9}$ Acm^{-2}		Max. at 620–570 nm	4059
$I_{Phot} \cong 22 \times 10^{-9}$ Acm^{-2}			4059
$I_{Phot} \cong 10^4 I_{Dark}$		Sublinear to light, ohmic below 0.1 mV cm^{-1}	4048
$I_{Phot} \cong 100 I_{Dark}$		10^{16} Photons cm^{-2} sec^{-1}; non-linear	4023, 4031
$I_+ \cong 8 \times 10^{-10}$ A $I_- \cong 1.5 \times 10^{-10}$ A	0.08–0.1	I_+ linear to field	4064
Quant. yield ~ 0.01			2232
4.3×10^{-7} electrons/photon; I_{Phot} Max. = 5 $\times 10^{-13}$ A			4053
Quant. yield $\propto$ light$^{1.5}$			4100
Quantum yield ~ 10^{-3}; $I_{Phot} \cong 10^{-13}$ A	0.172	$I_{Phot} = aE + bE^2$ $=$ field	4016, 2014

TABLE 8.15A2 (*Continued*)

Substance	Purifi-cation	Cell Type	Threshold	Other experimental conditions
Tetraphenylporphine	S	Sublimed film on Au	>700 nm	At 450 nm, aqu. electrode
Thiacyanine		SW	400	Monolayer on In_2O_3; 30 nm, photoelectro-chem.
Thiapyrylium Dye (Aryl Subst.) + Triphenylmethane (dialkylamino) substituted	Chromat.	SW, transp. Ni elect.	~730 nm	Dispersed in poly(iso-propylidenediphenyl-carbonate) polymer as 2:1 complex of dye and polymer
Trinitrofluorene: Polyvinylcarbazole (1:1)CTC		SW, film		23°C
2,4,6 Triphenyl Thio-pyrylium$^+$/ClO_4 (CTC)	R	SW		At 500 nm
CTC's Pyrylium on Thiopyrylium$^+$/ polycyanide$^-$	R	SW	~900 nm	SC, Ag electrodes

Magnitude of effect	Thermal activation energy (eV) (E/kT)	Remarks	Ref.
I_{Phot} Max ~ 10^{-8} Acm^{-2}			4055
$I_{Phot} \cong 10^{-7}$ Acm^{-2}			4099
$I_{Phot} \cong 1000 I_{Dark}$ for 10^{14} Photons cm^{-2} sec.$^{-1}$ Quant. yield = 0.5 at 1 mVcm^{-1}	0.3	$I_{Phot} \propto$ light$^{0.7}$ 2.4×10^3 Vcm^{-1}	2315, 4030
Quant. yield ≤ 0.23			4042
	1.2		4000
In visible, $I_{Phot} \cong 10 I_{Dark}$	0.5 to 0.8	Sublinear to light intensity $\rho_{Dark} \cong 10^{10}$ to 10^{12} Ωcm; ohmic $E_{Dark}(E/kT)$ 0.4 to 0.9 eV	4000

TABLE 8.16A1 Photo-Effects. See also Text Tables 22.5 and 22.6.

Substance	Purification	Substrate	Cell type	Second contact	Polarity of Substrate
Anthracene		Ag	Subl. film		
Chlorophyll-a		Al	SW cast film	Hg Au Al	−
β-Carotene Glass	Dried	Au, Cr, In$_2$O$_3$	Sw	Au, Cr, In$_2$O$_3$	
Chlorophyll + Acceptor		Hg	SW	Al	
Chlorophyll + Ubiquinone or Plastoquinone Doped		Hg	SW	Al	
Erythrosine B(ZnO sensitized)					
Fluorescein					
Hexylmethacrylate-Polystyrene copolymers		Au/Se	SW	Au	
Metal-free PHTHC		Ag	Subl. film		
Metal-free PHTHC (x-form) in PVK binder		Al	SW	SnO$_2$	
PHTHC	S, 3 times	Au	SW	Au	+
PHTHC	S, 3 times	Al	SW	Au	+
Fe-PHTHC		Ag	Subl. film		
Ni-PHTHC		Ag	Subl. film		
Zn-PHTHC		Ag	Subl. film		

Temp. (°C)	Maximum photo-effect (eV) (on open circuit)	Sign of majority carrier	Remarks	Ref.
	0.8 µV		$\tau = 5.6$ msec threshold 8500Å	2327
	200–500 250 300		Power $\zeta \approx 10^{-3}$%. For other second contacts, e.g., Cr, Cu, Ni, Ag, Ga, Hg, In(60–40%), see original ref.	2328
21	I_{Phot} on short circuit $\cong 10^{-11}$ 10^{-12} A/cm²			2313
21	150		Monolayer; $Q = 2 \times 10^{-4}$ e^-/hr; Conversion (power) efficiency 4×10^{-5}%	2329
21	270		$Q = 0.002$; Power conversion efficiency $= 4 \times 10^{-4}$%	2329
	250		Threshold 0.7 µm at 0.52 µm	2330
	up to 300		at 0.485 µm	2330
	50		at 5000Å, 10^{15} $h\nu$/cm² sec	2331
	10.6		$\tau = 25.3$ µsec threshold 4.0 eV	2332
	1100		Power $\zeta > 6$% at 0.06 W/m², 670 nm; at 1400 W/cm² ζ drops to 0.01%	2333
25	310	P		2047
25	390	P		2047
	0			2332
	11		$\tau = 43.2$ µsec threshold	2332
	0.01		$\tau = 18.4$ µsec	2332

TABLE 8.16A1 *Continued*

Substance	Purification	Substrate	Cell type	Second contact	Polarity of substrate
Zn-PHTHC		Au	SW	In	+
Pentacene		Au	SW		
Polyvinylcarbazole: trinitrofluorenone (1:1)	Au	SW cart film	SnO		
Polyvinylcarbazole: trinitrofluorenone (1:1)		Au	SW cast film	SnO	
Polyvinylcarbazole: I_2 Complex		SnO	SW	SnO	
Polysulfurnitride		GaAs	SW		
Rose bengal					
Tetracene		Al	Thin film SW	Au	—
Tetracene		Al	Thin film		
			SW		
Tetramethyl-*p*-phenylenediamine in CH_3CN (Complex)			SW		

Temp. (°C)	Maximum photo-effect (mV) (on open circuit)	Sign of majority carrier	Remarks	Ref.
20	Quantum yield = 14% $\zeta < 0.023$ % 800		at 6328 Å	2334
	17.5 μV		$\tau = 20$ msec threshold 8300 Å	2327
	80 $I_{sh} = 8 \times 10^{-11}$ A/5 cm²		ζ at 6250 Å $< 10^{-3}$ %	2335
20	5 to 15		Linear to 140 mW/cm² above this $\propto \sqrt{\text{Illum.}}$	2079
	700			2135
	up to 300		at 0.565 μm	2330
	600 $I_{sh} = 7 \times 10^{-8}$ A $I_{sh} \propto$ illum.$^{0.6}$ to ^{1}	$\mu = 0.5$ cm²/Vsec	Max power $\zeta = 10^{-4}$ %	2336
	12 μV		$\tau = 10$ msec threshold 5200 Å	2327
	$I_{Phot} \cong 10^{-6}$ A/cm²; 40			2337

TABLE 8.16A2 Photo-EMF Supplementary Data. *See also Text Tables 22.5 and 22.6*

Substance	Purification	Substrate	Cell type	Second contact	Polarity of substrate
Anthracene Sol. in Tetrahydrofuran, 10^{-1} M.	Z, R	SnO			
Bacterio-Rhodopsin on Collodion/Phospholipid			SW film		Rhodopsin +
Bisphenol-A-Polycarbonate, N-Isopropylcarbazole. Doped (40%).	R	Ni	SW, film	Ni	
Chlorophyll-a or -b		SnO_2	Photo electrochem.		
Chlorophyll a: Stearic Acid		SnO_2	SW		
Mg-Chlorophyll		Pt	Photo electrochem.		+
Mn-Chlorophyll		Pt	Photo electrochem.		—
Mg-Chlorophyll		Pt	Photo-electrochem.		+
Mn-Chlorophyll		Pt	Photo-electrochem.		—
Chloroplast extract plus biliprotein on black lipid membrane.			Photo electrochem.		
Chloroplast extract plus biliprotein on black lipid membrane phylocyanine doped.			Photo electrochem.		
Crystalviolet-CdS junction.			SU		
Cytochrome c/ Membrane/ Ubiquinone 0-Photosystem.		~10^{11} reaction centers per cm^2	SW, aq. solutions Light ⊥ to membrane		Cytochrome is + terminal
Dyes, Photoreducible + high valency metal ion			Photo-electrochem.		

Temp. (°C)	Maximum photo-effect (mV) (on open circuit)	Sign of majority carrier	Remarks	Ref.
	3	p	Anthracene positive ions, Dember Effect.	4034
			570 nm	4077, 4082
	yield ⩽ 0.2		Threshold 360 nm	4027
	12 mV at 415 nm 8 mv at 675 nm		Monolayer $I_{sh} = 7 \times 10^{-9}$ Acm^{-2} at 675 nm	4084
	⩽ 18 Quant. yield- ⩽ 15.7%		Mixed monolayer	4084
	118, $I_{sh} \cong 0.2$ μAcm^{-2}		Thin layer on liq. cryst. attached to Pt electrode	4094
	62			
	52		No liq. crystal	
	5			
	12.6			4061
	22.5			4061
	0.16			4036
25	50; $I_{sh} \cong 50 \times 10^{-9}$ A/cm^2		800 nm; synthetic lipid bilayer membrane	4019 2329
	79		at 53.3 mWcm^{-2} illum. at 147 mWcm^{-2} illum.	4072

TABLE 8.16A2 *Continued*

Substance	Purification	Substrate	Cell type	Second contact	Polarity of Substrate
Fe-Thiazine		Pt	Photo electro-chem.	SnO_2	
Furan-Quinone yellow pigment	S	Al	SW	NESA	+
Hydroxysquarylium	S	Ga or Hg	SW, film = 100″ 1000 Å	In_2O_3	
Lipid (artificial) membrane dye-sensitized			Photo-electro-chem.		1
Merocyanine		Al_2O_3	SW, CF 100–500 Å	Al	
Merocyanine	Sublimed film	Ag	SW	Al Ag	
Mesochlorine eb + Phenyl Hydrazine RedOx Agent		ZnO	Photo-electro-chem.	Pt	
Mesochlorine eb + Phenyl Hydrazine RedOx Agent, Chlorophyll doped		Zn	Photo-electro-chem.	Pt	
Methylene blue: Ni(II) Dithiolate 1:1 complex		Brass Ag	CP, SW	SnO_2	Illum. electrode-negative
Napthalene	S	Aquadag	SU	Aquadag	+ or − depending on conditions or prehistory
Octaethyl-porphine	Chromat.	Al	SW	Ag	−
Tetraphenyl porphine	Chromat.	Al	SW film 260 nm	Ag	−

Temp. (°C)	Maximum photo-effect (mV) (on open circuit)	Sign of majority carrier	Remarks	Ref.
	Power $\eta = 0.036\%$			4073
	I_{sh} 10^{-11} to 10^{-12} Acm^{-2}		Sublimed film	
	450; power conv. efficiency $= 0.1\%$ $I_{sh} \leqslant 1$ mA/cm^2		0.14 mWcm^{-2}; incident light. Quant. yield 2.3%	4029
	$\leqslant 25$		Light changes RedOx potential by 0.4 V. Light/dark internal resist. $\cong$ 2.5 to 2.8	4058
	1200; $I_{sh} = 1.8$ A/cm^2		Power conv. eff. $=$ 0.7%, sunlight. Quant. yield 90% at 420 nm	4088
	1.2, $I_{sh} = 1.8$ mA/cm^2 I_{Phot} max. at $\lambda >$ Absorpt. Max.		Power conv. eff. $=$ 0.7% at 78 mWcm^{-2} sunlight; threshold 650 nm	2336
	$I_{sh} = 3$ μAcm^{-2}		Quant. yield $\cong$ 0.1; at 680 nm	4063
	$I_{sh} = 5.8 \times 10^{-8}$ Acm^{-2}			
25	80	—	Dember voltage resistivity $=$ 10^9 Ωcm	4076
22	Quant. yield $= 10^{-7}$ electrons/$h\nu$ ~20		SC, light beam in, ab plane, sample ∥ to ab plane, peak at 410 nm	4009
20	600; $I_{sh} = 0.25$ μA/cm^{-2}		$R_{Dark} = 29.5$ MΩcm^{-2} (sunlight)	4080; 4083
20	600; $I_{sh} = 0.1$ μA/cm^{-2}		$R_{Dark} = 13.4$ MΩcm^{-2} (sunlight)	

TABLE 8.16A2 *Continued*

Substance	Purification	Substrate	Cell Type	Second contact	Polarity of Substrate
Pentacene		SnO_2	SW, Sublimed film from SC	Au	
Pheophytins			Photoelectrochem.		
Phthalocyanine, metal-free, doped, with excess phthalocyanine	Dispersed in polymer, washed	NESA	SW		
Cu-Phthalocyanine				Cu	+
Cu-Phthalocyanine	S	SnO_2	Photoelectrochem.		+
Cu-Phthalocyanine, O_2 doped			Photoelectrochem.		
Mg-Phthalocyanine		Al	SW	Ag	
Mg-Phthalocyanine		Ag		Al	
Mg-Phthalocyanine, O_2 doped	R, S Chroma.	Al, Ag	SW, 2500 Å film	Ag, Al	+
Phylocyanine			Photoelectrochem.		
Pinacyanole-CdS Junction			SU		

602

Temp. (°C)	Maximum photo-effect (mV) (on open circuit)	Sign of majority carrier	Remarks	Ref.
	>100	p	Threshold 2.2 eV	4010
	100, Quant. yield 0.05 at 750 nm		White light	4068
	630, carrier generation efficiency 16-39%			4007
	Quantum yield ~0.05		For 16 eV photons, electron free path ~11Å	4025
	30; $I_{Phot} \cong 10^{-7}$ Acm^{-2}	p	30Å film	4098
	$I_{sh} \cong 25\ \mu Acm^{-2}$		+600 mV vs. SCE at pH 1.2	2092
	0.7, Quantum eff. = 0.01%	n	Schottky barrier 0.6 eV and 250Å at 690 nm; = 0.1 cm^2	4018
	0.85, Quantum yield $\cong$ 0.01, $I_{sh} \sim 9 \times 10^{-8}$ A/cm^2		Non-linear at 690 nm	4028
20	700: I_{sh} = 0.15 to 0.20 A/w		At 670 nm; n-intrinsic-p-function	2086
	6.1			4061
	0.25 0.55			4036

TABLE 8.16A2 *Continued*

Substance	Purification	Substrate	Cell type	Second contact	Polarity of substrate
Polyacetylene		Cu	Photo-electro-chem. Free-standing film	Pt	
Poly(β-N Carbazo-ethyl)-L		In$_2$O$_3$	SW	Au	−
Polyquinoxaline, sens. W.					
Polyvinylcarbazole	R	Au	SW	SnO$_2$	+
Porphyrin Mn(III0		Pt	EF, Photo-electro-chem.		
Proflavine		Pt	Photo-electro-chem. Redox agent is ethylene diamine-tetra-acetic acid	Pt	
Pyrylium (type dye salt in insol. binder					
Rhodamine-B		Pt	Photo-electro-chem. Redox agent is ethylene diamine-tetra-acetic acid	Pt	

Temp. (°C)	Maximum photo-effect (mV) (on open circuit)	Sign of majority carrier	Remarks	Ref.
	300; $I_{sh} \cong 40\ \mu Acm^{-2}$		Na-Polysulfide electrolyte. Sunlight. Quant. yield 1% for 2.4 eV quanta.	4088
	$I_{sh} \sim 10^{-8}\ Acm^{-2}$, A/W of incident radiation	n	at 700 nm	4045
	Linear currents		ohmic	4066
	$I_{sh} \sim 5 \times 10^{-3}\ Acm^2$	p	EF 15 μ thick	4079
	$I_{Phot} \leq 2 \times 10^{-7}\ A/cm^2$. Power conv. $\eta < 0.01\%$ at 620 to 650 nm		10^{15} molecules on 0.2 cm^2 Pt aq sol. in tetraethylammonium perchlorate	4071
	190, $I_{sh} = 12\ \mu Acm^{-2}$		pH = 6; 500 W Xenon lamp	4089
	$\cong <800$, $I_{sh} \leq 1\ mAcm^2$		Power conv. eff. = 0.2%	4065
	28 $I_{sh} = 0.092\ \mu Acm^{-2}$		pH = 6; 500 W Xenon lamp	4089

TABLE 8.16A2 *Continued*

Substance	Purification	Substrate	Cell type	Second contact	Polarity of substrate
Rhodamine-B absorbed on SC ab-plane of: Perylene Anthracene Chrysene Phenanthrene Naphthalene		Electrolytic contact	SW	Electrolytic contact.	+
Rhodamine:Hydroquinone			Photoelectrochem. Redox agent: Ethylene-Diaminetetraacetic acid		
Riboflavine		Pt		Pt	
Rose-Bengal		Pt	Photoelectrochem. Redox agent: Ethylene-Diaminetetraacetic acid	Pt	
Tetracene		Al	SW	Au	
Tetraphenyl-porphine	S	Au	SW, Sub-film	aqueous	+
Zn-Tetraphenyl porphine		Pt	Photoelectrochem.	added acceptors, e.g. O_2	+
Zn-Tetraphenyl porphine			Photoelectrochem.		

Temp. (°C)	Maximum photo-effect (mV) (on open circuit)	Sign of majority carrier	Remarks	Ref.
	Quant, yield 7.5 × 10^{-3}, Light 3 µW		½ Monolayer absorbed 568.2 nm peak, 5 × 10^5 Vcm^{-1}	4095
	Quant, yield ~5 × 10, Light 1 mW			
	$I_{sh} \cong 10^{-6}$ to 10^{-8} Acm^{-2}		Peak 560 nm; 2000 W lamp	4093
	720 $I_{sh} = 58$ µAcm^{-2}		pH = 6; 500 W Xenon lamp	4089
	19 $I_{sh} = 0.025$ µAcm^{-2}		pH = 6; 500 W Xenon lamp	4089
			Schottky barrier 1.2 eV, threshold at 2640 Å	4017
	35		at 560 nm	4055
	530, $I_{sh} = 40$ µAcm^{-2}	p	0.1 N HCl, Quant. yield at 435 nm = 2%	4050 4050
	1100			4056

TABLE 8.16A2 *Continued*

Substance	Purification	Substrate	Cell type	Second contact	Polarity of substrate
Thionine: Alkylane Polyamine polymers					
Thionine: Fe II/III		Pt	Photo-electro-chem.	SnO_2	+
Thionine: Fe II/III Gel		Pt	Photo-electro-chem.	SnO_2	+
Thionine: Fe II/III + Na Acetate			Photo-electro-chem.		
2,4,7 Trifluorenone					
Trinitrofluorenone: Polyvinyl-carbazole (1:0.4)		Al		Se	$\mu_+ \cong \mu_- \cong 10^{-7}$ cm^2/Vsec
Triphenylamine dispersed in polycarbonate	Z, R	Ni	SW	Se	

Temp. (°C)	Maximum photo-effect (mV) (on open circuit)	Sign of majority carrier	Remarks	Ref.
	180			4078
	50 to 100 $I_{sh} \cong 10\ \mu Acm^{-2}$			4090
	163 $I_{sh} = 4.7\ \mu Acm^{-2}$		100 mWcm^{-2} light; pH = 2	4091
	230		pH = 4	4096
	at low illumin.		ohmic	4066
25	Prim. Quantum		3×10^{14} photons cm^{-2} sec^{-1}, 345 nm, 1 mV/cm	4026
−45 to 50	Quantum yield 0.7 at 0.64 mVcm^{-1}	p	$\mu_{+} \cong f(\text{field}) \cong 10^{-5}$ to 10^{-7} cm^2/Vsec. Se is sensitizer.	4030

TABLE 8.20A Glass Transition Temperatures Evaluated from Glow Curves Compared to Literature Values. *After J. van Turnhout, Polymer J., 2, 173 (1971).*

The experimental values were obtained dilatometrically. For a discussion of glow curves and their application to the location of transition temperatures T_g, see Sections 10.3, 16.4 and 19.3.

Polymer	T_g (°C)	Literature Values (°C)
PcHMA	83	90
PEMA	66	65
PMMA	106	105
PPhMA	105	105
PET	88	81
PC-n	152	149
89 TFE co 11 HFP	75	77
PVC	69	68

Names of polymer used

PEMA Polyethylmethacrylate
PTFE Polytetrafluoroethylene
PHFP Polyhexylfluorophthalate
For others see Table 10.3A

TABLE 8.21A Donicities DN and Electron Acceptance Numbers AN of Organic Compounds. *After V. Gutmann, Chimia,* **31**, *1 (1977).*

Solvent	DN	AN
Acetic Acid	—	52.9
Acetone (AC)	17.0	12.5
Acetonitrile (An)	14.1	19.3
Acetylchloride	0.7	—
Benzene	0.1	8.2
Benzonitrile (BN)	11.9	15.5
Benzoylchloride	2.3	—
Carbontetrachloride	—	8.6
Chloroform	—	23.1
Diethylether	19.2	3.9
Dichloroethylene carbonate	3.2	16.7
Diglyme	≈ 24.0	10.2
Dimethylacetamide (DMA)	27.3	13.6
Dimethylformamide (DMF)	24.0	16.0
Dimethylsulfoxide (DMSO)	29.8	19.3
Dioxane	14.8	10.8
Ethanol	19.0	37.1
Ethylenesulfite (ES)	15.3	—
Hexamethylphosphorotriamide (HMPA)	38.8	10.6
Methanol	20.0	41.3
Nitrobenzene	4.4	14.8
Nitromethane	2.7	20.5
Propanol	18.0	33.5
Propylene carbonate (PDC)	15.3	18.3
Pyridine (py)	33.1	14.2
Tetrahydrofuran	20.0	8.0
Tributylphosphate	23.7	—
Water (gas)	18	54.8

TABLE 8.21B Supplementary Table of Selected Donicities *DN* vs. Antimonypentachloride in 1,2-Dichloroethane as Solvent. *After V. Gutmann, Chimia, 31, 1 (1977)*

Compound	DN
Ammonia	59.0
Benzylcyanide	15.1
iso-Butyronitrile	15.4
t-Butyronitrile	57.0
n-Butyronitrile	16.6
1,2-Dichlorethane	0.1
N,N-Diethylacetamide	32.2
N,N-Diethylformamide	30.9
Dimethoxyethane	24
Diphenylphosphonic chloride	22.4
Ethylamine	55.5
Ethylacetate	17.1
Ethylenecarbonate	16.4
Ethylenediamine	55
Hydrazine	44
Methylacetate	16.5
N-methyl-ϵ-caprolactame	27.1
N-methyl-2-pyrrolidone	27.3
Phenylphosphonic dichloride	18.5
Phenylphosphonic difluoride	16.4
Propionitrile	16.1
Sulpholane	14.8
Sulfurylchloride	0.1
Tetrachloroethylene carbonate	0.8
Thionylchloride	0.4
Trimethylphophate	23.0
Triethylamine	61.0
Water (Liquid)	33

TABLE 8.21C Supplementary Table of Electron Acceptance Numbers *AN* vs. Hexane as the Reference Solvent. *After V. Gutmann, Chimia,* **31**, *1 (1977)*

Solvent	AN
CF_3COOH	105.3
CF_3SO_3H	129.1
CH_3SO_3H	126.1
Diethylether	3.9
Dioxane	10.8
Formamide	39.8
Hexane (ref. solvent)	0
N-methylpyrrolidinone	13.3
$SbCl_5$ in Dichloroethane	100.0

TABLE 8.22A The Dielectric Constants and Conductivities of Homogeneous Solids Showing Nomadic Polarizations. After Pohl and Pollak.[2282]

	Dielectric constant ϵ_r	$T(°C)$	Pressure (kbars)	Frequency (Hz)	Resistivity (dc, Ω-cm)
Selenium (trigonal single crystal)	1 000	27	10^{-3}	1000	...
$Li_2SO_4 \cdot H_2O$	380; 105	20	10^{-3}	0.26; 10	$10^5 - 10^6$
$[Li^+NH_2-NH; \cdots SO_4^-]$	$10^3 - 10^4$	27	10^{-3}	10	
PAQR, (Acridone–PMA; JM-41)	70	25	6.3	100	1.1×10^7
PAQR, (Acridone–MTA; JM-61)	80	25	6.3	100	6.5×10^7
PAQR, (Thianthrene–PMA; JM-39)	1 900	25	6.3	100	9×10^4
PAQR, (Thianthrene–MTA; JM-51)	123	25	6.3	100	4×10^6
PAQR, (9-Thioxanthene–PMA; JM-48)	2 700	25	6.3	100	5.3×10^6
PAQR, (Carbazole–MTA; JM-65)	2 200	25	6.3	100	9×10^5
PAQR, (Carbazole–MTA; JM-65)	300 000	101	6.3	30	9×10^5
PAQR, (Xanthene–MTA; JM-64)	190	25	6.3	100	6×10^7

PAQR, (Anthraquinone–PMA; DPIA)	5 030	23	3.4	300	7.1×10^4
$(CS_2)_z$ (Type B)	20 000	27	10^{-3}	100	5×10^2 to $\sim 7 \times 10^4$
Poly[Cu(II)–N,N'-dimethylrubeanate]	1 520	27	1.37	1000	2×10^4
Poly[Cu(II–N,N'-di(2-bydroxyethyl)-rubeanate]	100	24	31.5	100	4.2×10^{10}
Quinazone polymer (α-form)	12	27	11	100	10^{10}
Quinazone polymer (β-form)	500	67	11	100	5×10^8
N-ethyl phenothiazine: I_2	500	50	10^{-3}	11.5 GHz	~ 2 (dc)
Pyropolymer (undoped, made at 600°C)	10 000	24	28	100	10^7
Pyropolymer (Ca-doped; made at 600°C)	630	24	4.2	1000	$\sim 2 \times 10^3$
Poly[Ni(II) quinoxaline-2,3 dithiol] complex	180	20	10^{-3}	100	10^8

TABLE 9.1A Minimum Resistivities of Phthalocyanine under Pressure. After A. Onodera, N. Kawai and T. Kobayashi, Solid State Commun., 17, 775 (1975)

Phthalocyanine	Minimum resistivity (ohm cm)	Pressure (kbar)	Deviation (Å)
VOPc	0.2	290^a	$<2.6^1$
ThPc$_2$	0.2	330^a	1.45^2
SnPc$_2$	2×10	360^a	1.35^3
PbPc	2	490^a	0.4^4
CuPc	4	620^b	
ZnPc	1×10	640^b	
H$_2$Pc	4×10	570^a	
CuPcCl$_{16}$	8×10^5	520^b	

Pc: Phthalocyanine
VO: Vanadyl
a: Pressure for Resistivity Minimum
b: Highest pressure applied
The deviation values refer to the departure of the central metal atom from the central plane of the molecule.

References:
1. J. M. Assour, J. Goldmacher and S. E. Harrison, J. Chem. Phys., 43, 159 (1965).
2. T. Kobayashi and N. Uyeda, Abst. Proc. Sympos. Struct. of Molecules, Sendai, Japan 1972, p. 119.
3. W. E. Bennett, D. E. Droberg and N. C. Baenziger, Inorg. Chem., 12, 930 (1973).
4. K. Ukey, Acta Cryst., B29, 2290 (1973).

TABLE 9.2A   Sensitization of Photoconductivity of Polynaphthalene, Polyvinylcarbazole and Polyacenaphthalene. *After Hayashi et al.*[4157]

$E_{1/2}$ refers to the electrographic sensitivity; it is termed the half decay exposure: the specimen is positively charged in the dark to a saturation potential V_0 from a corona discharge. Light is then applied and the loss of charge is measured in terms of the potential across the sample. $E_{1/2}$ is that exposure which produces a charge decay to 1/2 of the original saturation value.

	Sensitizations with a Dye and/or Lewis Acids	
	Without dye $E_{1/2}$ (lux. sec)	With crystal violet $E_{1/2}$ (lux. sec)
Acids and acid anhydrides		
Acetic acid	none	217
Monochloroacetic acid	10500	49
Dichloroacetic acid	5880	46
Trichloroacetic acid	3745	26
Monobromoacetic acid	11900	75
Tribromoacetic acid	13850	553
Benzoic acid	none	196
Tetrachlorophthalic acid	1595	112
Tetrachlorophthalic anhydride	3910	110
Succinic anhydride	39200	142
Maleic acid	7175	84
Maleic anhydride	11270	166
Terephthalic acid	none	350
Cinnamic acid	5250	143
Phthalic acid	4100	73
Phthalic anhydride	4230	85
Acetanilide	none	80
Benzenesulfonic acid	none	78
p-Toluenesulfonic acid	none	85
Phenol	none	2170
p-Nitrophenol	5950	85
Picric acid	650	36
Ketones		
Acetophenone	none	287
Benzoine	11600	245
Benzil	6800	270
Aldehydes		
p-Nitrobenzaldehyde	1030	287
o-Nitrobenzaldehyde	1350	325
2,4-Dichlorobenzaldehyde	none	441
p-Dimethylaminobenzaldehyde	none	280
p-Hydroxybenzaldehyde	none	860
Furfural	none	175
Bromal	900	77

TABLE 9.2A *Continued*

	Sensitizations with a Dye and/or Lewis Acids	
	Without dye $E_{1/2}$ (lux. sec)	With crystal violet $E_{1/2}$ (lux. sec)
Quinones		
p-Quinone	2850	1103
p-Chloranil	1100	560
α-Naphthoquinone	1295	297
Anthraquinone	577	175
1-Aminoanthraquinone	1190	332
2-Aminoanthraquinone	490	280
1,4-Diaminoanthraquinone	9325	9975
1-Nitroanthraquinone	297	140
1,5-Dichloroanthraquinone	402	241
2-Chloroanthraquinone	595	260

	Sensitizations by Dyes	
	V_0, (V)	$E_{1/2}$ (lux. sec)
Phenylmethanes		
Victoria pure blue	1660	6500
Victoria blue	1540	>100000
Crystal violet	1520	>100000
Brilliant green	1050	20000
Acid violet 6G	1720	73000
Fuchsin	2280	27000
Basic cyanine BX	1760	11000
Malachite green	1720	9000
Auramine	1520	73000
Naphthalene green	1520	93000
Xanthenes		
Rhodamine 6G	940	6000
Rhodamine B extra	1250	22000
Fluorescein	2140	40000
Sulforhodamine B	1590	73000
Eosine A	2330	61000
Eosine S	1950	49000
Phloxine	1840	64000
Rose Bengale	1080	>100000
Azines		
Neutral red	1400	46000
Phenosafranine	2200	38000
Pinacryptol green	1900	62000
Selestine blue	1720	>100000

TABLE 9.2A *Continued*

Sensitization by Dyes (*Continued*)		
	V_0, (V)	$E_{1/2}$ (lux. sec)
Acridines		
Acridine yellow	1840	76000
Acridine orange	1680	81000
Pinacryptol yellow	2080	>100000
Thiazines		
Methylene blue	1400	1500
New methylene blue	1340	14000
Cyanines		
Garocyanine	1800	>100000
Chinolin blue	1920	67000
Pinacyanole	450	>100000
Azo dyes		
Oill red B	1640	>100000
Chrom blue black RC	2080	54000
Butter yellow	1840	41000
Fast light yellow G	1680	74000

Sensitization of Polyacenaphthalene by Lewis Acids		
	V_0, (V)	$E_{1/2}$ (luc. sec)
Acids and acid anhydrides		
Acetic acid	1535	>100000
Monochloroacetic acid	1550	>100000
Dichloroacetic acid	1040	>100000
Trichloroacetic acid	1340	11000
Monobromoacetic acid	1020	>100000
Tribromoacetic acid	1360	21000
Phenol	1245	>100000
p-Nitrophenol	1000	6000
Picric acid	1320	6000
Hydroquinone	1380	27000
Tetrachlorophthalic acid	960	7000
Tetrachlorophthalic anhydride	1310	7000
Benzoic acid	1500	82000
2,4,6-Trinitrobenzoic acid	1350	4000
Telephthalic chloride	1070	82000
Phthalic anhydride	1500	75000
Cinnamic acid	1630	80000
Acetanilide	1520	>100000
Oxalic acid	1120	92000

TABLE 9.2A *Continued*

Sensitization of Polyacenaphthalene by Lewis Acids (*Continued*)

	V_0, (V)	$E_{1/2}$ (lux. sec)
Acids and acid anhydrides (*cont.*)		
Succinic anhydride	1350	>100000
Maleic acid	1190	60000
Maleic anhydride	830	56000
Fumaric acid	1110	>100000
o-Toluic acid	1535	>100000
p-Aminobenzoic acid	1400	>100000
Aldehydes and Ketones		
p-Hydroxybenzaldehyde	1480	>100000
2,4-Dichlorobenzaldehyde	1480	>100000
p-Nitrobenzaldehyde	1560	98000
p-Dimethylaminobenzaldehyde	1510	>100000
Bromal	1400	>100000
o-Nitrobenzaldehyde	1535	>100000
Acetophenone	1680	>100000
Benzophenone	1550	>100000
Benzil	1480	>100000
Benzoin	1610	>100000
Quinones		
α-Naphthoquinone	850	18500
2,3-Dichloronaphthoquinone	1090	1400
1-Nitoroanthraquinone	1420	1200
2,6-Dichloro-*p*-benzoquinone	1390	4000
2-Chloroanthraquinone	1390	1900
2-Methylanthraquinone	1110	11000
p-Benzoquinone	1320	36000
p-Chloranil	1520	1800
Anthraquinone	1210	8800
1,5-Dichloroanthraquinone	1100	1500
Anthraquinone-2-carboxilic acid	970	4000
Benzanthron	1470	>100000
α-Benzanthraquinone	1600	5000
1-Aminoanthraquinone	1330	>100000
2-Aminoanthraquinone	350	>100000
1,4-Diaminoanthraquinone	950	>100000
Alizarine	1350	>100000
Quinizarine	1670	4100
Others		
5-Nitroacenaphthylene	1535	>100000
Trinitrobenzene	1170	2400

TABLE 9.2B Sensitization of the Photoconductivity of Polyvinylcarbazole. After Hayashi et al.[4157]

	Sensitization with a Dye and/or Lewis Acids	
	Without dye $E_{1/2}$ (lux. sec)	With crystal violet $E_{1/2}$ (lux. sec)
Acids and acid anhydrides		
Acetic acid	none	217
Monochloroacetic acid	10500	49
Dichloroacetic acid	5880	46
Trichloroacetic acid	3745	26
Monobromoacetic acid	11900	75
Tribromoacetic acid	13850	553
Benzoic acid	none	196
Tetrachlorophthalic acid	1595	112
Tetrachlorophthalic anhydride	3910	110
Succinic anhydride	39200	142
Maleic acid	7175	84
Maleic anhydride	11270	166
Telephthalic acid	none	350
Cinnamic acid	5250	143
Phthalic acid	4100	73
Phthalic anhydride	4230	85
Acetanilide	none	80
Benzenesulfonic acid	none	78
p-Toluenesulfonic acid	none	85
Phenol	none	2170
p-Nitrophenol	5950	85
Picric acid	650	36
Ketones		
Acetophenone	none	287
Benzoine	11600	245
Benzil	6800	270
Aldehydes		
p-Nitrobenzaldehyde	1030	287
o-Nitrobenzaldehyde	1350	325
2,4-Dichlorobenzaldehyde	none	441
p-Dimethylaminobenzaldehyde	none	280
p-Hydroxybenzaldehyde	none	860
Furfural	none	175
Bromal	900	77
Quinones		
p-Quinone	2850	1103
p-Chloranil	1100	560
α-Naphthoquinone	1295	297

TABLE 9.2B *Continued*

Sensitization with a Dye and/or Lewis Acids (*Cont.*)

	Without dye $E_{1/2}$ (lux. sec)	With crystal violet $E_{1/2}$ (lux. sec)
Quinones (Cont.)		
Anthraquinone	577	175
1-Aminoanthraquinone	11190	332
2-Aminoanthraquinone	490	280
1,4-Diaminoanthraquinone	9325	9975
1-Nitroanthraquinone	297	140
1,5-Dichloroanthraquinone	402	241
2-Chloroanthraquinone	595	260
2-Methylanthraquinone	621	186
2,6-Dichloro-*p*-benzoquinone	665	224
1,2-Benzanthraquinone	297	200
Anthraquinone-2-carboxylic acid	630	149
Anthraquinone-2-sulfonic acid	none	190
Polynitrate		
Trinitrobenzene	435	252

Sensitization with Dyes

Dye	Sensitivity, $E_{1/2}$ (lux. sec)	
Triphenyl Methanes		
Victoria pure blue		1137
Victoria blue	612	612
Crystal violet		350
Briliant green		420
Methyl violet		1015
Acid violet 6B		10850
Fuchsine		none
Basic cyanine BX		612
Basic cyanine 6G		4550
Malachite green		490
Auramine		13300
Napthalene green		none
Xanthenes		
Rhodamine 6G		657
Acid rhodamine G		none
Rhodamine B extra		725
Fluorescein		none
Rose bengale		none
Sulfurhodamine B		4290
Eosine A		6300
Eosine S		5250
Phloxine		15750

TABLE 9.2B *Continued*

Sensitizations with Dyes *Continued*	
Dye	Sensitivity, $E_{1/2}$ (lux. sec)
Azines	
Phenosafranine	6650
Pinacryptol green	13300
Selestine blue	none
Acridines	
Pinacryptol yellow	2800
Acridine orange	3355
Tripafravine	5900
Triazines	
Methylene blue	1750
New methylene blue	402
Cyanines	
Chinoline blue	42000
Pinacyanole	none
1,1'-Diethyl-4,4'-quinocarbocyanine iodide	40000
1,1'-Diethyl-4,4'-quinocyanine iodide	22000
3,3'-Diethyl-thiacarbocyanine iodide	8250
3,3'-Diethyl-2,2'-thiacyanine iodide	15500
2-(*p*-Dimethylaminostyryl)-benzothiazole-ethiodide	4650
1,1'-Diethyl-2,4'-quinocyanine iodide	16000
Azo dyes	
Oil red B	14000
Chrom blue black RC	none
Butter yellow	14000
Fast light yellow G	none
Neutral red	8205
Anthraquinones	
Quinizarin	1260
Alizarin	9800

TABLE 10.2A Piezoelectric Substances. *With permission, D. Vasilescu et al., Nature (London), 225, 635 (1970)*

The + sign indicates that the compound exhibits piezoelectricity, and the − sign that it is non-piezoelectric.

Comparison of Experimental and Theoretical Piezoelectricity for Amino-Acids

Amino-acid	Spatial group	Crystalline class	Theoretical piezoelectricity	Experimental piezoelectricity
L-Alanine	$P2_1 2_1 2_1$	222	+	+
D-Alanine				+
DL-Alanine	Pba	mm2	+	+
β-Alanine	Pbca	mmm	−	−
L-Arginine·$2H_2O$	$P2_1 2_1 2_1$	222	+	−
D-Arginine·HCl				+
L-Asparagine·H_2O	$P2_1 2_1 2_1$	222	+	+
L-Aspartic acid	$P2_1$	2	+	+
DL-Aspartic acid	I2/a	2/m	−	−
L-Cysteine	$P2_1$	2	+	+
L-Cystine	$P6_1 22$	622	+	+
DL-Cystine				+
L-Glutamine	$P2_1 2_1 2_1$	222	+	−
L-Glutamic acid	$P2_1 2_1 2_1$	222	+	−
D-Glutamic acid				−
Glycine				+
L-Histidine				+
DL-Histidine				−
L-Hydroxyproline	$P2_1 2_1 2_1$	222	+	+
L-Isoleucine				+
D-Isoleucine				+
L-Leucine				+
D-Leucine	$P2_1 2_1 2_1$	222	+	−
DL-Leucine	P1	1	+	−
L-Lysine·HCl·$2H_2O$	$P2_1$	2	+	+
L-Methionine				−
DL-Methionine				−
L-Phenylalanine				−
D-Phenylalanine				−
L-Proline	$P2_1 2_1 2_1$	222	+	+
DL-Proline				−
L-Serine				+
DL-Serine	$P2_1/a$	2/m	−	+
L-Threonine	$P2_1 2_1 2_1$	222	+	+
DL-Threonine		mm2	+	+
L-Tryptophan				−
DL-Tryptophan				−
L-Tyrosine				+
DL-Tyrosine				+
L-Valine	$P2_1$	2	+	−
D-Valine				+
DL-Valine	P1 or P$\bar{1}$	1 or $\bar{1}$	+ or −	−

TABLE 10.2B Piezoelectric Nucleic Acids. *D. Vasilescu, Private Communication (1971)*

Substance	Piezoelectric coefficient
Adenosine	-320×10^{-6}
Cytidine	-350×10^{-6}
Cytidylic Acid	-156×10^{-6}
Desoxyadenosine	-540×10^{-6}
Deoxyuridine	-610×10^{-6}
DNA (calf thymus)	-49×10^{-6}
RNA (E. Coli)	-16×10^{-6}
Thymidine	-280×10^{-6}
Uridine	-250×10^{-6}

TABLE 10.2C Piezoelectric Temperature Coefficients of Various Substances. *After F. W. Cope, J. Biol. Phys., 3, 11, (1975)*

Sample	$\Delta(\times 10^{-6})(°C)^{-1}$	Reference
Quartz	6	1
Zinc sulfide	46	1
Potassium sodium tartrate	400	1
1-Chloramphenicol	100	2
DNA (thymus)	50	3
DNA (sperm)	56	3
DNA (yeast)	18	1
RNA (yeast)	6	1
Myosin	85	3
Actomyosin	100	3
Collagen	47	3

The constant Δ is characteristic of the material and is calculated from measurements of the frequency shifts of piezoelectric resonance absorption peaks as a function of temperature.

1. S. V. Toulsky, A. K. Kuhushkin and L. A. Blumenfeld, "Molecular Biophysics," G. M. Frank, ed., Nauk, Moscow, 1965, p. 41 ff.
2. J. Duchesne and A. Monfils, *J. Chem. Phys.*, 23, 762 (1955).
3. J. Duchesne and A. Monfils, *Nature* (London), 188, 405 (1960).

TABLE 10.3A Reduced Heterocharge for Various Polymer Electrets.
After J. van Turnhout, Polymer J., 2, 173 (1971).

The charges are given in terms of "reduced density" $\sigma/\epsilon_0 E$; where σ is the observed charge density, ϵ_0 the low frequency permittivity and E the polarization field.

Homo-polymer	$\sigma/\epsilon_0 E$	Copolymer	$\sigma/\epsilon_0 E$
PcHMA	1.8	80 MMA co 20 AN	14.4
PMMA	4.0	70 MMA co 30 AN	26
PPhMA	3.1	60 MMA co 40 tbMAm	12.3
PC-n	0.8	60 MMA co 40 cHMA	9.9
PET	1.0	80 MMA co 20 diMit	14.3
PVC	16.1	50 MMA co 50 PhMA	4.3
		60 MMA co 40 S	13.8
		80 S co 20 AN	82

Names of the Polymers used	
PAN	polyacrylonitrile
PtBMAm,	poly (t-butylmethacrylamide)
PC-n,	polycarbonate (Makrofol-n)
PET,	poly (ethylene terephthalate) (Mylar)
PcHMA,	poly (cyclohexyl methacrylate)
PdiMIt,	poly (dimethyl itaconate)
PMMA,	poly (methyl methacrylate)
PPhMA,	poly (phenyl methacrylate)
PS,	polystyrene
PVC,	poly (vinyl chloride)

TABLE 10.4A Dielectric Constants and Conductivities of Organic Homogeneous Solids Showing Nomadic Polarization. After H. A. Pohl, "Organic Semiconductors," Quantum Theoretical Research Group, Oklahoma State University, Stillwater, Okl. Res. Note 35, April 1973, p. 43; see also J. Biol. Phys., 2, 169 (1974).

	Dielectric constant ϵ_r	$T(°C)$	Pressure (kbars)	Frequency (Hz)	Resistivity (dc, Ω-cm)
PAQR, (Acridone-PMA[a]; JM-41)	70	25	6.3	100	1.1×10^7
PAQR, (Acridone-MTA[b]; JM-61)	80	25	6.3	100	6.5×10^7
PAQR, (Thianthrene-PMA; JM-39)	1900	25	6.3	100	9×10^4
PAQR, (Thianthrene-MTA; JM-51)	123	25	6.3	100	4×10^6
PAQR, (9-Thioxanthene-PMA; JM-48)	2700	25	6.3	100	5.3×10^6
PAQR, (Carbazole-MTA; JM-65)	2200	25	6.3	100	9×10^5
PAQR, (Carbazole-MTA; JM-65)	300,000	101	6.3	30	9×10^5
PAQR, (Xanthene-MTA; JM-64)	190	25	6.3	100	6×10^7
PAQR, (Anthraquinone-PMA; DPIA)	5030	23	3.4	300	7.1×10^4

Poly[Cu(II)-N,N-dimethyl rubeanate]	1520	27	1.37	1000	2×10^4
Poly[Cu(II)-N,N'-di(2-hydroxyethyl)-rubeanate]	100	24	31.5	100	4.2×10^{10}
Quinazone polymer (α-form)	12	27	11	100	10^{10}
Quinazone polymer (β-form)	500	67	11	100	5×10^8
Pyropolymer (undoped, made at 600°C)	10,000	24	28	100	10^7
Pyropolymer (Ca-doped; made at 600°C)	630	24	4.2	1000	$\sim 2 \times 10^3$
Poly[Ni(II)quinoxaline-2,3 dithiol] complex	180	20	10^{-3}	100	10^8

[a] PMA-pyromellitic dianhydride
[b] MTA-mellitic trianhydride

TABLE 10.5A   List of all Tables in Parts A and B

Quanty Listed	Table	Part A Page
Adsorption, effects of	3.7 to 3.9	201, 207 209
Band widths	4.4	239, 241
Band splitting	4.4	240
Biological materials, conductivity data	8.11	758
Carrier concentrations, calculated from steady state assumptions	6.12	380
Charge transfer complexes, conductivity data	5.2, 8.5	667, 720
Chromatographic methods, substances suitable for purification by	3.3	148
Complex metal compounds, conductivity data	8.4	718
Conductivity data	3.5, 3.6, 8.2 to 8.8, 8.11 to 8.13	174, 178, 706, 758
Conductivity parameters, typical values	8.1	449
Conductivity to mobility ratios	6.11	361
Contact potentials	8.17	796
Diaminodurene-chloranil complex, conductivity properties of single crystal and microcrystalline samples of	3.5	174
Dipole induction, work reqired	6.2	340
DPPH, resistance data for single crystals and films of	3.6	178
Dye films, properties	8.10	490
Dye conductivity data	8.8	750
Electret-forming substances	10.3	618
Electron affinities, absolute	A_G, 6.8; A_G, 6.10	698, 704
Electron affinities ΔA_G, relative to benzene	6.9	703
Energies of lowest optically observed excited states	5.1, 5.2	653, 667
Ferroelectric materials, properties of various	10.1	607
Free radicals and radical salts, conductivity data	8.6	732
Glasses, conductivity data	8.12	766
Impurity determinations, estimated sensitivities of different methods	3.1	127
Intermolecular resonance integrals	4.2, 4.3	237
Ionization energies I_G, I_c	6.6, 6.7	669, 693
Liquids, conductivity data	8.12, 8.13	766, 771
Long chain compounds, conductivity data	8.7	736
Melts, conductivity data	8.13	771
Mobility as a function of pressure, anthracene	4.9	253

TABLE 10.5A (*Continued*)

Quantity Listed	Table	Part A Page
Mobilities in		
Anthracene	4.7	250
Naphthalene	4.7	250
Other organic solids	4.8	251
Molecular crystals, conductivity data on		
metal-free	8.2	706
Organic materials, mean values of data	8.18	797
Overlap integrals	4.1	232
Phthalocyanine, metal-free β-, conductivity data	8.3	717
Photoconductivity data for		
Anthracene	6.13	397
Organic materials	8.15	776
Photoconductivity		
Results of Simpson's model	6.14	400
Results of Murrell's nonexciton model	6.15	401
Results of Murrell's exciton model	6.16	403
Photoconductors, oxygen and vacuum	8.9	487
Photo-emf's	8.16	790
Photoemission thresholds I_c	6.7	693
Piezoelectrics, list of	10.2	614
Polarizabilities	6.1	339
Polarization energy, calculated values of		
terms	6.4	342
Polarization energy, equated to $I_G - I_c$	6.5	345
Polarization energy, individual molecule		
contribution for naphthalene crystals	6.3	341
Polymers, conductivity data	8.7	736
Seebeck coefficients	8.14	772
Sublimation methods, substances purified by	3.4	153
Velocity component functions	4.5, 4.6	241
Zone refining, substance suitable for	3.2	135

Quantity Listed	Table	Part B Page
Acceptor-doped metal-free phthalocyanine	22.3	435
Activation energies of organic substances	4.9A	492
Affinities, proton	8.5D	530
Anion radical salts derived from tetrabromo-p-		
diphenoquinone	8.6C	543
Bandwidths vs pressure	15.3	129
Biological materials	8.11A	556
Biological materials, supplemental	8.11B	563
Complexes, charge transfer	8.5A	520

TABLE 10.5A (*Continued*)

Quantity Listed	Table	Part B Page
Complexes, some metal	8.2B	520
Complexes, proton and protonated complexes	8.5C	529
Complexes, ternary and inclusion complexes	8.5B	526
Conduction and dielectric activation energies	15.8	150
Dielectric constants and conductivities of organic homogenous solids showing nomadic polarization	10.4A	628
Dielectric constants and conductivities of homogenous solids showing nomadic polarization	8.22A	614
Donicities and electron acceptance numbers of organic compounds	8.21A	611
Donicities and electron acceptance numbers of organic compounds, supplementary	8.21B	612
Electrical conductivity of quasi-one-dimensional organic conductors	21.1	402–411
Electrical properties of semiconducting anion-radical salts derived from tetrachloro-*o*-diphenoquinone	8.6B	542
Electrical properties of $7,7',8,8'$-tetracyanoquinodimethane salts of ionene polymers and their model compounds	8.7F	552
Electrical resistivity, tetrathiotetracene	12.1	51
Electrical resistivity of phthalocyanines under pressure	19.1	362
Electrical room temperature resistivities of PAQR polymers and copolymers	8.7D	550
Electrical resistivities and activation energies for ten PAQR polymers after precompression to 32 kbar at 450 K, evaluated at 1.82 kbar and 300 K	8.7E	551
Electron affinities of organic materials, absolute	6.8A	504
Electron acceptance numbers vs hexane as the reference solvent, supplemental	8.21C	613
Electron affinities of acceptors, and CT band maxima of methiodides	6.8F	510
Electron affinities of acceptors, and excited state energies of their atoms	6.20A	516
Electron affinity, group contributions to	6.8D	519
Electron affinity of organic compounds	6.10A	513
Electron affinities of the elements	6.17A	513
Electron affinities of organic molecules	6.8C	506

TABLE 10.5A (*Continued*)

Quantity Listed	Table	Part B Page
Electron affinities, polarographic	6.8E	519
Electron affinities, relative and absolute electron	6.9A	510
Electron affinities of substituents	6.18A	514
Electron transfer integrals for perylene and anthracene	15.7	143
Energy gaps and related quantities in organic solids	15.12	161
Energy parameters of aromatic hydrocarbons	22.7	452
Energy transfer, triplet-triplet to napthalene	16.2	230
Glass transition temperatures evaluated from flow curves compared to literature values	8.20A	610
Intermolecular resonance integrals for perylene	15.6	142
Intermolecular resonance integrals at different pressures	15.2	128
Ionization energies and work functions of organic molecules	6.6A1	494
Ionization energies and work functions of organic molecules, supplemental	6.6A2	498
Ionization energies, gas phase	14.1	94
Ionization potentials of various molecules and free radicals	6.6B	499
Ionization energies, experimental data for photo-emission threshold and molecular	6.7A1	500
Ionization energies, experimental data for photo-emission threshold and molecular ionization energies, supplemental	6.7A2	503
Ionization potentials of anthracene	14.2	95
Ionization potentials and electron affinities of purines, pyrimidines, and poly-pyrroles, calculated	6.19A	514
Long-chain compounds and polymers	8.7C	544
Metal-free molecular crystals	8.2A	518
Minimum resistivities of phthalocyanine under pressure	9.1A	616
Mobilities, effective	4.7D	480
Mobilities of molecular crystals	4.7C	478
Mobility, and its activation energy for organic liquids and melts	4.7E1	481
Mobility, and its activation energy for organic liquids and melts, supplemental	4.7E2	482
Mobility, and its activation energy for anthracene	4.7A	476

TABLE 10.5A (*Continued*)

Quantity Listed	Table	Part B Page
Mobility, and its activation energy for napthalene	4.7B	478
Mobility, and its activation energy for organic substance other than anthracene and naphthalene	4.8A1	484
Mobility, and its activation energy for organic substances other than anthracene and naphthalene, supplemental	4.8A2	490
Mobility (drift) of anthracene, comparison of theoretical and experimental anisotropy under atmospheric pressure	15.4	129
Mobility (drift) of anthracene under 3 kbar pressure	15.5	130
Mobilities (drift, Hall) for napthalene	15.11	161
Mobility, temperature dependence	15.1	126
Organic dyes	8.8A	555
Photocurrent/applied electric field relations for low density polyethylene	22.2	533
Photocurrents for several polymers	22.1	531
Photoconductivity data	8.15A1	576
Photoconductivity data, supplemental	8.15A2	582
Photo-effects	8.16A1	594
Photo-effects, supplemental	8.16A2	596
Photovoltaic properties of phthalocyanine	22.5	442
Photovoltaic, output of organic cells	22.6	449
Piezoelectric substances	10.2A	624
Piezoelectric nucleic acids	10.2B	625
Piezoelectric temperature coefficients of various substances	10.2C	625
Polarization and activation energy for conductivity, change with pressure	19.2, 19.3	369, 370
Polarization energies, calculated for selected hydrocarbons	14.4	103
Polarization energies, calculated for molecular complexes	14.5	104
Polarization and adiabatic ionization energies of organic solids	6.8B	505
Quantum yields, expected behavior	14.3	99
Rate constants for monomolecular and biomolecular exciton recombination	13.2	73
Reduced heterocharge for various polymer electrets	10.3A	627
Relaxation processes of excited molecules	13.1	62

TABLE 10.5A (Continued)

Quantity Listed	Table	Part B Page
Semiconducting free radicals and radical salts: TCNQ as acceptor	8.6A	532
Sensitization of photoconductivity of polynaphthalene, polyvinylcarbazole and polyacenaphthalene	9.2A	617
Sensitization of the photoconductivity of polyvinylcarbazole	9.2B	621
Sensitizers for TiO_2	16.1	224
Solutes for hole injection into anthracene	22.4	440
Solvent effect on the charge transfer adsorption band of the acenaphthenetetrachlorophthalic anhydride complex at $20°C$	5.4A	493
Thermo-emf (Seebeck coefficients) of organic semiconductors	8.14A	564
Thermoelectric power data for conductive PAQR polymers	8.14B	575
Trap models	15.9, 15.10	154, 155
Values of constants	20.1	380

References to Addenda Tables

2001 J. H. T. Kho and H. A. Pohl, *J. Polymer Sci.*, A-1, (7) 139 (1969).
2002 H. A. Pohl, *J. Biol. Phys.*, 2, 142 (1974).
2003 P. Dupuis, D. SC. Thesis, Université De Nancy, 1971.
2004 H. A. Pohl and J. R. Wyhof, *J. Non-Cryst. Solids*, 11, 137 (1972).
2005 H. A. Pohl, "Organic Semiconductors," Research Note 35, Oklahoma State University, April 1973.
2006 J. H. T. Kho and H. A. Pohl, *J. Polymer Sci.*, A-1,(7)139 (1969).
2007 W. Bücker, Dr. Ing. Thesis, T. H. Aachen (W. Germany), 1972.
2008 J. H. Ranicar, Ph.D. Thesis, Monash University, Melbourne, 1972.
2009 J. P. Farges, D. Sc. Thesis, Université de Nice (France), 1974.
2010 A. Brau, D. Sc. Thesis, Université de Nice, 1976.
2011 J. Heim and L. Libera, *Kristall und Technik*, 2, 459 (1967).
2012 S. N. Bhat and C. N. R. Rao, *Canad. J. Chem.*, 47, 3899 (1969).
2013 A. R. Blythe, M. R. Boon and P. G. Wright, *Disc. Faraday Soc.*, 51, 110 (1971).
2014 I. F. Shchegolev, *Phys. Stat. Solidi*, 12A, 9 (1972).
2015 I. Shirotani, H. Inokuchi and S. Akimoto, *Bull. Chem. Soc. Japan*, 40, 2277 (1967).
2016 H. Inokuchi et al., *Bull. Chem. Soc. Japan*, 40, 2695 (1967).
2017 F. Gutmann and H. Keyzer, *J. Chem. Phys.*, 46, 1969 (1967).
2018 J.-P. Farges et al., *Phys. Stat. Solidi*, 37, 745 (1970).

2019 J.-P. Farges, A. Brau and F. Gutmann, *J. Phys. Chem. Solids,* **33**, 1723 (1972).
2020 F. Gutmann, A. M. Herman and A. Rembaum, *Nature* (Lond.), **221**, 1237 (1969).
2021 N. Wakayama and H. Inokuchi, *J. Catalysis,* **11**, 143 (1968).
2022 M. Ohmasa et al., *Bull. Chem. Soc. Japan,* **41**, 1998 (1968).
2023 M. Ohmasa, M. Kinoshita and H. Akamatu, *Bull. Chem. Soc. Japan,* **44**, 395, (1971).
2024 K. D. Straub, Ph.D. Thesis, Duke University, 1968.
2025 M. E. Burnell, D. D. Eley and V. Subramangan, *Ann. N.Y. Acad. Sci.,* **158**, 191 (1969).
2026 D. Vasilescu in *Physico-Chemical Properties of Nucleic Acids.,* J. D. Duchesne, ed., Academic Press, N.Y., 1973.
2027 J. Altieri and J. E. Krizan, *J. Biol. Phys.,* **3**, 110 (1975).
2028 E. Subertova, V. Prosser and J. Drobnik, *Biopolymers,* **8**, 421 (1969).
2029 R. Pethig, *J. Biol. Phys.,* **1**, 193 (1973).
2030 E. M. Trukhan, *Biofizika,* **11**, 412 (1966).
2031 D. D. Eley and R. Pethig, *Conduction in Low Mobility Materials,* Taylor and Francis, London, 1971, p. 397.
2032 R. Bogomolni and M. P. Klein, Proc. Int. Conf. Photosynthetic Unit, Gotlinburg, Tenn., p. 20, 1970.
2033 E. M. Trukhan et al., *Biofizika,* **15**, 1052 (1970).
2034 R. F. Kopczewski and H. S. Cole, in "Electrophotography," Supplement 3 to *Appld. Optics,* J. N. Howard, ed., 1969, p. 156.
2035 S. N. Bhat and C. N. R. Rao, *Canad. J. Chem.,* **47**, 3299 (1969).
2036 S. Koizumi and Y. Matsunaga, *Bull. Chem. Soc. Japan,* **45**, 423 (1972).
2037 I. Schopov, *Polymer Lett.,* **4**, 1023 (1966).
2038 I. Schopov and C. Vodenicharov, *J. Macromolec. Sci. Chem.,* **A-4**, 1627 (1970).
2039 A. M. Hermann and A. Rembaum, *J. Polymer Sci.,* Pt. C, No. 17, 107 (1967).
2040 A. M. Hermann and A Rembaum, *J. Appl. Phys.,* **37**, 3642 (1966).
2041 N. Petruzella, S. Takeda and R. C. Nelson, *J. Chem. Phys.,* **47**, 4247 (1967).
2042 V. Hadek, *J. Chem. Phys.,* **49**, 5202 (1968).
2043 N. Geacintov, M. Pope and H. Kallmann, *J. Chem. Phys.,* **45**, 2639 (1966); *ibid;* 3884.
2044 T. Tanaka and Y. Inuishi, *Japan J. Appl. Phys.,* **5**, 974 (1966).
2045 L. E. Lyons and K. A. Milne, *J. Chem. Phys.,* **65**, 1474 (1976).
2046 E. Krikorian and R. J. Sneed, *J. Appl. Phys.,* **40**, 2306 (1969).
2047 K. J. Hall, J. S. Bonham and L. E. Lyons, in press, *Aust. J. Chem.,* 1978.
2048 V. Hadek, H. Noguchi and A. Rembaum, *Macromolecules,* **4**, 494 (1971).
2049 C. I. Buravov et al., *Soviet Phys. JETP Lett.,* **12**, 99 (1970).
2050 R. B. Somoano et al., *Phys. Rev.,* **B-15**, 595 (1977).
2051 J. Ferraris et al., *J. Amer. Chem. Soc.,* **95**, 948 (1973).
2052 A. Q. Kulshreshtha and T. Mookherji, *Molec. Cryst.,* **10**, 75 (1970).
2053 W. Klöpfer and H. Rabenhorst, *J. Chem. Phys.,* **46**, 1362 (1967).
2054 Y. Matsunaga and K. Shono, *Bull. Chem. Soc. Japan,* **43**, 2007 (1970).
2055 M. Ieda et al., Proc. Internatl. Micro-Symp., Polarization, Conduction in Insulating Polymers, Bratislava (CSR) 1972.
2056 J. H. Ranicar et al., *Aust., J. Phys.,* **24**, 325 (1971); J. H. Ranicar, Ph.D. Thesis, Monash Univ., Melbourne, 1972.
2057 M. Prince and B. Bremer, *J. Polymer Sci.,* **B-5**, 843 (1967).

2058 F. Sandrolini et al., *Polymer Lett.*, **8**, 749 (1970).
2059 E. L. Frankevich and E. I. Balabanov, *Sov. Phys.-Solid State*, **7**, 570 (1965).
2060 I. Schopov and C. Vodeicharov, *J. Macromolec. Sci.-Chem.*, **A-4**, 1627 (1970).
2061 I. Schopov, *Polymer Lett.*, **4**, 1023 (1966).
2062 K. Okamoto et al., *Bull. Chem. Soc. Japan*, **46**, 2324 (1973).
2063 W. Hanue, *Z. Anorg. u. Allgem. Chem.*, **347**, 67 (1966).
2064 I. F. Shchegolev et al., *Zh. ETF. Pis. RED.*, **8**, 353 (1968).
2065 D. Zosel et al., *Phys. Stat. Sol.*, **32**, K-75 (1969).
2066 K. D. Straub, Ph.D. Thesis, Duke University, 1968.
2067 L. Alcacer and A. U. Maki, *J. Phys. Chem.*, **78**, 215 (1974); **80**, 1912 (1976).
2068 A. Kofler, *Z. Electrochem.*, **50**, 200 (1944); Y. Matsunaga and G. Saito, *Bull. Chem. Soc. Japan*, **45**, 963 (1972).
2069 G. Saito and Y. Matsunaga, *Bull. Chem. Soc. Japan*, **46**, 714, 1609 (1973).
2070 P. Stilbs and R. Olofsson, *Acta Chem. Scand.*, **A-28**, 647 (1974); G. Olofsson and I. Olofsson, *Tetrahedron*, **29**, 1711 (1973).
2071 I. Agranat and S. Cohen, *Bull. Chem. Soc. Japan*, **47**, 723 (1974).
2072 J. Nishijo, *Bull. Chem. Soc. Japan*, **47**, 1539 (1974).
2073 N. Inoue and Y. Matsunaga, *ibid.*, **45**, 3478 (1972).
2074 Y. Matsunaga and R. Osawa, *ibid.*, **47**, 1589 (1974).
2075 Y. Matsunaga and G. Saito, *ibid.*, **47**, 2873 (1974).
2076 T. J. Sonnonstine and A. M. Hermann, *J. Chem. Phys.*, **60**, 1335 (1974).
2077 T. J. Sonnonstine, A. Wiglesworth and A. M. Hermann, *J. Chem. Phys.*, **59**, 3865 (1973).
2078 W. Mey and A. M. Hermann, *Phys. Rev.*, **B-7**, 1652 (1973).
2079 A. M. Hermann and A. Rembaum, *Polymer Lett.*, **5**, 445 (1967).
2080 R. Somoano et al., *J. Chem. Phys.*, **62**, 4970 (1975).
2081 R. B. Somoano et al., *J. Chem. Phys.*, **63**, 4970 (1975).
2082 A. M. Hermann and L. G. Wilhite, *Proc. Louisiana Acad. Sci.*, **32**, 126 (1969).
2083 H. A. Pohl et al., Res. Note 77, Sept. 1978, Quantum Theor. Res. Group, Oklahoma State University, Stillwater, Ok. 74074.
2084 J. H. T. Kho and H. A. Pohl, *J. Polymer Sci.*, Pt. A-1, **7**, 139 (1969).
2085 H. A. Pohl, *J. Biol. Phys.*, **2**, 113 (1974).
2086 C. G. Norrell and H. A. Pohl, *J. Polymer Sci.*, Polymer Phys. Ed., **12**, 913 (1974).
2087 S. G. Burnay and H. A. Pohl, *J. Non-Cryst. Solids*, in press, 1979.
2088 R. Somoano et al., *Polymer Lett.*, **8**, 467 (1970).
2089 S. K. Khanna et al., *Phys. Rev.*, **B-15** (1979), in print.
2090 A. Rembaum et al., *J. Am. Chem. Soc.*, **93**, 2532 (1971).
2091 S. K. Khanna et al., *Solid State Commun.*, **18**, 1405 (1976).
2092 R. Williams et al., *Phys. Rev., 1979*, in print.
2093 R. B. Somoano et al., *Phys. Stat. Solidi*, **81**, 281 (1977); **82**, K117 (1977).
2094 E. Ehrenfreund et al., *Solid State Commun.*, **22**, 139 (1977).
2095 A. Rembaum, *J. Polymer Sci.*, C. No. 29, 157 (1970).
2096 A. Kh. Yusupbekov et al., *Uzb. Khim. Zh.*, **1976**, 50.
2097 M. Kajiwara and H. Saito, *Polymer.*, **17**, 1013 (1976); *Bull. Chem. Soc. Japan*, **50**, 1023 (1977).
2098 L. A. Akopyan et al., *Vysokomol. Soedin* Ser. A, **19**, 271 (1977).
2099 G. Kossmehl and M. Rohde, *Macromolec. Chem.*, **178**, 715 (1977).
2100 M. Prasad and B. D. Choubey, *J. Indian Chem. Soc.*, **54**, 499 (1977).
2101 I. Lupu, *Bull. Univ. Brasou*, Ser. C, **18**, 93 (1976).
2102 M. H. Cohen et al., U.S. Pat., 4,026,903, 31 May, 1977.

2103 A. I. Levchenko et al., *Khim. Geterotskil. Soedin,* 598 (1978).
2104 L. Hellner and L. Wsieck, *J. Res. Natl. Bur. Stand.,* **A-75**, 487 (1971).
2105 V. Dycsmons et al., *Chem. Phys. Lett.,* **5**, 361 (1970).
2106 H. Klotz et al., *Z. Phys. Chem.* (Leipzig), **237**, 305 (1968).
2107 S. Mima, *Osaka Kagyo Gijutsu Shikensho Kiho,* **14**, 13, 25, 30 (1963) via *Nuclear Sci. Abst., Japan,* **2** (3) 0002730 (1963).
2108 A. C. Kopkinson and I. G. Czimadia, *Theor. Chim. Acta,* **31**, 83 (1973).
2109 L. D. Betowski et al., *Inorg. Chem.,* **11**, 424 (1972).
2110 Sh.-L. Chong and J. L. Franklin, *J. Am. Chem. Soc.,* **84**, 6630 (1972).
2111 F. Sersen et al., *Collect. Czech. Chem. Comm.,* **42**, 2173 (1977).
2112 B. A. Zhubavov et al., *Dokl. Akad. Nauk SSR,* **229**, 875 (1976).
2113 R. Sellam et al., *Bull. Inst. Chem. Res.,* Kyoto Univ., **56**, 1 (1978).
2114 M. S. B. Munson, *J. Am. Chem. Soc.,* **87**, 3294 (1965).
2115 F. W. Lampe and J. H. Futrell, *Trans. Faraday Soc.,* **59**, 1957 (1963).
2116 M. A. Haney and J. L. Franklin, *Trans. Faraday Soc.,* **65**, 1794 (1969); *J. Phys. Chem.,* **73**, 4328 (1969).
2117 D. H. Ave et al., *J. Am. Chem. Soc.,* **94**, 4726 (1972).
2118 T. C. Waddington, *Adv. Inorg. Chem. Radiochem.,* **1**, 157 (1959).
2119 R. V. Hodges and J. L. Beauchamp, *Inorg. Chem.,* **14**, 2887 (1975).
2120 W. L. Jolly and D. N. Hendrickson, *J. Am. Chem. Soc.,* **92**, 1863 (1970).
2121 K. M. A. Rafacy and W. A. Chupka, *J. Chem. Phys.,* **48**, 5205 (1967).
2122 P. Kriemler and S. E. Buttrill, Jr., *J. Am. Chem. Soc.,* **92**, 1123 (1970).
2123 T. Takahashi et al., *Solid State Comm.,* **26**, 705 (1978).
2124 G. Teodorescu, *Bull. Inst. Politech.,* "Gheorghe, Gheorghi-Dej," Buceresti, Ser. Chim. Metal, **40**, 3 (1978).
2125 L. C. Isett, *Phys. Rev.,* **B-18**, 439 (1978).
2126 J. Altieri and J. E. Krizan, *J. Biol. Phys.,* **3**, 103 (1975).
2127 C. K. Chiang et al., *J. Chem. Phys.,* **69**, 5098 (1978); *J. Am. Chem. Soc.,* **100**, 1013 (1978); *Phys. Rev. Lett.,* **39**, 1098 (1978).
2128 H. Shirakawa et al., *Die Makromolekulare Chemie,* **179**, 1565 (1978); *Chem. Comm.,* 1978, 578.
2129 T. Hibma et al., *J. Chem. Phys.,* **49**, 4755 (1968).
2130 V. Hadek, *J. Chem. Phys.,* **49**, 5202 (1968).
2131 J. Petersen et al., *J. Am. Chem. Soc.,* **99**, 286 (1977).
2132 A. G. MacDiarmid and A. J. Heeger, Symp. Struct. and Highly Conducting Polymers and Graphite, IBM, San Jose, Calif., March, 1979, Abst. 1 and 15.
2133 A. Dall'Olio et al., *C. R. Acad Sci.* (Paris), **433**, 267c (1968).
2134 A. F. Diaz, IBM, San Jose, Calif., Private Comm. 1979.
2135 M. J. Cohen and J. S. Harris, Symp. Struct. and Prop. Highly Conducting Polymers and Graphite, IBM, San Jose, Calif., March, 1979; Abst., No. 27.
2136 I. B. Goldberg et al., *J. Chem. Phys.,* **70**, 1132 (1979).
2137 P. Politzer and K. C. Daiker, *J. Chem. Phys.,* **68**, 5289 (1978).
2138 E. A. Baum et al., *J. Phys.,* **D-11**, 703 (1978).
2139 D. J. Sandman et al., *J. Am. Chem. Soc.,* **100**, 202 (1978).
2140 M. Suzuki et al., *Japan J. Appl. Phys.,* **14**, 741 (1975); *Kanazawa Daigaku Kogakubu Kiyo,* **6**, 149, 159 (1971).
2141 C. D. Cooper et al., *J. Chem. Phys.,* **69**, 2367 (1978).
2142 V. Kampars and O. Neilands, *Russ. Chem. Revs.,* **46**, 503 (1977).
2143 Z. Burshtein and D. F. Williams, *Mol. Cryst. Liqu. Cryst.,* **43**, 1 (1977).
2144 R. C. Hughes, *J. Chem. Phys.,* **58**, 2212 (1973).

2145 S. L. Hsu et al., *J. Chem. Phys.*, **69**, 106 (1978).
2146 R. T. McIver and J. Scott Miller, *J. Am. Chem. Soc.*, **96**, 4324 (1974).
2147 L. C. Isett, *Phys. Rev.*, **B-18**, 439 (1978).
2148 E. A. Perez-Albuerne, *J. Chem. Phys.*, **55**, 1549 (1971).
2149 B. Reimer and H. Bässler, *Chem. Phys. Lett.*, **43**, 81 (1976).
2150 R. W. Munn and W. Siebrand, *J. Chem. Phys.*, **52**, 47, 6391 (1970).
2151 W. M. Tratt et al., **69**, 3150 (1978).
2152 P. W. Tiedemann et al., "Engergies of Molecule Ions by the Molecular Beam Photoionization Method," 175th ACS Natl. Meeting, Anaheim, Calif., March 1978.
2153 V. K. Potapov et al., *Dokl. Akad. Nauk SSSR.*, **180**, 398 (1968).
2154 K. Watanabe et al., *J. Quant. Spectroscop. Rad. Transfer*, **2**, 369 (1962).
2155 R. Hernandez et al., *J. Elektron. Spectroscop. Related Phenom.*, **10**, 333 (1977).
2156 K. Kimura et al., *ibid.*, **6**, 41 (1975).
2157 C. R. Brundle et al., *J. Am. Chem. Soc.*, **94**, 1451 (1972).
2158 R. A. Arndt and A. C. Damask, *J. Chem. Phys.*, **45**, 4627 (1966).
2159 K. E. Murphy and T. B. Flanagan, *J. Solid State Chem.*, **22**, 367 (1977).
2160 W. Mey et al., *J. Chem. Phys.*, **58**, 2542 (1973).
2161 V. Gaidelis et al., *Thin Solid Films*, **38**, 9 (1976).
2162 Z. Burshtein and D. F. Williams, *Phys. Rev.*, **B-15**, 5769 (1977).
2163 M. D. Tabak, *J. Non-Cryst. Solids*, **6**, 357 (1971).
2164 R. A. Holroyd et al., *J. Phys. Chem.*, **83**, 435 (1979).
2165 S. Bagchi and M. Chowdury, *J. Phys. Chem.*, **83**, 629 (1979).
2166 A. G. MacDiarmid and A. J. Heeger, *Proceedings of the NATO ASI on Molecular Metals*, Les Arcs, France, Plenum Press, N.Y., 1979; C. K. Chiang et al., *J. Chem. Phys.*, **69**, 5098 (1978); G. B. Street and T. C. Clarke, *J. Am. Chem. Soc.*, Advances in Chem., in press (1979); S. L. Hsu et al., *J. Chem. Phys.*, **69**, 1106 (1978).
2167 D. F. Barbe and C. R. Westgate, *Solid State Comm.*, **7**, 563 (1969).
2168 M. Audenaert et al., *Solid State Comm.*, **30**, 797 (1979).
2169 I. Schopov and M. Vodenicharova, *Makromolec. Chem.*, **179**, 63 (1978).
2170 S. D. Phadke, *Thin Solid Film*, **55**, 391 (1978).
2171 A. Dall'Olio et al., *C. R. Acad. Sci.* (Paris), **267C**, 433 (1968).
2172 J. Guillet, *Angew. Makromolek. Chem.*, **68**, 163 (1978).
2173 C. J. Norrell et al., *J. Polymer Sci.*, **12**, 913 (1974).
2174 A. Mierzejewski et al., *Pr. Nauk Inst. Chem. Org. Fiz. Politech. Wroclaw*, **16**, 245 (1978).
2175 G. Delacote et al., *Solid State Comm.*, **4**, 137 (1966).
2176 Z. Burshtein and D. F. Williams, *Phys. Rev.*, **B-15**, 5769 (1977).
2177 Y. Kakamura et al., *Chem. Sett.*, **1974** 301.
2178 Proc. Int. Conf. Conduct. Breakdown Dielect. Liqu., 6th (1978) Editions Front, Dreux, France, 1978.
2179 W. Döldissen et al., see ref., 2178.
2180 D. H. Spielberg et al., *Phys. Rev.*, **B-3**, 2012 (1971).
2181 J. B. Webb and D. F. Williams, *J. Phys.*, **C-11**, 3245 (1978).
2182 S. G. Burnway and M. Pohl, *J. Non-Cryst. Solids*, in press 1979.
2183 M. Schadt and D. F. Williams, *Phys. Stat. Solidi*, **39**, 223 (1970).
2184 Y. Maruyama et al., *Molec. Cryst. and Liqu. Cryst.*, **20**, 373 (1973).
2185 A. L. Farragher and F. M. Page, *Trans. Faraday Soc.*, **63**, 2369 (1967).
2186 T. L. Kunii and H. Kuroda, *Theor. Chim. Acta*, **11**, 97 (1968).
2187 W. E. Wentworth, E. Chen and J. E. Lovelock, *J. Phys. Chem.*, **70**, 445 (1966).
2188 P. R. Hammond, *J. Chem. Phys.*, **55**, 3468 (1971).

2189 W. F. Kuhn et al., *J. Chem. Phys.*, 49, 5551 (1977).
2190 S. R. Rafikov et al., *Zh. Obshch. Khim.*, 47, 2610 (1977).
2200 R. A. Holroyd and M. Allen, *J. Chem. Phys.*, 54, 5014 (1971).
2201 M. Kotani and H. Akamatu, *Bull. Chem. Soc., Japan*, 43, 120 (1970).
2202 T. Hiruoka et al., *ibid.*, 42, 1481 (1969).
2203 G. Eckert and F. Gutmann, *Electroanalyt. Chem. Interfacial Electrochem.*, 62 267 (1975).
2204 F. M. Page and G. C. Goode, *Negative Ions and the Magnetron*, Wiley-Interscience, N.Y. and Lond., 1969, p. 137-144.
2205 T. L. Kunii and H. Kuroda, *Theor. Chim. Acta*, 11, 97 (1968).
2206 C. D. Cooper, W. T. Naff and R. N. Compton, *J. Chem. Phys.*, 63, 2752 (1975).
2207 R. J. Celotta, R. A. Bennett and J. L. Hall, *J. Chem. Phys.*, 60, 1740 (1979).
2208 H. Hotop et al., *J. Chem. Phys.*, 58, 2373 (1973).
2209 R. H. Young et al., *J. Chem. Phys.*, 70, 443 (1979).
2210 L. B. Schein, *Phys. Rev.*, B-15, 1024 (1977).
2211 A. Suzuki, *Bull. Chem. Soc. Japan*, 49, 3347 (1976).
2212 T. Kayiwara et al., *Bull. Chem. Soc. Japan*, 40, 1055 (1967).
2214 O. Dresner, *Phys. Rev.*, 143, 558 (1966).
2215 J. M. Thomas and J. O. Williams, *Prog. Solid State Chem.*, 6, 119 (1971).
2216 Proc. 6th., Molecular Crystal Symp., 1973, Sehlen Elmer, W. Germany; W. Mey et al., *J. Chem. Phys.*, 58, 2542 (1973).
2217 E. H. Martin and J. Hirsch, *J. Non-Cryst. Solids*, 4, 133 (1970).
2218 V. Adamec, *Kolloid Z. and Z. Polymere*, 249, 1085 (1971).
2219 W. Tantraporn, *J. Appl. Phys.*, 39, 2012 (1968).
2220 E. H. Martin and J. Hirsch, *J. Non-Cryst. Solids*, 4, 133 (1970); Kodak Ltd., Harrow, U.K., Interim Rept., 1971, Pt. 19.
2221 D. A. Seanor, "Elect. Prop. of Polymers," in *Polymer Sci.*, Ch. 17; A. D. Jenkins, ed., Nth. Holland. Publ., N.Y., 1972.
2222 R. Pethig and K. Morgan, *Nature* (London), 214, 266 (1967).
2223 C. R. Westgate and G. Warfield, *J. Chem. Phys.*, 46, 94 (1967).
2224 H. Meler and W. Albrecht, *Z. Naturforschg.*, 24-A, 257 (1969).
2225 A. M. Hermann and J. S. Ham, *Rev. Sci. Inst.*, 36, 1553 (1965).
2226 W. Bückner, Dr. Ing. Thesis, T. H. Aachen, W. Germany, 1972, p. 77.
2227 S. S-S. Huang and C. R. Freeman, *J. Chem. Phys.*, 69, 1585 (1978).
2228 M. Campos and G. A. Giacometti, *Appl. Phys.*, 32, 794 (1978).
2229 K. Seki et al., *Bull. Chem. Soc. Japan*, 49, 904 (1976).
2230 K. Ishii et al., *Chem. Phys.*, 41, 154 (1976).
2231 M. Batley and L. E. Lyons, *Mol. Cryst.*, 3, 357 (1968).
2232 S. Hino et al., *Chem. Phys.*, 36, 335 (1975).
2233 H. Inokuchi, *Faraday Soc. Disc.*, 51, 183 (1971).
2234 K. Seki et al., *Bull. Chem. Soc. Japan*, 49, 904 (1976).
2235 T. Hiruoka et al., *Chem. Phys. Lett.*, 18, 930 (1973).
2236 T. Hiruoka, Ph. D. Thesis, Univ. of Tokyo, 1973.
2237 S. Hino, K. Seki and H. Inokuchi, *Chem. Phys. Lett.*, 36, 335 (1975).
2238 K. Ishii et al., *Chem. Phys. Lett.*, 41, 154 (1976).
2239 P. Nielsen, A. J. Epstein and D. J. Sandman, *Solid State Comm.*, 15, 53 (1974); References cites therein.
2240 J. J. Ritsko et al., *J. Chem. Phys.*, 67, 687 (1977).
2241 K. Seki and H. Inokuchi, *Chem. Phys. Lett.*, 65, 158 (1979).
2242 S. Hino et al., *Chem. Phys. Lett.*, 36, 335 (1975).

REFERENCES TO ADDENDA TABLES

2243 H. Kawamura and H. Inokuchi, *Bull. Chem. Soc. Japan,* **45**, 710 (1972).
2244 A. Nakajima and H. Akamatu, *Bull. Chem. Soc. Japan,* **42**, 3030 (1969).
2245 T. Kitagawa, *J. Molec. Spectroscopy,* **26**, 1 (1968).
2246 R. Boschi, J. N. Murell and W. Schmidt, *Disc. Faraday Soc.,* **54**, 116 (1972).
2247 P. A. Clark, F. Brogli and E. Heilbronner, *Helv. Chim. Acta,* **55**, 1415 (1972).
2248 R. Boschi and W. Schmidt, *Tetrahedron Lett.,* **1972**, 2557.
2249 H. Inokuchi, H. Kuroda and H. Akamatu, *Bull. Chem. Soc. Japan,* **34**, 749 (1961).
2250 A. A. Pankratov and B. P. Bespalov, "Tezist Dokl. Vses. Soveshch Kompleksan Perenosom Zaryada Ion-Radikalnym Solyam," 3rd. 1975 (Publ. 1976) *Zinante* Riga USSR, 1976, p. 157.
2251 R. Edgell et al., *Chem. Phys. Lett.,* **33**, 600 (1975).
2252 D. J. Sandman et al., *J. Chem. Phys.,* **70**, 305 (1973).
2253 J. Berkowitz, *J. Chem. Phys.,* **70**, 2819 (1979).
2254 M. Kotani and A. Akamatu, *Bull. Chem. Soc. Japan,* **43**, 120 (1970).
2255 Y. Marada and H. Inokuchi, *Bull. Chem. Soc. Japan,* **39**, 1443 (1966).
2256 M. Kochi, Y. Harada and H. Inokuchi, *Bull. Chem. Soc. Japan,* **40**, 531 (1967).
2257 T. Hiruoka et al., *Bull. Chem. Soc. Japan,* **42**, 1481 (1969).
2258 M. Kotani and A. Akamatu, *Faraday Soc. Disc.,* **51**, 124 (1971).
2259 S. Hino, T. Hiruoka and H. Inokuchi, *Bull. Chem. Soc., Japan,* **48**, 1133 (1974).
2260 F. T. Viselov et al., *Sov. Phys. Solid State,* **11**, 2775 (1970).
2261 D. K. Davies, Proc. Conf. Static. Electrification, London, 1967, p. 29.
2262 J. Aihara and H. Inokuchi, *Bull. Chem. Soc., Japan,* **43**, 1265 (1970).
2263 B. Mallik et al., *J. Biochem. Biophys.,* **15**, 233 (1978).
2264 K. Ohno et al., *Bull. Chem. Soc., Japan,* **46**, 2353 (1973).
2265 J. I. Aihara and H. Inokuchi, *Chem. Lett.,* **1973**, 421.
2266 G. W. Mines and H. W. Thompson, *Spectrochim. Acta,* **A-31**, 137 (1975).
2267 K. Seki et al., *Bull. Chem. Soc. Japan,* **47**, 1608 (1974).
2268 S. Hino and H. Inokuchi, *Chem. Lett.,* **1974**, 365.
2269 K. Kimura et al., *J. Am. Chem. Soc.,* **100**, 6564 (1978).
2270 J. M. Williams and W. H. Hamill, *J. Chem. Phys.,* **49**, 4467 (1968).
2271 A. Fulton and L. E. Lyons, *Aust. J. Chem.,* **21**, 873 (1968).
2272 G. R. Johnston and L. E. Lyons, *Chem. Phys. Lett.,* **2**, 489 (1968); *Phys. Stat. Solidi,* **37**, K75 (1970).
2273 Yu. A. Berlin et al., *J. Chem. Phys.,* **69**, 2401 (1978).
2274 I. Kalinowski et al., Proc. Tihany Symp. Radiat. Chem., 1976, Publ. 1977, **4**, 171 (1977).
2275 L. Nyikos et al., Proc. Tihany Symp. Radiat. Chem., 1976, publ. 1977, **4**, 179 (1977).
2276 S. S-S. Huang and G. R. Freeman, *J. Chem. Phys.,* **70**, 1538 (1979).
2277 E. Sacher, Amer. Rep. Conf. Elect. Insul. Dielect. Phenomena, 1976, publ., 1978, p. 33.
2278 L. Brehner et al., *Phys. Stat. Solidi,* **A-50**, K239 (1978).
2279 C. K. Chang, *Bull. Amer. Phys. Soc.,* **24**, (3) 293 (1979).
2280. Y. Nakato et al., *Bull. Chem. Soc. Japan,* **45**, 1299 (1972).
2281 Y. W. Park et al., *Bull. Amer. Chem. Soc.,* **24**, 328 (1979).
2282 H. A. Pohl and M. Pollak, *J. Chem. Phys.,* **66**, 4031 (1977).
2283 J. Langer, *Pr. Nauk Inst. Chem. Org. Fiz. Politech. Wroclaw* (Poland), **16**, 221 (1978).
2284 R. C. Nelson, *J. Chem. Phys.,* **47**, 4451 (1967).

2285 L. E. Lyons, Priv. Commun., 1980.
2286 I. Klinowski et al., Conduct. and Breakdown of Dielect. Liqu. Proc. Inst. Conf., 5th, 1975, J. M. Goldschwartz, ed., Delft Univ. Press, 1975, p. 11.
2287 A. A. Balakin et al., *Khim Vys. Energ.*, **12**, 210 (1978).
2288 W. F. Schmidt et al., *Z. Naturforsch.*, **33A**, 1383 (1978).
2289 Y. Matsunaga and Y. Suzuki, *Bull. Chem. Soc. Japan*, **45**, 3375 (1972).
2290 S. Doi et al., *Bull. Chem. Soc. Japan*, **50**, 837 (1977); K. Kan and Y. Matsunaga, *ibid.*, **45**, 2096 (1972).
2291 G. Dix, *Phys. Stat. Solidi*, **A24**, 139 (1974).
2292 L. I. Buravov et al., *J. Chem. Commun.*, 1976, 20; *cf.* also L. C. Isetti and E. A. Perez-Albuerne, *Solid State Commun.*, **21**, 433 (1977).
2293 Y. Yamamoto et al., Technol. Rep., Osaka University, **28**, 1430 (1978).
2294 A. K. Kostin and A. V. Vannikov, *Fiz. Tverd. Tela*, **20**, 3407 (1978).
2295 A. Mierzejewski et al., *Pr. Nauk Inst. Chem. Org. Fiz. Politech.*, Wroclaw (Poland), **16**, 295 (1978).
2296 K. Morokuma, *Acc. Chem. Res.*, **10**, 294 (1977).
2297 J. K. Lau and P. Kebarle, *J. Am. Chem. Soc.*, **98**, 7452 (1976).
2298 H. Umeyama and K. Morokuma, *J. Am. Chem. Soc.*, **98**, 4400 (1976).
2299 Z. Zboinski, *Phys. Stat. Solidi*, **B74**, 561 (1976).
2300 C. Hamann, *Phys. Stat. Solidi*, **A10**, 509 (1972).
2301 T. G. Abd-El-Malik and G. A. Cox, *J. Phys. Chem. Solid State Phys.*, **10**, 63 (1977).
2302 I. Vaisnys and R. Kirk, *Phys. Rev.*, **141**, 641 (1966).
2303 J. P. Farges et al., *J. Phys. Chem. Solids*, **33**, 1723 (1972).
2304 I. F. Schegdlev et al., *JETP Lett.*, **8**, 218 (1968); L. I. Buravov et al., *ibid.*, **12**, 99 (1970); **32**, 612 (1971).
2305 A. J. Epstein et al., *Solid State Commun.*, **9**, 1803 (1971); *cf.*, also, *ibid.*, **23** (6), 335 (1977).
2306 A. Rembaum et al., *J. Phys. Chem.*, **73**, 513 (1969).
2307 P. Dupuis and J. Neel, *C. R. Acad. Sci.* (Paris), **265-C**, 777, 1297 (1967).
2308 R. Pethig and A. Szent-Györgyi, in *Bioelectrochemistry*, Proc. Australia-USA Joint Seminar on Bioelectrochemistry, Pasadena 1979, H. Keyzer and F. Gutmann, eds., Plenum, N.Y., 1980.
2309 Y. Nakamara et al., *Chem. Lett.*, 1979, 877.
2310 Y. Nakamara et al., *Chem. Phys. Lett.*, **47**, 251 (1977).
2311 A. Fort and A. Coret, *Phys. Stat. Solidi*, **B85**, 599 (1977).
2312 P. Schlotter et al., *Chem. Phys.*, **19**, 353 (1977).
2313 Y. Hoshino and K. Tateishi, *J. Phys. Soc. Japan*, **37**, 1024 (1979); **41**, 1625 (1976).
2314 G. A. Corker and I. Lundström, *J. Appl. Phys.*, **49**, 686 (1978).
2315 P. M. Borsenberger et al., *J. Appl. Phys.*, **49**, 5555 (1978); W. J. Dulmage et al., *ibid.*, 5543; P. M. Borsenberger and A. Chowdry, *J. Appl. Phys.*, **49**, 273 (1978); *cf.* also W. Mey, *ibid.*, **50**, 8090 (1979).
2316 M. Akiyama, *J. Phys. Soc. Japan*, **31**, 148 (1971).
2317 V. A. Benderskii and N. N. Usov, *Dokl. Akad. Nauk SSSR*, **167**, 898 (1966).
2318 K. Lochner et al., *Phys. Stat. Solidi*, **B76**, 533 (1976); **B88**, 653 (1978).
2319 R. R. Chance et al., *Chem. Phys.*, **13**, 181 (1976); *cf.* also G. C. Stevens et al., *ibid.*, **28**, 399 (1978).
2320 B. Reimer and H. Bässler, *Phys. Stat. Solidi*, **B-85**, 145 (1978).

2321 P. M. Borsenberger and A. I. Ateya, *J. Appl. Phys.*, **49**, 4035 (1978).
2322 A. D. Adler et al., *Porphyrins*, **5**, Pt. C, 483 (1978).
2323 E. L. Frankevich et al., *Phys. Stat. Solidi*, **B77**, 265 (1976).
2324 J. Gonzalez-Basurto et al., *Molec. Cryst. Liqu. Cryst.*, **51**, 303 (1979).
2325 H. Yasunaga et al., *J. Phys. Soc. Japan*, **37**, 1024 (1974); **46**, 839 (1974).
2326 Y. Aoyagi et al., *J. Phys. Soc. Japan*, **31**, 164 (1971); *J. Appl. Phys.*, **43**, 249 (1972).
2327 S. C. Dahlberg and M. E. Musser, *J. Chem. Phys.*, **71**, 2806 (1979).
2328 C. W. Tang and A. C. Albrecht, *J. Chem. Phys.*, **62**(6), 2139 (1975).
2329 A. F. Janzen and J. R. Bolton, *J. Am. Chem. Soc.*, **101**, 6342 (1979).
2330 J. Lagowski and H. C. Gates, *J. Appl. Phys.*, **49**, 2821 (1978).
2331 K. Okumura, *J. Appl. Phys.*, **45**, 5317 (1974).
2332 S. C. Dahlberg and M. E. Musser, *J. Chem. Phys.*, **71**, 2806 (1979).
2333 R. O. Loutfy and J. H. Sharp, *J. Chem. Phys.*, **71**, 1211 (1979).
2334 F. R. Fan and L. R. Faulkner, *J. Chem. Phys.*, **69**, 3341, 3334 (1978).
2335 P. J. Reucroft et al., *J. Appl. Phys.*, **46**, 5218 (1975); *Appl. Phys. Lett.*, **25**, 664 (1974).
2336 A. K. Gosh and T. Feng, *J. Appl. Phys.*, **44**, 2781 (1973).
2337 T. Imura, *Bull. Chem. Soc. Japan*, **46**, 2075 (1973).

4000 T. Tamura et al., *Bull. Chem. Soc. Japan*, **47**, 448 (1974).
4001 Y. Sakai et al., *Bull. Chem. Soc. Japan*, **47**, (1974).
4002 K. Kato et al., *Bull. Chem. Soc. Japan*, **47**, (1974).
4003 C. L. Braun and G. M. Dobbs, *J. Chem. Phys.*, **53**, 2718 (1970).
4004 A. Bromberg et al., *J. Chem. Phys.*, **60**, 4059 (1974).
4005 J. O. M. Bockris, in *Bioelectrochemistry*, H. Keyzer and F. Gutmann, eds., Plenum Press, N.Y., 1980, p. 20.
4006 T. Miyasaka et al., *J. Am. Chem. Soc.*, **100**, 6657 (1978).
4007 E. R. Menzel and R. O. Loutfy, *Chem. Phys. Lett.*, **72**, 522 (1980).
4008 G. R. Johnston and L. E. Lyons, *Chem. Phys. Lett.*, **2**, 489 (1968).
4009 A. Dias Tavares, *J. Chem. Phys.*, **53**, 2520, (1970).
4010 E. A. Silinsh et al., *Phys. Stat. Solidi*, **A25**, 339 (1974).
4011 K. Kato and C. L. Braun, *J. Chem. Phys.*, **72**, (1980).
4012 M. Soma, *J. Chem. Phys.*, **52**, 6042 (1970).
4013 C. F. Hackett, *J. Chem. Phys.*, **55**, 3178 (1971).
4014 S. S. Takeda et al., *J. Chem. Phys.*, **54**, 3195 (1971).
4015 D. F. Barbe and C. R. Westgate, *J. Chem. Phys.*, **52**, 4046 (1970).
4016 N. Houser and R. C. Jarnagin, *J. Chem., Phys.*, **52**, 1069 (1970).
4017 A. J. Twarowski and A. C. Albrecht, *J. Chem. Phys.*, **70**, 2255 (1979).
4018 A. K. Ghosh et al., *J. Appl. Phys.*, **45**, 230 (1974).
4019 M. Schonfeld et al., Proc. Nat. Acad. Sci. (USA), **76**, 6351 (1979); N. K. Packham et al., *FEBS Lett.*, **110**, 101 (1980); Editorial, *Physics Today*, **33**, No. 9, 19 (1980).
4020 I. Lundström et al., *J. Appl. Phys.*, **49**, 701 (1978).
4021 G. A. Corker and I. Lundström, *J. Appl. Phys.*, **48**, 686 (1978).
4022 N. Oron et al., *Molec. Cryst.*, **6**, 415 (1970); *J. Appl. Phys.*, **41**, 2649 (1970).
4023 A. Prock and R. Zahradnik, *J. Chem. Phys.*, **49**, 3204 (1968).
4024 P. H. Fang et al., *Japan J. Appl. Phys.*, **11**, 1298 (1972).
4025 W. Pong and J. A. Smith, *J. Appl. Phys.*, **44**, 174 (1973).

4026 J. Mort and R. L. Emerald, *J. Appl. Phys.*, **45**, 175 (1974).
4027 P. M. Borsenberger et al., *J. Appl. Phys.*, **50**, 914 (1979).
4028 A. K. Ghosh et al., *J. Appl. Phys.*, **45**, 230 (1974).
4029 V. Y. Merritt and J. H. Hovel, *Appl. Phys. Lett.*, **29**, 414 (1978).
4030 P. M. Borsenberger et al., *J. Appl. Phys.*, **49**, 5555, 5543, 273 (1978); W. Mey et al., *ibid.*, 3607.
4031 A. Prock, *Bull. Chem. Soc. Japan*, **40**, 2452 (1964).
4032 R. R. Chance and R. H. Baugham, *J. Chem. Phys.*, **64**, 3889 (1976).
4033 M. Yokoyama et al., *J. Chem. Phys.*, **67**, 1742 (1977).
4034 A. Bergman et al., *J. Chem. Phys.*, **65**, 1184 (1976).
4035 Z. D. Popovic and J. H. Sharp, *J. Chem. Phys.*, **66**, 5076 (1977).
4036 N. Petruzella et al., *J. Chem. Phys.*, **50**, 3527 (1969).
4037 N. Geacintov and M. Pope, *J. Chem. Phys.*, **50**, 814 (1969).
4038 Y. Maruyama and K. Funabashi, *J. Chem. Phys.*, **56**, 2342 (1972).
4039 T. Huang et al., *J. Chem. Phys.*, **56**, 4702 (1972).
4040 S. E. Harrison, *J. Chem. Phys.*, **51**, 465 (1969).
4041 H. Bässler et al., *J. Chem. Phys.*, **51**, 3695 (1969).
4042 P. J. Melz, *J. Chem. Phys.*, **57**, 1694 (1972).
4043 G. Pfister et al., *J. Chem. Phys.*, **57**, 2979 (1972).
4044 R. G. Williams and B. A. Lowry, *J. Chem. Phys.*, **56**, 5736 (1972).
4045 K. Tanikawa et al., *J. Appl. Phys.*, **48**, 2424 (1977).
4046 S. Tutihashi, *J. Appl. Phys.*, 3097, 3104 (1972).
4047 V. Tulagin, *J. Opt. Soc. Amer.*, **59**, 328 (1969).
4048 A. I. Lakatos, *J. Appl. Phys.*, **46**, 1744(1975).
4049 N. V. Joshi and M. Castillon, *Chem. Phys. Lett.*, **46**, 317 (1977).
4050 T. Kawai et al., *Chem. Phys. Lett.*, **56**, 541 (1978).
4051 H. Müller et al., *Chem. Phys. Lett.*, **50**, 22 (1977).
4052 H. Killesreiter and S. Schneider, *Phys. Chem. Lett.*, **52**, 191 (1977); *cf.* also Ref. 4081.
4053 Y. Kamura et al., *Chem. Phys. Lett.*, **46**, 356 (1977).
4054 B. Reimer and H. Bässler, *Chem. Phys. Lett.*, **55**, 315 (1978).
4055 M. Soma, *Chem. Phys. Lett.*, **50**, 93 (1977).
4056 J. H. Wang, *Proc. Natl. Acad. Sci.*, USA, **62**, 653 (1969).
4057 N. V. Joshi and M. Castillon, *Chem. Phys. Lett.*, **41**, 490 (1976).
4058 J. M. Mountz and H. Ti. Tien, *Photochem. Photobiol.*, **29**, 93 (1979).
4059 H. Meier et al., *Photochem. Photobiol.*, **16**, 353 (1972).
4060. G. Briere and F. Gaspard, *Chem. Phys. Lett.*, **7**, 537 (1970).
4061 D. S. Berns, *Photochem. Photobiol.*, **24**, 117 (1976).
4062 N. Bauser and W. Klöpfer, *Chem. Phys. Lett.*, **7**, 134 (1970).
4063 H. Tributsch, *Photochem. Photobiol.*, **16**, 261 (1972); H. Tributsch and M. Calvin, *ibid.*, **14**, 95 (1971).
4064 P. C. Lianos, *Diss. Abst. Inst.*, 839 (8) 3656 (1979).
4065 C. W. Tang et al., *Res. Discl.*, **173**, 73 (1978).
4066 H. H. Hoerhold et al., *Faserforsch., Textiltech.*, **29**, 571 (1978).
4067 E. L. Frankevich et al., *Pr. Nauk Inst. Chem. Org. Fiz. Politech.*, Wroclaw, **16**, 109 (1978).
4068 D. C. Brune et al., *Natl. Bur. Stand. (USA) Specl. Publ.*, **526**, 204 (1978).
4069 M. R. Padhye et al., *Proc. Nucl. Phys. Solid State Phys. Symp.*, **20C**, 622 (1977).
4070 S. D. Padke et al., *Proc. Nucl. Phys. Solid State Phys. Symp.*, **20C**, 146 (1977).
4071 Y. Umezawa and T. Yamamura, *J. Electrochem. Soc.*, **126**, 705 (1979).

REFERENCES TO ADDENDA TABLES 645

4072 Japan Pat. Kokai, 78,147,926 (23 Dec., 1978).
4073 N. N. Lichtin et al., *Adv. Chem. Ser.*, 173, 296 (1979).
4074 Z. Burshtein et al., *J. Phys. Chem. Sol.*, 39, 1125 (1978).
4075 S. S. Chen, *J. Membr. Biol.*, 47, 113 (1979).
4076 J. R. Rosseinsky and R. E. Malpas, *J. Electroanal. Chem.*, 68, 120 (1976); 89, 433 (1978).
4077 V. P. Skulachev, Proc. FEBS Meetg., 1977, Part 45, Membrane Proteins, publ. 1978, p. 49; *cf.*, also N. G. Abdulaev et al., *FEBA Lett.*, 90, 190 (1978).
4078 M. Kaneko et al., *Makromolec. Chem.*, 179, 2431 (1978).
4079 K. Okamoto et al., *Bull. Chem. Soc. Japan*, 46, 1948 (1973).
4080 F. J. Kampars and M. Gouterman, *J. Phys. Chem.*, 81, 690 (1977); see also Ref. 2142.
4081 H. Killesreiter et al., *Ber. Bunsen Ges. Phys. Chem.*, 82, 503 (1978).
4082 L. Packer et al., *Proc. Em. Conf. Living Syst. Energy Conv.*, 1976, R. Buvet et al., ed., Nth. Holland, Amsterdam, 1977, p. 119; T. R. Herrmann and G. W. Rayfield, *Biophys. J.*, 21, 111 (1978).
4083 I. I. Dilung and E. I. Kapinus, *Russ. Chem. Res.*, 47, 43 (1978).
4084 T. Yiyasaka et al., *J. Am. Chem. Soc.*, 100, 6657 (1978).
4085 H. Yasunaga et al., *J. Phys. Soc. Japan*, 46, 839 (1979).
4086 M. I. Fedorov and V. A. Benderskii, *Soc. Phys. Semicond.*, 4, 1198, 1720 (1971).
4087 D. L. Morel et al., *Appl. Phys. Lett.*, 32, 495 (1978).
4088 S. N. Chen et al., *Appl. Phys. Lett.*, 36, 96 (1980).
4089 H. Tsubomura et al., *J. Phys. Chem.*, 83, 2103, (1979).
4090 W. D. Clark and J. A. Eckert, *Solar Energy*, 17, 147 (1975).
4091 K. Shigehara et al., *Electrochim. Acta*, 23, 855 (1978).
4092 S. Magner and M. Savy, *Electrochim. Acta*, 23, 669 (1978); T. Imura et al., Int. Conf. Solar Energy, Univ. of Ontario, London, Ont., Canada, 1976 Proc.
4093 J. Nasielski et al., *Electrochim. Acta*, 23, 605 (1978).
4094 M. Aizawa et al., *Electrochim. Acta*, 23, 1185 (1978); 24, 89 (1979).
4095 F. Willig et al., *Electrochim. Acta*, 24, 463 (1979); H. Gerischer and F. Willig, in *Topics in Current Chemistry*, F. L. Boscke, ed., 61, 31 (1976) Springer, Berlin, 1976.
4096 H. Ti Tien and J. M. Mountz, *J. Electrochim. Soc.*, 125, 885 (1978).
4097 G. Kossmehl, *Ber. Bunsen Ges. Phys. Chem.*, 83, 417 (1979).
4098 N. Minami, *Ber. Bunsen Ges. Phys. Chem.*, 83, 476 (1979).
4099 W. Arden and P. Fromherz, *Ber. Bunsen Ges. Phys. Chem.*, 82, 868 (1978).
4100 S. Arnold et al., *Phys. Stat. Solidi*, B-94, 263 (1979).
4101 D. A. Sell and A. Kuppermann, *J. Chem. Phys.*, 71, 4703 (1979).
4102 J. Berkowitz, *J. Chem. Phys.*, 70, 2819 (1979).
4103 P. L. Krowick, *J. Appl. Phys.*, 39, 5806 (1968).
4104 M. J. Dewar and S. D. Worley, *J. Chem. Phys.*, 51, 263 (1969).
4105 S. C. Khandelwal and J. L. Rocker, *Chem. Phys. Lett.*, 34, 355 (1975).
4106 E. Amouyal et al., *Photochem. Photobiol.*, 29, 1071 (1979).
4107 D. C. Frost et al., *Chem. Phys. Lett.*, 2, 663 (1969).
4108 K. Ishii et al., *Chem. Phys. Lett.*, 41, 154 (1976).
4109 N. Sato et al., *Annual Rev. Met. Molec. Sci.*, Okazaki, Japan, 1979, p. 97.
4110 K. Seki and H. Inokuchi, *Chem. Phys. Lett.*, 65, 158 (1979).
4111 J. Lin et al., *J. Amer. Chem. Soc.*, 102, 4627 (1980).
4112 C. Colbourne, *Chem. Phys. Lett.*, 72, 247 (1980).
4113 K. Kimura et al., *J. Amer. Chem. Soc.*, 100, 6564 (1978).

4114 Y. Nakato et al., *Bull. Chem. Soc. Japan,* **47**, 3001 (1974).
4115 A. K. Kostin and A. V. Vannikov, *Fiz. Tverd. Tela* (Leningrad), **20**, 3407 (1978).
4116 J. Mort, *Phys. Rev.,* **B5**, 3329 (1972).
4117 R. L. Emerald and J. Mort, *J. Appl. Phys.,* **45**, 3943 (1974).
4118 J. Dresner, *J. Chem. Phys.,* **52**, 6343 (1970).
4119 W. F. Schmidt and A. J. Allen, *J. Chem. Phys.,* **50**, 5037 (1969).
4120 A. V. Vannikov, *Russ. Chem. Revs.,* **44**, 906 (1975).
4121 J. M. L. Engels and A. J. M. Van Kimmenade, *Chem. Phys. Sett.,* **42**, 250 (1976).
4122 W. Döldissen et al., *Chem. Phys. Let.,* **56**, 347 (1978).
4123 E. E. Conrad and J. Silverman, *J. Chem. Phys.,* **51**, 450 (1969).
4124 J.-P. Dodelet et al., *J. Chem. Phys.,* **59**, 1293 (1973).
4125 N. E. Cipollini and A. O. Allen, *J. Chem. Phys.,* **67**, 131 (1977).
4126 T. G. Ryan and G. R. Freeman, *J. Chem. Phys.,* **68**, 5144 (1978).
4127 H. Sato et al., *J. Appl. Phys.,* **45**, 1675 (1974).
4128 E. I. Walker et al., *J. Chem. Phys.,* **68**, 4134 (1968).
4129 J. B. Webb et al., *Solid State Comm.,* **31**, 905 (1979).
4130 Z. Burshtein and D. F. Williams, *J. Chem. Phys.,* **67**, 3592 (1977).
4131 J. B. Webb et al., *J. Phys.,* **C11**, 3245 (1978).
4132 N. Cipollini et al., *Chem. Phys. Lett.,* **53**, 404 (1978).
4133 A. Szymanski and M. M. Labes, *J. Chem. Phys.,* **50**, 1898 (1969).
4134 B. Fitton, *J. Phys. Chem. Solids,* **30**, 211 (1969).
4135 G. H. Richardson et al., *J. Chem. Phys.,* **63**, 74 (1975).
4136 M. T. Jones and M. Josita Feighan, *J. Chem. Phys.,* **49**, 5549 (1968).
4137 V. Kampars and O. Neilands, *Zh. Obshch. Khim.,* **47**, 150 (1977).
4138 J. H. Richardson et al., *J. Chem. Phys.,* **59**, 5068 (1973).
4139 B. K. Janousek and J. I. Bravnan, *J. Chem. Phys.,* **72**, 694 (1980).
4140 K. Kimura et al., *J. Chem. Phys.,* **70**, 3317 (1979).
4141 P. Baraldi et al., *J. Electrochem. Soc.,* **126**, 1207 (1979).
4142 S. Takagi and K. Kawabe, *J. Phys. Soc. Japan,* **46**, 440 (1979).
4143 W. Hwang and K. C. Kao, *J. Chem. Phys.,* **58**, 3521 (1973).
4144 J.-P. Dodelet and G. R. Freeman, *J. Chem. Phys.,* **59**, 1293 (1973).
4145 M. M. Nicholson and F. A. Pizzarello, *J. Electrochem. Soc.,* **126**, 1490 (1979).
4146 A. Abou El Hela and H. H. Afifi, *J. Phys. Chem. Solids,* **40**, 257 (1979).
4147 D. L. Morel and H. Berger, *J. Appl. Phys.,* **46**, 863 (1975).
4148 M. I. Fedorov and V. A. Benderskii, *Soviet Phys. Semicond.,* **4**, 1198 (1970).
4149 H. Yasunaga et al., *J. Phys. Soc. Japan,* **33**, 1024 (1974).
4150 D. Vasilescu, Private Communication, 1971.
4151 G. Kossmehl, *Ber. Bunsen Ges. Phys. Chem.,* **83**, 417 (1979).
4152 R. C. Wheland and J. L. Gillson, *J. Amer. Chem. Soc.,* **98**, 3916 (1976).
4153 J. B. Torrence et al., *J. Amer. Chem. Soc.,* **101**, 4748 (1979).
4154 Y. Nakahara et al., *Chem. Phys. Lett.,* **47**, 250 (1977).
4155 Y. Nakahara, K. Kimura, H. Inokuchi and T. Yagi, *Chem. Phys. Lett.,* **73**,
4156 K. Kimura et al., *J. Amer. Chem. Soc.,* **100**, 6564 (1978); *J. Chem. Phys.,* **70**, 3317 (1979).
4157. Y. Hayashi et al., *Bull. Chem. Soc. Japan,* **39**, 1664 (1966).

25

Post-scriptum

An annotated and selected supplemental bibliography covering mainly the period 1981-1982.

25.1 Supplemental to Chapter 12 "Experimental and Methodology"

A critical re-examination of the application of van der Pauw's arrangement: C. Albers et al., *Cryst. Res. Technol.*, **17**, (5) 585 (1982).

A TTF/TCNQ electrode in an organic solvent is said to provide quasi-reversible electron transfer: D. J. Sandman et al., *Synth. Met.*, **4**, (3) 249 (1982).

A study of the a.c. conductance of organic solid thin films; the dependence of the conductance on the square of the frequency is said to be a contact effect: R. J. Quian et al., *Wu ki Hsueh Pao*, **29**, (8) 992 (1980).

An a.c. method for conductance measurements: R. N. Gupta and M. Misra, *Indian J. Pure and Appl. Phys.*, **19**, (12) 1151 (1981).

Several models for the a.c. conductance of disordered solids, below 10^{11} Hz, are scrutinized and discussed: U. Strom and K. L. Ngai, *J. Phys. Colloq.*, C4, Pt. 1, p. 123 (1981).

The effect pf doping and of impurities on carrier transport in organic molecular crystals, especially in ultrapure anthracene, is discussed: N. Karl, *Chem. Scr.*, **17**, (1-5) 201 (1981).

Hall effect measurements in samples exhibiting a van der Pauw geometry. An empirical averaging technique is described said to avoid errors otherwise liable to occur and reaching up to 600%: D. M. Boerger et al., *J. Appl. Phys.*, **52**, (1) 269 (1981).

An a.c. Hall effect measuring system is described in which the currents for the sample, the magnet and the reference are all derived from one digital oscillator: D. C. Abbas and D. J. Phelps, *J. Phys.*, **E-14**, (9) 1078 (1981).

The measurement of the conductivity of thin films in a microwave cavity is described: Y. Le Cleach, *Revs. Phys. Appl.* (France), **17**, (8) 481 (1982).

Field modulated electro-reflectance studies: L. Sebastian and G. Weiser, *Chem. Phys.*, **62**, (3) 447 (1981).

Experimental studies of electro-optic and ferro-electric phenomena in organic solids: G. F. Lipscomb, *Dissert. Abst. Int. B.*, **41**, (10) 3822 (1981).

Reviews: Hall effect measurements in polycrystalline and in powdered samples: J. W. Orton and M. J. Powell, *Rep. Progr. Phys.*, **43**, (11) 1263 (1980).

A critical discussion of methodology: R. D. Gould, *J. Appl. Phys.*, **53**, (4) 3353 (1982).

25.2 Supplemental to Chapter 13 "Excited States"

The migration of mobile excitons is discussed in terms of a percolation model and applied to isotropic naphthalene: ACS Symp. Ser., 1981, **162**, p. 57.

The participation of exciplexes formed by the interaction of excited singlet states of a sensitizing dye with the chromophore of a polymer is said to be involved in the generation of free charge carriers: B. M. Rumyansev et al, *Vysokomol. Soedin* ser. A., **22**, (11) 2545 (1980).

Dipole moments of excited states and of photochemical transients obtained from microwave dielectric measurements are reported: R. W. Fessenden et al., *J. Phys. Chem.*, **86**, (19) 3803 (1982).

Ground state energies of exciton complexes bound to a donor molecule are reported: K. K. Bajaj et al., Proc. Int. Conf. Phys. Semicond., 15th **1980** p. 457; S. Tanaka and Y. Toyozawa, eds.

The role of bi-excitons as charge transfer bound states for exciton interaction: G. Gumbs and C. Mavroyannis, *Phys. Rev.*, **B-24**, (12) 7258 (1981); *Solid State Commun.*, **41**, (3) 237 (1982). See also T. D. Clark et al., *Phys. Status Solidi*, **B-110**, (1) 341 (1982).

Excitons, polarons and bi-polar entities in conducting polymers: S. A. Brazovskii and N. N. Kirova, *Pisma Zh. Eksp. Teor. Fiz.*, **33**, (1) 6 (1981).

Theoretical studies of charged defects in doped polyacetylenes indicate polarons with a binding energy of about 0.05 eV producing charged solitons. The appearance of doubly charged defects *viz.* bipolarons is proposed: J. Bredas et al., *Mol. Cryst. Liq. Cryst.*, **77** (1-4) 319 (1981); *cf.*, also M. Dugay and J. Rouston, *ibid.*, p. 333.

Exicton self-trapping has been discussed: D. P. Craig et al., *Chem. Phys.*, **46**, 87 (1980).

Self-trapping of excimers in dichloroanthracene has been studied: U. Mayer et al., *Chem. Phys.*, **59**, 449 (1981); *cf.*, also P. Krauss et al., *Acta Cryst.*, B-35, 1419 (1979).

25.3 Supplemental to Chapter 14: "Ionized States"

Calculations of ionization potentials of conjugated polymers including polyacetylenes: J. L. Bredas et al., *J. Chem. Phys.*, **76**, (7) 3673 (1982).

Charge transfer states in anthracene: W. Siebrand and P. J. Bounds, *Mater. Sci.*, **7**, (1) 67 (1981).

Calculation of charge transfer energies in molecular crystals: J. P. Bounds, *Mater. Sci.*, **7**, (2-3) 107 (1981).

Review: Ionization or appearance potentials for over 2000 inorganic and organic crystals are listed and the subject is reviewed, 668 references: H. M. Rosenstock et al., Natl. Stand. Ref. Data Ser. (U.S. Natl. Bur. Stand.), 1980, NSRDS-NBS 66 Pt., 374 pp. It is derived theoretically that the Verdet and the Hall coefficients change sign when the chemical potential of the system attains zero value; then, in a narrow band semiconductor, conductivity changes from n- to p-type: I. P. Kogotyuk and F. V. Skripnik, Spektrosk. Mol. Krist. Mater. Resp. Shk.-Semin., 4th 1979; *Izd. Nauk., Dumka Kiev*, **1**, 212; M. T. Shpak, ed., Publ. 1981.

25.4 Supplemental to Chapter 15 "Theories of Charge Transfer"

The band model is said to be generally applicable to electron transport in metal phthalocyanines, but holds for hole transport only in the b-direction: S. Mitra, *Indian J. Phys.*, A-56, (3) 197 (1982).

A theoretical study of energy and electron transport in disordered solids: R. Silbey, *Mater. Sci.*, **7**, (1) 77 (1981).

Hopping charge transfer in disordered semiconductors: M. van der Meer et al., *Phys. Status Solidi*, **110**, (2) 571 (1982).

Resonance integrals of sixteen dye molecules are listed; mainly cyanine dyes: Tang, Ying Wu et al., *Qinghua Daxue Xuebao,* **21**, (4) 71 (1981).

In disordered structures such as glasses, self-trapping of like carriers (two electrons OR two holes) is possible; thus bipolarons are produced. They may exhibit a negative correlation energy, of the order of 1 eV or less, leading to an effective attraction between carrier pairs involving anomalously high dielectric susceptibilities of the disordered atomic subsystem: M. I. Klinger and V. G. Karpov, *Zh. Eksp. Teor. Fiz.,* **82**, (5) 1787 (1982).

The carrier mobility in polycrystalline semiconductors is treated *via* a grain boundary model: K. R. Kumar and M. Satyam, *Appl. Phys. Lett.,* **39**, (11) 898 (1981).

Carrier drift and diffusion in materials containing traps: A. I. Rudenko and V. I. Arkhipov, *Phil. Mag.,* **B-45**, (2) 177 (1982); see also *ibid.,* pp. 189 and 209.

The Hall mobility is derived from Hall effect measurements on polycrystalline semiconductor films: J. W. Orton, *Thin Solid Films,* **86**, (4) 351 (1981).

Effect of surface space charge layers on Hall- and Seebeck-effects in semiconducting films: S. K. Datta and A. K. Chaudhuri, *J. Phys.,* **D-14**, (9) L-141 (1981).

Correlations between the activation energy and the pre-exponential factor and variable range hopping conductivity: A. Kurobe and H. Kamimura, *J. Phys. Soc. Jpn.,* **51**, (6) 1904 (1982); see also A. E. Pochtenni and E. V. Ratnikov, *Dokl. Akad. Nauk. BSSR,* **25**, (10) 896 (1981); see also Yu. A. Vidadi, *Zh. Fiz. Khim.,* **55**, (9) 2413 (1981).

Percolation treatment of conductivity in the small polaron-high temperature hopping regime in disordered systems: G. P. Triberis and L. R. Friedman, *J. Phys.,* **C-14**, (31) 4631 (1981).

25.5 Supplemental to Chapter 16 "Review of Published Data"

25.5.1 (General): Macroscopic aspects of carrier transport in thin films in a direction perpendicular to the planes of the faces. If the film is not trap-free, three characteristic lengths determine the measured properties: 1. free carrier Debye length which governs screening; 2. diffusion length which governs carrier survival against geminate recombination; 3. film thickness. In the presence of traps, the Debye length exceeds the screening length: H. K. Henish and J. C. Manifacier, *Thin Solid Films,* **89**, (1) 3 (1982).

SUPPLEMENTAL TO CHAPTER 16 "REVIEW OF PUBLISHED DATA"

Conductivity, mobility and Seebeck coefficient measurements for tetrathiotetracene:I_3 charge transfer complexes are reported for varying degrees of disorder: S. K. Khanna et al., *Phys. Rev.*, B-24, (6) 2958 (1981).

Supersensitization of electron transfer in monolayer assemblies due to the dye molecules becoming organized into aggregates and undergoing a charge transfer interaction with the supersensitizing agent. That agent acts as electron donor and does not itself lead to an observable (photo)-current: Th. L. Penner and D. Möbius, Amer. Chem. Soc. Abst. of papers, 21st Aug. 1982–Coll. #41.

25.5.2 (Phthalocyanines): Refractive index and apparent density of evaporated films of Cu-phthalocyanine: Qian Renyuan et al., *Kexue Tongbao*, 26, (6) 522 (1981); Shi Zurong et al., *Huaxue Tongbao*, 2, 24 (1982); Wang Dianxun et al., *Huaxue Tongbao*, 11, 655 (1980).

Bulk trapping states in beta-zinc-phthalocyanine: a trap level of 0.37 eV below the edge of the conduction band, and with an activation energy of 1.67 eV is reported: T. G. Abdel-Malik et al., *Phys. Stat. Solidi*, A-72, (1) 99 (1982).

Conductivity and Seebeck effect data on beta-Cu-phthalocyanine; residual impurities exert a strong effect. Conductivity is ascribed to hopping: A. N. Asanov and A. T. Vartanyan, *Zhur. Fiz. Khim.*, 56, (1) 205 (1982).

The d.c. and a.c. conductivity of Cu, Co and Pb phthalocyanine have been determined: W. Waclawec et al., *Mater. Sci.*, 7, (2-3) 385 (1981).

The molecular positions in lattice images as well as distortions due to the appearance of high-energy sites or to boundaries are located by electron microscopy: J. R. Fryer, Electron Microsc. Proc. Eur. Congr., 7th 1980, 2, p. 676, P. Bredero, ed.

The values of depth profiles at Fe/Cu-phthalocyanine interfaces has been determined as about 300 to 400 Å using Auger electron spectroscopy: E. R. Davis and L. R. Faulkner, *J. Electrochem. Soc.*, 128, (6) 1349 (1981).

The conductivity of Ti-phthalocyanine is said to increase by 10^8 Scm^{-1} under high pressure due to the increase in intermolecular conjugation: Yu. A. Berlin et al., *Zh. Fiz. Khim.*, 56, (2) 499 (1982).

Co and Zn phthalocyanines exhibit electrochromism: V. I. Gavrilov et al., *Elektrokhimiya*, 16, (10) 1611 (1980).

Poly-Fe-phthalocyanine deposited from sulfuric acid is said to be active as an electro-catalyst: L. Kreja and A. Plewka, *Electrochim. Acta*, 27, (2) 251 (1982); cf., also J. Zagal et al., *J. Electroanal. Chem. Interfacial Electrochem.*, 135 (2) 343 (1982).

Electrocatalytical activity is also reported for graphite electrodes coated with sulfonated Fe,Co,Cu,Ni phthalocyanine; the Fe-substituted compound has the highest activity: M. E. Munoz et al., *Bul. Soc. Chil. Quimic.*, **27** (1) 121 (1982); *cf.*, also J. H. Zagal, *Contrib. Cient. Tecnol.*, (Univ. Tec. Estado Santiago Chile), **48**, 39, 47 (1981).

Monoclinic Pb-phthalocyanine is suggested to behave as a quasi one-dimensional semi-metal; it exhibits switching between semiconducting and metallic behaviour and a characteristic order of molecular dipoles.: F. Przyborowski and C. Hamann, *Cryst. Res. Technol.*, **17**, (8) 1041 (1982).

Review: "Metalloid porphyrins and phthalocyanines,": P. Sayer et al., *Acc. Chem. Res.*, **15**, (3) 73 (1982).

25.5.3 (Mobilities): Mobilities in organic molecular crystals: L. B. Schein and D. W. Brown, *Mol. Cryst. Liqu. Cryst.*, **87**, (1-2) 1 (1982), gives list of mobility values. The hole mobility of iodoform is reported as anisotropic with values of 0.05 to 0.7 cm^2/Vsec over the range of 200 to 340 K, using the time-of-flight method: J. Giermanska et al., *Mater. Sci.*, **7**, (2-3) 153 (1981).

The conductivity of anilin-black has been studied: J. J. Langer, *Mater. Sci.*, **7**, (2-3) 223 (1981).

25.5.4 (Polymers): Poly(phenylene sulfide) doped with H_2SO_4 or SO_3 is reported to exhibit high conductivity: K. K. Showa Denko Jpn. Pat. 82 21456 (1982).

The conductivity of high density polyethylene is said to be independent of the nature of the metal contact(s); there is no charge injection: V. Adamec et al., *Konf. Cesk. Fyz.* (Sb. Prednasek) 7th, 1981, **1** (1) paper No. 213.

The conductivities of polypyrrole and of polybiphenylene have been studied: M. Tanaka et al., *Mo. Cryst. Liq. Cryst.*, **83**, (1-4) 1309 (1982).

The conductivities of polymeric azomethines with azobenzene units are reported: G. Kossmehl and J. Wallis, *Makromol. Chem.*, **183**, (2) 347; *cf.*, also *ibid.*, p. 331.

The anisotropy of the conductivity in oriented polymers: S. Hirota, *J. Appl. Phys.*, **53**, (5) 3792 (1982).

Essentially non-conducting polymers such as poly-urethane can be made reasonably conducting by admixing an organic semiconductor: Matsushita Elect. Indust. Co. Ltd. Jpn. Pat. 30 98407 (26th July 1980).

The volume v fraction of poly-conjugated regions in polyacrylonitrile-based semiconductors, the average distance between them as well as the conductivity increase with rising tempurature; v being given by

$$v = \frac{m - m''}{m' - m''}$$

where, m' and m'' refer to the semiconductor densities at a temperature T, at 200 °C and at 1100 °C, respectively: U. Abdurakhmanov and M. A. Magrupov, Dokl. Nauk Uzb. SSR, (1980) (3) 45.

Review: The energy migration in polymers is reviewed: several papers by several authors in *Macromolecules,* **15** (3) (1982).

25.5.5 (Seebeck Coefficients): Theory of the Seebeck effect in polycrystalline semiconductors: J. Jerhot and J. Vleck, *Thin Solid Films,* **92**, (3) 259 (1982), also D. Deschat et al., *Phys Status Solidi,* **A-71** (2) K-205 (1982).

The Seebeck and the Hall coefficients in percolation theory, considering a two-component disordered system, are theoretically discussed: A. S. Skal et al., *Phil. Mag.,* **B-45**, (3) 323, 335 (1982).

25.5.6 (Liquids): Electrical conductivity in undehydrated hydrocarbons in low electric fields; charge transfer occurs mainly *via* the water molecules and is strongly affected by convection effects: K. Yatsuzuka and A. Watanabe, *Jpn. J. Appl. Phys.,* Pt. 1, **21**, (6) 920 (1982).

Cyclohexane and methylene chloride exhibit an increase in their polarization charges during the passage of a current: L. S. Kazatskaya, *Elektron. Obrab. Mater.,* (1982) (3) 27.

The Hall effect in organic liquids such as dimethylformamide is studied; mobility values as high as 200 cm^2/Vsec are reported. L. S. Kazatskaya, *Elektron. Obrab. Mater.,* (1981) (1) 52.

In dimethylether, it is said that the energy abstracted from the applied electric field is dissipated mainly *via* inelastic collisions, the electrons going into relatively stable localized states: N. Gee and G. R. Freeman, *Canad. J. Chem.,* **60**, (8) 1034 (1982).

In organic liquids, residual charges are retained over long periods of time because of the tendency of the molecules not to give up charges acquired during passage of current: L. S. Kazatskaya, *Elektron. Obrab. Mater.,* (1982), (2) 24.

The conductivity of, and space charges in, bromobenzene solutions in dimethylformamide were studied: L. S. Kazatskaya et al., *Elektron. Obrab. Mater.,* (1981), (5) 31.

The conductivity of molten methylpyridinium halides: D. S. Newman et al., *J. Electrochem. Soc.*, **128**, (11) 2331 (1981).

The conductivity of fused polymers, especially low and high density polyethylene, of polystyrene and of poly(vinylidenefluoride) is reported to be severaly affected by impurities, the d.c. current dropping with time to a final, steady, state: M. Takeuchi and H. Nagasaka, *J. Elec rostat.*, **12**, 189 (1982).

Review: Liquid semiconductors are reviewed: A. F Regel and A. Andreev, *Porbl. Sovrem. Fiz.*, **1980**, p. 162.

25.6 Supplemental to Chapter 17 "Charge Transfer Complexes"

So many of the semi-infinite number of possible adducts between donors and acceptors have been reported that only those compounds are here referred to which exhibit some extra features of interest apart from their semi-conductivity.

25.6.1 (Charge Transfer Complexes Involving TCNQ): Photoelectron spectra of TCNQ:Cu are reported: R. S. Potember et al., *Chem. Ser.*, **17** (1-5) 219 (1981).

TCNQ adducts of pyridine derivatives have been studied: A. Graja et al., *Mol. Cryst. Liq. Cryst.*, **85**, (1-4) 1647 (1982).

The UV/spectra, ESR and electrical conductivities of N-alkyl-2-aminopyridinium ion radical type adducts with TCNQ have been studied; the size of the alkyl groups considerably affects the conductivity of the complex. 1:1 as well as 1:2 stoichiometries were obtained with 1:2 formed in acetonitrile: Bi Xian-Tong et al., *Acta Chim. Sinica*, **40**, 32 (1982).

The Tetramethylbenzidine:TCNQ complex: N. S. Hemamalini et al., Proc. Nucl. Phys. Solid State Phys. Symp., 1981, Publ. 1982, 24-C 489 (1982).

Charge transport in single crystals of TCNQ complexed with methyl derivatives of pyridinium. The low temperature behavior is said to be controlled by deep-lying electrically active impurity levels (about 5×10^6 cm^{-3}) which are partially compensated donors: M. Przybylki and A. Graja, *Physica B & C* (Amsterdam), **104**, (3) 278 (1981).

In Cu-TCNQ films it is suggested that the complex salt is only produced in the presence of an applied field because TCNQ° becomes evident from IR reflectance spectroscopy, Auger and X-ray photoelectron spectroscopy, only after application of an electric field: R. S. Potember et al., *Chem. Scr.*, **17**, (1-5) 219 (1981).

In TTF:TCNQ adducts, the degree of charge transfer is said to be about 0.5 as ascertained from Raman spectroscopy; M. Tokumoto et al., *Mol. Cryst. Liq. Cryst.*, **85**, (1-4) 1585 (1982).

TCNQ adducts of Cu-chelates are reported to exhibit a room temperature conductivity of 1.9 Siemens cm^{-1}; M. Inoue et al., *Mol. Cryst. Liq. Cryst.*, **86**, (1-4) 1879 (1982).

A study of TCNQ complexes at pressures up to 80 kBars: A. K. Bandyopadhyay et al., *Mater. Sci.*, **7**, (2-3) 97 (1981); cf., also *J. Phys. Chem.*, **13**, (29) L-803 (1980).

Vapor deposited TT:TCNQ thin films are reported to have a room temperature conductivity of 65 Siemens cm^{-1} and an activation energy of 0.02 eV; charge transport is ascribed to linear hopping: C. Hamann et al., *Mater. Sci.*, **7**, (2-3) 181 (1981).

Charge transfer complexes from five ring donors, e.g., 1,2 dithiole derivatives, 1:1 and 1:2 have been obtained: J. Amzil et al., *Chem. Scr.*, **17**, (1-5) 65 (1981).

The 4,4' dithiopyranylidene:TCNQ complex is said to show a room temperature conductivity of 30 Siemens cm^{-1}: D. J. Sandman et al., *J. Chem. Soc*, Perkin Trans., II; **1980**, (11) 1578.

The 1,3(di(1-methyl-4 pyridinium)propane)$^{2+}$:TCNQ$_4$ adduct is reported to show a room temperature conductivity of about 50 Siemens/cm^{-1} if hydrated, the activation energy being 0.05 eV; G. J. Ashwell, *Chem. Scr.*, **17**, (1-5) 107 (1981).

The TTF:TCNQ complex has been studied: S. Etemad, *Phys. Rev.*, **B-24**, (9) 4959 (1981).

The electrical properties of TCNQ complexes with ionenes have been investigated; their conductivity rises as the number of CH$_2$ groups per polymer unit increases: W. Ciesielski et al., *Acta Polym.*, **33**, (5) 318 (1982).

The crystal structure and electronic properties of TTF:TCNQ complexes are discussed: S. Kagoshima, *Busei Kenkyu*, **34**, (1) 1 (1980).

Reviews: V. A. Starodub and I. V. Krivoshei, *Usp. Khim.*, **51** (5) 764 (1982).

K. Pigon and H. Chojnacki, *Mol. Interact.*, **2**, 451 (1981); S. Ikeda, *Kagaku Zokan* (Kyoto), **87**, 7 (1980); P. Delhaes, et al., *J. Chim. Phys. Phys.-Chim. Biol.*, **79**, (4) 299 (1982).

25.6.2 (Charge Transfer Complexes Not Involving TCNQ): Transition-metal complexes of macrocyclic tetrapyrroles: Y. Matsuda et al., *Inorg. Chem.*, **20**, (7) 2239 (1981).

Perylene:Chloranil adducts were studied over the frequency range of 10^{-5} to 10^{+5} Hz. At very low frequency, charge transport is said to be by activated hopping between trapping sites: R. Pethig and S. Bone, *Mater. Sci.*, **7**, (2-3) 291 (1981).

The conductivity of evaporated films of *p*-terphenyl:tetracyanoethylene complexes was investigated over the range of 160 to 300 K: A. Lipinski et al., *Mater. Sci.*, **7**, (2-3) 231 (1981).

Adducts between rhodamine B deposited on single crystals of anthracene were studied: G. Papier et al., *Ber. Bunsen Ges. Phys. Chem.*, **86**, (7) 670 (1982).

The electrical conductivity and the thermoelectric properties of dimethylpyridinium:TCNQ adducts have been studied: W. Przybylski and A. Graja, *Mater. Sci.*, **7**, (2-3) 321 (1981).

A metal-to-semiconductor transition in TTF:TCNQ is said to occur when measurements of, *e.g.*, thermoelectric powers are effected along the b-axis at 56 K: B. P. Chandra, Proc. Nucl. Phys. Solid State Phys. Symp., 1981, publ. 1982, **24C**, 617 (1982).

The electrical properties of metalloid porphyrin- and phthalocyanine-adducts are reviewed: P. Sayer et al., *Acc. Chem. Res.*, **15**, (3) 73 (1982).

The pressure dependence of the Seebeck coefficient of TTF:TCNQ adducts has been measured at room temperatures as -10 μV/K at 15 kBar and as -30 μV/K at 1 bar. The sign of the Seebeck coefficient changes at the metal to semiconductor transition: C. Weyl et al., *J. Phys.* (France), **43** (7) 1167 (1982).

The role of water in metachromatic complexes of methylene blue was studied: J. B. Lawton and G. O. Phillips, *Makromol. Chem.*, **183** (6) 1497 (1982).

Polyelectrolyte charge transfer complexes between glycol chitosan and polysaccharides are reported: R. Srinivasan and R. Kamalam, *Biopolymers*, **21**, (2) 265 (1982).

Alkali metal:aromatic complexes are catalytically active in the para/ortho hydrogen conversion process: M. Sano et al., *Mater. Sci.*, **7**, (2-3) 347 (1981).

A ternary intercalation complex K-graphite-dimethylsulfoxide exhibits its highest conductivity along the layer plane: N. Okuyama et al., *Physica B & C*, **105**, (1-3) 298 (1981).

25.6.3 (Polymeric Charge Transfer Complexes): The arsenic pentafluoride: poly(2,5-thiophenediyl) adduct has a conductivity at room temperatures of 0.02 S cm^{-1} and an activation energy of 0.04 eV: G. Kossmehl and G. Chatlitheodorou, *Mol. Cryst. Liqu. Cryst.*, **83**, (1-4) 1323 (1982).

Semiconducting complexes between poly(1,1'-ferroceneylene) and iodine or TCNQ: K. Sanechika et al., *Polymer J.* (Tokyo), **13**, (3) 255 (1981).

In polyvinylcarbazole:TCNQ complexes, conductivity is said to involve Poole-Frenkel electron transfer; the activation energy is 0.37 eV: A. Kuczkowski, *Eur. Polymer J.*, **18**, (2) 109 (1982).

Acceptor doped polyphenylene complexes are reported: H. Eckhardt et al., *Mater. Sci.*, **7**, (2-3) 121 (1981).

Poly(*p*-phenylene sulfide):hexafluoroarsenate adducts are *p*-type with a conductivity of 1-10 S cm^{-1}: T. C. Clarke et al., *Chem. Scr.*, **17**, (1-5) 161 (1981).

The semiconducting polymides doped with metal compounds exhibit good high temperature adhesive properties: L. T. Taylor et al., *Org. Coat. Plast. Chem.*, **43**, 635 (1980).

Polymeric ionene:TCNQ complexes are reported: E. N. Varakina et al., *Vysokomol Soedin*, Ser. B, **23**, (10) 740 (1981).

Doped polyphenylenes are said to show conductivities in excess of 0.01 S cm^{-1}: V. Muench et al., Eur. Pat. Appl. EP 44 935 (3rd Feb. 1982).

Reviews: R. B. Seymour, *Polymer News*, **8**, (2) 41 (1982); C. B. Duke and H. Gibson, in *Kirk-Othmer Encycl. Chem. Technol.*, 3rd ed. 1982; M. Grayson and D. Eckroth, eds., Wiley, N.Y.

25.7 Supplemental to Chapter 18 "Biological Materials"

Anhydrous Films of tetrahemoprotein cytochrome C3 exhibit good conductivity at 292 K: K. Ichimura et al., *Chem. Lett.*, **(1982)**, (1) 19.

It is shown that radicals and/or radical ions move relatively freely along hydrocarbon chains, especially when any unsaturation permits the formation and isomerization of allyl radicals, giving rise to electrical conductivity. The trans-membrane charge transfer time is reported as $10^{-9} - 10^{-11}$ sec. in spite of low carrier mobility because of the high field developed across the thin membrane: I. I. Ivanov, *Biofizika*, **27**, (2) 326 (1982).

The conductivity of bilipid membranes rises upon addition of oxidants and drops upon addition of reductants. The stability of the system, however, exhibits the converse behavior: M. Dely et al., *Stud. Biophys.*, **85**, (2) 115 (1981).

Charge transfer complexes between amino-acids and paraaminobenzoquinone are reported: A. Chattopadhay and S. C. Lamiri, *J. Indian Chem. Soc.*, **57**, (6) 604 (1980).

Insect cuticula transversely cut emit electrons under solar irradiation, effect even visible microscopically; E. Morgenstern, in *Electromagnetic Bio-Information*, F. A. Popp, ed., Urban & Schwarzenberg, Munich-Vienna-Baltimore, 1979, p. 62.

Organic semiconductors are suggested as model systems for primary photochemical processes such as photosynthesis: E. L. Frankevich and M. M. Tribel, *Latvian PSR Zinat Akad. Vestis, Fiz. Teh. Ser.*, **(1981)** (6) p. 40.

The conductivity of, and effect of oxygen on, microcrystalline films of chlorophyll-*a* has been studied: O. Inganaes and I. Lumdstroem, *Thin Solid Films*, **85**, (2) 129 (1981).

The charge transport in compressed hydrated bovine serum albumen is ascribed to multiple hopping *via* a series of traps; the very low mobility is inversely proportional to the applied electric field: P. G. Lederer et al., *J. Chem. Soc. Faraday Trans. I*, **77**, (12) 2989 (1981).

The role of bacterio-rhodopsin in the proton pumping activity of halobacteria is investigated: C. Bagyinka et al., *Acta Biol. Acad. Sci. Hung.*, **32**, (3-4) 311 (1981).

The role of excited states in biological phenomena is discussed in very many contributions; *e.g.*, Excited states and carcinogenesis, F. A. Popp, *Strahlen Therapie*, **144**, 208 (1972); *Z. Naturf.*, **28C**, 517 (1973); F. Kaiser, *ibid.*, **33A**, 294 (1978); F. A. Popp, *Bio-Photon Phys.*, **1**, 1 (1978); **2**, 15 (1978).

The trans-membrane conductance is modified by the incorporation of alamethicin: G. Baumann and G. S. Easton, *J. Membrane Biol.*, **52**, 237 (1980); *cf.*, also M. P. Borisova et al., *Biochim. Biophys. Acta*, **553**, 450 (1979); also G. Ehrenstein et al., *Biophys. Sci. Abst.*, **97a**, TPM-A10 (1982).

Light transduction by pigmented bi-lipid membranes, especially the purple membrane of H.halobium, has been studied by H. Ti Tien: *Bioelectrochem. Bioenerget.*, **5**, 318 (1978).

Molecular information transfer in doped monolayers has been studied: D. Möbius and G. Debuch, *Ber. Bunsen Ges. Phys. Chem.*, **80**, 1180 (1976).

Refractive index and reflectance measurements indicate that black lipid bilayers have a permittivity near to two: R. Waldbillig, *Biophys. J.*, **33**, 160a (1981).

Charge transfer interactions between heparin and aminoglycoside antibiotics have been demonstrated: G. M. Eckert et al., *J. Biol. Phys.*, **10**, 51 (1982).

Charge transfer complexes involving heparin have been studied: G. Eckert and F. Gutmann, *Electroanal. Chem. Interfacial Electrochem.*, **62**, 267 (1975); *cf.*, also F. Gutmann and J. P. Farges in *Modern Aspects of Electrochemistry*, No. 13, B. E. Conway and J. O'M. Bockris, eds., Plenum Press, N.Y., 1979, p. 380.

25.8 Supplemental to Chapter 19 "Structure"

Effect of hetero and peripheral substitutions on quantum yield in tetraphenyl-dithiaporphyrins: K. Yamashita et al., *Chem. Lett.*, **(1982)** (6) 807.

Aromaticity and electrical properties of liquids: S. Yasufuku and Y. Inuishi, *J. Electrostat.*, **12**, 601 (1982).

Carrier transport in the van der Waal's solid diphenyl (*p*-chlorophenyl)-pyrazoline is studied in single crystals, in the glassy and in the supercooled liquid states. The hole mobility in the single crystal is reported as

0.01 cm²/Vsec while under the other two conditions it is about a 1000 times less: H. Katayama et al., *Mol. Cryst. Liqu. Cryst.*, **69**, (3-4) 257 (1981).

Stacking disorder in layered semiconductors causes the localization of electronic wave functions over a finite number of layers; charge transport then is determined by hopping from layer to layer: K. Maschke, *Cryst. Res. Technol.*, **16**, (2) 265 (1981).

Intrachain interactions and conformational aspects of polymers, and structural aspects of dopant-polymer interactions are studied: The latter interactions may become considerable. Intercalation of a dopant is facilitated by rotations about C-O or C-S bonds and by stretching of the C-O-C or C-S-C bond angles. Such intercalation raises the conductivity *via* interchain transitions: S. Tripathy, *Mo. Cryst. Liq. Cryst.*, **83**, (1-4) 1271 (1982).

The non-ohmic effects of Anderson localization in disordered systems are discussed; no metallic conductivity is said to be possible for a disordered two-dimensional system but if the dimensionality exceeds the value of two, a mobility edge exists at which the conductivity tends to zero: E. Abrahams et al., *Phil. Mag.*, **B-42**, (6) 827 (1980).

Percolation is said to be a non-linear effect in disordered semiconductors: A. Ya. Vinnikov et al., *Fiz. Tverd. Tela* (Leningrad), **24**, (5) 1338 (1982).

The conductivity of pyrolyzed polyacrylonitrile is discussed in terms of percolation theory: M. A. Magrupov et al., *Pisma Zh. Tekh. Fiz.*, **7**, (20) 1244 (1981).

Reviews: Disordered two-dimensional systems: N. F. Mott, *J. Phys.*, **C-14**, (8) L-183 (1981).

A discussion of the Anderson localization: J. L. Pichard and G. Sarma, *J. Phys. Colloq.*, **(1981)**, C4, Pt. 1, p. 37.

The structural basis for semiconductivity and metallic behavior in polymer-dopant systems: R. H. Baughman et al., *Chem. Revs.*, **82**, (2) 209 (1982).

The relationship between conductivity and aromaticity: F. Wudl, *Pure Appld. Chem.*, **54**, (5) 1051 (1982).

Structure and conductivity of organic semiconductors: H. Shimoda and M. Sukigara, *Nippon Sharkin Gakkaishi*, **44**, (3) 201.

25.9 Supplemental to Chapter 20 "Space Charge Effects"

25.9.1 (Space Charge Limited Currents): Theory of space charge limited currents in quasi-one dimensional solids: E. G. Wilson, *Chem. Phys. Lett.*, **90**, (3) 221 (1982).

From an exact solution of the nonlinear partial integro-differential equation describing the motion of space charge broadened charge packets in the time-of-flight method, values for the carrier mobility are derived: D. F. Nelson, *Phys. Rev.*, **B-25**, (8) 5267 (1982).

25.9.2 (Glow Curves): Glow curves obtained from polymers are discussed: J. van der Schueren et al., *IEEE Trans. Electro. Insul.*, **E-I-17**, (3) 189 (1982).

Glow curves for polyvinylcarbazole: J. Plans and F. J. Balta-Calleja, *Colloid. Polymer Sci.*, **260**, (3) 258 (1982).

Resolution of complex retardation modes in polymers: C. La Cabanne et al., *Org. Coat. Plast. Chem.*, **43**, 550 (1980).

It is suggested to use the second derivative of the glow curve to obtain useful information: E. Klier, *Phys. Status Solidi*, **A-67**, (2) K-151 (1981).

Electrode effects and carrier injection into polyvinylchloride are studied by means of glow curves for Al,Sn,Ag,Cu electrodes: Sh. Behari and I. M. Talwar, *Acta Polym.*, **33**, (5) 314 (1982).

The space charge polarization in pyrene doped polystyrene films is strongly dependent on the dopant concentration: A. R. Tiwari et al., *Thin Solid Films*, **88**, (2) 121 (1982).

25.9.3 (Electrets): Polar polymer forming electrets may be protected against ingress of humidity by coating both sides with a nonpolar layer: A. Mishra, *J. Appl. Polymer Sci.*, **27**, (6) 1967 (1982).

Photoelectrets may be formed from poly(ethylene naphthalate) by the development of carrier traps; the trap depth, from glow curves, is said to be 0.1 to 0.5 eV; K. Kenzo et al., *Japan J. Appl. Phys.*, Pt. 1, **21**, (7) 1025 (1982).

Magneto-electrets are reported formed from polypropylene. Its dielectric properties are described: R. K. Srivastara and C. S. Bhatnarar, *Indian J. Pure Appl. Phys.*, **18**, (11) 845 (1980).

p-Alkoybenzylidine-p'-amino-2-chloropropyl cinnamates are reported to exhibit large permittivities and to exhibit large spontaneous polarizations: K. Yoshuo et al., *Technol. Rep. Osaka Univ.*, **30**, (1051-1582) 483 (1980).

Photoelectrets are reported from polyimide films: J. K. Quamara et al., *Acta Polym.*, **33**, (8) 501 (1982).

Photoelectrets may be formed from pure as well as from Cu-phthalocyanine doped polystyrene: S. K. Shrivastava et al., *Br. Polymer J.*, **13**, (4) 151 (1981); *cf.*, also *ibid.*, p. 147.

Thermoelectrets from polyvinylchloride are reported: T. Gancheva and P. Dinev, *Elektron. Obrab. Mater.*, **(1982)**, (1) 49.

Electrets from polypropylene are reported: K. M. Falck et al., *Phys. Rev.*, **B-25**, (8) 5509 (1982).

Pyroelectrets are reported in poled beta-crystalline poly(vinylidene fluoride): N. Nobukazu et al., *Jpn. J. Appl. Phys.*, Pt. 1, **21**, (5) 706 (1982).

25.9.4 (Pyroelectric and Piezoelectric Effects): Pyroelectricity is reported from poly(ethylene terephthalate) with a pyroelectric coefficient of 4 nC m^{-2} K^{-1}: H. J. Wintle and J. R. Turlo, *J. Appl. Phys.*, **50**, (11 pt. 1) 7128 (1979).

Strong piezoelectricity is said to occur in crystalline films of vinylidene fluoride-trifluoroethylene copolymers, cast from the melt. The films are also ferroelectric, and may be poled to form electrets usable for microphones: T. Yamada, *Jpn. J. Appl. Phys.*, **20**, (Suppl. 20-4) 121 (1981).

The piezo and pyroelectric properties of poly(vinylidenefluoride) are discussed: W. T. Chen et al., *J. Macromolec. Sci. Phys.*, **B-21**, (3) 397 (1982); *cf.*, also A. G. Kolbeck, *Polymer Eng. Sci.*, **22**, (7) 444 (1982).

Biphenyl derivatives are reported to be pyroelectric: D. V. Belous et al., USSR Pat. 779380 (1980).

Reviews: Piezoelectric thin polymeric films: A. Kawabata, *Denshi Tsushin Gakkaishi*, **65**, (2) 132 (1982).

Piezoelectricity in polymers: R. G. Kepler and R. A. Anderson, *CRC Crit. Rev. Solid State Mater. Sci.*, **9**, (4) 399 (1980).

Pyroelectricity and piezoelectricity in vertebrates: H. Athenstaedt, *Ann. N.Y. Acad. Sci.*, **238**, 68 (1974).

Experimental studies of electro-optic and ferro-electric phenomena in organic solids: G. F. Lipscomb, *Diss. Abst. Int. B., 1981*, **41** (10) 3822, Order #8107772 Univ. Microfilm Int.

A bibliography to 1980 is available: S. B. Lang, *Ferroelectrics*, **34**, (4) 239 (1981).

A literature guide to pyroelectricity has been published: S. B. Lang, *Ferroelectrics*, **43**, (3-4) 251 (1982).

25.10 Supplemental to Chapter 21
"Quasi-One-Dimensional Organic Conductors"

25.10.1 (Polyacetylenes):

25.10.1.1 General: Physical properties of polyacetylene films doped with MoCl$_5$ and WCl$_6$: M. Rolland et al., *Polymer*, **23**, (6) 834 (1982).

Synthesis of polyacetylene: H. Shirakawa, *Kagaku, Zokan* (Kyoto), **87**, 15 (1980).

One dimensional order-disorder transition effects are studied in polydiacetylene films: G. N. Patel and Y. P. Khanna, *J. Polymer Sci. Polymer Phys. Ed.*, **20**, (6) 1029 (1982).

Defects in single crystals of polydiacetylene: R. S. Young and J. Peterman, *J. Polymer Sci. Polymer Phys. Ed.*, **20**, (6) 961 (1982).

Diffusion profiles of impurities (dopants) in polymers, *e.g.*, of iodine in polydiacetylene. Iodine penetrates through interfibril spaces and simultaneously undergoes a 1st order reaction on the fibril surfaces: J. P. Louboutin and F. Beniere, *J. Phys. Chem. Solids,* **43**, (3) 233 (1983).

Fluctuation induced tunneling between fibers in iodine doped polyacetylene in the metallic regime: W. Mayr et al., *Solid State Commun.,* **43**, (2) 117 (1982).

Polydiacetylenes, doped with, *e.g.*, chlorosulfuric acid: M. Masashi et al., *Chem. Scr.,* **17**, (1-5) 131 (1981); *cf.,* also A. J. Epstein et al., *ibid.,* p. 135.

The conditions for n-type doping of polyacetylenes are studied; from ESR and IR spectroscopy it is deduced that the dopant concentrates on the fibril surface before penetrating inside the fibril: B. Francois et al., *J. Chem. Phys.,* **75**, (8) 4142 (1981); *cf.,* also F. Beniere et al., *J. Phys. Chem. Solids,* **42**, (8) 649 (1981).

Ultraviolet irradiation of polyacetylene films treated with diaryliodonium or triarylsulfonium salts leads to doping of the exposed areas and improves control of the process: T. C. Clarke et al., *J. Chem. Soc. Chem. Commun.,* **1981**, (8) 384 (1981).

A discussion of excited states in polydiacetylene: U. Dinur and M. Karplus, *Chem. Phys. Lett.,* **88**, (2) 171 (1982); *cf.,* also B. Gazdy, *ibid.,* p. 220.

Impurity states in doped trans-polyacetylene are said to modify severely the structure of the soliton distortions in that the most stable distortion is a charged polaron: G. W. Bryant and A. J. Glick, *J. Phys.,* **C-15**, (13) L-391 (1982).

Polyacetylene treated with very strong acids or their esters becomes a p-type semiconductor with a conductivity of in excess of 100 S/cm: K. K. Showa Denko Jpn Pat. 36,285 March 1979; *cf.,* also Jpn. Pat. 81,145,928 (1981); 80,129,426 Oct. 1980 and 80,129,424 Oct. 1980.

Grown films of polyacetylene are reported to be anisotropic while shear-flow, *in situ,* polymerized films are isotropic: H. Kiell et al., *Mol. Cryst. Liq. Cryst.,* **77**, (1-4) 147 (1981).

25.10.1.2 Electrical Properties: The Hall mobility magnetoresistance and the nonlinear current voltage characteristic of doped polyacetylene are reported: K. Seeger et al., *Chem. Scr.,* **17**, (1-5) 129 (1981).

The conductivity of Br-doped polyacetylene in the form of a 1000 Å film is ascribed to 3-dimensional disordered hopping; doping is said to destroy the crystallinity: J. Kanicki et al., *Thin Solid Films,* **93**, (3) 243 (1982).

The electrical conductivity of polyacetylene is said to be in agreement with a study of its optical properties: reported in 5 papers in Kyoto Daigaku Genshiro Jikkensho (Tech. Rept) DURRI-TR 217 (1982); *cf.,* also M. Tanaka et al., *Mol. Cryst. Liq. Cryst.,* **83**, (1-4) 1107 (1982).

Impedance spectra of trans-polyacetylene have been obtained; the impedance consists of 2 components, one associated with intrinsic conductivity showing classical semiconductor behavior, and the other associated with inhomogeneities: C. K. Chiang and A. D. Franklin, *Solid State Commun.*, **40**, (8) 775 (1981).

The a.c. conductivity of iodine doped polyacetylene is reported to exceed greatly the d.c. conductivity, at low doping but at high dopant levels both are about equal: D. M. Hoffmann et al., *Mol. Cryst. Liq. Cryst.*, **83**, (1-4) 1175 (1982).

It is the size and the shape of the dopant molecules which control the conductivity rather than the doping conditions: E. K. Sichels et al., *Phys. Rev.*, **B-25**, (8) 5574 (1982).

The carrier mobility in polydiacetylene-toluene sulfate is reported to have the extraordinarily high value of greater than 20×10^4 cm^2/Vsec: K. J. Donovan and E. G. Wilson, *Phil. Mag.*, **B-(1981)**, p. 449. The validity of these results has been questioned: A. S. Siddiqui, *J. Phys.*, **C-15** (9), L-263 (1982).

Doping of polyacetylene with H_2IrCl_6 results in a conductivity increase by a factor of 10^7 to 2000 S/cm with a dopant concentration below 0.01 mole-%: M. Rubner et al., *J. Chem. Soc. Chem. Commun.*, **(1982)**, (9) 507 (1982).

The a.c. conductivity of semiconducting trans-polyacetylene has been studied over the frequency range of 10 Hz to 10 MHz. The complex conductance is said to be frequency independent for all doping levels and all types of contacts employed: P. M. Grant and M. Krounbi, *Solid State Commun.*, **36**, (4) 291 (1980).

The measured Seebeck coefficient of -850 μV/K of polyacetylene near the compensation point is the same for lightly doped p-type as it is for the n-type material, agreeing with a soliton hopping mode of charge transfer: Y. W. Park et al., Rept. TR-80-4 (1980) ex. Gov. Announ. Index (USA), **80**, 23 4944.

25.10.1.3 Magnetic Properties:
NMR studies on polyacetylenes: H. Thoman et al., *Mol. Cryst. Liq. Cryst.*, **83**, (1-4) 1065 (1982).

The magnetic properties of polydiacetylenes are reported: M. Peo et al., *Springer Ser. Solid State Sci.*, **23**, (Phys. One Dimens) 218 (1981); *cf.*, also B. R. Weinberger et al., *Phys. Rev.*, **B-20**, 223 (1979) and I. B. Goldberg et al., *J. Chem. Phys.*, **70**, 1132 (1979).

25.10.1.4 Dielectric Properties:
The dielectric properties of polyacetylene are treated: M. Gamoudi et al., *J. Phys.* (France), **43**, (6) 953 (1982).

The momentum-dependent dielectric function of trans-polyacetylene is studied: J. J. Ritsko, *Phys. Rev.*, **B-26**, (4) 2192 (1982).

25.10.1.5 Theory, Solitons: Soliton diffusion in polydiacetylene is studied by means of NMR; the spin diffusion in trans-polyacetylene is described in terms of neutral domain walls: F. Devreux et al., *Springer Ser. Solid State Sci.*, **23**, (Phys. One Dimens.) 194 (1981).

Neutral solitons were observed during the thermal isomerization of cis- to trans-polyacetylene. The solitons are mobile along the chain resulting in their annihilation by an intramolecular process or else by an intermolecular event cross linking two polymer chains. The effect of doping is discussed: J. C. W. Chien, *J. Polymer Sci. Polymer Lett. Ed.*, **19**, (5) 249 (1981).

The dominant conduction mechanism in polyacetylene is said to be phonon assisted hopping between soliton bound states, *viz.*, intersoliton hopping. Solitons are characterized by a) no increase in spin concentration upon doping, b) a highly one dimensional spin mobility and c) strongly affected by cis to trans isomerization. Hopping, in turn involves a strongly field and frequency dependent conductivity: S. Kivelson, *Mol. Cryst. Liq. Cryst.*, **77**, (1-4) 65 (1981). For experimental verification, see A. J. Epstein, *ibid.*, p. 81.

It is shown theoretically that in quasi-one dimensional conductors with a commensurability index defined as (distortion period/lattice spacing) $= n$, excitations exist determined by $\pm 2e/n$ or $2e/n \pm e$, where e stands for the electronic charge, being kinks in the order parameter: J. R. Schrieffer, *Mol. Cryst.*, **77**, (1-4) 209 (1981).

Fractionally charged solitons are discussed: M. J. Rice and E. J. Mele, *Mol. Cryst. Liq. Cryst.*, **77**, (1-4) 223 (1981).

The role of Peierls instabilities in doped polyacetylene is discussed: M. J. Rice and E. J. Mele, *Chem. Scr.*, **17**, (1-5) 121 (1981).

It is suggested that doped polyacetylene contains soliton droplets each involving about 3% doped concentration as long as the disorder and any pinning of dopant ions may be neglected. Charged solitons attract each other if their formation energy exceeds $E/\sqrt{2}$ where $2E$ stands for the semiconductor band gap: B. Horovitz, *Chem. Scr.*, **17**, (1-5) 127 (1981).

The evolution of, *e.g.*, polyacetylene from insulator to semi-metal is characterized by 4 regimes: a) accomodation of excess carriers in solitons pinned at impurity sites, as long as the dopant concentration remains low; b) at slightly higher dopant concentrations, statistical fluctuations give rise to a form of quasi-metallic isolated sub-regions; c) at still higher dopant levels, a dense band of localized states is formed *via* a charge density wave and d) at very high dopant concentrations, bond alternation becomes suppressed in the presence of the now prevailing high degree of disorder: J. E. Mele and M. J. Rice, *Phys. Rev.*, **B-23**, (10) 5397 (1981).

Kinks and polaron-solitons are predicted from theory: A. R. Bishop et al., *Molec. Cryst. Liq. Cryst.*, **77**, (1-4) 235 (1981); *cf.*, also: Th. Holstein, *ibid.*, p. 235; *cf.*, also Ch. Jouanin and J. P. Albert, *ibid.*, p. 349; *cf.*, also G. Lieser et al., *ibid.*, p. 169; *cf.*, also W. H. Meier, *ibid.*, p. 137.

Solitons in polyacetylene are discussed: K. Maki, *Molec. Cryst. Liq. Cryst.*, **77**, (1-4) 277 (1981); *cf.*, also Y. Tomkiewicz et al., *Phys. Rev.*, **B-24**, (8) 4348 (1981).

Soliton excitations in polyacetylene are discussed in terms of a topological entity, a moving domain wall: W. P. Su et al., *Physical Rev.*, **B-22**, (4) 2099 (1980).

25.10.1.6 Morphology and Structure: A calendered film of polyacetylene is shown to exhibit considerable anisotropy: Showa Denko., Jpn. Pat. 82 53 324 (1982).

The crystal structure of iodine doped polyacetylene has been studied with reference to the pristine compound: C. W. Chien et al., *Macromolecules*, **15**, (4) 1012 (1982); *cf.*, also K. Miyasaka, Kyoto Daisaku Genshiro Jikkensho (Tech. Rept) KURRI-TR-217, 14, (1982).

The electronic structure of polyacetylenes has been investigated: E. Tokomoto et al., Kyoto Daigaku Genshiro Jikkenso (Tech. Rept.) KURRI-TR-217, 45 (1982); *cf.*, also T. Yamabe, *ibid.*, p. 48; *cf.*, also H. Fukudome and Y. Sasai, *ibid.*, p. 52; *cf.*, also K. Harrada and M. Tasumi, *ibid.*, p. 59; *cf.*, also K. Yoshino et al., *ibid.*, p. 28.

The morphology and dopant distribution in polyacetylenes has been studied; several fibrillar and rod-like forms are retained upon doping: A. J. Epstein et al., *Polymer*, **23**, (8) 1211 (1982); *cf.*, also V. N Spector et al., *Vysokomol. Soedin*, **A-23**, (9) 2128 (1981), *cf.*, also H. Rommelmann et al., *Mol. Cryst. Liq. Cryst.*, **77**, (1-4) 177 (1981); *cf.*, also A. Janossy et al., *ibid.*, p. 185; *cf.*, also S. Pekker and A. J. Janossy, *ibid.*, p. 77.

The morphology of polyacetylene films has been investigated: M. Rolland et al., *J. Microsc. Spectroscop. Electron.* (France), **7**, (1) 21 (1982); *cf.*, also *Nature* (London), **294**, (5836) 60 (1981).

In iodine doped polyacetylene, π-orbitals are said to develop in a direction perpendicular to the polymer zig-zag layer and interacting with I_x; for I_3: polyacetylene, the interchain distance is obtained as about 7.6 Å, *i.e.*, twice the 3.8 Å lattice spacing: K. Shimamura et al., *Makromol. Chem. Rapid Commun.*, **3**, (5) 269 (1982).

The electronic structure of polyacetylene, as measured by electron loss spectroscopy, is reported to involve wide valency and conduction bands; also an excitonic or indirect absorption edge at 1.42 eV: J. J. Ritsko, *Mater. Sci.*, **7**, (2-3) 337 (1981).

A high strength polyacetylene membrane with a conductivity of about 1100 S/cm and a tensile strength of 6.17 kg/mm^2 has been developed; the density of a 0.08 mm thick membrane is 0.69 g/ml: Toray Indust. Jpn. Pat. 120715 (1981).

Experiments on iodine doped polyacetylenes for a dopant concentration exceeding 1% molar, including measurements of the Seebeck coefficient and of conductivities, are reported: M. Audenaert et al., *Phys. Rev.*, **B-24**, (12) 7380 (1981); *cf.*, also J. Kanicki et al., *J. Chem. Soc. Trans. Faraday, II*, **77**, (12) 2157 (1981).

The sharpness of the semiconductor-metal transition in doped polyacetylene is enhanced the more uniform is the dopant distribution: D. Moses et al., *Phys. Rev.*, **B-25**, (12) 7652 (1982).

The photoconductivity, dark conductivity, capacitance and nonlinear current voltage data are given for iodine and for AsF_5 doped polyacetylenes: E. T. Kang et al., *J. Polymer Sci. Polymer Lett. Ed.*, **20**, (3) 143 (1982).

The swelling of iodine doped polyacetylene is reported to amount to about 1% and rises with increasing dopant concentration to a volume expansion of about 3%: B. Francois et al., *Synthet. Met.*, **4**, (2) 131 (1981).

Doping of polyacetylene films to the metallic state by means of xenon-fluorides or with iodine pentafluoride is reported: H. Selig et al., *J. Chem. Soc. Chem. Commun.*, **(1981)**, (24) 1288.

Ferric, Al and Zr chlorides are said to be active dopants for polyacetylene: I. Kulszewski et al., *Mol. Cryst. Liq. Cryst.*, **83**, (1-4) 1191 (1982).

The chemistry of polyacetylene doping is studied: T. C. Clarke et al., *Mol. Cryst. Liq. Cryst.*, **83**, (1-4) 1033 (1982).

25.10.1.8 Reviews: A. J. Heeger, *Comments Solid State Phys.*, **10**, (2) 53 (1981); S. Etemad, *Mol. Cryst. Liq. Cryst.*, **77**, (1-4) 43 (1981); J. C. W. Chien, *Macromolec. Main Lect. Int. Symp. 27th, 1981*, Pergamon, Oxford (England) 1982; H. Benoit and P. Rempp, eds.; P. Bernier et al., *Port. Phys.*, **11**, (3-4) 183 (1980).

Crystal energy, cohesion and ionicity: R. M. Metzger, *Top. Curr. Phys.*, **26**, 80 (1981).

Solitons: A. J. Heeger et al., *Springer Ser. Solid State Sci.*, **23**, (Phys. one dimens.) 179 (1981); *cf.*, also B. Horovitz, *ibid.*, p. 212; *cf.*, also Y. Tomkiewicz, *ibid.*, p. 214.

The metal-insulator transition: N. F. Mott, *Proc. Royal Soc. London*, **A-382**, (1782) 1 (1982).

25.10.2 (Other Organic Semi-Metals): Perylene$_2$ $(AsF_6)_{0.75}$ $(PF)_{0.35}$.0.85 CH_2Cl_2 is reported to exhibit metallic conductivity above 200 K: P. Koch et al., *Mol. Cryst. Liq. Cryst.*, **86**, (1-4) 1827 (1982).

(Tetramethyl-diselenol)$_2$.ClO$_4$ is said to be metallic above 1.07 K: K. Murata et al., *J. Phys. Soc. Japan*, **51**, (6) 1817 (1982).

Doped dialkylated phenazines are reported to show a conductivity of 10 to 100 S/cm at 300 K: K. Dietz et al., *Chem. Scr.*, **17**, (1-5) 93 (1981).

It is proposed that metallic conductivity in an organic crystal may not require separate donor and acceptor stacks nor incommensurability between the donor lattice and the charge density waves within the charged ion structure: F. Wudl, *Mol. Cryst. Liq. Cryst.*, **79**, 423 (1982).

Ditetramethyl-tetraselena fulvalinium fluorosulfonate compounds are reported to exhibit a metal-to-insulator transition at 86 K which is not driven by spin waves: F. Wudl et al., *J. Chem. Phys.*, **76**, (11) 5497 (1982).

Metallic iodine is reported for 235 kBar at 225 K: N. Nobuko et al., *J. Phys. Soc. Japan*, **51**, (6) 1811 (1982).

Polythiazyl-(SN)$_x$ is compared to polyacetylenes; exhibits metallic conductivity: G. S. Street and T. C. Clarke, *Adv. Chem. Ser.*, **186**, 177 (1980).

Madagascar natural graphite plus an electron donor such as LiC$_6$ or LiC$_{12}$ is swaged in a tube of 70% Cu/30% Ni to exhibit metallic conductivity: D. Billand et al., *Mater. Sci. Eng.*, **47**, (2) 137 (1981).

The electrochemical oxidation at a Pt electrode of heterocyclic compounds at 300 K, in, *e.g.*, acetonitrile is reported to yield organic semimetals; indole exhibits a conductivity of 200 S/cm and azulene of 100 S/cm: G. Tourillon and F. Garner, *J. Electroanal. Chem. Interfacial Electrochem.*, **135**, (1) 173 (1982).

PF$_6$ naphthaceno-dithiols form charge transfer complexes exhibiting metallic conductivity above 220 K with a 300 K conductivity of 762 S/cm: H. Tanaka et al., *Chem. Lett.*, (**1982**), (5) 727.

Bis(ethylene dithiolo) TTF perchlorate shows metallic behavior down to 1.4 K and acts as a two-dimensional system: G. Saito et al., *Solid State Commun.*, **42**, (8) 557 (1982).

The metallic state in TCNQ-based organic semi-metals is stabilized and extended to very low temperatures by irradiation at low dosages with X-rays, sufficient to damage about 0.2% of the molecules: L. Forro et al., *J. Phys.* (France), **43**, (6) 977 (1982).

Bis(tetramethyl tetraselena fulvalenium) perrhenate 2:1 charge transfer salts are said to show metallic behavior above 182 K and to behave as nonmagnetic semiconductors below that temperature: C. S. Jacobsen et al., *J. Phys.*, C-15, (12) 2651 (1982).

Tetrathiatetracene$_2$:I$_{3+\delta}$ are reported to show metallic conductivity: B. M. Gorelov et al., *Chem. Scr.*, **17**, (1-5) 23 (1981); *cf.*, also P. Delhaes et al., *ibid.*, p. 41 who report a conductivity of 1000 to 10000 S/cm in the temperature range 0.1 to 300 K; *cf.*, also M. Kaveh, *ibid.*, p. 49 and E. M. Conwell, *ibid.*, p. 69.

Magnetoresistance measurements on organic semimetals are reported: P. M. Chaikin et al., *Mol. Cryst. Liq. Cryst.,* **79**, 435 (1982).

Peryleniumyl hexafluorophosphates are reported to behave as one-dimensional semimetals between 200 and 300 K: H. J. Keller et al., *Chem. Scr.,* **17**, (1-5) 107 (1981).

The semi-metallic conductivity of dibenzotetrathiafulvalene complexes with the acceptor DDQ (2,-3 dichloro-5,6-dicyano-p-benzoquinone) has been measured as 8 S/cm: J. J. Mayerle and J. B. Torrance, *Bull. Chem. Soc. Jpn,* **54**, (10) 3170 (1981).

N-methyl-N-ethylmorpholinium:TCNQ is said to exhibit an electronic Peierls transition at 335 K, which temperature drops with increasing pressure to reach 155 K at 3.5 kBar; the compound is semi-metallic above and semiconducting below the transition: S. Kagoshima, *Bussei Kerkyu,* **34** (1) (1980).

Oxidatively stabilized polyacrylonitrile, annealed at 710 to 950 K *in vacuo,* has a fibrous strcuture containing, at room temperatures, metallical conducting pathways in an impurity semiconductor matrix: R. L. Lerner, *J. Appl. Phys.,* **52**, (11) 6757 (1981).

Ultrapure pyrolysates of polyacrylonitrile films exhibit a dramatic conductivity increase during pyrolysis, at 390 to 435 °C showing a final conductivity of higher than 5 S/cm. The maximum value of the conductivity is said to be rather independent of molecular weight. The conductivity increase during pyrolysis is ascribed to the formation of conjugated C=C and C=N bonds: H. Tech et al., *Mol. Cryst. Liq. Cryst.,* **83**, (1-4) 1329 (1982).

Reviews: R. H. Baughman et al., *Org. Coat. Plast. Chem.,* **43**, 762 (1980); D. Jerome and H. J. Schulz, *Springer Ser. Solid State Sci.,* **23**, (Phys. One Dimens.) 239 (1981); T. J. Marks et al., *Ext. Linear Chain Cpds.,* **1**, 197 (1982); K. K. Kanazawa et al., *Synth. Met.,* **4**, (2) 119 (1981); Mainly theory: G. Grüner, *Chem. Scr.,* **17**, (1-5) 207 (1981); for other reviews *cf.,* Section 25.10.1.8.

25.11 Supplemental to Chapter 22,"Photo Effects"

25.11.1 (Photovoltaic Phenomena): Poly(vinylpyrolidone):iodine complexes between two dissimilar electrodes exhibit spontaneous dark- and photovoltages: P. K. C. Pillai and B. Vasanta, Proc. Nucl. Phys. Solid State Phys. Symp. New Delhi, 1981 publ. 1982, 1981.

Photovoltages in the sandwich cell In/polyacetylene/other metal were obtained: J. Kanieki et al., *Mol. Cryst. Liq. Cryst.,* **83**, (1-4) 1351 (1982).

Treatment with surfactants is said to yield an 18% quantum yield in metal/phthalocyanine sandwich cells at 638 nm dropping with time: J. P. Dodelet et al., *J. Appl. Phys.,* **53**, (6) 4270 (1982).

Thin films of microcrystalline metal-free phthalocyanine dispersed in a polymeric binder are painted onto a conducting substrate which acts as a photocathode in contact with an aqueous electrolyte containing a RedOx couple, *e.g.*, *p*-benzoquinone. A linear photovoltaic response is obtained amounting to 250 mV at a mA/cm^2 level and said to be stable with time. The photocurrent is ascribed to excitonic reactions at the semiconductor/ electrolyte interface: R. O. Loutfy and C. M. McIntyre, *Sol. Energy Mat.*, **6**, (4) 467 (1982); *cf.*, also R. O. Loutfy et al., *J. Appl. Phys.*, **52**, (8) 5218 (1981) for the tin oxide/phthalocyanine/In system yielding an open circuit voltage of 0.45 V and a short circuit current of 0.2 mA/cm^2.

Photovoltaic sandwich cells of merocyanine thin films with ZnO and Ag electrodes are reported: K. Judo et al., *Jpn. J. Appl. Phys.*, **20**, (Suppl. 20-2) 135 (1981); *cf.*, also Ph. Yamin et al., *J. Phys. Chem.*, **86**,(19) 3796 (1982). who report an open circuit voltage of 1.1 V and a quantum yield of 15 to 30% replacing In by Al/Al$_2$O$_3$ top, semi-transparent electrodes.

The violanthrone-metal system is reported to be photovoltaic due to exciton dissociation at the contact interface: K. Roy et al., *Indian J. Pure Appl. Phys.*, **19**, (11) 1092 (1981).

Review: V. M. Fridkin et al., *Ferroelectrics*, **43**, (3-4) 99 (1982).

25.11.2 (Other Photo-Effects): Photocurrent spectra in polycrystalline tetracene films are reported: R. Signeski and J. Kalmourski, *Thin Solid Films*, **75**, (2) 151 (1981).

Indigo coated polystyrene films are said to exhibit photoconductivity in the visible region: Y. K. Kulshrestha and A. Srivastava, *Indian J. Pure Appl. Phys.*, **19**, (5) 478 (1981).

The photoconductivity of metal-free phthalocyanine in the near infrared is ascribed to carrier generation *via* surface charge transfer states: R. O. Loutfy, *J. Phys. Chem.*, **86**, (17) 3302 (1982).

A photo-stimulated charge transfer in the pyranthrone:dodecane system has been studied: R. C. Ahuja and K. Hauffe, *J. Chem. Soc. Faraday Trans. II*, **78**, (6) 941 (1982).

The photoconductivity of poly(*p*-xylene) has been studied: Y. Takai et al., *J. Phys.*, **D-15**, (5) 917 (1982).

The dark- and photo-conductivity of polymeric cyanines has been investigated: G. Kossmehl and P. Bocionek, *Makromolek. Chem.*, **182**, (12) 3445 (1981).

The mechanism of photocurrent generation in single crystalline anthracene is discussed: P. J. Bounds, *Chem. Phys. Lett.*, **89**, (1) 1 (1982).

In the dye-sensitized photoconductivity of polymeric semiconductors, it is suggested theoretically that the free carriers arise from the dissociation of exciplexes: B. M. Rumyansev et al., *Vysokomol. Soedin,* Ser. **A-22**, (11) 2545 (1980).

The frequency dependent photoconductivity in organic molecular crystals is discussed: R. Kuehne et al., *Z. Phys.*, **B-41**, (2) 181 (1981).

The relation between the Onsager thermalization distance and the energy of the incident photon has been investigated by W. A. Silinsh et al., Int. Conf. Defects in Insul. Cryst. Proc. Riga, USSR, May 1981.

Reviews: V. S. Mylnikov, *Usp. Khim.*, **50** (10) 1972 (1981).

25.12 Supplemental to Chapter 23 "Retrospect, Outlook and Speculations"

25.12.1 (Superconductivity): (Tetramethyl-diselena)$_2$.ClO$_4$ is said to exhibit superconductivity below 1.07 K; above it is a semi-metal: K. Murata et al., *J. Phys. Soc. Jpn.*, **51**, (6) 1817 (1982); *cf.*, also D. U. Gubser et al., *Phys. Rev.*, **B-24**, (1) 478 (1981).

(Tetramethyl-tetraselena-fulvalene)$_2$PF$_6$ exhibits superconductivity: D. Jerome, *Chem. Soc.*, **17**, (1-5) 13 (1981); *cf.*, also C. More et al., *J. Phys. Lett.* (France), **42**, (13) 313 (1981).

Tetramethyl-tetraselena-fulvalenium-hexafluorophosphate is said to form a quasi-one-dimensional superconductor: H. J. Schulz, *Mol. Cryst. Liq. Cryst.*, **79**, (1-4) 199 (1982); *cf.*, also B. Horovitz et al., *ibid.*, p. 235.

(Tetramethyl-tetraselena-fulvalene)$_2$AsF$_6$ is reported to become superconductive at low temperatures: R. Brusetti et al., *J. Phys.* (France), **43**, (5) 801 (1982); *cf.*, also R. L. Greene et al., *Mol. Cryst. Liq. Cryst.*, **79**, (1-4) 183 (1982).

Considerable variations in the conditions of preparation, including the purity of the constituents, of tetramethyl-tetraselena-fulvalene superconductors are said not to seriously affect the low temperature superconductivity of such compounds: E. M. Engler et al., *Mol. Cryst. Liq. Cryst.*, **79**, (1-4) 15 (1982).

An excitonic mechanism is suggested for the possible superconductivity in polycytosine: J. Ladik et al., *Phys. Lett.*, **A-81**, (8) 488 (1981).

Magnetic field effects in organic superconductors have been studied: L. N. Bulaevskii and A. I. Buzdin, *Chem. Soc.*, **17**, (1-5) 67 (1981).

Bipolaronic superconductivity is suggested to arise *via* a strong electron-phonon interaction in the form of localized non-overlapping Cooper pairs giving rise to molecular superconductivity: A. Alexander and J. Ranninger, *Phys. Rev.*, **B-24**, (3) 1164 (1981).

It is suggested that excitonic superconductivity on the lines proposed by Little may be possible along dislocations where linear pathways are formed: H. Fukuyama, *J. Phys. Soc. Jpn.*, **51**, (6) 1709 (1982).

A deformation induced attraction between electrons is suggested to lead to localized electron pairs, *vis.*, bipolarons, which are suggested to act like itinerant Cooper pairs: E. K. Chakraverty, *J. Phys.* (France), **42**, (9) 1351 (1981).

The transition from Peierls to TCS superconductivity occurs as the back scattering electron-phonon interaction-coupling is reduced to a critical value. Formation of Cooper pairs has to compete with the generation of charge density waves: H. Gutfreund and M. Weger, *Chem. Scr.*, **17**, (1-5) 53 (1981).

Superconductivity in graphite-potassium intercalation compounds is experimentally explored: Y. Korke et al., *J. Phys. Chem. Solids,* **41**, (10) 1111 (1980).

A critical temperature of 1.4 K is reported for the superconductivity of graphite-alkali metal amalgam intercalation compounds: Y. Iye and S. Tanuma, *Phys. Rev.*, **B-25**, (7) 4583 (1982).

The low temperature Hall mobility of the electrons in organic superconductors has been found to exceed 10^6 cm^2/Vsec: E. M. Conwell and N. C. Banik, *Mol. Cryst. Liq. Cryst.*, **79**, 451 (1982).

The superconductivity of tetramethyltetraselena fulvalene adducts with various electron donors has been discussed: S. S. P. Parkin et al., *J. Phys.*, **C-14**, (15) L-445 (1981); *cf.*, also K. Bechgaard et al., *Mol. Cryst. Liq. Cryst.*, **79**, (1-4) 271 (1982); *cf.*, also R. Gubser et al., *ibid.*, p. 225; *cf.*, also J. F. Kwak et al., *ibid.*, p. 111.

The Meissner effect in deuterated tetramethyl-tetraselena-fulvalene has been studied: H. Schwenk et al., *Mol. Cryst. Liq. Cryst.*, **79**, (1-4) 277 (1982).

Reviews: The properties of tetramethyltetraselenafulvalene are reviewed: Yu. A. Bogod et al., *Physica,* **B** and **C**, (Amsterdam)-**108**, (1-3) 905 (1981). F. L. Vogel et al., *Springer Ser. Solid State Sci.*, **38**, (Phys. Intercalation Cpds) 288 (1981); K. Norbert, *Phys. Bl.*, **37**, (1) 11 (1981).

25.12.2 (Devices): A monolayer of chlorophyll on Cd-palmitate is used as a solar cell, based on photovoltaics: Jpn. Pat. 82-06231 (1982).

Photocells are formed from cross-linked polymeric dyes: H. Tuchida, Jpn. Pat., 81-54671; -54670; -54669 (1981).

A photochromic transition between red and blue forms of crystalline films of Zn-dithizonate, as well as photovoltaic activity, is reported: S. C. Dahlberg and C. B. Reinganum, *J. Chem. Phys.*, **76**, (11) 5515 (1982).

Doped polyacetylenes are reported to form useful solar cells: Y. Matsumura et al., Eur. Pat. Appl. 22,271,14 Jan., 1981; Jpn. Pat. Appl. 79-86,402, 10th July, 1979.

A Schottky barrier diode is formed from a polyacetylene film on a suitable substrate: Toray Indust. Jpn. Pat. 81-129,370 (1981).

(Polycarbazolyl-ethyl)-2-glutamate, which is p-type, surface doped with trinitrofluorenone forms a p-n rectifying junction: M. Hatano and K. Tanikawa, *Org. Coat. Plast. Chem.*, **40**, 288 (1979).

A Schottky diode rectifier has been produced from polyacetylene films on Al,Ga,In: Showa Denko K. K. Jpn. Pat. 81,147,487 (1981).

Rectifying tunnel diode junctions have been developed as Al-Al$_2$O$_3$ doped with thiourea-Pb M. K. Konkin and J. G. Adler, *J. Appl. Phys.*, **53**, (7) 5057 (1982); *Surf. Sci.*, **118**, (1-2) 303 (1982).

Phthalocyanine deposited on Pb as a film less than 1 μ thick acts as a Schottky barrier exhibiting a strong photovoltaic effect suitable for solar energy conversion; the currents reported are of the order of mA/W incident radiation. The barrier height is said to decrease in the order Al-SnO$_2$-Au-InO$_3$: S. Kanayama et al., *Jpn. J. Appl. Phys.*, **20**, (Suppl. 20-2) 141 (1981).

The stabilization of n-Si polycrystalline photoelectrochemical cells by coating of the surface with polypyrrole is reported: J. A. Frank, *Mol. Cryst. Liq. Cryst.*, **83**, (1-4) 1373 (1982); *cf.*, also T. Skotheim et al., *ibid.*, p. 1361.

Doping of polystyrene with anthraquinone, thioindigo, or perinone shifts the wavelength of incident light to longer wavelengths, *e.g.*, from 550 nm to 590 nm: Teijin Ltd. Jpan. Pat. 82-38844 (1982).

La-diphthalocyanine is reported to be electrochromic, showing a reversible color change as a function of the applied electric field: M. M. Nicholson, *Gov. Rep. Announce. Index*, (USA) **82**, (10) 1901 (1982) Avail. NTIS.

A photo-induced memory effect associated with the presence of deep traps is reported for polyvinylcarbazole:tetranitrofluorenone systems: Y. Nishio and E. Inoue, *Photogr. Sci. Eng.*, **25**, (1) 35 (1981).

Polytetrafluorethylene electrets are said to be suitable for condenser microphones: B. Lowkis and E. Motyl, *Mater. Sci.*, **7**, (2-3) 251 (1981).

The application of pyroelectric triglycine sulfate as infrared sensors is described: R. L. Kroes and D. Reiss, *NASA Tech. Memo.* 1981, NASA Tm-82394; Avail. NTIS; Sci. Tech. Aerospace Rep., *19* (9), Abstract No. N81-18852.

Thiophene-iodine charge transfer complexes are proposed as active cathodes *vs.* Li in solid state batteries: Jpn. Pat. 82-40870 (1982).

Solid state batteries based on iodine containing charge transfer complexes are investigated: A. V. Di Stefano et al., *Solid State Ionics,* **5**, 693 (1981).

Cu-Cu TCNQ films are said to act as switching devices, the two states involving the Cu TCNQ ⇌ TCNQ forms: E. I. Kamitsos et al., *Solid State Commun.*, **42**, (8) 561 (1982).

Secondary batteries based on electrochemically doped poly-(p-phenylene) forming n- and p-type complexes with Li are reported: L. W. Shacklette et al., *J. Chem. Soc. Chem. Commun.*, (1982), (6) 361.

Polyacetylene may be controllably doped p-type or n-type resulting in a light-weight rechargeable battery of high power density: P. J. Nigrey et al., *Mol. Cryst. Liq. Cryst.*, **83**, (1-4) 1341 (1982).

Fibrillar, doped, polyacetylene charge transfer complexes are proposed for use as flexible heat and electricity conductors for photoelectric devices: S. Ikeda et al., (K. K. Showa Denko Ltd) Jpn. Pat. Int. Appl., 8002,146 (Oct. 16th, 1980).

Flexible, stretchable and electrically conducting sheets are reported formed from thermoplastic polymer mixtures such as butyl-rubber co-polymers, heated with an organic peroxide and then mixed with carbon powder: Dainepon Jushi Kenkyusho, Jpn. Pat. 81-65,035 (June 2nd, 1981).

Pyrene derivatives plus iodine and/or TCNQ are proposed as thermistors or heat sensors: Mitsubishi Chem. Ind. Co. Ltd., Jpn. Pat., 81-120 649 (1981).

Phosphazene polymers are proposed as thermistors: Otsuka Pharm. Co. Jpn. Pat. 82-52,102 (1982).

Thermistors in sheet form are reported for TCNQ adducts: Katsushita Electric Indust. Co. Ltd.: Jap. Pat. 80-109,313 (1980).

Switching devices formed from standing stacks of Pb-phthalocyanine on a suitable substrate are proposed: C. Hamann et al., *Mater. Sci.*, **7**, (2-3) 181 (1981).

Semiconductors with a dumb-bell shaped structure are said to exhibit spontaneous oscillations in the current saturation region: H. Kuwano et al., *Jpn. J. Appl. Phys.*, Pt. I, **21**, (6) 896 (1982).

Author Index

Abbas, D. C., 608
Abate, L., 304
Abbi, J. C., 250
Abbi, S. C., 84, 132, 189, 307, 375
Abd-El-Malik, T. G., 642, 651
Abdulaev, N. G., 339
Abdurakchamno, B. M., 59
Abdurakhamanov, U., 653
Abkowitz, M., 256
Abou El Hela, A., 646
Abraham, R. J., 316
Abrahams, E., 60, 374, 659
Abrahams, S. C., 394
Abrahart, E. N., 251
Abu-Bakr, A., 186, 249
Adamec, V., 248, 250, 395, 640
Adamowski, J., 86
Adams, R. N., 316
Adkhamov, A. A., 192
Adler, A. D., 643
Adler, J. G., 672
Adrian, G. S., 341
Afifi, H. H., 646
Agranat, I., 311, 636, 637
Agranovich, V. M., 86
Ahmed,L. I., 58, 40, 40, 41, 310
Ahrland, S., 309
Ahuja, R. C., 669

Ashwell, G. J., 654
Aidelis, V. G., 455
Aihara, J. P., 95, 106, 641
Aikara, J., 304
Aikawa, Y., 372
Aizawa, M., 637
Akamatu, A., 641
Akamatu, M., 314
Akimoto, S., 635
Akimov, I. A., 252
Akiyama, I., 96, 107
Akiyama, M., 642
Akopyan, L. A., 637
Albers, C., 648
Albers, R. C., 373
Alberty, R. A., 58
Albery, W. J., 106, 468
Albrecht, A. C., 339, 437, 456, 643
Albrecht, W., 55, 253, 456, 462, 467
Alcazer, L., 316, 269, 313, 637
Aldrich, C., 85
Aleksandrov, S. B., 106
Alexander, A., 670
Alexander, S., 85, 250
Alfimov, M. V., 252
Algie, J. E., 59
Alkatis, S. A., 310, 310
Allen, M., 54, 15

AUTHOR INDEX

Almgren, M., 311, 282
Altieri, J., 320, 337, 636, 638
Altwegg, L., 83
Ambegaokar, V., 186
Ametov, K. K., 395
Amouyal, E., 645
Amov, I., 52
Amov, P., 248
Amritkar, R. E., 85
Amsler, P. E., 307, 308
Amzil, J., 655
Andersen, J. R., 414, 415, 417
Andersen, N. H., 415
Anderson, J. E., 315
Anderson, P. W., 249, 311, 372, 374
Anderson, R. A., 386, 394, 661
Andre, J. J., 60
Andreev, A. A., 60, 654
Andrews, E. T., 469
Andriessen, H. J. M., 305
Andrieux, A., 417
Andryushin, E. A., 86
Anex, B. J., 313
Angell, C A., 250
Aniansson, E. A. G., 310
Ansari, A. A., 455
Anthonsen, J. W., 313
Antula, J., 189
Aoyagi, Y., 642
Apgar, P. A., 417
Appel, O. J., 84
Applequist, J. R. C., 307
Appleton, G., 310, 312
Aprilesi, G., 454
Arapov, Yu. G., 56
Archer, M. D., 106, 468
Arden, W., 254, 454, 645
Argyrakis, A., 189
Argyrakis, P., 83
Arimatsu, S., 306
Aris, F. C., 188
Arkhipov, V. I., 650
Arnard, R., 315
Arndi, R., 317
Arndt, R. A., 395, 639
Arnett, E. M., 311, 284
Arnold, S., 645
Asanov, A. N., 651
Asbrink, L., 93, 105
Ashbaugh, A. L., 316
Ashley, D. J. B., 469

Asmolov, G. N., 310, 281
Assour, J. M., 376
Ateya, A. I., 637, 643
Athenstaedt, H., 340, 661
Atherton, N. M., 315
Atkins, P. W., 315
Atzmuller, H., 83
Auch, W., 316, 300
Audenaert, M., 249, 636, 666
Augustynski, J., 60
Ausloos, P., 304, 311
Aust, R. G., 376
Ave, D. H., 638
Avrami, M., 333, 340

Baba, H., 315
Babler, F., 306
Bachman, C. H., 336, 340
Backstrom, G., 249
Baenziger, N. C., 376
Bagchi, J., 54
Bagchi, S., 105, 313, 314, 317
Bagley, B. G., 146, 188
Bagyinka, C., 658
Bajaj, K. K., 85, 647
Bak, P., 252
Baker, B. G., 60
Balabanov, E. I., 52, 304, 637
Balakin, A. A., 255, 642
Balcerowicz, K., 55
Balode, D., 393, 428, 455
Balsenc, L. L., 60
Balta-Calleja, F. J., 660
Bandyopadhyay, A. K., 655
Banik, N. C., 671
Banks, R., 414
Bannister, C. E., 311
Banville, B. M., 82, 84, 396
Banyai, P., 58
Barach, J. P., 341
Baraldi, P., 646
Barancock, D., 340
Barbe, D. F., 639, 643
Barber, J., 87
Barbieri, M., 341
Bard, A. J., 253, 467
Bardeen, J., 188, 416
Bargeman, D., 305
Bari, R. A., 256
Barigandi, M., 316
Barkalov, I. M., 468

AUTHOR INDEX

Barkenov, V. K., 365, 375
Barker, G. C., 58
Barrell, P., 455
Bartczak, W. M., 187
Barthel, J., 60, 255
Bartlet, H., 316
Barton, A. F. M., 191
Bartzack, W. M., 255
Basov, N. G., 457
Bassett, C. A. L., 340, 341, 469
Bässler, H., 74, 83, 86, 171, 192, 312, 373, 434, 455, 639, 642, 644
Basu, S., 308
Bates, D. R., 54, 105
Batley, M., 94, 95, 98, 99, 102, 103, 104, 106, 107, 191, 367, 369, 370, 376, 456, 640
Batra, I. P., 185
Batt, R. H., 100, 101, 106, 426, 455
Batt, R. J., 189
Bau, L., 340
Bauer, D., 305, 315
Bauer, H. H., 315
Baughman, R. H., 312, 363, 374, 417, 470, 644, 659, 668
Baum, E. A., 638
Baumann, G., 658
Baumgartner, W., 374
Bauser, H., 73, 248
Bauser, N., 644
Baxt, L. M., 396
Bayless, J. H., 311, 283
Beauchamp, J. L., 311, 283
Bechgaard, K., 414, 415, 417, 671
Beck, A., 192
Beck, F., 254, 468
Beck, J., 56
Becker, R. O., 336, 340, 469
Bederson, B., 54, 105
Bednar, J., 252
Bednarek, S., 86
Beeby, J. L., 4, 52
Beens, H., 86, 315
Beer, A. C., 457
Beeson, K. W., 339
Beevers, M. S., 55
Behari, Sh., 660
Behr, J. P., 309
Beitz, J. V., 94, 105, 145, 189, 255
Belkind, A. E., 96, 106, 452, 453, 457

Belkind, P. E., 20, 54
Belmont, M. R., 254
Belov, G. P., 56
Beltran, D., 56
Benderskii, V. A., 339, 435, 436, 455, 456, 463, 468, 642
Bendig, J., 191
Bendiz, I., 308
Benedek, G., 417
Beni, G., 12, 53, 256, 416
Beniere, F., 662
Bennet, C. A. L., 336, 340
Bennet, W. E., 376
Bennett, R. A., 105, 316
Benoit, H., 666
Bentley, W. H., 376
Bercher, D. E., 340
Berets, D. J., 41, 58
Berg, C., 414
Berg, W. F., 252, 457
Berger, H. 646
Berger, P. A., 415
Bergman, A., 644
Bergman, J. G., 467
Bergmann, G., 53
Bergmann, J. G., 313
Berkovich, L. A., 59, 249
Berkowitz, J., 645
Berlin, K. D., 372
Berlin, Yu. A., 192, 255
Berlinsky, A. J., 60, 416
Bernard, C., 417
Bernardi, F., 305
Bernas, A., 54
Bernasconi, J., 4, 60, 132, 189, 249, 250
Bernier, P., 666
Berns, D. S., 644
Bernstein, H. J., 315
Berry, K., 468
Berry, R. S., 185
Berry, W., 20, 54, 106
Bertin, J., 308
Bespalov, B. P., 304, 313, 641
Besson, J. M., 371, 376
Bethe, H., 189
Bethea, C. G., 313
Bethner, C. G., 467
Betowski, L. D., 638
Bhat, S. N., 289, 313, 635
Bhatnagar, C. S., 390, 395

Bhatnarar, C. S., 660
Bichov, G., 395
Bickley, R. I., 457
Bieber, A., 60
Bieler, L., 469
Billand, D., 667
Binks, A. E., 434, 455
Birks, J. B., 185, 187, 313
Bischof, P., 341
Bishop, A. R., 417, 665
Bi Xian-Tong, 654
Bizzaro, W., 74, 83
Blackhurst, A. J., 315
Blakely, J. M., 57
Blanc, G., 58
Blanc, R., 315
Blank, M., 341
Blank, M. L., 341
Blasczkiewicz, B., 314
Blavrock, A. E., 339
Blinov, L. M., 307
Bloch, A. N., 413, 414, 415, 417
Block, H., 313
Block, J. H., 57
Bloom, H., 58, 455, 457, 468
Bloor, J. E., 468
Blount, G. H., 53
Blume, H., 86
Blythe, A. R., 53, 248, 250, 635
Blyumenfeld, L. A., 339
Borstel, G., 87
Braundmeier, A. J., 86
Bixon, M., 106
Boardman, N. K., 330, 339
Bobyl, K. G., 255
Bobyl, V. C., 384, 394
Bocionek, P., 669
Bockris, J. O'M., 53, 56, 58, 60, 86, 191, 253, 255, 304, 309, 311, 316, 339, 340, 438, 456, 468, 643, 658
Bode, U., 457
Boerger, D. M., 647
Boettger, H., 10, 53, 186, 360, 374, 428, 455
Boeyens, J. C. A., 375
Bogod, Yu. A., 671
Bogomolni, R. A., 338, 339, 469, 636
Boguslavsky, L. I., 42, 57, 58, 454
Boishchev, V. S., 249
Bokestein, G., 306

Bolard, J., 315
Bondar, V. V., 85
Bone, S., 9, 52, 53, 327, 338, 655
Bongers, P. F., 55
Bonham, J. S., 170, 190, 378, 380, 381, 382, 383, 393, 442, 445, 456, 636
Bonnier, J. M., 315 Bonsignore, A., 3
Bonsignore, A., 338
Boon, M. R., 635
Bordu, G., 52
Borisova, T., 249
Bornzin, G. A., 390, 395
Borodko, Yu. G., 416
Borovikov, A. Ya., 295, 314
Borsenberger, P. M., 427, 455, 642, 643, 644
Borster, G., 84
Boschi, R., 641
Boscke, F. L., 645
Boto, K. G., 316
Bouchez, P., 152, 189
Boulton, A. J., 250
Bounds, P. J., 649, 669
Bourg, C. M. A., 309
Bourret, L. A., 341
Bowmaker, G. A., 316
Bowman, R. W., 312
Brachman, M. K., 393
Bradley, R. S., 376
Branacomb, L. M., 105
Brau, A., 55, 192, 256, 305, 314, 372, 468, 635, 636
Braun, C. L., 100, 101, 102, 106, 107, 189, 252, 426, 455
Braun, D., 309, 312
Braun, R., 86
Braunstein, J., 255
Bravnan, J. I., 646
Brazovskii, S. A., 649
Breckenridge, W. C., 341
Bredas, J., 649
Bredas, J. L., 649
Bredero, P., 651
Bree, A., 454
Brehner, L., 641
Bremer, B., 636
Bremmer, C., 248
Brenig, W., 186, 187, 188, 373, 374
Brenner, A., 32, 56
Bretschneider, H., 308, 314

AUTHOR INDEX

Breyer, B., 315, 393
Bridgeman, P. W., 367, 368, 376
Briegleb, G., 53, 55, 304, 313
Bright, A. A., 55, 313, 417
Briere, G., 644
Bright, A. A., 55, 313, 417
Brighton, C. T., 469
Brill, J. W., 417
Brillante, A., 74, 83, 313
Brillson, L. J., 190
Brinen, J. S., 314
Brittain, J. O., 250
Broadhurst, M. G., 386, 394, 396
Broberg, J. E., 376
Brocklehurst, B., 147, 187, 188
Brocklehurst, J. R., 187
Broglie, F., 641
Brom, H. B., 256
Bromberg, A., 329, 339, 643
Brook, R. J., 190
Brophy, J., 58
Broslow, W., 191
Brossat, T., 395
Brot, J. M., 253, 467
Broude, V. L., 85
Brout, R., 84
Brown, D. W., 653
Brown, G. H., 249
Brown, N. M. D., 189
Brown, R. G., 255
Bruce, J. M., 312
Bruckenstein, S., 187
Brugg, C. E., 309
Bruggemann, R., 316
Brundle, C. R., 54, 106, 639
Brune, D. C., 644
Bruselfi, R., 670
Bruylants, A., 309
Bryant, G. W., 662
Bryskin, V. V., 10, 53, 186, 360, 374, 428, 455
Bube, R. H., 10, 53
Bucher, H., 192
Buchner, W., 393
Buck, M., 306
Bücker, W., 190, 248, 635, 640
Buckingham, A. D., 254, 309
Budniko, G. K., 375
Budzinski, E. E., 315
Buettner, H., 84, 85

Buguslavskii, L. I., 59
Bulaevskii, I. N., 60, 416, 670
Bunen, G. V., 311
Bunton, C. A., 311
Buravov, C. I., 636
Buravov, L. J., 29, 56
Burland, D. M., 4, 52
Burnay, S. G., 160, 188, 637
Burnell, M. E., 636
Burnway, S. G., 639
Burnett, G. M., 190
Burshtein, Z., 116, 186, 455, 638, 645
Bush, J. H., 312
Buttner, H., 86
Buttrill, S. E. Jr., 638
Buvet, R., 645
Buzdin, A. I., 670
Byakov, V. M., 255
Byrd, N. R., 249

Cahen, D., 59
Caillon, P., 383, 393
Calderwood, J. H., 248, 250
Caldorford, J., 315
Callcott, T. A., 415
Calvin, M., 339, 341, 437, 455, 469, 644
Camochan, P., 338
Campbell, C. K., 161, 189
Campillo, A. C., 74, 83
Campos, M., 32, 52, 188, 388, 640
Cann, J. R., 468
Capek, V., 102, 107
Cardew, M. H., 324, 338
Cardon, F., 58
Cardona, M., 457
Careem, M., 59, 189
Carmin, P. C., 314
Carneiro, K., 416
Carr, C. J., 468
Carrion, J. P., 55, 313
Carruthers, P., 84
Carruthers, T., 414
Case, B., 309
Casey, H. C., 457
Caspar, W., 316
Castonguay, J., 312
Castro, G., 414, 417
Cava, M. P., 415
Caywood, J. M., 54, 187
Cellota, R. C., 105

Celotta, R. J., 14, 54
Chaiken, P. M., 414, 416, 466
Chaiken, R. F., 427, 455
Chaikin, P. H., 53
Chaikin, P. M., 13, 256, 415, 417, 668
Chakraverly, E. K., 67
Chalozonites, N., 340
Chalvet, O., 191
Chamay, A., 469
Chan, C. K., 187
Chan, S., 56
Chan, W. S., 251
Chance, B., 187, 249
Chance, R. R., 101, 102, 106, 107, 189, 252, 312, 417, 426, 455, 642, 644
Chandler, D., 254
Chandross, E. A., 306
Chandra, B. P., 656
Chang, C. K., 641
Chang, J. C., 81, 87
Chang, J. Y., 376
Chang, Y. C., 20, 54, 106
Charle, K. P., 57
Charlton, H. M., 469
Chatlitheodoru, G., 656
Chattopadhay, A., 657
Chaudhuri, N. K., 374
Chaudhuri, P., 307, 375, 414
Chem, I., 192
Chen, E., 191
Chen, E. C. M., 16, 54
Chen, I., 33, 57
Chen, S. N., 645
Chen, S. S., 645
Chepel, V. F., 387, 395
Cherry, R. J., 339
Chew, M., 310
Chiang, C. K., 187, 250, 313, 417, 638, 663
Chibisov, A., 339
Chien, C. L., 192
Chien, C. W., 665
Chien, J. C. W., 664
Chikayzova, E. G., 316
Chistayokov, E. A., 456
Choi, S. I., 120, 185, 187
Chojnacki, H., 655
Chong, Sh-L., 638
Choubey, B. D., 637
Chow, W. S., 339

Chowdhury, M., 54, 105, 313, 314, 317, 639
Chowdry, A., 642
Christophorou, L. G., 53, 255
Christov, J. W., 340
Christov, S. G., 145, 146, 164, 168, 188, 189, 190, 249, 311, 330, 339
Chu, C. W., 413
Chu, N. Y. C., 86
Chupka, W. A., 638
Cignitti, M., 337
Ciesielski, W., 655
Cilento, G., 308
Cipollini, N. E., 255, 646
Clark, A., 57
Clark, D. T., 83
Clark, J. M., 60
Clark, P. A., 641
Clark, W. D. K., 253, 467, 468, 645
Clarke, T. C., 87, 249, 250, 252, 417, 657, 662, 666, 667
Clark, T. D., 648
Clement, R., 308
Cohen, J., 394
Cohen, J. A., 416
Cohen, M., 30, 56, 340
Cohen, M. D., 86
Cohen, M. H., 53, 111, 186, 346, 347, 348, 349, 372, 373, 637
Cohen, M. J., 413, 414, 417, 638
Cohen, M. L., 58
Cohen, S., 307, 311, 637
Colaccico, G., 469
Colbourne, C., 645
Cole, H. S., 636
Coleman, L. B., 1, 51, 59, 413, 414, 416
Coleman, R. V., 60
Coll, C. F., 53
Collins, Th. C., 83
Colson, R., 11, 52
Colson, S. D., 85, 189
Combescot, H., 77, 83
Comisarow, M. B., 55
Conrad, E. E., 646
Conway, B. E., 53, 56, 58, 60, 86, 191, 253, 255, 304, 309, 311, 316, 468, 658
Conwell, E. M., 416, 470
Cook, J. W., 413

Cooke, S. P., 29. 56
Cooper, C. D., 92, 105, 638, 640
Cooper, J. R., 417
Cooper, L. N., 416
Cooper, M. S., 466, 470
Cooper, R. R., 52
Cooper, W. F., 252
Cope, F. W., 325, 331, 332, 337, 338, 339, 340, 387, 395
Cope, J. W., 467
Coppage, F. N., 106
Corfield, P. W. R., 415
Corker, G. A., 642, 643
Cornillon, R., 395
Corrsin, L., 254
Costa, S. M. De B., 77, 87
Cot, R. J., 252
Cotman, C., 341
Cottles, V. M., 12, 53
Cottrell, G. A., 33, 55
Courtens, E., 107
Coutts, T. J., 360, 374
Covington, A. K., 309, 316
Cowan, D. O., 413, 414, 415, 417
Cox, G. A., 394, 642
Cox, R. J., 253
Craig, D. P., 649
Cramer, F., 308, 309
Crane, G. R., 313, 467
Craven, R. A., 413, 414
Cresp, H., 339
Crider, C. A., 59
Croft, R. C., 308
Croitorou, Z., 55, 315
Crooks, J. E., 311
Crossley, J., 55, 60
Crump, R. A., 315
Csavinszky, P., 186
Cu, A., 312
Cuatrecasas, P., 59
Currey, J. D., 469
Cutler, M., 254
Czimadia, I. G., 638

Dahlberg, S. C., 643
Dahm, D. J., 414, 415
Daiker, K. C., 638
Dainton, F. S., 255
Dakin, H. D., 338
Dalal, E. N., 250

Dall'Olio, A., 638, 639
Damadian, R., 332, 340
Damask, A. C., 395, 639
Daniels, F., 58
Danilevich, O. I., 53
Danno, T., 187
Datta, S. K., 650
Datta, T., 415
David, C., 248, 254
Davidson, R. S., 87
Davies, D. K., 641
Davies, E. A., 106
Davies, H. W., 376
Davies, M., 53
Davis, E. A., 189, 191, 373, 374, 425, 455
Davis, E. E., 308
Davis, E. R., 651
Davis, G. T., 394
Davis, H. T., 255
Davis, K. M. C., 191
Davydov, A. S., 56, 83, 85, 87, 251
Davydov, B. L., 313, 467
Dayhurst, G., 316
Debies, T. P., 54
Debuch, G., 658
Debye, P., 52
Deck, R., 415
DeFelice, L. G., 58
de Gruyter, W., 457
Deguchi, K., 309, 310, 312
De Haas, M. P., 255
Deitsch, A. J., 468
de Jonge, J., 454
Delacote, G., 124, 188, 639
Delannoy, P., 377, 378
Delhaes, P., 667
Dell, E. M., 106
Delsaut, A., 316
Del Sole, R., 84, 310
Deltowi, R., 311
de Lucia, J., 393
De Mey, G., 1, 53
Dempster, C. J., 306
DePasquali, G., 413
Deperasinska, I., 305, 314
Deschat, D., 653
Deutch, B., 469
Deutsch, J. W., 341
DeVault, D., 187

De la Vega, J. R., 312
Devlin, J. P., 59, 304, 375
Devreux, F., 84, 664
Devyatkov, N. D., 470
Dexter, R. L., 85, 188
Dewar, M. J., 645
Diaconu, I., 396
Dias Tavares, A., 643
Diaz, A. F., 37, 58, 249, 638
Dietz, F., 251
Dietz, K., 667
Digby, P. S. B., 338
Dikarev, B. M., 256
Dilung, I. I., 645
Dimitrescu, S., 396
Dimitrova, I., 314
Dinev, P., 660
DiSalvo, F. W., 415
Dissado, L. A., 74, 83
Di Stefano, A. V., 672
Dix, G., 642
Dlott, D. D., 83, 85
Doane, L. M., 28, 56
Doblhofer, K., 57, 59
Dobrianski, B. J., 376
Dobrovolskii, V. N., 53
Dobson, C. M., 338
Dodelet, J.-P., 5, 52, 255, 646, 668
Doehler, J., 83
Dogonadze, R. R., 57, 311
Doi, S., 268, 306, 642
Dolan, G. J., 249
Döldissen, W., 255, 639, 646
Donati, D., 83, 252, 254
Donavan, K. J., 663
Dosfale, T., 190
Dresner, J., 54, 377, 393, 394, 646
Dresner, O., 640
Dresselhaus, M. S., 308
Dresvyannikov, V. G., 396
Dreyfus, G., 386, 395
Drickamer, H. G., 367, 375, 376
Druger, S. D., 185
Drumbell, J. E., 251
Dryden, J. S., 395
Dube, C. B., 55
Duchesne, J. D., 636
Dudkowski, S. J., 252
Dudley, H. W., 338
Duerksen, W. K., 304

Dugay, M., 649
Duhaj, P., 340
Duic, L., 469
Duke, C. B., 186, 187, 188, 249, 256, 422, 657
Dulmage, W. J., 642
Dulov, A. A., 186
Dumai, J. M., 311
Duniec, J. T., 339
Dunitz, D., 375
Dupey, R. P., 374
DuPre, D. B., 249
Dupuis, P., 372, 635, 642
Durand, P., 55, 314
Duroure, C., 417
Duschek, D., 304
Dvey-Akharon, H., 345, 372
Dyankov, S., 308
Dycsmons, V., 638
Dynes, R. C., 83
Dzhaparidze, K. G., 467

Eagen, C. E., 84, 251
Eastman, J. W., 314
Easton, G. S., 658
Ebenhahn, M., 414
Eckardt, A., 468
Eckart, C., 145, 146, 188
Eckert, G., 314, 316, 465, 468, 469, 640, 658
Eckert, H., 310, 455, 469
Eckert, J. A., 467, 468, 645
Eckert, J. C., 417
Eckhard, J. J., 313
Eckhardt, H., 656
Economou, E. N., 374
Edelman, S., 394
Eden, J., 324, 338
Edgell, R., 641
Efrima, Sh., 186, 192
Efros, A. L., 355, 374
Ehrenfreund, E., 414, 637
Ehrenreich, H., 456
Ehrenstein, G., 658
Eidus, Ya. T., 191
Eiermann, H., 59
Einstein, T. L., 310
Eirich, F. R., 310
Eisenberg, A., 374
Eisner, M., 469

El-Sayed, M. A., 85
El-Wahaidy, E. F., 393, 383
Elenbaumer, R., 470
Eley, D. D., 13, 29, 31, 53, 56, 190, 324, 338, 636
Elli, A., 83
Ellis, J. D., 55
Elnahwy, S., 127, 128, 129, 130, 160, 187
Elving, P. J., 316
Emerald, R. L., 644, 646
Emergy, J., 252
Emert, J., 282, 310
Emery, V., 416
Emilsson, A., 341
Emin, D., 131, 151, 186, 188, 256
Emin, J., 160, 189
Enck, R. C., 106
Engberts, J. B. F. N., 307
Engelking, P. C., 54
Engels, J. M. L., 646
Engelsma, G., 456
Engler, E. M., 414, 415, 417, 670
Enomoto, T., 372
Epros, A. L., 373
Epstein, A. F., 251
Epstein, A. J., 106, 187, 416, 642, 664
Erdey-Fruz, T., 309, 311
Eremenko, O. N., 415
Eremenko, V. V., 107
Ermakova, V. D., 251, 254
Ernst, L., 463, 468
Ernst, N., 57
Ershov, B. G., 255
Estigneev, V. B., 338, 339
Etemad, S., 252, 414, 415, 416, 417, 655, 666
Euler, K. J., 58
Evans, D. F., 255
Evans, M., 53
Evans, S. M., 336, 340
Ewald, A. H., 375
Exner, O., 171, 190
Eyring, H., 57, 256
Fabian, J., 374
Fabish, T. J., 33, 55
Fabre, J. M., 417
Falck, K. M., 660
Falge, H. J., 87

Fan, F.-R., 190, 374, 643
Fang, P. H., 643
Farges, J. P., 23, 42, 53, 55, 56, 58, 60, 86, 191, 192, 253, 256, 304, 305, 310, 372, 468, 469, 635, 636, 642
Farragher, A. L., 639
Fauchais, P., 312
Faulkner, L. R., 190, 374, 456, 643
Fawcett, W. R., 305
Fayer, M. D., 55
Feakins, F., 309
Fedder, A. P., 192
Fedder, P. A., 360, 374
Fedorov, M. I., 59, 456, 645, 646
Fedorov, V. I., 57
Fedutin, D. N., 416
Feher, G., 58
Feigina, M. Yu., 339
Felix, G. N., 311, 312
Fendler, J. H., 310
Feng, T., 455
Ferguson, J., 306
Ferrari, J., 306
Ferraris, J. P., 413, 414, 636
Fessenden, R. W., 648
Feuerbacher, B., 457
Fialkov, Yu. Ya., 295, 314
Filimonov, A. A., 396
Fincher, C. R., 250, 313, 417
Finnegan, T. F., 413
Finsy, R., 60
Fintelmann, D., 468
Firth, D. R., 58
Fisch, K., 360, 374
Fischer, B., 86
Fischer, J. E., 308, 371, 375
Fischer, P., 248
Fischer, S. F., 150, 188
Fishman, H., 58
Fishman, M. L., 310
Fisum, O. I., 396
Fivac, R. C., 371, 376
Flanders, D. C., 59
Fleming, R. J., 250, 384, 394
Fleming, R. M., 417
Focke, W., 305
Foder, G., 338
Fogel, M. B., 87
Folkesson, B., 310

Fomin, A., 59, 249
Fong, F. K., 85
Forrest, I., 56, 306, 310, 455, 468, 469
Forro, L., 667
Forster, E. O., 123, 186, 314
Forster, F., 313
Forster, R., 375
Forster, Th., 85
Fort, A., 642
Foster, R., 86, 304, 306, 309, 312, 313, 316, 317, 341, 468
Fournie, R., 55, 314
Foweraker, A. R., 186
Fowler, W. B., 188
Fraenkel, G., 316
Franceschetti, D. R., 4, 52
Francis, A. H., 85
Francois, B., 662, 666
Frank, A. J., 310
Frank, C. W., 254
Frank, F., 309
Frank, G. W., 78, 84
Frank, H. J., 311
Frankevich, E. L., 52, 637, 643, 644, 657
Franklin, A. D., 663
Franklin, J. L., 638
Fredin, L., 83, 306, 309, 375
Freedman, G. R., 256
Freeman, G. R., 5, 52, 59, 646, 653
Freeman, H. C., 361, 374
Freeman, J. A., 341
Freeman, R., 55
Frei, R. W., 55
Freimans, J., 307
Frenkel, J., 189
Fridkin, V. M., 394, 448, 456
Friedenberg, Z. B., 469
Friedman, L., 11, 53, 159, 160, 188, 650
Friemans, J., 274
Friend, R. H., 52, 60, 416, 417
Frisch, K. C., 308
Fritzsche, H., 256, 374
Frohlich, H., 466. 469
Frohlich, P., 86, 306
Fromherz, P., 254, 454
Frosh, R. P., 85
Frost, D. C., 645
Frost, H. M., 340
Frumkin, A. N., 21, 55
Fryer, J. R., 651

Fuchs, R., 84
Fuhrmann, F., 249
Fujihara, M., 305
Fujishima, A., 339
Fukada, E., 340, 394, 395
Fukuda, Y., 308
Fukudome, H., 665
Fukui, M., 84
Fukayama, H., 670
Fulton, A., 106, 468, 641
Fumi, J. G., 373
Funabashi, K., 644
Funatsu, A., 305
Funfschilling, J., 393
Fung, K. K., 307
Funi, F. G., 55, 317
Fuoni, J. G., 373
Furutsuka, T., 338
Futrell, J. H., 638
Fyfe, C. A., 316, 468

Gabay, S., 375
Gabes, W., 55, 313
Gabler, R., 337
Gachkovskii, V. F., 304
Gadzuk, J. W., 187, 189
Gaehas, H. J., 187
Gagara, L. S., 350, 373
Gagarin, S. G., 310
Gagliardi, L. F., 311, 312
Gaidelis, V., 639
Gailis, A., 428, 455
Galanin, M. D., 83
Gallagher, T. J., 254
Gallo, C. F., 55
Galstyan, G., 396
Gamble, F. R., 308, 376
Gamoudi, M., 663
Gancheva, T., 660
Gardini, G. P., 250
Garito, A. F., 413, 414, 415, 416, 417
Garland, J. C., 249
Garner, D. P., 312
Garner, F., 667
Garofano, T., 454
Gascoyne, P. R. C., 324, 338, 470
Gashgari, M. A., 254
Gaspard, F., 644
Gates, H. C., 643
Gau, S. C., 417

AUTHOR INDEX

Gaul, S., 303, 316
Gause, E. M., 315
Gautier, F., 60
Gavrilov, V. I., 651
Gaylord, G., 56
Gaylord, N. G., 310
Gazdy, B., 662
Geacintov, N., 100, 101, 106, 107, 189, 428, 455, 456, 636
Gebicki, J., 308
Gee, N., 653
Geiger, R. S., 341
Geiger, W. E., 189
Geil, W., 58
Geis, M. W., 59
Geiss, R. H., 417
Gelsdorf, J., 254
Gemert, M. J. V. C., 52
Gemmer, R., 413, 414
Gerard, P., 161, 192
Gerischer, H., 8, 39, 57, 254, 645
Gerson, F., 315
Gerulaitis, N. Y., 467
Geserich, H. P., 417
Gestblom, B., 52
Geusekens, G., 248, 254
Ghosh, A. K., 455, 643, 644
Giacoletto, L. J., 58
Giacometti, G. A., 640
Giacometti, J. A., 32, 52, 188
Giaguinta, P. V., 52
Giardino, N., 249
Gibbons, D. J., 393
Gibbs, D. F., 376
Gibson, H., 657
Giera, J., 305
Giermanska, J., 652
Gilbert, B. C., 314
Gilbert, J. P., 253
Gill, W. D., 139, 153, 186, 187, 188, 190, 248, 249, 252, 313, 417
Gillbanks, D. N., 314, 315
Gillson, J. L., 413, 646
Ginsberg, A. P., 187
Ginter, M. L., 304
Giovanelli, K. H., 307
Giral, L., 417
Girvin, S. M., 53
Glaser, R. M., 185
Glazov, V. M., 255, 266, 305

Glick, A. J., 662
Glocker, D. A., 413
Gludau, W., 417
Gochev, A. D., 330, 339
Godfray, G., 393
Godik, E. E., 55
Godlewski, J., 56, 380, 393
Goel, M., 395
Gofman, J. W., 337
Goldberg, I. B., 663
Goldenberg, I. B., 84
Goldenberg, M., 274, 307
Goldfein, S., 340
Goldmacher, J., 376
Goldschwartz, J. M., 254, 255, 642
Golub, A. M., 308
Gombos, G., 341
Gonzalez-Basurto, J., 643
Goode, G. C., 54
Goodenow, J. M., 304
Goodgame, D. M. L., 313
Goodings, E. P., 254
Goodman, A. M., 54
Goodman, J. W., 457
Goodman, L., 127
Gooley, C. M., 468
Gor'kov, L. P., 374
Gordina, T. A., 312
Gordon, D., 306
Gordon, M., 86, 306
Gorelov, M., 667, 668
Gorenbein, A. Ya., 316
Gorodetskaya, A. V., 55
Gorodyskii, V. A., 305, 314
Gorshkov, V. T., 311
Gosar, P., 120
Gosh, A. K., 643, 644
Goswami, D. N., 57
Gotov, A., 311
Gould, R. D., 648
Gourlay, P., 468
Gouterman, M. P., 455, 645
Grabowski, Z. R., 307
Grace, J. D., 376
Gracey, J. P. V., 304
Grady, P. L., 13, 53
Graja, A., 654, 656
Grant, A. J., 308, 373, 374
Grant, P. M., 417, 663
Grass, F., 316

Grassi, H., 53, 60, 372
Gratzel, M., 310, 311
Grawaz, B. M. A., 390, 395
Grayson, M., 657
Greatbach, W., 467
Greatorex, D., 59
Grechischkins, R. V., 315
Grechishkin, V. S., 306
Grechov, V. V., 106
Green, J. R., 376
Greene, R. L., 414, 415, 417, 670
Greenstein, M., 83
Greiner, A. C., 468
Grensing, D., 359, 374
Griffith, P. R., 55
Griffith, R. W., 348, 373
Grimley, T. B., 310
Groff, R. P., 413
Groll, P., 316
Gross, B., 390, 395
Grover, M., 85
Grundnes, J., 304
Grüner, G., 415, 416, 668
Gubernatis, J. E., 373
Gubser, D. U., 670
Gubser, R., 671
Gudbjarnason, A., 341
Guesten, H., 86
Guidotti, D., 372
Guillet, J., 255, 634
Gumbs, G., 648
Gupta, A., 415
Gupta, C. L., 59
Gupta, R. N., 647
Guruswamy, V., 438, 456
Gusman, G., 311
Gutfreund, H., 416, 671
Gutmann, F., 27, 42, 52, 53, 56, 58, 84, 86, 130, 150, 253, 294, 295, 304, 305, 306, 308, 310, 314, 315, 316, 337, 338, 339, 340, 341, 393, 455, 457, 463, 465, 467, 468, 469, 635, 636, 642, 658
Gutmann, V., 183, 184, 263, 304, 305, 309, 314

Haarer, D., 54, 305
Hacobian, S., 315, 316
Hackett, C. F., 643

Hadek, V., 59, 312, 364, 372, 375, 415, 636
Hadni, A., 396
Haen, P., 417
Haglund, H., 59
Haken, H., 74, 86, 469
Halbritter, J., 31, 56
Hale, J. M., 35, 57
Hall, B., 375
Hall, D. E., 467
Hall, D. G., 86
Hall, K. J., 440, 441, 442, 445, 447, 456, 636
Hallenga, K., 60
Halperin, B. I., 373, 374
Ham, J. S., 53, 640
Haman, J. P., 83
Hamann, C., 467, 642, 652, 655, 672
Hamann, D. R., 190
Hambury, H. J., 469
Hamill, W. H., 86, 457
Hammer, A., 73
Hammond, P. R., 93, 105, 467, 639
Hampe, A., 385, 394
Hanamura, G., 84, 85
Hancock, F. E., 314
Haneman, D., 456
Haney, M. A., 638
Hang, H., 85
Hanke, W., 83, 182, 191, 464, 468
Hanna, M. W., 316
Hannay, N. B., 60
Hanscomb, J. R., 250
Hansma, P. K., 189
Hanson, D. M., 84, 85, 106, 304, 307, 384, 396, 457
Hanson, P., 304, 468
Hantzsch, A., 313
Hanue, W., 637
Harada, Y., 376, 641
Harbour, J. R., 182, 191, 254
Hardwicke, A. M., 469
Hardy, W. N., 31, 56
Haroniec, M., 58
Harper, R. A., 3, 52
Harrada, K., 665
Harrah, L. A., 78, 84
Harris, A. B., 360, 374
Harris, C. B., 85
Harris, J., 375, 638

Harris, W. A., 314
Harrison, R., 341
Harrison, S. E., 376
Hartman, R. D., 387, 395
Hartmann, A. K., 389, 395
Hartmann, G. C., 454, 457
Hartmann, K. W., 311
Hartmann, N., 311
Hartmann, R. D., 249, 250, 375
Hartmann, T. E., 57
Hase, H., 255
Hashimoto, M., 313
Hassan, A. R., 84
Hassel, D., 375
Hatano, M., 55, 372, 672
Hatch, G. F., 85
Hattori, K., 250
Hauffe, K. H., 252, 253, 306, 437, 455, 457, 669
Haug, H., 84
Hauser, J. J., 374
Hayakawa, S., 83
Hayashi, K., 468
Hayashi, Y., 252, 646
Haydon, S. C., 251
Hayes, T. L., 337
Hayes, T. M., 4, 52
Hayon, E., 305
Heathcote, J. G., 313
Hedvig, P., 395
Heeger, A. J., 252, 413, 414, 416, 417, 470, 638, 639, 666
Heesemann, J., 251
Heikes, K., 185
Heiland, G., 252
Heilbronner, E., 641
Heim, J., 635
Heinekamp, S., 84
Heinrichs, J., 85
Helene, C., 304
Heleskivi, J., 251
Hellner, L., 638
Hemming, R., 457
Hemamalini, N. S., 638
Henderson, D., 53
Henderson, E., 455
Henderson, R., 339
Hendrickson, D. N., 638
Henish, H. K., 53, 190, 650
Henry, A., 376, 467

Hensel, J. C., 83
Henson, J. R., 312
Herbstein, F. H., 375
Herbstein, I. H., 375
Herchenroeder, P., 345, 372
Hermann, A. M., 5, 12, 52, 53, 114, 115, 116, 185, 249, 308, 312, 366, 372, 375, 393, 415, 467, 636, 637, 640
Herold, A., 417
Herrmann, T. R., 645
Hersh, S. P., 13, 53
Herspring, A., 190, 248
Hertz, J. A., 256
Herzberg, G., 105
Hess, K., 56
Hetzler, H., 250
Hetzler, U., 192
Heublein, G., 310
Heyn, M. P., 339
Heywang, W., 394
Hibma, T., 638
Hicks, J. C., 396
Hidefumi, H., 55
Higasi, K., 315
Hilger, A., 309
Hill, E. B., 313
Hill, G. J., 376
Hill, H., 55
Hill, N. E., 8, 52
Hill, R. M., 189, 373
Hilti, B., 415
Hindley, N. K., 8, 52
Hino, S., 20, 54, 106, 450, 451, 453, 455, 456, 640
Hirano, F., 461, 467
Hiraoka, K., 86
Hirayama, H., 306
Hirohashi, R., 249, 455
Hirota, S., 652
Hirsch, J., 52, 249, 640
Hiruoka, T., 456, 457, 640
Hisa, M., 190
Hochstrasser, R. M., 67, 86, 87
Hociuc, M., 187
Hodges, R. V., 638
Hodgkin, A. L., 332, 340
Hoegl, H., 186, 456
Hoerhogd, H., 252
Hoerhold, H. H., 644
Hofberger, W., 59

AUTHOR INDEX

Hoffmann, H., 310
Hoffmann, D. M., 663
Hogarth, C. A., 57
Hoijtink, G. J., 191
Holczer, K., 257, 416
Holland, L. R., 3, 51
Holm, R. T., 84, 310
Holmes, P. J., 339
Holroyd, R. A., 15, 16, 54, 105, 255, 639, 640
Holstein, T., 187, 665
Homer, J., 316
Honda, K., 339
Hong, H. K., 85
Hong, J., 250
Hooge, F. N., 58
Hooper, H. O., 316
Hoover, W. H., 469
Hopf, H., 464
Hopfield, J. J., 87, 188, 339
Hopfinger, A. J., 372
Hoppenbrouwers, A. M. M., 58
Hora, H., 456
Horgan, A. M., 254
Horiuti, J., 310
Horn, P. M., 372, 413
Hornig, J. F., 100, 101, 106, 189, 426, 455
Horowitz, B., 416, 664, 670
Hoshen, J., 85
Hoshino, J., 373
Hoshino, Y., 642
Hosseini, N. M., 58
Hotop, H., 54, 105, 640
Hottman, S. D., 59
Houle, C. R., 341
Houser, N., 643
Housman, D. L., 81, 87
Hove, M. J., 306
Hovel, J. H., 644
Howard, J. N., 636
Howell, F. L., 251
Howie, A., 372
Hoyland, J. R., 127, 187
Hsieh, E. T., 256
Hsu, S. L., 252, 639
Hu, C., 414, 417
Huang, S. S-S., 59, 256, 640, 641
Huang, T., 644
Huebner, K., 365, 375

Huffman, W. A., 253
Hughes, R. C., 84, 98, 106, 188, 248, 638
Huml, J., 375
Humphrey, C. E., 649
Hunt, B. G., 341
Hush, N. S., 178, 191
Huxley, A. L., 332, 340
Huyskens, P. C., 311, 312
Hwang, S. B., 339
Hwang, W., 378, 393, 646

Ibers, J. Z., 375
Ibragimova, R. Kh., 309
Ichikawa, M., 470
Ichimura, K., 657
Ichimura, Sh., 80, 84, 86
Ideka, M., 253
Ido, Y., 456
Ieda, M., 430, 431, 432, 455, 636
Igbal, T., 57
Ihaya, Y., 313
Iida, Y., 305, 309, 374
Ikeda, S., 249, 250, 655, 673
Iliev, V., 308
Ilten, D. F., 469
Imamura, H., 470
Imura, I., 307
Imura, T., 182, 191, 338, 339, 454, 645
Inacker, O., 251
Inganaes, O., 657
Ingham, J. D., 191
Ingram, D. J., 315
Inkeno, Sh., 52
Inkson, J. C., 190
Inokuchi, H., 18, 54, 55, 56, 59, 95, 97, 106, 187, 252, 308, 375, 376, 453, 455, 456, 635, 636, 640, 641
Inoshita, I., 308
Inoue, A., 86
Inoue, E., 672
Inoue, H., 307
Inoue, M., 655
Inoue, N., 311, 637
Insepov, Z. A., 86, 308
Inuishi, Y., 255, 434, 456, 457, 636, 658
Ionescu, N. I., 58
Iqubal, Z., 312
Iseng, Y. W., 339
Isett, L. C., 189, 251, 415, 638, 639

Ishihara, Y., 59
Ishii, K., 264, 265, 305, 313, 640, 641
Ishii, Y., 85
Ishizuka, Y., 58
Islam, N., 250
Itaya, K., 307
Ito, I., 372
Ito, T., 249
Itoh, M., 86, 306, 307
Ivanov, I. I., 657
Ivory, D. M., 250
Ivove, E., 253
Iwasawa, Y., 305
Iye, Y., 671
Izumida, T., 312

Jacobsen, C. S., 414, 415, 667
Jacquemin, J. L., 52
Jahnke, H., 468
Jain, K. M., 58, 306
Jain, V. J., 394
Jajinski, R. J., 55
Jakai, Y., 455
Janda, M., 187, 190
Jannakovdakis, D., 314
Jano, I., 191
Janossy, A., 415, 416, 665
Janousek, B. K., 646
Jansen, L., 309
Jarnagin, R. C., 57, 427, 455, 643
Jaros, M., 383, 393
Jarrigeon, M., 250
Jarvis, D. H., 378, 383, 393, 456
Jaud, J., 256
Jaworski, M., 56
Jayanty, R. K. M., 457
Jayathirta, Y., 305
Jeffries, C. D., 83
Jenkins, A. D., 83, 248
Jennings, B. R., 186
Jerhot, J., 653
Jerome, D., 16, 60, 372, 375 668
Jeudy, M. J., 467
Jhon, M. S., 256
Joannopoulos, J. D., 190
Johansson, L. Y., 56, 254, 464, 468
Johari, G. P., 250
John, W., 356, 373, 374
Johns, H. E., 63, 84

Johnson, D., 52
Johnson, E. C., 307
Johnson, G. R., 174, 176, 413, 415
Johnson, H., 415
Johnson, J. W., 468
Johnson, M. R., 307
Johnson, O., 310
Johnston, G. R., 2, 7, 60, 107, 190
Johnston, L., 191
Jolly, W. L., 638
Jones, B. K., 58
Jones, J. B., 315
Jones, M., 415
Jones, M. T., 646
Jones, T. E., 414
Jonscher, A. K., 8, 52, 53, 251, 388, 395, 455
Jorgenson, T. E., 340
Jortner, J., 53, 83, 100, 106, 187, 373
Joshi, N. V., 644
Josita Feighan, M., 646
Jouanin, C., 665
Judo, K., 669

Kaahwa, Y., 250
Kadhim, A. H., 366, 375
Kagiya, T., 309
Kagoshima, S., 252, 655, 668
Kainer, H., 314
Kaiser, F., 658
Kajiwara, T., 187
Kakamura, Y., 639
Kalinowski, I., 641
Kalinowski, J., 380, 393
Kalis, R. K., 84
Kalishnikov, S. G., 57
Kallman, H., 57, 86, 455, 636
Kalmourski, J., 669
Kalnins, K., 311
Kamalam, R., 656
Kamaras, K., 60, 415
Kamimura, H., 650
Kaminski, J., 254
Kaminskii, V. F., 415, 646
Kamitsos, E. I., 672
Kampars, V., 92, 105
Kampars, F. J., 455, 645
Kamura, Y., 52, 59, 371, 375, 376, 644
Kan, K., 306
Kanayama, S., 672

Kanazawa, K. K., 37, 57, 58, 249, 668
Kanda, S., 375, 395
Kaneko, M., 468, 645
Kaneto, K., 456
Kang, E. T., 666
Kanicki, J., 662, 666
Kanieki, J., 668
Kanizawa, T., 191
Kano, K., 310
Kao, K. C., 255, 378, 393, 646
Kapinus, E. I., 645
Kaplan, J. L., 316
Kaplan, M. L., 415
Kaplunov, M. G., 416
Karbainon, Yu. A., 316
Karl, N., 54, 100, 106, 150, 188, 189, 253, 266, 305, 373, 426, 455, 647
Karmilov, A. Yu., 316
Karpin, G. M., 305
Karplus, M., 662
Karpov, V. G., 650
Karreman, G., 340
Kartheuser, E., 74, 86
Kasica, H., 383, 393
Kastening, B., 316
Katayama, H., 659
Kates, M., 339
Kato, K., 643
Katritzky, A. R., 250
Katusin-Razem, B., 310
Katz, J. I., 188
Kaufman, L., 341
Kavesh, S., 395
Kawabata, A.. 661
Kawabe, K., 338, 339, 646
Kawada, K., 375
Kawai, N., 362, 374, 375
Kawai, T., 644
Kawaki, N., 375
Kawamura, H., 54, 106, 253, 456, 641
Kawski, A., 254
Kay, E., 192, 250
Kayakawa, R., 394
Kayiwara, T., 640
Kazandzan, V. I., 314
Kazatskaya, L. S., 256, 653
Kearns, D. R., 427, 455
Kebarle, P., 642
Kedzia, J., 394
Keenan, R. W., 341

Keidekel, M. L., 416
Kein, N., 373
Keiper, R., 85, 428, 455
Keldysh, L. V., 314
Keller, H. J., 416, 668
Kelly, A. R., 339
Kelly, R. G., 341
Kemeny, G., 174, 190
Kennedy, T. A., 56
Kenzo, K., 660
Kepler, R. G., 5, 60, 100, 106, 185, 186, 187, 248, 386, 390, 394, 396, 467
Kertesz, M., 416
Kestner, N. R., 255
Kestyuki, D. G., 337
Keszthelі, C. P., 253
Kevan, L., 256
Keyes, T., 189
Keyzer, H., 27, 28, 52, 56, 191, 250, 294, 295, 305, 306, 308, 310, 314, 315, 337, 338, 339, 340, 341, 468, 469, 635, 642, 643
Khairutdinov, R. F., 187, 188, 191, 255,
Khammond, P. R., 191
Khanna, S. K., 8, 30, 52, 56, 252, 257, 414, 415, 417, 637, 651
Khanna, Y. P., 661
Khannay, N. B., 340
Kharadze, G. A., 374
Khare, M. L., 395
Kharkyanen, V. N., 469
Khidekel, M. L., 415
Kho, J. H. T., 59, 375, 635, 637
Khourkyamen, V. N., 465
Khvorov, M. M., 249
Kiaz, A. F., 57
Kibblewhite, J. F. J., 310, 311
Kiefer, J., 86
Kieffer, F., 255
Kiell, H., 662
Kigashimura, T., 255
Kikuchi, M., 253
Killesreiter, H., 86, 645
Kimura, K., 54, 56, 253, 338, 341, 641, 645, 646
King, D. L., 86
King, G. I., 339
Kinnunen, P. K. S., 333, 334, 341
Kinoshita, M., 316, 636
Kirby, D. R. S., 469

Kirk, R., 642
Kirzan, J. E., 320, 337
Kiselev, A. V., 339
Kistenmacher, J. Th., 341
Kistenmacher, T., 414
Kitaeva, T. A., 416
Kitagawa, T., 641
Kitaigorodsky, A. I., 188
Kittel, C., 58, 84, 190
Kivelson, H., 141, 187
Kivits, P., 385, 394
Klaffer, J., 83
Klafter, A. J., 373
Klasnic, L., 54
Klason, K., 58
Klein, D. J., 60, 375
Klein, G., 83
Klein, M. P. 636
Klier, E., 660
Kliewer, K. L., 84
Klindukhov, V. P., 317
Klinger, M. I., 650
Klinowski, I., 255, 642
Klöpffer, W., 73, 84, 86, 248, 311, 312, 636, 644
Klotz, H., 638
Knibbe, H., 306
Knights, J. C., 106, 425, 455
Knotek, M. L., 374, 457
Knox, R. S., 85
Kobayashi, A., 374
Kobayashi, H., 374
Kobayashi, T., 362, 374, 376
Kobinata, S., 305, 315
Koch, F., 190
Kochi, M., 19, 54, 98, 456, 641
Kocki, M., 106
Kodaira, T., 468
Koenig, J. L., 55
Kofler, A., 310, 311, 637
Kogotyuk, I. P., 649
Kohl, P. A., 467
Koizumi, S., 17, 54, 93, 105, 305, 309, 374, 636
Kok, J. A., 59
Kokado, H., 252
Kol, G. Ya., 53
Kolbanovskii, Yu. A., 310
Kolbasov, G. Y., 58
Kolbeck, A. G., 661

Kolendritskii, D. D., 84
Kollman, P., 304 311, 312
Kolomeits, B. T., 254
Kommandeur, J., 268, 305
Kondratov, V. K., 268, 305
Konig, R., 251
Konkin, M. K., 187, 190, 672
Konobeev, Yu. V., 86
Konzelmann, U., 4, 52
Koops, C. G., 52
Kopczewski, R. F., 636
Kopelman, R., 83, 85, 189
Kopf, H., 26, 56, 192, 468
Kopkinson, A. C., 638
Koppelman, R., 76, 84
Kopylov, V. V., 305
Kopylova, T. N., 307
Korb, J., 255
Korke, Y., 671
Korn, A., 160, 188
Korolev, A. I., 467
Kortum, G., 339
Kosar, J., 251
Kosche, H. H., 457
Kosower, E. M., 54, 307, 313
Kossmehl, G., 248, 249, 631, 645, 646
Kossut, J., 59
Kostin, A. K., 642, 646, 652, 669
Kotani, M., 640, 641
Kotov, A. I., 415, 416
Koutecky, J., 191
Kovalskii, P. N., 390, 395
Kovshev, E. I., 256
Koziol, J., 311
Kral, K., 257
Kramer, P., 59
Kranck, H., 58
Krans, W., 340
Krasnikov, N. N., 255
Kraus, W., 469
Krauss, P., 649
Krausse, H., 254
Kraut, W., 387, 396
Kreja, L., 651
Kreysig, D., 191
Kriemler, P., 638
Krikorian, E., 59, 190, 252, 636
Krishina, A. D., 253
Krishnan, V., 305
Krishtalik, G. I., 311

Krizan, J. E., 636, 638
Kroes, R. L., 672
Kroleveta, A. N., 53
Krolikowski, W. F., 55
Kroll, M., 304
Kronberger, H., 376
Krounbi, M., 663
Kronganz, V. A., 251, 256
Krug, W., 414
Krumhansl, J. A., 417
Krygowski, T. M., 305
Krylov, O. V., 281, 310
Kryszewski, M., 57, 250, 395
Kubaj, J., 58
Kubarev, S. I., 185
Kubovy, A., 187, 190
Kuczkowski, A., 656
Kuder, J. E., 105, 139, 140, 147, 187
Kuehne, R., 670
Kuhn, H., 192, 251
Kuhn, W. F., 640
Kukhtim, R. E., 306
Kulevsky, N., 309
Kulshreshtha, A. Q., 636
Kulshreshtha, Y. K., 669
Kulszewski, I., 666
Kumar, N., 85
Kunii, T. L., 105, 305, 639, 640
Kuntz, I. D., 339
Kurik, M. V., 59, 85
Kurobe, A., 650
Kuroda, H., 105, 305, 314, 639, 640
Kurracose, J. C., 252
Kurstein, G., 311
Kusabayashi, S., 455
Kushawa, S. C., 339
Kushelevsky, A. P., 313
Kuwano, H., 673
Kuznetsov, A. I., 55
Kuznetsov, A. M., 311
Kwak, J. F., 13, 53, 256, 313, 414, 415, 416, 417, 671
Kydd, R. A., 454

Labbe, J., 249
Labes, M. M., 248, 249, 308, 417, 466
Labos, E., 468
La Cabanne, C., 660
Lachish, W., 317
Ladik, J., 338, 465, 670

Lagowski, J. J., 58, 309, 643
Laguis, J., 86
Laibowitz, R. B., 414
Lakatos, A. I., 54, 186, 248, 644
Lakdar, T. B., 315
Lakes, R. S., 3, 52
Lakik, J., 469
Lakin, W. D., 192, 420, 457
Lal, K., 3, 52
Lalancette, J. M., 308
Lama, W. L., 55
Lamin, S. C., 657
Lamiri, S. C., 657
Lampe, F. W., 638
Lampert, M. A., 393,
Lamsweerde-Gallez, D. V., 316
Landar, T. B., 55
Landau, C. D., 256
Landau, L. D., 416
Landel, R. F., 375
Lane, J. E., 59
Lang, F. T., 304
Lang, R. P., 291, 306, 314
Lang, S., 394
Lang, S. B., 661
Langan, J. D., 189
Lange, H., 310
Langer, J., 251, 641, 652
Langler, E. M., 372
Langmuir, I., 21, 55
LaPlace, S. J., 414, 415, 417
Lardon, M., 248
Larson, D. C., 466
Larsson, R., 468
Latimer, M., 191
Latour, M., 395
Lau, J. K., 642
Lauchlan, L., 417
Laurentlev, V. V., 387, 395
Lavine, L. S., 340, 469
Lawrence, J. W., 337
Lawson, D. D., 191
Lawton, J. B., 656
Lax, M., 138, 141, 188, 308, 348, 373
Leal Ferreira, G., 395
Lebev, V. P., 250
LeBlanc, O. H., 5, 60
Lebovitz, Z., 84
Lechner, F., 340, 469
LeCleach, Y., 648
LeComber, P. G., 188, 372, 373

Ledwith, A., 341
Lederer, P. G., 658
Lee, K. O., 175, 190
Lee, P. A., 417
Lee, R., 341, 470
Lee, S. I., 249, 374
Lee, V. Y., 415
LeFur, D., 417
Le Gloan, A. C., 250
Lehn, J. M., 309
Leismann, H., 308
Lelauvain, M., 417
Leloir, L. F., 341
Lemonon, C., 395
Lenchenko, V. M., 253
Lengyel, G., 190
Lennarz, W. J., 341
Leonhardt, H., 86
Lerner, R. L., 668
Le Sar, R., 76, 84
Leupold, D., 251
Levchenko, A. I., 638
Levi, L., 28, 56
Levich, V. G., 57
Levine, B. F., 313, 467
Levy, D. D., 469
Lewiner, J., 386, 395
Lewis, T. J., 186, 250, 325, 327, 338
Ley, L., 457
Li, P. C., 59, 304
Lianos, P. C., 644
Lias, S. G., 304
Libby, W. F., 187
Libera, L., 314, 635
Liboff, A. R., 469
Licea, I., 254
Lichtenecker, K. Z., 175, 190
Lichtin, N. N., 645
Lifshitz, E. M., 416
Lilly, A. C., 57
Lim, E. C., 83
Lima, C., 372
Lin, C. T., 253
Lin, J., 645
Lin, S. F., 54, 106
Lindau, I., 54, 106
Ling, G. N., 332, 340
Linkens, A., 394
Lipatov, Yu. S., 250

Lipinski, A., 59, 656
Lipscomb, G. F., 648, 661
Lipskis, K., 10, 13, 53
Liptuga, A. I., 55
Lisitsa, M. P., 83
Litt, M. H., 305, 312, 361, 374, 387, 395
Little, W. A., 417
Litvinenko, V. Yu., 158, 188
Liu, S. F., 457
Liv, L. J., 310
Livingston, R., 339
Llenado, R. A., 316
Llewellyn, J. P., 55
Lochner, K., 642
Louie, S. G., 190
Lovelock, J. E., 191Lob
Lobanov, N. A., 339
Lobo, R., 87
Lochner, K., 312, 345, 372, 642
Logan, B. F., 1, 51
Lohmann, F., 34, 35, 57, 393
Lomova, L. G., 396
Long, J. P., 32, 57
Longworth, J. W., 307
Lonngren, K., 87
Loon, R. V., 60
Lorand, L., 326, 338
Loreitye, J. N., 58
Louboutin, J. P., 662
Louie, S. G., 57
Louis, E. J., 417
Loutfy, R. O., 643, 669
Louzos, J. K., 467
Lovelock, J. E., 639
Lowdin, P. O., 315
Lowkis, B., 672
Lozhkin, B. T., 57, 59, 454
Lozier, R. H., 339, 469
Lu, P., 187
Ludmer, Z., 86
Ludwig, J., 305
Luksha, E., 467
Lumb, M. D., 457
Lumdstroem, I., 657
Ludström, L., 642, 643
Lupu, I., 637
Lustrin, I., 469
Lutskii, A. E., 274, 307
Lux, F., 269, 306

Lykox, P. G., 315
Lyons, L. E., 1, 2, 7, 14, 15, 17, 34, 36, 54, 57, 58, 68, 92, 94, 95, 97, 101, 102, 105, 106, 107, 152, 170, 174, 176, 178, 179, 180, 182, 183, 189, 190, 191, 251, 252, 254, 308, 329, 338, 367, 369, 370, 376, 422, 423, 424, 426, 442, 444, 446, 447, 454, 455, 456, 468, 636, 641, 642, 643
Lyubovskii, R. B., 415

MacDiarmid, A. G., 252, 313, 417, 638, 639
MacDonald, J. R., 4, 52, 393
MacNeil, J. D., 55
Mackay, R. A., 54
Mackenzie, G. A., 416
Madek, V., 190
Maeda, H., 249
Magner, S., 645
Magrupov, M. A., 653, 659
Mah, C. Lowe, 415
Mahan, G. D., 53
Maher, J., 256
Mahood, J. S., 468
Mair, S. L., 307
Makarov-Mironov, A. M., 385, 394
Maki, A. H., 269, 313, 316
Maki, A. U., 637
Maki, K., 665
Maksimychev., A. V., 52
Maksyutin, Yu, K., 317
Malikov, B. F., 360, 362, 374
Mallet, G., 395
Mallik, B., 641
Malpas, R. E., 645
Malyutenko, V. K., 55
Mamann, S. D., 376
Manassen, J., 183, 191
Mandelkorn, L., 309, 312
Mandliv, A., 84
Mangin, J., 396
Mangini, A., 305
Manifacier, J. C., 10, 53, 190, 650
Mann, B., 192
Mantione, M. J., 305
Manzoli-Guidotti, L., 341
Manzoli, F. A., 341
Marada, Y., 641
Marchi, R. P., 256

Marcus, R. A., 57
Marello, V., 83
Margilli, C. G., 341
Margulis, V. B., 42, 58, 454
Mark, Ch., 314
Mark, H. B., Jr., 57
Mark, P., 393
Marklet, J. L., 316
Marshall, A. G., 55
Marshall, E. J., 255
Martin, E. H., 641
Martin, E. M., 52, 249
Martin, G. C., 304
Martin, J. H., 456
Martinov, Yu. I., 312
Marton, J. P., 337
Marty, J., 372
Martynov, I. Yu., 311, 312
Maruchin, J., 455
Marugima, Sh., 253
Maruyama, Y., 141, 142, 143, 186, 253, 376, 639, 644
Masashi, M., 662
Mascarenhas, S., 387, 396, 395
Maschke, K., 317, 371, 376, 659
Massey, H. S. W., 54, 105
Masuda, K., 313
Masumar, M., 317
Mataga, N., 306
Matsen, F. A., 178, 191
Matsuda, A., 55
Matsuda, Y., 655
Matsui, A., 85
Matsui, M., 395
Matsumura, M., 252
Matsumura, Y., 671
Matsunaga, Y., 11, 17, 54, 55, 59, 93, 105, 268, 283, 304, 306, 307, 308, 309, 310, 313, 316, 317, 636, 637, 642
Maurer, W., 56, 191
Mavroyannis, C., 648
Maxia, V., 385, 394
Maxwell, C., 395
Mayer, C. W., 415
Mayer, U., 304, 309, 649
Mayer, V., 191
Mayerle, J. J., 414, 415, 417, 668
Mayoh, B., 305
Mayoral, J., 396
Mayr, W., 662

Mazaud, A., 417
Mazur, K., 396
Mazur, P., 309
McAlear, J. H., 460, 470
McAuliffe, C. A., 470
McCafferty, E. M., 52
McCapra, F., 55
McCartin, P. J., 339
McCormick, D. B., 372
McDaniel, E. W., 105
McDiarmid, A. G., 470
McDowell, C. A., 53
McDowell, J. R., 57
McDowell, M. R. C., 105
McGhie, A. R., 185, 186, 192
McGill, T. C., 57
McGlynn, S. P., 305
McGovern, I. T., 308
McGregor, K. G., 36, 57, 440, 456
McGroddy, J., 83
McIntyre, C. M., 669
McIver, R. T., 639
McKinney, J. E., 394
McLaughlin, J., 338
McLevige, W. V., 59
McMullan, J. T., 1, 51
McWhan, D. B., 417
Mead, C. A., 187, 189
Meakins, R. J., 395
Meaudre, M., 52
Meaudre, R., 52, 377
Medvedev, V. S., 107
Meerschaut, A., 417
Meessen, A., 316
Meguro, K., 309, 310, 312
Mehendru, P. C., 252, 254
Mehl, W., 35, 57, 393
Meier, H., 55, 60, 83, 252, 253, 449, 456, 461, 462, 467, 644
Meier, W. H., 665
Meilanov, Y. S., 329, 339
Meister, T. G., 312
Melby, L. R., 413
Mele, E. J., 190, 664
Maler, H., 640
Melo, E. C. C., 77, 87
Melz, P. J., 644
Memming, R., 152, 187, 189, 311
Mendenhall, G. D., 24, 55
Menefee, E., 395

Menzel, E. R., 643
Menzel, R., 457
Meredith, G. R., 87
Merkel, P. B., 457
Merritt, V. Y., 644
Merski, J., 313
Meshkov, A. M., 455
Mesnard, G., 377
Metiu, H., 186
Meton, M., 161, 192
Metzger, R. M., 666
Mey, W., 114, 115, 116, 186, 186, 637, 638
Meyer, R. J., 186, 338
Meyer, W., 171, 190
Michaelides, J. P. O. M., 58
Michel, Bayerle, M. E., 84
Micheron, F., 394, 395
Mierzejewski, A., 158, 189, 642
Mihaly, G., 372, 415, 416
Mikawa, H., 455
Mikhailov, I. D., 185
Mikulski, C. M., 417
Miles, M. G., 413, 415
Miljak, M., 417
Miller, I., 340
Miller, I. F., 390, 396, 395
Miller, J. R., 94, 105, 145, 188, 189, 255
Miller, J. S., 251, 416
Milne, K. A., 101, 105, 106, 107, 153, 168, 189, 192, 252, 422, 423, 424, 426, 455, 636
Mima, S., 638
Mimura, T., 86
Minami, N., 645
Mines, G. W., 641
Minkin, C., 469
Minkoff, L., 340
Minomura, S., 375, 376
Mirovich, L. V., 316
Mishra, A., 66
Misra, M., 647
Misra, T. N., 314
Misurkin, I., 248
Mitchell, E. J., 284, 311
Mitchell, P. R., 307, 308, 309
Mitra, S., 649
Miyagawa, M., 254
Miyajima, N., 313
Miyakawa, T., 84

Miyasaka, K., 665
Mizuno, M., 415
Mizutami, T., 455
Moan, J., 54
Möbius, D., 251, 254, 651
Mobley, M. J., 375
Moehl, A., 341
Mohwald, M., 308
Moldavan, A., 60
Molinie, P., 417
Moller, C. K., 313
Mollers, F., 152, 187, 189, 311
Molnar, B., 56
Monberg, E. M., 85
Monceau, P., 417
Moncton, D. E., 417
Montenay-Garestier, T., 304
Montfoort, A., 341
Montgomery, H. C., 1, 51
Montroll, E., 5, 52, 138, 141, 155, 186, 187
Mookhemji, T., 636
Mooney, E. F., 315
Moore, W. J., 332, 340, 465, 466, 469
Morachevskii, A. A., 305, 314
Morawitz, H., 56
Morchardt, I. G., 3, 51
Mordclewski, J., 189
More, C., 670
Morel, D. L., 455, 645, 646
Morgan, A. E., 57
Morgan, I. G., 341
Morgan, K., 53, 640
Morimoto, K., 253, 312, 462, 467
Morokuma, K., 283, 311, 642
Morris, C. G., 83
Morris, G. C., 17, 54, 60, 85, 191
Morrison, S. R., 55, 253, 468
Mort, G., 373
Mort, J., 52, 54, 136, 137, 138, 186, 188, 248, 252, 253, 313, 454, 644, 646
Mortensen, K., 415
Morton, N., 118, 187
Mose, S. D., 666
Moss, M. L., 469
Mott, N. F., 53, 65, 85, 131, 186, 191, 254, 351, 352, 353, 355, 357, 373, 374, 428, 455, 659, 666
Motyl, E., 672
Moumtzis, J., 314

Mountz, J. M., 644, 645
Mouross, F., 84
Mozunder, A., 106, 255
Mracek, J., 8, 52
Mrha, J., 468
Muchnik, T. L., 85
Muench, V., 657
Mujasaka, T., 339
Mujunder, R., 338
Mulder, B. J., 83, 86, 421, 454
Mulder, G. L., 454
Muller, E., 256
Müller, H., 644
Muller, N., 339
Mulliken, R. S., 267, 304, 307, 313
Munits, I. N., 59
Munn, R. W., 83, 85, 118, 119, 123, 127, 185, 186, 187, 191, 639
Munoz, M. E., 652
Munro, D. C., 376
Munson, M. S. B., 638
Murakai, Y., 248
Murakami, Y., 253, 312, 462, 467
Murano, K., 330, 341
Murasaki, N., 395
Murata, K., 670
Murgich, J., 317
Murray, J. J., 308
Murrell, J. N., 314, 641
Muruyama, Y., 125
Musser, M. E., 643
Musuda, K., 305
Mutai, K., 307
Muus, L., 315
Mylinkov, V. S., 186

Naegele, D., 345, 372
Naff, W. T., 640
Nagaev, E. I., 188
Nagaev, E. L., 160
Nagakura, S., 305, 315
Nagao, M., 372
Nagasaka, H., 654
Nagy, J. B., 309
Nagy, N., 316
Nagy, O. B., 309
Nakada, I., 59
Nakahara, Y., 322, 338, 646
Nakajima, A., 641
Nakamura, A., 310

Nakamura, N., 316
Nakamura, T., 40, 58
Nakata, J., 339
Nakato, Y., 256, 641, 646
Nasielski, J., 645
Nasimov, I. V., 339
Naumann, C. F., 307, 308, 366, 376, 375
Naundorf, G., 73
Nava, F., 454
Neel, J., 642
Negita, H., 309
Nehl, W., 393
Neilands, D., 92, 105
Neilands, O., 646
Neilsen, P., 456, 457
Nelander, B., 83, 306, 309, 375
Neldel, H., 171, 190
Nelson, D. F., 660
Nelson, R. C., 251, 253, 329, 339, 641
Nesmeyanov, A. N., 308
Nespurek, S., 190, 393
Netschey, A., 468
Nettel, J. J., 84
Neuberg, C., 338
Newman, B. A., 394
Newman, D. S., 255, 654
Newman, K. E., 309, 316
Newman, M., 456
Newman, O. M. G., 455
Ngai, K. L., 346, 373, 647
Nicholson, J. R., 191
Nicholson, M. M., 646, 672
Nicodemo, L., 249
Nicolet, M. A., 377, 378
Nicolson, G. A., 468
Nielsen, P., 54, 106
Nieman, G. C., 85
Nigrey, P. J., 673
Niguchi, H., 386, 394
Niki, K., 338
Nishide, H., 468
Nishijo, H., 311, 637
Nishikawa, H., 468
Nishio, Y., 672
Nixon, D. E., 308
Nixon, W. B., 312, 314
Nobe, K., 255
Nobukazu, N., 660
Noda, S., 54, 256
Noguchi, H., 372, 636

Noguchi, K., 469
Noolandi, J., 454, 457
Norbert, K., 671
Nordhage, F., 249
Noreland, E., 52
Norman, G. E., 86, 308
Norrell, C. J., 372, 637, 639
Norton, L. A., 341
Novikov, Yu. N., 308
Novotny, M., 415
Noyes, W. A., 34, 57
Noyima, H., 59
Nozieret, P., 83
Nozoe, T., 308
Nurmukhametov, R. N., 374
Nutt, G. F., 251
Nuyts, R., 83
Nyikos, L., 255, 641

O'Brien, J. S., 341
O'Brien, R. M., 389
O'Conner, B. T., 469
O'Connor, D. V., 308
O'Reilly, T. J., 393
Ochs, F. W., 85
Oesterhelt, P., 339
Offen, H. W., 366, 375
Ohmasa, M., 309, 312, 313, 314, 636
Ohnishi, R., 55
Ohnishi, T., 468
Ohno, K., 56, 641
Ohno, M., 97, 107
Ohrbach, N., 306
Okamoto, K., 84, 429, 455, 637, 645
Okamoto, Y., 367, 376
Okeke, C. E., 55, 253
Okuhara, T., 55
Okuyara, K., 643
Okuyama, N., 656
Olofsson, G., 310, 312, 637
Olofsson, I., 310, 312, 637
Olofsson, R., 637
Olson, J. M., 254, 456
Olson, M. L., 307
Ondreijka, A., 340
Ong, N. P., 32, 56, 414, 417
Ong, P. H., 388, 395
Onishi, R., 192
Ono, K., 338
Onodera, A., 362, 371, 374, 375

Onsager, L., 100, 106, 423, 426, 455
Orchinnokov, Yu., 339
Orlov, I. G., 304
Ormancey, G., 393
Ormitoto, K. M., 248
Oron, N., 643
Orton, J. W., 648, 650
Osawa, E., 296, 314
Osawa, R., 311, 637
Oshiki, M., 394
Osipov, V. V., 308
Osterhoff, D. D., 249
Ostertag, E., 84
Ostrowski, M. A., 340
Ottavi, H., 60
Otto, A., 85
Otto, P., 326, 338
Ottolenghi, M., 306
Ovchinnikov, A. A., 248
Overhof, H., 371, 376
Ovshinsky, S. R., 374
Owen, A. E., 373
Owen, G. P., 190
Ozaki, M., 417

Padhye, M. R., 455, 644
Padke, S. D., 644
Padova, J.I., 309
Pae, K. D., 394
Page, F. M., 54, 639, 640
Pahwa, D. R., 3, 52
Pai, D. H., 253
Pai, D. M., 106, 248, 454
Pajak, Z., 314
Pal, L., 416
Palik, E. D., 84, 310
Palmer, L. D., 14, 54, 92, 105
Pampallona, M., 467
Pan, D. S., 86
Pandey, K. C., 190
Pandey, R. K., 59
Panish, M. B., 457
Pankratov, A. A., 641
Papier, G., 656
Paritskii, L. G., 378, 393
Park, Y. W., 250, 417, 641, 663
Parker, G. M., 187, 189
Parkhurst, L. J., 313
Parkin, S. S. P., 671
Parkinson, G. M., 188

Parkyns, N. D., 304
Parodi, A. J., 341
Parry, G. S., 308
Pasman, P., 308
Patel, G. N., 661
Patel, V. V., 414
Pawelka, Z. B., 251
Pawley, G. S., 59, 373
Pawluk, R. J., 340, 341, 469
Pearl, B. W., 469
Pearson, J. M., 312, 341
Pedersen, H. J., 415
Peebles, D., 417
Peierls, R. E., 399, 400, 401, 416
Pekker, S., 665
Pellizza, F., 304
Pemsler, J. P., 255
Penn, J. C., 187, 189
Penner, T. L., 651
Penney, T., 414
Pee, M., 663
Peover, M. E., 191, 316
Percel, V., 250, 313
Peredereeva, S. I., 304, 312, 313
Perewoschikof, N. F., 340
Perez-Albuerne, E. A., 189, 251, 415, 639
Perlman, M. M., 394, 395, 396, 467
Perlstein, J. H., 416
Perlstein, J. M., 60
Perluzzo, G., 53
Person, W. B., 304, 307, 313
Peshev, O., 310
Petelenz, P., 63, 76, 83, 85, 86, 308
Petersen, A. S., 416
Peterman, J., 661
Petersen, J., 638
Pethig, R., 9, 52, 53, 56, 250, 315, 320, 323, 324, 326, 327, 337, 338, 636, 640, 642, 655
Petkov, J., 374
Petree, M. C., 250
Petrov, A. A., 304
Petruzella, N., 251, 636, 644
Pfister, G., 52, 134, 135, 155, 157, 162, 186, 187, 189, 248, 249, 252, 254, 312, 373, 644
Phadke, S. D., 249, 639
Phariseau, P., 83
Phelps, D. J., 648
Philips, J. C., 249, 365, 375

Phillips, G. D., 656
Phillips, P., 250
Phillips, T., 414
Phillips, T. E., 2, 51
Phillips, W. A., 52, 374
Philpott, M. R., 84, 85, 313
Pichard, J. L., 659
Piciulo, P., 310
Pickard, W. F., 58
Pigon, K., 57, 345, 372
Pilla, A. A., 340, 341, 469
Pillai, P. K. C., 668
Pilling, M. J., 255
Pilling, M. M., 187
Pintschovius, L., 417
Pissanetzky, S., 317
Pistoulet, B., 373
Pivovarov, A. P., 307
Plans, J., 660
Plewka, A., 651
Plummer, E. W., 187, 189
Poberezhets, I. I., 59
Pochorov, J., 305
Pochtenni, A. E., 650
Poehler, T., 415
Poehler, T. O., 413, 414
Pohl, H. A., 53, 59, 132, 160, 186, 188, 189, 248, 249, 250, 251, 304, 337, 341, 343, 363, 367, 372, 375, 376, 466, 467, 469, 470, 635, 637, 641
Pohl, M., 639
Pohl, R. D., 374
Pokhodnya, K. I., 416
Pokrovskii, Ya. E., 83
Polanco, J. E., 308
Polansky, O. E., 374
Poleshuk, O. Kh., 317
Polis, B. D., 339
Politzer, P., 638
Pollak, M., 11, 53, 157, 159, 186, 188, 189, 250, 372, 373, 374, 641
Pollmann, J., 85, 86
Polymeropoulos, E. E., 251, 254, 454
Pong, W., 20, 54, 96, 106, 453, 457, 643
Pope, M., 57, 100, 106, 107, 189, 455, 456, 636, 644
Popescu, P., 190
Popot, J.-L., 341
Popov, B. M., 394
Popov, V. G., 39, 58

Popovic, Z. D., 457, 644
Popp, F. A., 657, 658
Poradowska, H., 305
Porter, G., 339
Portis, A. M., 32, 56
Potapov, V. K., 639
Potasek, M. J., 188, 339
Potember, R. S., 654
Poulton, B. R., 469
Powell, C. J., 106
Powell, M. J., 648
Powell, R. C., 69, 85, 252, 308
Powles, J. G., 254
Pragst, F., 307
Prakash, J., 313
Prasad, M., 637
Prasad, P. N., 85
Pratt, S., 189
Pravednikov, A. N., 374
Prigogine, I., 255
Prince, M., 636
Probst, K. H., 253
Prochorow, J. P., 304, 306, 314, 372
Prock, A., 643, 644
Prokofev, A. K., 307
Prosser, V., 636
Prout, C. K., 305
Prutton, M., 57
Przybylki, M., 654, 656
Pulfrey, D. L., 57
Pullman, B., 59, 304
Putley, E. H., 192
Pyle, R. E., 413, 414

Qian Renyan, 651
Quamara, J. K., 660
Quayum, A., 314
Quian, R. J., 647
Quickenden, T. I., 106
Qureshi, M. S., 390, 395

Raaen, V. F., 413
Rabelais, J. W., 54
Rabenhorst, H., 636
Rabolt, J. F., 87
Radushkevich, L. V., 467
Rafacy, K. M. A., 638
Rafikov, S. R., 640
Raghavan, V., 340
Rajeshwar, K., 252

Ramaley, L. R., 303, 316
Ramamurthy, M. V., 314
Ramdall, S., 252
Randles, J. E., 315
Ranicar, J. H., 186, 190, 249, 250, 394, 384, 635, 636
Ranninger, J., 670
Rantor, M., 376
Rao, C. N. R., 636
Rashba, E. I., 64, 83
Ratajcak, H., 311
Rather, H., 85
Ratnikov, E. V., 650
Ravenz, J., 394
Rayfield, G. W., 645
Rayment, T., 373
Reboul, J.-P., 383, 393
Redfield, D., 186
Redi, M., 339
Ree, R. S., 256
Ree, T., 256
Regel, A. R., 654
Regensburger, P. J., 248
Reimer, B., 312, 313, 434, 639, 642
Reinganum, C. B., 671
Reiss, D., 672
Rembaum, A., 53, 248, 249, 250, 312, 315, 372, 376, 374, 375, 467, 636, 637
Rempp, P., 666
Rettig, M. F., 105
Reucroft, P. J., 643
Reynolds, D. C., 83
Rhoderick, E. H., 192
Phodes, C. K., 457
Rhodes, L. M., 312
Ribault, M., 417
Rice, M. J., 4, 60, 87, 249, 256, 311, 312, 416, 417, 664
Rice, R. M., 84
Rice, S. A., 187, 255
Rice, T. M., 417
Richardson, G. H., 646
Richardson, L. M., 186
Richardson, T., 341
Richez, J., 469
Rieckhoff, K. I., 375
Rimai, D., 413
Rinaldi, R. A., 469
Rindorf, G., 415
Ritsko, J. J., 640, 664, 665
Ritvay-Emandity, K., 416
Robert, C., 417
Roberts, G. G., 190, 308
Robertson, W., 306
Robillard, J. J., 253, 310, 462, 467
Robin, M. B., 457
Robinson, B. H., 311
Robinson, G. W., 85
Rock, P. A., 340
Rocker, J. L., 645
Rodan, G. A., 337, 341
Rodgers, D. H., 314
Rohde, M., 249, 637
Rolland, M., 661
Romaczewski, Z., 56
Romanets, R. G., 186, 254
Romanowsky, E., 255
Rommelman, H., 665
Romming, C., 375
Roos, J., 388, 395
Rose, J. H., 84, 313
Rosen, R., 250
Rosenberg, B., 171, 172, 173, 174, 190, 314, 324, 338
Ross, M., 457
Rosseinsky, D. R., 33, 57
Rosseinsky, J. R., 645
Rossiter, B. W., 192, 251
Rossler, N., 311
Roth, W. L., 311, 312
Rothenberg, S., 311
Rouser, G., 341
Rouston, J., 649
Rouxel, J., 417
Rowe, C. A., 455
Rowe, J. E., 190
Rowley, B. A., 469
Roy, J. K., 306
Roy, K., 669
Royer, M., 395
Rozenberg, L. P., 415
Rozenshtein, L. D., 456
Rozental, A. I., 378, 393
Rubner, M., 663
Ruby, R. H., 339
Rudd, E. J., 316
Rudenko, A. I., 650
Rudyak, V. M., 394
Rudzinski, W., 58

Rule, N. G., 254
Rumyansev, B. M., 648, 670
Runyan, W. R., 57
Rupp, W., 415
Ruppel, W., 183, 191
Russell, A. A., 414

Sacher, E., 641
Sadaoka, Y., 372, 376, 393, 383, 467
Saels, J., 7, 52
Sagin, J., 254
Sagiv, J., 251
Saha, K., 132, 189, 250, 375
Saito, G., 310, 311, 637, 667
Saito, H., 637
Saito, S., 250
Saitoh, M., 373
Saitt, A. I., 55
Saitto, A. I., 315
Sakai, Y., 186, 372, 376, 383, 393, 467
Sakoda, Sh., 76, 84
Sakurai, T., 252
Salamon, M. B., 413, 414
Salkind, A., 340
Salkov, E. A., 83
Salo, T., 251
Samara, G. A., 367, 376, 375
Samoc, A., 254, 384, 394, 385
Samorjai, G. A., 57
Sampson, E. L., 341
Samson, S., 415
Samulski, E. T., 249
Sanche, L., 86
Sand, H., 106
Sandman, D. J., 106, 416, 641, 647, 655
Sandrolini, F., 637
Sanechika, K., 656
Sano, M., 308, 313, 316, 656
Santangelo, M., 454
Sapar, A., 393
Sapru, K., 374
Saraki, Y., 374
Sarma, G., 659
Sarrabayrouse, G., 187, 190
Sasaki, A., 83, 307
Sata, T., 469
Sato, H., 248, 304
Sato, M., 56
Sato, N., 107
Sato, T., 256

Satomura, M., 253
Satyam, M., 650
Savage, J. W., 417
Sawoda, M., 253
Sawyer, P. N., 469
Sayed, M. M., 31, 56
Sayer, P., 652
Scarfone, L. M., 186
Sceats, M. G., 60, 85
Scott Miller, J., 639
Schaadt, M., 160, 188
Schadt, M., 393, 639
Schafer, D. E., 413, 415
Schafer, F. P., 251
Schaffert, R. M., 253
Schaffman, M. J., 414
Schatz, E. R., 185
Schechtman, B. H., 106, 449, 456
Schegdlev, I. F., 642
Schein, L. B., 52, 106, 122, 123, 126,
   185, 186, 192, 420, 454, 640
Scheinbein, J. T., 394
Scher, H., 5, 45, 52, 134, 135, 138, 141,
   155, 157, 162, 186, 187, 188, 189,
   308, 348, 373
Scherer, G., 60
Schermann, W., 312
Schimmel, P. R., 312
Schindler, P. W., 309
Schklovskii, B. I., 189, 192, 355, 356, 373, 374
Schlotter, P., 18, 54, 84
Schmickler, W., 187, 190
Schmid, P. E., 371, 376
Schmidlin, F. W., 192
Schmidt, P. P., 57
Schmidt, V. H., 251
Schmidt, W., 641
Schmidt, W. F., 642, 646
Schmugge, K., 59
Schneider, A. D., 390
Schneider, S., 644
Schneider, W. G., 31, 56, 315
Schoenborn, B. P., 339
Schold, D. M., 254
Schonborn, M., 468
Schopov, I. S., 186, 190, 249, 636, 639
Schramm, C. J., 305
Schreiber, H., 373
Schreiber, J., 356, 374

Schriefer, J. R., 310, 416, 417, 664
Schroder, V., 83
Schroff, I., 304, 307
Schroff, L. G., 307
Schuchard, R., 85, 428, 455
Schuhmann, D., 191
Schultz, A. J., 414, 416, 417
Schultz, J., 185
Schultz, J. M., 186, 248, 396, 395
Schultz, T. D., 414
Schulz, G. J., 86
Schulz, H. J., 668, 670
Schulz, T. D., 252
Schumaker, R., 414, 417
Schuster, H. G., 416
Schwab, G. M., 310
Schwab, H. P., 191
Schwartz, G., 315
Schwarz, M., 376
Schwarzer, E., 74, 86
Schwenk, H., 671
Schwob, H. P., 393
Scott, A., 87
Scott, A. C., 341
Scott, B. A., 414, 415
Scudder, J. A., 375
Seal, E. H., 469
Seanor, D. A., 62, 83, 248, 640
Sebastian, L., 648
Sedlak, W., 337
Seeger, K., 31, 56, 191, 662
Seely, G. A., 339
Seghatchian, N. J., 468
Seiden, D. M., 414
Seifert, K., 339
Seitz, F., 84
Seitz, W. R., 55
Seki, H., 146, 147, 188, 248
Seki, K., 18, 54, 106, 452, 457, 640, 645
Selig, H., 666
Seliger, H. H., 83
Sergeev, G. B., 309
Sersen, F., 306
Sessler, G. M., 32, 57, 387, 396, 395, 467
Setchell, F., 468
Setser, D. W., 86
Sever, R. J., 339
Seymour, R. B., 312
Shacklette, L. W., 672
Shamos, M. H., 340, 469

Shamos, M. I., 340
Sharma, D. L., 394
Sharp, J. H., 644
Shaw, R. E., 455
Shchegolev, I. F., 12, 29, 53, 56, 185, 191, 256, 304, 415, 416, 635, 637
Sheng, S. J., 384, 396
Shibaeva, R. P., 415
Shigehara, K., 467, 645
Shihara, H., 316
Shimada, M., 306
Shimamura, K., 665
Shimoda, H., 659
Shipman, L. L., 81, 87
Shirakawa, H., 249, 250, 374, 417, 661
Shirota, Y., 253
Shirotani, I., 107, 371, 376, 374, 375, 635
Shi Zurong, 651
Shlyapinktokh, V. Ya., 55
Shneider, A. D., 395
Shockley, W., 188
Shold, M., 464, 468
Shono, K., 268, 317, 636
Shore, M. B., 84
Shostokoskii, M. F., 316
Shpak, M. T., 649
Shramm, H., 250
Shrivastava, S. K., 660
Shu, P., 414
Shuvaev, V. P., 57
Sichels, E. K., 663
Sidaravichyus, I. B., 467
Siddigui, A. S., 663
Siebrand, W., 85, 118, 119, 123, 127, 185, 186, 187, 191, 639, 649
Siegel, H., 307, 308, 309, 366, 372, 376, 375
Siegel, M., 54
Siegman, A. E., 251, 457
Signeski, R., 669
Silbey, R., 83, 85, 86, 185, 188, 649
Silin, A. P., 86
Silinsh, E., 190, 252, 328, 336, 338, 341, 393, 643
Silinsh, W. A., 670
Silver, M., 186, 305, 313, 350, 373, 427, 455
Silver, R. N., 83
Silverman, B. D., 304, 309, 312, 375, 415

Silverman, H., 340
Silverman, J., 646
Simionescu, C. I., 313, 337
Simmons, J. G., 154, 155, 188
Simonsen, M. G., 58, 60
Simov, D., 307
Simpson, W. T., 313
Singh, R. A., 289, 313
Singh, V., 317
Siracusa, G., 304
Sivyer, A., 469
Skal, A. S., 653
Skeda, M., 248
Skilandat, H., 316
Skotheim, T., 672
Skove, M. J., 413
Skripnik, F. V., 649
Skulachev, V. P., 645
Skulski, L., 314
Slater, E. C., 330, 339
Slavinski, J., 55
Slifkin, M. A., 259, 304, 306, 307, 309, 313, 314, 315, 316, 468
Sligh, J. L., 32, 56
Sliva, P. O., 253
Slobodyanik, V. V., 306
Slocum, D. W., 304
Sluckin, T. J., 372
Smith, D., 315
Smith, D. E., 311
Smith, D. S., 41, 58
Smith, H. I., 59
Smith, J. A., 20, 54, 96, 106, 457
Smith, L. M., 341
Smith, R. A., 188
Smith, S. D., 340
Smith, S. O., 469
Smith, V. H., 85
Smolenski, B. I., 469
Smolenskii, G. A., 394
Smolyanskaya, A. Z., 470
Smyth, C. P., 52
Smyth, D. M., 457
Smith, J. A., 453
Sneed, R. J., 59, 190, 252, 636
Snyder, F., 341
Soares de Campos, M., 395, 396
Sobczyk, L., 251
Sokolova, E. B., 160, 188
Soling, H., 415

Solyom, J., 416
Somerton, U. W., 315
Sommer, G., 100, 106, 189, 426, 455
Sommerford, A., 189
Somoano, R. B., 55, 250, 313, 372, 392, 396, 415, 417, 636, 637
Somogyi, I. K., 374
Soni, V., 315
Sonnonstine, T. J., 5, 52, 185, 637
Soos, Z. G., 60, 85, 252, 375, 416
Soran, M. S., 417
Spadaro, J. A., 469
Spange, S., 310
Spannring, W., 83, 171, 192
Spear, W. E., 52, 111, 186, 252, 373, 374
Spector, V. N., 665
Spicer, W. E., 54, 55, 106, 449, 456
Spielberg, D. H., 160, 161, 188, 372
Spitzler, M., 437, 455
Spivey, D., 323, 338
Sponable, E. I., 467
Spong, P. L., 393
Srinivasan, R., 656
Srinivasan, S., 374, 469
Srivastara, R. K., 660
Srivastava, A., 669
St. Onges, H., 250
Staab, H. A., 306, 307, 276
Stahl, A., 87
Stanby, J., 59
Stanche, A., 249
Starodub, V. A., 655
Starzak, M. E., 316
Staudinger, A., 374
Stedule, W., 84
Stefoni, S., 341
Steinbach, F., 26, 56, 192, 464, 468
Steiner, B. W., 105
Steinfeld, J. I., 251
Stern, M. B., 341
Sternlicht, H., 85
Sterzel, H. J., 309, 312
Steudle, W., 84
Stevens, G. C., 642
Stewart, F., 469
Stilbs, P., 310, 312, 637
Stiles, P. J., 383, 393
Stock, J. T., 28, 56
Stoeckenius, W., 339, 469
Stokes, G., 59

Stolka, M., 307
Stone, F. G., 305
Stoneham, A. M., 85
Stoneham, M. E., 341, 470
Stoyanov, S., 307
Stradins, J., 316
Stradowski, C. Z., 255
Straley, J. P., 373
Strassler, S., 416
Stratton, R., 189
Straub, K. D., 325, 338, 636, 637
Straw, H., 314
Street, G. B., 249, 252, 417
Street, G. S., 667
Street, R. A., 374
Streitweiser, A., 182, 191
Strohbusch, F., 312
Strome, F. C., 102, 107
Strong, R. L., 304
Stuart, M., 57
Stubb, T., 251
Stucky, G. D., 60, 251, 414, 416
Studke, J., 188
Stufkens, D. J., 55, 313
Stuke, J., 186, 187, 373, 374
Styrov, U. V., 57
Su, W. P., 87, 417, 665
Subbotin, G. I., 315
Subertova, E., 636
Suchet, J. P., 373
Sugi, M., 187, 257
Sukhov, N. L., 255
Sukigara, M., 659
Sumi, H., 115, 125, 132, 186, 189, 192
Sumita, I., 394
Summers, J. W., 361, 374
Sundberg, K. R., 307
Sutherland, I. O., 307
Sutin, N., 253
Suzi, M., 192
Suzuki, A., 640
Suzuki, M., 188, 638
Suzuki, Y., 316
Suzumura, M., 386, 394
Swalen, J. D., 84
Swart, P. L., 161, 189
Swenberg, C., 189
Swiatek, J., 59, 186, 381, 384, 393, 394
Swiatkiewicz, J., 345, 372
Swift, J., 23, 55

Sworakowski, J., 57, 378, 393, 460, 467
Sybesma, C., 66, 79, 87, 338
Szent-Györgyi, A., 323, 337, 338, 341, 465, 466, 469, 470, 642
Szymanski, A., 57, 248, 466, 646

Tabak, M. D., 639
Tabor, D., 57
Takagi, H., 313
Takahashi, K., 60
Takahashi, T., 344, 372, 638
Takai, Y., 669
Takeda, S., 84, 251, 636, 643
Takemoto, K., 306
Takenaka, I., 333, 340
Takeuchi, M., 654
Takizawa, T., 56, 251, 253
Talwar, I. M., 394, 660
Tamai, T., 57
Tamaki, T., 307
Tamamura, J., 363, 375
Tamamura, T., 306, 434, 455
Tamamushi, R. 60
Tamaru, K., 56, 470
Tamres, M., 304, 306
Tamura, T., 252, 643
Tan, F. R., 456
Tanaka, H., 667
Tanaka, K., 55, 56
Tanaka, M., 55, 417, 652, 662
Tanaka, T., 434, 457, 636
Tanase, S., 372, 648
Tang, C. W., 339, 437, 456, 643
Tang, Ying Wu., 650
Taniguchi, Y., 306
Tanikawa, K., 644, 672
Tanikawa, M., 460, 466
Tannhauser, D. S., 373
Tantraporn, W., 52, 640
Tanuma, S., 671
Tappel, A. L., 341
Taranko, A. R., 414
Tarvin, R. F., 468
Tatekawi, M., 40
Tatewaki, M., 58
Tauc, J., 311, 372, 373
Tauchert, W., 54
Taylor, L. T., 657
Taylor, P. C., 373
Taylor, P. L., 373

Taylor, P. R., 87
Tech, H., 668
Teitelbaum, R. C., 305, 268
Tench, A. J., 310, 311
Teo, Boon-Keng, 187
Teodorescu, G., 388, 395, 638
Terenin, A., 306
Terentev, V. A., 84
Testa, A. C., 312
Thakton, G. D., 188
Thayer, J. S., 305
Thewalt, U., 309
Thiele, D., 314
Thoman, H., 663
Thomas, G. F., 316
Thomas, G. A., 413
Thomas, J. K., 310
Thomas, J. M., 384, 394
Thomas, M., 372
Thomas, R., 415
Thomas, Y. M., 310
Thompson, A. H., 308
Thompson, T. E., 308, 371, 375
Thorne, S. W., 339
Thorua, P., 373
Thorup, N., 415
Thouless, D. J., 249
Tick, P. A., 7, 52
Tiedemann, P. W., 639
Tien, H. Ti., 337, 644, 658
Tissier, B., 417
Titov, V. V., 304, 313
Titvinenko, L. M., 308
Tiwari, A. R., 660
Tokiwa, F., 310
Tokomuto, M., 654
Toler, J., 7, 52
Tollin, G., 182, 191, 254
Tomasik, P., 311
Tomioka, K., 84
Tomkiewicz, Y., 252, 315, 317, 414, 665
Tong, B. Y., 320, 337
Toombs, G. A., 60, 416
Toropova, V. F., 316
Torrance, J. B., 257, 309, 312, 375, 414, 415
Tosatti, E., 84, 86
Toshima, Sh., 307
Tossatti, E., 310
Toureille, A., 383, 393

Tourillok, G., 667
Toyoda, K., 305, 394
Teyozawa, Y., 648
Tramer, A., 304, 372
Tratt, W. M., 639
Treinin, A., 305
Treusch, J., 417
Trevoy, D. J., 415
Tribel, M. M. 657
Triberis, G. P., 650
Tributsch, H., 644
Tripathy, S., 659
Trofinchek, A. K., 316
Trukhan, E. M., 29, 31, 56, 253, 340, 636
Trullinger, S. E., 417
Truro, N., 282, 310
Tschirwitz, U., 456, 467
Tsong, T. T., 190
Tsubomura, H., 252, 305, 306, 645
Tsuchida, E., 56, 468
Tsuchiya, S., 470
Tsuda, M., 56
Tsuji, K., 310, 315
Tuchida, H., 671
Tulagin, V., 644
Turcel, M., 85
Turlet, J. M., 84, 310
Tumo, J. R., 661
Turner, D. R., 339
Turner, R. E., 85
Turner, S. R., 307, 341
Turovskii, I. V., 307
Tutihasi, S., 644
Twaroswski, A. J., 643
Twiselton, D. R., 316
Tyagai, V. A., 53, 58
Tyczkowski, J., 190, 249
Tyurin, Yu. I., 57
Tyutyulkov, N., 86, 306, 315, 374

Ubbelohde, A. R., 304, 308
Uberle, A., 314
Ueba, H., 80, 84, 86
Ueno, N., 257
Ukei, K., 376
Ulbert, K., 177, 190, 364, 375
Ulbrich, R. G., 83
Ulstrup, J., 57, 59
Umeyama, H., 642

Umezawa, Y., 254, 456, 463, 467, 468, 644
Underhill, A. E., 416
Underhill, C., 308
Unwin, P. N. T., 339
Uosaki, K., 456
Ursu, I., 315
Usdin, E., 310, 455, 468, 469
Usov, N. N., 435, 436, 455, 642
Usui, Y., 311
Uth, H. J., 252, 253
Uyeda, N., 376

Vaisnys, I., 642
Vakser, A. I., 385, 396
van Beek, L. K. H., 251
van Deenen, L. L. M., 341
van der Meer, M., 649
Vanderschueren, J., 394
Van Est-Stammer, R., 307
van Golde, L. M. G., 341
van Heuvelen, A., 150, 188
Vannikov, A. V., 52, 253, 454, 642, 646
van Reuth, E. C., 249
van Ruyen, L. J., 59
Van Schooten, J., 191
van Turnhout, J., 384, 388, 394, 395
Varakina, E. N., 657
Varma, C. M., 374
Vartanyan, A. T., 251, 651
Vasanta, B., 668
Vashishta, P., 84
Vasilensko, N. A., 374
Vasilescu, D., 53, 58, 43, 387, 395, 636, 646
Vasilev, G. P., 58
Vaughan, W. E., 305, 315
Veda, H., 311
Veno, N., 457
Verhoven, J. W., 54
Vermeulen, G., 454
Vezzetti, C. F., 394
Vidadi, Y. A., 383, 393, 456, 650
Vijh, A. K., 183, 191
Vilenskaya, R. L., 470
Vilesov, F. I., 106, 456
Vilfau, I., 120, 185, 372
Virtanen, J. A., 333, 334, 341
Viselov, F. T., 641
Vlasova, R. M., 256

Vleck, J., 653
Vodeicharov, C., 637
Vodenicharov, C., 186, 189, 190, 249, 636
Vodenicharova, M., 52, 169, 170, 189, 190, 248, 249, 253
Vogel, F. L., 308, 371, 375, 671
Voigt, E. M., 375
Volger, J., 10, 53
Volkenshtein, Th., 182, 191
Volkenstein, M. V., 337
Volpin, M. E., 191, 308
von Baltz, R., 387, 396
von Goldammer, E., 316
Von Zelewski, A., 306
Voronyuk, P. I., 12, 53
Vozzhennikov, V. M., 374

Waclawec, W., 315, 394, 651
Wada, Y., 248, 250, 394, 467
Wadda, Y., 53
Waddington, T. C., 638
Waechter, C. J., 341
Wagner, K. W., 395
Wagner, M., 469
Wakayama, N. I., 58, 636
Wakill, M., 309
Walatka, V. V., 413, 414
Waldbillig, R., 658
Walker, E. I., 646
Walker, S., 314
Wallace, S. C., 310
Wallis, J., 652
Walmsey, R. H., 304, 307
Walsh, W. M., 415
Wang, J. H., 644
Wang, T. T., 345, 372
Ward, L., 314
Ware, W. R., 86, 306, 308
Warfield, G., 640
Warfield, R. W., 250
Warmack, R. J., 413, 415
Warren, L. J., 191, 254
Warta, W., 189
Watanabe, A., 653
Waszczak, J. V., 415
Watanabe, K., 253
Watanabe, M., 373
Watanabe, T., 56, 329, 339
Watkins, D. M., 416

AUTHOR INDEX

Watson, C. R., 415
Watson, J., 469
Watson, P. K., 250
Wauk, M. T., 85
Waysand, G., 417
Webb, J. B., 186, 639, 646
Webb, S. J., 337, 341, 466, 470
Weber, B., 55
Weber, G., 396
Weber, W. H., 84, 251
Webman, I., 53, 255
Webmann, I., 12
Weger, M., 52, 371, 375, 416, 417
Wegner, G., 312
Wehry, E. L., 86, 306
Wei, L. Y., 187
Weigl, J., 252
Weigl, J. W., 254
Weikowitsch, C. E., 305
Weiser, G., 248, 648
Weiss, J. J., 376
Weissberger, A., 192, 251
Weissman, M., 58
Weisz, S. Z., 456
Welber, B., 390, 396, 415
Weller, A., 86
Wellinghoff, J., 305
Wendin, G., 97, 107
Wentworth, W. E., 16, 54, 191, 639
Werner, H., 309
Werner, H. W., 57
West, J. E., 32, 57, 387, 396, 395, 467
West, M. A., 84
West, R., 305
Westbrook, R. D., 51
Westgate, C. R., 31, 56, 192, 639, 640
Weyl, C., 417, 656
Wheeler, D. R., 376
Wheeler, P. C., 469
Whelan, T. D., 87
Wheland, R. C., 185, 192, 413, 646
Whillans, D. W., 63, 84
White, C. T., 346, 373
White, D. R., 188, 248, 254
Whiteman, J. D., 86
Whitten, D. G., 306
Whitten, W. B., 456
Wigglesworth, D., 637
Wikswo, J. P., 341
Wilhite, L. G., 249, 637

Willardson, R. K., 457
Willersinn, H., 312
William, J. D., 52
Williams, D. F., 191, 455, 639
Williams, D. J., 254, 317, 469
Williams, D. R., 160, 186, 188, 191
Williams, G., 315
Williams, I. F., 393
Williams, I. M., 254
Williams, J. D., 84, 457
Williams, J. M., 416
Williams, J. O., 83, 113, 182, 185, 188,
    191, 248, 249, 252, 253, 254, 394,
    460, 467
Williams, J. P., 310, 316, 303
Williams, D. F., 116, 185
Williams, R., 54, 415, 637
Williams, R. J. P., 188
Williams, W. G., 377, 393, 394
Williamson, S. J., 341
Willig, C., 187
Willig, F., 56, 57, 60, 251, 253
Willig, H. G., 56, 251, 253
Wilson, E. G., 659
Wilson, J. D., 413, 414, 415
Wilson, N. K., 305
Wing, M., 105
Wintle, H. J., 455, 661
Wisdom, N. E., 314
Witkiewicz, Z., 304
Witt, H. T., 339
Wlodarski, W., 59
Woehrle, D., 252
Wohrle, D., 253
Wolcott, L. E., 469
Wolf, H. C., 73
Wolfe, J., 316
Wolfe, J. P., 83
Wood, D. J., 416
Woodruff, D. P., 457
Woods, C., 469
Woolam, J. A., 415
Wooten, F., 55
Worcester, D. L., 337
Worlock, J. M., 83
Woronkow, M. G., 468
Wright, P. G., 53, 635
Wrixon, A. D., 55, 315
Wubbles, G. G., 312
Wudl, F., 413, 415, 659, 667

Wuertz, D., 373
Wyhof, J. R., 250, 251, 363, 372, 375, 635

Yagi, T., 646
Yagubskii, E. B., 415, 416
Yahabe, T., 312
Yahia, J., 53
Yakusheva, O. B., 252
Yamabe, T., 665
Yamada, A., 468
Yamada, T., 661
Yamagishi, F. G., 414
Yamaguchi, O., 309
Yamaguchi, T., 188
Yamamoto, N., 147, 192
Yamamoto, O., 372
Yamamoto, Y., 461, 467, 642
Yamamura, T., 254, 456, 463, 467, 468, 644
Yamane, T., 363, 375
Yamaoka, Y., 251
Yamashita, K., 658
Yamashita, S., 253
Yamin, P., 669
Yardley, J. T., 185
Yarwood, J., 317
Yashiro, D., 311
Yasuda, I., 340
Yasufuku, S., 658
Yasunaga, H., 436, 455, 643, 645, 646
Yasuniwa, T., 304
Yata, S. M., 396
Yatsimirskii, V. K., 178, 191
Yatsuo, T., 310
Yatsuzuka, K., 653
Yayakawa, R., 467
Yen, S. P. S., 250, 372, 375, 415
Yim-Tan, G. K., 106
Yiyasaka, T., 645
Yoda, C. H., 394
Yokomoto, Y., 253
Yokoyama, M., 78, 84, 86, 644

Yokoyama, T., 462, 467
Yomosa, S., 306, 314
Yoneyama, M., 372
Yoon, D. Y., 345, 372
Yoshida, Z., 296, 314
Yoshihara, K., 72, 73, 86
Yoshino, K., 250, 456
Yoshuo, K., 660
Young, R. H., 5, 6, 52, 640
Young, R. S., 661
Yukhnovskii, I., 314
Yunaki, T., 313
Yurovskaya, M. A., 307, 308
Yusuda, I., 469
Yusupbekov, A. Kh., 637

Zablowska, M., 384, 394
Zagal, J., 651, 652
Zagrubskii, A. A., 106, 456
Zahradnik, A., 643
Zallen, R., 373
Zalukaev, P., 361, 374
Zamaraev, K. I., 187, 191, 255
Zander, M., 59
Zboinski, Z., 384, 394, 642
Zeichmann, W., 309
Zeit, W., 336, 340
Zeller, H. R., 192, 304
Zeller, R. C., 374
Zellweger, U., 60
Zereva, G. I., 415
Zheludev, I. S., 394
Ziegler, J., 54, 305, 266
Ziel, V. D., 58
Zimin, A. B., 64, 83
Zololukhin, S. P., 415
Zosel, D., 53, 256, 637
Zrtamonov, B. P., 60
Zschokke-Granacher, I., 393
Zuev, V. A., 58, 39
Zvargkina, A. V., 415
Zverena, G. L., 415

Subject Index

Abnormal bonding model, 350
Absolute gap, 353
ac Conductivity, 3, 211, 243
Acenaphthene-tetrachlorophthalic anhydride, 279
Acetone, 277
Acetonitrile, 277, 295
Acetophenone, 360
Acetylcholine, 296, 299
ac Hall technique, 10
ac Kerr effect, 25
Acoustic phonon scattering, 124
Acridine, 69, 301, 420
Acridine orange, 283
Acrylamide, 462
Action of the acceptor, 436
Action spectrum, 425
Activated auto-complexes, 274
Activated charge transfer complexes, 269
Activated hopping, 7
Activated ternary complex, 278
Activation energies for the photocurrent, 430
Activation energy of photoconductivity, 215
Active space, 298
Activity, 302
Adiabatic ionization energy, 94

Adsorbate, 280
Adsorbed dye monolayers, 217
Adsorbent, 280
Adsorption, 39, 280, 420
Adsorption bonds, 454
Aerosol filtration, 461
Affinity chromatography, 44
Aggregates with a ternary structure, 427
Alpha-donors and acceptors, 260
Al-polyvinylcarbazole copper, 362
Alkanes, 234
Aluminum contacts, 425
Amaranth, 296
Ambient gases, 96, 436
Amino acid-chloranil, 300
Amino-acid residues, 320
Aminoacids, 261, 387
Amorphous solid, 11, 232, 345, 347
Ampholytes, 44
Amphoteric compounds, 273
Anharmonicity effects, 2
Anhydrides decreasing stability, 268
Anhydrous cytochromes, 321
Aniline black, 219
Anisotropic permittivity, 345
Anisotropy, 1, 10, 23, 195, 243, 344, 434, 652
Annealing, 203

SUBJECT INDEX

Anomalous fluorescence, 217
Anthracene, 64–183
Anthraquinone, 195
Anthraquinone anthrahydroquinone complexes, 282
Aplysia photoneuron, 331
Aprotic solvents, 277
Arachidic acid, 225
Arginine, 326, 335
Aromatic polyimides, 288
Ascorbic acid, 327
Asymmetric carrier propagation, 134
Attenuation length, 20, 98
Auger, 39
Auramine, 344
Auto-complexes, 236, 270, 273, 274, 290, 465
Auto-ionization, 76, 94, 97, 110, 215, 450
Average hopping energy, 359
Axon, 335
Azido group, 93
Azine, 420
Azo-dyes, 420
Azulene, 348

Back-bond states, 39
Background conductivity, 293
Bacterial photosynthesis, 330
Bacteriorhodopsin, 329
Bond model, 109, 434
Barrier layer photocells, 448
Barrier tunnelling, 148
Benzaldehyde, 77
Benzene, 77, 131, 148, 233, 278, 366, 384
Benzene:Br_2, 291
Benzene:Cl_2 291
Benzene:TCNQ complex, 289
Benzene and olefins, 270
Benzidine:TCNQ complex, 277
Benzonitrilerubidium, 96
Benzophene aromatic amines, 272
Benzophenone, 195
Benzoporphyrazines, 260
Benzoquinone, 435
Benzyl radicals, 284
Benzyl-phenyl sulfones, 274
Bi-exciton, 75, 76, 236, 648, 649
Bi-softaron, 82, 105, 113, 360

Bioamines, 465
Bipolymers, 13
Bipyridine, 345
Biradical complex, 325
Bis(cyclopentadienylcobalt)cyclo-octatetracene, 150
1,8-bis(dimethylamino)naphthalene, 285
Bix(o-phenylenediamido)Ni complexes, 269
Bisphenol-polycarbonate polymer, 427
Bistable switching, 461
Blocking, 425
Blue shift, 366
Bone, 336, 390
Bosons, 64
Bromanil, 367, 435
Bulk polaritons, 80
Bulk recombination, 426

CPZ:I_2, 303
CPZ:dilantin, 303
CT stabilization energies, 267
Camphor, 345, 368
Cancer, 337
Cancer prevention, 465
Caoutchouc, 383
Capacitance maximum, 298 ff
Capture coefficient of electrons, 383
Carbon tetrachloride, 429
Carbon-filled polymers, 123
Carbonyl, 326
Carcinosarcoma, 465
Cardiac pace-maker, 462
Carnaube wax, 389
Carrier generation, 97, 111, 368, 422
Carrier mobilities for durene, 116
Casein, 326
Casein-methylglyoxal, 324
Catalysis, 25, 228, 280, 464
Catalysis on phthalocyanines, 464
Catalytic activity, 182
Cation mobilities, 233
Cell reproduction, 337
Cellular electric environment, 465
Centers of dissociation, 436
Cetyltrimethyl-ammonium-bromide, 282
Chain flexibility, 201
Chains, 397
Characteristic photocurrents, 423
Charge density, 364

SUBJECT INDEX

Charge separation distances, 336
Charge transfer, 21, 269
Charge transfer complex, 329, 325, 336, 363, 364, 654
Charge transfer excitons, 64
Charge transfer optical absorption, 183
Charge transfer states, 269, 328, 436
Charge transport, 275
Charge-dipole, 363
Charge-dipole interactions, 268
Charge-transfer absorptions, 17
Charge-transfer excitation energy, 266
Charge-transfer states, 428
Chelate metal complexes, 319
Chemical shift, 300
Chemiluminescence, 23, 24, 25, 224
Chemisorption, 39, 40
Chick epiphysis, 337
Chloranil, 277, 366, 435
Chloranil adenine, 271
Chlorobiphenyl complexes, 299
Chlorodicyanoquinone, 286
Chloroform, 301
Chlorophyll, 67, 81, 225, 328, 437
Chloroplasts, 330
Chloropromazine, 261, 271, 281, 296, 299, 365, 436, 465
Cholesterol, 173, 332
Cholesteryl-laureate, 148
Christov characteristic temperature, 145, 164
Chromophores, 230
Chymotrypsin, 326
Chymotrypsinogen, 326
Circular dichroism, 25
Clausius-Mosotti equation, 297
Co-solubility of donors and acceptors, 266
Co-sublime, 44
Coal, 277
Coaxial transmission cavity, 31
Coehn's Rule, 21
Coherence length, 400
Coherence length of the wave function, 111
Cohesive energy density, 184
Cole-Cole distribution of relaxation times, 8, 199
Colicin synthesis, 466
Collagen, 281, 387, 390
Collagen complex, 327

Collision-induced dipole moments, 299
Collodial charge transfer complexes, 281
Compensated impurity semiconductors, 356
Compensation Law, 171
Competitive interaction, 275
Complete charge transfer, 269
Complex conductivity, 212
Complex formation, 261
Complex permittivity, 212
Complexes based on I_2 as acceptor, 268
Conducting pathway in neurons, 333
Conduction and biological structure, 320
Conduction data, 319
Conductivity, 367
Conductivity maximum, 293
Conductivity measurements, 1
Conductivity titrations, 277, 291, 296
Conjugation, 334
α-conglutin, 324
Contact energy barrier, 420
Contact potential, 21, 32, 420, 437
Contact-less methods, 32
Contacts, 420
Continuous Time Random Walk, 155
Controlled addition of traps, 195
Cooperative electric dipole effects, 466
Coordinate bond energies, 263
Copolymers of styrol, 288
Coronene, 131, 182, 372
Cotton effect, 25, 290
Cottrell filter, 46
Coulomb interaction, 400
Counterions, 283
Coupled system, 421
Critical micelle concentration, 281
Crown ethers, 260, 274
Crystal violet, 219
Crystalline transitions, 345
Cu-phthalocyanine, 96, 384, 461
Cu-polyvinylpyridine complex, 464
Cube-power law, 19
Current peak, 133
Current responses, 163
Cyanine, 420
Cyanine dyes, 215
Cyclic voltammetry, 303
α-cyclodextrin, 278
Cyclohexane, 282
Cyclophanes, 260

Cytochrome oxidase, 325
Cytochrome-c, 322, 326, 337
Cytotonous theory of Damadian, 332

Dangling bonds, 38
Dangling bonds, 38
Dark conductance, 1
Dead molecule, 326
Debye length, 1, 356
Debye temperature, 115, 132, 174
Debye theory, 8
Decay mechanism, 61
Deep traps, 383
Density fluctuations, 236
Depolarization current spectra, 389
Depolarization factor, 31
Desulfovibrio vulgaris, 322
Diamagnetic susceptibility, 301
Diamino-anthraquinone, 436
1,6-diaminopyrene chloranil complex, 277
Diamond, 368
Diazabicyclo-(2,2,2)octane, 269
Dibromonaphthalene, 72
Dichloroanthracene, 65
2,3-Dichloro-5,6-dicyanobenzoquinone, 271
Dichlorodicyanoquinone, 361
Dichloroethane, 262, 272, 285
Dielectric Time Domain Spectroscopy, 8
Dielectric absorption frequency peak, 200, 299
Dielectric dispersion, 7, 325
Dielectric liquids, 234
Dielectric measurement, 324
Dielectric polaron, 65
Dielectric relaxation effects, 149, 299
Diethylether, 263
Diffusion, 377
Diffusion and drift in disordered systems, 350
Diffusion coefficients, 233
Dimethylacetamide, 279
Dimethylalloxazine, 260
Dimethylamine, 282
Dimethylaniline, 270
Dimethylaniline-pyrene, 282
Dimethylformanide, 277, 279
Dimethylsulfoxide, 277, 279, 285
Dinitrogen-metal complexes, 280
Diphenyl, 182

Diphenyl-p-phenylenediamine, 364
Diphenyl-p-phenylenediamine I_2, 364
Diphenylcyclopropenone, 284
Diphenyldiamine derivatives, 266
Diphenylmethanes, 274
Dipolar plasmons, 82
Dipolar volume polarization, 385
Dipole moments, 267, 297
Dipole-dipole, 363
Dipole-dipole (Forster) transfer, 80, 268, 270, 282, 363
Direct (elastic) tunnelling, 143
Direct one-photon, intrinsic photoconduction, 102
Direct pyroelectric effect, 386
Disorder energy, 156, 353
Disordered solids, 3, 647, 649
Dispersive plasmas, 392
Dispersive transport, 136
Disproportionation reaction, 439
Distribution of hopping times, 156
2,3-dithia-(4,5)-spirodecane bromine complex, 272
d-Lactic acid, 326
DNA, 320, 387
Dolichol, 333, 334
Domains, 209 ff
Donicity, 259, 262
Donor, 302
Donor surface centers, 436
Donor-acceptor complexes, 62
Donor-acceptor-micelle, 283
Doped CTC, 275
Doped ZnO, 437
Doped glasses, 234
Doped phthalocyanines, 435
Doped polyacetylenes, 136, 288
Doped polymers, 136, 228
Doping, 219, 222, 272, 286, 421, 460
Double injection, 158, 382
Double layer, 299
Drift mobility, 5, 127, 158, 194, 233
Drude-Lorentz model, 391
Drug interactions, 296, 464
Dry-state conductivity, 323
Durene, 422
Dye reductant, 463
Dye-lasers, 453
Dyes, 214, 282, 448
Dynamic bridge, 7

SUBJECT INDEX

ESCA, 39
ESR absorption, 267
Effect of anion solvation on complex formation, 279
Effective mass, 360, 392
Effective mass of charge carrier, 4
Effective mobilities, 158
Effective thickness, 378, 381
Elastomeric ionenes, 344
Electret, 387, 388, 390, 428, 461, 627, 660
Electret microphone, 461
Electric neutrality, 328
Electrical heating tape, 461
Electro-catalysis, 464
Electro-chromism, 218
Electro-optic, 299
Electro-reflectance, 228
Electrocatalysis, 26
Electrochemical energy gap correlations, 178
Electrochromism, 273
Electrolytic contacts, 33, 421
Electrolytic double layer, 299, 381
Electron acceptance number, 260, 262
Electron affinities, methiodides, 93
Electron affinity, 22, 42, 89, 93, 178, 280, 519, 510, 514
Electron beam induced drift mobility transients, 158
Electron capture method, 15
Electron gas model, 12
Electron injection, 421
Electron plasma, 289
Electron scavenging, 461
Electron spin resonance, 300
Electron superconductive tunnelling, 332
Electron transfer in micellar solutions, 282
Electron-attenuation length, 20
Electron-hole droplets, 77
Electron-phonon interaction, 116, 119, 120, 121, 399, 400
Electronic absorption spectra, 290
Electronic charge density wave, 399
Electronic conduction in amorphous semiconductors, 350
Electronic desaturatio of proteins, 465
Electronic droplets, 356
Electronic interchain coupling, 400

Electronic polaron, 352
Electronic spin concentrations, 361
α-Electron resonance, 320
Electrophotography, 227, 228, 287, 462
Electroreflectance, 23
Ellipsoid of revolution, 29
Elovich equation, 331
Emulsion polymerization, 464
Energy conversion, 462
Energy gap, 97, 110, 179
Energy level diagram, 113
Energy parameters of aromatics, 452
Energy transfer distance, 275
Energy-limited, 355
Entropy, 278
Entropy change, 261
Entropy of fusion, 234
Eosin, 271, 282
Equilibrium constant, 296, 302
Escape depth, 96, 452
Ethanol, 277
Ethanol glasses, 234
Ethylchlorophyllides-a, 329
Ethylenic double bonds, 335
Ethylensulfite, 279
Excess electrons, 243, 235
Excimer, 79, 142, cf also 441, 444
Excimer decay, 78, 439, 440
Excimer fluorescence, 231
Excimer formation, 78, 439, 440
Excimers, 62, 77, 270, 271, 419, 440
Exciplex, 77, 269, 270, 271, 282, 430
Excitation activated complexes, 269
Excited dimers, 270
Excited dye, 215, 420
Excited precursor, 272
Excited states, 222
Exciton, 63, 68, 79, 70, 71, 122, 209, 275, 419, 436, 439, 443, 450, 451, 648
Exciton band, 61
Exciton clusters, 75, 76
Exciton conduction bandwidths, 74
Exciton diffusion length, 429
Exciton dynamics, 65
Exciton migration, 71, 274
Exciton percolation, 72
Exciton quenching, 75
Exciton reactions, 74
Exciton transfer, 74, 236

Exciton trapping, 67, 275
Exciton-induced injection, 445
Exciton-phonon scattering, 67
Exciton-photon interaction, 222
Excitonic insulator, 63
Exciplex, 274
Extended field emission, 166
Extended thermionic emission region, 167
Extrinsic photoconduction, 423

Faraday rotation, 25
Fatty acid monolayers, 217
Fatty acids doped with all-trans -carotene, 147
Fermi levels, 21, 22
Fermi sea, 356
Ferricytochrome-c, 321
Ferro-electricity, 385
Ferrocene-ferricenium, 260, 367, 463
Ferrocytochrome-c, 321
Ferroelectric semiconductors, 448
Ferroin, 439
Fibrinogen, 326
Field dependence of the mobility, 153
Field-emitted ions, 32
Field-induced bistable structures, 372
Fischer-Tropsch synthesis, 26
Flash photoconductivity, 221
Flavine-adenine dinuclectides, 273
Fluorenone doped polyvinylcarbazole, 1 139, 147
Fluorescein, 282
Fluorescence, 222, 298, 422
Fluorescence quenchers, 424
Fluorobenzene, 77
Fluorocarbons, 388
Formaldehyde-phenol copolymer, 388
Formamide, 295
Formation of ion pairs, 290
Formation of photo-oxides, 272
Forster mechanism, 282
Forster-Dexter interaction, 72
Four point probe, 1, 3, 647, 648
Fourier transformation, 22
Fowler-Nordheim (field) emission, 32
Free charge density, 379, 382
Free electron mass, 4
Free radical decay, 331
Free-electron plasma, 392
Frenkel excitons, 63, 64, 67, 69, 210

Frequency dispersion of the ac conductivity, 249
Frohlich sliding mode, 401
Fuel cells, 228, 464
Furan, 77

Galvenomagnetic effects, 11
Gas phase ionization, 18
Gaussian transport, 156
Geminate recombination, 97, 99, 234, 426
Germanium, 327
Giant polarizations, 461
Glass transition temperature, 204, 384, 428, 610
Glasses, 205, 236
Glowcurves, 204, 384, 660
Glow discharges, 32
Glycine, 320
Glyoxalase, 326
Graphite, 276, 371
Graphitic carbon, 168

H. halobium, 330
Hadamard transform, 22
Half-wave potential, 302
Hall effect, 9, 29, 122, 159, 160, 648
Hall mobilities, 159, 194, 288
Halobacterium halobium, 326, 329
Heats, 284
Hemes in cytochrome-c3, 323
α-helix, 320, 344
Hemoglobin, 173
Heparin, 464
Heptane, 278
Heterocharges, 388
Heterogeneous electron transfer, 152
Heterojunction, 223
Hexaiodobenzene, 105, 367
Hexamethylbenzene, 182, 365, 366
Hexamethylene-tetramine, 301
High voltage cables, 281
Highly conducting polymer, 207, 412
Highly conjugated polymers, 46
Hindered rotation, 148
Hodgkin-Huxley theory, 332
Hole, 98
Hole injection, 35
Homocharges, 388
Homoconjugated system, 333

SUBJECT INDEX

Hopping, 3, 112, 119, 130, 146, 360, 386, 428, 649
Hopping distances, 327
Hopping energy, 357
Hopping frequency, 131
Hopping length, 357, 359
Hopping probability, 357
Hopping rate, 354
Host matrix, 5
Hot electron scattering, 96
Hubbard mode, 1, 351, 359
Hydrated electrons, 283
Hydrates, 62
Hydration, 321, 323
Hydration isotherms, 323
Hydrocarbon solvents, 270
Hydrogen bonded complexes, 284
Hydrogen bonding, 268, 284, 285
Hydrogen bonding interactions, 363
Hydrogen bonded network, 321
Hydroxy-dinitro pyridines, 283
6-Hydroxydopamine, 261, 272, 296
8-Hydroxyquinoline, 361
Hyperelectronic polarization, 207, 363
Hyperprotonic polarization, 209, 213
Hypothermia, 281
Hysteresis, 296
Hysteris of conductivity, 371

Ideal contact, 161
Igepal, 282
Iminophenanthrene, 78
Imizadole, 435
Impurity center, 61
Inclusion adducts, 274, 277
Incommensurate-commensurate transition transitions, 412
Indentification of charge transfer complexes, 288
Index of Interface Behavior, 169
Indirect charge transfer, 276
Indole, 261, 271
Indole exciplexes, 269
Indophenine, 174
Induced circular dichronism, 290
Induced polarization, 267
Inelastic-electron-tunnelling spectroscopy, 39
Injected plasmas, 382
Injecting electrode, 377

Injection efficiency, 272, 420. 439
Injection into traps, 445
Injection of electrons, 461
Injection plasma, 158
Instabilities, 392
Insulator, 378
Interband transitions, 391
Intercalation complex, 275, 371
Intercalation ternaries, 276
Interchain transport, 385
Interfacial polarizations, 7
Interfacial potential wells, 350
Intermolecular charge transfer band, 290
Intermolecular resonance integrals, 125, 127
Interplanar distance for solid charge transfer complexes, 363
Interrupted strand model, 200, 288
Intra-site interaction, 239
Intrachain CTC, 288
Intramolecular aromatic stacking interaction, 365
Intramolecular charge transfer, 270, 298, 363
Intramolecular complexes, 272
Intramolecular excimers, 282
Intramolecular proton transfer, 285
Intramolecular stacking, 365
Intrinsic photoconductivity, 97, 100, 424, 441
Inverse piezoelectric effect, 386
Iodanil, 367
Iodine-naphthalene, 293
Ion exchange resin, 332
Ion migration in liquids, 233
Ion pair production efficiency, 422
Ion radicals, 18, 270
Ion-molecule equilibrium, 261
Ionene polymer, 286, 287, 362
Ionic charge transfer complexes, 284
Ionic conductivity, 196
Ionic contributions to conduction, 323
Ionization energies, 498, 499, 500, 503, 514
Ionic solvation in non-aqueous solvents, 279
Ionization cross-section, 94
Ionization potentials, 16, 17, 18, 94, 178, 280
Irreversible protonation, 296

Iso-octane, 236
Isobestic, 290
Isoelectric focussing, 44
Isoelectric point, 44, 45
Isotopic exchange, 272

Johnson noise, 42
Josephson effect, 332

K-edge, 22
Keratin, 47
Kerr effect, 25, 299, 344

LEED, 39
Lamellar compounds, 273, 275, 365
Langmuir-Blodgett technique, 148
Layered materials, 371
Liftoff, 232
Ligand mediated charge transfer, 149
Light activated charge transfer complex, 269
Light harvesting molecules, 330
Lignocain, 281
Limb regeneration, 465
Linear chains, 74
Linear free-energy relationship, 175
Linear organics, 365
Linear polyacenes, 328
Ling's assocation-induction hypothesis, 332
Liquid electrodes, 35
Liquid hydrocarbons, 233, 236
Liquid semiconductors, 231
Liquids, 233
Living molecule, 326
Local level cross section, 383
Localization, 111
Localization radii, 139
Localized excited state, 466
Localized hops, 325
Localized quasi-particles, 232
Localized states, 50
London dispersion, 363
Long range electron transfer, 234
Long range order, 111, 336, 349
Long range percolation, 73
Long range tunnelling, 144, 147
Long-lived triplet state, 216
Long-range hopping, 324
Lorentz force, 11

Lower Hubbard band, 351
Luminescence, 25, 453
Lyons equation, 102, 114
Lysine, 326
Lysozyme, 326

MIM, 440
Macromolecule, 44
Macroscopic mobilities, 233
Magnesium phthalocyanine molecule, 260
Magnetic fields of the action potential, 335
Magneto-electret, 390
Magnetron triode, 89
Maleonitrile dithiol ligand, 152
Markovian process, 347
Matrix-element-limited, 355
Maximum power conversion efficiency, 446
Maxwell-Wagner effect, 6, 213, 389
Meen Field (MF) theory, 400
Mean drift mobility electrons, 194
Mean free path, 98, 111
Mechanical roughening, 384
Melanin, 269, 271, 300, 331, 461
Melting point, 232
Membrane potential, 332
Membranes, 337
Memory, 465
Merocyanine dyes, 218
Metachromasia, 219
Metal chelate, 261
Metal-free polyphthalocyanine, 160
Metal-ion bridged CTC, 365
Metal-phthalocyanine, 261, 361
Metal-to-insulator phase transition, 399
Metal-to-nonmetal transition, 400
Metallic behavior, 398
Metallo-organics, 361
Methanol, 282
Methiodide complexes, 290
Methyl viologen, 330
2-methyl tetrahydrofuran-pyromellitic anhydride complexes, 271
Methylalcohol, 279
Methylene blue, 271
Methylglyoxal, 326, 465
Methylglyoxal complexes, 325
Methylglyoxal molecular dipole moments, 325

Methylglyoxal-ascorbic acid-protein interaction, 327
3-Methylpentane, 390
2-Methyl-tetrahydrofuran glasses, 94, 145
Meyer-Neldel rule, 171
Mg-phthalocyanine, 280
Micelles, 282
Microdielectrophoresis, 466
Microdomain model, 365
Microscopic mobilities, 158, 195, 233
Microwave Hall measurements, 31
Microwave method, 8, 29
Minority carrier injection, 170
Mitochondria, 334, 335
Mixed ligand complexes, 365
Mixed monolayers, 217, 421
Mixing rule, 175
Mobility, 112, 116, 140, 154, 235, 476, 478, 480, 481, 482, 484, 490, 650, 652
Mobility components in naphthalene, 114
Mobility edge, 348
Mobility gap, 348
Mobility studies, 6
Molecular electron pump, 227
Molecular ions, 273
Molecular polarization, 297
Molecularly doped polymers, 229
Monolayer assemblies, 225
Monte Carlo techniques, 350
Mott excitons, 210
Multi-electron hops, 355
Multi-exciton complexes, 77
Multiple charge transfer bands, 290
Multiple trapping, 155
Munn-Siebrand formalism, 118
Muscle stretch receptors, 332
Myelin, 333

2-N-arylamino-6-naphthalene sulfonates, 274
N-dimethylaniline, 275
N-isopropyl carbazole, 137
N-methylimidazole, 285
N-methylphenothiazine, 268
N-purinoylaminopyrimidine, 274
NMR chemical shift, 262, 267
Naphthacene, 96, 367, 383, 452
Naphthalene, 69, 74, 93, 116, 132, 136, 182, 230, 364, 366, 388, 390

Naphthalene-sensitized charge transfer, 429
α-Naphthylamine-picric acid, 284
n-donors, 260
Nearest neighbor hopping, 112, 132, 350
Nearest-neighbor cage, 266
Negative photoconductivity, 454
Neomycin, 296
Neopentane, 236
Nernst, Rhighi-Leduc and Ettinghausen effect, 11
Nerve cell membrane, 333
Nerve conduction, 331
Nerve impulses, 335
Nerve superconduction, piezoelectricity, pyroelectricity, 332
Nervonic acid, 333, 334
Net polarization, 267
Neural processes, 332
Neuronal thresholds, 465
Nitrodiphenylmethanes, 273
Nitroethanol amine, 296
Nitromethane, 279
Noise, 24
Noise measurements, 42
Noise mobility, 42
Nomadic polarization, 209, 343
Non-Gaussian transport, 136
Non-coherent charge transfer, 110
Non-ionic conduction, 323, 324
Non-vibronic polaritons, 80
Nondative ground state, 264
Nonlinear vibrational excitation, 392
Normal hopping probability, 353
Nuclear acids, 387
Nuclear magnetic resonance, 300
Nuclear quadrupole resonance, 301
Nucleic acid-protein complex, 365
Nucleic acids, 296

Ohmic contact, 161
One dimensional metal, 399
One-dimensional conductivity, 428
Onsager critical distance, 427
Onsager geminate recombination theory, 101 ff, 423
Onsager radius, 65
Optical reflectance spectroscopy, 23, 390
Optimum hopping distance, 353
Organic chain conductor, 363

Organic charge transfer salts, 271
Organic dyes, 555
Organic glasses, 136
Organic photovoltaic systems, 440
Organic semi-metals, 4, 276, 398
Organic switching, 460
Organometallic pseudohalides, 266
Organotellurides, 360
Ortho-para hydrogen conversion, 26, 228
Osteogenesis, 465
Oxalic acid, 284
Oxidation potentials, 178
Oxidized cholesterol, 173
Oxygen, 325
Pair
Pair production, 450
Palisade type of molecular arrangement, 281
PAQR polymers, 207
Paracyclophane-quinhydrones, 273, 276
Paramagnetic complex, 301
Paramagnetic susceptibility, 301
Parinaric acid, 333
Peierls instability, 400
Peierls transition temperature, 400
Pentacene, 74, 215, 336, 428, 448
Percolation, 143, 152, 275, 347, 350, 356, 360, 648, 650, 659
Percolation fraction, 350
Perdeuterated anthracene, 116
Periodic organization of molecules, 319
Permittivity, 6, 244
Permittivity tetrations, 298
Perylene, 78, 96, 131, 142, 182, 372, 452
Perylene chloranil, 325
Perylene metal dithiolate complexes, 269, 301
Perylene-cesium-hydrogen, 276
Perylene-iodine, 177
Phase photons, 244
Phenanthrene, 195, 345, 390
Phenazine, 270, 301, 384
Phenolformaldehyde, 199
Phenothiazines, 269, 281, 285, 300
Phenylenedialdehydes, 362
Phenylmethane, 420
Phenytoin, 281
Pheophytin, 225
Phonon scattering, 74
Phonon wind, 77

Phonon-exciton scattering, 70
Phosphate esters, 461
Phosphorescent state, 298
Photoactivated charge transfer, 214
Photobiology, 327
Photocatalysis, 26, 454
Photochemical cycles, 330
Photoconduction, 265, 266, 329
Photoconduction quantum yields, 422
Photoconductive transients, 434
Photoconductivity, 229, 271, 273, 422, 428, 434, 617
Photoconductivity activation energy, 434
Photoconductivity in charge transfer complex salts, 434
Photoconductivity transient methods, 23
Photo-crosslinkable polymers, 227
Photocurrent, 229, 271, 531, 476, 582
Photocurrent spectra, 290
Photodetachment, 13
Photodetrapping, 159
Photodielectric effects, 25
Photo-dissociation of CTC, 271
Photo-electrets, 390
Photoeffects, 22, 419, 594, 596
Photoelectric memory, 466
Photoelectrochemical cells, 463
Photoelectron spectra, 449
Photoelectron spectroscopy, 96
Photoemission, 15, 16, 264, 328, 449, 450, 452
Photoexcitation of adsorbed dyes, 228
Photofluorescence, 443
Photogeneration of carriers, 5, 171, 420, 437
Photoinduced-RedOx reactions, 463
Photoinjection, 163, 437
Photoionization, 94, 235, 265, 281
Photomagnetic effects, 419, 453
Photon energy, 17
Photopolymerization, 464
Photoreduction processes, 285
Photoresponse, 436
Photo-sensitizer, 223
Photosynthesis, 329, 448
Photosynthetic bacteria, 67
Photovoltaic effects, 273, 419, 437, 438, 442, 445, 449
Photovoltaic sandwich cells, 441
Phthalocyanine, 182, 327, 367, 435, 437,

439, 441, 448, 449, 453, 454, 463, 651
Picoline, 301
Pi-donors and acceptors, 260
Piezo-electric effect, 386
Piezo-electric metalloorganic polymers, 387
Piezoconduction, 371
Piezoelectric effects in bioorganic compounds, 387
Piezoelectricity, 332, 336, 345, 624, 625
Pinacyanol, 215
Pinned Frohlich mode, 412
Piperidinoxyl radical, 275
Planar dyes, 219
Plasma frequency, 390
Plasmas, 390
Plasmons, 82, 392, 452
β-Pleated sheets, 320
p-n Junctions, 223, 438, 448, 460, 464
Polaritons, 79, 80
Polarizability, 288
Polarizable centers, 211
Polarization currents, 2
Polarization energies, 102, 369, 370, 505
Polarization waves, 82
Polarized dark currents, 386
Polarized light, 23
Polarized reflectivity, 391
Polarized single-crystal spectroscopic, 288
Polarographic RedOx (half-wave) potentials, 182
Polarography, 301
Polarons, 6, 63, 119, 124, 209, 420, 649
Poly(2-vinylpyridine), 131
Poly(4-vinylpyridine) complexes, 287
Poly(N-epoxypropyl-carbazole), 430
Poly(N-vinyl carbazole), 344, 429
Poly(alkylmethacrylate), 231
Poly(arylenevinylenes), 131, 197
Poly(ethylene-terephthalate), 196
Poly(pyromellitimide), 214
Poly--naphthol, 383
Poly-(1-naphthylmethyl)-L-glutamate, 344
Poly-vinyl-naphthalene, 286
Polyacenaphthalene, 225
Polyacenaphthylene, 286, 420
Polyacetylenes, 201, 207, 223, 231, 363, 412, 460, 661 ff
Polyacrylonitrile, 160

Polycarbonates, 204, 231, 384, 388
Polydiacetylene, 208, 345, 434
Polyenes, 361
Polyesters, 231, 387
Polyethylene, 281, 360, 364, 385, 434
Polyethyleneterphthalate, 384
Polyimides, 203, 384, 387
Polyindophenines, 197
Polyisoprene, 195
Polymalonitrile, 203
Polymers, 430, 652
Polymeric binders, 231
Polymeric charge-transfer complexes, 286
Polymeric ionene-TCNQ, 243
Polymeric structure determination, 344
Polymethylmethacrylates, 273, 364, 384, 388
Polymethylvinylketone, 230
Polynaphthalene, 420
Polynaphthylene, 203
Polypeptide chains, 320 ff
Polyphenylvinylketone, 230
Polyproline, 234
Polypropylene, 386, 388
Polypyrimidines, 203
Polypyrroles, 201, 207
Polyquinoline-carboxylic acid, 203
Polyquinoline-carboxylic acid films, 169
Polyquinolinecarboxylic acid, 199
Polysarcosine chains, 288
Polysilazane, 203
Polystyrene, 202, 286, 365, 384, 386
Polysulfurnitride, 412
Polytetrafluorethylene, 384
Polythene, 202
Polyvinyl chloride, 383, 388
Polyvinylacetate, 231, 385
Polyvinylalcohol, 424
Polyvinylbenzophenone, 230
Polyvinylcarbazole, 77, 131, 196, 287, 420
Polyvinylcarbazole trinitrofluorenone, 434
Polyvinylchloride, 131, 384
Polyvinylidene fluoride, 345, 386
Polyvinylpyridine-iodine-lithium, 462
Poole-Frenkel effect, 152, 164, 202, 231
Poole-Frenkel effects, 101
Porphyrins, 437, 439, 463
Positional degeneracy, 232

Potential energy curve of a CTC, 261
Potentiometric titration, 303
Power conversion efficiency, 438, 463
Pressure, 366, 367, 412
Pressure dependence of mobility, 127
Pressure dependent electrical properties, 371
Primary and secondary acceptors, 330
Primary cells, 42, 462
Primary process in photosynthesis, 329
Probability of carrier separation, 426
Probability of complexation, 269
Proline, 320
Protein, 173, 296, 320, 323
Protein primary sorption, 323
Protein-complexes, 324
Protein-methylglyoxal complexes, 324, 326
Proton affinity, 283
Proton complexes, 286
Proton delocalization, 284
Proton sponges, 285
Proton transfer, 272, 284
Proton transfer complexes, 283, 284, 285, 296
Proton tunnelling, 146, 284
Protonated acceptor site, 229
Protonation, 284
Protonic bond, 285
Protonic conductors, 212
Protonic interaction, 285
Protoporphyrin, 449
Pseudo-bridge, 3
Pseudo-gaps, 347
Pseudo-polymer, 333
Pseudo-ternary complexes, 278
Pseudopotential, 38
Psychotropic drugs, 465
Pulse technique, 425
Pulsed photoconductivity, 427
Pumping of electrons, 334
Pure field emission, 167
Pure thermionic emission, 167
Purification, 44
Purine-indole, 365
Purines, 260
Purple membrane, 329
Pyrene, 72, 152, 182, 260, 281, 282, 364, 366, 448

Pyrene dimethylamine micellar adducts, 282
Pyrene-naphthalene, 267
Pyridine, 77, 285, 296, 300, 301
Pyridine iodine complex, 267, 303
Pyrimidine, 62
Pyro-electric phenomena, 332, 345, 385, 387
Pyropolymers, 208, 363

Q-switched, giant-impulse laser, 453
Quadrature balance, 298
Quadrupole moment, 301
Quantity of charge without recombination, 328
Quantum efficiencies, 98, 329, cf also 115
Quantum yield, 16, 17, 427, 435, 449, 452
Quasi-crystalline globules, 236
Quasi-one dimensional compounds, 397
Quasi-one dimensional crystal structure, 398
Quasicontinuum, 72
Quenching fluorescence, 270
Quinone, 435

RNA, 320
Radial p-n junction, 2
Radiation-induced conductivity, 203
Radius of the localized state, 354
Raman spectroscopy, 291
Random phase model, 112
Random walk model, 65, 222
Randomly oreinted polymer chain, 197
Reaction zone, 299
Real solid, 350
Recombination, 98
Recombination centers, 434
Recombination rate, 426
Rectifying p-n junction, 460
Red shift, 366
RedOx systems, 421, 439
Reduction of oxygen, 464
Reduction potentials, 178
Reflectance spectra, 244, 289
Refractive index, 288
Regeneration region, 426
Regenerative photoelectrochemical cells, 438
Relative drift velocities, 98

Relative electron affinities, 91
Relaxation phenomena in polymers, 384
Relaxations, 7
Residual polarization, 386
Residual potentials, 430
Resistivity minima, 367
Resistivity tensor, 1
Resolved Picosecond Spectroscopy, 22
Resonance tunnelling, 36, 143, 144
Retinals, 173, 326
Reverse charge-transfer complexes, 260
Reversible crystallization, 345
Rhodamine, 437
Rhodopsin, 326
Rojinski-Zeldovich equation, 331
Rose bengal, 271
Rosenmund-von Braun nitrile synthesis, 26
Rotational dispersion, 25
Rotational energy barrier, 148
Rubbery state, 206
Ruthenocene, 429

SH-glutathione, 465
Saccharose, 387
Salicylic acid esters, 285
Sample thickness, 388
Sandwich configurations, 270
Sandwich-type cell, 429
Sarcoma, 337, 466
Saturable optical absorptivity, 454
Saturation photocurrent, 441
Saturation polarization, 390
Scaling law, 164, 377, 382
Scattering, 125
Schottky (thermionic) transfer, 32
Schottky barrier, 168, 223, 438
Schottky effect, 101, 164, 231
Second order optical effects, 290
Second-order, non-linear optical coefficients, 461
Secondary hydration sites, 323
Sedation of mice, 281
Seebeck coefficients, 12, 236, 237, 287, 564, 575, 651, 653
Segmental mobility, 288
Segregated stacks, 363
Selective long-range interactions, 337
Self-complexing, 270, 272
Self-trapped polaron, 65

Semiconducting TCNQ free radical salts, 243
Semiconducting gap, 245
Semiconductor-liquid-electrolyte interfaces, 438
Semiconjugated system, 333, 334
Semimetals, 31
Semi-methylene blue, 224
Sensitization, 214, 420, 651
Sensitization of photoconductivity, 215, 617, 621
Sensitized photoconductive organic layers, 462
Sensors for gases, 372
Serotonine, 296
Serotonine-picrate monohydrate, 277
Serum albumin, 324, 326
Shock-induced polarization, 213
Short-lived ion pairs, 270
Short-range order, 232
Siliconmonoxide, 389
Simultaneous short-range hops, 352
Single crystal, 116
Singlet exciton fission, 105
Singlet exciton oxidation, 74
Singlet excitons, 75, 422
Singlet-triplet separation energy, 271
Sodium lauryl sulfate, 282
Soft phonon, 400
Softaron, 82, 360
Softons, 82, 360
Solar conversion efficiencies, 438, 448, 463
Solid emulsion, 461
Solid state physical theory of Cope, 332
Solid-liquid interface electronic reactions, 331
Solitons, 82
Soluble proteins, 321
Solution RedOx couples, 438
Solvate complex, 279
Solvation, 278, 367
Solvation and association effects, 267
Solvation of the acceptor, 266
Solvatochromism, 290
Solvent dipoles, 277
Solvent inclusion complexes, 285
Solvent interactions, 295, 493
Solvent shared ion pairs, 271
Space charge, 133, 659

Space charge limited currents, 35, 164, 377, 383, 447
Space charge polarization, 388
Spatial extension of the wave function, 356
Spectral dependence of the quantum yield, 98
Spectral shift, 421
Spiraloconjugated system, 334
Spirochromenes, 462
Spiropyrans, 236
Spivey equation, 323
Stabilization energy of the charge-transfer interaction, 265
Stacks, 397
Stearic acid, 329, 389
Stern model, 381
Stilbene, 388
Stochastic hopping theory, 138
Stochastic transport, 346
Stoichiometric CTC, 272, 293
Storage materials, 463
Strained bonds, 346
Stress-induced orientation, 366
Stretched polymers, 366
Strong charge transfer complexes, 266
Structural proteins, 321
Structural traps, 158
Structure factor, 38
Structure-electrical property relationships, 343
Structured instabilities, 319
Substituted benzenes, 364
Super-exchange, 72
Supercells, 38
Superconductivity, 32, 64, 335, 412
Superconductivity fluctuations, 363
Superexchange, 74
Superlinearity of the current, 430
Surface charge transfer complexes, 280
Surface ejection, 427
Surface excitons, 79, 280
Surface generation, 422
Surface ionization, 89
Surface plasmons, 80
Surfaces, 37
Surfactants, 282
Switching devices, 360
Switching effects, 372
Synaptic vesicles, 334

Synaptosomes, 334
Synthetic metals, 276
System, 397

TCNE, 158
TCNQ, 398, 450, 460
TCNQ complexes, 131
TCNQ-Polymer, 361
TTF-TCNQ, 5, 96, 453
TTF-TCNQ, 5, 96, 201, 453
$TaSi_2$ (pyridine), 276
Tailing of states, 346
Ternary complex, 225, 275, 278, 290, 365
Ternary metal ion complex, 365
Ternary surface complexes, 277
Ternary, protonic, surface complexes, 286
Tetracene, 69, 78, 79, 179, 383, 428, 429, 437, 448, 460
Tetrachlorobenzene, 72
Tetrachloroquinone, 286
Tetracyanobenzene, 72
Tetracyanoethylene, 89, 260, 275, 286, 361, 366, 435
Tetracyanoquinodimethane, 398, 450, 460
Tetrafluoroethylene, 149
Tetrafluoroethylene films, 200
Tetramethylbenzidine, 281
Tetramethylsulfone, 279
Tetranitrobiphenyl-4,4'-diol, 284, 285
Tetranitrofluorenone, 361
Tetraselenafulvalene TCNQ complex, 344
Tetraselenonaphthacene, 367, 371
Tetrathiafulvalene, 398
Tetrathionaphthacene, 367, 371
Tetrathiotetracene, 223, 363, 383
Thermal activation energies of the dark electrical conductivity, 110
Thermal activation energy of photoconductivity, 182
Thermalization of the host electron, 425
Thermally activated hopping, 201
Thermally erasable ESR signals, 466
Thermally stimulated current, 384
Thermochromism, 290
Thermoelectric devices, 461
Thermoelectric power, 237, 240
Thermoelectric voltage, 11
Thermoluminescent peaks, 384, 385

SUBJECT INDEX

Thermomagnetic effects, 11
Thiazine, 420
Thiazine derivatives, 8, 299
Thin films, 49, 371
Thiocarbazones, 365
Thiophene, 77
Thiotetracene, 428
Threshold energy, 13
Three-dimensional Peierls distortion, 344
Three-dimensional ordering, 344
Three-dimensional phase transitions, 47
Thrombosis, 465
Thylakoids, 330
Time of flight method, 5, 133
Time-dependent conductivities, 296
Titration apparatus, 26
Transfer integrals, 142
Transformation kinetics of inner-outer complex transitions, 296
Transient charging currents, 202
Transient current, 134, 377
Transient photoconductivity, 384, 420
Transit time, 136
Transmission of nerve signals, 332
Trap density, 379, 441
Trap distribution, 377
Trap-controlled hopping, 157, 229
Trap-controlled mobility, 220
Trap-density, 378
Trap-free hopping, 157
Trap-free mobility, 116
Trap-limited transport, 152
Trap-to-trap exciton transfer, 72
Trapped carriers, 426
Trapped space charges, 377
Trapping center, 170
Travelling space charge, 136
Tri-octylphosphine, 290
Triarylcarbonium, 462
Trichloroacetic acid, 286
Tricyanovinyl derivatives, 273
Triethylamine-iodine complex, 267
Triglycine sulfate, 387
2,4,6-Trinitrochlorobenzene, 366
Trinitrofluorenone, 460, 462
Trinitrofluorenone doped polyesters, 137
Trinitrophenylamines, 273

Triplet exciplexes, 77
Triplet emission processes, 290
Triplet exciton diffusion, 74
Triplet excitons, 75, 422
Tropyllium tetrafluoroborate, 366
Tunnel-junction, 460
Tunnelling, 50, 141, 143, 357, 360
Tunnelling mobility, 141
Tunnelling via localized states, 143
Two-channel hopping, 125
Two-dimensional soliton, 82
Two-dimensional space charge layers, 383
Two-dimensional superconductivity, 276
Two-photon absorption, 422
Two-photon ion pair yield, 422
Two-trap level models, 154

Ultrasound, 463
Uncompensated space charge, 328
Unsaturated, uncharged π-electron acceptors, 93
Ureanitrate, 344

Valerolactones, 362
Variable range hopping, 110, 112, 132, 275, 351, 352, 356
Vidicon type detector, 22
Voids, 281
Voltage-shortened compaction, 49
Voltammetry, 301
Volume polarization, 388
Vorton, 82

Wannier excitons, 63, 76
Weak charge transfer complexes, 264
Wet solar cell, 463
Wien mass filter, 14
Work function, 15
Wound healing, 465

Xanthene, 420
Xerographic discharge, 5
Xylene, 278

Zero current transport, 12
Zone refining, 44